A GARDEN OF PLEASANT FLOWERS

(Paradisi in Sole: Paradisus Terrestris)

by

JOHN PARKINSON

Dover Publications, Inc.

New York

PUBLISHER'S NOTE

One of the great seventeenth-century English botanists was John Parkinson (1567–1650), apothecary to King James I and, after the publication in 1629 of the *Paradisus Terrestris* (dedicated to Queen Henrietta Maria), "first royal botanist" to Charles I. He is considered as the last Englishman "who belonged to the true lineage of herbalists." His largest work was the 1640 *Theatrum Botanicum. The Theater of Plantes, or An Universall and Complete Herball* (it was the premature announcement of the *Theatrum* that spurred Thomas Johnson's competitive revision of John Gerard's *Herball* in 1633). There has been much controversy over the value of the *Theatrum* and its alleged plagiarisms from Lobelius (Mathias de l'Obel), but no one has disputed the charm of Parkinson's first book, the *Paradisus Terrestris,* and its pioneering importance for English (and, later, American) horticulture, with special emphasis on the flower garden. The original 1629 edition is reprinted here in its entirety; the posthumous 1656 edition, though labeled as "corrected and enlarged," was substantially identical.

The woodcut illustration on the title page depicts Adam and Eve in Paradise (etymologically an Iranian/Greek word for "park" or "garden"). At the top of the page is the chief Hebrew name of God. At the very bottom is a four-line French poem meaning: "Whoever wishes to compare art with nature and our parks with Eden, indiscreetly measures the stride of the elephant by the stride of the mite and the flight of the eagle by that of the gnat." The main cartouche contains the punning Latin title (to be rendered "Park in Sun's [= Parkinson's] Earthly Paradise") and a long, fully descriptive English title. All the woodcuts in the volume, including the title page and the portrait, were done in England by the German artist C. Switzer (either Christoph or his son Christopher).

Following the dedication is Parkinson's epistle to the reader, in which he clearly states the aim, approach and limits of the work, and promises a future fourth part on simples (this eventually developed into the *Theatrum*). The sixteenth-century predecessors he mentions are Clusius (Charles de l'Ecluse) from Artois, William Turner (the "father of English botany"), the Fleming Dodonaeus (Rembert Dodoens) and Gerard.

Next come Latin compliments from friends and colleagues of Parkinson: a prose letter (dated Oct. 1, 1629) by the great physician Sir Theodore Turquet de Mayerne, who served James I, Charles I and Charles II (Mayerne apologizes for his inability to versify); and poems by the physician Othowell Meverall, William Atkins (a printer's error for the court physician Henry Atkins?), William "Brodus" and the above-mentioned Thomas Johnson.

The woodcut portrait of Parkinson bears a Latin legend meaning: "Portrait of John Parkinson, London apothecary, at 62 years of age, 1629 A.D."

Parkinson's main text, in easy, readable English, discusses nearly 1000 plants from all over the world that could then be grown in England. These fall under the three headings announced on his title page: the "garden of pleasure," i.e., flowers and fragrant herbs (this section alone comprises three-fourths of the entire work); the kitchen garden (culinary herbs and vegetables); and the orchard (trees, shrubs and vines that produce edible fruit). Some 812 plants are illustrated on 108 full-page plates. Some of these renderings were original with Switzer; many were adapted from the works of Lobelius (1570) and Clusius (1601), others from the *Hortus Floridus* of Crispin de Passe (Utrecht, 1614) and elsewhere. There are a few other drawings of garden plans and tools. Each of the three divisions of the book begins with an essay on "ordering" (preparing) the respective type of garden, but there are no specific growing instructions in the descriptions of the individual plants (in which the heading "vertues" means "medicinal properties").

The work concludes with an index of Latin plant names, an English plant name index, a table of medicinal properties, a list of errata, and the colophon of the printers.

Published in Canada by General Publishing Company, Ltd., 30 Lesmill Road, Don Mills, Toronto, Ontario. Published in the United Kingdom by Constable and Company, Ltd., 10 Orange Street, London WC2H 7EG.

This Dover edition, first published in 1976, is an unabridged republication of the work originally printed by Humphrey Lownes and Robert Young, London, in 1629. A new Publisher's Note has been written for the present edition.

International Standard Book Number: 0-486-23392-8
Library of Congress Catalog Card Number: 76-15697

Manufactured in the United States of America
Dover Publications, Inc.
180 Varick Street
New York, N.Y. 10014

PARADISI IN SOLE
Paradisus Terrestris.
or
A Garden of all sorts of pleasant flowers which our
English ayre will permitt to be noursed up:
with
A Kitchen garden of all manner of herbes, rootes, & fruites,
for meate or sause used with us,
and
An Orchard of all sorte of fruitbearing Trees
and shrubbes fit for our Land
together
With the right orderinge planting & preseruing
of them and their uses & vertues
Collected by John Parkinson
Apothecary of London
1629

Qui veut parangonner L'artifice a Nature,   Le pas de L'elephant par le pas du ciron,
Et nos parcs à L'Eden indiscret il mesure.   Et de L'Aigle le vol par cil du moucheron,

TO
THE QVEENES
MOST EXCELLENT
MAIESTIE.

Madame,

Knowing your Maiestie so much delighted with all the faire Flowers of a Garden, and furnished with them as farre beyond others, as you are eminent before
them; this my Worke of a Garden, long
before this intended to be published, and
but now only finished, seemed as it were
destined, to bee first offered into your
Highnesse hands, as of right challenging the proprietie of
Patronage from all others. Accept, I beseech your Maiestie,
this speaking Garden, that may informe you in all the particulars of your store, as well as wants, when you cannot see
any of them fresh vpon the ground : and it shall further en-
courage him to accomplish the remainder; who, in praying
that your Highnesse may enioy the heauenly Paradise, after
the many yeares fruition of this earthly, submitteth to be

Your Maiesties

in all

humble deuotion,

IOHN PARKINSON.

TO THE COVRTEOVS
READER.

Lthough the ancient Heathens did appropriate the first inuention of the knowledge of Herbes, and so consequently of Physicke, some vnto Chiron the Centaure, and others vnto Apollo or Æculapius his sonne; yet wee that are Christians haue out of a better Schoole learned, that God, the Creator of Heauen and Earth, at the beginning when he created Adam, inspired him with the knowledge of all naturall things (which successiuely descended to Noah afterwardes, and to his Posterity): for, as he was able to giue names to all the liuing Creatures, according to their seuerall natures ; so no doubt but hee had also the knowledge, both what Herbes and Fruits were fit, eyther for Meate or Medicine, for Vse or for Delight. And that Adam might exercise this knowledge, God planted a Garden for him to liue in, (wherein euen in his innocency he was to labour and spend his time) which hee stored with the best and choysest Herbes and Fruits the earth could produce, that he might haue not onely for necessitie whereon to feede, but for pleasure also; the place or garden called Paradise importing as much, and more plainly the words set downe in Genesis the second, which are these ; Out of the ground the Lord God made to grow euerie tree pleasant to the sight and good for meate; and in the 24. of Numbers, the Parable of Balaam, mentioning the Aloe trees that God planted; and in other places if there were neede to recite them. But my purpose is onely to shew you, that Paradise was a place (whether you will call it a Garden, or Orchard, or both, no doubt of some large extent) wherein Adam was first placed to abide; that God was the Planter thereof, hauing furnished it with trees and herbes, as well pleasant to the sight, as good for meate, and that hee being to dresse and keepe this place, must of necessity know all the things that grew therein, and to what vses they serued, or else his labour about them, and knowledge in them, had been in vaine. And although Adam lost the place for his transgression, yet he lost not the naturall knowledge, nor vse of them: but that, as God made the whole world, and all the Creatures therein for Man, so hee may vse all things as well of pleasure as of necessitie, to bee helpes vnto him to serue his God. Let men therefore, according to their first institution, so vse their seruice, that they also in them may remember their seruice to God, and not (like our Grand-mother Eve) set their affections so strongly on the pleasure in them, as to deserue the losse of them in this Paradise, yea and of Heauen also. For truly from all sorts of Herbes and Flowers we may draw matter at all times not only to magnifie the Creator that hath giuen them such diuersities of formes, sents and colours, that the most cunning

Worke-

Worke-man cannot imitate, and such vertues and properties, that although wee know many, yet many more lye hidden and vnknowne, but many good instructions also to our selues: That as many herbes and flowers with their fragrant sweete smels doe comfort, and as it were reuiue the spirits, and perfume a whole house; euen so such men as liue vertuously, labouring to doe good, and profit the Church of God and the Common wealth by their paines or penne, doe as it were send forth a pleasing sauour of sweet instructions, not only to that time wherein they liue, and are fresh, but being drye, withered and dead, cease not in all after ages to doe as much or more. Many herbes and flowers that haue small beautie or sauour to commend them, haue much more good vse and vertue: so many men of excellent rare parts and good qualities doe lye hid vnknown and not respected, vntill time and vse of them doe set forth their properties. Againe, many flowers haue a glorious shew of beauty and brauery, yet stinking in smell, or else of no other vse: so many doe make a glorious ostentation, and flourish in the world, when as if they stinke not horribly before God, and all good men, yet surely they haue no other vertue then their outside to commend them, or leaue behind them. Some also rise vp and appear like a Lilly among Thornes, or as a goodly Flower among many Weedes or Grasse, eyther by their honourable authoritie, or eminence of learning or riches, whereby they excell others, and thereby may doe good to many. The frailty also of Mans life is learned by the soone fading of them before their flowring, or in their pride, or soone after, being either cropt by the hand of the spectator, or by a sudden blast withered and parched, or by the reuolution of time decaying of it owne nature: as also that the fairest flowers or fruits first ripe, are soonest and first gathered. The mutabilitie also of states and persons, by this, that as where many goodly flowers & fruits did grow this yeare and age, in another they are quite pulled or digged vp, and eyther weedes and grasse grow in their place, or some building erected thereon, and their place is no more known. The Ciuill respects to be learned from them are many also: for the delight of the varieties both of formes, colours and properties of Herbes and Flowers, hath euer beene powerfull ouer dull, vnnurtured, rusticke and sauage people, led only by Natures instinct; how much more powerfull is it, or should be in the mindes of generous persons? for it may well bee said, he is not humane, that is not allured with this obiect. The study, knowledge, and trauel in them, as they haue been entertained of great Kings, Princes and Potentates, without disparagement to their Greatnesse, or hinderance to their more serious and weighty Affaires: so no doubt vnto all that are capable thereof, it is not onely pleasant, but profitable, by comforting the minde, spirits and senses with an harmelesse delight, and by enabling the iudgement to conferre and apply helpe to many dangerous diseases. It is also an Instructer in the verity of the genuine Plants of the Ancients, and a Correcter of the many errours whereunto the world by continuance hath bin diuerted, and almost therein fixed, by eradicating in time, and by degrees, the pertinacious wilfulnesse of many, who because they were brought vp in their errours, are most vnwilling to leaue them without consideration of the good or euill, the right or wrong, they draw on therewith. And for my selfe I may well say, that had not mine owne paines and studies by a naturall inclination beene more powerfull in mee then any others helpe (although some through an euill disposition and ignorance haue so far traduced me as to say this was rather another mans worke then mine owne, but I leaue them to their folly) I had neuer done so much as I here publish; nor been fit or prepared for a larger, as time may suddenly (by Gods permission) bring to light, if the maleuolent dispositions of degenerate spirits doe not hinder the accomplishment.

But

But perſwading my ſelfe there is no ſhowre that produceth not ſome fruit, or no word but worketh ſome effect, eyther of good to perſwade, or of reproofe to euince; I could not but declare my minde herein, let others iudge or ſay what they pleaſe. For I haue alwaies held it a thing vnfit, to conceale or bury that knowledge God hath giuen, and not to impart it, and further others therewith as much as is conuenient, yet without oſtentation, which I haue euer hated. Now further to informe the courteous Reader, both of the occaſion that led me on to this worke, and the other occurrences to it. Firſt, hauing peruſed many Herbals in Latine, I obſerued that moſt of them haue eyther neglected or not knowne the many diuerſities of the flower Plants, and rare fruits are known to vs at this time, and (except Cluſius) haue made mention but of a very few. In Engliſh likewiſe we haue ſome extant, as Turner and Dodonæus tranſlated, who haue haue ſaid little of Flowers, Gerard who is laſt, hath no doubt giuen vs the knowledge of as many as he attained vnto in his time, but ſince his dates we haue had many more varieties, then he or they euer heard of, as may be perceiued by the ſtore I haue here produced. And none of them haue particularly ſeuered thoſe that are beautifull flower plants, fit to ſtore a garden of delight and pleaſure, from the wilde and vnfit: but haue enterlaced many, one among another, whereby many that haue deſired to haue faire flowers, haue not known either what to chooſe, or what to deſire. Diuers Bookes of Flowers alſo haue been ſet forth, ſome in our owne Countrey, and more in others, all which are as it were but handfuls ſnatched from the plentifull Treaſury of Nature, none of them being willing or able to open all ſorts, and declare them fully ; but the greateſt hinderance of all mens delight was, that none of them had giuen any deſcription of them, but the bare name only. To ſatisfie therefore their deſires that are louers of ſuch Delights, I took vpon me this labour and charge, and haue here ſelected and ſet forth a Garden of all the chiefeſt for choyce, and faireſt for ſhew, from among all the ſeuerall Tribes and Kindreds of Natures beauty, and haue ranked them as neere as I could, or as the worke would permit, in affinity one vnto another. Secondly, and for their ſakes that are ſtudious in Authors, I haue ſet down the names haue bin formerly giuen vnto them, with ſome of their errours, not intending to cumber this worke with all that might bee ſaid of them, becauſe the deciding of the many controuerſies, doubts, and queſtions that concerne them, pertaine more fitly to a generall Hiſtory : yet I haue beene in ſome places more copious and ample then at the firſt I had intended, the occaſion drawing on my deſire to informe others with what I thought was fit to be known, reſeruing what elſe might be ſaid to another time & worke; wherein (God willing) I will inlarge my ſelfe, the ſubiect matter requiring it at my hands, in what my ſmall ability can effect. Thirdly, I haue alſo to embelliſh this Worke ſet forth the figures of all ſuch plants and flowers as are materiall and different one from another : but not as ſome others haue done, that is, a number of the figures of one ſort of plant that haue nothing to diſtinguiſh them but the colour, for that I held to be ſuperfluous and waſte. Fourthly, I haue alſo ſet down the Vertues and Properties of them in a briefe manner, rather deſiring to giue you the knowledge of a few certaine and true, then to relate, as others haue done, a needleß and falſe multiplicitie, that ſo there might as well profit as pleaſure be taken from them, and that nothing might be wanting to accompliſh it fully. And ſo much for this firſt part, my Garden of pleaſant and delightfull Flowers. My next Garden conſiſteth of Herbes and Rootes, fit to be eaten of the rich and poor as nouriſhment and food, as ſawce or condiment, as ſallet or refreſhing, for pleaſure or profit; where I doe as well play the Gardiner, to ſhew you (in briefe, but not at large) the times

and

and manner of sowing, setting, planting, replanting, and the like (although all these things, and many more then are true, are set down very largely in the severall bookes that others haue written of this subiect) as also to shew some of the Kitchen vses (because they are Kitchen herbes &c.) although I confesse but very sparingly, not intending a treatise of cookery, but briefly to giue a touch thereof; and also the Physicall properties, to shew somewhat that others haue not set forth; yet not to play the Empericke, and giue you receipts of medicines for all diseases, but only to shew in some sort the qualities of Herbes, to quicken the minds of the studious. And lastly an Orchard of all sorts of domesticke or forraine, rare and good fruits, fit for this our Land and Countrey, which is at this time better stored and furnished then euer in any age before. I haue herein endeauoured, as in the other Gardens, to set forth the varieties of euery sort in as briefe a manner as possibly could be, without superfluous repetitions of descriptions, and onely with especiall notes of difference in leaues, flowers and fruits. Some few properties also are set downe, rather the chiefest then the most, as the worke did require. And moreouer before euery of these parts I haue giuen Treatises of the ordering, preparing and keeping the seuerall Gardens and Orchard, with whatsoeuer I thought was conuenient to be known for euery of them.

Thus haue I shewed you both the occasion and scope of this Worke, and herein haue spent my time, paines and charge, which if well accepted, I shall thinke well employed, and may the sooner hasten the fourth Part, A Garden of Simples; which will be quiet no longer at home, then that it can bring his Master newes of faire weather for the iourney.

Thine in what he may,

IOHN PARKINSON.

Ioanni

Ioanni Parkinſono *Pharmacopoeo Londinenſi ſolertiſ-*
ſimo Botanico conſummatiſſimo
T.D.M. S.P.D.

POema panegyricum Opus tuum indefeſſi laboris, vtili-
tatis eximiæ poſtulat, & meriti iure à me extorqueret
(mi Parkinſone) ſi fauentibus Muſis, & ſecundo Apol-
line in bicipiti ſomniare Parnaſſo,& repentè Poetæ mihi
prodire liceret. In fœtus tui bonis auibus in lucem editi,
& prolixiorem nepotum ſeriem promittentis laudes, alii
Deopleni Enthouſiaſtæ carmine ſuos pangant elenchos;
quos ſub figmentis ampullata hyperbolicarum vocum mulcedine, vates
ferè auribus mentibuſue inſinuant. Veritas nuditatis amans, fuco natiuum
candorem obumbranti non illuſtranti perpetuum indixit bellum : In ſim-
plicitate,quam aſſertionum neruoſa breuitas exprimit, exultat. Audi quid
de te ſentiam, Tu mihi ſis in poſterum Crateuas Brittannus; inter omnes,
quotquot mihi hic innotuerunt, peritiſsimus, exercitatiſsimus, oculatiſ-
ſimus, & emunctiſsimæ naris Botanicus : Cuius opera in fortunata hac
Inſula rem herbariam tractari, emendari, augeri, & popularibus tuis ver-
naculo ſermone ad amuſsim tradi, non decentiæ modo, ſed etiam neceſsi-
tatis eſt. Macte tua ſedulitate (Vir optime) neque te laborum tam arduis
lucubrationibus datorum hactenus pœniteat, vel deinceps impendendo-
rum pigeat.Difficilia quæ pulchra.Leniet debitæ laudis dulcedo vigiliarum
acerbitatem, & Olympicum ſtadium cito pede, à carceribus ad metas ala-
criter decurrentem nobile manet βραβειον. Sed memento Artem longam,
Vitam eſſe breuem. Μηδὲν ἀναβαλλόμενΘ. Vide quid ad antiquum illum, cuius
ſi non animam, ſaltem genium induiſti, Crateuam ſcribat Hippocrates,
Τέχνης πάσης ἀλλότριον ἀναβολὴ ἰητρικῆς δὲ κ̀ πάνυ; ἐν ᾗ ψυχῆς κίνδυνΘ ἢ ὑπέρθεσις. Nobiliſsimam
Medicinæ partem Botanicam eſſe reputa. Floræ nunc litaſti & Pomonæ,
Apollini vt audio propediem HORTO MEDICO facturus.Amabò integræ
Veſtæ ſacra conficito, eiuſque variegatum multis ſimplicium morbifugo-
rum myriadibus ſinum abſolutè pandito, quem ſine velo nobis exhibeas.
Nulla dies abeat ſine linea. Sic tandem fructus gloriæ referes vberrimos,
quos iuſtè ſudoribus partos, vt in cruda & viridi ſenectute decerpas diu,
iiſque longum fruaris opto. *Vale. Datum Londini Calendas Octobris anno*
ſalutis 1629.

Theodorus de Mayerne *Eques aurat. in Aula*
Regum Magnæ Britanniæ Iacobi & Caroli
P. & F. Archiatrorum Comes.

Ad eximium arte & vſu Pharmacopæum & Botanographum *I. Parkinſonum.*

Gu. Turne-
rus. M. D.

Io. Gerar-
dus Chirur-
gus.

Erbarum vires, primus te (*magne Britannæ*)
   Edocuit medicas, inclytus arte ſophus.
Atque cluens herbis alter, Chironis alumnus,
   Deſcripſit plantas, neu cadat vlla ſalus.
Fortunate ſenex, ſis tu nunc tertius Heros
Hortos qui reſeras, deliciaſque ſoli,
Et flores Veneris lætos, herbaſque virentes,
Arboreos fætus, pharmacum & arte potens.
Poſteritas iuſtos poſthac tibi ſolvet honores,
   Laudabitque tuæ dexteritatis opus.

Ottuellus Meuerell. D. M. & Collegiæ
Med. Lond. ſocius.

Amico ſuo *Ioanni Parkinſono.*

Xtollunt alij quos (Parkinſone) labores
   Da mihi iam veniam comminuiſſe tuos.
Extremos poteris credi migraſſe per Indos:
   Cum liber haud aliud quam tuus hortus hic eſt:
Ipſe habitare Indos tecum facis, haud petis Indos
   I nunc, & tua me comminuiſſe refer.
Eſt liber Effigies, tuus hic qui pingitur hortus,
   Digna manu facies hæc, facieque manus!
Vidi ego ſplendentem varigatis vndique gemmis
   Vna fuit Salomon, turba quid ergo fuit?
Vt vario ſplendent Pallacia regia ſumptu,
   Et Procerum turbis Atria tota nitent:
Tunc cum feſta dies veniam dedit eſſe ſuperbis
   Quoſque ficus texit, nunc tria rura tegunt:
Plena tuo pariter ſpectatur Curia in Horto,
   Hic Princeps, Dux hic, Sponſaque pulchra Ducis.
Quæque dies eſt feſta dies, nec parcius vnquam
   Luxuriant, lauta hæc; Quotidiana tamen.
Ecce velut Patriæ Paradiſi haud immemor Exul,
   Hunc naturali pingit amore ſibi.
Pingit & ad vivum ſub eodem nomine, & hic eſt
   Fronticuli ſudor quem cerebrique dedit:
Aſtat Adam medius Paradiſo noſter in iſto
   Et ſpecies nomen cuique dat ipſe ſuum.
Hos cape pro meritis, qui florem nomine donas
   Æternum florens tu tibi Nomen habe.

Guilielmus Atkins.

Ad Amicum *Ioannem Parkinsonum* Pharmacopæum, & Archibotanicum Londinensem.

Africa quas profert Plantas, quas India mittit,
  Quas tua dat tellus, has tuus hortus habet:
Atque harum Species, florendi tempora, vires,
  Et varias formas iste libellus habet:
Nescio plus librum talem mirabor, an hortum
Totus inest horto mundus ; at iste libro.
Parkinsone tuus liber, & labor, & tua sit laus,
Herbas dum nobis das ; datur herba tibi.

Guilielmus Brodus Pharmacopæus
ac Philobotanicus Londinensis.

Ad Amicum *Ioannem Parkinsonum* Pharmacopæum & Botanicum insignem. Carmen.

Quam magno pandis Floræ penetralia nixu
  Atque facis cœlo liberiore frui?
Omnibus vt placeas, ô quam propensa voluntas,
  Solicitusque labor nocte dieque premit?
Quam magno cultum studio conquirere in hortum
  Herbarum quicquid mundus in orbe tenet,
Immensus sumptus, multosque extensus in annos
  Te labor afficiunt? & data nulla quies.
Talia quærenti, surgit novus ardor habendi,
  Nec tibi tot soli munera magna petis ;
Descriptos vivâ profers sub imagine flores,
  Tum profers mensæ quicquid & hortus alit,
Laudatos nobis fructus & promis honores,
  Profers, quas celebrant nullibi scripta virum,
Herbarum species, quibus est quoque grata venustas :
  Sic nos multiplici munere, Amice, beas.
Hoc cape pro meritis, florum dum gratia floret,
  Suntque herbis vires ; en tibi Nomen erit.
In serum semper tua gloria floreat ævum,
  Gloria quæ in longum non peritura diem.

Thomas Iohnson vtriusque
Societatis consors.

THE ORDERING OF THE
GARDEN OF PLEASVRE.

CHAP. I.

The situation of a Garden of pleasure, with the nature of soyles, and how to amend the defects that are in many sorts of situations and grounds.

HE seuerall situations of mens dwellings, are for the most part vnauoideable and vnremoueable; for most men cannot appoint forth such a manner of situation for their dwelling, as is most fit to auoide all the inconueniences of winde and weather, but must bee content with such as the place will afford them; yet all men doe well know, that some situations are more excellent than others: according therfore to the seuerall situation of mens dwellings, so are the situations of their gardens also for the most part. And although diuers doe diuersly preferre their owne seuerall places which they haue chosen, or wherein they dwell; As some those places that are neare vnto a riuer or brooke to be best for the pleasantnesse of the water, the ease of transportation of themselues, their friends and goods, as also for the fertility of the soyle, which is seldome bad neare vnto a riuers side; And others extoll the side or top of an hill, bee it small or great, for the prospects sake; And againe, some the plaine or champian ground, for the euen leuell thereof: euery one of which, as they haue their commodities accompanying them, so haue they also their discommodities belonging vnto them, according to the Latine Prouerbe, *Omne commodum fert suum incommodum.* Yet to shew you for euerie of these situations which is the fittest place to plant your garden in, and how to defend it from the iniuries of the cold windes and frosts that may annoy it, will, I hope, be well accepted. And first, for the water side, I suppose the North side of the water to be the best side for your garden, that it may haue the comfort of the South Sunne to lye vpon it and face it, and the dwelling house to bee aboue it, to defend the cold windes and frosts both from your herbes, and flowers, and early fruits. And so likewise I iudge for the hill side, that it may lye full open to the South Sunne, and the house aboue it, both for the comfort the ground shall receiue of the water and raine descending into it, and of defence from winter and colds. Now for the plaine leuell ground, the buildings of the house should be on the North side of the garden, that so they might bee a defence of much sufficiency to safeguard it from many iniurious cold nights and dayes, which else might spoyle the pride thereof in the bud. But because euery one cannot so appoint his dwelling, as I here appoint the fittest place for it to be, euery ones pleasure thereof shall be according to the site, cost, and endeauours they bestow, to cause it come nearest to this proportion, by such helpes of bricke or stone wals to defend it, or by the helpe of high growne and well spread trees, planted on the North side thereof, to keepe it the warmer. And euery of these three situations, hauing the fairest buildings of the house facing the garden in this manner before specified, besides the benefit of shelter it shall haue from them, the buildings and roomes abutting thereon, shall haue reciprocally the beautifull prospect into it, and haue both sight and sent of whatsoeuer is excellent, and worthy to giue content out from it, which is one of the greatest pleasures a garden can yeeld his Master. Now hauing shewed you the best place where this your

A   garden

garden should be, let me likewise aduise you where it should not be, at least that it is the worst place wherein it may be, if it be either on the West or East side of your house, or that it stand in a moorish ground, or other vnwholsome ayre (for many, both fruits, herbes, and flowers that are tender, participate with the ayre, taking in a manner their chiefest thriuing from thence) or neare any common Lay-stalles, or common Sewers, or else neare any great Brew-house, Dye-house, or any other place where there is much smoake, whether it be of straw, wood, or especially of sea-coales, which of all other is the worst, as our Citie of London can giue proofe sufficient, wherein neither herbe nor tree will long prosper, nor hath done euer since the vse of sea-coales beganne to bee frequent therein. And likewise that it is much the worse, if it bee neare vnto any Barnes or Stackes of corne or hey, because that from thence will continually with the winde bee brought into the garden the strawe and chaffe of the corne, the dust and seede of the hey to choake or pester it. Next vnto the place or situation, let mee shew you the grounds or soyles for it, eyther naturall or artificiall. No man will deny, but the naturall blacke mould is not only the fattest and richest, but farre exceedeth any other either naturall or artificiall, as well in goodnesse as durability. And next thereunto, I hold the sandy loame (which is light and yet firme, but not loose as sand, nor stiffe like vnto clay) to be little inferiour for this our Garden of pleasure; for that it doth cause all bulbous and tuberous rooted plants to thriue sufficiently therein, as likewise all other flower-plants, Roses, Trees, &c. which if it shall decay by much turning and working out the heart of it, may soone be helped with old stable manure of horses, being well turned in, when it is old and almost conuerted to mould. Other grounds, as chalke, sand, grauell, or clay, are euery of them one more or lesse fertill or barren than other; and therefore doe require such helpes as is most fit for them. And those grounds that are ouer dry, loose, and dustie, the manure of stall fedde beasts and cattell being buried or trenched into the earth, and when it is thorough rotten (which will require twice the time that the stable soyle of horses will) well turned and mixed with the earth, is the best soyle to temper both the heate and drinesse of them. So contrariwise the stable dung of horses is the best for cold grounds, to giue them heate and life. But of all other sorts of grounds, the stiffe clay is the very worst for this purpose; for that although you should digge out the whole compasse of your Garden, carry it away, and bring other good mould in the stead thereof, and fill vp the place, yet the nature of that clay is so predominant, that in a small time it will eate out the heart of the good mould, and conuert it to its owne nature, or very neare vnto it: so that to bring it to any good, there must bee continuall labour bestowed thereon, by bringing into it good store of chalke, lime, or sand, or else ashes eyther of wood or of sea-coales (which is the best for this ground) well mixed and turned in with it. And as this stiffe clay is the worst, so what ground soeuer commeth nearest vnto the nature thereof, is nearest vnto it in badnesse, the signes whereof are the ouermuch moysture thereof in Winter, and the much cleauing and chapping thereof in Summer, when the heate of the yeare hath consumed the moysture, which tyed and bound it fast together, as also the stiffe and hard working therein: but if the nature of the clay bee not too stiffe, but as it were tempered and mixed with sand or other earths, your old stable soyle of horses will helpe well the small rifting or chapping thereof, to be plentifully bestowed therein in a fit season. Some also do commend the casting of ponds and ditches, to helpe to manure these stiffe chapping grounds. Other grounds, that are ouermoist by springs, that lye too neare the vpper face of the earth, besides that the beds thereof had need to be laid vp higher, and the allies, as trenches and furrowes, to lye lower, the ground it selfe had neede to haue some good store of chalke-stones bestowed thereon, some certaine yeares, if it may be, before it be laid into a Garden, that the Winter frosts may breake the chalke small, and the Raine dissolue it into mould, that so they may bee well mixed together; than which, there is not any better manure to soyle such a moist ground, to helpe to dry vp the moysture, and to giue heate and life to the coldnesse thereof, which doth alwayes accompany these moist grounds, and also to cause it abide longer in heart than any other. For the sandy and grauelly grounds, although I know the well mollified manure of beasts and cattell to be excellent good, yet I know also, that some commend a white Marle, and some a clay to be well spread thereon, and after turned thereinto: and for the chalkie ground, *è conuerso,* I commend fatte clay to helpe it. You must vnderstand, that the lesse rich or more barren that your ground is, there needeth

deth the more care, labour, and coſt to bee beſtowed thereon, both to order it rightly, & ſo to preſerue it from time to time : for no artificiall or forc't ground can endure good any long time, but that within a few yeares it muſt be refreſhed more or leſſe, according as it doth require. Yet you ſhall likewiſe vnderſtand, that this Garden of pleaſure ſtored with theſe Out-landiſh flowers ; that is, bulbous and tuberous rooted plants, and other fine flowers, that I haue hereafter deſcribed, and aſſigned vnto it, needeth not ſo much or ſo often manuring with ſoyle, &c. as another Garden planted with the other ſorts of Engliſh flowers, or a Garden of ordinary Kitchin herbes doth. Your ground likewiſe for this Garden had neede to bee well cleanſed from all annoyances (that may hinder the well doing or proſpering of the flowers therein) as ſtones, weedes, rootes of trees, buſhes, &c. and all other things cumberſome or hurtfull ; and therefore the earth being not naturally fine enough of it ſelfe, is vſed to bee ſifted to make it the finer, and that either through a hurdle made of ſticks, or lathes, or through ſquare or round ſieues plat-ted with fine and ſtrong thin ſtickes, or with wyers in the bottome. Or elſe the whole earth of the Garden being courſe, may be caſt in the ſame manner that men vſe to try or fine ſand from grauell, that is, againſt a wall ; whereby the courſer and more ſtony, fal-ling downe from the fine, is to be taken away from the foote of the heape, the finer ſand and ground remaining ſtill aboue, and on the heape. Or elſe in the want of a wall to caſt it againſt, I haue ſeene earth fined by it ſelfe in this manner : Hauing made the floore or vpper part of a large plat of ground cleane from ſtones, &c. let there a reaſonable round heape of fine earth be ſet in the midſt thereof, or in ſtead thereof a large Garden flower-pot, or other great pot, the bottome turned vpwards, and then poure your courſe earth on the top or head thereof, one ſhouell full after another ſomewhat gently, and thereby all the courſe ſtuffe and ſtones will fall downe to the bottome round about the heape, which muſt continually be carefully taken away, and thus you may make your earth as fine as if it were caſt againſt a wall, the heape being growne great, ſeruing in ſtead there-of. Thoſe that will not prepare their grounds in ſome of theſe manners aforeſaid, ſhall ſoone finde to their loſſe the neglect thereof : for the traſh and ſtones ſhall ſo hinder the encreaſe of their roots, that they will be halfe loſt in the earth among the ſtones, which elſe might be ſaued to ſerue to plant whereſoeuer they pleaſe.

CHAP. II.

The frame or forme of a Garden of delight and pleaſure, with the ſeuerall varieties thereof.

ALthough many men muſt be content with any plat of ground, of what forme or quantity ſoeuer it bee, more or leſſe, for their Garden, becauſe a more large or conuenient cannot bee had to their habitation : Yet I perſwade my ſelfe, that Gentlemen of the better ſort and quality, will prouide ſuch a parcell of ground to bee laid out for their Garden, and in ſuch conuenient manner, as may be fit and anſwerable to the degree they hold. To preſcribe one forme for euery man to follow, were too great preſumption and folly : for euery man will pleaſe his owne fancie, according to the extent he deſigneth out for that purpoſe, be it orbicular or round, triangular or three ſquare, quadrangular or foure ſquare, or more long than broad. I will onely ſhew you here the ſeuerall formes that many men haue taken and delighted in, let euery man chuſe which him liketh beſt, or may moſt fitly agree to that proportion of ground hee hath ſet out for that purpoſe. The orbicular or round forme is held in it owne proper exiſtence to be the moſt abſolute forme, containing within it all other formes whatſoeuer ; but few I thinke will chuſe ſuch a proportion to be ioyned to their habitation, being not accep-ted any where I thinke, but for the generall Garden to the Vniuerſity at Padoa. The tri-angular or three ſquare is ſuch a forme alſo, as is ſeldome choſen by any that may make another choiſe, and as I thinke is onely had where another forme cannot be had, neceſ-ſitie conſtraining them to be therewith content. The foure ſquare forme is the moſt vſu-ally accepted with all, and doth beſt agree to any mans dwelling, being (as I ſaid before) behinde the houſe, all the backe windowes thereof opening into it. Yet if it bee longer than the breadth, or broader than the length, the proportion of walkes, ſquares, and knots may be ſoon brought to the ſquare forme, and be ſo caſt, as the beauty thereof may

A 2 be

bee no leffe than the foure fquare proportion, or any other better forme, if any be. To forme it therfore with walks, croffe the middle both waies, and round about it alfo with hedges, with fquares, knots and trayles, or any other worke within the foure fquare parts, is according as euery mans conceit alloweth of it, and they will be at the charge: For there may be therein walkes eyther open or clofe, eyther publike, or priuate, a maze or wilderneffe, a rocke or mount, with a fountaine in the midft thereof to conuey water to euery part of the Garden, eyther in pipes vnder the ground, or brought by hand, and emptied into large Cifternes or great Turkie Iarres, placed in conuenient places, to ferue as an eafe to water the neareft parts thereunto. Arbours alfo being both gracefull and neceffary, may be appointed in fuch conuenient places, as the corners, or elfe where, as may be moft fit, to ferue both for fhadow and reft after walking. And becaufe many are defirous to fee the formes of trayles, knots, and other compartiments, and becaufe the open knots are more proper for thefe Out-landifh flowers; I haue here caufed fome to be drawne, to fatisfie their defires, not intending to cumber this worke with ouer ma-nie, in that it would be almoft endleffe, to expreffe fo many as might bee conceiued and fet downe, for that euery man may inuent others farre differing from thefe, or any other can be fet forth. Let euery man therefore, if hee like of thefe, take what may pleafe his mind, or out of thefe or his own conceit, frame any other to his fancy, or caufe others to be done as he liketh beft, obferuing this *decorum*, that according to his ground he do caft out his knots, with conuenient roome for allies and walkes; for the fairer and larger your allies and walkes be, the more grace your Garden fhall haue, the leffe harme the herbes and flowers fhall receiue, by paffing by them that grow next vnto the allies fides, and the better fhall your Weeders cleanfe both the beds and the allies.

Chap. III.

The many forts of herbes and other things, wherewith the beds and parts of knots are bordered to fet out the forme of them, with their commodities and difcommodities.

IT is neceffary alfo, that I fhew you the feuerall materials, wherewith thefe knots and trayles are fet forth and bordered; which are of two forts: The one are liuing herbes, and the other are dead materials; as leade, boords, bones, tyles, &c. Of herbes, there are many forts wherewith the knots and beds in a Garden are vfed to bee fet, to fhew forth the forme of them, and to preferue them the longer in their forme, as alfo to be as greene, and fweete herbes, while they grow, to be cut to perfume the houfe, keeping them in fuch order and proportion, as may be moft conuenient for their feuerall natures, and euery mans pleafure and fancy: Of all which, I intend to giue you the knowledge here in this place; and firft, to begin with that which hath beene moft anci-ently receiued, which is Thrift. This is an euerliuing greene herbe, which many take to border their beds, and fet their knots and trayles, and therein much delight, becaufe it will grow thicke and bufhie, and may be kept, being cut with a paire of Garden fheeres, in fome good handfome manner and proportion for a time, and befides, in the Summer time fend forth many fhort ftalkes of pleafant flowers, to decke vp an houfe among o-ther fweete herbes: Yet thefe inconueniences doe accompany it; it will not onely in a fmall time ouergrow the knot or trayle in many places, by growing fo thicke and bufhie, that it will put out the forme of a knot in many places: but alfo much thereof will dye with the frofts and fnowes in Winter, and with the drought in Summer, whereby many voide places will be feene in the knot, which doth much deforme it, and muft therefore bee yearely refrefhed: the thickneffe alfo and bufhing thereof doth hide and fhelter fnayles and other fmall noyfome wormes fo plentifully, that Gilloflowers, and other fine herbes and flowers being planted therein, are much fpoyled by them, and cannot be helped without much induftry, and very great and daily attendance to deftroy them. Germander is another herbe, in former times alfo much vfed, and yet alfo in many pla-ces; and becaufe it will grow thicke, and may be kept alfo in fome forme and proportion with cutting, and that the cuttings are much vfed as a ftrawing herbe for houfes, being pretty and fweete, is alfo much affected by diuers: but this alfo will often dye and grow out of forme, and befides that, the ftalkes will grow too great, hard and ftubby, the rootes doe fo farre fhoote vnder ground, that vpon a little continuance thereof, will

fpread

spread into many places within the knot, which if continually they be not plucked vp, they will spoile the whole knot it selfe; and therefore once in three or foure yeares at the most, it must be taken vp and new set, or else it will grow too roynish and cumbersome. Hyssope hath also been vsed to be set about a knot, and being sweet, will serue for strewings, as Germander: But this, although the rootes doe not runne or creep like it, yet the stalkes doe quickly grow great aboue ground, and dye often after the first yeares setting, whereby the grace of the knot will be much lost. Marierome, Sauorie, and Thyme, in the like manner being sweete herbes, are vsed to border vp beds and knots, and will be kept for a little while, with cutting, into some conformity; but all and euery of them serue most commonly but for one yeares vse, and will soone decay and perish: and therefore none of these, no more than any of the former, doe I commend for a good bordering herbe for this purpose. Lauander Cotton also being finely slipped and set, is of many, and those of the highest respect of late daies, accepted, both for the beauty and forme of the herbe, being of a whitish greene mealy colour, for his sent smelling somewhat strong, and being euerliuing and abiding greene all the Winter, will, by cutting, be kept in as euen proportion as any other herbe may be. This will likewise soone grow great and stubbed, notwithstanding the cutting, and besides will now and then perish in some places, especially if you doe not strike or put off the snow, before the Sunne lying vpon it dissolue it: The rarity & nouelty of this herbe, being for the most part but in the Gardens of great persons, doth cause it to be of the greater regard, it must therfore be renewed wholly euery second or third yeare at the most, because of the great growing therof. Slips of Iuniper or Yew are also receiued of some & planted, because they are alwayes green, and that the Iuniper especially hath not that ill sent that Boxe hath, which I will presently commend vnto you, yet both Iuniper and Yew will soon grow too great and stubbed, and force you to take vp your knot sooner, than if it were planted with Boxe. Which lastly, I chiefly and aboue all other herbes commend vnto you, and being a small, lowe, or dwarfe kinde, is called French or Dutch Boxe, and serueth very well to set out any knot, or border out any beds: for besides that it is euer greene, it being reasonable thicke set, will easily be cut and formed into any fashion one will, according to the nature thereof, which is to grow very slowly, and will not in a long time rise to be of any height, but shooting forth many small branches from the roote, will grow very thicke, and yet not require so great tending, nor so much perish as any of the former, and is onely receiued into the Gardens of those that are curious. This (as I before said) I commend and hold to bee the best and surest herbe to abide faire and greene in all the bitter stormes of the sharpest Winter, and all the great heates and droughts of Summer, and doth recompence the want of a good sweet sent with his fresh verdure, euen proportion, and long lasting continuance. Yet these inconueniences it hath, that besides the vnpleasing sent which many mislike, and yet is but small, the rootes of this Boxe do so much spread themselues into the ground of the knot, and doe draw from thence so much nourishment, that it robbeth all the herbes that grow neare it of their sap and substance, thereby making all the earth about it barren, or at least lesse fertile. Wherefore to shew you the remedy of this inconuenience of spreading, without either taking vp the Boxe of the border, or the herbes and flowers in the knot, is I thinke a secret knowne but vnto a few, which is this: You shall take a broad pointed Iron like vnto a Slise or Chessill, which thrust downe right into the ground a good depth all along the inside of the border of Boxe somewhat close thereunto, you may thereby cut away the spreading rootes thereof, which draw so much moisture from the other herbes on the inside, and by this meanes both preserue your herbes and flowers in the knot, and your Boxe also, for that the Boxe will be nourished sufficiently from the rest of the rootes it shooteth on all the other sides. And thus much for the liuing herbes, that serue to set or border vp any knot. Now for the dead materials, they are also, as I said before diuers: as first, Leade, which some that are curious doe border their knots withall, causing it to be cut of the breadth of foure fingers, bowing the lower edge a little outward, that it may lye vnder the vpper crust of the ground, and that it may stand the faster, and making the vpper edge either plain, or cut out like vnto the battlements of a Church: this fashion hath delighted some, who haue accounted it stately (at the least costly) and fit for their degree, and the rather, because it will be bowed and bended into any round square, angular, or other proportion as one listeth, and is not much to be misliked, in that the Leade

doth

doth not eafily breake or fpoile without much iniury, and keepeth vp a knot for a very long time in his due proportion : but in my opinion, the Leade is ouer-hot for Summer, and ouer-cold for Winter. Others doe take Oaken inch boords, and fawing them foure or fiue inches broad, do hold vp their knot therewith : but in that thefe boordes cannot bee drawne compaffe into any fmall fcantling, they muft ferue rather for long outright beds, or fuch knots as haue no rounds, halfe rounds, or compaffings in them. And befides, thefe boordes are not long lafting, becaufe they ftand continually in the weather, efpecially the ends where they are faftned together will fooneft rot and perifh, and fo the whole forme will be fpoyled. To preuent that fault, fome others haue chofen the fhanke bones of Sheep, which after they haue beene well cleanfed and boyled, to take out the fat from them, are ftucke into the ground the fmall end downewards, and the knockle head vpwards, and thus being fet fide to fide, or end to end clofe together, they fet out the whole knot therewith, which heads of bones although they looke not white the firft yeare, yet after they haue abiden fome frofts and heates will become white, and prettily grace out the ground : but this inconuenience is incident to them, that the Winter frofts will raife them out of the ground oftentimes, and if by chance the knockle head of any doe breake, or be ftrucke off with any ones foot, &c. going by, from your ftore, that lyeth by you of the fame fort, fet another in the place, hauing firft taken away the broken peece: although thefe will laft long in forme and order, yet becaufe they are but bones many miflike them, and indeed I know but few that vfe them. Tyles are alfo vfed by fome, which by reafon they may bee brought compaffe into any fafhion many are pleafed with them, who doe not take the whole Tyle at length, but halfe Tyles, and other broken peeces fet fomewhat deepe into the ground, that they may ftand faft, and thefe take vp but little roome, and keepe vp the edge of the beds and knots in a pretty comely manner, but they are often out of frame, in that many of them are broken and fpoiled, both with mens feete paffing by, the weather and weight of the earth beating them downe and breaking them, but efpecially the frofts in Winter doe fo cracke off their edges, both at the toppes and fides that ftand clofe one vnto another, that they muft bee continually tended and repaired, with frefh and found ones put in the place of them that are broken or decayed. And laftly (for it is the lateft inuention) round whitifh or blewifh pebble ftones, of fome reafonable proportion and bigneffe, neither too great nor too little, haue beene vfed by fome to be fet, or rather in a manner but laide vpon the ground to fafhion out the traile or knot, or all along by the large grauelly walke fides to fet out the walke, and maketh a pretty handfome fhew, and becaufe the ftones will not decay with the iniuries of any time or weather, and will be placed in their places againe, if any fhould be thruft out by any accident, as alfo that their fight is fo confpicuous vpon the ground, efpecially if they be not hid with the ftore of herbes growing in the knot; is accounted both for durability, beauty of the fight, handfomneffe in the worke, and eafe in the working and charge, to be of all other dead materials the chiefeft. And thus, Gentlemen, I haue fhewed you all the varieties that I know are vfed by any in our Countrey, that are worth the reciting (but as for the fafhion of Iawe-bones, vfed by fome in the Low-Countries, and other places beyond the Seas, being too groffe and bafe, I make no mention of them) among which euery one may take what pleafeth him beft, or may moft fitly be had, or may beft agree with the ground or knot. Moreouer, all thefe herbes that ferue for borderings, doe ferue as well to be fet vpon the ground of a leuelled knot; that is, where the allies and foot-pathes are of the fame leuell with the knot, as they may ferue alfo for the raifed knot, that is, where the beds of the knot are raifed higher than the allies : but both Leade, Boordes, Bones, and Tyles, are onely for the raifed ground, be it knot or beds. The pebble ftones againe are onely for the leuelled ground, becaufe they are fo fhallow, that as I faid before, they rather lye vpon the earth than are thruft any way into it. All this that I haue here fet downe, you muft vnderftand is proper for the knots alone of a Garden. But for to border the whole fquare or knot about, to ferue as a hedge thereunto, euery one taketh what liketh him beft; as either Priuet alone, or fweete Bryer, and white Thorne enterlaced together, and Rofes of one, or two, or more forts placed here and there amongft them. Some alfo take Lauander, Rofemary, Sage, Southernwood, Lauander Cotton, or fome fuch other thing. Some againe plant Cornell Trees, and plafh them, or keepe them lowe, to

forme

forme them into an hedge. And fome againe take a lowe prickly fhrubbe, that abideth alwayes greene, defcribed in the end of this Booke, called in Latine *Pyracantha*, which in time will make an euer greene hedge or border, and when it beareth fruit, which are red berries like vnto Hawthorne berries, make a glorious fhew among the greene leaues in the Winter time, when no other fhrubbes haue fruit or leaues.

Chap. IV.

The nature and names of diuers Out-landifh flowers, that for their pride, beauty, and earlineffe, are to be planted in Gardens of pleafure for delight.

Hauing thus formed out a Garden, and diuided it into his fit and due proportion, with all the gracefull knots, arbours, walkes, &c. likewife what is fit to keepe it in the fame comely order, is appointed vnto it, both for the borders of the fquares, and for the knots and beds themfelues; let vs now come and furnifh the inward parts, and beds with thofe fine flowers that (being ftrangers vnto vs, and giuing the beauty and brauery of their colours fo early before many of our owne bred flowers, the more to entice vs to their delight) are moft befeeming it: and namely, with Daffodils, Fritillarias, Iacinthes, Saffron-flowers, Lillies, Flowerdeluces, Tulipas, Anemones, French Cowflips, or Beares eares, and a number of fuch other flowers, very beautifull, delightfull, and pleafant, hereafter defcribed at full, whereof although many haue little fweete fent to commend them, yet their earlineffe and exceeding great beautie and varietie doth fo farre counteruaile that defect (and yet I muft tell you with all, that there is among the many forts of them fome, and that not a few, that doe excell in fweetneffe, being fo ftrong and heady, that they rather offend by too much than by too little fent, and fome againe are of fo milde and moderate temper, that they fcarce come fhort of your moft delicate and dantieft flowers) that they are almoft in all places with all perfons, efpecially with the better fort of the Gentry of the Land, as greatly defired and accepted as any other the moft choifeft, and the rather, for that the moft part of thefe Out-landifh flowers, do fhew forth their beauty and colours fo early in the yeare, that they feeme to make a Garden of delight euen in the Winter time, and doe fo giue their flowers one after another, that all their brauery is not fully fpent, vntil that Gilliflowers, the pride of our Englifh Gardens, do fhew themfelues: So that whofoeuer would haue of euery fort of thefe flowers, may haue for euery moneth feuerall colours and varieties, euen from Chriftmas vntill Midfommer, or after; and then, after fome little refpite, vntill Chriftmas againe, and that in fome plenty, with great content and without forcing; fo that euery man may haue them in euery place, if they will take any care of them. And becaufe there bee many Gentlewomen and others, that would gladly haue fome fine flowers to furnifh their Gardens, but know not what the names of thofe things are that they defire, nor what are the times of their flowring, nor the fkill and knowledge of their right ordering, planting, difplanting, tranfplanting, and replanting; I haue here for their fakes fet downe the nature, names, times, and manner of ordering in a briefe manner, referring the more ample declaration of them to the worke following. And firft of their names and natures: Of Daffodils there are almoft an hundred forts, as they are feuerally defcribed hereafter, euery one to be diftinguifhed from other, both in their times, formes, and colours, fome being eyther white, or yellow, or mixt, or elfe being fmall or great, fingle or double, and fome hauing but one flower vpon a ftalke, others many, whereof many are fo exceeding fweete, that a very few are fufficient to perfume a whole chamber, and befides, many of them be fo faire and double, eyther one vpon a ftalke, or many vpon a ftalke, that one or two ftalkes of flowers are in ftead of a whole nofe-gay, or bundell of flowers tyed together. This I doe affirme vpon good knowledge and certaine experience, and not as a great many others doe, tell of the wonders of another world, which themfelues neuer faw nor euer heard of, except fome fuperficiall relation, which themfelues haue augmented according to their owne fanfie and conceit. Againe, let me here alfo by the way tell you, that many idle and ignorant Gardiners and others, who get names by ftealth, as they doe many other things, doe call

some

some of these Daffodils Narcisses, when as all know that know any Latine, that Narcissus is the Latine name, and Daffodill the English of one and the same thing; and therefore alone without any other Epithite cannot properly distinguish seuerall things. I would willingly therefore that all would grow iudicious, and call euery thing by his proper English name in speaking English, or else by such Latine name as euery thing hath that hath not a proper English name, that thereby they may distinguish the seuerall varieties of things and not confound them, as also to take away all excuses of mistaking; as for example: The single English bastard Daffodill (which groweth wilde in many Woods, Groues, and Orchards in England.) The double English bastard Daffodill. The French single white Daffodill many vpon a stalke. The French double yellow Daffodill. The great, or the little, or the least Spanish yellow bastard Daffodill, or the great or little Spanish white Daffodill. The Turkie single white Daffodill, or, The Turkie single or double white Daffodill many vpon a stalke, &c. Of Fritillaria, or the checkerd Daffodill, there are halfe a score seuerall sorts, both white and red, both yellow and blacke, which are a wonderfull grace and ornament to a Garden in regard of the Checker like spots are in the flowers. Of Iacinthes there are aboue halfe an hundred sorts, as they are specified hereafter; some like vnto little bells or starres, others like vnto little bottles or pearles, both white and blew, sky-coloured and blush, and some starlike of many pretty various formes, and all to giue delight to them that will be curious to obserue them. Of Crocus or Saffron flowers, there are also twenty sorts; some of the Spring time, others flowring onely in the Autume or Fall, earlier or later than another, some whereof abide but a while; others indure aboue a moneth in their glorious beauty. The Colchicum or Medowe Saffron, which some call the sonne before the father, but not properly, is of many sorts also; some flowring in the Spring of the yeare, but the most in Autume, whereof some haue faire double flowers very delightfull to behold, and some party coloured both single and double so variable, that it would make any one admire the worke of the Creatour in the various spots and stripes of these flowers. Then haue wee of Lillies twenty seuerall sorts and colours, among whom I must reckon the Crowne Imperiall, that for his stately forme deserueth some speciall place in this Garden, as also the Martagons, both white and red, both blush and yellow, that require to be set by themselues apart, as it were in a small round or square of a knot, without many other, or tall flowers growing neare them. But to tell you of all the sorts of Tulipas (which are the pride of delight) they are so many, and as I may say, almost infinite, doth both passe my ability, and as I beleeue the skill of any other. They are of two especiall sorts, some flowring earlier, and others later than their fellowes, and that naturally in all grounds, wherein there is such a wonderfull variety and mixture of colours, that it is almost impossible for the wit of man to descipher them thoroughly, and to giue names that may be true & seuerall distinctions to euery flower, threescore seuerall sorts of colours simple and mixed of each kind I can reckon vp that I haue, and of especiall note, and yet I doubt not, but for euery one of them there are ten others differing from them, which may be seen at seuerall times, and in seuerall places: & besides this glory of variety in colors that these flowers haue, they carry so stately & delightfull a forme, & do abide so long in their brauery (enduring aboue three whole moneths from the first vnto the last) that there is no Lady or Gentlewoman of any worth that is not caught with this delight, or not delighted with these flowers. The Anemones likewise or Windeflowers are so full of variety and so dainty, so pleasant and so delightsome flowers, that the sight of them doth enforce an earnest longing desire in the minde of any one to be a possessour of some of them at the least: For without all doubt, this one kinde of flower, so variable in colours, so differing in forme (being almost as many sorts of them double as single) so plentifull in bearing flowers, and so durable in lasting, and also so easie both to preserue and to encrease, is of it selfe alone almost sufficient to furnish a garden with their flowers for almost halfe the yeare, as I shall shew you in a fit and conuenient place. The Beares eares or French Cowslips must not want their deserued commendations, seeing that their flowers, being many set together vpon a stalke, doe seeme euery one of them to bee a Nosegay alone of it selfe: and besides the many differing colours that are to be seene in them, as white, yellow, blush, purple, red, tawney, murrey, haire colour, &c. which encrease much delight in all sorts of the Gentry of the Land, they are not vnfurnished with a pretty sweete sent,

which

which doth adde an encrease of pleasure in those that make them an ornament for their wearing. Flowerdeluces also are of many sorts, but diuided into two especiall kindes; the one bearing a leafe like a flagge, whose rootes are tuberous, thicke and short (one kinde of them being the Orris rootes that are sold at the Apothecaries, whereof sweete powders are made to lye among garments) the other hauing round rootes like vnto Onions, and narrow long leaues somewhat like grasse : Of both these kindes there is much variety, especially in their colours. The greater Flagge kinde is frequent enough and disperfed in this Land, and well doth serue to decke vp both a Garden and House with natures beauties : But the chiefe of all is your Sable flower, so fit for a mourning habit, that I thinke in the whole compasse of natures store, there is not a more patheticall, or of greater correspondency, nor yet among all the flowers I know any one comming neare vnto the colour of it. The other kinde which hath bulbous or Onion like rootes, diuersifieth it selfe also into so many fine colours, being of a more neate shape and succinct forme than the former, that it must not bee wanting to furnish this Garden. The Hepatica or Noble Liuerwoort is another flower of account, whereof some are white, others red, or blew, or purple, somewhat resembling Violets, but that there are white threads in the middest of their flowers, which adde the more grace vnto them; and one kinde of them is so double, that it resembleth a double thicke Dasie or Marigold, but being small and of an excellent blew colour, is like vnto a Button : but that which commendeth the flower as much as the beauty, is the earlinesse in flowring, for that it is one of the very first flowers that open themselues after Christmas, euen in the midst of Winter. The Cyclamen or Sowebread is a flower of rare receipt, because it is naturally hard to encrease, and that the flowers are like vnto red or blush coloured Violets, flowring in the end of Summer or beginning of Autumne : the leaues likewise hereof haue no small delight in their pleasant colour, being spotted and circled white vpon greene, and that which most preferreth it, is the Physicall properties thereof for women, which I will declare when I shall shew you the seuerall descriptions of the varieties in his proper place. Many other sorts of flowers there are fit to furnish this Garden, as Leucoium or Bulbous Violet, both early and late flowring. Muscari or Muske Grape flower. Starre flowers of diuers sorts. Phalangium or Spiderwort, the chiefe of many is that sort whose flowers are like vnto a white Lilly. Winter Crowfoote or Wolfes bane. The Christmas flower like vnto a single white Rose. Bell flowers of many kindes. Yellow Larkes spurre, the prettiest flower of a score in a Garden. Flower-gentle or Floramour. Flower of the Sunne. The Maruaile of Peru or of the world. Double Marsh Marigold or double yellow Buttons, much differing and farre exceeding your double yellow Crowfoote, which some call Batchelours Buttons. Double French Marigolds that smell well, and is a greater kinde than the ordinary, and farre surpasseth it. The double red Ranunculus or Crowfoote (farre excelling the most glorious double Anemone) and is like vnto our great yellow double Crowfoote. Thus hauing giuen you the knowledge of some of the choisest flowers for the beds of this Garden, let me also shew you what are fittest for your borders, and for your arbours. The Iasmine white and yellow. The double Honysockle. The Ladies Bower, both white, and red, and purple single and double, are the fittest of Outlandish plants to set by arbours and banqueting houses, that are open, both before and aboue to helpe to couer them, and to giue both sight, smell, and delight. The sorts of Roses are fittest for standards in the hedges or borders. The Cherry Bay or Laurocerasus. The Rose Bay or Oleander. The white and the blew Syringa or Pipe tree, are all gracefull and delightfull to set at seuerall distances in the borders of knots; for some of them giue beautifull and sweete flowers. The Pyracantha or Prickly Corall tree doth remaine with greene leaues all the yeare, and may be plashed, or laid downe, or tyed to make a fine hedge to border the whole knot, as is said before. The Wilde Bay or Laurus Tinus, doth chiefly desire to be sheltered vnder a wall, where it will best thriue, and giue you his beautifull flowers in Winter for your delight, in recompence of his fenced dwelling. The Dwarfe Bay or Mesereon, is most commonly either placed in the midst of a knot, or at the corners thereof, and sometimes all along a walke for the more grace. And thus to fit euery ones fancy, I haue shewed you the variety of natures store in some part for you to dispose of them to your best content.

CHAP.

CHAP. V.

The nature and names of those that are called vsually English flowers.

THofe flowers that haue beene vfually planted in former times in Gardens of this Kingdome (when as our forefathers knew few or none of thofe that are recited before) haue by time and cuftome attained the name of Englifh flowers, although the moft of them were neuer naturall of this our Land, but brought in from other Countries at one time or other, by thofe that tooke pleafure in them where they firft faw them : and I doubt not, but many other forts than here are fet downe, or now knowne to vs, haue beene brought, which either haue perifhed by their negligence or want of skill that brought them, or elfe becaufe they could not abide our cold Winters ; thofe onely remaining with vs that haue endured of themfelues, and by their encreafing haue beene diftributed ouer the whole Land. If I fhould make any large difcourfe of them, being fo well knowne to all, I doubt I fhould make a long tale to fmall purpofe : I will therefore but briefly recite them, that you may haue them together in one place, with fome little declaration of the nature and quality of them, and fo paffe to other matters. And firft of Primrofes and Cowflips, whereof there are many prettie varieties ; fome better knowne in the Weft parts of this Kingdome, others in the North, than in any other, vntill of late being obferued by fome curious louers of varieties, they haue been tranfplanted diuerfly, and fo made more common : for although we haue had formerly in thefe parts about London greene Primrofes vfually, yet we neuer faw or heard of greene Cowflips both fingle and double but of late dayes, and fo likewife for Primrofes to be both fingle and double from one roote, and diuers vpon one ftalke of diuers fafhions, I am fure is not vfuall : all which defire rather to bee planted vnder fome hedge, or fence, or in the fhade, than in the Sunne. Single Rofe Campions, both white, red, and blufh, and the double red Rofe Campion alfo is knowne fufficiently, and will abide moderate Sunne as well as the fhade. The flower of Briftow or None-fuch is likewife another kinde of Campion, whereof there is both white flowring plants and blufh as well as Orange colour, all of them being fingle flowers require a moderate Sunne and not the fhadow : But the Orange colour Nonefuch with double flowers, as it is rare and not common, fo for his brauery doth well deferue a Mafter of account that will take care to keepe and preferue it. Batchelours Buttons both white and red, are kindes of wilde Campions of a very double forme, and will reafonably well like the Sunne but not the fhade. Wall flowers are common in euery Garden, as well the ordinary double as the fingle, and the double kinde defireth no more fhade than the fingle, but the greater kindes both double and fingle muft haue the Sunne. Stock-Gilloflowers likewife are almoft as common as Wall-flowers, efpecially the fingle kindes in euery womans Garden, but the double kindes are much more rare, and poffeffed but of a few, and thofe onely that will bee carefull to preferue them in Winter ; for befides that the moft of them are more tender, they yeeld no feede as the fingle kindes doe to preferue them, although one kinde from the fowing of the feed yeeld double flowers : They will all require the comfort of the Sunne, efpecially the double kindes, and to be defended from cold, yet fo as in the Summer they doe not want water wherein they much ioy, and which is as it were their life. Queenes Gilloflowers (which fome call Dames Violets, and fome Winter Gilloflowers, are a kinde of Stock-Gilloflower) planted in Gardens to ferue to fill vp the parts thereof for want of better things, hauing in mine opinion neither fight nor fent much to commend them. Violets are the Springs chiefe flowers for beauty, fmell, and vfe, both fingle and double, the more fhadie and moift they ftand the better. Snapdragon are flowers of much more delight, and in that they are more tender to keep, and will hardly endure the fharpe Winters, vnleffe they ftand well defended, are fcarce feene in many Gardens. Columbines fingle and double, of many forts, fafhions, and colours, very variable both fpeckled and party coloured, are flowers of that refpect, as that no Garden would willingly bee without them, that could tell how to haue them, yet the rarer the flowers are, the more trouble to keepe ; the ordinary forts on the con-

trary

trary part will not be loft, doe what one will. Larkes heeles, or fpurres, or toes, as in feuerall Countries they are called, exceed in the varietie of colours, both fingle and double, any of the former times; for vntill of late dayes none of the moft pleafant colours were feene or heard of: but now the fingle kindes are reafonable well difperft ouer the Land, yet the double kindes of all thofe pleafant colours (and fome other alfo as beautifull) which ftand like little double Rofes, are enioyed but of a few: all of them rife from feed, and muft be fowne euery yeare, the double as well as the fingle. Panfyes or Hartes eafes of diuers colours, and although without fent, yet not without fome refpect and delight. Double Poppies are flowers of a great and goodly proportion, adorning a Garden with their variable colours to the delight of the beholders, wherein there is fome fpeciall care to be taken, left they turne fingle; and that is, if you fee them grow vp too thicke, that you muft pull them vp, and not fuffer them to grow within leffe than halfe a yard diftance, or more one from another. Double Daifies are flowers not to be forgotten, although they be common enough in euery Garden, being both white and red, both blufh and fpeckled, or party coloured, befides that which is called Iacke an Apes on horfebacke, they require a moift and fhadowie place; for they are fcorched away, if they ftand in the Sunne in any dry place. Double Marigolds alfo are the moft common in all Gardens. And fo are the French Marigolds that haue a ftrong heady fent, both fingle and double, whofe glorious fhew for colour would caufe any to beleeue there were fome rare goodneffe or vertue in them. Thefe all are fometimes preferued in the Winter, if they bee well defended from the cold. But what fhall I fay to the Queene of delight and of flowers, Carnations and Gilloflowers, whofe brauery, variety, and fweete fmell ioyned together, tyeth euery ones affection with great earneftneffe, both to like and to haue them? Thofe that were knowne, and enioyed in former times with much acceptation, are now for the moft part leffe accounted of, except a very few: for now there are fo many other varieties of later inuention, that troubleth the other both in number, beauty, and worth: The names of them doe differ very variably, in that names are impofed and altered as euerie ones fancy will haue them, that carryed or fent them into the feuerall Countries from London, where their trueft name is to be had, in mine opinion. I will here but giue you the names of fome, and referre you to the worke enfuing for your further knowledge. The red and the gray Hulo. The old Carnation, differing from them both. The Gran Pere. The Camberfiue. The Sauadge. The Chriftall. The Prince. The white Carnation, or Delicate. The ground Carnation. The French Carnation. The Douer. The Oxford. The Briftow. The Weftminfter. The Daintie. The Granado, and many other Gilloflowers too tedious to recite in this place, becaufe I haue amply declared them in the booke following. But there is another fort of great delight and varietie, called the Orange tawny Gilloflower, which for the moft part hath rifen from feed, and doth giue feed in a more plentifull manner than any of the former forts, and likewife by the fowing of the feed there hath been gained fo many varieties of that excellent worth and refpect, that it can hardly be expreffed or beleeued, and called by diuers names according to the marking of the flowers; as The Infanta. The Stript Tawny. The Speckled Tawny. The Flackt Tawny. The Grifeld Tawny, and many others, euery one to bee diftinguifhed from others: Some alfo haue their flowers more double and large than others, and fome from the fame feed haue fingle flowers like broad fingle Pinkes: the further relation of them, *viz.* their order to fowe, encreafe, and preferue them, you fhall haue in the fubfequent difcourfe in a place by it felfe. Pinkes likewife both fingle and double are of much variety, all of them very fweete, comming neare the Gilloflowers. Sweete Williams and Sweete Iohns, both fingle and double, both white, red, and fpotted, as they are kindes of wilde Pinkes, fo for their grace and beauty helpe to furnifh a Garden, yet defire not to ftand fo open to the Sunne as the former. Double and fingle Peonies are fit flowers to furnifh a Garden, and by reafon of their durability, giue out frefh pleafure euery yeare without any further trouble of fowing. And laftly, Hollihocks both fingle and double, of many and fundry colours, yeeld out their flowers like Rofes on their tall branches, like Trees, to fute you with flowers, when almoft you haue no other to grace out your Garden: the fingle and double doe both yeeld feed, and yet doe after their feeding abide many yeares. Thus haue I fhewed you moft of the Englifh, as well as (I did before) the Outlandifh

landifh flowers, that are fit to furnifh the knots, trailes, beds, and borders of this Garden. Rofes onely, as I faid before, I referue to circle or encompaffe all the reft, becaufe that for the moft part they are planted in the outer borders of the quarters, and fometimes by themfelues in the middle of long beds, the forts or kindes whereof are many, as they are declared in their proper place : but the White Rofe, the Red, and the Damaske, are the moft ancient Standards in England, and therefore accounted naturall.

Chap. VI.

The order and manner to plant and replant all the forts of Out-landifh flowers fpoken of before, as well thofe with bulbous rootes, as others with ftringie rootes.

WHereas it is the vfuall cuftome of moft in this Land, to turne vp their Gardens, and to plant them againe in the Spring of the yeare, which is the beft time that may bee chofen for all Englifh flowers, yet it is not fo for your Out-landifh flowers. And herein indeede hath beene not onely the errour of a great many to hinder their rootes from bearing out their flowers as they fhould, but alfo to hinder many to take delight in them, becaufe as they fay they will not thriue and profper with them, when as the whole fault is in the want of knowledge of the fit and conuenient time wherein they fhould bee planted. And beeaufe our Englifh Gardiners are all or the moft of them vtterly ignorant in the ordering of thefe Out-landifh flowers, as not being trained vp to know them, I haue here taken vpon mee the forme of a new Gardiner, to giue inftructions to thofe that will take pleafure in them, that they may be the better enabled with thefe helpes I fhall fhew them, both to know how they fhould be ordered, and to direct their Gardiners that are ignorant thereof, rightly to difpofe them according to their naturall qualities. And I doe wifh all Gentlemen and Gentlewomen, whom it may concerne for their owne good, to bee as carefull whom they truft with the planting and replanting of thefe fine flowers, as they would be with fo many Iewels ; for the rootes of many of them being fmall, and of great value, may be foone conueyed away, and a cleanly tale faire told, that fuch a roote is rotten, or perifhed in the ground if none be feene where it fhould be, or elfe that the flower hath changed his colour, when it hath been taken away, or a counterfeit one hath beene put in the place thereof ; and thus many haue been deceiued of their daintieft flowers, without remedy or true knowledge of the defect. You fhall therefore, if you will take the right courfe that is proper for thefe kindes of flowers, not fet or plant them among your Englifh flowers ; for that when the one may be remoued, the other may not be ftirred : but plant thofe rootes that are bulbous, or round like Onions, eyther in knots or beds by themfelues which is the beft, or with but very few Englifh or Out-landifh flower plants that haue ftringie rootes : For you muft take this for a generall rule, that all thofe rootes that are like Lillies or Onions, are to bee planted in the moneths of Iuly or Auguft, or vnto the middle or end of September at the furtheft, if you will haue them to profper as they fhould, and not in the Spring of the yeare, when other gardening is vfed. Yet I muft likewife giue you to vnderftand, that if Tulipas, and Daffodils, and fome other that are firme and hard rootes, and not limber or fpongie, being taken vp out of the ground in their fit feafon, that is, in Iune, Iuly, and Auguft, and likewife kept well and dry, may bee referued out of the ground vntill Chriftmas or after, and then (if they could not be fet fooner) being fet, will thriue reafonable well, but not altogether fo well as the former, being fet long before : but if you fhall remoue thefe bulbous rootes againe, either prefently after their planting hauing fhot their fmall fibres vnder the round rootes, and fprung likewife vpwards, or before they be in flower at the fooneft (yet Tulipas, Daffodils, and many other bulbous, may be fafely remoued being in flower, and tranfplanted into other places, fo as they be not kept too long out of the ground) you fhall much endanger them either vtterly to perifh, or to be hindered from bearing out their flowers they then would haue

B   borne,

borne, and for two or three years after from bearing flowers againe. For the order of their planting there are diuers wayes, some whereof I will shew you in this place: Your knot or beds being prepared fitly, as before is declared, you may place and order your rootes therein thus, Eyther many rootes of one kind set together in a round or cluster, or longwise crosse a bed one by another, whereby the beauty of many flowers of one kinde being together, may make a faire shew well pleasing to many ; Or else you may plant one or two in a place dispersedly ouer the whole knot, or in a proportion or diameter one place answering another of the knot, as your store will suffer you, or your knot permit : Or you may also mingle these rootes in their planting many of diuers sorts together, that they may giue the more glorious shew when they are in flower; and that you may so doe, you must first obserue the seuerall kindes of them, which doe flower at one and the same time, and then to place them in such order and so neare one vnto another, that their flowers appearing together of seuerall colours, will cause the more admiration in the beholders : as thus, The Vernall Crocus or Saffron flowers of the Spring, white, purple, yellow, and stript, with some Vernall Colchicum or Medow Saffron among them, some Deus Caninus or Dogges teeth, and some of the small early Leucoium or Bulbous Violet, all planted in some proportion as neare one vnto another as is fit for them, will giue such a grace to the Garden, that the place will seeme like a peece of tapestry of many glorious colours, to encrease euery ones delight : Or else many of one sort together, as the blew, white and blush Grape flowers in the same manner intermingled, doe make a maruellous delectable shew, especially because all of them rise almost to an equall height, which causeth the greater grace, as well neare hand as farre of. The like order may be kept with many other things, as the Hepatica, white, blew, purple, and red set or sowne together, will make many to beleeue that one roote doth beare all those colours : But aboue and beyond all others, the Tulipas may be so matched, one colour answering and setting of another, that the place where they stand may resemble a peece of curious needle-worke, or peece of painting : and I haue knowne in a Garden, the Master as much commended for this artificiall forme in placing the colours of Tulipas, as for the goodnesse of his flowers, or any other thing. The diuers sorts and colours of Anemones or Winde-flowers may be so ordered likewise, which are very beautifull, to haue the seuerall varieties planted one neare vnto another, that their seuerall colours appearing in one place will be a very great grace in a Garden, or if they be dispersed among the other sorts of flowers, they will make a glorious shew. Another order in planting you may obserue ; which is this, That those plants that grow low, as the Aconitum Hyemale or Winter-wolues bane, the Vernall Crocus or Saffron flowers of diuers sorts, the little early Leucoium or Bulbous Violet, and some such other as rise not vp high, as also some Anemones may be very well placed somewhat neare or about your Martagons, Lillies, or Crownes Imperiall, both because these little plants will flower earlier than they, and so will bee gone and past, before the other greater plants will rise vp to any height to hinder them ; which is a way may well be admitted in those Gardens that are small, to saue roome, and to place things to the most aduantage. Thus hauing shewed you diuers wayes and orders how to plant your rootes, that your flowers may giue the greater grace in the Garden, let mee shew you likewise how to set these kindes of rootes into the ground ; for many know not well eyther which end to set vpwards or downewards, nor yet to what depth they should be placed in the ground. Daffodils if they be great rootes, will require (as must bee obserued in all other great plants) to bee planted somewhat deeper then the smaller of the same kinde, as also that the tops or heads of the rootes be about two or three fingers breadth hid vnder ground. The Tulipas likewise if you set them deepe, they will be the safer from frosts if your ground be cold, which will also cause them to be a little later before they be in flower, yet vsually if the mould be good, they are to be set a good hand breadth deep within the ground, so that there may be three or foure inches of earth at the least aboue the head, which is the smaller end of the roote: for if they shall lye too neare the vpper face or crust of the earth, the colds & frosts will pierce and pinch them the sooner. After the same order and manner must Hyacinthes, whether great or small, and other such great rootes be planted. Your greater rootes, as Martagons, Lillies, and Crownes Imperiall, must be set much deeper then any other bulbous roote, because they are greater rootes then others, and by themselues also, as is moſt vſuall either in ſome ſquare, round, triangle, or other ſmall part in the Garden, becauſe they ſpread and take vp a very great deale of ground. All of them likewiſe are to be ſet with the broad end of the roote downewards, and the ſmall end vpwards, that is, both Lillies, Daffodils, Hyacinthes, and Tulipas, and all other ſorts of round rootes, which ſhew one end to bee ſmaller than another. But the Colchicum or Medow-Saffron onely requireth an exception to this generall rule, in regard the roote thereof hath a ſmall eminence or part on the one ſide thereof, which muſt bee ſet or planted downeward, and not vpward; for you ſhall obſerue, if the roote lye a little moiſt out of the ground, that it will ſhoote fibres out at the ſmall long end thereof, although you may perceiue when you take it vp, that the fibres were at the other broad end or ſide of the roote. As for the Crowne Imperiall, which is a broad round roote and flat withall, hauing a hole in the middle, for the moſt part quite thorow, when it is taken vp in his due time out of the ground, you ſhall perceiue the ſcales or cloues of the rootes to bee a little open on the vpperſide, and cloſe and flat on the vnderſide, which will direct you which part to ſet vpward, as alſo that the hole is bigger aboue then it is below. The Perſian Lilly is almoſt like vnto the Crowne Imperiall, but that the roote thereof is not ſo flat, and that it hath a ſmaller head at the one part, whereby it may be diſcerned the plainer how to be ſet. The Fritillaria is a ſmall white root diuided as it were into two parts, ſo that many haue doubted, as formerly in the Crowne Imperiall, what part to ſet vppermoſt; you ſhall therefore marke, that the two parts of the roote are ioyned together at the bottome, where it ſhooteth out fibres or ſmall ſtringie rootes, as all other ſorts of bulbous rootes doe, and withall you ſhall ſee, that betweene the two parts of the roote a ſmall head will appeare, which is the burgeon that will ſpring vp to beare leaues and flowers. In the rootes of Anemones there are ſmall round ſwelling heads, eaſie enough to be obſerued if you marke it, which muſt be ſet vpwards. All other ſorts of ſtringie rooted plants (and not bulbous or tuberous rooted) that loſe their greene leaues in Winter, will ſhew a head from whence the leaues and flowers will ſpring, and all others that keepe their greene leaues, are to bee planted in the ſame manner that other herbes and flower-plants are accuſtomed to be. But yet for the better thriuing of the ſtringie rooted plants, when you will plant them, let me informe you of the beſt way of planting, and the moſt ſure to cauſe any plant to comprehend in the ground without failing, and is no common way with any Gardiner in this Kingdome, that euer I heard or knew, which is thus : Preſuming that the ſtringie rooted plant is freſh and not old gathered, and a plant that being remoued will grow againe, make a hole in the ground large enough where you meane to ſet this roote, and raiſe the earth within the hole a little higher in the middle then on the ſides, and ſet the roote thereon, ſpreading the ſtrings all abroad about the middle, that they may as it were couer the middle, and then put the earth gently round about it, preſſing it a little cloſe, and afterwards water it well, if it be in Summer, or in a dry time, or otherwiſe moderately : thus ſhall euery ſeuerall ſtring of the roote haue earth enough to cauſe it to ſhoote forth, and thereby to encreaſe farre better than by the vſuall way, which is without any great care and reſpect to thruſt the rootes together into the ground. Diuers other flower plants are but annuall, to bee new ſowne euery yeare ; as the Maruaile of the world, the Indian Creſſes, or yellow Larkes heeles, the Flower of the Sunne, and diuers other : they therefore that will take pleaſure in them, that they may enioy their flowers the earlier in the yeare, and thereby haue ripe ſeede of them while warme weather laſteth, muſt nurſe vp their ſeedes in a bed of hot dung, as Melons and Cowcumbers are, but your bed muſt be prouided earlier for theſe ſeeds, than for Melons, &c. that they may haue the more comfort of the Summer, which are to be carefully tended after they are tranſplanted from the hot bed, and couered with ſtraw from colds, whereby you ſhall not faile to gaine ripe ſeed euery yeare, which otherwiſe if you ſhould miſſe of a very kindly & hot Summer, you ſhould neuer haue. Some of theſe ſeedes neede likewiſe to be tranſplanted from the bed of dung vnder a warme wall, as the Flower of the Sunne, and the Maruaile of the world, and ſome others, and that for a while after their tranſplanting, as alſo in the heate of Summer, you water them at the roote with water that hath ſtood a day or two in the Sunne, hauing firſt laid a round wiſpe of hay or ſuch other thing round about the roote, that ſo all helpes may further their giuing of ripe ſeede. One or two rules more I will giue you concerning

theſe

thefe dainty flowers, the firft whereof is this, That you fhall not bee carefull to water any of your bulbous or tuberous rooted plants at any time; for they all of them do better profper in a dry ground than in a wet, onely all forts of tuberous rooted Flowerdeluces vpon their remouall had neede of a little water, and fome will doe fo alfo to fuch Tulipas and other bulbous rootes as they tranfplant, when they are in flower, and this is I grant in fome fort tolerable, if it bee not too much, and done onely to caufe the ftalke and flower to abide fometime the longer before they wither, but elfe in no other cafe to be permitted. The fecond rule is, That I would aduife you to water none of your dainty flowers or herbes, with any water that hath prefently before been drawne out of a well or pumpe, but onely with fuch water that hath ftood open in the Sunne in fome cifterne, tubbe, or pot for a day at the leaft, if more the better: for that water which is prefently drawne out of a well, &c. is fo cold, that it prefently chilleth & killeth any dainty plant be it younger or elder grown, wherof I haue had fufficient proofe: and therfore I giue you this caution by mine own experience. Thus haue I directed you from point to point, in all the particulars of preparing & planting that belong to this Garden, fauing only that yet I would further enforme you, of the time of the flowring of thefe Out-landifh plants, according to the feuerall monethe in the yeare, that euery one may know what flowers euery moneth yeeldeth, and may chufe what them liketh beft, in that they may fee that there is no moneth, but glorieth in fome peculiar forts of rare flowers. I would likewife rather in this place fhew you, the true and beft manner & order to encreafe and preferue all forts of Gilloflowers & Carnations, then ioyne it with the Chapter of Gilloflowers in the worke following, becaufe it would in that place take vp too much roome. And laftly, I muft of neceffity oppofe three fundry errours, that haue poffeffed the mindes of many both in former and later times, which are, that any flower may be made to grow double by art, that was but fingle before by nature: And that one may by art caufe any flower to grow of what colour they will: And that any plants may be forced to flower out of their due feafons, either earlier or later, by an art which fome can vfe. All which being declared, I then fuppofe enough is fpoken for an introduction to this worke, referring many other things to the feuerall directions in the Chapters of the booke.

Chap. VII.

The feuerall times of the flowring of thefe Out-landifh flowers, according to the feuerall monethe of the yeare.

I Intend in this place onely to giue you briefly, the names of fome of the chiefeft of thefe Out-landifh flowers, according to the feuerall monethe of the yeare wherein they flower, that euery one feeing what forts of flowers euery moneth yeeldeth, may take of them which they like beft. I begin with Ianuary, as the firft moneth of the yeare, wherein if the frofts be not extreme, you fhall haue thefe flowers of plants; the Chriftmas flower or Helleborus niger verus, Winter wolues bane or Aconitum hyemale, Hepatica or Noble Liuer wort blew and red, and of fhrubbes, the Laurus Tinus or Wilde Bay tree, and Mefereon or the dwarfe Bay: but becaufe Ianuarie is oftentimes too deepe in frofts and fnow, I therefore referre the Hepaticas vnto the moneth following, which is February, wherein the weather beginneth to be a little milder, and then they will flower much better, as alfo diuers forts of Crocus or Saffron flower will appeare, the little early Summer foole or Leucoium bulbofum, and towards the latter end thereof the Vernall Colchicum, the Dogges tooth Violet or Deus Caminus, and fome Anemones, both fingle and double, which in fome places will flower all the Winter long. March will yeeld more varieties; for befides that it holdeth fome of the flowers of the former moneth, it will yeeld you both the double blew Hepatica, and the white and the blufh fingle: then alfo you fhall haue diuers other forts of Crocus or Saffron flowers, Double yellow Daffodils, Orientall Iacinths and others, the Crowne Imperiall, diuers forts of early Tulipas, fome forts of French Cowflips, both tawney, murry, yellow, and blufh, the early Fritillaria or checkerd Daffodill,

dill, and some other sorts of early Daffodils, and many sorts of Anemones. In Aprill commeth on the pride of these strangers; for herein you may behold all the sorts of Auricula Vrsi or Beares Eares, many sorts of Anemones, both single and double, both the sorts of Tulipas, the earlier vntill the middle of the moneth, and the later then beginning; which are of so many different colours, that it is almost impossible to expresse them, the white, red, blacke, and yellow Fritillarias, the Muscari or Muske Grape flower, both ash colour and yellow. Diuers other sorts of Iacinths and Daffodils, both single and double, the smaller sorts of Flowerdeluces, the Veluet Flowerdeluce and double Honysuckles, with diuers others. May likewise at the beginning seemeth as glorious as Aprill, although toward the end it doth decline, in regard the heate of the Sunne hath by this time drawne forth all the store of natures tenderest dainties, which are vsually spent by the end of this moneth, and then those of stronger constitution come forward. Herein are to bee seene at the beginning the middle flowring Tulipas, and at the end the later sort: some kindes of Daffodils, the Day Lillies, the great white Starre flower, the Flowerdeluce of Constantinople or the mourning Sable flower, the other sorts of Flowerdeluces. Single and double white Crowfoote, and single and double red Crowfoot, the glory of a Garden: the early red Martagon, the Persian Lilly, the yellow Martagon, the Gladiolus or Corne flagge, both white, red, and blush: the double yellow Rose, and some other sorts of Roses. In Iune doe flower the white and the blush Martagon, the Martagon Imperiall, the mountaine Lillies, and the other sorts of white and red Lillies, the bulbous Flowerdeluces of diuers sorts, the red flowred Ladies bower, the single and double purple flowred Ladies bower, the white Syringa or Pipe tree, for the blew Pipe tree flowreth earlier, the white and the yellow Iasmin. Iuly holdeth in flower some of the Ladies bowers and Iasmines, and besides doth glory in the Female Balsame apple, the Indian Cresses or yellow Larkes spurres, the purple Flower-gentle and the Rose Bay. In August begin some of the Autumne bulbous flowers to appeare, as the white and the purple Colchicum or Medow Saffron, the purple mountaine Crocus or Saffron flower, the little Autumne Leucoium and Autumne Iacinth, the Italian Starrewort, called of some the purple Marigold, the Meruaile of Peru or of the world, the Flower of the Sunne, the great blew Bell-flower, the great double French Marigold. September flourisheth with the Flower of the Sunne, the Meruaile of the world, the purple Marigold, and blew Bell-flower spoken of before, and likewise the other sorts of Medow Saffron, and the double kinde likewise, the siluer Crocus, the Autumne yellow Daffodill, Cyclamen also or Sowbread shew their flowers in the end of this moneth. October also will shew the flowers of Cyclamen, and some of the Medow Saffrons. In Nouember, as also sometimes in the moneth before, the party coloured Medow Saffron may bee seene, that will longest hold his flower, because it is the latest that sheweth it selfe, and the ash coloured mountaine Crocus. And euen December it selfe will not want the true blacke Hellebor or Christmas flower, and the glorious shew of the Laurus Tinus or wilde Bay tree. Thus haue I shewed you some of the flowers for euery moneth, but I referre you to the more ample declarion of them and all the others, vnto the work following.

Chap. VIII.

The true manner and order to encrease and preserue all sorts of Gilloflowers, as well by slippes as seedes.

BEcause that Carnations and Gilloflowers bee the chiefest flowers of account in all our English Gardens, I haue thought good to entreate somewhat amply of them, and that a part by it selfe, as I said a little before, in regard there is so much to be said concerning them, and that if all the matters to be entreated of should haue beene inserted in the Chapter of Gilloflowers, it would haue made it too tedious and large, and taken vp too much roome. The particular matters whereof I mean in this place to entreate are these: How to encrease Gilloflowers by planting and by sowing, and how to preserue them being encreased, both in Summer from noysome and hurtfull vermine that destroy them, and in Winter from frosts, snowes, and windes, that spoile them. There are two wayes of planting, whereby to encrease these faire flowers; the one is by slipping, which is the old and ready vsuall way, best knowne in this Kingdome; the other is more sure, perfect, ready, and of later inuention, *videlicet*, by laying downe the branches. The way to encrease Gilloflowers by slipping, is so common with all that euer kept any of them, that I thinke most persons may thinke me idle, to spend time to set downe in writing that which is so well known vnto all: Yet giue me leaue to tell them that so might imagine, that (when they haue heard or read what I haue written thereof, if they did know fully as much before) what I here write, was not to informe them, but such as did not know the best, or so good a way as I teach them: For I am assured, the greatest number doe vse, and follow the most vsuall way, and that is not alwaies the best, especially when by good experience a better way is found, and may be learned; and therefore if some can doe a thing better than others, I thinke it is no shame to learne it of them. You shall not then (to take the surest course) take any long spindled branches, nor those branches that haue any young shootes from the ioynts on them, nor yet sliue or teare any slippe or branch from the roote; for all these waies are vsuall and common with most, which causeth so many good rootes to rot and perish, and also so many slippes to be lost, when as for the most part, not the one halfe, or with some, not a third part doth grow and thriue of those slippes they set. And although many that haue store of plants, doe not so much care what hauocke they make to gaine some, yet to saue both labour and plants, I doe wish them to obserue these orders: Take from those rootes from whence you intend to make your encrease, those shootes onely that are reasonable strong, but yet young, and not either too small and slender, or hauing any shootes from the ioynts vpon them; cut these slippes or shootes off from the stemme or roote with a knife, as conueniently as the shoote or branch will permit, that is, either close vnto the maine branch, if it be short, or leauing a ioynt or two behinde you, if it be long enough, at which it may shoote anew: When you haue cut off your slippes, you may either set them by and by, or else as the best Gardiners vse to doe, cast them into a tubbe or pot with water for a day or two, and then hauing prepared a place conuenient to set them in, which had neede to bee of the finest, richest, and best mould you can prouide, that they may thriue therein the better, cut off your slippe close at the ioynt, and hauing cut away the lowest leaues close to the stalke, and the vppermost euen at the top, with a little sticke make a little hole in the earth, and put your slippe therein so deep, as that the vpper leaues may be wholly aboue the ground, (some vse to cleaue the stalke in the middle, and put a little earth or clay within the cleft, but many good and skilfull Gardiners doe not vse it); put the earth a little close to the slippe with your finger and thumbe, and there let it rest, and in this manner doe with as many slippes as you haue, setting them somewhat close together, and not too farre in sunder, both to saue ground and cost thereon, in that a small compasse will serue for the first planting, and also the better to giue them shadow: For you must remember in any case, that these slippes new set, haue no sight of the Sunne, vntill they be well taken in the ground, and shot aboue ground, and also that they want not water, both vpon the new planting and after. When these slippes are well growne vp, they must be transplanted into such other places as you thinke meete; that is, either into the ground in beds, or otherwise, or into pots, which that you may the more safely doe, after you haue well watered the ground, for halfe a day before you intend to transplant them, you shall separate them seuerally, by putting down a broad pointed knife on each side of the slippe, so cutting it out, take euery one by it selfe, with the earth cleauing close vnto the root, which by reason of the moisture it had formerly, and that which you gaue presently before, will be sufficient with any care had, to cause it to hold fast vnto the roote for the transplanting of it: for if the earth were dry, and that it should fall away from the roote in the transplanting, it would hazzard and endanger the roote very much, if it did thriue at all. You must remember also, that vpon the remouing of these slips, you shadow them from the heate of the Sunne for a while with some straw or other thing, vntill they haue taken hold in their new place. Thus although it bee a little more labour and care than the ordinary way is, yet it is surer, and will giue you plants that

will

will be so ftrongly growne before Winter, that with the care hereafter fpecified, you shall haue them beare flowers the next yeare after, and yeeld you encreafe of slippes alfo. To giue you any fet time, wherein thefe slippes will take roote, and begin to shoote aboue ground, is very hard to doe; for that euery slip, or yet euery kinde of Gilloflower is not alike apt to grow; nor is euery earth in like manner fit to produce and bring forward the slippes that are fet therein : but if both the slippe be apt to grow, and the earth of the beft, fit to produce, I thinke within a fortnight or three weekes, you shall fee them begin to put forth young leaues in the middle, or elfe it may be a moneth and more before you shall fee any fpringing. The beft time likewife when to plant, is a fpeciall thing to be knowne, and of as great confequence as any thing elfe : For if you slippe and fet in September, as many vfe to doe, or yet in Auguft, as fome may thinke will doe well, yet (vnleffe they be the moft ordinary forts, which are likely to grow at any time, and in any place) the moft of them, if not all, will either affuredly perifh, or neuer profper well : for the more excellent and dainty the Gilloflower is, the more tender for the moft part, and hard to nurfe vp will the slippes be. The beft time therefore is, that you cut off fuch slippes as are likely, and fuch as your rootes may fpare, from the beginning of May vntill the middle of Iune at the furtheft, and order them as I haue shewed you before, that fo you may haue faire plants, plenty of flowers, and encreafe fufficient for new fupply, without offence or loffe of your ftore. For the enriching likewife of your earth, wherein you shall plant your slippes, that they may the better thriue and profper, diuers haue vfed diuers forts of manure; as ftable foyle of horfe, beafts or kine, of sheepe, and pigeons, all which are very good when they are thoroughly turned to mould, to mixe with your other earth, or being fteeped in water, may ferue to water the earth at times, and turned in with it. And fome haue likewife proued Tanners earth, that is, their barke, which after they haue vfed, doth lye on heapes and rot in their yards, or the like mould from wood-ftackes or yards; but efpecially, and beyond all other is commended the Willow earth, that is, that mould which is found in the hollow of old Willow trees, to be the moft principall to mixe with other good earth for this purpofe. And as I haue now giuen you directions for the firft way to encreafe them by slipping, fo before I come to the other way, let mee giue you a caueat or two for the preferuing of them, when they are beginning to runne vtterly to decay and perifh : The one is, that whereas many are ouer greedy to haue their plants to giue them flowers, and therefore let them runne all to flower, fo farre fpending themfelues thereby, that after they haue done flowring, they grow fo weake, hauing out fpent themfelues, that they cannot poffibly be preferued from the iniuries of the fucceeding Winter; you shall therefore keepe the kinde of any fort you are delighted withall, if you carefully looke that too many branches doe not runne vp and fpindle for flowers, but rather either cut fome of them downe, before they are run vp too high, within two or three ioynts of the rootes; or elfe plucke away the innermoft leaues where it fpringeth forwards, which you fee in the middle of euery branch, before it be runne vp too high, which will caufe them to breake out the fafter into slips and fuckers at the ioynts, to hinder their forward luxurie, and to preferue them the longer : The other is, If you shall perceiue any of your Gilloflower leaues to change their naturall frefh verdure, and turne yellowifh, or begin to wither in anie part or branch thereof, it is a fure figne that the roote is infected with fome cancker or rottenneffe, and will foone shew it felfe in all the reft of the branches, whereby the plant will quickly be loft : to preferue it therefore, you shall betime, before it be runne too farre, (for otherwife it is impoffible to faue it) either couer all or moft of the branches with frefh earth, or elfe take the faireft slippes from it, as many as you can poffibly, and caft them into a pot or tubbe with water, and let them there abide for two or three daies at the leaft : the firft way hath recouered many, being taken in time. Thus you shall fee them recouer their former ftiffeneffe and colour, and then you may plant them as you haue beene heretofore directed; and although many of them may perifh, yet shall you haue fome of them that will grow to continue the kinde againe. The other or fecond way to encreafe Gilloflowers by planting, is, as I faid before, by in-laying or laying downe the branches of them, and is a way of later inuention, and as frequently vfed, not onely for the tawney or yellow Gilloflower, and all the varieties therof, but with the other kinds of Gilloflowers, whereof experience hath shewed

that

that they will likewise take if they be so vsed, the manner whereof is thus : You must choose out the youngest, likeliest, and lowest branches that are nearest the ground (for the vpper branches will sooner breake at the ioynt, than bend downe so low into the earth, without some pot with earth raised vp vnto them) and cut it on the vnderside thereof vpwards at the second ioynt next vnto the roote, to the middle of the branch, and no more, and not quite thorough in any case, and then from that second ioynt vnto the third, slit or cut the branch in the middle longwise, that so it may be the more easily bended into the ground, the cut ioynt seeming like the end of a slippe, when you haue bended downe the branch where it is cut into the ground (which must bee done very gently for feare of breaking) with a little sticke or two thrust slopewise, crosse ouer it, keepe it downe within the earth, and raise vp sufficient earth ouer it, that there it may lye and take roote, which commonly will be effected within sixe weekes or two moneths in the Summer time, and then (or longer if you doubt the time too short for it to take sufficient roote) you may take or cut it away, and transplant it where you thinke good, yet so as in any case you shadow it from the heate of the Sunne, vntill it haue taken good hold in the ground. The other way to encrease Gilloflowers, is by sowing the seede : It is not vsuall with all sorts of Gilloflowers to giue seede, but such of them as doe yeeld seede may be encreased thereby, in the same manner as is here set downe. The Orange tawney Gilloflower and the varieties thereof is the most vsuall kinde, (and it is a kinde by it selfe, how various soeuer the plants be that rise from the seede) that doth giue seede, and is sowne, and from thence ariseth so many varieties of colours, both plaine and mixt, both single and double, that one can hardly set them downe in writing : yet such as I haue obserued and marked, you shall finde expressed in the Chapter of Gilloflowers in the worke following. First therefore make choise of your seede that you intend to sowe (if you doe not desire to haue as many more single flowers as double) that it bee taken from double flowers, and not from single, and from the best colours, howsoeuer some may boast to haue had double and stript flowers from the seede of a single one · which if it were so, yet one Swallow (as we say) maketh no Summer, nor a thing comming by chance cannot bee reckoned for a certaine and constant rule ; you may be assured they will not vsually doe so : but the best, fairest, and most double flowers come alwaies, or for the most part, from the seede of those flowers that were best, fairest, and most double ; and I doe aduise you to take the best and most double : for euen from them you shall haue single ones e-now, you neede not to sowe any worser sort. And againe, see that your seede bee new, of the last yeares gathering, and also that it was full ripe before it was gathered, lest you lose your labour, or misse of your purpose, which is, to haue faire and double flowers. Hauing now made choise of your seede, and prepared you a bedde to sowe them on, the earth whereof must be rich and good, and likewise sifted to make it the finer ; for the better it is, the better shall your profit and pleasure bee : hereon, being first made leuell, plaine, and smooth, sowe your seede somewhat thinne, and not too thicke in any case, and as euenly as you can, that they be not too many in one place, and too few in another, which afterwards couer with fine sifted earth ouer them about one fingers thicknesse ; let this be done in the middle of Aprill, if the time of the yeare be temperate, and not too cold, or else stay vntill the end of the moneth : after they are sprung vp and growne to be somewhat bigge, let them bee drawne forth that are too close and neare one vnto another, and plant them in such place where they shall continue, so that they stand halfe a yard of ground distance asunder, which after the planting, let be shadowed for a time, as is before specified ; and this may bee done in the end of Iuly, or sooner if there be cause. I haue not set downe in all this discourse of planting, transplanting, sowing, setting, &c. any mention of watering those slips or plants, not doubting but that euery ones reason will induce them to thinke, that they cannot prosper without watering : But let this Caueat be a sufficient remembrance vnto you, that you neuer water any of these Gilloflowers, nor yet indeede any other fine herbe or plant with cold water, such as you haue presently before drawne out from a pumpe or Well, &c. but with such water as hath stood open in the aire in a cisterne, tubbe, or pot, for one whole day at the least ; if it be two or three daies it will be neuer the worse, but rather the better, as I haue related before : yet take especiall heede that you doe not giue them too much to ouer-glut them at any time, but temperately to ir-

rorate

rorate, bedew or sprinkle them often. From the seedes of these Gilloflowers hath risen both white, red, blush, stamell, tawny lighter and sadder, marbled, speckled, striped, flaked, and that in diuers manners, both single and double flowers, as you shall
see them set downe in a more ample manner in the Chapter of Gilloflowers. And
thus much for their encrease by the two wayes of planting and sowing : For as for a
third way, by grafting one into or vpon another, I know none such to be true, nor to
be of any more worth than an old Wiues tale, both nature, reason, and experience, all
contesting against such an idle fancy, let men make what ostentation they please. It
now resteth, that we also shew you the manner how to preserue them, as well in Summer from all noysome and hurtfull things, as in the Winter and Spring from the sharp
and chilling colds, and the sharpe and bitter killing windes in March. The hurtfull
things in the Summer are especially these, too much heate of the Sunne which scorch-
eth them, which you must be carefull to preuent, by placing boughes, boords, clothes
or mats, &c. before them, if they bee in the ground; or else if they bee in pots, to
remoue into them into the shadow, to giue them refreshing from the heate, and giue
them water also for their life : too much water, or too little is another annoyance,
which you must order as you see there is iust cause, by withholding or giuing them water gently out of a watering pot, and not cast on by dishfuls : Some also to water their
Gilloflowers, vse to set their pots into tubbes or pots halfe full of water, that so the
water may soake in at the lower holes in each flower pot, to giue moisture to the roots
of the Gilloflowers onely, without casting any water vpon the leaues, and assuredly it
is an excellent way to moisten the rootes so sufficiently at one time, that it doth saue a
great deale of paines many other times. Earwickes are a most infestuous vermine, to
spoyle the whole beauty of your flowers, and that in one night or day; for these creatures delighting to creepe into any hollow or shadowie place, doe creepe into the
long greene pods of the Gilloflowers, and doe eate away the white bottomes of their
leaues, which are sweete, whereby the leaues of the flowers being loose, doe either
fall away of themselues before, or when they are gathered, or handled, or presently
wither within the pods before they are gathered, and blowne away with the winde.
To auoide which inconuenience, many haue deuised many waies and inuentions to
destroy them, as pots with double verges or brimmes, containing a hollow gutter betweene them, which being filled with water, will not suffer these small vermine to
passe ouer it to the Gilloflowers to spoile them. Others haue vsed old shooes, and such
like hollow things to bee set by them to take them in : but the best and most vsuall
things now vsed, are eyther long hollow canes, or else beasts hoofes, which being
turned downe vpon stickes ends set into the ground, or into the pots of earth, will
soone draw into them many Earwickes, lying hid therein from sunne, winde, and
raine, and by care and diligence may soone bee destroyed, if euery morning and euening one take the hoofes gently off from the stickes, and knocking them against the
ground in a plain allie, shake out all the Earwicks that are crept into them, which quickly with ones foot may be trode to peeces. For sodain blasting with thunder and lighte-
ning, or fierce sharpe windes, &c. I know no other remedy, vnlesse you can couer
them therefrom when you first foresee the danger, but patiently to abide the losse,
whatsoeuer some haue aduised, to lay litter about them to auoide blasting; for if any
shall make tryall thereof, I am in doubt, he shall more endanger his rootes thereby, being the Summer time, when any such feare of blasting is, than any wise saue them from
it, or doe them any good. For the Winter preseruation of them, some haue aduised to
couer them with Bee-hiues, or else with small Willow stickes, prickt crossewise into
the ground ouer your flowers, and bowed archwise, and with litter laid thereon,
to couer the Gilloflowers quite ouer, after they haue beene sprinkled with sope ashes
and lyme mixt together : and this way is commended by some that haue written thereof, to be such an admirable defence vnto them in Winter, that neither Ants, nor
Snailes, nor Earwickes shall touch them, because of the sope ashes and lyme, and neyther frosts nor stormes shall hurt them, because of the litter which so well will defend
them; and hereby also your Gilloflowers will bee ready to flower, not onely in the
Spring very early, but euen all the Winter. But whosoeuer shall follow these directions, may peraduenture finde them in some part true, as they are there set downe for
the Winter time, and while they are kept close and couered; but let them bee assured,
                                                                              that

that all such plants, or the most part of them, will certainely perish and dye before the Summer be at an end : for the sope ashes and lyme will burne vp and spoile any herbe; and againe, it is impossible for any plant that is kept so warme in Winter, to abide eyther the cold or the winde in the Spring following, or any heate of the Sun, but that both of them will scorch them, and carry them quite away. One great hurt vnto them, and to all other herbes that wee preserue in Winter, is to suffer the snow to lye vpon them any time after it is fallen, for that it doth so chill them, that the Sunne afterward, although in Winter, doth scorch them and burne them vp : looke therefore vnto your Gilloflowers in those times, and shake or strike off the snow gently off from them, not suffering it to abide on them any day or night if you can; for assure your selfe, if it doth not abide on them, the better they will be. The frosts likewise is another great annoyance vnto them, to corrupt the rootes, and to cause them to swell, rot, and breake : to preuent which inconuenience, I would aduise you to take the straw or litter of your horse stable, and lay some thereof about euery roote of your Gilloflowers (especially those of the best account) close vnto them vpon the ground, but be as carefull as you can, that none thereof lye vpon the greene leaues, or as little as may be, and by this onely way haue they been better defended from the frosts that spoile them in Winter, then by any other that I haue seen or knowne. The windes in March, and Sunneshine dayes then, are one of the greatest inconueniences that happeneth vnto them : for they that haue had hundreds of plants, that haue kept faire and greene all the Winter vntill the beginning or middle of March, before the end thereof, haue had scarce one of many, that either hath not vtterly perished, or been so tainted, that quickly after haue not been lost; which hath happened chiefly by the neglect of these cautions before specified, or in not defending them from the bitter sharpe windes and sunne in this moneth of March. You shall therefore for their better preseruation, besides the litter laid about the rootes, which I aduise you not to remoue as yet, shelter them somewhat from the windes, with eyther bottomlesse pots, pales, or such like things, to keep away the violent force both of windes and sun for that moneth, and for some time before & after it also : yet so, that they be not couered close aboue, but open to receiue ayre & raine. Some also vse to wind withes of hey or straw about the rootes of their Gilloflowers, and fasten them with stickes thrust into the ground, which serue very well in the stead of the other. Thus haue I shewed you the whole preseruation of these worthy and dainty flowers, with the whole manner of ordering them for their encrease : if any one haue any other better way, I shall be as willing to learne it of them, as I haue beene to giue them or any others the knowledge of that I haue here set downe.

Chap. IX.

That there is not any art whereby any flower may be made to grow double, that was naturally single, nor of any other sent or colour than it first had by nature ; nor that the sowing or planting of herbes one deeper than other, will cause them to be in flower one after another, euery moneth in the yeare.

THe wonderfull desire that many haue to see faire, double, and sweete flowers, hath transported them beyond both reason and nature, feigning and boasting often of what they would haue, as if they had it. And I thinke, from this desire and boasting hath risen all the false tales and reports, of making flowers double as they list, and of giuing them colour and sent as they please, and to flower likewise at what time they will, I doubt not, but that some of these errours are ancient, and continued long by tradition, and others are of later inuention : and therefore the more to be condemned, that men of wit and iudgement in these dayes should expose themselues in their writings, to be rather laughed at, then beleeued for such idle tales. And although in the contradiction of them, I know I shall vndergoe many calumnies, yet notwithstanding, I will endeauour to set downe and declare so much, as I hope may by reason

perswade

perſwade many in the truth, although I cannot hope of all, ſome being ſo ſtrongly wedded to their owne will, and the errours they haue beene bred in, that no reaſon may alter them. Firſt therefore I ſay, that if there were any art to make ſome flowers to grow double, that naturally were ſingle, by the ſame art, all ſorts of flowers that are ſingle by nature, may be made to grow double : but the ſorts of flowers that are ſingle by nature, whereof ſome are double, were neuer made double by art ; for many ſorts abide ſtill ſingle, whereof there was neuer ſeene double : and therefore there is no ſuch art in any mans knowledge to bring it to paſſe. If any man ſhall ſay, that becauſe there are many flowers double, whereof there are ſingle alſo of the ſame kinde, as for example, Violets, Marigolds, Daiſyes, Daffodils, Anemones, and many other, that therefore thoſe double flowers were ſo made by the art of man : *viz.* by the obſeruation of the change of the Moone, the conſtellations or coniunctions of Planets, or ſome other Starres or celeſtiall bodies. Although I doe confeſſe and acknowledge, that I thinke ſome conſtellations, and peraduenture changes of the Moone, &c. were appointed by the God of nature, as conducing and helping to the making of thoſe flowers double, that nature hath ſo produced ; yet I doe deny, that any man hath or ſhall euer be able to proue, that it was done by any art of man, or that any man can tell the true cauſes and ſeaſons, what changes of the Moone, or conſtellations of the Planets, wrought together for the producing of thoſe double flowers, or can imitate nature, or rather the God of nature, to doe the like. If it ſhall bee demanded, From whence then came theſe double flowers that we haue, if they were not ſo made by art? I anſwer, that aſſuredly all ſuch flowers did firſt grow wilde, and were ſo found double, as they doe now grow in Gardens, but for how long before they were found they became double, no man can tell ; we onely haue them as nature hath produced them, and ſo they remaine. Againe, if any ſhall ſay, that it is likely that theſe double flowers were forced ſo to be, by the often planting and tranſplanting of them, becauſe it is obſerued in moſt of them, that if they ſtand long in any one place, and not be often remoued, they will grow ſtill leſſe double, and in the end turne ſingle. I doe confeſſe, that *Facilior eſt deſcenſus quàm aſcenſus,* and that the vnfruitfulneſſe of the ground they are planted in, or the neglect or little care had of them, or the growing of them too thicke or too long, are oftentimes a cauſe of the diminiſhing of the flowers doubleneſſe ; but withall you ſhall obſerue, that the ſame rootes that did beare double flowers (and not any other that neuer were double before) haue returned to their former doubleneſſe againe, by good ordering and looking vnto : ſingle flowers haue only beene made ſomewhat fairer or larger, by being planted in the richer and more fruitfull ground of the Garden, than they were found wilde by nature ; but neuer made to grow double, as that which is naturally ſo found of it ſelfe : For I will ſhew you mine owne experience in the matter. I haue been as inquiſitiue as any man might be, with euery one I knew, that made any ſuch report, or that I thought could ſay any thing therein, but I neuer could finde any one, that could aſſuredly reſolue me, that he knew certainly any ſuch thing to be done : all that they could ſay was but report, for the obſeruation of the Moone, to remoue plants before the change, that is, as ſome ſay, the full of the Moone, others the new Moone, whereupon I haue made tryall at many times, and in many ſorts of plants, accordingly, and as I thought fit, by planting & tranſplanting them, but I could neuer ſee the effect deſired, but rather in many of them the loſſe of my plants. And were there indeed ſuch a certaine art, to make ſingle flowers to grow double, it would haue beene knowne certainly to ſome that would practiſe it, and there are ſo many ſingle flowers, whereof there were neuer any of the kinde ſeene double, that to produce ſuch of them to be double, would procure both credit and coyne enough to him that ſhould vſe it ; but *Vltra poſſe non eſt eſſe* : and therefore let no man beleeue any ſuch reports, bee they neuer ſo ancient ; for they are but meere tales and fables. Concerning colours and ſents, the many rules and directions extant in manie mens writings, to cauſe flowers to grow yellow, red, greene, or white, that neuer were ſo naturally, as alſo to be of the ſent of Cinamon, Muske, &c. would almoſt perſwade any, that the matters thus ſet downe by ſuch perſons, and with ſome ſhew of probability, were conſtant and aſſured proofes thereof : but when they come to the triall, they all vaniſh away like ſmoake. I will in a few words ſhew you the matters and manners of their proceedings to effect this purpoſe : Firſt (they ſay) if you ſhall ſteepe

yous

your seedes in the lees of red Wine, you shall haue the flowers of those plants to be of a purple colour. If you will haue Lillies or Gilloflowers to be of a Scarlet red colour, you shall put Vermillion or Cynaber betweene the rinde and the small heads growing about the roote : if you will haue them blew, you shall dissolue Azur or Byse between the rinde and the heads : if yellow, Orpiment : if greene, Vardigrease, and thus of any other colour. Others doe aduise to open the head of the roote, and poure into it any colour dissolued, so that there be no fretting or corroding thing therein for feare of hurting the roote, and looke what colour you put in, iust such or neare vnto it shall the colour of the flower bee. Some againe doe aduise to water the plants you would haue changed, with such coloured liquor as you desire the flower to be of, and they shall grow to be so. Also to make Roses to bee yellow, that you should graft a white Rose (some say a Damaske) vpon a Broome stalke, and the flower will be yellow, supposing because the Broome flower is yellow, therefore the Rose will be yellow. Some affirme the like, if a Rose be grafted on a Barbery bush, because both the blossome and the barke of the Barbery is yellow, &c. In the like manner for sents, they haue set downe in their writings, that by putting Cloues, Muske, Cinamon, Benzoin, or any other such sweete thing, bruised with Rose water, between the barke and the body of trees, the fruit of them will smell and taste of the same that is put vnto them ; and if they bee put vnto the toppe of the rootes, or else bound vnto the head of the roote, they will cause the flowers to smell of that sent the matter put vnto them is of : as also to steep the seeds of Roses, and other plants in the water of such like sweet things, and then to sowe them, and water them morning and euening with such like liquor, vntill they be growne vp ; besides a number of such like rules and directions set downe in bookes, so confidently, as if the matters were without all doubt or question : whenas without all doubt and question I will assure you, that they are all but meere idle tales & fancies, without all reason or truth, or shadow of reason or truth : For sents and colours are both such qualities as follow the essence of plants, euen as formes are also, and one may as well make any plant to grow of what forme you will, as to make it of what sent or colour you will; and if any man can forme plants at his will and pleasure, he can doe as much as God himselfe that created them. For the things they would adde vnto the plants to giue them colour, are all corporeall, or of a bodily substance, and whatsoeuer should giue any colour vnto a liuing and growing plant, must be spirituall : for no solide corporeall substance can ioyne it selfe with the life and essence of an herbe or tree, and the spirituall part of the colour thereof is not the same with the bodily substance, but is a meere vapour that riseth from the substance, and feedeth the plant, whereby it groweth, so that there is no ground or colour of reason, that a substantiall colour should giue colour to a growing herbe or tree : but for sent (which is a meere vapour) you will say there is more probability. Yet consider also, that what sweete sent soeuer you binde or put vnto the rootes of herbes or trees, must be either buried, or as good as buried in the earth, or barke of the tree, whereby the substance will in a small time corrupt and rot, and before it can ioyne it selfe with the life, spirit, and essence of the plant, the sent also will perish with the substance : For no heterogeneall things can bee mixed naturally together, as Iron and Clay; and no other thing but homogeneall, can be nourishment or conuertible into the substance of man or beast : And as the stomach of man or beast altereth both formes, sents, and colours of all digestible things ; so whatsoeuer sent or colour is wholsome, and not poysonfull to nature, being receiued into the body of man or beast, doth neither change the bloud or skinne into that colour or sent was receiued : no more doth any colour or sent to any plant; for the plants are onely nourished by the moisture they draw naturally vnto them, be it of wine or any other liquor is put vnto them, and not by any corporeall substance, or heterogeneall vapour or sent, because the earth like vnto the stomach doth soone alter them, before they are conuerted into the nature and substance of the plant. Now for the last part I vndertooke to confute, that no man can by art make all flowers to spring at what time of the yeare hee will; although, as I haue here before shewed, there are flowers for euery moneth of the yeare, yet I hope there is not any one, that hath any knowledge in flowers and gardening, but knoweth that the flowers that appeare and shew themselues in the seuerall moneths of the yeare, are not one and the same, and so made to flower by art ; but that they are seuerall sorts of plants, which

will

will flower naturally and conſtantly in the ſame moneths one yeare, that they vſe to doe in another, or with but little alteration, if the yeares proue not alike kindly: As for example, thoſe plants that doe flower in Ianuary and February, will by no art or induſtry of man be cauſed to flower in Summer or in Autumne; and thoſe that flower in Aprill and May, will not flower in Ianuary or February; or thoſe in Iuly, Auguſt, &c. either in the Winter or Spring: but euery one knoweth their owne appointed naturall times, which they conſtantly obſerue and keepe, according to the temperature of the yeare, or the temper of the climate, being further North or South, to bring them on earlier or later, as it doth with all other fruits, flowers, and growing greene herbes, &c. except that by chance, ſome one or other extraordinarily may be hindered in their due ſeaſon of flowring, and ſo giue their flowers out of time, or elſe to giue their flowers twice in the yeare, by the ſuperaboundance of nouriſhment, or the mildneſſe of the ſeaſon, by moderate ſhowers of raine, &c. as it ſometimes alſo happeneth with fruits, which chance, as it is ſeldome, and not conſtant, ſo we then terme it but *Luſus naturæ :* or elſe by forcing them in hot ſtoues, which then will periſh, when they haue giuen their flowers or fruits. It is not then, as ſome haue written, the ſowing of the ſeedes of Lillies, or any other plants a foote deepe, or halfe a foote deepe, or two inches deepe, that will cauſe them to be in flower one after another, as they are ſowne euery moneth of the yeare; for it were too groſſe to thinke, that any man of reaſon and iudgement would ſo beleeue. Nor is it likewiſe in the power of any man, to make the ſame plants to abide a moneth, two, or three, or longer in their beauty of flowring, then naturally they vſe to doe; for I thinke that were no humane art, but a ſupernaturall worke. For nature ſtill bendeth and tendeth to perfection, that is, after flowring to giue fruit or ſeede; nor can it bee hindered in the courſe thereof without manifeſt danger of deſtruction, euen as it is in all other fruit-bearing creatures, which ſtay no longer, then their appointed time is naturall vnto them, without apparent damage. Some things I grant may be ſo ordered in the planting, that according to that order and time which is obſerued in their planting, they ſhall ſhew forth their faire flowers, and they are Anemones, which will in that manner, that I haue ſhewed in the worke following, flower in ſeuerall moneths of the yeare; which thing as it is incident to none or very few other plants, and is found out but of late, ſo likewiſe is it knowne but vnto a very few. Thus haue I ſhewed you the true ſolution of theſe doubts : And although they haue not beene amplified with ſuch Philoſophicall arguments and reaſons, as one of greater learning might haue done, yet are they truely and ſincerely ſet downe, that they may ſerue *tanquam galeatum,* againſt all the calumnies and obiections of wilfull and obdurate perſons, that will not be reformed. As firſt, that all double flowers were ſo found wilde, being the worke of nature alone, and not the art of any man, by planting or tranſplanting, at or before the new or full Moone, or any other obſeruation of time, that hath cauſed the flower to grow double, that naturally was ſingle : Secondly, that the rules and directions, to cauſe flowers to bee of contrary or different colours or ſents, from that they were or would be naturally, are meere fancies of men, without any ground of reaſon or truth. And thirdly, that there is no power or art in man, to cauſe flowers to ſhew their beauty diuers moneths before their naturall time, nor to abide in their beauty longer then the appointed naturall time for euery one of them.

C   THE

THE GARDEN
OF
PLEASANT FLOWERS.

CHAP. I.

Corona Imperialis. The Crowne Imperiall.

Ecaufe the Lilly is the more ftately flower among manie : and amongft the wonderfull varietie of Lillies,
knowne to vs in thefe daies, much more then in former
times, whereof fome are white, others blufh, fome purple, others red or yellow, fome fpotted, others without fpots, fome ftanding vpright, others hanging or
turning downewards, The Crowne Imperiall for his
ftately beautifulnefs, deferueth the firft place in this our
Garden of delight, to be here entreated of before all o-
ther Lillies : but becaufe it is fo well knowne to moft
perfons, being in a manner euery where common, I fhall
neede onely to giue you a relation of the chiefe parts
thereof (as I intend in fuch other things) which are thefe : The roote is yellowifh on
the outfide, compofed of fewer, but much thicker fcales, then any other Lilly but the
Perfian, and doth grow fometimes to be as great as a pretty bigge childes head, but
fomewhat flat withall, from the fides whereof, and not from the bottome, it fhooteth
forth thicke long fibres, which perifh euery yeare, hauing a hole in the midft thereof,
at the end of the yeare, when the old ftalke is dry and withered, and out of the which
a new ftalke doth fpring againe (from a bud or head to be feen within the hollowneffe
on the one fide) the yeare following : the ftalke then filling vp the hollowneffe, rifeth
vp three or foure foote high, being great, round, and of a purplifh colour at the bottome, but greene aboue, befet from thence to the middle thereof with many long and
broad greene leaues, very like to the leaues of our ordinary white Lilly, but fomewhat
fhorter and narrower, confufedly without order, and from the middle is bare or naked without leaues, for a certaine fpace vpwards, and then beareth foure, fixe, or tenne
flowers, more or leffe, according to the age of the plant, and the fertility of the foyle
where it groweth : The buddes at the firft appearing are whitifh, ftanding vpright a-
mong a bufh or tuft of greene leaues, fmaller then thofe below, and ftanding aboue
the flowers, after a while they turne themfelues, and hang downewards euerie
one vpon his owne footeftalke, round about the great ftemme or ftalke, fometimes of
an euen depth, and other while one lower or higher than another, which flowers are
neare the forme of an ordinary Lilly, yet fomewhat leffer and clofer, confifting of
fixe leaues of an Orange colour, ftriped with purplifh lines and veines, which adde
a great grace to the flowers : At the bottome of the flower next vnto the ftalke, euery

leafe thereof hath on the outſide a certaine bunch or eminence, of a darke purpliſh colour, and on the inſide there lyeth in thoſe hollow bunched places, certaine cleare drops of water like vnto pearles, of a very ſweete taſte almoſt like ſugar : in the midſt of each flower is a long white ſtile or pointell, forked or diuided at the end, and ſixe white chiues tipt with yellowiſh pendents, ſtanding cloſe about it : after the flowers are paſt, appeare ſixe ſquare ſeede veſſels ſtanding vpright, winged as it were or welted on the edges, yet ſeeming but three ſquare, becauſe each couple of thoſe welted edges are ioyned cloſer together, wherein are contained broad, flat, and thinne ſeedes, of a pale browniſh colour, like vnto other Lillies, but much greater and thicker alſo. The ſtalke of this plant doth oftentimes grow flat, two, three, or foure fingers broad, and then beareth many more flowers, but for the moſt part ſmaller then when it beareth round ſtalkes. And ſometimes it happeneth the ſtalke to be diuided at the top, carrying two or three tufts of greene leaues, without any flowers on them. And ſometimes likewiſe, to beare two or three rowes or crownes of flowers one aboue another vpon one ſtalke, which is ſeldome and ſcarce ſeene, and beſides, is but meere accidentall : the whole plant and euery part thereof, as well rootes, as leaues and flowers, doe ſmell ſomewhat ſtrong as it were the ſauour of a Foxe, ſo that if any doe but come neare it, he cannot but ſmell it, which yet is not vnwholſome.

I haue not obſerued any variety in the colour of this flower, more then that it will be fairer in a cleare open ayre, and paler, or as it were blaſted in a muddy or ſmoakie ayre. And although ſome haue boaſted of one with white flowers, yet I could neuer heare that any ſuch hath endured in one vniforme colour.

The Place.

This plant was firſt brought from Conſtantinople into theſe Chriſtian Countries, and by the relation of ſome that ſent it, groweth naturally in Perſia.

The Time.

It flowreth moſt commonly in the end of March, if the weather be milde, and ſpringeth not out of the ground vntill the end of February, or beginning of March, ſo quicke it is in the ſpringing : the heads with ſeed are ripe in the end of May.

The Names.

It is of ſome called *Lilium Perſicum*, the Perſian Lilly : but becauſe wee haue another, which is more vſually called by that name, as ſhall be ſhewed in the next Chapter, I had rather with Alphonſus Pancius the Duke of Florence his Phyſitian, (who firſt ſent the figure thereof vnto Mr. Iohn de Brancion) call it *Corona Imperialis*, The Crowne Imperiall, then by any other name, as alſo for that this name is now more generally receiued. It hath been ſent alſo by the name *Tuſai*, and *Tuſchai*, and *Turſani*, or *Turſanda*, being, as it is like, the Turkiſh names.

The Vertues.

For any Phyſicall Vertues that are in it, I know of none, nor haue heard that any hath been found out : notwithſtanding the ſtrong ſent would perſwade it might be applyed to good purpoſe.

CHAP. II.
Lilium Perſicum. The Perſian Lilly.

THe roote of the Perſian Lilly is very like vnto the root of the Crowne Imperiall, and loſing his fibres in like maner euery yeare, hauing a hole therin likewiſe where the old ſtalke grew, but whiter, rounder, and a little longer, ſmaller, and not ſtinking at all like it, from whence ſpringeth vp a round whitiſh greene ſtalke, not

much

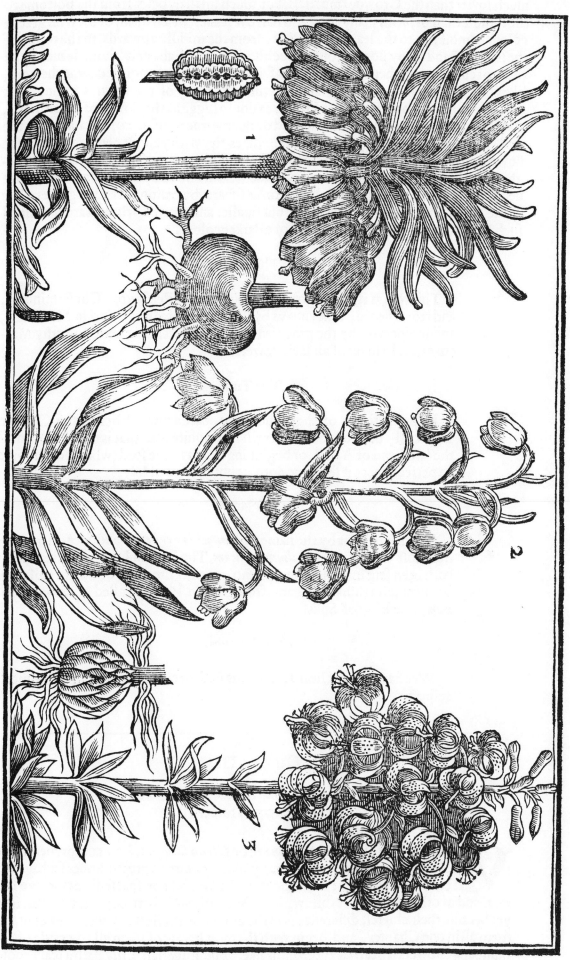

1 *Corona Imperialis.* The Crowne Imperiall.   2 *Lilium Persicum.* The Persian Lilly.
3 *Martagon Imperiale.* The Martagon Imperiall.

much

much lower than the Crowne Imperiall, but much smaller, beset from the bottome to the middle thereof, with many long and narrow leaues, of a whitish or blewish greene colour, almost like to the leafe of a Tulipa : from the middle vpwards, to the toppe of the stalke, stand many flowers one aboue another round about it, with leaues at the foote of euery one of them, each whereof is pendulous or hanging downe the head, like vnto the Crowne Imperiall, and not turning vp any of the flowers againe, but smaller than in any other kinde of Lilly, yea not so bigge as the flower of a Fritillaria, consisting of sixe leaues a peece, of a dead or ouerworne purplish colour, hauing in the midst a small long pointell, with certaine chiues tipt with yellow pendents : after the flowers are past (which abide open a long time, and for the most part flower by de-grees, the lowest first, and so vpwards) if the weather be temperate, come sixe square heads or seede vessels, seeming to be but three square, by reason of the wings, very like to the heads of the Crowne Imperiall, but smaller and shorter, wherein are contained such like flat seed, but smaller also, and of a darker colour.

The Place.

This was, as it is thought, first brought from Persia vnto Constantinople, and from thence, sent vnto vs by the meanes of diuers Turkie Merchants, and in especiall, by the procurement of M^r. Nicholas Lete, a worthy Mer-chant, and a louer of all faire flowers.

The Time.

It springeth out of the ground very neare a moneth before the Crowne Imperiall, but doth not flower till it bee quite past (that is to say) not vntill the latter end of Aprill, or beginning of May : the seed (when it doth come to perfection, as it seldome doth) is not ripe vntill Iuly.

The Names.

It hath been sent by the name of *Pennachio Persiano*, and wee thereupon doe most vsually call it *Lilium Persicum*, The Persian Lilly. Clusius saith it hath been sent into the Low-Countries vnder the name of Susam giul, and he thereupon thinking it came from Susis in Persia, called it *Lilium Susia-num*, The Lilly of Susis.

The Vertues.

Wee haue not yet heard, that this hath beene applyed for any Physicall respect.

Chap. III.

Martagon Imperiale, siue Lilium Montanum maius,
The Martagon Imperiall.

Vnder this title of *Lilium Montanum*, or *Lilium Siluestre*, I do comprenend only those kindes of Lillies, which carry diuers circles of greene leaues set togethei at certaine distances, round about the stalke, and not sparsedly as the two for-mer, and as other kindes that follow, doe. And although there bee many of this sort, yet because their chiefest difference is in the colour of the flower, wee will containe them all in one Chapter, and begin with the most stately of them all, because of the number of flowers it beareth vpon one stalke. The Imperiall Lilly hath a scaly roote, like vnto all the rest of the Lillies, but of a paler yellow colour, closely compact or set together, being short and small oftentimes, in comparison of the greatnesse of the
                                                                                    stemme

ſtemme growing from it. The ſtalke is browniſh and round at the bottome, and ſometimes flat from the middle vpwards, three foote high or more, beſet at certaine diſtances with rondles or circles of many broad leaues, larger and broader for the moſt part than any other of this kinde, and of a darke green colour: It hath two or three, and ſometimes foure of theſe rondles or circles of leaues, and bare without any leafe betweene; but aboue toward the tops of the ſtalkes, it hath here and there ſome leaues vpon it, but ſmaller than any of the other leaues: at the toppe of the ſtalke come forth many flowers, ſometime three or foure ſcore, thicke thruſt, or confuſedly ſet together, and not thinne or ſparſedly one aboue another, as in the leſſer of this kinde of Mountaine Lilly. It hath been ſometimes alſo obſerued in this kinde, that it hath borne manie flowers at three ſeuerall ſpaces of the ſtalke, one aboue another, which hath made a goodly ſhew; each flower whereof is pendulous, hanging downe, and each leafe of the flower turning vp againe, being thicke or fleſhy, of a fine delayed purple colour, ſpotted with many blackiſh or browniſh ſpots, of a very pleaſant ſweet ſent, which maketh it the more acceptable: in the middle of the flower hangeth downe a ſtile or pointell, knobbed or buttoned at the end with ſixe yellow chiues, tipt with looſe pendents of an Orient red or Vermillion colour, which will eaſily ſticke like duſt vpon any thing that toucheth them: the heads or ſeede veſſels are ſmall and round, with ſmall edges about them, wherein is contained flat browne ſeede like other Lillies, but leſſer. The root is very apt to encreaſe or ſet of, as we call it, wherby the plant ſeldome commeth to ſo great a head of flowers, but riſeth vp with many ſtalkes, and then carry fewer flowers.

Of this kinde there is ſometimes one found, that beareth flowers without any ſpots: the leaues whereof and ſtalke likewiſe are paler, but not elſe differing.   *Martagon Imperiale flore non punctato.*

Martagon flore albo. The White Martagon.

We haue alſo ſome other of this kind, the firſt wherof hath his ſtalke & leafe greener than the former, the ſtalke is a little higher, but not bearing ſo thicke a head of flowers, although much more plentifull than the leſſer Mountaine Lilly, being altogether of a fine white colour, without any ſpots, or but very few, and that but ſometimes alſo: the pendents in the middle of this flower are not red, as the former, but yellow; the roote of this, and of the other two that follow, are of a pale yellow colour, the cloues or ſcales of them being brittle, and not cloſely compact, yet ſo as if two, and ſometimes three ſcales or cloues grew one vpon the head or vpperpart of another; which difference is a ſpeciall note to know theſe three kindes, from any other kinde of Mountaine Lilly, as in all old rootes that I haue ſeene, I haue obſerued, as alſo in them that are reaſonably well growne, but in the young rootes it is not yet ſo manifeſt.

Martagon flore albo maculato. The White ſpotted Martagon.

The ſecond is like vnto the firſt in all things, ſaue in this, that the flowers hereof are not altogether ſo white, and beſides hath many reddiſh ſpots on the inſide of the leaues of the flower, and the ſtalke alſo is not ſo greene but browniſh.

Martagon flore carneo. The bluſh Martagon.

A third ſort there is of this kinde, whoſe flowers are wholly of a delayed fleſh colour, with many ſpots on the flowers, and this is the difference hereof from the former.

Lilium Montanum ſiue ſilueſtre minus. The leſſer Mountaine Lilly.

The leſſer Mountaine Lilly is ſo like in root vnto the greater that is firſt deſcribed, that it is hard to diſtinguiſh them aſunder; but when this is ſprung vp out of the ground, which is a moneth after the firſt: it alſo carrieth his leaues in rondles about the ſtalke, although not altogether ſo great nor ſo many. The flowers are more thinly ſet on the ſtalkes one aboue another, with more diſtance betweene each flower than the former, and are of a little deeper fleſh colour or purple, ſpotted in the ſame manner. The buds

or

1 *Martagon flore albo.* The white Martagon.　　2 *Martagon siue Lilium Canadense maculatum.* The spotted Martagon, or Lilly of Canada.　　3 *Martagon Pomponeum.* The Martagon Pompony, or early red Martagon.

or heads of flowers, in some of these before they be blowne, are hoary white, or hairie, whereas in others, there is no hoarinesse at all, but the buddes are smooth and purplish : in other things this differeth not from the former.

Of this sort also there is one that hath but few spots on the flowers, whose colour is somewhat paler than the other. *Lilium Montanum non maculatum.*

Martagon Canadense maculatum. The spotted Martagon of Canada.

Although this strange Lilly hath not his flowers hanging downe, and turning vp againe, as the former kinds set forth in this Chapter; yet because the green leaues stand at seuerall ioynts as they do, I must needs insert it here, not knowing where more fitly to place it. It hath a small scaly roote, with many small long fibres thereat, from whence riseth vp a reasonable great stalke, almost as high as any of the former, bearing at three or foure distances many long and narrow greene leaues, but not so many or so broad as the former, with diuers ribbes in them : from among the vppermost rundle of leaues breake forth foure or fiue flowers together, euery one standing on a long slender foote stalke, being almost as large as a red Lilly, but a little bending downewards, and of a faire yellow colour, spotted on the inside with diuers blackish purple spots or strakes, hauing a middle pointell, and sixe chiues, with pendents on them.

The Place.

All these Lillies haue been found in the diuers Countries of Germany, as Austria, Hungaria, Pannonia, Stiria, &c. and are all made Denisons in our London Gardens, where they flourish as in their owne naturall places. The last was was brought into France from Canada by the French Colonie, and from thence vnto vs.

The Time.

They flower about the later end of Iune for the most part, yet the first springeth out of the ground a moneth at the least before the other, which are most vsually in flower before it, like vnto the Serotine Tulipas, all of them being early vp, and neuer the neere.

The Names.

The first is vsually called *Martagon Imperiale*, the Imperiall Martagon, and is *Lilium Montanum maius*, the greatest Mountaine Lilly; for so it deserueth the name, because of the number of flowers vpon a head or stalke. Some haue called it *Lilium Sarasenicum*, and some *Hemerocallis*, but neither of them doth so fitly agree vnto it.

The second is *Lilium Montanum maius flore albo*, and of some *Martagon Imperiale flore albo*, but most vsually *Martagon flore albo*, the white Martagon. The second sort of this second kinde, is called *Martagon flore albo maculato*, the spotted white Martagon. And the third, *Martagon flore carneo*, the blush Martagon.

The third kinde is called *Lilium Montanum*, the Mountaine Lilly, and some adde the title *minus*, the lesser, to know it more distinctly from the other. Some also *Lilium Siluestre*, as Clusius, and some others, and of Matthiolus *Martagon*. Of diuers women here in England, from the Dutch name, Lilly of Nazareth. The last hath his title *Americanum & Canadense*, and in English accordingly.

CHAP. IV.

CHAP. IV.

1. *Martagon Pomponeum ſiue Lilium rubrum præcox, vel Lilium Macedonicum.*
The early red Martagon, or Martagon Pompony.

AS in the former Chapter we deſcribed vnto you ſuch Lillies, whoſe flowers being pendulous, turne their leaues backe againe, and haue their greene leaues, ſet by ſpaces about the ſtalke : ſo in this wee will ſet downe thoſe ſorts, which carry their greene leaues more ſparſedly, and all along the ſtalke, their flowers hanging downe, and turning vp againe as the former, and begin with that which is of greateſt beauty, or at leaſt of moſt rarity.

1. Martagon Pomponeum anguſti folium præcox.

1. This rare Martagon hath a ſcaly root cloſely compact, with broader and thinner ſcales than others, in time growing very great, and of a more deepe yellow colour then the former, from whence doth ſpring vp a round greene ſtalke in ſome plants, and flat in others, two or three foote high, bearing a number of ſmall, long, and narrow greene leaues, very like vnto the leaues of Pinkes, but greener, ſet very thicke together, and without order about the ſtalke, vp almoſt vnto the toppe, and leſſer by degrees vpwards, where ſtand many flowers, according to the age of the plant, and thriuing in the place where it groweth ; in thoſe that are young, but a few, and more ſparſedly, and in others that are old many more, and thicker ſet : for I haue reckoned threeſcore flowers and more, growing thicke together on one plant with mee, and an hundred flowers on another : theſe flowers are of a pale or yellowiſh red colour, and not ſo deep red as the red Martagon of Conſtantinople, hereafter ſet down, nor fully ſo large : yet of the ſame faſhion, that is, euery flower hanging downe, and turning vp his leaues againe. It is not ſo plentifull in bearing of ſeede as the other Lillies, but when it doth, it differeth not but in being leſſe.

2. Martagon anguſti folium magis ſerotinum.
3. Martagon Pomponeum latifolium pracox.

There is another, whoſe greene leaues are not ſo thicke ſet on the ſtalke, but elſe differeth not but in flowring a fortnight later.

There is another alſo of this kind, ſo like vnto the former in root, ſtalk, flower, & maner of growing, that the difference is hardly diſcerned ; but conſiſteth chiefly in theſe two points : Firſt, that the leaues of this are a little broader and ſhorter then the former ; and ſecondly, that it beareth his flowers a fortnight earlier than the firſt. In the colour or forme of the flower, there can no difference bee diſcerned, nor (as I ſaid) in any other thing. All theſe Lillies doe ſpring very late out of the ground, euen as the yellow Martagons doe, but are ſooner in flower then any others.

4. Martagon flore phæniceo.

A fourth kinde hereof hath of late been knowne to vs, whoſe leaues are broader and ſhorter then the laſt, and the flowers of a paler red, tending to yellow, of ſome called a golden red colour : but flowreth not ſo early as they.

2. *Lilium rubrum Byzantinum, ſiue Martagon Conſtantinopolitanum.*
The red Martagon of Conſtantinople.

1. The red Martagon of Conſtantinople is become ſo common euery where, and ſo well knowne to all louers of theſe delights, that I ſhall ſeeme vnto them to loſe time, to beſtow many lines vpon it ; yet becauſe it is ſo faire a flower, and was at the firſt ſo highly eſteemed, it deſerueth his place and commendations, howſoeuer encreaſing the plenty hath not made it dainty. It riſeth out of the ground early in the ſpring, before many other Lillies, from a great thicke yellow ſcaly root, bearing a round browniſh ſtalke, beſet with many faire greene leaues confuſedly thereon, but not ſo broad as the common white Lilly, vpon the toppe whereof ſtand one, two, or three, or more flowers, vpon long footeſtalkes, which hang downe their heads, and turne vp their leaues againe, of an excellent red crimſon colour, and ſometimes paler, hauing a long pointell in the middle, compaſſed with ſixe whitiſh chiues, tipt with looſe yellow pendents, of a reaſonable good ſent, but ſomewhat faint. It likewiſe beareth ſeede in heads, like vnto the other, but greater.

Martagon Constantinopolitanum maculatum.
The red spotted Martagon of Constantinople.

We haue another of this kinde, that groweth somewhat greater and higher, with a larger flower, and of a deeper colour, spotted with diuers blacke spots, or strakes and lines, as is to be seene in the Mountaine Lillies, and in some other hereafter to be described; but is not so in the former of this kinde, which hath no shew of spots at all. The whole plant as it is rare, so it is of much more beauty then the former.

2. *Martagon Pannonicum, siue Exoticum flore spadicee.*
The bright red Martagon of Hungarie.

Although this Martagon or Lilly bee of another Countrey, yet by reason of the neerenesse both in leafe and flower vnto the former, may more fitly be placed next vnto them, then in any other place. It hath his roote very like the other, but the leaues are somewhat larger, and more sparsedly set vpon the stalke, else not much vnlike: the flowers bend downe, and turne vp their leaues againe, but somewhat larger, and of a bright red, tending to an Orenge colour, that is, somewhat yellowish, and not crimson, like the other.

3. *Martagon Luteum punctatum.* The Yellow spotted Martagon.

1. This Yellow Martagon hath a great scaly or cloued roote, and yellow, like vnto all these sorts of turning Lillies, from whence springeth vp a round greene strong stalke, three foote high at the least, confusedly set with narrow long greene leaues, white on the edges vp to the very toppe thereof almost, hauing diuers flowers on the head, turning vp againe as the former doe, of a faint yellowish, or greenish yellow colour, with many blacke spots or strakes about the middle of the leafe of euery flower, and a forked pointell, with sixe chiues about it, tipt with reddish pendents, of a heauie strong smell, not very pleasant to many. It beareth seede very plentifully, in great heads, like vnto the other former Lillies, but a little paler.

2. *Martagon Luteum non maculatum.* The Yellow Martagon without spots.

The other yellow Martagon differeth in no other thing from the former, but onely that it hath no spots at all vpon any of the leaues of the flowers; agreeing with the former, in colour, forme, height, and all things else.

3. *Martagon Luteum serotinum.* The late flowring Yellow Martagon.

There is yet another yellow Martagon, that hath no other difference then the time of his flowring, which is not vntill Iuly, vnlesse in this, that the flower is of a deeper yellow colour.

The Place.

The knowledge of the first kindes of these early Martagons hath come from Italy, from whence they haue bin sent into the Low-Countries, and to vs, and, as it seemeth by the name, whereby they haue bin sent by some into these parts, his originall should be from the mountaines in Macedonia.

The second sort is sufficiently knowne by his name, being first brought from Constantinople, his naturall place being not farre from thence, as it is likely. But the next sort of this second kinde, doth plainly tell vs his place of birth to be the mountaines of Pannonia or Hungarie.

The third kindes grow on the Pyrenæan mountaines, where they haue been searched out, and found by diuers louers of plants, as also in the Kingdome of Naples.

The

The Time.

The firſt early Martagons flower in the end of May, or beginning of Iune, and that is a moneth at the leaſt before thoſe that come from Conſtantinople, which is the ſecond kinde. The two firſt yellow Martagons flower ſomewhat more early, then the early red Martagons, and ſometimes at the ſame time with them. But the third yellow Martagon, as is ſaid, flowreth a moneth later or more, and is in flower when the red Martagon of Conſtantinople flowreth. And although the early red and yellow Martagons, ſpring later then the other Martagons or Lillies, yet they are in flower before them.

The Names.

The firſt early red Lillies or Martagons haue beene ſent vnto vs by ſeuerall names, as *Martagon Pomponeum*, and thereafter are called Martagon of Pompony, and alſo *Lilium* or *Martagon Macedonicum*, the Lilly or Martagon of Macedonia. They are alſo called by Cluſius *Lilium rubrum præcox*, the one *anguſtiore folio*, the other *latiore folio*. And the laſt of this kinde hath the title *flore phæniceo* added or giuen vnto it, that is, the Martagon or Lilly of Macedonia with gold red flowers.

The Martagons of Conſtantinople haue beene ſent by the Turkiſh name *Zuſiniare*, and is called *Martagon*, or *Lilium Byzantinum* by ſome, and *Hemerocallis Chalcedonica* by others ; but by the name of the Martagon of Conſtantinople they are moſt commonly receiued with vs, with the diſtinction of *maculatum* to the one, to diſtinguiſh the ſorts. The laſt kinde in this *claſsis*, hath his name in his title, as it hath been ſent vnto vs.

The Yellow Martagons are diſtinguiſhed in their ſeuerall titles, as much as is conuenient for them.

CHAP. V.

Lilium Aureum & Lilium Rubrum. The Gold and Red Lillies.

THere are yet ſome other kindes of red Lillies to bee deſcribed, which differ from all the former, and remaine to be ſpoken of in this place. Some of them grow high, and ſome lowe, ſome haue ſmall knots, which wee call bulbes, growing vpon the ſtalkes, at the ioynts of the leaues or flowers, and ſome haue none : all which ſhall be intreated of in their ſeuerall orders.

Lilium pumilum cruentum. The dwarfe red Lilly.

The dwarfe red Lilly hath a ſcaly roote, ſomewhat like vnto other Lillies, but white, and not yellow at all, and the cloues or ſcales thicker, ſhorter, and fewer in number, then in moſt of the former : the ſtalke hereof is not aboue a foote and a halfe high, round and greene, ſet confuſedly with many faire and ſhort greene leaues, on the toppe of which doe ſtand ſometimes but a few flowers, and ſometimes many of a faire purpliſh red colour, and a little paler in the middle, euery flower ſtanding vpright, and not hanging downe, as in the former, on the leaues whereof here and there are ſome blacke ſpots, lines or markes, and in the middle of the flower a long pointell, with ſome chiues about it, as is in the reſt of theſe Lillies.

Lilium rubrum multiplici flore. This kinde is ſometimes found to yeeld double flowers, as if all the ſingle flowers ſhould grow into one, and ſo make it conſiſt of many leaues, which notwithſtanding

his

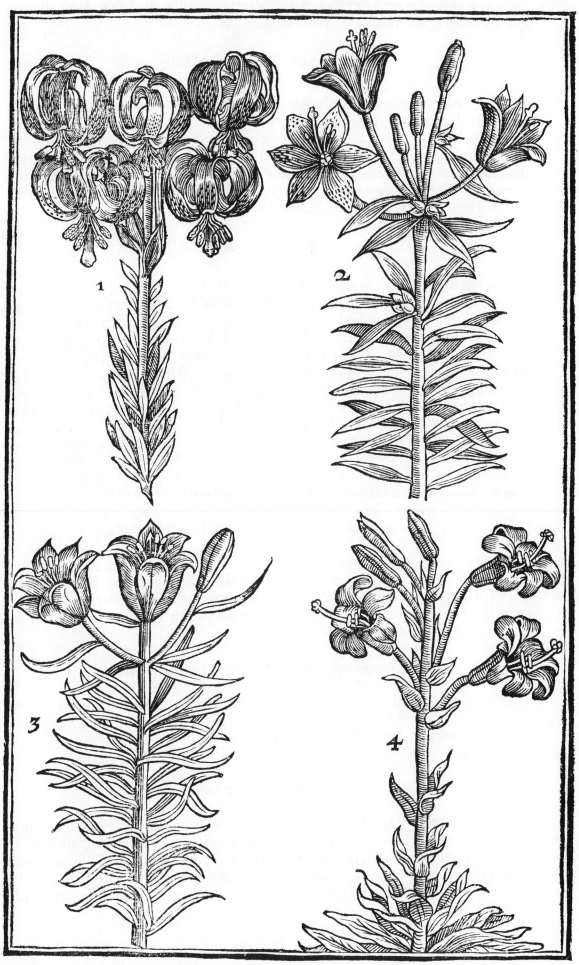

1 *Martagon rubrum sine luteum.* The red or the yellow Martagon. 2 *Lilium Bulbiferum.* The red bulbed Lilly. 3 *Lilium aureum.* The gold red Lilly. 4 *Lilium album.* The white Lilly.

his so continuing sundry yeares, vpon transplanting, will *redire ad ingenium*, that is, quickly come againe to his old byas or forme.

Lilium Aureum. The Gold red Lilly.

The second red Lilly without bulbes groweth much higher then the first, and almost as high as any other Lilly : the roote hereof is white and scaly, the leaues are somewhat longer, and of a darke or sad greene colour; the flowers are many and large, standing vpright as all these sorts of red Lillies doe, of a paler red colour tending to an Orenge on the inside, with many blacke spots, and lines on them, as in the former, and more yellow on the outside : the seede vessels are like vnto the roundish heads of other Lillies, and so are the seedes in them likewise.

1. *Lilium minus bulbiferum.* The dwarfe bulbed Lilly.

The first of the Lillies that carrieth bulbes on the stalke, hath a white scaly roote like the former; from whence riseth vp a small round stalke, not much higher then the first dwarfe Lilly, seeming to be edged, hauing many leaues thereon of a sad greene colour set about it, close thrust together : the greene heads for flowers, will haue a kind of woollinesse on them, before the flowers begin to open, and betweene these heads of flowers, as also vnder them, and among the vppermost leaues, appeare small bulbes or heads, which being ripe if they be put into the ground, or if they fall of themselues, will shoote forth leaues, and beare flowers within two or three yeares like the mother plant, and so will the bulbes of the other hereafter described : the flowers of this Lilly are of a faire gold yellow colour, shadowed ouer with a shew of purple, but not so red as the first, or the next to bee described. This Lilly will shoote strings vnder ground, like as the last red Lilly will doe also, whereat will grow white bulbed roots, like the rootes of the mother plant, thereby quickly encreasing it selfe.

2. *Lilium Cruentum bulbiferum.* The Fierie red bulbed Lilly.

The second bulbed Lilly riseth vp with his stalke as high as any of these Lillies, carrying many long and narrow darke greene leaues about it, and at the toppe many faire red flowers, as large or larger then any of the former, and of a deeper red colour, with spots on them likewise, hauing greater bulbes growing about the toppe of the stalke and among the flowers, then any else.

Lilium Cruentum flore pleno. The Fierie red double Lilly.

The difference of this doth chiefly consist in the flower, which is composed of manie leaues, as if many flowers went to make one, spotted with black spots, and without any bulbes when it thus beareth, which is but accidentall, as the former double Lilly is said to be.

3. *Lilium maius bulbiferum.* The greater bulbed red Lilly.

The third red Lilly with bulbes, riseth vp almost as high as the last, and is the most common kinde we haue bearing bulbes. It hath many leaues about the stalke, but not of so sad a greene colour as the former : the flowers are of as pale a reddish yellow colour as any of the former, and comming neerest vnto the colour of the Gold red Lilly. This is more plentifull in bulbes, and in shooting strings, to encrease rootes vnder ground, then the others.

The Place.

These Lillies doe all grow in Gardens, but their naturall places of growing is the Mountaines and the Vallies neere them in Italy, as Matthiolus
saith :

faith : and in many Countries of Germany, as Hungarie, Auftria, Stiria, and Bohemia, as Clufius and other doe report.

The Time.

They flower for the moft part in Iune, yet the firft of thefe is the earlieft of all the reft.

The Names.

All thefe Lillies are called *Lilia Rubra*, Red Lillies : Some call them *Lilium Aureum, Lilium Purpureum*, *Lilium Puniceum*, *& Lilium Cruentum*. Some alfo call them *Martagon Chimiftarum*. Clufius calleth thefe bulbed Lillies *Martagon Bulbiferum*. It is thought to be *Hyacinthus Poetarum*, but I referre the difcuffing thereof to a fitter time. Wee haue, to diftinguifh them moft fitly (as I take it) giuen their proper names in their feuerall titles.

CHAP. VI.

Lilium Album. The White Lilly.

NOw remaineth onely the White Lilly, of all the whole family or ftocke of the Lillies, to bee fpoken of, which is of two forts. The one is our common or vulgar White Lilly ; and the other, that which was brought from Conftantinople.

Lilium Album vulgare. The ordinary White Lilly.

The ordinary White Lilly fcarce needeth any defcription, it is fo well knowne, and fo frequent in euery Garden ; but to fay fomewhat thereof, as I vfe to doe of euery thing, be it neuer fo common and knowne ; it hath a cloued or fcaly roote, yellower and bigger then any of the red Lillies : the ftalke is of a blackifh greene colour, and rifeth as high as moft of the Lillies, hauing many faire, broad, and long greene leaues thereon, larger and longer beneath, and fmaller vpon the ftalke vpwards ; the flowers are many or few, according to the age of the plant, fertility of the foile, and time of ftanding where it groweth : and ftand vpon long greene footftalkes, of a faire white colour, with a long pointell in the middle, and white chiues tipt with yellow pendents about it ; the fmell is fomewhat heady and ftrong.

Lilium Album Byzantinum. The White Lilly of Conftantinople.

The other White Lilly, differeth but little from the former White Lilly, either in roote, leafe, or flower, but only that this vfually groweth with more number of flowers, then euer we faw in our ordinary White Lilly : for I haue feene the ftalke of this Lilly turne flat, of the breadth of an hand, bearing neere two hundred flowers vpon a head, yet moft commonly it beareth not aboue a dozen, or twenty flowers, but fmaller then the ordinary, as the greene leaues are likewife.

The Place.

The firft groweth onely in Gardens, and hath not beene declared where it is found wilde, by any that I can heare of. The other hath beene fent from Conftantinople, among other rootes, and therefore is likely to grow in fome parts neere thereunto.

The Time.

They flower in Iune or thereabouts, but fhoote forth greene leaues in

Autumne,

Autumne, which abide greene all the Winter, the ftalke fpringing vp betweene the lower leaues in the Spring.

The Names.

It is called *Lilium Album*, the White Lilly, by moft Writers; but by Poets *Rofa Iunonis*, Iuno's Rofe. The other hath his name in his title.

The Vertues.

This Lilly aboue all the reft, yea, and I thinke this onely, and none of
the reft is vfed in medicines now adayes, although in former times Empe-
ricks vfed the red; and therefore I haue fpoken nothing of them in the end
of their Chapters, referuing what is to be faid in this. This hath a mollifying, digefting, and cleanfing quality, helping to fuppurate tumours, and to
digeft them, for which purpofe the roote is much vfed. The water of the
flowers diftilled, is of excellent vertue for women in trauell of childe bearing, to procure an eafie deliuery, as Matthiolus and Camerarius report. It
is vfed alfo of diuers women outwardly, for their faces to cleanfe the skin,
and make it white and frefh. Diuers other properties there are in thefe
Lillies, which my purpofe is not to declare in this place. Nor is it the fcope
of this worke; this that hath been faid is fufficient: for were it not, that I
would giue you fome tafte of the qualities of plants (as I faid in my Preface)
as I goe along with them, a generall worke were fitter to declare them
then this.

CHAP. VII.

Fritillaria. The checkerd Daffodill.

ALthough diuers learned men do by the name giuen vnto this delightfull plant,
thinke it doth in fome things partake with a Tulipa or Daffodill, and haue
therefore placed it betweene them; yet I, finding it moft like vnto a little Lilly,
both in roote, ftalke, leafe, flower, and feede, haue (as you fee here) placed it next
vnto the Lillies, and before them. Hereof there are many forts found out of late, as
white, red, blacke, and yellow, befides the purple, which was firft knowne; and of
each of them there are alfo diuers forts: and firft of that which is moft frequent, and
then of the reft, euery one in his place and order.

1. *Fritillaria vulgaris.* The common checkerd Daffodill.

The ordinary checkerd Daffodill (as it is vfually called, but might more properly
be called the fmall checkerd Lilly) hath a fmall round white roote, and fomewhat
flat, made as it were of two cloues, and diuided in a maner into two parts, yet ioyning
together at the bottome or feate of the roote, which holdeth them both together:
from betweene this cleft or diuifion, the budde for the ftalke &c. appeareth, which in
time rifeth vp a foote, or a foote and a halfe high, being round and of a brownifh
greene colour, efpecially neere vnto the ground, whereon there ftandeth difperfedly
foure or fiue narrow long and greene leaues, being a little hollow: at the toppe of the
ftalke, betweene the vpper leaues (which are fmaller then the loweft) the flower fheweth it felfe, hanging or turning downe the head, but not turning vp againe any of his
leaues, as fome of the Lillies before defcribed doe; (fometimes this ftalke beareth
two flowers, and very feldome three) confifting of fixe leaues, of a reddifh purple colour, fpotted diuerfly with great fpots, appearing like vnto fquare checkers, of a deeper colour; the infide of the flower is of a brighter colour then the outfide, which
hath fome greenneffe at the bottome of euery leafe: within the flower there appeare

fixe

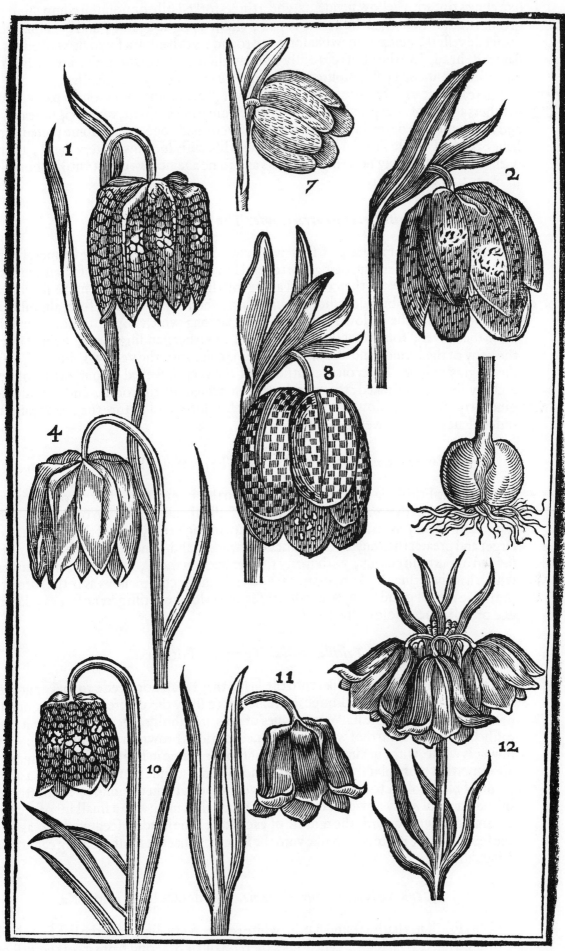

1 *Fritillaria vulgaris.* The common Fritillaria. 2 *Fritillaria flore atrorubente.* The darke red Fritillaria. 4 *Fritillaria alba.* The white Fritillaria. 7 *Fritillaria lutea punctata.* The yellow checkerd Fritillaria. 8 *Fritillaria lutea Italica.* The great yellow Italian Fritillaria. 10 *Fritillaria lutea Lusitanica.* The small yellow Fritillaria of Portugall. 11 *Fritillaria Pyrenæa.* The blacke Fritillaria. 12 *Fritillaria umbellifera.* The Spanish blacke Fritillaria.

fixe chiues tipt with yellow pendents, and a three-forked ftile or pointell compaffing a greene head, which when the flower is paft, rifeth vpright againe, and becommeth the feede veffell, being fomewhat long and round, yet hauing a fmall fhew of edges, flat at the head, like the head of a Lilly, and without any crowne as the Tulipa hath, wherein is contained pale coloured flat feede, like vnto a Lilly, but fmaller.

*Fritillaria vul-
garis pallidior,
præcox, & fe-
rotina.*

There is fome variety to be feene in this flower; for in fome the colour is paler, and in others againe of a very high or deepe colour: fometimes alfo they haue eight leaues, and fometimes ten or twelue, as if two flowers were made one, which fome thereupon haue called a Double Fritillaria. Some of them likewife doe flower very early, euen with or before the early flowring Tulipas; and fome againe flower not vntill a moneth or more after the former.

2. *Fritillaria flore atrorubente.* The bloud red Fritillaria.

The roote of this Fritillaria is fomewhat rounder and clofer then the former, from whence the ftalke rifeth vp, being fhorter and lower then in any other of thefe kindes, hauing one or two leaues thereon, and at the top thereof two or three more fet clofer together, which are broader, fhorter, and whiter then any of them before, almoft like vnto the leaues of the yellow Fritillaria, from among which toppe leaues commeth forth the flower, fomewhat bending downe, or rather ftanding forth, being larger then any of the former, and almoft equall in bigneffe vnto the yellow Fritillaria, of a duskie gray colour all ouer on the outfide, and of a very darke red colour on the infide, diuerfly fpotted or ftraked: this very hardly encreafeth by the roote, and as feldome giueth ripe feede, but flowreth with the other firft forts, and before the blacke, and a-bideth leffe time in flower then any.

3. *Fritillaria maxima purpurea fiue rubra.* The great purple or red Fritillaria.

This great Fritillaria hath his roote equall to the bigneffe of the reft of his parts, from whence rifeth vp one, & oftentimes two ftalks, hauing one, two or three flowers a peece on them, as nature and the feafons are fitting: euery one of thefe flowers are larger and greater then any of the former defcribed, and pendulous as they are, of a fad red or purplifh colour, with many thwart lines on them, and fmall long markes, which hardly feeme checkerwife, nor are fo eminent or confpicuous as in the former: the ftalke is ftrong and high, whereon are fet diuers long whitifh greene leaues, larger and broader then thofe of the former.

4. *Fritillaria alba.* The white Fritillaria.

The white Fritillaria is fo like vnto the firft, that I fhall not neede to make another defcription of this: it fhall (I hope) be fufficient to fhew the chiefe differences, and fo proceed to the reft. The ftalke and leaues of this are wholly greene, whereby it may eafily be knowne from the former, which, as is faid, is brownifh at the bottome. The flower is white, without almoft any fhew of fpot or marke in it, yet in fome the markes are fomewhat more plainly to be feene, and in fome againe there is a fhew of a faint kinde of blufh colour to be feene in the flower, efpecially in the infide, the bottomes of the leaues of euery flower fometimes are greenifh, hauing alfo a fmall lift of greene, comming downe towards the middle of each leafe: the head or feede veffell, as alfo the feede and the roote, are fo like vnto the former, that the moft cunning cannot di-ftinguifh them.

5. *Fritillaria flore duplici albicante.* The double blufh Fritillaria.

This Fritillaria hath a round flattifh white roote, very like vnto the laft Fritillaria, bearing a ftalke with long greene leaues thereon, little differing from it, or the firft or-dinary Fritillaria: the flower is faid to be conftant, compofed of many leaues, being ten at the leaft, and moft vfually twelue, of a pale whitifh purple colour, fpotted like vnto the paler ordinary Fritillaria that is early, fo that one would verily thinke it were but an accidentall kinde thereof, whereas it is (as is faid before) held to bee conftant, continuing in this manner.

6. *Fritillaria flore luteo puro.* The pure yellow Fritillaria.

The pure yellow Fritillaria hath a more round, and not fo flat a whitifh roote as the former kindes, and of a meane bigneffe; from the middle rifeth vp a ftalke a foote and a halfe high, and fometimes higher, whereon are fet without order diuers long and fomewhat broad leaues of a whitifh greene colour, like vnto the leaues of the blacke Fritillaria, but not aboue halfe fo broad: the flower is fomewhat fmall and long, nor much vnlike to the blacke for fhape and fafhion, but that the leaues are fmaller and rounder pointed, of a faint yellowifh colour, without any fhew of fpots or checkers at all, eyther within or without the flower, hauing fome chiues and yellow pendents in the middle, as is to be feene in all of them: the feede is like the firft kinde.

7. *Fritillaria flore luteo vario fiue punctato.* The checkerd yellow Fritillaria.

This Fritillaria groweth not much lower then the former, and brownifh at the ri-fing vp, hauing his leaues whiter, broader, and fhorter then it, and almoft round poin-ted. The flower is greater, and larger fpread then any other before, of a faire pale yel-low colour, fpotted in very good order, with fine fmall checkers, which adde a won-derfull pleafing beauty thereunto: it hath alfo fome lifts of greene running downe the backe of euery leafe. It feldome giueth feede; the roote alfo is like the other, but not fo flat.

8. *Fritillaria lutea maxima Italica.* The great yellow Italian Fritillaria.

This kinde of Fritillaria rifeth vp with a round and browne greene ftalke, whereon are fet diuers leaues fomewhat broad and fhort, which compaffe the ftalke at the bot-tome of them, of a darke greene colour; at the toppe of the ftalke, which bendeth a little downewards, doe moft vfually ftand three or foure leaues, betweene which com-meth forth moft vfually but one flower, which is longer then the laft, hanging downe the head as all the others doe, confifting of fixe leaues, of a darke yellowifh purple colour, fpotted with fome fmall red checkers. This kinde flowreth late, and not vn-till all the reft are paft.

9. *Fritillaria Italorum polyanthos flore paruo.* The fmall Italian Fritillaria.

This fmall Italian Fritillaria carrieth more ftore of flowers on the ftalke, but they are much fmaller, and of a yellowifh greene colour, fpotted with long and fmall darke red checkers or markes: the ftalke hath diuers fmall fhort greene leaues thereon, vnto the very toppe.

10. *Fritillaria lutea Iuncifolia Lufitanica.* The fmall yellow Fritillaria of Portugall.

The leaues of this Fritillaria are fo fmall, narrow and long, that it hath caufed them to take the name of rufhes, as if you fhould call it, The rufh leafed Fritillaria, which ftand on a long weake round ftalke, fet without order: the flower is fmall and yellow, but thicker checkerd with red fpots then any of the other yellow Fritillaria's; the ftalk of the flower, at the head thereof, being alfo of a yellowifh colour.

11. *Fritillaria Pyrenæa fiue Apenninea.* The blacke Fritillaria.

The roote of this kinde doth often grow fo great, that it feemeth like vnto the roote of a fmall Crowne Imperiall: the ftalke is ftrong, round, and high, fet without order, with broader and whiter greene leaues then any of the former, bearing one, two, or three flowers; fometimes at the toppe, being not fo large as thofe of the ordinary pur-ple Fritillaria, but fmaller, longer, and rounder, fometimes a little turning vp the brims or edges of the leaues againe, and are of a yellowifh fhining greene colour on

the

the inside, sometimes spotted with red spots almost through the whole inside of the flower, vnto the very edge, which abideth of a pale yellow colour, and sometimes there are very few spots to be seene, and those from the middle onely on the inside (for on the outside there neuer appeareth any spots at all in this kinde) and sometimes with no shew of spots at all, sometimes also of a more pale greene, and sometime of a more yellow colour: the outside of the flowers doe likewise vary, for in some the outside of the leaues are of a darke sullen yellow, &c. else more pale yellow, and in other of a darke purplish yellow colour, which in some is so deepe, and so much, that it rather seemeth blacke then purple or yellow, and this especially about the bottome of the flower, next vnto the stalke, but the edges are still of a yellowish greene: the head of seede, and the seede likewise is like vnto the former, but bigger in all respects.

12. *Fritillaria Hispanica vmbellifera.* The Spanish blacke Fritillaria.

This Fritillaria is no doubt of kindred to the last recited, it is so like, but greater in all parts thereof, as if growing in a more fruitfull soile, it were the stronger and lustier to beare more store of flowers: the flowers grow foure or fiue from the head together, hanging downe round about the stalke, like vnto a Crowne Imperiall, and are of a yellowish greene colour on the inside, spotted with a few red spots, the outside being blackish as the former.

The Place.

The first of these plants was first brought to our knowledge from France, where it groweth plentifully about Orleance; the other sorts grow in diuers other Countries, as some in Portugall, Spaine, Italy, &c. as their names doe import, and as in time they haue been obserued by those that were curious searchers of these rarities, haue been sent to vs.

The Time.

The early kindes doe flower in the beginning of Aprill or thereabouts, according to the mildenesse or sharpenesse of the precedent Winter. The other doe flower after the first are past, for a moneths space one after another, and the great yellow is very late, not flowring vntill about the middle or end of May.

The Names.

This hath receiued diuers names: some calling it *Flos Meleagridis*, the Ginny Hen Flower, of the variety of the colours in the flower, agreeing with the feathers of that Bird. Some call it *Narcissus Caparonius*, of the name of the first inuentor or finder thereof, called Noel Caperon, an Apothecary dwelling in Orleance, at the time he first found it, and was shortly after the finding thereof taken away in the Massacre in France. It is now generally called *Fritillaria*, of the word *Fritillus*, which diuers doe take for the Chesse borde or table whereon they play, whereunto, by reason of the resemblance of the great squares or spots so like it, they did presently referre it. It is called by Lobel *Lilionarcissus purpureus variegatus, & tessulatus*, making it a kinde of Tulipa; but as I said in the beginniug of the Chapter, it doth most neerely resemble a small pendulous Lilly, and might therefore rightly hold the name of *Lilium variegatum*, or in English, the checkerd Lilly. But because the errour which first referred it to a Daffodill, is growne strong by custome of continuance, I leaue to euery one their owne will, to call it in English eyther Fritillaria, as it is called of most, or the checkerd Daffodill, or the Ginnie Hen flower, or, as I doe, the checkerd Lilly. I shall not neede in this place further to explaine the seuerall names of euery of them, hauing giuen you them in their titles.

The

The Vertues.

I haue not found or heard by any others of any property peculiar in this plant, to be applied either inwardly or outwardly for any difease : the chiefe or onely vfe thereof is, to be an ornament for the Gardens of the curious louers of thefe delights, and to be worne of them abroad, which for the gallant beauty of many of them, deferueth their courteous entertainment, among many other the like pleafures.

Chap. VIII.

Tulipa. The Turkes Cap.

NExt vnto the Lillies, and before the Narciffi or Daffodils, the difcourfe of Tulipas deferueth his place, for that it partaketh of both their natures ; agreeing with the Lillies in leaues, flowers, and feede, and fomewhat with the Daffodils in rootes. There are not onely diuers kindes of Tulipas, but fundry diuerfities of colours in them, found out in thefe later dayes by many the fearchers of natures varieties, which h aue not formerly been obferued : our age being more delighted in the fearch, curiofity, and rarities of thefe pleafant delights, then any age I thinke before. But indeede, this flower, aboue many other, deferueth his true commendations and acceptance with all louers of thefe beauties, both for the ftately afpect, and for the admirable varietie of colours, that daily doe arife in them, farre beyond all other plants that grow, in fo much, that I doubt, although I fhall in this Chapter fet downe the varieties of a great many, I fhall leaue more vnfpoken of, then I fhall defcribe ; for I may well fay, there is in this one plant no end of diuerfity to be expected, euery yeare yeelding a mixture and variety that hath not before been obferued, and all this arifing from the fowing of the feede. The chiefe diuifion of Tulipas, is into two forts : *Præcoces*, early flowring Tulipas, and *Serotinæ*, late flowring Tulipas. For that fort which is called *Mediæ* or *Dubiæ*, that is, which flower in the middle time betweene them both, and may be thought to be a kinde or fort by it felfe, as well as any of the other two : yet becaufe they doe neerer participate with the *Serotinæ* then with the *Præcoces*, not onely in the colour of the leafe, being of the fame greenneffe with the *Serotinæ*, and moft vfually alfo, for that it beareth his ftalke and flower, high and large like as the *Serotinæ* doe ; but efpecially, for that the feede of a *Media Tulipa* did neuer bring forth a *Præcox* flower (although I know Clufius, an induftrious, learned, and painfull fearcher and publifher of thefe rarities, faith otherwife) fo farre as euer I could, by mine owne care or knowledge, in fowing their feede apart, or the affurance of any others, the louers and fowers of Tulipa feede, obferue, learne, or know : and becaufe alfo that the feede of the *Serotinæ* bringeth forth *Medias*, and the feede of *Medias Serotinæ*, they may well bee comprehended vnder the generall title of *Serotinæ* : But becaufe they haue generally receiued the name of *Mediæ*, or middle flowring Tulipas, to diftinguifh betweene them, and thofe that vfually doe flower after them ; I am content to fet them downe, and fpeake of them feuerally, as of three forts. Vnto the place and ranke likewife of the *Præcoces*, or early flowring Tulipas, there are fome other feuerall kinds of Tulipas to be added, which are notably differing, not onely from the former *Præcox Tulipa*, but euery one of them, one from another, in fome fpeciall note or other : as the *Tulipa Boloniensis flore rubro*, the red Bolonia Tulipa. *Tulipa Boloniensis flore luteo*, the yellow Bolonia Tulipa. *Tulipa Perfica*, the Perfian Tulipa. *Tulipa Cretica*, the Candie Tulipa, and others : all which fhall bee defcribed and entreated of, euery one apart by it felfe, in the end of the ranke of the *Præcoces*, becaufe all of them flower much about their time. To begin then with the *Præcox*, or early flowring Tulipas, and after them with the *Medias* and *Serotines*, I fhall for the better method, diuide their flowers into foure primary or principall colours, that is to fay, White, Purple, Red, and Yellow, and vnder euery one of thefe colours, fet downe the feuerall varieties

ties

ties of mixtures we haue seene and obserued in them, that so they may be both the better described by me, and the better conceiued by others, and euery one placed in their proper ranke. Yet I shall in this, as I intend to doe in diuers other plants that are variable, giue but one description in generall of the plant, and then set downe the varietie of forme or colour afterwards briefly by themselues.

Tulipa præcox. The early flowring Tulipa.

The early Tulipa (and so all other Tulipas) springeth out of the ground with his leaues folded one within another, the first or lowest leafe riseth vp first, sharpe pointed, and folded round together, vntill it be an inch or two aboue the ground, which then openeth it selfe, shewing another leafe folded also in the bosome or belly of the first, which in time likewise opening it selfe, sheweth forth a third, and sometimes a fourth and a fifth: the lower leaues are larger then the vpper, and are faire, thicke, broad, long, and hollow like a gutter, and sometimes crumpled on the edges, which will hold water that falleth thereon a long time, of a pale or whitish greene colour, (and the *Mediæ* and *Serotinæ* more greene) couered ouer as it were with a mealinesse or hoarinesse, with an eye or shew of rednesse towards the bottome of the leaues, and the edges in this kinde being more notable white, which are two principall notes to know a *Præcox Tulipa* from a *Media* or *Serotina*: the stalke with the flower riseth vp in the middle, as it were through these leaues, which in time stand one aboue another, compassing it at certaine vnequall distances, and is often obserued to bend it selfe crookedly downe to the ground, as if it would thrust his head thereinto, but turning vp his head (which will be the flower) againe, afterwards standeth vpright, sometimes but three or foure fingers or inches high, but more often halfe a foote, and a foot high, but the *Mediæ*, and *Serotinas* much higher, carrying (for the most part) but one flower on the toppe thereof, like vnto a Lilly for the forme, consisting of sixe leaues, greene at the first, and afterwards changing into diuers and sundry seuerall colours and varieties, the bottomes likewise of the leaues of these sometimes, but most especially of the *Mediæ*, being as variable as the flower, which are in some yellow, or green, or blacke, in others white, blew, purple, or tawnie; and sometimes one colour circling another: some of them haue little or no sent at all, and some haue a better then others. After it hath been blowne open three or foure dayes or more, it will in the heate of the Sunne spread it selfe open, and lay it selfe almost flat to the stalke: in the middle of the flower standeth a greene long head (which will be the seed vessell) compassed about with sixe chiues, which doe much vary, in being sometimes of one, and sometimes of another colour, tipt with pendents diuersly varied likewise: the head in the middle of the flower groweth after the flower is fallen, to be long, round, and edged, as it were three square, the edges meeting at the toppe, where it is smallest, and making as it were a crowne (which is not seen in the head of any Lilly) and when it is ripe, diuideth it selfe on the inside into sixe rowes, of flat, thinne, brownish, gristly seede, very like vnto the seede of the Lillies, but brighter, stiffer, and more transparent: the roote being well growne is round, and somewhat great, small and pointed at the toppe, and broader, yet roundish at the bottome, with a certaine eminence or seate on the one side, as the roote of the Colchicum hath; but not so long, or great, it hath also an hollownesse on the one side (if it haue borne a flower) where the stalke grew, (for although in the time of the first springing vp, vntill it shew the budde for flower, the stalke with the leaues thereon rise vp out of the middle of the roote; yet when the stalke is risen vp, and sheweth the budde for flower, it commeth to one side, making an impression therein) couered ouer with a brownish thin coate or skin, like an Onion, hauing a little woollinesse at the bottome; but white within, and firme, yet composed of many coates, one folding within another, as the roote of the Daffodils be, of a reasonable good taste, neyther very sweete, nor yet vnpleasant. This description may well serue for the other Tulipas, being *Mediæ* or *Serotinas*, concerning their springing and bearing, which haue not any other great variety therein worth the note, which is not expressed here; the chiefe difference resting in the variety of the colours of the flower, and their seuerall mixtures and markes, as I said before: sauing onely, that the flowers of some are great and large, and of others smaller, and the leaues of some long

and

Tulipa præcox alba sine rubra, &c. vnius coloris. The early white or red Tulipa, &c. being of one colour.
2 *Tulipa præcox purpurea oris albis.* The early purple Tulipa with white edges, or the Prince. 3 *Tulipa præcox variegata.* The early stript Tulipa. 4 *Tulipa præcox rubra oris luteis.* The early red Tulipa with yellow edges, or the Duke.

and pointed, and of others broad and round, or bluntly pointed, as shall bee shewed in the end of the Chapter : I shall therefore onely expresse the colours, with the mixture or composure of them, and giue you withall the names of some of them, (for it is impossible I thinke to any man, to giue seuerall names to all varieties) as they are called by those that chiefly delight in them with vs.

Tulipa præcox Alba.	The early White Tulipa.
1 *Niuea tota interdum purpureis staminibus, vel saltem luteis, fundo puro kaua luteo.*	1 The flower whereof is either pure snow white, with purple sometimes, or at least with yellow chiues, without any yellow bottome.
2 *Alba siue niuea fundo luteo.*	2 Or pure white with a yellow bottome.
3 *Albida.*	3 Or milk white that is not so pure white.
4 *Alba, venis cæruleis in dorso.*	4 White with blew veines on the outside.
5 *Alba purpureis oris.* ⎫ *Harum flores vel*	5 White with purple edges. ⎫ Some of these a-
6 *Alba carneis oris.* ⎬ *constantes, vel*	6 White with blush edges. ⎬ biding constant, & others sprea-
7 *Alba sanguineis oris.* ⎭ *dispergentes.*	7 White with red edges. ⎭ ding or running.
8 *Alba oris magnis carneis, & venis intro respicientibus.*	8 White with great blush edges, and some strakes running from the edge inward.
9 *Alba extra, carnei vero coloris intus, oras habens carneas saturatiores.*	9 White without, and somewhat blush within, with edges of a deeper blush.
10 *Albida, oris rubris, vel oris purpureis.*	10 Whitish, or pale white with red or purple edges.
11 *Albida purpurascentibus maculis extra, intus vero carnei vinacissimi.*	11 Whitish without, with some purplish veins & spots, & of a liuely blush within.
12 *Alba, purpureis maculis aspersa extra, intus vero alba purpurantibus oris.*	12 White without, spotted with small purple spots, and white within with purple edges.
13 *Dux Alba, i. e. coccineis & albis variata flammis, à medio ad oras intercursantibus.*	13 A white Duke, that is, parted with white & crimson flames, from the middle of each leafe to the edge.
14 *Princessa, i.e. argentei coloris maculis purpurascentibus.*	14 The Princesse, that is, a siluer colour spotted with fine deepe blush spots.
15 *Regina pulcherrima, albis & sanguineis aspersa radijs & punctis.*	15 The Queen, that is, a fine white sprinkled with bloud red spots, and greater strakes.

Tulipa præcox purpurea.	The early purple Tulipa.
1 *Purpurea satura rubescens, vel violacea.*	1 A deep reddish purple, or more violet.
2 *Purpurea pallida, Columbina dicta.*	2 A pale purple, called a Doue colour.
3 *Persici coloris saturi.*	3 A deep Peach colour.
4 *Persici coloris pallidioris.*	4 A paler Peach colour.
5 *Pæoniæ floris coloris.*	5 A Peony flower colour.
6 *Rosea.*	6 A Rose colour.
7 *Chermesina peramæna.*	7 A Crimson very bright.
8 *Chermesina parum striata.*	8 A Crimson stript with a little white.

9 Princeps,

9 *Princeps, i.e. purpurea saturatior vel dilu-*
tior, oris albis magnis vel paruis, fundo lu-
teo, vel albo orbe, quæ multum variatur,
& colore, & oris, ita vt purpurea ele-
gans oris magnis albis, dicta est, Princeps
excellens, &

10 *Princeps Columbina, purpurea diluti-*
or.

11 *Purpurea Ghermesiua, rubicundioris colo-*
ris, albidis vel albis oris.

12 *Purpurea, vel obsoleta albidis oris Prin-*
ceps Brancion.

13 *Purpurea diluta, oris dilutioris purpurei*
coloris.

14 *Purpurea in exterioribus, carnei vero ad*
medium intus, oris albis, fundo luteo.

15 *Purpurea albo plumata extra, oris albis,*
purpurascens intus, fundo luteo, vel orbe
albo.

16 *Aliæ, minus elegans plumata, minoribusq́;*
oris albidis.

9 A Prince or Bracklar, that is, a deepe or pale purple, with white edges, greater or smaller, and a yellow bottome, or circled with white, which varieth much, both in the purple & edges, so that a faire deep purple, with great white edges, is called, The best or chiefe Prince, and

10 A paler purple with white edges, called a Doue coloured Prince.

11 A Crimson Prince or Bracklar.

12 A Brancion Prince, or purple Brancion.

13 A purple with more pale purple edges.

14 Purple without, and blush halfe way within, with white edges, and a yellow bottome.

15 Purple feathered with white on the out side, with white edges, and pale purple within, the ground being a little yellow, or circled with white.

16 Another very neere vnto it, but not so fairely feathered, being more obscure, and the edges not so great or whitish.

Tulipa præcox rubra.

The early red Tulipa.

1 *Rubra vulgaris fundo luteo, & aliquando*
nigro.

2 *Rubra satura oris luteis paruis, dicta*
Roan.

3 *Baro, i.e. rubra magis intensa, oris luteis*
paruis.

4 *Dux maior & minor, i.e. rubra magis aut*
minus elegans satura, oris luteis maximis
vel minoribus, & fundo luteo magno. Alia
alijs est magis amœna, in alijs etiam fundo
nigro vel obscuro viridi.

5 *Ducissa, i.e. Duci similis, at plus lutei*
quàm rubri, oris magnis luteis, & rubore
magis aut minus intus in gyrum acto, fundo
item luteo magno.

6 *Testamentum Brancion, i.e. rubra sangui-*
nea satura, aut minus rubra, oris pallidis,
magnis vel paruis: alia alijs magis aut mi-
nus elegans diuersimodo.

1 An ordinary red, with a yellow, & sometimes a blacke bottome.

2 A deep red, with a small edge of yellow, called a Roane.

3 A Baron, that is, a faire red with a small yellow edge.

4 A Duke, a greater and a lesser, that is, a more or less faire deep red, with greater or lesser yellow edges, and a great yellow bottome. Some of this sort are much more or lesse faire then others, some also haue a blacke or darke greene bottome.

5 A Dutchesse, that is like vnto the Duke, but more yellow then red, with greater yellow edges, and the red more or lesse circling the middle of the flower on the inside, with a large yellow bottome.

6 A Testament Brancion, or a Brancion Duke,

7 *Flambans, ex rubore & flauedine radiata, vel striata fundo luteo.*

8 *Mali Aurantij coloris, ex rubore, & flauedine integrè, non separatim mixta, oris luteis paruis, vel absq́, oris.*

9 *Minij, siue Cinabaris coloris, i.e. ex purpurea, rubedine, & flauedine radiata, vnguibus luteis, & aliquando oris.*

10 *Rex Tuliparum, i.e. ex sanguineo & aureo radiatim mixta, à flammea diuersa, fundo luteo, orbe rubro.*

11 *Tunica Morionis, i.e. ex rubore & aureo separatim diuisa.*

Duke, that is, a faire deepe red, or lesse red, with a pale yellow or butter coloured edge, some larger others smaller: and some more pleasing then others, in a very variable manner.

7 A Flambant, differing from the Dutchesse; for this hath no such great yellow edge, but streaks of yellow through the leafe vnto the very edge.

8 An Orenge colour, that is, a reddish yellow, or a red and yellow equally mixed, with small yellow edges, and sometimes without.

9 A Vermillion, that is, a purplish red, streamed with yellow, the bottome yellow, and sometimes the edges.

10 The Kings flower, that is, a crimson or bloud red, streamed with a gold yellow, differing from the Flambant, the bottome yellow, circled with red.

11 A Fooles coate, parted with red and yellow guardes.

Tulipa præcox lutea.

1 *Lutea siue flaua.*
2 *Pallida lutea siue straminea.*
3 *Aurea, oris rubicundis.*
4 *Straminea, oris rubris.*
5 *Aurea, rubore persusa extra.*
6 *Aurea, vel magis pallida, rubore in gyrum acta simillima Ducissa, nisi minus rubedinis habet.*
7 *Aurea, extremitatibus rubris, dici potest, Morionis Pileus præcox.*

The early yellow Tulipa.

1 A faire gold yellow without mixture.
2 A strawe colour.
3 A faire yellow with reddish edges.
4 A strawe colour, with red edges.
5 A faire yellow, reddish on the out side onely.
6 A gold or paler yellow, circled on the inside a little with red, very like the Dutchesse, but that it hath lesse red therein.
7 A gold yellow with red toppes, and may be called, The early Fooles Cap.

Tulipa de Caffa. The Tulipa of Caffa.

There is another fort or kinde of early Tulipa, differing from the former, whofe pale greene leaues being as broad and large as they, and fometimes crumpled or waued at the edges, in fome haue the edges onely of the faid leaues for a good breadth, of a whitifh or whitifh yellow colour, and in others, the leaues are lifted or parted with whitifh yellow and greene : the ftalke rifeth not vp fo high as the former, and beareth a flower at the toppe like vnto the former, in fome of a reddifh yellow colour, with a ruffet coloured ground or bottome, and in others, of other feuerall colours : the feede and roote is fo like vnto others of this kinde, that they cannot be diftinguifhed.

There is (as I doe heare) of this kinde, both *Præcoces*, and *Serotina*, early flowring, and late flowring, whereof although wee haue not fo exact knowledge, as of the reft, yet I thought good to fpeake fo much, as I could hitherto vnderftand of them, and giue others leaue (if I doe not) hereafter to amplifie it.

Tulipa Boloniensis, fiue Bombycina flore rubro major.
The greater red Bolonia Tulipa.

There are likewife other kindes of early Tulipas to bee fpoken of, and firft of the red Bolonia Tulipa ; the roote whereof is plainly difcerned, to be differing from all others : for that it is longer, and not hauing fo plaine an eminence at the bottome thereof, as the former and later Tulipas, but more efpecially becaufe the toppe is plentifully ftored with a yellowifh filke-like woollineffe : the outfide likewife or skinne is of a brighter or paler red, not fo eafie to be pilled away, and runneth vnder ground both downeright and fidewife (efpecially in the Countrey ground and ayre, where it will encreafe aboundantly, but not either in our London ayre, or foreʼt grounds) fomewhat like vnto the yellow Bolonia Tulipa next following. It fhooteth out of the ground with broad and long leaues, like the former ; but neither fo broad, nor of fo white or mealy a greene colour as the former, but more darke then the late flowring Tulipa, fo that this may bee eafily difcerned by his leafe from any other Tulipa aboue the ground, by one that is skilfull.   It beareth likewife three or foure leaues vpon the ftalke, like the former, and a flower alfo at the toppe of the fame fafhion, but that the leaues hereof are alwayes long, and fomewhat narrow ; hauing a large blacke bottome, made like vnto a cheuerne, the point whereof rifeth vp vnto the middle of the leafe, higher then any other Tulipa ; the flower is of a pale red colour, nothing fo liuely as in the early or late red Tulipas, yet fweeter for the moft part then any of them, and neereft vnto the yellow Bolonia Tulipa, which is much about the fame fent.

Tulipa pumilio rubra, fiue Bergomenfis rubra media & minor.
The dwarfe red Bergomo Tulipa, a bigger and a leffer.

There are two other forts hereof, and becaufe they were found about Bergomo, do carry that name, the one bigger or leffer then another, yet neither fo great as the former, hauing very little other difference to bee obferued in them, then that they are fmaller in all parts of them.

Tulipa Boloniensis flore luteo. The yellow Bolonia Tulipa.

The roote of this Tulipa may likewife bee knowne from the former red (or any other Tulipa) in that it feldome commeth to bee fo bigge, and is not fo woolly at the toppe, and the skinne or outfide is fomewhat paler, harder, and fharper pointed : but the bottome is like the former red, and not fo eminent as the early or late Tulipas. This beareth much longer and narrower leaues then any (except the Perfian & dwarfe yellow Tulipas) and of a whitifh greene colour : it beareth fometimes but one flower on a ftalke, and fometimes two or three wholly yellow, but fmaller, & more open then the other kinds, and (as I faid) fmelleth fweete, the head for feede is fmaller then in others, and hath not that crowne at the head thereof, yet the feed is like, but fmaller.

E 2   *Tulipa*

Tulipa Narbonensis, siue Monspeliensis vel pumilio.
The French or dwarfe yellow Tulipa.

This Tulipa is very like vnto the yellow Bolonia Tulipa, both in roote, leafe, and flower, as also in the colour thereof, being yellow: the onely difference is, that it is in all things lesser and lower, and is not so apt to beare, nor so plentifull to encrease by the roote.

Tulipa Italica maior & minor. The Italian Tulipa the greater and the lesser.

Both these kindes of Tulipas doe so neere resemble the last kinde, that I might almost say they were the same, but that some difference which I saw in them, maketh mee set them apart; and consisteth in these things, the stalkes of neither of both these rise so high, as of the first yellow Bolonia Tulipa: the leaues of both sorts are writhed in and out at the edges, or made like a waue of the sea, lying neerer the ground, and the flower being yellow within, is brownish or reddish on the backe, in the middle of the three outer leaues the edges appearing yellow. Both these kindes doe differ one from the other in nothing, but in that one is bigger, and the other smaller then the other which I saw with Iohn Tradescante, my very good friend often remembred.

Tulipa Lusitanica, siue pumilio versicolor. The dwarfe stript Tulipa.

This dwarfe Tulipa is also of the same kindred with the three last described; for there is no other difference in this from them, then that the flower hath some red veins running in the leaues thereof.
There are two other sorts of dwarfe Tulipas with white flowers, whereof Lobel hath made mention in the Appendix to his *Aduersaria*; the one whereof is the same that Clusius setteth forth, vnder the title of *Pumilio altera*: but because I haue not seen either of them both, I speake no further of them.

Tulipa pumilio alba. The white dwarfe Tulipa.

But that white flower that Iohn Tradescante shewed me, and as hee saith, was deliuered him for a white Pumilio, had a stalke longer then they set out theirs to haue, and the flower also larger, but yet had narrower leaues then other sorts of white Tulipas haue.
Tulipa Bicolor. The small party coloured Tulipa.

Vnto these kindes, I may well adde this kinde of Tulipa also, which was sent out of Italy, whose leaues are small, long, and narrow, and of a darke greene colour, somewhat like vnto the leaues of an Hyacinth: the flower is small also, consisting of sixe leaues, as all other Tulipas doe, three whereof are wholly of a red colour, and the other three wholly of a yellow.

Tulipa Persica. The Persian Tulipa.

This rare Tulipa, wherewith we haue beene but lately acquainted, doth most fitly deserue to be described in this place, because it doth so neerely participate with the Bolonia and Italian Tulipas, in roote, leafe, and flower: the roote hereof is small, couered with a thicke hard blackish shell or skinne, with a yellowish woollinesse both at the toppe, and vnder the shell. It riseth out of the ground at the first, with one very long and small round leafe, which when it is three or foure inches high, doth open it selfe, and shew forth another small leafe (as long almost as the former) breaking out of the one side thereat, and after it a third, and sometimes a fourth, and a fift; but each shorter then other, which afterwards be of the breadth of the dwarfe yellow Tulipa, or somewhat broader, but much longer then any other, and abiding more hollow, and of the colour of the early Tulipas on the inside: the stalke riseth vp a foot and a halfe
high

Tulipa Bombycina flore vnbre. The red Bolonia Tulipa. 2 Tulipa Boloniensis flore luteo. The yellow Bolonia Tulipa. 3 Tulipa pumilio rubra
ſiue lutea. The red or yellow dwarfe Tulipa. 4 Folium Tulipa de Caſſa per totum ſtriatum. The leafe of the Tulipa of Caſſa ſtriped throughout
the whole leaſe. 5 Folium Tulipa de Caſſa per oras ſtriatum. The leafe of the Tulipa of Caſſa ſtriped at the edges oncly. 6 Tulipa Perſica. The
Peſſian Tulipa. 7 Tulipa Cretica. The Tulipa of Candie. 8 Tulipa Armeniaca. The Tulipa of Armenia.

high fometimes, bearing one flower thereon, compofed of fixe long and pointed leaues of the forme of other fmall Tulipas, and not fhewing much bigger then the yellow Italian Tulipa, and is wholly white, both infide and outfide of all the leaues, except the three outtermoft, which haue on the backe of them, from the middle toward the edges, a fhew of a brownifh blufh, or pale red colour, yet deeper in the midft, and the edges remaining wholly white : the bottomes of all thefe leaues are of a darke or dun tawnie colour, and the chiues and tippes of a darkifh purple or tawnie alfo. This doth beare feed but feldome in our Country, that euer I could vnderftand, but when it doth, it is fmall like vnto the Bolonia or dwarfe yellow Tulipas, being not fo plentifull alfo in parting, or fetting of by the roote as they, and neuer groweth nor abideth fo great as it is brought vnto vs, and feldome likewife flowreth after the firft yeare : for the rootes for the moft part with euery one grow leffe and leffe, decaying euery yeare, and fo perifh for the moft part by reafon of the frofts and cold, and yet they haue been fet deepe to defend them, although of their owne nature they will runne downe deep into the ground.

Tulipa Byzantina duobus floribus Clufij. The fmall Tulipa of Conftantinople.

The fmall Tulipa of Conftantinople, beareth for the moft part but two leaues on the ftalke, which are faire and broad, almoft like vnto the Candy Tulipa, next hereunto to be defcribed : the ftalke it felfe rifeth not aboue a foote high, bearing fometimes but one flower, but moft commonly two thereon, one below another, and are no bigger then the flowers of the yellow Bolonia Tulipa, but differing in colour ; for this is on the outfide of a purplifh colour, mixed with white and greene, and on the infide of a faire blufh colour, the bottome and chiues being yellow, and the tippes or pendents blackifh : the roote is very like the yellow Bolonia Tulipa.

Tulipa Cretica. The Tulipa of Candie.

This Tulipa is of later knowledge with vs then the Perfian, but doth more hardly thriue, in regard of our cold climate ; the defcription whereof, for fo much as wee haue knowledge, by the fight of the roote and leafe, and relation from others of the flower, (for I haue not yet heard that it hath very often flowred in our Country) is as followeth. It beareth faire broad leaues, refembling the leaues of a Lilly, of a greenifh colour, and not very whitifh : the ftalke beareth thereon one flower, larger and more open then many other, which is eyther wholly white, or of a deepe red colour, or elfe is variably mixed, white with a fine reddifh purple, the bottomes being yellow, with purplifh chiues tipt with blackifh pendents : the roote is fmall, and fomewhat like the dwarfe yellow Tulipa, but fomewhat bigger.

Tulipa Armeniaca. The Tulipa of Armenia.

This fmall Tulipa is much differing from all the former (except the fmall or dwarfe white Tulipas remembred by Lobel and Clufius, as is before fet downe) in that it beareth three or foure fmall, long, and fomewhat narrow greene leaues, altogether at one ioynt or place ; the ftalke being not high, and naked or without leaues from them to the toppe, where it beareth one fmall flower like vnto an ordinary red Tulipa, but fomewhat more yellow, tending to an Orenge colour with a blacke bottome : the roote is not much bigger then the ordinary yellow Bolonia Tulipa, before fet downe.
And thefe are the forts of this firft *Claffu* of early Tulipas.

Tulipa media. The meaner or middle flowring Tulipa.

For any other, or further defcription of this kinde of Tulipa, it fhall not neede, hauing giuen it fufficiently in the former early Tulipa, the maine difference confifting firft in the time of flowring, which is about a moneth after the early Tulipas, yet fome more fome leffe : for euen in the *Præcoces*, or early ones, fome flower a little earlier, and later then others, and then in the colours of the flowers ; for wee haue obferued many
colours,

colours, and mixtures, or varieties of colours in the *Medias*, which we could neuer fee in the *Præcoces*, and fo alfo fome in the *Præcoces*, which are not in the *Medias* : yet there is farre greater varieties of mixture of colours in thefe *Medias*, then hath been obfer-ued in all the *Præcoces*, (although Clufius faith otherwife) eyther by my felfe, or by any other that I haue conuerfed with about this matter, and all this hath happened by the fowing of the feede, as I faid before. I will therefore in this place not trouble you with any further circumftance, then to diftinguifh them, as I haue done in the former early Tulipas, into their foure primary colours, and vnder them, giue you their feuerall varieties and names, for fo much as hath come to my knowledge, not doubting, but that many that haue trauelled in the fowing of the feed of Tulipas many yeares, may obferue each of them to haue fome variety that others haue not: and therefore I thinke no one man can come to the knowledge of all particular diftinctions.

Tulipa media alba.	The white meane flowring Tulipa.
1 *Niuea, fundo albo vel luteo.*	1 A fnow white, with a white or yellow bottome.
2 *Argentea, quafi alba cineracea fundo lute-fcente, purpureis ftaminibus.*	2 A filuer colour, that is, a very pale or whitifh afhe colour, with a yellowifh bottome and purple chiues.
3 *Margaritina alba, carneo dilutifsima.*	3 A Pearle colour, that is, white, with a wafh or fhew of blufh.
4 *Alba, fundo cæruleo vel nigro.*	4 A white, with a blew or black bottome.
5 *Albida.*	5 A Creame colour.
6 *Alba, oris rubris.*	6 A white, with red edges.
7 *Alba, purpureis oris.*	7 A white, with purple edges.
8 *Alba, oris coccineis.*	8 A white, with crimfon edges.

Hæc tria genera in aliquibus conftanter tenent oras, in a'y dilpergunt.

Thefe three forts doe hold their edges conftant in fome, but well fpread in others.

9 *Albidi primum, deinde albidior, oris pur-pureis, & venis intrò refpicientibus, dicta nobis Hackquenay.*

9 A pale or whitifh yellow, which after a few dayes groweth more white, with purplifh red edges, and fome ftreakes running inward from the edge, which we call an Hackney.

10 *Alba, fanguineo colore variata, fundo vel albifsimo, vel alio.*

10 A white mixed with a bloud red very variably, and with a pure white, or o-ther coloured bottome.

11 *Alba, radiatim difpofita flammis, & ma-culis coccineis.*

11 A white, ftreamed with crimfon flames, and fpots through the whole flower.

12 *Alba, purpurea rubedine plumata, diuer-farum fpecierum, quæ cum fuperiore, vel albo, vel luteo, vel paruo cæruleo conftant fundo, quæ conftanter tenent punctatos co-lores, & non difpergunt, fed poft trium aut quatuor dierum fpatium pulchriores appa-rent.*

12 A white, fpeckled with a reddifh pur-ple, more or leffe, of diuers forts, with white, yellow, or blew bottomes, all which doe hold their markes conftant, and doe not fpread their colours, but fhew fairer after they haue ftood blown three or foure dayes.

13 *Panni argentei coloris, i.e. alba, plumata, punctata, ftriata, vel diuerfimodè variata, rubedine dilutiore, vel faturatiore purpu-rea, interius vel exterius, vel vtrinq, diuerfarum fpecierum.*

13 A cloth of filuer of diuers forts, that is, a white fpotted, ftriped, or otherwife marked with red or purple, in fome pa-ler, in fome deeper, either on the infide, or on the outfide, or on both.

14 *Tunica morionis alba varia, i.e. ex albo & purpureo ftriata diuerfimodè, fundo albo vel alio.*

14 A white Fooles coate of diuers forts, that is, purple or pale crimfon, and white, as it were empaled together, ey-ther with a white ground or other, whereof there is great variety.

15 *Holias alba vel albida, abfq; fundo, vel fundo purpureocæruleo, vel cæruleo albo cir-cundato, diuersè fignata, vel variata intus ad medietatem foliorum, furfum in orbem vt plurimum, vel ad oras pertingens am-plas & albas. Hæ fpecies tantoperè multi-plicantur, vt vix fint explicabiles.*

15 A white Holias, that is, a faire white, or paler white, eyther without a bottome, or with a blewifh purple bottome, or blew and white circling the bottome,

and

and from the middle vpwards, speckled and straked on the inside for the most part, with bloud red or purplish spots and lines vnto the very edges, which abide large and white. Of this kinde there are found very great varieties, not to be expressed.

Of this sort there is so much variety, some being larger or fairer marked then others, their bottomes also varying, that it is almost impossible to express them.

Tanta est huius varietas, vel multitudine, vel striarum paucitate & distinctione, vel fundis variantibus, vt ad tædium esset perscribere.

Tulipa media purpurea.	The meane flowring purple Tulipa.
1 *Purpurea satura.*	1 A faire deep purple.
2 *Purpurea dilutior, diuersarum specierum, quarum Rosea vna, Carnea sit altera.*	2 A paler purple, of many sorts, whereof a Rose colour is one, a Blush another.
3 *Persici coloris, duarum aut trium specierum.*	3 A Peach colour of two or three sorts.
4 *Chermesina, obscura, aut pallida.*	4 A Crimson, deepe, or pale.
5 *Stamela, intensior aut remissior.*	5 A Stamell, darke or light.
6 *Xerampelina.*	6 A Murrey.
7 *Purpurea, striata.*	7 A purple, stript and spotted.
8 *Persici saturi, vel diluti coloris, vndulata, vel radiata.*	8 A Peach colour, higher or paler, waued or stript.
9 *Columbina, oris & radijs albis.*	9 A Doue colour, edged and straked with white.
10 *Purpurea rubra, oris albis, similis Præcoci, dicta Princeps.*	10 A faire red purple, with white edges, like vnto the early Tulipa, called a Prince
11 *Chermesina, vel Heluola, lineis albis in medio, & versus oras, fundo cæruleo, vel albo, item q̃, albo orbe.*	11 A faire Crimson, or Claret wine colour, with white lines both in the middle, and towards the edges, most haue a blew bottome, yet some are white, or circled with white.
12 *Purpurea remissior, aut intensior, oris albis, paruis aut magnis, vt in Princspe præcoci, fundo vel cæruleo orbe albo, vel albo orbe cæruleo amplo.*	12 A light or deepe purple, with white edges, greater or smaller, like the early Prince, the bottomes eyther blew circled with white, or white circled with a large blew.
13 *Holias Heluola, sanguineis guttis intus à medio sursum in orbem, fundo cæruleo.*	13 A purple Holias, the colour of a pale Claret wine, marked and spotted with bloud red spots, round about the middle of each leafe vpward on the inside onely, the bottome being blew.
14 *Tunica Morionis purpurea rubra satura, albido striata, quam in alba saturatior, fundo ex cæruleo & albo.*	14 A Crimson Fooles Coate, a darke crimson, and pale white empaled together, differing from the white Fooles Coate, the bottome blew and white.
15 *Purpurea rubra satura vel diluta, albo vel albedine, punctata vel striata diuersimodè, dicta Cariophyllata.*	15 A deeper or paler reddish purple, spotted or striped with a paler or purer white, of diuers sorts, called the Gilloflower Tulipa.

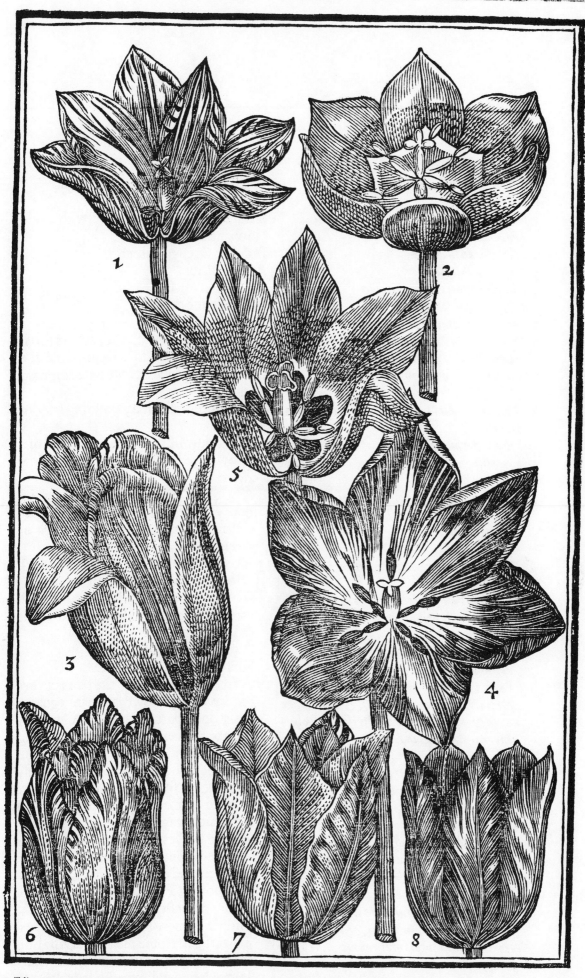

1 *Tulipa rubra & lutea varia* The Fooles Coate red and yellow. 2 *Tulipa Holeas alba absǫ; fundo*. The white Holeas without a bottome. 3 *Tulipa argentea, vel punctata, &c.* The cloth of siluer, or other spotted Tulipa. 4 *Tulipa alba flammis coccineis.* The white Fooles Coate. 5 *Tulipa Holeas alba, &c. fundo purpureo, &c.* A white Holeas, &c. with a purple bottome, &c. 6 *Tulipa rubra & lutea flammea, &c.* A red and yellow flamed Tulipa, &c. 7 *Tulipa alba striata & punctata*. A white striped and spotted Tulipa. 8 *Tulipa altera variata, &c.* Another variable Tulipa.

Tulipa media rubra.

1 *Rubra communis, fundo luteo, vel nigro.*
2 *Mali Aurantij coloris.*
3 *Cinabaris coloris.*
4 *Lateritij coloris.*
5 *Rubra, luteo aspersa.*
6 *Rubra, oris luteis.*
7 *Testamentum Brancion rubra satura, oris pallidis, diuersarum specierum, rubore variantium, & orarum amplitudine.*

8 *Cinabaris radiata, magis aut minus serotina.*
9 *Rubra purpurascens obsoleta, exterioribus folijs, perfusa luteo intus, oris pallidis luteis.*
10 *Rubra purpurascens elegans extra, & intus lutescens, oris pallidis luteis, fundo luteo vel viridi.*

11 *Rubra flambans coccinea, crebris maculis luteis absq; fundo.*
12 *Flambans elegantior rubra, i.e. radijs luteis intercursantibus ruborem*

13 *Flambans remissior vtroq; colore.*
14 *Panni aurei coloris.*
15 *Tunica Morionis verior, seu Palto du Sot. optima, tenijs amplis amœnis & crebris, ex rubro & flauo separatim diuisis & excurrentibus, flos constans.*

16 *Tunica Morionis altera, tenijs minoribus & minus frequentibus, magis aut minus alia alijs inconstans.*

17 *Tunica Morionis pallida, i.e. tenijs vel strijs frequentioribus in vtroq; colore pallidis, flos est constans & elegans.*
18 *Pileus Morionis, radijs luteis, in medio foliorum latis, per ruborem excurrentibus, fundo luteo, apicibus luteis, & tribus exterioribus folijs luteis oris rubris, vel absq; oris.*

The meane flowring red Tulipa.

1 A faire red which is ordinary, with a yellow or blacke bottome.
2 A deepe Orenge colour.
3 A Vermillion.
4 A pale red, or Bricke colour.
5 A Gingeline colour.
6 A red with small yellow edges.
7 A Testament Brancion of diuers sorts, differing both in the deepnesse of the red, and largenesse of the pale coloured edges.

8 A Vermillion flamed, flowring later or earlier.
9 A dead purplish red without, and of a yellowish red within, with pale yellow edges.
10 A bright Crimson red on the outside, more yellowish on the inside, with pale yellow edges, and a bottome yellow or greene.

11 A red Flambant, spotted thicke with yellow spots without any bottome.
12 A more excellent red Flambant, with flames of yellow running through the red.
13 A pale coloured Flambant.
14 A cloth of gold colour.
15 A true Fooles Coate, the best is a faire red & a faire yellow, parted into guards euery one apart, varied through euery leafe to the very edge, yet in most abiding constant.
16 Another Fooles Coate, not so fairely marked, nor so much, some of these are more or lesse constant in their marks, & some more variable then others.
17 A pale Fooles Coate, that is, with pale red, and pale yellow guardes or stripes very faire and constant.
18 A Fooles Cappe, that is, with lists or stripes of yellow running through the middle of euery leafe of the red, broader at the bottome then aboue, the bottome being yellow, the three outer leaues being yellow with red edges, or without.

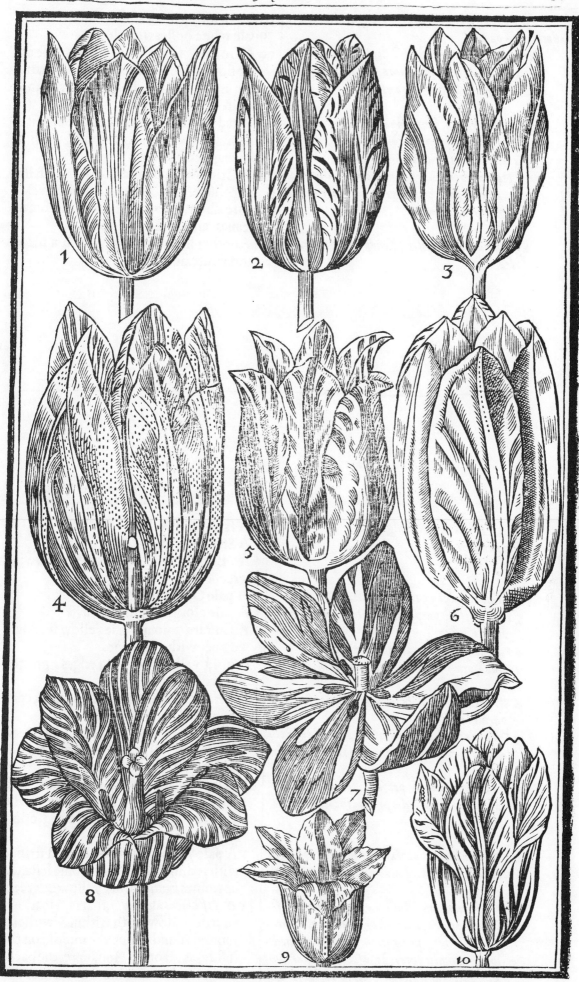

1 *Tulipa tricolor.* A Tulipa of three colours. 2 *Tulipa Macedonica, siue de Caffa varia.* The Tulipa of Caffa purple, with pale white stripes. 3 *Tulipa Heluola charmesina versicolor.* A pure Claret wine colour variable. 4 *Tulipa Caryophyllata Wilmeri.* Mr. Wilmers Gilloflower Tulipa. 5 *Tulipa Chermesina flammis albis.* A Crimson with white flames. 6 *Tulipa Goliah.* A kind of Zwitter called Goliah. 7 *Tulipa le Zwisse.* A Tulipa called the Zwitler. 8 *Tulipa alba flammis coccineis.* Another white Flambant or Fooles Coate. 9 *Tulipa Cinnaberina albo flammata.* The Vermillion flamed. 10 *Tulipa plumata rubra & lutea.* The feathered Tulipa red and yellow.

19 *Le Suiſſe, teniis radiata magnis ex rubore & pallore.*

20 *Altera dicta Goliah à floris magnitudine, teniis radiata ſimillima le Suiſſe, niſi rubor & albedo ſint elegantiores.*

21 *Holias rubra, i.e. ſanguinea argenteis radiis, & guttis in orbem diſpoſitis, praſertim interiùs, fundo viridi ſaturo.*

22 *Holias coccinea, rubra coccinea, albo radiata in orbem, circa medium foliorum interiùs, fundo albo.*

23 *Alia huic ſimilis, fundo albo & caruleo.*

19 A Swiſſe, paned with a faire red and pale white or ſtrawe colour.

20 A Goliah, ſo called of the bigneſſe of the flower, moſt like to the Swiſſe in the marks and guardes, but that the red and white is more liuely.

21 A red Holias. A bloud red ſtript with ſiluer white veines and ſpots, with a darke green bottome.

22 A Crimſon red Holias, that is, a faire purpliſh red, ſpotted with white circlewiſe about the middle of the inner leaues, and a white bottome.

23 Another like thereunto, with a blew and white bottome.

Tulipa media lutea.

1 *Lutea, ſiue Aurea vulgaris.*
2 *Straminea.*
3 *Sulphurea.*

4 *Mali Aurantij pallidi coloris.*
5 *Lutea dilutè purpurea ſtriata, aurei panni pallidi inſtar,*
6 *Pallidè lutea fuſcedine adumbrata.*
7 *Flaua, oris rubris magnis, aut paruis.*

8 *Straminea oris rubris magnis intenſis, vel paruis remiſſis.*
9 *Obſcura & fuliginoſa lutea, inſtar Folij decidui, ideoq; Folium mortuum appellatur.*
10 *Flaua, rubore perfuſa, etiamque ſtriata per totum, dorſo coccineo, oris pallidis.*

11 *Pallidè lutea, perfuſa & magis aut minus rubore ſtriata, fundo vel luteo, vel viridi.*
12 *Teſtamentum Cluſij, i.e. lutea pallida fuligine obfuſca, exteriùs & interiùs ad oras vſq; pallidas, per totum vero floris medium, maculis interiùs aſperſa inſtar omnium aliarum Holias, dorſo obſcurior, fundo viridi.*

The meane flowring yellow Tulipa.

1 A faire gold yellow.
2 A Strawe colour.
3 A Brimſtone colour pale yellowiſh greene.
4 A pale Orenge colour.
5 A pale cloth of gold colour.
6 A Cuſtard colour a pale yellow ſhadowed ouer with a browne.
7 A gold yellow with red edges, greater or ſmaller.
8 A Strawe colour with red edges, deeper or paler, greater or ſmaller.
9 A ſullen or ſmoakie yellow, like a dead leafe that is fallen, and therefore called, *Fueille mort.*
10 A yellow ſhadowed with red, and ſtriped alſo through all the leaues, the backſide of them being of a red crimſon, and the edges pale.
11 A pale yellow, ſhadowed and ſtriped with red, in ſome more in ſome leſſe, the bottomes being either yellow or green.
12 A *Teſtamentum Cluſij*, that is, a ſhadowed pale yellow, both within & without, ſpotted round about the middle on the inſide, as all other Holias are, the backe of the leaues being more obſcure or ſhadowed with pale yellow edges, and a greene bottome.

13 *Flambans lutea, diuersimodè intus magis aut minus striata, vel in alijs extra maculata rubore, fundo vt plurimum nigro, vel in alijs luteo.*

13 A yellow Flambant of diuers sorts, that is, the whole flower more or lesse streamed or spotted on the inside, and in some on the outside with red, the bottome in most being blacke, yet in some yellow.

14 *Flambans pallidior & elegantior.*

14 A paler yellow Flambant more beautifull.

15 *Holias lutea intensior vel remissior diuersimodè, in orbem radiata interius, rubris maculis ad supremas vsq̃; oras, aliquoties crebrè, alias parcè, fundo viridi, vel tanetto obscuro.*

15 A yellow Holias, paler or deeper yellow very variable, spotted on the inside round about the middle, with red sometimes plentifully, or else sparingly, with a green or dark tawny bottome.

16 *Holias straminea rubore striata & punctata, instar alba Holias.*

16 A strawe coloured Holias, spotted and streamed with red, as is to bee seene in the white Holias.

17 *Tunica Morionis lutea, alijs dicta Flammea, in qua color flavius magis & conspicuus rubore, diuersimodè radiata.*

17 A yellow Fooles coate, of some called a flame colour, wherein the yellow is more then the red, diuersly streamed.

Huc reddenda esset viridium Tuliparum classis, quæ diuersarum etiam constat specierum. Vna viridis intensior, cuius flos semper ferè semiclausus manet staminibus fimbriatis. Altera remissior, instar Psittaci pennarum viridium, luteo variata oris albis. Tertia adhuc dilutiori viriditate oris purpureis. Quarta, cujus folia æqualiter purpura diluta, & viriditate diuisa sunt. Quinta, folijs longissimis stellamodo expansis, ex rubore & viriditate coacta.

Vnto these may be added the greene Tulipa, which is also of diuers sorts. One hauing a great flower of a deepe green colour, seldome opening it selfe, but abiding alwaies as it were halfe shut vp and closed, the chiues being as it were feathered. Another of a paler or yellowish green, paned with yellow, and is called, The Parret, &c. with white edges. A third of a more yellowish gieen, with red or purplish edges. A fourth, hath the leaues of the flower equally almost parted, with greene and a light purple colour, which abiding a long time in flower, groweth in time to be fairer marked: for at the first it doth not shew it selfe so plainely diuided. Some call this a greene Swisser. A fifth hath the longest leaues standing like a starre, consisting of greene and purple.

Tulipa Serotina. The late flowring Tulipa.

The late flowring Tulipa hath had his description expressed in the precedent discourse, so that I shall not neede to make a repetition of what hath already beene set downe. The greatest matter of knowledge in this kinde is this, That it hath no such plentifull variety of colours or mixtures in his flowers, as are in the two former sorts, but is confined within these limits here expressed, as farre as hath come to our knowledge.

Tulipa Serotina.	The late flowring Tulipa.
Rosea intensior, aut remissior.	A Rose colour deeper or paler.
Rubra vulgaris, aut saturatior, & quasi nigricans, fundo luteo vel nigro, vel nigro orbe, aureo incluso, dicta Oculus Solis.	An ordinary red, or else a deeper red like blacke bloud, with a blacke or yellow bottome, or blacke circled with yellow, called the Suns eye.
Lutea communis.	An ordinary yellow.
Lutea oris rubris.	A yellow with red edges.
Lutea guttis sanguineis, fundo nigro vel vario.	A yellow with red spots and veines, the bottome blacke or discoloured.

F

There

There yet remaine many obferuations, concerning thefe beautifull flowers, fit to be knowne, which could not, without too much prolixity, be comprehended within the body of the defcription of them; but are referued to bee intreated of a part by them-felues.

All forts of Tulipas beare vfually but one ftalke, and that without any branches: but fometimes nature is fo plentifull in bearing, that it hath two or three ftalkes, and fometimes two, or more branches out of one ftalke (euery ftalke or branch bearing one flower at the toppe) but this is but feldome feene; and when it doth happen once, it is hardly feene againe in the fame roote, but is a great figne, that the roote that doth thus, being an old roote, will the fame yeare part into diuers rootes, whereof euery one, being of a reafonable greatneffe, will beare both his ftalke and flower the next yeare, agreeing with the mother plant in colour, as all the of-fets of Tulipas doe for the moft part: for although the young of-fets of fome doe vary from the maine roote, euen while it groweth with them, yet being feparated, it will bee of the fame colour with the mother plant.

There groweth oftentimes in the *Medias*, and fometimes alfo in the *Præcoces*, but more feldome, a fmall bulbe or roote, hard aboue the ground, at the bottome of the ftalke, and betweene it and the lower leafe, which when the ftalke is dry, and it ripe, being put into the ground, will bring forth in time a flower like vnto the mother plant, from whence it was taken.

The flowers alfo of Tulipas confift moft commonly of fixe leaues, but fometimes they are feene to haue eight or tenne, or more leaues; but vfually, thofe rootes beare but their ordinary number of fixe leaues the next yeare: the head for feede then, is for the moft part foure fquare, which at all other times is but three fquare, or when the flower wanteth a leafe or two, as fometimes alfo it doth, it then is flat, hauing but two fides.

The forme of the flower is alfo very variable; for the leaues of fome Tulipas are all fharpe pointed, or all blunt and round pointed, and many haue the three outer leaues fharpe pointed, and the three inner round or pointed, and fome contrariwife, the three outermoft round pointed, and the three inner fharpe pointed. Againe, fome haue all the leaues of the flowers long and narrow, and fome haue them broader and fhorter. Some *Præcoces* alfo haue their flowers very large and great, equall vnto eyther the *Media*, or *Serotina*, which moft commonly are the largeft, and others haue them as fmall as the Bolonia Tulipa.

The bottomes of the leaues of the flowers are alfo variably diuerfified, and fo are both the chiues or threeds that ftand vp about the head, and the tips or pendents that are hanging loofe on the toppes of them; and by the difference of the bottomes or chiues, many flowers are diftinguifhed, which elfe are very like in colour, and alike alfo marked.

For the fmell alfo there is fome diuerfity; for that the flowers of fome are very fweete, of others nothing at all, and fome betweene both, of a fmall fent, but not offen-fiue: and yet fome I haue obferued haue had a ftrong ill fent; but how to fhew you to diftinguifh them, more then by your owne fenfe, I cannot: for the feedes of fweete fmelling Tulipas doe not follow their mother plant, no more then they doe in the colour.

And laftly, take this, which is not the leaft obferuation, worth the noting, that I haue obferued in many: When they haue beene of one entire colour for diuers yeares, yet in fome yeare they haue altered very much, as if it had not beene the fame, *viz.* from a purple or ftamell, it hath beene variably either parted, or mixed, or ftriped with white, eyther in part, or through the whole flower, and fo in a red or yellow flower, that it hath had eyther red or yellow edges, or yellow or red fpots, lines, veines, or flames, running through the red or yellow colour, and fometimes it hath happened, that three leaues haue been equally parted in the middle with red and yel-low, the other three abiding of one colour, and in fome the red had fome yellow in it, and the yellow fome red fpots in it alfo; whereof I haue obferued, that all fuch flow-ers, not hauing their originall in that manner, (for fome that haue fuch or the like markes from the beginning, that is, from the firft and fecond yeares flowring, are con-ftant, and doe not change) but as I faid, were of one colour at the firft, doe fhew the

weakneffe

weakneſſe and decay of the roote, and that this extraordinary beauty in the flower, is but as the brightneſſe of a light, vpon the very extinguiſhing thereof, and doth plainly declare, that it can doe his Maſter no more ſeruice, and therefore with this iollity doth bid him good night. I know there is a common opinion among many (and very confidently maintained) that a Tulipa with a white flower, hath changed to beare a red or yellow, and ſo of the red or yellow, and other colours, that they are likewiſe inconſtant, as though no flowers were certaine : but I could neuer either ſee or heare for certaine any ſuch alteration, nor any other variation, but what is formerly expreſſed. Let not therefore any iudicious be carried away with any ſuch idle conceit, but rather ſuſpect ſome deceit in their Gardeners or others, by taking vp one, and putting in another in the place, or elſe their owne miſtaking.

Now for the ſowing, planting, tranſplanting, choiſe, and ordering of Tulipas, which is not the leaſt of regard, concerning this ſubiect in hand, but (as I think) would be willingly entertained ; What I haue by my beſt endeauours learned, by mine owne paines in almoſt forty yeares trauell, or from others informations, I am willing here to ſet downe ; not doubting, but that ſome may adde what hath not come to my knowledge.

Firſt, in the ſowing of ſeedes of Tulipas, I haue not obſerued (whatſoeuer others haue written) nor could of certainty learne of others, that there doth ariſe from the ſeedes of *Præcoces* any *Medias* or *Serotine* Tulipas, (or but very ſeldome) nor am certainly aſſured of any : but that the ſeedes of all *Præcoces* (ſo they be not doubtfull, or of the laſt flowring ſorts) will bring *Præcoces* : And I am out of doubt, that I neuer ſaw, nor could learne, that euer the ſeede of the *Medias* or *Serotines* haue giuen *Præcoces* ; but *Medias* or *Serotines*, according to their naturall kinde. But if there ſhould bee any degeneration, I rather incline to thinke, that it ſooner commeth to paſſe *(à meliore ad peius*, for *facilis eſt deſcenſus*, that is) that *Præcoces* may giue *Medias*, then that *Medias* or *Serotines* ſhould giue *Præcoces*.

For the choiſe of your ſeede to ſowe. Firſt, for the *Præcoces*, Cluſius ſaith, that the *Præcox Tulipa*, that beareth a white flower, is the beſt to giue the greateſt variety of colours. Some among vs haue reported, that they haue found great variety riſe from the ſeede of the red *Præcox*, which I can more hardly beleeue : but Cluſius his experience hath the greater probability, but eſpecially if it haue ſome mixture of red or purple in it. The purple I haue found to be the beſt, next thereunto is the purple with white edges, and ſo likewiſe the red with yellow edges, each of them will bring moſt of their owne colours. Then the choiſe of the beſt *Medias*, is to take thoſe colours that are light, rather white then yellow, and purple then red ; yea white, not yellow, purple, not red : but theſe againe to be ſpotted is the beſt, and the more the better ; but withall, or aboue all in theſe, reſpect the ground or bottome of the flower, (which in the *Præcox Tulipa* cannot, becauſe you ſhall ſeldome ſee any other ground in them but yellow) for if the flower be white, or whitiſh, ſpotted, or edged, and ſtraked, and the bottome blew or purple (ſuch as is found in the Holias, and in the Cloth of ſiluer, this is beyond all other the moſt excellent, and out of queſtion the choiſeſt of an hundred, to haue the greateſt and moſt pleaſant variety and rarity. And ſo in degree, the meaner in beauty you ſowe, the leſſer ſhall your pleaſure in rarities be. Beſtowe not your time in ſowing red or yellow Tulipa ſeede, or the diuers mixtures of them ; for they will (as I haue found by experience) ſeldome be worth your paines. The *Serotina*, or late flowring Tulipa, becauſe it is ſeldome ſeene, with any eſpeciall beautifull variety, you may eaſily your ſelues gheſſe that it can bring forth (euen as I haue alſo learned) no raritie, and little or no diuerſity at all.

The time and manner to ſowe theſe ſeedes is next to be conſidered. You may not ſowe them in the ſpring of the yeare, if you hope to haue any good of them ; but in the Autumne, or preſently after they be thorough ripe and dry : yet if you ſowe them not vntill the end of October, they will come forward neuer the worſe, but rather the better ; for it is often ſeene, that ouer early ſowing cauſeth them to ſpring out of the ground ouer early, ſo that if a ſharpe ſpring chance to follow, it may goe neere to ſpoile all, or the moſt of your ſeede. Wee vſually ſowe the ſame yeares ſeede, yet if you chance to keepe of your owne, or haue from others ſuch ſeed, as is two years old, they will thriue and doe well enough, eſpecially if they were ripe and well gathered :

You muſt not ſowe them too thicke, for ſo doing hath loſt many a pecke of good ſeede, as I can tell; for if the ſeede lye one vpon another, that it hath not roome vpon the ſprouting, to enter and take roote in the earth, it periſheth by and by. Some vſe to tread downe the ground, where they meane to ſowe their ſeede, and hauing ſowne them thereon, doe couer them ouer the thickneſſe of a mans thumbe with fine ſifted earth, and they thinke they doe well, and haue good reaſon for it: for conſidering the nature of the young Tulipa rootes, is to runne downe deeper into the ground, euery yeare more then other, they thinke to hinder their quicke deſcent by the faſtneſſe of the ground, that ſo they may encreaſe the better. This way may pleaſe ſome, but I doe not vſe it, nor can finde the reaſon ſufficient; for they doe not conſider, that the ſtiffeneſſe of the earth, doth cauſe the rootes of the young Tulipas to bee long before they grow great, in that a ſtiffe ground doth more hinder the well thriuing of the rootes, then a looſe doth, and although the rootes doe runne downe deeper in a looſe earth, yet they may eaſily by tranſplanting be holpen, and raiſed vp high enough. I haue alſo ſeene ſome Tulipas not once remoued from their ſowing to their flowring; but if you will not loſe them, you muſt take them vp while their leaſe or ſtalke is freſh, and not withered: for if you doe not follow the ſtalke downe to the roote, be it neuer ſo deepe, you will leaue them behinde you. The ground alſo muſt be reſpected; for the finer, ſofter, and richer the mould is, wherein you ſowe your ſeede, the greater ſhall be your encreaſe and varietie: Sift it therefore from all ſtones and rubbiſh, and let it be either fat naturall ground of it ſelfe, or being muckt, that it bee thoroughly rotten: but ſome I know, to mend their ground, doe make ſuch a mixture of grounds, that they marre it in the making.

After the ſeede is thus ſowne, the firſt yeares ſpringing bringeth forth leaues, little bigger then the ordinary graſſe leaues; the ſecond yeare bigger, and ſo by degrees euery yeare bigger then other. The leaues of the *Præcoces* while they are young, may be diſcerned from the *Medias* by this note, which I haue obſerued. The leaues of them doe wholly ſtand vp aboue the ground, ſhewing the ſmall footſtalkes, whereby euerie leaſe doth ſtand, but the leaues of the *Medias* or *Serotines* doe neuer wholly appeare out of the ground, but the lower part which is broad, abideth vnder the vpper face of the earth. Thoſe Tulipas now growing to bee three yeares old, (yet ſome at the ſecond, if the ground and ayre be correſpondent) are to bee taken vp out of the ground, wherein yee ſhall finde they haue runne deepe, and to be anew planted, after they haue been a little dryed and cleanſed, eyther in the ſame, or another ground againe, placing them reaſonable neare one vnto another, according to their greatneſſe, which being planted and couered ouer with earth againe, of about an inch or two thickneſſe, may be left vntaken vp againe for two yeare longer, if you will, or elſe remoued euery yeare after, as you pleaſe; and thus by tranſplanting them in their due ſeaſon (which is ſtill in the end of Iuly, or beginning of Auguſt, or thereabouts) you ſhall according to your ſeede and ſoyle, haue ſome come to bearing, in the fifth yeare after the flowring, (and ſome haue had them in the fourth, but that hath beene but few, and none of the beſt, or in a rich ground) ſome in the ſixth and ſeuenth, and ſome peraduenture, not vntill the eighth or tenth yeare: but ſtill remember, that as your rootes growe greater, that in re-planting you giue them the more roome to be diſtant one from another, or elſe the one will hinder, if not rot the other.

The ſeede of the *Præcoces*, doe not thriue and come forward ſo faſt as the *Medias* or *Serotines*, nor doe giue any of-ſets in their running downe as the *Medias* doe, which vſually leaue a ſmall roote at the head of the other that is runne downe euery yeare; and beſides, are more tender, and require more care and attendance then the *Medias*, and therefore they are the more reſpected.

This is a generall and certaine rule in all Tulipas, that all the while they beare but one leaſe, they will not beare flower, whether they bee ſeedlings, or the of-ſets of elder rootes, or the rootes themſelues, that haue heretofore borne flowers; but when they ſhew a ſecond leaſe, breaking out of the firſt, it is a certaine ſigne, that it will then beare a flower, vnleſſe ſome caſualty hinder it, as froſt or raine, to nip or ſpoile the bud, or other vntimely accident befall it.

To ſet or plant your beſt and bearing Tulipas ſomewhat deeper then other rootes, I hold it the beſt way; for if the ground bee either cold, or lye too open to the cold

Northerne ayre, they will be the better defended therein, and not fuffer the frofts or cold to pierce them fo foone: for the deepe frofts and fnowes doe pinch the *Præcoces* chiefly, if they bee too neare the vppermoft cruft of the earth; and therefore many, with good fucceffe, couer ouer their ground before Winter, with either frefh or old rotten dung, and that will maruelloufly preferue them. The like courfe you may hold with feedlings, to caufe them to come on the forwarder, fo it bee after the firft yeares fowing, and not till then.

To remoue Tulipas after they haue fhot forth their fibres or fmall ftrings, which grow vnder the great round rootes, (that is, from September vntill they bee in flower) is very dangerous; for by remouing them when they haue taken faft hold in the ground, you doe both hinder them in the bearing out their flower, and befides, put them in hazzard to perifh, at leaft to bee put backe from bearing for a while after, as oftentimes I haue proued by experience: But when they are now rifen to flower, and fo for any time after, you may fafely take them vp if you will, and remoue them without danger, if you haue any good regard vnto them, vnleffe it be a young bearing roote, which you fhall in fo doing much hinder, becaufe it is yet tender, by reafon it now beareth his firft flower. But all Tulipa roots when their ftalke and leaues are dry, may moft fafely then be taken vp out of the ground, and be fo kept (fo that they lye in a dry, and not in a moift place) for fixe moneths, without any great harme: yea I haue knowne them that haue had them nine moneths out of the ground, and haue done reafonable well, but this you muft vnderftand withall, that they haue not been young but elder rootes, and they haue been orderly taken vp and preferued. The dryer you keep a Tulipa roote the better, fo as you let it not lye in the funne or winde, which will pierce it and fpoile it.

Thus Gentlewomen for your delights, (for thefe pleafures are the delights of leafure, which hath bred your loue & liking to them, and although you are herein predominant, yet cannot they be barred from your beloued, who I doubt not, wil fhare with you in the delight as much as is fit) haue I taken this paines, to fet downe, and bring to your knowledge fuch rules of art, as my fmall skill hath enabled mee withall concerning this fubiect, which of all other, feemed fitteft in this manner to be enlarged, both for the varietie of matter, and excellency of beautie herein, and alfo that thefe rules fet forth together in one place, might faue many repetitions in other places, fo that for the planting and ordering of all other bulbous rootes, and the fowing the feedes of them, you may haue recourfe vnto thefe rules, *(tanquam ad normam & examen)* which may ferue in generall for all other, little diuerfitie of particulars needing exception.

The Place.

The greater Tulipas haue firft beene fent vs from Conftantinople, and other parts of Turkie, where it is faid they grow naturally wilde in the Fields, Woods, and Mountaines; as Thracia, Macedonia, Pontus about the Euxine Sea, Cappadocia, Bithynia, and about Tripolis and Aleppo in Syria alfo: the leffer haue come from other feuerall places, as their names doe defcipher it out vnto vs; as Armenia, Perfia, Candye, Portugall, Spaine, Italy, and France. They are all now made Denizens in our Gardens, where they yeeld vs more delight, and more encreafe for their proportion, by reafon of the culture, then they did vnto their owne naturals.

The Time.

Thefe doe flower fome earlier, fome later, for three whole moneths together at the leaft, therein adorning out a Garden moft glorioufly, in that being but one kinde of flower, it is fo full of variety, as no other (except the Daffodils, which yet are not comparable, in that they yeeld not that alluring pleafant variety) doe the like befides. Some of the *Præcoces* haue beene in flower with vs, (for I fpeake not of their owne naturall places, where the Winters are milder, and the Spring earlier then ours) in the moneth of Ianuary, when the Winter before hath beene milde, but many in February,

and

and all the *Præcoces*, from the beginning to the end of March, if the yeare be kindly : at what time the *Medias* doe begin, and abide all Aprill, and part of May, when the *Serotines* flower and fade ; but this, as I said, if the yeare be kindly, or else each kinde will be a moneth later. The seede is ripe in Iune and Iuly, according to their early or late flowring.

The Names.

There haue beene diuers opinions among our moderne Writers, by what name this plant was knowne to the ancient Authors. Some would haue it be *Cosmosandalos*, of the Ancient. Dodonæus referreth it to πυπῶν of Theophraftus, in his seuenth Booke and thirteenth Chapter : but thereof he is so briefe, that besides the bare name, wee cannot finde him to make any further relation of forme, or quality. And Bauhinus, vpon Matthiolus Commentaries of Dioscorides, and in his Pinax also, followeth his opinion. Camerarius in his Hortus Medicus is of opinion, it may be referred to the Helychryfum of Crateua. Gefner, as I thinke, first of all, and after him Lobel, Camerarius, Clufius and many others, referre it to the Satyrium of Dioscorides : and surely this opinion is the most probable for many reafons. Firft, for that this plant doth grow very frequent in many places of Greece, and the lesser Afia, which were no doubt sufficiently knowne both to Theophraftus, and Dioscorides, and was accounted among bulbous rootes, although by fundry names. And fecondly, as Dioscorides fetteth forth his Satyrium, so this most commonly beareth three leaues vpon a ftalke (although fometimes with vs it hath foure or fiue) like vnto a Lilly, whereof fome are often feen to be both red, in the firft fpringing, and also vpon the decaying, efpecially in a dry time, and in a dry ground : the flower likewife of fome is white, and like a Lilly ; the roote is round, and as white within as the white of an egge, couered with a browne coate, hauing a fweetifh, but not vnpleafant tafte, as any man without danger many try. This defcription doth fo liuely fet forth this plant, that I thinke wee fhall not neede to be any longer in doubt, where to finde Dioscorides his Satyrium Triphyllum, feeing wee haue fuch plenty growing with vs. And thirdly, there is no doubt, but that it hath the fame qualities, as you fhall hereafter heare further. And laftly, that plant likewife that beareth a red flower, may very well agree with his Erythronium ; for the defcriptions in Dioscorides are both alike, as are their qualities, the greateft doubt may be in the feede, which yet may agree vnto Lin or Flaxe as fitly, or rather more then many other plants doe, in many of his comparifons, which yet wee receiue for currant. For the feede of Tulipas are flat, hard, and fhining as the feede of *Linum* or Flaxe, although of another colour, and bigger, as Dioscorides himfelfe fetteth it downe. But if there fhould be a miftaking in the writing of λίνν for λέαν in the Greeke Text, as the flippe is both eafie and likely, it were then out of all queftion the fame : for the feede is very like vnto the feede of Lillies, as any man may eafily difcerne that know them, or will compare them. It is generally called by all the late Writers, *Tulipa*, which is deriued from the name *Tulpan*, whereby the Turkes of *Dalmatia* doe entitle their head Tyres, or Caps ; and this flower being blowne, laide open, and inuerted, doth very well refemble them. We haue receiued the early kinde from Conftantinople, by the name of *Cafa lale*, and the other by the name of *Cauala lale*. Lobel and others doe call it *Lilio-narciffus*, becaufe it doth refemble a Lilly in the leafe, flower, and feede, and a Daffodill in the roote. We call it in Englifh the Turkes Cap, but moft vfually Tulipa, as moft other Chriftian Countries that delight therein doe. Dalefchampius calleth it Oulada.

The Vertues.

Dioscorides writeth, that his firft Satyrium is profitable for them that haue

haue a convulsion in their necke, (which wee call a cricke in the necke) if it be drunke in harsh (which we call red) wine.

That the roots of Tulipas are nourishing, there is no doubt, the pleasant, or at least the no vnpleasant taste, may hereunto perswade; for diuers haue had them sent by their friends from beyond Sea, and mistaking them to bee Onions, haue vsed them as Onions in their pottage or broth, and neuer found any cause of mislike, or any sense of euill quality produced by them, but accounted them sweete Onions.

Further, I haue made tryall of them my selfe in this manner. I haue preserued the rootes of these Tulipas in Sugar, as I haue done the rootes of Eringus, Orchis, or any other such like, and haue found them to be almost as pleasant as the Eringus rootes, being firme and sound, fit to be presented to the curious; but for force of Venereous quality, I cannot say, either from my selfe, not hauing eaten many, or from any other, on whom I haue bestowed them : but surely, if there be any speciall propertie in the rootes of Orchis, or some other tending to that purpose, I thinke this may as well haue it as they. It should seeme, that Dioscorides doth attribute a great Venereous faculty to the seede, whereof I know not any hath made any especiall experiment with vs as yet.

Chap. IX.

Narcissus. The Daffodill.

THere hath beene great confusion among many of our moderne Writers of plants, in not distinguishing the manifold varieties of Daffodils; for euery one almost, without consideration of kinde or forme, or other speciall note, giueth names so diuersly one from another, that if any one shall receiue from seuerall places the Catalogues of their names (as I haue had many) as they set them down, and compare the one Catalogue with the other, he shall scarce haue three names in a dozen to agree together, one calling that by one name, which another calleth by another, that very few can tell what they meane. And this their confusion, in not distinguishing the name of *Narcissus* from *Pseudonarcissus*, is of all other in this kinde the greatest and grossest errour. To auoide therefore that gulfe, whereof I complaine that so manie haue bin endrenched; and to reduce the Daffodils into such a methodicall order, that euery one may know, to what *Classis* or forme any one doth appertaine, I will first diuide them into two principall or primary kindes : that is, into *Narcissos*, true Daffodils, and *Pseudonarcissos*, bastard Daffodils: which distinction I hold to be most necessarie to be set downe first of all, that euery one may be named without confusion vnder his owne primary kind, and then to let the other parts of the subdiuision follow, as is proper to them, and fittest to expresse them. Now to cause you to vnderstand the difference betweene a true Daffodill and a false, is this; it consisteth onely in the flower, (when as in all other parts they cannot bee distinguished) and chiefly in the middle cup or chalice; for that we doe in a manner onely account those to bee *Pseudonarcissos*, bastard Daffodils, whose middle cup is altogether as long, and sometime a little longer then the outter leaues that doe encompasse it, so that it seemeth rather like a trunke or long nose, then a cup or chalice, such as almost all the *Narcissi*, or true Daffodils haue; I say almost, because I know that some of them haue their middle cup so small, that we rather call it a crowne then a cup; and againe, some of them haue them so long, that they may seem to be of the number of the *Pseudonarcissi*, or bastard Daffodils: but yet may easily be knowne from them, in that, although the cup of some of the true Daffodils be great, yet it is wider open at the brim or edge, and not so long and narrow all alike as the bastard kindes are; and this is the chiefe and onely way to know how to seuer these kindes, which rule holdeth certaine in all, except that kinde which is called *Narcissus Iuncifolius reflexo flore*, whose cup is narrow, and as long as the leaues that turne vp againe.

Secondly,

Secondly, I will subdiuide each of these again apart by themselues, into foure sorts; and first the *Narcissos*, or true Daffodils into

Latifolios, broad leafed Daffodils.

Angustifolios, narrow leafed Daffodils.

Iuncifolios, Rushe Daffodils, and

Marinos, Sea Daffodils.

These sorts againe doe comprehend vnder them some other diuisions, whereby they may the better be distinguished, and yet still bee referred to one of those foure former sorts: as

Monanthos, that is, Daffodils that beare but one flower, or two at the most vpon a stalke, and

Polyanthos, those that beare many flowers together vpon a stalke: as also

Simplici flore, those that beare single flowers, and

Multiplici flore, or *flore pleno*, that is, haue double flowers.

Vernales, those that flower in the Spring, and among them some that are earlier; and therefore called

Præcoces, early flowring Daffodils, and

Autumnales, those that flower in Autumne onely.

And lastly, with the *Pseudonarcissos*, or bastard Daffodils, I will keepe the same order, to distinguish them likewise into their foure seuerall sorts; and as with the true Daffodils, so with these false, describe vnder euery sort: first, those that beare single flowers, whether one or many vpon a stalke; and then those that beare double flowers, one or many also. As for the distinctions of *maior* and *minor*, greater and lesser, and of *maximus* and *minimus*, greatest and least, they doe not onely belong to these Daffodils; and therefore must be vsed as occasion permitteth, but vnto all other sort of plants. To begin therefore, I thinke fittest with that stately Daffodill, which for his excellency carrieth the name of None such.

1. *Narcissus latifolius omnium maximus, amplo calice flauo, siue Nompareille.*
The great None such Daffodill, or Incomparable Daffodill.

This *Narcissus Nompareille* hath three or foure long and broad leaues, of a grayish greene colour, among which riseth vp a stalke two foote high at the least, at the toppe whereof, out of a thinne skinnie huske, as all Daffodils haue, commeth forth one large single flower, and no more vsually, consisting of sixe very pale yellow large leaues, almost round at the point, with a large cuppe in the middle, somewhat yellower then the leaues, the bottome whereof next vnto the stalke is narrow and round, rising wider to the mouth, which is very large and open, and vneuenly cut in or indented about the edges. The cup doth very well resemble the chalice, that in former dayes with vs, and beyond the Seas is still vsed to hold the Sacramentall Wine, that is with a narrower bottome, and a wide mouth. After the flower is past, sometimes there commeth (for it doth not often) a round greene head, and blacke round seede therein, like vnto other Daffodils, but greater. The roote is great, as other Daffodils that beare large flowers, and is couered ouer with a brownish coate or skinne. The flower hath little or no sent at all.

Flore geminato This doth sometimes bring forth a flower with ten or twelue leaues, and a cup much larger, as if it would be two, euen as the flower seemeth.

2. *Narcissus omnium maximus flore & calice flauo.*
The great yellow Incomparable Daffodill.

This other kinde differeth neither in forme, nor bignesse of leafe or flower from the former, but in the colour of the circling leaues of the flower, which are of the same yellow colour with the cup.

Flore geminato. This doth sometimes degenerate and grow luxurious also, bringing forth two flowers vpon a stalke, each distinct from other, and sometimes two flowers thrust together, as if they were but one, although it be but seldome; for it is not a peculiar kinde that is constant, yearly abiding in the same forme.

3. *Narciſſus maximus griſeus calice flauo.* The gray Peerleſſe Daffodill.

This Peerleſſe Daffodill well deſerueth his place among theſe kindes, for that it doth much reſemble them, and peraduenture is but a difference raiſed from the ſeede of the former, it is ſo like in leafe and flower, but that the leaues ſeeme to be ſomewhat greater, and the ſixe outer leaues of the flower to be of a gliſtering whitiſh gray colour, and the cup yellow, as the former, but larger.

4. *Narciſſus latifolius flauo flore amplo calice, ſiue Matteneſſe.*
The leſſer yellow Nompareille, or the Lady Matteneſſes Daffodill.

The leaues of this Daffodill, are ſomewhat like vnto the leaues of the firſt kind, but not altogether ſo long or broad : the ſtalke likewiſe riſeth not vp fully ſo high, and beareth one flower like the former, but leſſer, and both the cuppe and the leaues are of one colour, that is, of a pale yellow, yet more yellow then in the former : the cup of this alſo is leſſer, and a little differing ; for it is neither fully ſo ſmall in the bottome, nor ſo large at the edges, nor ſo crumpled at the brimmes, ſo that all theſe differences doe plainly ſhew it to be another kinde, quite from the former.

The Place.

The places of none of theſe are certainly knowne to vs where they grow naturally, but we haue them onely in our Gardens, and haue beene ſent, and procured from diuers places.

The Time.

They flower ſometimes in the end of March, but chiefly in Aprill.

The Names.

The firſt and ſecond haue been ſent vs by the name of *Narciſſe Nompareille*, as it is called in French ; and in Latine, *Narciſſus omnium maximus amplo calice flauo*, and *Narciſſus Incomparabilis*, that is, the Incomparable Daffodill, or the greateſt Daffodill of all other, with a large yellow cuppe : but aſſuredly, although this Daffodill doth exceed many other, both in length and bigneſſe, yet the great Spaniſh baſtard Daffodill, which ſhall be ſpoken of hereafter, is in my perſwaſion oftentimes a farre higher and larger flower ; and therefore this name was giuen but relatiuely, we may call it in Engliſh, The great None ſuch Daffodill, or the Incomparable Daffodill, or the great Peerleſſe Daffodill, or the Nompareille Daffodill, which you will : for they all doe anſwer either the French or the Latine name ; and becauſe this name *Nompareille* is growne currant by cuſtome, I know not well how to alter it. The third kinde may paſſe with the title giuen it, without controule. The laſt is very well knowne beyond the Seas, eſpecially in the Low Countries, and thoſe parts, by the Lady Matteneſſe Daffodill, becauſe Cluſius receiued it from her. We may call it in Engliſh, for the correſpondency with the former, The leſſer yellow Nompareille, or Peerleſſe Daffodill, or the Lady Matteneſſe Daffodill, which you will.

Narciſſus Indicus flore rubro, dictus Iacobæus.
The Indian Daffodill with a red flower.

This Indian Daffodill is ſo differing, both in forme, not hauing a cuppe, and in colour, being red, from the whole Family of the Daffodils (except the next that followeth, and the Autumne Daffodils) that ſome might iuſtly queſtion the fitneſſe of his place here. But becauſe as all the plants, whether bulbous or other, that come from the

the Indies, either Eaſt or Weſt (although they differ very notably, from thoſe that grow in theſe parts of the world) muſt in a generall ſuruey and muſter be ranked euery one, as neere as the ſurueiours wit will direct him, vnder ſome other growing with vs, that is of neereſt likeneſſe ; Euen ſo vntill ſome other can direct his place more fitly, I ſhall require you to accept of him in this, with this deſcription that followeth, which I muſt tell you alſo, is more by relation then knowledge, or ſight of the plant it ſelfe. This Daffodill hath diuers broad leaues, ſomewhat like vnto the common or ordinary white Daffodill, of a grayiſh greene colour ; from the ſides whereof, as alſo from the middle of them, riſe vp ſometimes two ſtalkes together, but moſt vſually one after another (for very often it flowreth twice in a Summer) and often alſo but one ſtalke alone, which is of a faint reddiſh colour, about a foote high or more, at the toppe whereof, out of a deepe red ſkinne or huſke, commeth forth one flower bending downewards, conſiſting of ſixe long leaues without any cup in the middle, of an excellent red colour, tending to a crimſon; three of theſe leaues that turne vpwards, are ſomewhat larger then thoſe three that hang downewards, hauing ſixe threads or chiues in the middle, tipt with yellow pendents, and a three forked ſtile longer then the reſt, and turning vp the end thereof againe : the roote is round and bigge, of a browniſh colour on the outſide, and white within. This is ſet forth by Aldinus, Cardinall Farneſius his Phyſitian, that at Rome it roſe vp with ſtalkes of flowers, before any leaues appeared.

The Place, Time, and Names.

This naturally groweth in the Weſt Indies, from whence it was brought into Spaine, where it bore both in Iune and Iuly, and by the Indians in their tongue named AZCAL XOCHITL, and hath beene ſent from Spaine, vnto diuers louers of plants, into ſeuerall parts of Chriſtendome, but haue not thriued long in theſe tranſalpine colder Countries, ſo far as I can heare.

Narciſſus Trapezunticus flore luteo præcociſsimus.
The early Daffodill of Trebizond.

Becauſe this Daffodill is ſo like in flower vnto the former, although differing in colour, I thought it the fitteſt place to ioyne it the next thereunto. This early Daffodill hath three or foure ſhort very greene leaues, ſo like vnto the leaues of the Autumne Daffodill, that many may eaſily bee deceiued in miſtaking one for another, the difference conſiſting chiefly in this, that the leaues of this are not ſo broad or ſo long, nor riſe vp in Autumne : in the midſt of theſe leaues riſeth vp a ſhort green ſtalke, an handfull high, or not much higher vſually, (I ſpeake of it as it hath often flowred with mee, whether the cauſe be the coldneſſe of the time wherein it flowreth, or the nature of the plant, or of our climate, I am in ſome doubt ; but I doe well remember, that the ſtalkes of ſome plants, that haue flowred later with me then the firſt, haue by the greater ſtrength, and comfort of the Sunne, riſen a good deale higher then the firſt) bearing at the top, out of a whitiſh thinne ſkinne ſtriped with greene, one flower a little bending downewards, conſiſting of ſixe leaues, laid open almoſt in the ſame manner with the former Indian Daffodill, whereof ſome doe a little turne vp their points againe, of a faire pale yellow colour, hauing ſixe white chiues within it, tipt with yellow pendents, and a longer pointell : the roote is not very great, but blackiſh on the outſide, ſo like vnto the Autumne Daffodill, but that it is yellow vnder the firſt or outermoſt coate, that one may eaſily miſtake one for another.

The Place.

It was ſent vs from Conſtantinople among other rootes, but as wee may gheſſe by the name, it ſhould come thither from Trapezunte or Trebizond

The Time.

It flowreth ſometimes in December, if the former part of the Winter
haue

1 *Narciſſus Nonpareille.* The incomparable Daffodill. 2 *Narciſſus Matteneſe.* The leſſer yellow Nomparelle Daffodill. 3 *Narciſſus Iacobæus flore rubro.* The red Indian Daffodill. 4 *Narciſſus Trapezunticus.* The early Daffodill of Trabeſond. 5 *Narciſſus Montanus albus apophyſibus præditus.* The white winged Daffodill. 6 *Narciſſus Montanus, ſiue Nompareille totus albus.* The white Nompareille, or Peerleſſe Daffodill. 7 *Narciſſus albus oblongo calice.* The white Daffodill with a long cup.

haue been milde; but most vsually about the end of Ianuary, or else in Februarie the beginning or the end.

The Names.

Wee doe vsually call it from the Turkish name, *Narciſſus Trapezunticus*, and some also call it *Narciſſus vernus pracox*, as Cluſius doth, in Engliſh, The early Daffodill of Trebizond.

Narciſſus Montanus albus apophyſibus praditus.
The white Mountaine Daffodill with eares, or
The white winged Daffodill.

This Mountaine Daffodill riseth vp with three or foure broad leaues, somewhat long, of a whitiſh greene colour, among which riseth vp a stalke a foote and a halfe high, whereon standeth one large flower, and sometimes two, conſiſting of ſixe white leaues a peece, not very broad, and without any ſhew of yellowneſſe in them, three whereof haue vsually each of them on the backe part, at the bottome vpon the one ſide of them, and not on both, a little ſmall white peece of a leafe like an eare, the other three hauing none at all : the cup is almoſt as large, or not much leſſe then the ſmall Nompareille, ſmall at the bottome, and very large, open at the brimme, of a faire yellow colour, and sometimes the edges or brimmes of the cup will haue a deeper yellow colour about it, like as if it were diſcoloured with Saffron : the flower is verie ſweete, the roote is great and white, couered with a pale coate or ſkinne, not verie blacke, and is not very apt to encreaſe, ſeldome giuing of-ſets ; neither haue I euer gathered ſeede thereof, becauſe it paſſeth away without bearing any with me.

Narciſſus Montanus, ſiue Nompareille totus albus amplo calice.
The white Nompareille Daffodill.

This white Nompareille Daffodill, is in roote and leafe very like vnto the former mountain or winged Daffodill, but that they are a little larger : the stalke from among the leaues riseth vp not much higher then it, bearing at the top one large flower, compoſed of ſixe long white leaues, each whereof is as it were folded halfe way together, in the middle whereof standeth forth a large white cup, broader at the mouth or brims then at the bottome, very like vnto the leſſer Nompareille Daffodill before remembred, which hath cauſed it to be ſo entituled : the ſent whereof is no leſſe ſweete then the former.

The Place.

The naturall places of theſe Daffodils are not certainly knowne to vs; but by the names they carry, they ſhould ſeeme to bee bred in the Mountaines.

The Time.

Theſe flower not ſo early as many other kindes doe, but rather are to bee accounted among the late flowring Daffodils ; for they ſhew not their flowers vntill the beginning of May, or the latter end of Aprill, with the ſooneſt.

The Names.

The names ſet downe ouer the heads of either of them be ſuch, whereby they are knowne to vs : yet ſome doe call the firſt *Narciſſus auriculatus*, that is to ſay, The Daffodill with eares : and the other, *Narciſſus Nompareille totus albus*, that is to ſay, The white Nompareille, or Peerleſſe Daffodill.

1. *Narciſſus*

1. *Narcissus albus oblongo calice luteo præcox minor.*
The small early white Daffodill with a long cup.

The leaues of this early Daffodill are broad, very greene, and not whitish as others, three or foure standing together, about a foote long or better, among which riseth vp a greene stalke, not full so high as the leaues, bearing one flower at the toppe thereof of a reasonable bignesse, but not so great as the later kindes that follow are, consisting of six whitish leaues, but not perfect white, hauing a shew of a Creame colour appearing in them; in the middle is a long round yellow cup, about halfe an inch long or better. The smell of this flower is reasonable sweete, the roote is of a reasonable bignesse, yet lesser then the rootes of the later kindes.

2. *Narcissus pallidus oblongo calice flauo præcox.*
The early Strawe coloured Daffodill with a long cup.

The leaues of this Daffodill are as greene as the former, but much narrower; and the leaues of the flower are more enclining to yellow, but yet very pale, as if it were a light strawe colour, and seeme to bee a little more narrow and pointed then the former: the cup of this, is as long and yellow as the precedent. The smell whereof is very like the former, yet neither of them being so sweete as those that follow.

3. *Narcissus albus oblongo calice luteo serotinus maior.*
The great late flowring white Daffodill with a long cup.

This later flowring Daffodill hath his leaues somewhat narrow & long, of a grayish or whitish greene colour, among which the stalke riseth vp a foote and a halfe high, bearing one flower at the toppe, made of six white leaues, hauing the cup in the middle thereof as long as the former, and of a deepe yellow: the edges of this cuppe are sometimes plaine, and sometimes a little crumpled; they are often also circled at the brimmes with a Saffron colour, and often also without it, the smell whereof is very pleasant, and not heady: the roote hereof is reasonable bigge, and couered ouer rather with a pale then blackish skinne. This flower doth sometimes alter his forme into eight leaues, which being narrow and long, seeme like a white starre, compassing a yellow trunke.

4. *Narcissus totus pallidus oblongo calice serotinus minor.*
The late pale coloured Daffodill with a long cup.

There is another of this kinde, whose flower is wholly of a pale white, or yellowish colour, differing neither in leafe nor roote from the former.

5. *Narcissus pallidus oblongo calice flauo serotinus.*
The Strawe coloured late flowring Daffodill with a long yellow cup.

The chiefe difference of this Daffodill from the former, consisteth in the colour of the top of the flower, which is of a more yellow colour, and a little larger then the former, and the brimmes or edges of the cup of a deeper yellow, or Saffron colour. The smell of this is no lesse sweete then in the former.

6. *Narcissus albus oblongo calice flauo serotinus, duobus floribus in caule.*
The late white Daffodill with a long cup, and two flowers on a stalke.

This Daffodill is surely a kinde of it selfe, although it be so like the former, abiding constant in his forme and manner of flowring, vsually bearing without missing two flowers vpon a stalke, very like vnto the former great white kinde, that one cannot know any greater matter of difference betweene them, then that it beareth two flowers on a stalke: the cuppes whereof are seldome touched with any shew of Saffron colour on them at the brimmes or edges, as some of the former haue.

G The

The Place.

All thefe Daffodils doe grow on the Pyrenæan mountaines, and haue been fought out, and brought into thefe parts, by thofe curious or couetous fearchers of thefe delights, that haue made vs partakers of them.

The Time.

The former kindes flower earlier by a fortnight then the later, the one in the later end of March, and the other not vntill the middle of Aprill.

The Names.

Their names are giuen to euery one of them in their feuerall titles, as fitly as may beft agree with their natures ; and therefore I fhall not neede to fpeake any further of them.

Narciffus medioluteus vulgaris.
The common white Daffodill called Primrofe Peerleffe.

This Daffodill is fo common in euery Countrey Garden almoft through England, that I doubt I fhall but fpend my time in vaine, to defcribe that which is fo well knowne, yet for their fakes that know it not, I will fet downe the defcription of it in this manner. It hath long limber and broad leaues, of a grayifh greene colour, among which rifeth vp a ftalke, bearing at the toppe out of a skinnie huske fometimes but one flower, but moft commonly two flowers, and feldome three or more, but larger for the moft part, then any that beare many flowers vpon a ftalke, of a pale whitifh Creame colour, tending fomewhat neare vnto the colour of a pale Primrofe (which hath caufed our Countrey Gentlewomen, I thinke, to entitle it Primrofe Peerleffe) with a fmall round flat Crowne, rather then a cup in the middle, of a pale yellow colour, with fome pale chiues ftanding therein, being of a fweete, but ftuffing fent : the roote is reafonable great, and encreafing more then a better plant.

Narciffus mediocroceus ferotinus. The late flowring white Daffodill.

This Daffodill hath much fmaller leaues, and fhorter then the laft, the ftalke alfo rifeth not fo high by much, and beareth but one flower thereon, of a pure white colour, made of fix fmall leaues, and fomewhat narrow, ftanding feuerally one from another, and not fo clofe together as the former, but appearing like a ftarre : the cup is fmall and round, of a pale yellow colour, but faffrony about the brims, hauing fix fmall pale chiues in the middle, the fmell whereof is much fweeter then in the former.

The Place.

The firft is thought to grow naturally in England, but I could neuer heare of his naturall place. I am fure it is plentifull enough in all Country Gardens, fo that wee fcarce giue it place in our more curious parkes. The fecond liueth onely with them that delight in varieties.

The Time.

The firft Daffodill flowreth in the middle time, being neither of the earlieft, nor of the lateft ; but about the middle, or end of Aprill. The other flowreth with the lateft in May.

The Names.

I fhall not neede to trouble you with further repetitions of names, they hauing been fet downe in their titles, which are proper to them.

1 *Narciſſus vulgaris medio luteus.* The common white Daffodill, or Primroſe Peerleſſe. 2 *Narciſſus medio purpureus maximus.* The great white purple ringed Daffodill. 3 *Narciſſus medio purpureus præcox.* The early purple ringed Daffodill. 4 *Narciſſus medio purpureus ſtellatus.* The ſtarry purple ringed Daffodill. 5 *Narciſſus Perſicus.* The Perſian Daffodill. 6 *Narciſſus Autumnalis minor.* The leſſer Winter Daffodill. 7 *Narciſſus Autumnalis maior.* The greater Winter Daffodill.

1. *Narciſſus medio purpureus præcox.* The early purple ringed Daffodill.

This early Daffodill hath many long grayiſh greene leaues, ſomewhat narrower and ſtiffer then the former common white Daffodill, among which riſeth vp a long naked hollow ſtalke (as all other Daffodils haue) bearing at the toppe one flower, and ſeldome two, made of ſixe long white leaues, ſtanding cloſe together about the ſtalke; the cup is yellow, and ſo flat, that it might rather bee called a crowne: for it ſtandeth very cloſe to the middle, and very open at the brimmes, circled with a reddiſh or purple coloured ring, hauing certaine chiues in the middle of it alſo. The ſmell hereof is very ſweete, exceeding many other.

2. *Narciſſus medio purpureus ſerotinus.* The late purple ringed Daffodill.

The leaues of this Daffodill are alwayes broader then the former early one, and ſome are very neare twice as broad: the flower is very like the former, being large, and his leaues ſtanding cloſe one to the ſide of another; the ring likewiſe that compaſſeth the yellow coronet, is ſometimes of a paler reddiſh purple, and ſometimes as deepe a red as the former: ſo that it differeth not in any other materiall point, then that it flowreth not vntill the other is paſt and gone. The ſent of this is like the former, the roote hereof is greater, as well as the leafe and flower.

3. *Narciſſus medio purpureus maximus.*
The great white purple ringed Daffodill.

There is another kinde, whoſe flower (as well as leaues and rootes) is larger then any other of this kinde, which onely maketh it a diſtinct ſort from the other: it flowreth alſo with the later ſort of theſe purple ringed Daffodils.

4. *Narciſſus medio purpureus ſtellaris.* The ſtarry purple ringed Daffodill.

This Daffodill hath his leaues a little narrower and greener then the former ſorts, the flower alſo of this hath his ſixe white leaues not ſo broad, but narrower, and ſeeming longer then they, not cloſing together, but ſtanding apart one from another, making it ſeeme like a white ſtarre: it hath alſo a yellow coronet in the middle, circled about with purple, like the former. This doth ſmell nothing ſo ſweete as the firſt, but yet hath a good ſent.

The Place.

The firſt, third, and fourth of theſe Daffodils, haue alwayes beene ſent vs from Conſtantinople among other bulbous rootes, ſo that wee know no further of their naturall places.

The ſecond groweth in many places of Europe, both in Germany, France, and Italy, as Cluſius hath noted.

The Time.

The firſt flowreth very early in March, euen with the firſt Daffodils. The ſecond, third, and fourth, about a moneth after.

The Names.

The early and ſtarre Daffodils, haue been ſent vs by the Turkiſh name of *Deuebohini*, and *Serincade*. But their names, they haue receiued ſince, to bee endenizond with vs, are ſet downe in their ſeuerall titles.

Narciſſus Perſicus. The Perſian Daffodill.

This Perſian Daffodill differeth from all other kindes of Daffodils in his manner of
growing,

growing, for it neuer hath leaues and flowers at one time together, wherein it is like vnto a Colchicum, yet in roote and leafe it is a Daffodill. The roote is a little blackish on the outside, somewhat like the roote of the Autumne Daffodill, from whence riseth vp a naked foote stalke, bearing one pale yellow flower, breaking through a thinne skinne, which first enclosed it, composed of six leaues, the three outermost being a little larger then the rest, in the middle of the flower there are six small chiues, and a longer pointell. The whole flower is of an vnpleasant sent : After the flower is past, come vp the leaues, sometimes before Winter, but most vsually after the deepe of Winter is past with vs, in the beginning of the yeare, which are broad, long, and of a pale greene colour, like the leaues of other Daffodils, but not greene as the Autumne Daffodill is, and besides they doe a little twine themselues, as some of the Pancratium, or bastard Sea Daffodils doe.

Narciffus Autumnalis maior. The greater Autumne or Winter Daffodill.

The greater Autumne Daffodill riseth vp with three or foure faire broad and short leaues at the first, but afterwards grow longer, of a very deepe or darke greene colour, in the middle of which riseth vp a short, stiffe, round footestalke, bearing one faire yellow flower on the head thereof (inclosed at the first in a thinne skinne, or huske) and consisteth of six leaues as the former, with certaine chiues in the middle, as all or most other Daffodils haue, which passeth away without shew of any seed, or head for seed, although vnder the head there is a little greene knot, which peraduenture would beare seede, if our sharpe Winters did not hinder it. The roote is great and round, couered ouer with a blackish skinne or coate.

Narciffus Autumnalis minor. The lesser Autumne or Winter Daffodill.

Clusius setteth downe, that the manner of the flowring of this lesser Daffodill, is more like vnto the Persian Daffodill, then vnto the former greater Autumne kind; but I doe finde that it doth in the same sort, as the greater kinde, rise vp with his leaues first, and the flowers a while after : the flower of this is lesser, and a little paler then the flower of the greater kinde, but consisting in like sort of six leaues, narrow and sharpe pointed ; the greene leaues also are almost of as deepe a greene colour, as the greater kinde, but smaller and narrower, and a little hollow in the middle. The roote is also alike, but lesser, and couered with a blackish skinne as the former. This hath sometimes borne blacke round seede in three square heads.

The Place.

The Persian Daffodill hath beene sent sometimes, but very seldome, among other rootes from Constantinople, and it is probable by the name whereby it was sent, that it should naturally grow in Persia.

The other two haue likewise beene sent from Constantinople, and as it is thought, grow in Thracia, or thereabouts.

The Time.

They all doe flower much about one time, that is, about the end of September, and in October.

The Names.

The first hath been sent by the name of *Serincade Persiana*, and thereupon is called *Narciffus Persicus*, The Persian Daffodill.

The other two haue been thought by diuers to be Colchica, and so haue they called them, vpon no other ground, but that their flower is in forme and time somewhat like Colchicum, when as if they had marked them better, they might plainly discerne, that in all other things they did resemble Daffodils ; but now the names of *Colchicum luteum maius, & minus*, is quite

lost,

loit, time hauing worne them out, and they are called by moſt Herbariſts now adayes, *Narciſſus Autumnalis maior & minor*, The greater and the leſſer Autumne Daffodill.

Thus farre haue I proceeded with thoſe Daffodils, that hauing broad leaues, beare but one ſingle flower, or two at the moſt vpon a ſtalke : And now to proceed with the reſt, that haue broad leaues, and beare ſingle flowers, but many vpon a ſtalke.

Narciſſus Africanus aureus maior. The great yellow Daffodill of Africa.

This braue and ſtately Daffodill hath many very long and broad leaues, of a better greene colour, then many others that are grayiſh, among which appeareth a ſtalke, not riſing to the height of the leaues, bearing at the toppe out of a skinnie hoſe many faire, goodly, and large flowers, to the number of ten or twelue, if the roote bee well growne, and ſtand in a warme place, euery one being larger then any of the French, Spaniſh, or Turkie Daffodils, that beare many ſingle flowers vpon a ſtalke, and commeth neere vnto the bigneſſe of the Engliſh Daffodill, called Primroſe Peerleſſe, before deſcribed, or that French kinde hereafter deſcribed, that beareth the largeſt flowers, many vpon a ſtalke (which ſome would make to bee a kinde of that Engliſh Daffodill, but bearing more flowers) and of a faire ſhining yellow colour, hauing large, round, and open cups or boules, yellower then the outer leaues ; and is of ſo exceeding ſweete a ſent, that it doth rather offend the ſenſes by the aboundance thereof: the roote is great, and couered with a blackiſh browne coate or skinne.

Narciſſus Africanus aureus minor. The leſſer Barbary Daffodill.

This leſſer kinde is very neere the ſame with the former, but that it lacketh ſomewhat of his ſtatelineſſe of height, largeneſſe of flower and cup (being of a paler yellow) and beauty of colour, for it beareth neither of theſe equall vnto the former, but is in them all inferiour. And thus by this priuatiue, you may vnderſtand his poſitiue, and that ſhall be ſufficient at this time.

Narciſſus Byzantinus totus luteus. The yellow Turkie Daffodill.

Whereas the laſt deſcribed, came ſhort of the beauty of the former, ſo this lacketh of that beauty is in the laſt ; for this, although it haue very long leaues, and a high ſtalke, yet the flowers are neither ſo many, as not being aboue foure or fiue, nor ſo large, being not much greater then the ordinary French Daffodill hereafter deſcribed, nor the colour ſo faire, but much paler, and the cup alſo ſmaller ; and herein conſiſteth the chiefeſt differences betweene this, and both the other, but that the ſent of this is alſo weaker.

The Place.

The firſt and the ſecond grow in Barbary, about Argier, and Fez, as by the relation of them, that haue brought them into theſe parts, wee haue been enformed.

The laſt hath been often brought from Conſtantinople among other varieties of Daffodils, but from whence they receiued them, I could neuer learne.

The Time.

Theſe Daffodils do flower very early, euen with the firſt ſort of Daffodils, I meane after they haue been accuſtomed vnto our climate : for oftentimes vpon their firſt bringing ouer, they flower in Ianuary or February, eſpecially if they be preſerued from the froſts, and kept in any warme place; for they are very tender, and will ſoone periſh, being left abroad.

The Names.

The firſt is called by diuers in French, *Narciſſe d'Algiers*, and in many places

places of the Low Countries, *Narciſſen van Heck*, or *Narciſſus Heckius* ; by diuers others *Narciſſus Africanus aureus maior*, we may call it in Engliſh, The great African Daffodill, or the great Barbary Daffodill, or the great yellow Daffodill of Argiers, which you pleaſe.

The ſecond hath no other variation of name, then a diminutiue of the former, as is ſet downe in the title.

The third is no doubt the ſame, that Cluſius ſetteth downe in the twelfth Chapter of his ſecond Booke of the Hiſtory of more rare plants, and maketh the fourth ſort, which came from Conſtantinople, and may alſo be the ſame, which he maketh his fifth, which (as he ſaith) he receiued from Doctour *Simor Touar* of Seuill in Spaine. Wee call it, from the place from whence we receiued it, *Narciſſus Byzantinus*, with the addition of *totus luteus*, to put a diffrence from other ſorts that come from thence alſo : in Engliſh, The yellow ſingle Daffodill of Turkie.

Narciſſus Sulphureus maior. The greater Lemon coloured Daffodill.

The greater of theſe Daffodils, beareth three or foure greene and very long leaues, a foote and a halfe long at the leaſt, among which riſeth vp a round, yet creſted ſtalke, not ſo high as the leaues, bearing fiue or ſixe ſingle flowers thereon, euery one of them being greater then the ordinary French or Italian Daffodils, with many flowers vpon a ſtalke ; of a faint, but yet pleaſant yellow colour at the firſt, which after they haue been in flower a fortnight or thereabouts, change into a deeper, or more ſullen yellow colour : the cup in the middle is likewiſe larger, then in thoſe formerly named, and of a deeper yellow colour then the outer leaues, hauing onely three chiues within it. The ſmell is very pleaſant.

Narciſſus Sulphureus minor. The leſſer Lemon coloured Daffodill.

This leſſer Daffodill hath broader and ſhorter leaues then the former, of the colour of other Daffodils, and not greene like the former : the ſtalke of this riſeth vp higher then the leaues, bearing foure or fiue flowers vpon ſhorter footeſtalkes, and no bigger then the French Daffodill, of a pale yellow, which moſt doe call a Brimſtone colour, the cup or rather crowne in the middle, is ſmall, and broad open, of a little deeper yellow, hauing many chiues within it, and is as it were ſprinkled ouer with a kinde of mealineſſe. The ſmell of this is not full ſo pleaſant as the former.

The Place.

Both theſe haue been gathered on the Pyrenæan Mountaines, and both likewiſe haue been ſent out of Italy.

The Time.

They both flower in the middle time of the Daffodils flowring, that is, in Aprill.

The Names.

They haue their Latine names expreſſed in their titles, and ſo are their Engliſh alſo, if you pleaſe ſo to let them paſſe ; or elſe according to the Latine, you may call them, The greater and the leſſer Brimſtone coloured Daffodils ; ſome haue called them *Narciſſus Italicus*, but the Italians themſelues haue ſent them by the name of *Narciſſo Solfarigno*.

Narciſſus totus albus polyanthos. The milke white Daffodill many vpon a ſtalke.

The leaues of this Daffodill are of a meane ſize, both for length and breadth, yet ſomewhat greener then in the ordinary ſorts, that haue ſome whiteneſſe in them : the

flowers

flowers are many vpon the ftalke, as fmall for the moft part, as any of thefe kindes that beare many together, being wholly of a milke, or rather fnow white colour, both the cuppe, which is fmall, and the outer leaues that compaffe it; after which come fmall heads, wherein is contained round blacke feede, as all other Daffodils doe, although fome greater, and others leffer, according to the proportion of the plants: the roote is couered ouer with a blackifh skinne or coate; the fmell is very fweete.

There are two other forts more of this kinde, the differences whereof are, that the one hath his leaues fomewhat broader, and the flowers greater then the former: And the other fmaller leaues and flowers alfo, whofe cups being fmall, are neuer feene fully open, but as it were halfe clofed at the brimmes.

Narciffus latifolius totus albus, mediocri calice reflexus.
The milke white Daffodill with the great cup.

There is yet another fort of thefe milke white Daffodils, whofe leaues are as broad as any of the former, and whofe cup in the middle of the flower, is fomewhat larger then in any of the leffer forts, and leffer then in the greater kinde: but the leaues of the flowers doe a little turne themfelues vpwards, which maketh a chiefe difference.

The Place.

Thefe Daffodils grow in Spaine, from whence I receiued many that flourifhed a while, but perifhed by fome fierce cold Winters: they likewife grow in France, from whence many alfo haue been brought vnto vs. They haue likewife been fent from Conftantinople to vs, among other kindes of Daffodils.

The Time.

They that come from Conftantinople, for the moft part doe flower earlier then the other, euen after they are accuftomed to our ayre. Some of them flower notwithftanding in the end of March, the reft in Aprill.

The Names.

They are vfually called *Narciffus totus albus polyanthos*, adding thereunto the differences of *maior, medius,* and *minor*, that is, The milke white Daffo-dill, the greater, the middle, and the leffer; for fo fome doe diftinguifh them. The laft, for diftinction, hath his name in his title fufficient to ex-preffe him.

1. *Narciffus Narbonenfis, fiue medio luteus præcox,*
The early French Daffodill.

The leaues of this Daffodill, fpring vp out of the ground a moneth or two fome-times before the other of this kinde, that follow; being alfo fhorter, and narrower: the ftalke likewife is not very high, bearing diuers flowers at the top, breaking through a thinne skinne, as is vfuall with all the Daffodils, euery one whereof is fmall, confi-fting of fix white leaues, and a fmall yellow cup in the middle, which is of a prettie fmall fent, nothing fo ftrong as many others: the roote is great and round, and fel-dome parteth into of-fets, euen as all the other that follow, bearing many fingle flow-ers, doe.

2. *Narciffus Narbonenfis vulgaris.* The ordinary French Daffodill.

This Daffodill hath long and broad greene leaues, a little hollowifh in the middle, and edged on both fides; the ftalke is a foote and a halfe high, bearing at the toppe di-uers flowers, fomewhat larger then the former, confifting of fix white leaues, fome-what round; the cup is yellow in the middle, fmall and round, like vnto an Acorne cuppe, or a little fuller in the middle: this is the forme of that fort which was firft

brought

1 *Narcissus Africanus aureus maior.* The great yellow Daffodill of Africa. 2 *Narcissus Africanus luteus minor.* The lesser yellow Daffodill of Africa. 3 *Narcissus Narbonensis medio luteus.* The French Daffodill. 4 *Narcissus Pisanus, vel totus albus.* The Italian Daffodill, or the all white Daffodill. 5 *Narcissus Mussart.* Mussart his Daffodill. 6 *Narcissus Anglicus polyanthos.* The great English Daffodill.

brought vnto vs : But ſince there is found out ſome, whoſe cup is ſhorter, others flatter, ſome of a paler, others of a deeper yellow colour, and ſome that haue their cuppe longer then the reſt. The rootes of them all are couered with a blackiſh skin or coate.

3. *Narciſſus Narbonenſis maior amplo flore.*
The French Daffodill with great flowers.

The leaues of this Daffodill are ſomewhat like vnto the laſt, but not ſo broad, yet full as long, and ſpring ſooner out of the ground, yet not ſo early as the firſt of theſe kindes: the ſtalke hereof is flatter, and riſeth higher, bearing foure or fiue flowers, much larger then any of this kinde ; for euery one of them doth equall the Engliſh Daffodill, before deſcribed, but whiter then it, and the yellow cup larger, and more open then in any of the reſt. The roote of this is not ſo great, or round, as the former, but is more plentifull in of-ſets, then any other of theſe French, or Italian kindes.

4. *Narciſſus Piſanus.* The Italian Daffodill.

This Italian Daffodill hath his leaues as large, or larger then the ſecond French Daffodill, and his ſtalke ſomewhat higher, bearing many white flowers, very like vnto the common French Daffodill, but ſomewhat larger alſo; and the yellow cup in the middle likewiſe is larger, and rounder, then is vſually ſeen in any of the French kinds, except the laſt with the greateſt flowers.

5. *Narciſſus mediocroceus polyanthos.*
The French Daffodill with Saffron coloured cups.

This French Daffodill hath diuers leaues of a grayiſh greene colour, not ſo broad or long as the laſt recited Daffodill, but comming neerer vnto the ſecond French kinde, the flowers likewiſe are white, and many vpon a ſtalke, like thereunto, but the yellow cup is ſomewhat large, and circled with a Saffron like brimme or edge, which maketh the chiefeſt difference.

6. *Narciſſus mediocroceus alter, dictus Muſſart.* Muſſart his Daffodill.

The affinity between this & the laſt, (for it is not the ſame to be expreſſed vnder one title) hath made me ioyne it next vnto it, yet becauſe it hath a notable difference, it deſerueth a place by himſelfe. The leaues are large and long, and the flowers, being white, are larger alſo then in any other, except the greateſt, but the cup hereof is ſmall and ſhort, rather ſeeming a coronet then a cup, of a deepe Saffron colour all about the brimmes or edges.

7. *Narciſſus Anglicus polyanthos.* The great Engliſh Daffodill.

This Daffodill hath his leaues not much broader or longer, then the French kinde with great flowers, before deſcribed, the ſtalke with flowers riſeth not fully ſo high as it, bearing many flowers thereon, not altogether ſo white, yet whiter then the former Engliſh Daffodill, called Primroſe Peerleſſe, but nothing ſo large, and with ſhort, broad, and almoſt round leaues, ſtanding cloſe one vnto another : the yellow cup in the middle is bowle faſhion, being ſomewhat deeper then in any of the former kinds, but not much greater : the ſmell hereof is very ſweete and pleaſant.

8. *Narciſſus Narbonenſis, ſiue medio luteus ſerotinus maior.*
The greater late flowring French Daffodill.

The roote as well as the leaues of this Daffodill, are greater, larger, broader, and longer then in any other of the former French, or Italian kindes; the ſtalke is as high as any of them, bearing at the toppe fiue or ſixe white flowers, ſtanding open ſpread like a ſtarre, and not cloſe together, euery one whereof is large, and round pointed,

the

the cup is yellow, ſmall and ſhort, yet not lying flat to the flower, but a little ſtanding out with ſome threads in the middle, as all the former Daffodils haue. This is not ſo ſweete as the earlier kindes.

9. *Narciſſus medioluteus alter ſerotinus calice breui.*
The leſſer late flowring French Daffodill.

This Daffodill is of the ſame kinde with the laſt deſcribed, the onely difference is, that it is leſſer, and the yellow cuppe in the middle of the flower, is ſomewhat ſhorter then the former, although the former be ſhorter then many others, otherwiſe it differeth not, no not in time; for it flowreth late as the former doth.

The Place.

Theſe Daffodils haue been brought vs from diuers places: The firſt and ſecond grow naturally in many places of Spaine, that are open to the Sea: they grow likewiſe about Mompelier, and thoſe parts in France. They haue been likewiſe ſent among many other ſorts of Daffodils from Conſtantinople, ſo that I may thinke, they grow in ſome places neere thereunto.

The fourth groweth plentifully in Italy, about Piſa in Tuſcane, from whence we haue had plants to furniſh our Gardens.

The ſeuenth is accounted beyond Sea to be naturall of our Country, but I know not any with vs that haue it, but they haue had it from them.

The reſt haue been brought at diuers times, but wee know no further of their naturall places.

The Time.

The firſt flowreth earlier then any of the reſt by a moneth, euen in the beginning of March, or earlier, if the weather be milde. The other in Aprill, ſome a little before or after another. The late kinds flower not vntill May.

The Names.

There can be no more ſaid of the names of any of them, then hath beene ſet out in their titles; for they diſtinguiſh euery ſort as fitly as we can: onely ſome doe call the firſt two ſorts, by the name of *Donax Narbonenſis.*

After all theſe Daffodils, that hauing broad leaues beare ſingle flowers, either one or many vpon a ſtalke, I ſhall now goe on to ſet forth thoſe broad leafed Daffodils, that carry double flowers, either one or many vpon a ſtalke together, in the ſame order that we haue vſed before.

1. *Narciſſus albus multiplex.* The double white Daffodill.

The leaues of this Daffodill are not very broad, but rather of a meane ſize, being of the ſame largeneſſe with the leaues of the purple ringed Daffodill, the ſtalke riſeth vp to be a foote and a halfe high, bearing out of a thinne white ſkinne or hoſe, one flower and no more, conſiſting of many leaues, of a faire white colour, the flower is larger then any other double white Daffodill, hauing euery leafe, eſpecially the outermoſt, as large almoſt as any leafe of the ſingle Daffodill with the yellow cup, or purple ring. Sometimes it happeneth, that the flower is very little double, and almoſt ſingle, but that is either in a bad ground, or for that it hath ſtood long in a place without remouing; for then it hath ſuch a great encreaſe of rootes about it, that it draweth away into many parts, the nouriſhment that ſhould be for a few: but if you doe tranſplant it, taking away the of-ſets, and ſet his rootes ſingle, it will then thriue, and beare his flower as goodly and double, as I haue before deſcribed it: and is very ſweete.

2. *Narciſſus mediopurpureus multiplex.* The double purple ringed Daffodill.

There is little difference in the leaues of this kinde, from the leaues of the ſingle pur

ple

ple ringed Daffodill ; for it is probable it is of the same kinde, but by natures gift (and not by any humane art) made more plentifull, which abideth conftant, and hath not that dalliance, which oftentimes nature sheweth, to recreate the senses of men for the present, and appeareth not againe in the same forme : the chiefest difference is, that the flower (being but sometimes one on a ftalke, and sometimes two) consisteth of six white outer leaues, as large as the leaues of the single kinde, hauing many small yellow peeces, edged with purple circles round about them, instead of a cup ; and in the middle of these peeces, stand other six white leaues, lesser then the former, and a yellow cup edged with a purple circle likewise, parted into peeces, and they comprehend a few other white leaues, smaller then any of the other, hauing among them some broken peeces of the cup, with a few chiues also in the middle of the flower. The flower is very sweete.

There is of this kinde another, whose flower hath not so plaine a distinction, of a triple rowe of leaues in it : but the whole flower is confusedly set together, the outer leaues being not so large, and the inner leaues larger then the former ; the broken yellow cuppe, which is tipt with purple, running diuersly among the leaues ; so that it sheweth a fairer, and more double flower then the former, as it is indeed.

3. *Narcissus medioluteus corona duplici.*
The Turkie Daffodill with a double crowne.

This Daffodill hath three or foure leaues, as large and long almost, as the great double Daffodill of Conftantinople next following hath : the ftalke likewise is very neere as great, but as high altogether, bearing at the toppe foure or fiue flowers, the leaues whereof are as large, as of the first or second kinde of French Daffodils, before described, but not altogether of so pure a white colour; and being six in number, stand like the former single French Daffodils, but that the yellow cup in the middle of this is thicke and double, or as it were crumpled together, not standing very high to be conspicuous, but abiding lowe and short, so that it is not presently marked, vnlesse one looke vpon it precisely ; yet is exceeding sweete. The roote is like vnto the roote of the purple ringed Daffodill, or somewhat bigger.

4. *Narcissus Chalcedonicus flore pleno albo polyanthos.*
The double white Daffodill of Conftantinople.

This beautifull and goodly Daffodill (wherewith all Florists greatly desire to bee acquainted, as well for the beauty of his double flowers, as also for his superabounding sweete smell, one ftalke with flowers being instead of a nosegay) hath many very broad, and very long leaues, somewhat greener then gray, among which riseth vp a strong round ftalke, being sometimes almost flat, and ribbed, bearing foure or fiue, or more white flowers at the toppe, euery one being very great, large, and double, the leaues being confusedly set together, hauing little peeces of a yellow cup running among them, without any shew of that purple ring that is in the former, and fall away without bearing seed, euen as all, or most other double flowers doe : the smell is so exceeding sweet and strong, that it will soone offend the senses of any, that shall smell much vnto it : the roote is great and thicke, couered with a blackish coate.

5. *Narcissus Chalcedonicus fimbriatus multiplex polyanthos.*
The great double purple ringed Daffodill of Conftantinople.

This Daffodill differeth very little or nothing in leafe from the former, the onely difference is in the flowers, which although they bee double, and beare many vpon a ftalke, like vnto them, yet this hath the peeces of the yellow cuppes tipt with purple, as if they were shred or scattered among the white leaues, whereas the other hath only the yellow, without any shew of purple tips vpon them : the smell of this is as strong as of the other.

6. *Narcissus*

1 *Narciſſus albus multiplex.* The double white Daffodill. 2 *Narciſſus medioluteus corona duplici.* The Turkie Daffodill with a double crowne. 3 *Narciſſus mediopurpureus multiplex.* The double purple ringed Daffodill. 4 *Narciſſus Chalcedonicus flore plene albo polyanthos.* The double white Daffodill of Conſtantinople.

6. *Narcissus Cyprius flore pleno luteo polyanthos.*
The double yellow Daffodill of Cyprus.

The leaues of this Daffodill are almoſt as broad and long as the former, the ſtalke is a foot high and more, bearing foure or fiue flowers on the top, euery one very double, and of a fine pale yellow colour, of a ſtrong heady ſent. The root of this is alſo like the former.

The Place.

The firſt of theſe Daffodils, was firſt brought into England by Mr. Iohn de Franqueuille the elder, who gathered it in his owne Countrey of Cambray, where it groweth wilde, from whoſe ſonne, Mr. Iohn de Franqueuille, now liuing, we all haue had it. The reſt haue come from Conſtantinople at ſeuerall times ; and the laſt is thought to come from Cyprus. Wee haue it credibly affirmed alſo, that it groweth in Barbary about Fez and Argiers. Some of the double white kindes grow in Candy, and about Aleppo alſo.

The Time.

The Turkie kindes doe for the moſt part all flower early, in the end of March, or beginning of Aprill at the furtheſt, and the firſt double, about the middle or end of Aprill.

The Names.

All theſe Daffodils, except the firſt, haue had diuers Turkiſh names ſet vpon the packets, wherein they haue been ſent, but there is ſmall regard of certainty to be expected from them ; for that the name *Serincade*, without any more addition, which is a ſingle Daffodill, hath beene impoſed vpon that parcell of rootes, that haue borne moſt of them double flowers of diuers ſorts ; and the name *Serincade Catamer lale*, which ſignifieth a double flowred Daffodill, hath had many ſingle white flowers, with yellow cups, and ſome whoſe flowers haue been wholly white, cuppe and all, and ſome purple ringed, and double alſo among them. Their names, whereby they are knowne and called with vs, are, as fitly as may be, impoſed in their titles: And this I hope ſhall ſuffice, to haue ſpoken of theſe ſorts of Daffodils.

Hauing finiſhed the diſcourſe of the former ſort of broad leafed Daffodils, it is fit to proceede to the next, which are *Angustifolios Narcissos*, thoſe Daffodils that haue narrow leaues, and firſt to ſet downe thoſe that beare ſingle flowers, whether one or many flowers vpon a ſtalke, and then thoſe that beare double flowers in the ſame manner.

Narcissus Virgineus. The Virginia Daffodill.

This plant I thought fitteſt to place here in the beginning of this *Claſsis*, not finding where better to ſhroud it. It hath two or three long, and very narrow leaues, as greene as the leaues of the great *Leucoium bulboſum*, and ſhining withall, which grow ſometimes reddiſh, eſpecially at the edges : the ſtalke riſeth vp a ſpanne high, bearing one flower and no more on the head thereof, ſtanding vpright like a little Lilly or Tulipa, made of ſix leaues, wholly white, both within and without, except that at the bottome next to the ſtalke, and a little on the backſide of the three outer leaues, it hath a ſmall daſh or ſhew of a reddiſh purple colour : it hath in the middle a few chiues, ſtanding about a ſmall head pointed ; which head groweth to bee ſmall and long, containing ſmall blackiſh flat ſeede : the roote is ſmall, long, and round, a little blackiſh on the outſide, and white on the inſide.

The

The Place.

This bulbous plant was brought vs from Virginia, where they grow a-boundantly ; but they hardly thriue and abide in our Gardens to beare flowers.

The Time.

It flowreth in May, and feldome before.

The Names.

The Indians in Virginia do call it *Attamufco,* fome among vs do call it *Lilionarciffus Virginianus,* of the likeneffe of the flower to a Lilly, and the leaues and roote to a Daffodill. Wee for breuity doe call it *Narciffus Virgineus,* that is, The Daffodill of Virginia, or elfe you may call it according to the former Latine name, The Lilly Daffodill of Virginia, which you will ; for both names may ferue well to expreffe the plant.

Narciffus anguftifolius albidus præcox oblongo calice.
The early white narrow leafed Daffodill with a long cup.

This Daffodill hath three or foure narrow, long, and very greene leaues, a foote long for the moft part : the ftalke rifeth not vp fo high as the leaues, whereon ftandeth one flower, not altogether fo great as the late flowring Daffodill, with a long cuppe, defcribed before among the broad leafed ones, which confifteth of fix pale coloured leaues, not pure white, but hauing a wafh of light yellow among the white : the cuppe in the middle is round and long, yet not fo long as to bee accounted a baftard Daffodill, within which is a middle pointell, compaffed with fix chiues, hauing yellow mealy pendents.

The Place.

This Daffodill groweth with the other forts of broad leafed ones, on the Pyrenæan Mountaines, from whence they haue beene brought vnto vs, to furnifh our Gardens.

The Time.

It flowreth early, a moneth before the other forts of the fame fafhion, that is, in the beginning of March, if the time be milde, which the other before fpoken of doe not.

The Names.

It hath no other name that I know, then is expreffed in the title.

2. *Narciffus mediocroceus tenuifolius.*The fmall Daffodill with a Saffron crown.

This fmall Daffodill hath foure or fiue narrow leaues about a fpanne long, among which rifeth vp a ftalke fome nine inches high, bearing at the toppe one fmall white flower, made of fix leaues, with a fmall yellow cup in the middle, fhadowed ouer at the brimmes with a Saffron colour : the roote is fmall, round, and little long withall, couered with a blackifh skinne or coate.

3. *Narciffus minimus mediopurpureus.* The leaft purple ringed Daffodill.

This little Daffodill hath fmall narrow leaues, fhorter by much then any of the purple ringed Daffodils, before defcribed : the ftalke and flower keepe an equall proportion to the reft of the plant, being in forme and colour of the flower, like vnto the

H 2   Starre

Starre Daffodill before recited, but vnlike in the greatneſſe : this alſo is to bee obſerued, that the purple colour that circleth the brimmes of the cuppe, is ſo ſmall, that ſometimes it is not well perceiued.

4. *Narciſſus minimus Iuncifolŷ flore.* The leaſt Daffodill of all.

This leaſt Daffodill hath two or three whitiſh greene leaues, narrower then the two laſt recited Daffodils, and ſhorter by halfe, being not aboue two or three inches long, the ſtalke likewiſe is not aboue three or foure inches high, bearing one ſingle flower at the toppe, ſomewhat bigger then the ſmalneſſe of the plant ſhould ſeeme to beare, very like vnto the leaſt Ruſh Daffodill, and of the ſame bigneſſe, or rather ſomewhat bigger, being of a faint yellow colour, both leaues, and cup, or crowne, (if you pleaſe ſo to call it) ; for the middle part is ſpread very much, euen to the middle of the leaues almoſt, and lyeth flat open vpon the flower : the roote is ſmall, euen the ſmalleſt of any Daffodill, and couered with a blackiſh ſkinne or coate.

The Place.

The firſt of theſe Daffodils haue beene brought vs from the Pyrenæan Mountaines, among a number of other rare plants, and the laſt by a French man, called Francis le Veau, the honeſteſt roote-gatherer that euer came ouer to vs. The ſecond was ſent to Mr. Iohn de Franqueuille, before remembred, who imparted it to mee, as hee hath done many other good things; but his naturall place wee know not.

The Time.

They all flower about the latter end of Aprill.

The Names.

Being brought without names, wee haue giuen them their names according to their face and faſhion, as they are ſet downe in their titles.

Narciſſus Autumnalis minor albus. The little white Autumne Daffodill.

This little Autumne Daffodill riſeth with his flowers firſt out of the ground, without any leaues at all. It ſpringeth vp with one or two ſtalkes about a finger long, euery one bearing out of a ſmall huske one ſmall white flower, laid open abroad like vnto the Starre white Daffodill, before ſpoken of : in the middle of the flower is a ſmall yellow cup of a meane ſize, and after the flower is paſt, there commeth in the ſame place a ſmall head, containing ſmall, round, blacke ſeede, like vnto the Autumne Hyacinth: the leaues come vp after the ſeede is ripe and gone, being ſmall and narrow, not much bigger then the Autumne Hyacinth : the roote is ſmall and blackiſh on the outſide.

The Place.

This Daffodill groweth in Spaine, where Cluſius ſaw it, and brought it into theſe parts.

The Time.

It flowreth in the beginning of Autumne, and his ſeede is ripe in the end of October in thoſe hot Countries, but in ours it will ſcarce abide to ſhew a flower.

The Names.

The Spaniards, as Cluſius reporteth, call it *Tonada,* and he vpon the ſight thereof,

1 *Narciſſus Virgineus.* The Virginian Daffodill. 2 *Narciſſus minimus luncifoly flore.* The leaſt Daffodill of all. 3 *Narciſſus Autumnalis minor albus.*
The little white Autumne Daffodill. 4 *Narciſſus albus Autumnalis medio obſoletus.* The white Autumne Daffodill with a ſullen crown. 5 *Narciſſus*
luncifolius maximus amplo calice. The great Iunquilia with the Largeſt flower or cup. 6 *Narciſſus totus albus flore pleno Virginianus.* The double white
Daffodill of Virginia.

thereof, *Narciſſus Autumnalis minor albus*, and wee in Engliſh thereafter, The little white Autumne Daffodill.

Narciſſus albus Autumnalis medio obſoletus.
The white Autumne Daffodill with a ſullen crowne.

This Autumne Daffodill hath two or three leaues at the moſt, and very narrow, ſo that ſome doe reckon it among the Ruſh Daffodils, being ſomewhat broad at the bottome, and more pointed at the toppe, betweene theſe leaues commeth vp the ſtalke, bearing vſually two flowers and no more at the toppe, made of ſixe white leaues a peece, pointed and not round : the cup is ſmall and round, like vnto the cup or crowne of the leaſt Ruſh Daffodill, of a yellow colour at the bottome, but toward the edge of a dunne or ſullen colour.

Narciſſus anguſtifolius luteus ſemper florens Caccini.
The yellow Italian Daffodill of Caccini.

This Daffodill beareth a number of ſmall, long, narrow, and very greene leaues, broader then the leaues of any Ruſh Daffodill, among which riſe vp diuers ſtalkes, bearing at the head two or three flowers a peece, each of them being ſmall and yellow, the cup or crowne is ſmall alſo, of a deeper yellow then the flower. The Nobleman of Florence, who firſt ſent this plant to Chriſtian Porret at Leyden, after the death of Carolus Cluſius, writeth that euery ſtalke doth beare with him more ſtore of flowers, then are formerly ſet downe, and that it neuer ceaſeth to beare flowers, but that after one or moe ſtalkes haue been in flower together, and are paſt, there ſucceed other in their places.

The Place.

The firſt is naturall of Spain, the naturall place of the other is not known to vs.

The Time.

The times of the flowring, are ſet downe both in the title and in the deſcriptions ; the one to be in Autumne, the other to be all the Summer long.

The Names.

The Latine names are impoſed on them, as are fitteſt for them, and the laſt by that honourable man that ſent it, which is moſt fit to continue, and not to bee changed. But wee, to let it bee knowne by an Engliſh name to Engliſh people, haue entituled it, The yellow Italian Daffodill of Caccini: if any man can giue it a more proper name, I ſhall bee therewith right well content.

Narciſſus anguſtifolius, ſiue Iuncifolius maximus ample calice.
The great Iunquilia with the large flower or cup.

Although this Daffodill importeth by his name, not to be of this family, but of the next, conſidering it is ſo like vnto them, but bigger; yet I haue thought good to place it in the end of theſe narrow leafed Daffodils, as being indifferent, whether it ſhould bee referred to this or to that. For this carrieth diuers long greene leaues, like vnto the other Ruſh Daffodils, but thicker and broader, ſo that it may without any great errour, bee reckoned among theſe narrow leafed Daffodils, bearing at the toppe two or three very faire large flowers, with a large and more open cuppe, then in any other of the Ruſh Daffodils, both of them of a faire yellow colour, yet the cuppe a little deeper then the flower, and a little crumpled about the edges, and hath a pretty ſharpe ſent : the roote is greater and longer then the other Ruſh Daffodill, and couered likewiſe with a blackiſh coate.

The

The Place.

We haue this in Gardens onely, and haue not heard of his naturall place.

The Time.

It flowreth in Aprill.

The Names.

I leaue it indifferent, as I said, whether you will call it *Narciſſus anguſtifolius*, or *Iuncifolius magno calice*, or *maximus*, becauſe it is the greateſt of all the reſt of that kinde.

Narciſſus totus albus flore pleno Virginianus.
The double white Daffodill of Virginia.

The roote of this Daffodill, is very like vnto the former ſingle Virginia Daffodill, ſet forth in the firſt place of this ranke of narrow leafed Daffodils, but that it is a little bigger and rounder, being a little long withall, and blackiſh alſo on the outſide, as that is : from whence riſeth vp two leaues, ſomewhat broader then the former : but of a like greenneſſe : the ſtalke riſeth vp betweene theſe two leaues, about a ſpan high, or not much higher, bearing one faire double ſnow white flower, very like in the faſhion vnto the pale yellow double Daffodill, or baſtard Daffodill of Robinus, hereafter deſcribed : For it is in the like manner laid open flat, and compoſed of ſix rowes of leaues, euery rowe lying in order iuſt oppoſite, or one before another, whereof thoſe ſix leaues that make the firſt or outermoſt courſe, are the greateſt, and all the reſt lying, as I ſaid, one vpon or before another, are euery rowe ſmaller then others from the middle of this flower, thruſteth forth a ſmall long pointed forke or horne, white as the flower is.

The Place.

The place is named to be Virginia, but in what part it is not known to vs.

The Time.

It flowreth in the end of Aprill.

The Names.

It may be that this doth grow among the former ſingle kinde, and called by the ſame name Attamuſco, for that the plant is not much differing, yet hereof I am not certaine : But we, from the forme and countenance of the plant, doe call it *Narciſſus Virginianus*, The Virginian Daffodill, and becauſe it beareth a double flower, it hath the title of double added vnto it.

The third order of Daffodils, I ſaid in the beginning, was of *Iuncifolios*, Ruſh Daffodils, which are now next to be entreated of, I ſhall herein keepe the ſame order I vſed in the former ; but becauſe I finde none of this order, that beare but one flower vpon a ſtalke, I muſt begin with thoſe that beare many.

1. *Narciſſus Iuncifolius albus.* The white Iunquilia.

This white Ruſh Daffodill hath ſmall long leaues, a little broader, and of a whiter greene colour then the ordinary yellow Ruſh Daffodils : the ſtalke riſeth vp halfe a foote high or more, bearing two or three ſmall white flowers vpon a ſtalke, yet ſomewhat bigger then the common yellow Ruſh Daffodill, hauing a ſmall round cuppe in the middle, white alſo as the leaues are. The ſeede is ſmall,
blacke,

blacke, and round, as other feedes of Daffodils are : the roote is fmall and round, co-uered with a blackifh coate.

Narciſſus Iuncifolius albus magno calice. The white Iunquilia with a great cup.

There is of this kinde another fort, that hath the cup in the middle of the flower, a little larger then the other, but in all other things alike.

2. *Narciſſus Iuncifolius flore albo reflexo.*
The white turning Iunquilia, or Rufh Daffodill.

This turning white Daffodill hath foure or fiue long greene leaues, yet fhorter and broader then the ordinary yellow Iunquilia, and fully as greene alfo, from among which rifeth vp a flender greene ftalke, a foote high, bearing out of a thinne skinnie huske, three or foure, or more fnow white flowers, ftanding vpon long greene foot-ftalkes, euery flower hanging downe his head, and turning vp his fix narrow and long leaues, euen to the very foot-ftalke againe : from the middle of the flower hangeth downe a long round cuppe, as white as the leaues, within which are contained three fmall white chiues, tipt with yellow, and a fmall long pointell, thrufting out beyond the brimmes of the cup : after the flowers are paft, there come vp in their places fmall three fquare heads, wherein is contained very fmall, round, and blacke fhining feede : the roote is fmall, round, and a little long withall, couered with a blackifh browne coate or skin. The flower is quite without any good fent, or indeed rather none at all.

3. *Narciſſus Iuncifolius flore luteareflexo.*
The yellow turning Iunquilia, or Rufh Daffodill.

The leaues of this Rufh Daffodill are greater and longer then the former, and of a paler greene colour : the ftalke rifeth fomewhat higher, bearing two or three flowers thereon wholly of a gold yellow colour, both the cuppe and the leaues that turne vp againe.

4. *Narciſſus Iuncifolius calice albo reflexis folijs luteis.*
The yellow turning Iunquilia with a white cup.

This Daffodill hath his long rufh-like leaues ftanding vpright as the former, be-tweene which rifeth vp a greene ftalke, about a foote high or more, bearing two or three flowers thereon, whofe turning leaues are of a faire pale yellow, and the cuppe pale white, and not fo pure a white as the former.

5. *Narciſſus Iuncifolius calice luteo reflexis folijs albidis.*
The white turning Iunquilia with a yellow cup.

As the laft had the leaues of the flower that turne vp againe yellow, and the cuppe whitifh, fo this hath contrariwife the turning leaues of a whitifh yellow, and the long cup yellower, elfe in his long green leaues, or any other thing, there is fmall difference.

6. *Narciſſus Iuncifolius luteus magno calice.*
The Iunquilia, or Rufh Daffodill with a great cup.

This Rufh Daffodill hath bigger leaues, and longer then the ordinary yellow Rufh Daffodill, being a little flat on the one fide, and round on the other, but of the fame greenneffe with all the reft : the ftalke rifeth vp two foote high, bearing two, and fometimes three flowers thereon, being of a faire yellow colour, with a large open cup in the middle, of a little deeper yellow colour, like vnto the great Iunquilia with the large flower, before fet downe, whereof this is a kinde, no doubt; but that is larger and greater then this, both in leafe, flower, cup, &c. and this onely fomewhat leffe in all parts then that.

7. *Narciſſus*

1 *Narciſſus Iuncifolius albus.* The white Iunquilia. 2 *Narciſſus Iuncifolius flore albo reflexo.* The white turning Iunquilia. 3 *Narciſſus Iuncifolius calice luteo reflexis folijs albis.* The yellow turning Iunquilia. 4 *Narciſſus Iuncifolius luteus magno calice.* The yellow Iunquilia with a great cuppe. 5 *Narciſſus Iuncifolius luteus maior vulgaris.* The ordinary yellow Iunquilia. 6 *Narciſſus Iuncifolius Autumnalis flore viridi.* The greene Autumne Iunquilia. 7 *Narciſſus angustifolius aureus multiplex.* The golden double narrow leafed Daffodill. 8 *Narciſſus Iuncifolius flore plene.* The double Iunquilia.

7. *Narcissus Iuncifolius luteus vulgaris maior.*
The ordinary Iunquilia, or Rush Daffodill.

This ordinary Rush Daffodill hath foure or fiue long greene round leaues, like vnto Rushes, whereof it tooke the name : among these leaues riseth vp the stalke, round and greene, a foote and a halfe high very often, bearing at the toppe three or foure flowers all yellow, but much smaller then the last, and so is the cup also : the seede is small and blacke, inclosed in small cornered heads; the roote is blackish on the outside. The smell of the flower is very sweete in all these sorts of Rush Daffodils.

8. *Narcissus Iuncifolius luteus medius.* The smaller Iunquilia, or Rush Daffodill.

The leaues of this Daffodill are like vnto the former, but smaller and rounder, the stalke riseth not vp so high, nor are the flowers so great, but the leaues of the flower are a little rounder, and not so pointed as in the former, in all things else alike, sauing lesser.

9. *Narcissus Iuncifolius luteus minor.* The least Iunquilia, or Rush Daffodill.

This least Daffodill hath fiue or six small greene leaues, a little broader, and not so long as the last, among which riseth vp a stalke almost a foote high, bearing one or two small flowers at the toppe, of a paler yellow colour then the former, with a yellow open cuppe, or crowne rather in the middle, bigger then in either of the last two : the roote is very small and blacke, like vnto the last in roundnesse and colour.

10. *Narcissus Iuncifolius luteus albicantibus lineis distinctus.*
The yellow Iunquilia, or Rush Daffodill with white lines.

This Rush Daffodill hath round, greene, and long leaues, like vnto the ordinary Rush Daffodill, with a stalke bearing two or three yellow flowers, hauing leaues somewhat round at the point or end, with a line or strake of white in the middle of euerie one of them, the cup is short, and crowne fashion, a little crumpled about the brims : the seede, roote, or any thing else differeth not.

11. *Narcissus Iuncifolius Autumnalis flore viridi.*
The Autumne Rush Daffodill with a greene flower.

This strange Rush Daffodill (I call it strange, not onely because it differeth from all others of this kinde, but also because there are but few in these parts that haue had it, and fewer that doe still enioy it, in that it is perished withall that had it) hath but one onely leafe, very long, round, and greene, in all that euer I saw growing, which beareth no flower while that greene leafe is fresh, and to bee seene : but afterwards the stalke riseth vp, being like vnto the former greene leafe, round, naked, and greene vp to the toppe, where two or three flowers breake forth out of a small thin skinne, euery one consisting of six small and narrow greene leaues, very sharpe pointed at the end, and as it were ending in a small pricke or thorne : in the middle whereof is a small round cup, or rather crowne, of the same colour with the leaues and stalke, which flower smelleth very sweete, somewhat like vnto the rest of the Rush Daffodils : this sheweth not his flower vntill October, and the frosts quickly following after their flowring, cause them soone to perish.

12. *Narcissus angustifolius aureus multiplex.*
The golden double narrow leafed Daffodill.

The leaues of this Daffodill are very narrow, and of a whitish greene colour, not aboue foure or fiue inches long, from among which riseth vp a stalke about a foote high, bearing at the top one flower, consisting of some outer leaues, which are of a yellow

low colour, and of many other leaues in the middle being fmaller, and fet thicke and round together of a more yellow gold colour, but with fome whiter leaues among them, the middle part a little pointing forth : the flower ftandeth long before it doth perfect his colour, and abideth long in flower before the colour decay : the roote is in fafhion almoft like the ordinary Iunquilia, or Rufh Daffodill. I acknowledge this Daffodill hath not his proper place ; but becaufe the figure is fet in this table, let it thus paffe at this time.

13. *Narciſſus Iuncifolius luteus flore pleno.* The double Iunquilia, or Rufh Daffodill.

The double Rufh Daffodill hath his long greene leaues round, like the leaues of the common or ordinary Rufh Daffodill, and of the fame bigneffe, among which rifeth vp a long flender greene ftalke, bearing two or three, feldome more fmall flowers, yellow and double, that is, with diuers rowes of leaues, hauing the yellow cup fuch as is in the fingle flower, broken into fmall fhreads or peeces, running among the leaues of the flower, which peeces in fome flowers are not fo eafily feene, being fmaller then in others, this beareth no button or head vnder the flower for feede, his roote is round and blackifh, browne on the outfide, fo like vnto the common Rufh Daffodill, that it is almoft impoffible to know the one from the other.

There is another of this kinde, whofe flowers are fmaller, and not fo double, one, *Alter minori flore.* two, or three at the moft vpon a ftalke, and of leffe beauty by much.

The Place.

All thefe Rufh Daffodils, doe for the moft part grow in Spaine and France, and on the Pyrenæan Mountaines, which are betweene Spaine and France, which Mountains are the Nourferies of many of the fineft flowers, that doe adorne the Gardens of thefe louers of natures pride, and gathered in part by induftrious, learned, generous men, inhabiting neare thereunto, and in part by fuch as make a gaine of their labours, beftowed vpon thefe things. Onely that with the greene flower was gathered in Barbary, and imparted vnto vs from France.

The Time.

They flower in the Spring, that is, in March and Aprill, except fuch whofe time is fet downe to be in Autumne.

The Names.

Their names are fpecified in their titles, and therefore I fhall not need to fet downe any further repetitions.

To conclude therefore this difcourfe of true Daffodils, there remaineth to fpeake of the Sea Daffodils, which (as I faid in the beginning) is but one, that is frequent, and doth abide with vs. But there bee fome others found about the Cape of good Hope, and in the Weft Indies, and brought into thefe parts rather for oftentation, then continuance, where they haue flowred onely once (if peraduenture fo often) fo that being fuch ftrangers, of fo remote Countries, and of fo diuers natures, I fhall but fhew you fome of them, rather curforily then curioufly; and but onely for your fatisfaction, giue you knowledge of two or three of them, that there haue beene feene fuch in flower, and that they are fcarce to bee feene againe, except they bee fetcht a new euery yeare that they be feene.

Narciſſus Marinus, ſiue tertius Matthioli.
The great white Sea Daffodill, or Matthiolus his third Daffodill.

The roote of this Daffodill by long continuance, ftanding in one place without being remoued, groweth to be much greater and larger, then any other Daffodill whatfoeuer,

foeuer, and as bigge as any meane Squilla or Sea Onion roote, hauing many long, thicke, and white fibres, or long rootes, diuerfly branched, and fpread vnder the vpper part of the earth, befide fome others that grow downward, and perifh not euery yeare, as the fibres of all, or moft of the other Daffodils doe; and therefore this plant will not thriue, and beare flowers, if it be often tranfplanted, but rather defire to abide in one place without remouing, as I faid, and that not to be ouerfhadowed, or couered with other herbes ftanding too neare it, which then will flourifh, and beare aboundantly: from this roote, which is couered with many blackifh coates, arifeth fix or feuen, or more leaues, twice fo broad almoft, as any of the former Daffodils, but not fo long by halfe as many of them, being but fhort, in comparifon of the breadth, and of a white greene colour; from the middle of which leaues, as alfo from the fides fometimes, fpringeth vp one or two, or more ftalkes, roundifh and thicke, and fometimes a little flat and cornered, a foote high or fomewhat more, bearing at the toppe, out of a skinnie huske, eight, ten, twelue, or more very large flowers, confifting of fix white leaues a peece, fpread or laid open, with a white fhort cuppe or crowne in the middle, lying flat vpon the leaues, cut or diuided into fix corners (and not whole, as the cuppe or crowne of any other fingle Daffodill) from euery of which edges, or corners of this cup or crowne, ftandeth one white long thread, a little crooked or turning vp at the end, tipt with a yellow pendent, and fome other white threads tipt with yellow pendents, ftanding alfo in the middle: after the flower is paft, there come vp great three fquare heads, wherein the feede is contained, which is great, blacke, and round, like vnto the feede of other Daffodils, but greater: the flower hath a reafonable good fent, but not very ftrong.

The Place.

It was firft found by the Sea fide, in the Ifle of Sardinia, and on the high Mountaines alfo of the fame Ifle, where it hath borne by report, thirty fiue flowers vpon a ftalke: it groweth likewife about Illyricum, and in diuers other places.

The Time.

It fpringeth later out of the ground then any other Daffodill, that is to fay, not vntill the later end of March, or beginning of Aprill, and flowreth in the end of May, or the beginning of Iune: the feede is ripe in the end of Iuly, or beginning of Auguft.

The Names.

The firft that hath made mention of this Daffodill, was Matthiolus, who placed it in the third place among his Daffodils, and is moft vfually now a-dayes called, *Narciffus tertius Matthioli*, Matthiolus his third Daffodill, the rather, becaufe Clufius vpon a more mature deliberation, firft referred it thereunto, but called it at the firft, *Lilionarciffus Hemerocallidis facie*, and, as hee faith, Iacobus Plateau (who firft fent him the figure hereof, with the defcription) called it *Lilionarciffus Orientalis*, but Clufius vpon certaine information, that it grew in the places aforefaid, mifliked the name of *Orientalis*, and added *Hemerocallis*, which yet is not fit, for that his *Hemerocallis Valentina*, is a plaine Pancration or Sea baftard Daffodill, whofe middle cup is longer then the cup of any true Daffodill, which (as I faid in the beginning of this Chapter) is the chiefeft note of difference, betweene a true and a baftard Daffodill. I receiued the feede of this Daffodill among many other feedes of rare plants, from the liberality of Mr. Doctor Flud, one of the Phyfitians of the Colledge in London, who gathered them in the Vniuetfity Garden at Pifa in Italy, and brought them with him, returning home from his trauailes into thofe parts, by the name of *Martagon rarifsimum*, (and hauing fowne them, expected fourteene yeares, before I faw them beare a flower, which the firft yeare that it did flower, bore foure ftalkes of

flowers,

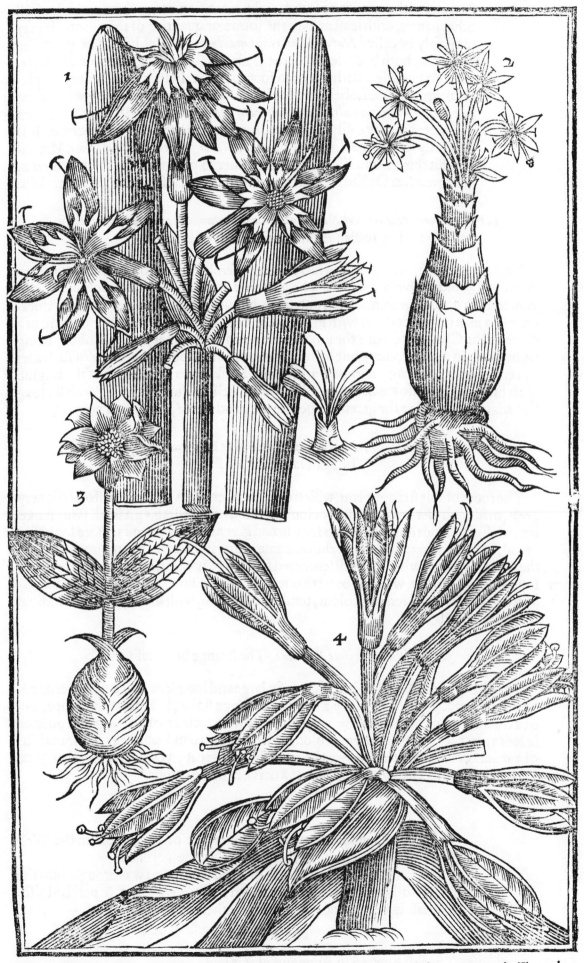

1 *Narcissus tertius Matthioli.* The great white Sea Daffodill. 2 *Narcissus Indicus Autumnalis.* The nd an
Autumne Daffodill. 3 *Narcissus marinus Africanus.* The Sea Daffodill of Africa. 4 *Narcissus marinus exoticus.* The strange Sea Daffodill.

flowers, with euery one of them eight or ten flowers on them) which of all other names, doth leaſt anſwer the forme or qualities of this plant. It may moſt fitly be called *Narciſſus marinus maximus*, in Engliſh, The great Sea Daffodill, both becauſe it is a true Daffodill, and the greateſt of all other, and alſo becauſe it hath not been found, but in Iſlands, or elſe in other places neare the Sea. Lobelius entituleth it *Pancratium Indicum alterum vernum, ſiue Narciſſus Indicus alter facie Pancratÿ Monſpeliaci*, but all this is wide from the matter, as may eaſily be known, by that that hath been ſaid before. It is generally (as I ſaid before) called of all *Narciſſus tertius Matthioli*, Matthiolus his third Daffodill, which may either ſo paſſe with vs, or as I called it, The great Sea Daffodill, which you will, & ſo Cluſius doth laſtly entitle it.

1. *Pancratium Indicum, aut Narciſſus Indicus Autumnalis quorundam Lobelÿ.*
The Indian Autumne Daffodill of Lobel.

This plant hath in my opinion, a farre nearer reſemblance vnto an Hyacinthus, then vnto any Daffodill: But becauſe Lobel hath ſo ſet it forth, I will ſo publiſh it vnto you, leauing it to iudgement. The roote is, as he ſaith, a ſpan long, and of the thickneſſe of a mans arme, couered with many white ſhells, whereof the outermoſt are of a darke red or Cheſnut colour: the flowers riſe vp in September, and October, being eight or ten in number, euery one by it ſelfe vpon a ſmall footſtalke, made of ſix leaues a peece, ſomewhat long, narrow, and pointed, like vnto the flowers of the Engliſh Colchicum, or Medowe Saffron, of a whitiſh yellow dunne colour, with ſix long threads in the middle: the greene leaues are long and broad, and broad pointed.

2. *Narciſſus Marinus Africanus, ſiue Exoticus Lobelÿ.*
The Sea Daffodill of Africa.

The roote of this ſtrange plant (which of ſome likeneſſe is called a Daffodill) is very great, made as it were of many ſcaly cloues, from whence riſeth vp a ſmall ſhort ſtalke, bearing hard aboue the ground two faire broad greene pointed leaues, more long then broad, ſo compaſſing the ſtalke at the bottome, that it ſeemeth to run through them: the ſtalke is ſpotted with diuers diſcoloured ſpots, and is bare or naked from theſe two leaues vnto the toppe, where it beareth one faire double flower, like vnto a double Auemone, of a delayed reddiſh colour, tending to a bluſh, with many threads ſet about the middle head.

3. *Narciſſus Marinus Exoticus.* The ſtrange Sea Daffodill

This ſtrange Sea Daffodill, hath fiue or ſix large and long leaues of a pale greene colour, from among which riſeth vp a ſtrong and bigge ſtalke, bearing at the toppe, out of a thinne hole or ſkinne, many very large flowers, made of ſix long and pointed leaues apeece, of a blewiſh purple colour, with a large round open cup in the middle, of a ſadder colour then the leaues: the roote is very great, yet like vnto other great Daffodils, the outer ſkins whereof are of a darke browne colour.

The Place.

The Indian Daffodils grew in the vpper part of Hiſpaniola in the Weſt Indies, and brought hither, where they all ſoone periſhed.

The other grew neare the Cape of good Hope, and was brought into the parts of Holland and thereabouts, from whence we had it, & periſhed alſo.

The laſt is vnknowne where it was gathered.

The Time.

The firſt flowred in Autumne, as it is ſaid.

The other in the firſt Summer of their bringing.

And ſo did the laſt, but the ſame rootes will not flower with vs againe.

The

The Names.

So much hath been said of their names in their titles, as hath come to our knowledge; and therefore let that suffice.

Thus hauing gone through the whole Family of the true Daffodils, (for so much as hath come to our knowledge) and set them downe euery one by his name, and in his order; it is fit that we speake of their bastard brethren, and shew you them also, in the same order held with the former, as neare as the plenty of variety herein, which is not the like with the former, will giue leaue, that when you know them both by face and name, you may the better know to place or distinguish of others, that haue not passed vnder this rod.

Pseudonarcissus aureus Hispanicus maximus.
The great yellow Spanish bastard Daffodill.

The roote of this kinde of Daffodill is reasonable great, and blackish on the out-side, desiring to be deepe in the ground; and therefore will runne downe, where it will then encrease into many of-sets, from whence rise vp many thicke, long, and stiffe leaues, of a grayish greene colour, among which riseth vp a round strong stalke, some-times three foote high or better, bearing at the toppe one onely faire great yellow flower, standing forth right, and not pendulous, consisting of six short and somewhat broad leaues, with a very great, large, and long trunke, of an equall largenesse, but open at the mouth, and turning vp the brimmes a little, which are somewhat crum-pled: after the flower is past, there commeth in the place a three square head, contai-ning round blacke seede, like vnto other Daffodils.

Pseudonarcissus Pyrenæus Hispanico & Anglico similis.
The Mountaine bastard Daffodill of diuers kindes.

There is much variety in this kinde of bastard Daffodill: For one sort hath verie broad and whitish greene leaues, somewhat short in comparison of others, that are of that breadth: the flower is wholly yellow, but a little paler then the former Spanish kinde, hauing the leaues of his flower long, and somewhat narrow, standing like wings about the middle trunke, which is as long as the leaues, and smaller then in many other of this kinde, but a little yellower then the wings. Another sort hath narrower green leaues then this last, and longer, the flower is all yellow, but the trunke is larger, wider, and more open at the mouth then the former, and almost as large as the former Spa-nish, but not so high as the last. A third hath the wings of the flower of a Strawe co-lour, but the trunke is long and narrow, of a faire yellow. A fourth hath such like flowers, but that it is shorter, both the wings and the trunke: Some likewise haue the wings of the flower longer, then the long trunke, and some shorter. Some also are all yellow, and some haue their wings onely a little more pale or white, like the English kinde: Some againe haue their trunkes long and narrow, others haue them larger and wider open, and crumpled at the brimmes; so that it is needlesse, to spend a great deale of time and labour vpon such smally respected flowers, but that in the beholding of them, we may therein admire the worke of the Creatour, who can frame such diuersity in one thing: But this is beside the text, yet not impertinent.

Pseudonarcissus pallidus præcox. The early Strawe coloured bastard Daffodill.

The leaues of this Daffodill are of a meane size, betweene the broadest and the nar-rower kindes, of a grayish greene colour, and not very long: the stalke riseth vp a foot high or more, whereon standeth one large great flower, equalling the greatest Spanish bastard Daffodill, before described, in the largenesse of his trunke, and hauing the brimmes turned vp a little, which maketh it seeme the larger: the wings or outer leaues are in a maner as short, as they are in the greatest Spanish kinde, (and not long flagging down, like vnto the Mountain kinds) and stand straight outright: all the whole flower is

I 2

of

of one euen colour, that is, of a fine pale yellow, somewhat like vnto the colour of a Lemon peele or rinde, but somewhat whiter, which vsually we call a Strawe colour: the greatnesse of the flower, the earlinesse of the flowring, and the difference of colour from all the rest of this kinde, hath made me entreate of it apart by it selfe, as being no lesse worthy.

Pseudonarcissus Hispanicus flore albo maior.
The great white Spanish bastard Daffodill.

This bastard Daffodill hath diuers leaues rising vp together, long and broad, somewhat like vnto the first Spanish kinde, but a little broader, and of a whiter greene colour, yet not so white, as in the lesser Spanish white kindes, hereafter described: among these leaues riseth vp a round strong stalke, about two foote high, bearing one white flower at the toppe, bending downe the head, as all these white kindes doe, but is not of so pure a white, as the lesser kindes that follow, yet whiter then the greatest white Spanish kinde, next of all to be described : the whole flower, as well trunke as wings, is much larger then the lesser white kindes, and almost equalling the first Spanish yellow, but a little longer and narrower, a little crumpled and turning vp at the brimmes: the head and seede are like the first ; the roote is greater and thicker then the first Spanish, and doth not encrease so much, nor is couered with a blacke, but rather with a whitish coate.

Pseudonarcissus Hispanicus maximus albidus.
The greatest Spanish white bastard Daffodill.

This kinde of bastard Daffodill is very like the last mentioned Daffodill, both in leaues and flowers, but larger in both : the flower of this is not full so white, but hath some shew of palenesse therein, and more vpon the first opening of the flower then afterwards, and is as great altogether, as the great Spanish yellow, at the least with a longer, and somewhat narrower trunke : the seede is like vnto the former, and so is the roote also, but greater, being white on the outside, and not blacke.

Pseudonarcissus Hispanicus flore albo medius & minor.
The two lesser white Spanish bastard Daffodils.

There are two other of these kindes of white Spanish Daffodils, one greater or lesser then the other, but neither of them so great as the former. The leaues of both are of a whitish greene colour, one a little broader then the other : the flowers of both are pure white, and bending downe the heads, that they almost touch the stalke againe, the greater flower hath the longer and narrower trunke ; and the lesser flower, the shorter and wider open, yet both a little crumpled at the edges or brimmes : the rootes of both are like one vnto another, but differ in the greatnesse. From the seede of these haue sprung much variety, few or none keeping either colour or height with the mother plants.

Pseudonarcissus Anglicus vulgaris. Our common English wilde bastard Daffodill.

This bastard Daffodill is so common in all England, both in Copses, Woods, and Orchards, that I might well forbeare the description thereof, and especially, in that growing wilde, it is of little respect in our Garden : but yet, lest I bee challenged of ignorance in common plants, and in regard of some variety therein worth the marking, I will set downe his description and variety as briefly as I may : It hath three or foure grayish greene leaues, long and somewhat narrow, among which riseth vp the stalke, about a span high or little higher, bearing at the toppe, out of a skinnie huske, as all other Daffodils haue, one flower (although sometimes I haue seene two together) somewhat large, hauing the six leaues that stand like wings, of a pale yellow colour, and the long trunke in the middle of a faire yellow, with the edges or brimmes a little crumpled or vneuen : after the flower is past, it beareth a round head, seeming three square, containing round blacke seede ; the roote is somewhat blackish on the outside.

But

1 *Pseudonarcissus Hispanicus maximus aureus.* The great yellow Spanish bastard Daffodill. 2 *Pseudonarcissus Pyreneus variformis.* The Mountaine bastard Daffodill of diuers kindes. 3 *Pseudonarcissus Hispanicus maior albus* The greater white Spanish bastard Daffodill. 4 *Pseudonarcissus Hispanicus minor albus.* The lesser Spanish white bastard Daffodill 5 *Pseudonarcissus tubo sexangulari.* The six cornered bastard Daffodill. 6 *Pseudonarcissus maximus aureus, siue Roseus Tradescanti.* Iohn Tradescants great Rose Daffodill. 7 *Pseudonarcissus aureus Anglicus maximus.* Master Wilmers great double Daffodill. 8 *Pseudonarcissus Hispanicus aureus flore pleno.* The double Spanish Daffodill, or Parkinsons double Daffodill. 9 *Pseudonarcissus Gallicus maior flore pleno.* The greater double French Daffodill. 10 *Pseudonarcissus Anglicus flore pleno.* The double English Daffodill, or Gerrards double Daffodill.

I 3

But there is another of this kinde like vnto the former, whose further description you haue here before ; the wings of which flower are much more white then the former, and in a manner of a milke white colour, the trunke remaining almost as yellow as the former, and not differing in any thing else.

Pseudonarcissus tubo sexangulari. The six cornered bastard Daffodill.

This kinde of Daffodill hath two or three long, and somewhat broader leaues then the last, between which commeth forth a stalke, bearing one flower somewhat large, hauing the six outer leaues of a pale yellow colour, and the long trunke plaited or cornered all along vnto the very edge into six parts, of a little deeper yellow then the wings.

The Place.

The first great Spanish kinde was brought out of Spaine. The rest from the Pyrenæan Mountaines, onely the last sauing one is plentifull in our owne Countrey, but the white sort of that kinde came with the rest from the same Mountaines.

The Time.

The pale or third kinde, and the English bee the most early, all the rest flower in Aprill, and the greatest yellow somewhat earlier, then the other greater or lesser white.

The Names.

Their seuerall names are expressed in their titles sufficient to distinguish them, and therefore there needeth no more to be said of them.

1. *Pseudonarcissus aureus maximus flore pleno, siue Roseus Tradescanti.*
The greatest double yellow bastard Daffodill, or
Iohn Tradescant his great Rose Daffodill.

This Prince of Daffodils (belongeth primarily to Iohn Tradescant, as the first founder thereof, that we know, and may well bee entituled the Glory of Daffodils) hath a great round roote, like vnto other Daffodils, couered with a brownish outer skinne or peeling, from whence riseth vp foure or fiue somewhat large and broad leaues, of a grayish greene colour, yet not fully so long and large as the next following Daffodill: from the middle whereof riseth vp a stalke almost as high and great as it, bearing at the toppe (out of a skinnie huske) one faire large great flower (the budde, before it breake open, being shorter and thicker in the middle, and ending in a longer and sharper point then any of the other Daffodils) very much spread open, consisting of smaller and shorter leaues then the next, but more in number, and thicker and rounder set together, making it seeme as great and double as any Prouince Rose, and intermixt with diuers yellow and pale leaues, as it were in rowes one vnder another. It abideth long in flower, and spreadeth, by standing long, to be the broadest in compasse of any of the Daffodils, but falleth away at the last without giuing any seede, as all double Daffodils doe.

2. *Pseudonarcissus aureus Anglicus maximus.* Mr. Wilmers great double Daffodill.

The other great double Daffodill doth so neare resemble our ordinary English double kinde, that I doe not finde therein any greater difference, then the largenesse both of leaues and flowers, &c. and the statelinesse of growth. It beareth three or foure large, long, and broad leaues, somewhat longer and broader then the former, and of a whitish greene colour : the stalke riseth to bee two foote high, growing (in a fruitfull and fat soyle) strong, and somewhat round, bearing at the toppe, out of a thin skinne, one great and faire double flower, each leafe whereof is twice as large and
broad

broad as the former, diuersly intermixt with a rowe of paler, and a rowe of deeper yellow leaues, wholly difperfed throughout the flower, the pale colour as well as the deeper yellow, in this as in the other fmall Englifh kinde, growing deeper by ftanding : fometimes the leaues hereof are fcattered, and fpread wholly, making it fhew a faire, broad, open flower : and fometimes the outer leaues ftand feparate from the middle trunke, which is whole and vnbroken, and very thicke of leaues : and fometimes the middle trunke will bee halfe broken, neither expreffing a full open double flower, nor a clofe double trunke, as it is likewife feene in the fmall Englifh kinde, as fhall bee declared in his place : this beareth no feede ; the roote hereof is thicke and great, and encreafeth as well as any other Daffodill.

3. *Pfeudonarciffus aureus Hifpanicus flore pleno.*
The great double yellow Spanifh baftard Daffodill, or Parkinfons Daffodill.

This double Spanifh Daffodill hath diuers leaues rifing from the roote, ftiffer, narrower, and not of fo whitifh a greene colour as the former, but more fullen or grayifh, plainely refembling the leaues of the fingle great kinde, from whence this hath rifen : the ftalke hereof likewife rifeth almoft as high as it, and neare the height of the laft recited double, bearing one double flower at the toppe, alwayes fpread open, and neuer forming a double trunke like the former, yet not fo faire and large as it, the outermoft leaues whereof being of a greenifh colour at the firft, and afterward more yellow, doe a little turne themfelues backe againe to the ftalke, the other leaues are fome of a pale yellow, and others of a more gold yellow colour, thofe that ftand in the middle are fmaller, and fome of them fhew as if they were hollow trunked, fo that they feeme to be greenifh, whitifh, yellow, and gold yellow, all mixed one among another : the root is great, round, and whitifh on the infide, couered with darke coloured skinnes or peelings. I thinke none euer had this kinde before my felfe, nor did I my felfe euer fee it before the yeare 1618. for it is of mine own raifing and flowring firft in my Garden.

4. *Pfeudonarciffus Gallicus maior flore pleno.*
The greater double French baftard Daffodill.

This greater double Daffodill, hath his whitifh greene leaues longer and broader then the fmaller French kinde, hereafter following, to bee defcribed, and broader, longer, and more limber then the double Englifh kinde : the ftalke rifeth vp not much higher, then the fmaller French kinde, but a little bigger, bearing at the top one great double flower, which when it is fully and perfectly blowne open (which is but feldome ; for that it is very tender, the leaues being much thinner, and thereby continually fubiect, vpon any little diftemperature of the time, to cleaue fo faft one vnto another, that the flower cannot blow open faire) is a faire and a goodly flower, larger by halfe then the fmaller kinde, and fuller of leaues, of the fame pale whitifh yellow, or Lemon colour, with the leffer, or rather a little whiter, and not fet in the fame order of rowes as it is, but more confufedly together, and turning backe the ends of the outermoft leaues to the ftalke againe, and hauing the bottome of the flower on the backfide fomewhat greene, neither of which is found in the leffer kinde : the roote is very like vnto the leffer kinde, but a little bigger and longer.

5. *Pfeudonarciffus Anglicus flore pleno.*
The double Englifh baftard Daffodill, or Gerrards double Daffodill.

The leaues of this double Daffodill are very like vnto the fingle kinde, being of a whitifh greene colour, and fomewhat broad, a little fhorter and narrower, yet ftiffer then the former French kinde : the ftalke rifeth vp about a foote high, bearing at the toppe one very double flower, the outermoft leaues being of the fame pale colour, that is to bee feene in the wings of the fingle kinde ; thofe that ftand next them, are fome as deepe a yellow as the trunke of the fingle, and others of the fame pale colour, with fome greene ftripes on the backe of diuers of the leaues : thus is the whole flower variably intermixt with pale and deepe yellow, and fome greene ftripes among them,
when

when it is fully open, and the leaues difperfed and broken. For fometimes the flower fheweth a clofe and round yellow trunke in the middle, feparate from the pale outer wings, which trunke is very double, fhewing fome pale leaues within it, difperfed among the yellow: And fometimes the trunke is more open, or in part broken, fhewing forth the fame colours intermixt within it: the flower paffeth away without giuing any feede, as all other bulbous rootes doe that beare double flowers: the roote is fmall, very like vnto the French double kindes, efpecially the leffer, that it is verie hard to know the one from the other.

The Place.

The firft and greateft kinde, we had firft from Iohn Tradefcante (as I faid before) whether raifed from feed, or gained from beyond Sea, I know not.

The fecond we firft had from Vincent Sion, borne in Flanders, dwelling on the Banke fide, in his liues time, but now dead; an induftrious and worthy louer of faire flowers, who cherifhed it in his Garden for many yeares, without bearing of any flowers vntill the yeare 1620. that hauing flowred with him, (and hee not knowing of whom hee receiued it, nor hauing euer feene the like flower before) he fheweth it to Mr. Iohn de Franqueuille, of whom he fuppofed he had receiued it, (for from beyond Sea he neuer receiued any) who finding it to bee a kinde neuer feene or knowne to vs before, caufed him to refpect it the more, as it is well worthy. And Mr. George Wilmer of Stratford Bowe Efquire, in his liues time hauing likewife receiued it of him (as my felfe did alfo) would needes appropriate it to himfelfe, as if he were the firft founder thereof, and call it by his owne name Wilmers double Daffodill, which fince hath fo continued.

The third is of mine owne foftering or raifing, as I faid before; for affuredly, it is rifen from the feede of the great Spanifh fingle kinde, which I fowed in mine owne Garden, and cherifhed it, vntill it gaue fuch a flower as is defcribed.

The fourth is not certainly knowne where his originall fhould be: Some thinke it to be of France, and others of Germany.

The laft is affuredly firft naturall of our owne Countrey, for Mr. Gerrard firft difcouered it to the world, finding it in a poore womans Garden in the Weft parts of England, where it grew before the woman came to dwell there, and, as I haue heard fince, is naturall of the Ifle of Wight.

The Time.

They doe all flower much about one time, that is, from the middle or end of March, as the yeare is forward, vnto the middle of Aprill.

The Names.

Vpon the three firft I haue impofed the names in Latine, as they are expreffed in their titles: and for the Englifh names, if you pleafe, you may let them paffe likewife as they are expreffed there alfo, that thereby euery one may be truely diftinguifhed, and not confounded. The fourth, befides the name in the title, is called of fome *Narciffus Germanicus*, which whether it be of Germany, or no, I know not; but that the name fhould import fo much. The laft doth vfually carry Mr. Gerrards name, and called Gerrards double Daffodill.

1. *Pfeudonarciffus anguftifolius flore flauefcente tubo quafi abfciffo.*
The narrow leafed baftard Daffodill with the clipt trunke.

This kinde of Daffodill hath long and narrow grayifh greene leaues, bearing one fingle flower at the toppe of his ftalke, like vnto the former fingle baftard kindes, before

fore specified, hauing his outer leaues of a pale yellow colour, and his trunke of a deeper yellow : the chiefe differences in this from the former, is in the leaues, being narrow, and then in the trunke of the flower, which is not crumpled or turned vp, as most of the other are; and that the brimmes or edges of the flower is as if it had beene clipt off, or cut euen.

2. *Pseudonarcissus Hispanicus medius & minor luteus.*
The two lesser Spanish yellow bastard Daffodils.

These two lesser kindes of Spanish Daffodils, doe but differ in greatnesse the one from the other, and not in any thing else; so that in declaring the one, you may vnderstand the other to bee a little greater. The lesser then hath three or foure narrow short whitish greene leaues, from among which commeth forth a short stalke, not aboue an hand breadth, or halfe a foote high, bearing one single flower, not fully standing outright, but a little bending downe, consisting of six small leaues, standing as wings about a small, but long trunke, a little crumpled at the brimmes : the whole flower, as well leaues as trunke, are of one deepe yellow colour, like vnto the great Spanish kinde : the roote is but small, and couered with a darkish coate. The other is in all parts greater, and (as I said) differeth not else.

3. *Pseudonarcissus Hispanicus luteus minimus.*
The least Spanish yellow bastard Daffodill.

The leaues of this small kinde are smaller and shorter then the former, seldome exceeding the length of three inches, and very narrow withall, but of the same grayish greene colour with the former : euery flower standeth vpon a small and short footestalke, scarce rising aboue the ground; so that his nose, for the most part, doth lye or touch the ground, and is made after the same fashion, and of the same colour with the former, but much smaller, as his roote is so likewise.

4. *Pseudonarcissus Gallicus minor flore pleno.*
The lesser French double bastard Daffodill.

The rootes of this lesser French kinde (if I may lawfully call it, or the greater kinde before specified, a bastard Daffodill; for I somewhat doubt thereof, in that the flower of either is not made after the fashion of any of the other bastard Daffodils, but doth more nearely resemble the forme of the double white Daffodill, expressed before among the true Daffodils) are like vnto the double English kinde, as also to the former double greater French kinde, and the leaues are of the same whitish greene colour also, but narrower and not longer : the stalke riseth a little higher then the English, and not fully so high as the greater French, bearing one faire double flower thereon, of a pale yellow or Lemon colour, consisting of six rowes of leaues, euery rowe growing smaller then other vnto the middle, and so set and placed, that euery leafe of the flower doth stand directly almost in all, one vpon or before another vnto the middle, where the leaues are smallest, the outermost being the greatest, which maketh the flower seeme the more beautifull : this and the greater kinde hath no trunke, or shew of any other thing in the middle, as all or most of the other former double bastard Daffodils haue, but are flowers wholly composed of leaues, standing double euen to the middle.

The Place.

The first is vndoubtedly a naturall of the Pyrenæan Mountaines.
The Spanish kindes grew in Spaine, and
The French double kinde about Orleance in France, where it is said to grow plentifully.

The Time.

The first flowreth at the end of March.

The

The Spanish kindes are the most early, flowring betimes in March. The French double doth flower presently after.

The Names.

More cannot bee said or added, concerning the names of any of these Daffodils, then hath been set downe in their titles : onely the French kinde is most vsually called *Robinus* his Daffodill.

Pseudonarcissus Iuncifolius albus. The white bastard Rush Daffodill, or Iunquilia.

This bastard Rush Daffodill hath two or three long and very greene leaues, very like vnto the small yellow Rush Daffodill, formerly described, but not altogether so round, among which riseth vp a short stalke, seldome halfe a foote high, bearing at the toppe, out of a small skinnie huske, one small white flower, sometime declining to a pale colour, hauing six small and short leaues, standing about the middle of the trunke, which is long, and much wider open at the mouth, then at the bottome : the small outer leaues or wings are a little tending to greene, and the trunke (as I said) is either white, or whitish, hauing the brimmes a little vneuen : the seede is small, blacke, and round, like vnto other Rush Daffodils, but smaller.

Pseudonarcissus Iuncifolius luteus maior.
The greater yellow Iunquilia, or bastard Daffodill.

The leaues of this greater kinde are longer, greater, and a little broader then the former ; the stalke also is higher, and the flower larger, more open at the mouth and crumpled, then the white, but wholly of a yellow colour : the seede and the roots are bigger, according to the proportion of the plant.

Pseudonarcissus Iuncifolius luteus minor. The lesser yellow bastard Iunquilia.

This is so like vnto the last in all things, that I shall not neede to trouble you with repetitions of the same things formerly spoken ; the chiefest difference is the smalnesse of the plant in all parts.

Pseudonarcissus Iuncifolius luteus serotinus. The late yellow bastard Iunquilia.

There is likewise a third kinde, as great as the greater yellow, and in all his parts expressing and equalling it, but is accounted the fairer, and flowreth somewhat later.

The Place.

The Pyrenæan Hils haue afforded vs all these varieties, and wee preserue them carefully ; for they are all tender.

The Time.

All these flower in Aprill, except the last, which is a moneth later.

The Names.

The French and Lowe-Countrey men call them *Trompettes*, that is, Trumpets, from the forme of the trunke ; wee sometimes call them also by that name, but more vsually bastard Iunquilia's.

Pseudonarcissus marinus albus, Pancratium vulgo.
The white Sea bastard Daffodill.

The Sea bastard Daffodill (to conclude this Chapter, and the discourse of Daffodils)

1 *Pseudonarcissus tubo quasi abscisso.* The bastard Daffodill with the clipt trunke. 2 *Pseudonarcissus Hispanicus minor.* The lesser Spanish bastard Daffodill.
3 *Pseudonarcissus Hispanicus minimus.* The least Spanish bastard Daffodill. 4 *Pseudonarcissus Gallicus minor flore pleno.* The lesser double French bastard Daffodil. 5 *Pancratium flore albo* The white Sea bastard Daffodil. 6 *Pseudonarcissus iuncifolius luteus maior.* The greater yellow bastard Iunquilia. 7 *Pseudonarcissus iuncifolius luteus minor.* The lesser yellow bastard Iunquilia 8 *Pseudonarcissus iuncifolius luteus serotinus.* The late yellow bastard Iunquilia 9 *Leucoium bulbosum præcox maius.* The great early bulbous Violet. † *Leucoium bulbosum præcox minus.* The lesser early bulbous Violet.
10 *Leucoium bulbosum autumnale.* The small Autumne bulbous Violet. 11 *Leucoium bulbosum maius serotinum,* The great late flowring bulbous Violet.

dils) hath diuers broad whitish greene leaues, but not very long, among which riseth vp a stiffe round stalke, at the top whereof breaketh out of a great round skinny huske, fiue or six flowers, euery one made somewhat of the fashion of the great bastard Rush Daffodill, but greater, and wholly white ; the six leaues, being larger and longer then in the Rush kinde, and extending beyond the trunke, are tipt with greene at the point of each leafe, and downe the middle likewise on the backside. The trunke is longer, larger, and wider open at the mouth, cut in or indented at the brims or edges, and small at the bottome, with diuers white threeds in the middle, and is very sweet : vnder the flower is a round greene head, which groweth very great, hauing within it, when it is ripe, flat and blacke seede : the roote is great and white.

Flore luteo, & flore rubro. It is reported, that there are found other sorts ; some that beare yellow flowers, and others that beare red: but we haue seene none such, and therefore I can say no more of them.

The Place.

This kinde groweth neare the Sea side, both in Spaine, Italy, and France, within the Straights, and for the most part, vpon all the Leuant shoare and Islands also, but will seldome either flower, or abide with vs in these colder Countries, as I haue both seene by those that I receiued from a friend, and heard by others.

The Time.

It flowreth in the end of Summer, that is, in August and September.

The Name.

Diuers doe call it *Pancratium,* as the learned of Mompeher, and others, with the addition of *flore Lily,* after they had left their old errour, in taking it to be *Scylla,* and vsing it for *Scylla,* in the *Trochisces* that go into Andromachus Treakle. The learned of Valentia in Spaine, as Clusius saith, doe call it *Hemerocallis,* thinking it to be a Lilly; and Clusius doth thereupon call it, *Hemerocallis Valentina :* but in my opinion, all these are deceiued in this plant ; for it is neither a Lilly, to haue the name of *Hemerocallis* giuen vnto it, nor *Scylla,* nor *Pancratium,* as many doe yet call it : for certainly this is a kinde of Daffodill; the forme both of roote, leafe, and flower, doth assure me that haue seene it, and not *Pancratium,* which (as Dioscorides testifieth) is a kinde of *Scylla,* and in his time called *Scylla,* with a red roote, and a leafe like a Lilly, but longer, and was vsed both with the same preparation and quantity, and for the same diseases that *Scylla* was vsed, but that his force was weaker: all which doth plainly shew the errours that many learned men haue been conuersant in, and that all may see how necessary the knowledge of Herbarisme is to the practice of Physicke ; And lest the roote of this Sea bastard Daffodill bee vsed in the stead of an wholsome remedy, which (as Clusius maketh mention) was deadly to him that did but cut his meate with that knife, which had immediately before cut this roote, and done in malice by him, that knew the force thereof, to kill his fellow, it working the more forceably by the euill attracting quality of the iron.

The Vertues of Daffodils in generall.

Howsoeuer Dioscorides and others, doe giue vnto some of them speciall properties, both for inward and outward diseases, yet know I not any in these dayes with vs, that apply any of them as a remedy for any griefe, whatsoeuer Gerrard or others haue written.

CHAP.

Chap. X.

Leucoium bulbosum. The bulbous Violet.

HAuing thus set downe the whole family, both of the true and bastard Daffodils, I should next set in hand with the Hyacinths ; but because *Leucoium bulbosum,* The bulbous Violet is a plant that doth challenge a place next vnto the Daffodils, as most nearly partaking with them, and a little with the Hyacinthes, I must of necessity interpose them, and shew their descriptions and differences, whereof some are early, of the first Spring, others later, and some of the Autumne.

Leucoium bulbosum præcox maius. The greater early bulbous Violet.

This bulbous Violet hath three or foure very greene, broad, flat, and short leaues, among which riseth vp a naked greene stalke, bearing out of a small skinny hose (as the former Daffodils doe) one white flower, hanging downe his head by a very small foot-stalke, made of six leaues, of an equall length, euery one whereof is tipt at the end with a small greenish yellow spot : after the flower is past, the head or seed-vessell groweth to be reasonable great, somewhat long and round, wherein is contained hard round seede, which being dry, is cleare, and of a whitish yellow colour : the roote is somewhat like a Daffodill roote, and couered with a blackish outside or skinne.

Leucoium bulbosum præcox minus. The lesser early bulbous Violet.

This lesser kinde riseth vp with two narrow grayish greene leaues, between which commeth forth the stalke, fiue or six inches high, bearing one small pendulous flower, consisting of three white leaues, which are small and pointed, standing on the outside, and hauing three other shorter leaues, which seeme like a cup in the middle, being each of them round at the ends, and cut in the middle, making the forme of an heart, with a greene tippe or spot at the broad end or edge : the seede is whitish, inclosed in long and round heads, like the former, but lesser : the roote is like a small Daffodill, with a blackish gray coate, and quickly diuideth into many of-sets.

There is another of this kinde, that came among other bulbous rootes from Constantinople, and differeth in nothing from it, but that it is a little greater, both in root, *Minus Byzantinum.* leafe, and flower.

The Place.

The two first are found in many places of Germany, and Hungary. The third, as I said, was brought from Constantinople.

The Time.

The two lesser sorts doe most commonly flower in February, if the weather be any thing milde, or at the furthest in the beginning of March, but the first is seldome in flower, before the other be well neare past, or altogether.

The Names.

Lobel and Dodonæus call the lesser kinde *Leucoium triphyllum,* and *Leuconarcissolirion triphyllum,* of the three leaues in the flower. Some doe call it *Viola bulbosa alba.* The first or greater kinde is called by Lobel, *Leuconarcissolirion paucioribus floribus* ; and by Dodonæus, *Leucoium bulbosum hexaphyllum.* We doe most vsually call them, *Leucoium bulbosum præcox maius, & minus,* The greater, or the lesser early bulbous Violet. In Dutch, *Somer Sottekens,* and not *Druifkens,* which are Grape-flowers, as some haue thought.

K *Leucoium*

1. *Leucoium bulbosum Vernum minimum.*
The small bulbous Violet of the Spring.

This small *Leucoium* sendeth forth his small and long greene leaues, like haires in Autumne, and before Winter, which abide greene vntill Aprill, and then wither away quite, and about May there arifeth vp a naked flender ftalke, at the toppe whereof breake forth two small white flowers, made of fix leaues a peece, hanging downe their heads, the three inner leaues being a little larger then the three outward, a little reddifh neare the ftalke, and very fweet : the root is fmall and round, and couered with a darke coate.

2. *Leucoium bulbosum Autumnale.* The small Autumne bulbous Violet.

As the former fmall *Leucoium* fprang vp with his leaues without flowers in Autumne, fo this contrariwife, rifeth vp with his flender brownifh ftalke of flowers in Autumne, before any greene leaues appeare, whereon ftand two or three very fmall fnow white pendulous flowers, confifting of fix leaues a peece, and a little reddifh at the bottome of the flower next vnto the ftalke, fo like vnto the former, that one would take them to be both one : after which, there grow fmall browne heads, containing fmall, blacke, round feed ; after the flower is paft, and the feede is ripening, and fometimes after the heads are ripe, the leaues begin to fpring vp, which when they are full growne, are long, greene, and as fmall, or fmaller then the leaues of the Autumne Hyacinth, which abide all the Winter, and Spring following, and wither away in the beginning of Summer : the roote is fmall, long, and white.

3. *Leucoium maius bulbosum ferotinum.*
The great late flowring bulbous Violet.

The late bulbous Violet hath three or foure broad flat greene leaues, very like vnto the firft, but longer, among which rifeth vp a flattifh ftalke, being thicker in the middle then at both edges, on the toppe whereof ftand three or foure flowers, hanging downe their heads, confifting of fix leaues a peece, all of an equall length and bigneffe, wholly white, except that each leafe hath a greene tippe at the end of them : the feede hereof is blacke and round ; the roote is reafonable great and white.

The Place.

The two former fmall ones were firft found in Spaine, and Portugall, and fent to me by Guillaume Boel ; but the firft was fo tender, that fcarce one of a fcore fprang with me, or would abide. The greateft haue beene found wilde in Germany and Auftria.

The Time.

The fmall ones haue their times expreffed in their titles and defcriptions, the laft flowreth not vntill May.

The Names.

Thefe names that are fet downe in their titles, doe paffe with all Herbarifts in thefe daies.

The Vertues.

Wee haue not knowne thefe plants vfed Phyfically, either inwardly or outwardly, to any purpofes in thefe dayes.

Снар.

Chap. XI.

Hyacinthus. The Hyacinth or Iacinth.

THe Iacinths are next to be entreated of, whereof there are many more kindes found out in thefe later times, then formerly were knowne, which for order and method fake, I will digeſt vnder feuerall forts, as neare as I can, that a-uoiding confufion, by enterlacing one among another, I may the better put euery fort vnder his owne kinde.

Hyacinthus Indicus maior tuberofa radice.
The greater Indian knobbed Iacinth.

I haue thought fitteſt to begin with this Iacinth, both becaufe it is the greateſt and higheſt, and alfo becaufe the flowers hereof are in fome likeneffe neare vnto a Daffo-dill, although his roote be tuberous, and not bulbous as all the reſt are. This Indian Iacinth hath a thicke knobbed roote (yet formed into feuerall heads, fomewhat like vnto bulbous rootes) with many thicke fibres at the bottome of them ; from the di-uers heads of this roote arife diuers ſtrong and very tall ſtalkes, befet with diuers faire, long, and broad leaues, ioyned at the bottome clofe vnto the ſtalke, where they are greateſt, and grow fmaller to the very end, and thofe that grow higher to the toppe, being fmaller and fmaller, which being broken, there appeare many threeds like wooll in them : the toppes of the ſtalkes are garnifhed with many faire large white flowers, each whereof is compofed of fix leaues, lying fpread open, as the flowers of the white Daffodill, with fome fhort threeds in the middle, and of a very fweete fent, or rather ſtrong and headie.

Hyacinthus Indicus minor tuberofa radice.
The fmaller Indian knobbed Iacinth.

The roote of this Iacinth is knobbed, like the roote of Arum or Wake Robin, from whence doe fpring many leaues, lying vpon the ground, and compaffing one another at the bottome, being long and narrow, and hollow guttered to the end, which is fmall and pointed, no leffe woolly, or full of threeds then the former : from the middle of thefe leaues rifeth vp the ſtalke, being very long and flender, three or foure foot long, fo that without it be propped vp, it will bend downe, and lye vpon the ground, where-on are fet at certaine diſtances many fhort leaues, being broad at the bottome, where they doe almoſt compaffe the ſtalke, and are fmaller toward the end where it is fharpe pointed: at the top of the ſtalke ſtand many flowers, with a fmall peece of a green leafe at the bottome of euery foot-ſtalke, which feeme to bee like fo many white Orientall Iacinths, being compofed of fix leaues, which are much thicker then the former, with fix chiues or threeds in the middle, tipt with pale yellow pendents.

The Place.

They both grow naturally in the Weſt Indies, from whence being firſt brought into Spaine, haue from thence been difperfed vnto diuers louers of plants.

The Time.

They flower not in thefe cold Countries vntill the middle of Auguſt, or not at all, if they bee not carefully preferued from the iniury of our cold Winters ; and then if the precedent Summer be hot, it may be flower a mo-neth fooner.

The Names.

Clufius calleth the leffer (for I thinke hee neuer faw the firſt) *Hyacinthus*

Indicus

Indicus tuberosa radice, that is in English, The Indian Iacinth with a tuberous roote : Some would call these *Hyacinthus Eriophorus Indicus*, that is, The Indian woolly Iacinth, because they haue much wooll in them when they are broken; yet some doe doubt that they are not two plants seuerall, as of greater and lesser, but that the greatnesse is caused by the fertility of the soyle wherein it grew.

1. *Hyacinthus Botroides maior Moschatus, siue Muscari flore flauo.*
The great yellow Muske Grape-flower, or yellow Muscari.

This Muske Iacinth or Grape-flower, hath fiue or six leaues spread vpon the ground in two or three heads, which at the first budding or shooting forth out of the ground, are of a reddish purple colour, and after become long, thicke, hollow, or guttered on the vpperside, of a whitish greene colour, and round and darke coloured vnderneath : in the middle of these heads of leaues, rise vp one or two hollow weake brownish stalkes, sometimes lying on the ground with the weight of the flowers, (but especially of the seede) yet for the most part standing vpright, when they are laden towards the toppe, with many bottle-like flowers, which at their first appearing, and vntill the flowers begin to blow open, are of a browne red colour, and when they are blowne, of a faire yellow colour, flowring first below, and so vpwards by degrees, euery one of these flowers is made like vnto a little pitcher or bottle, being bigge in the belly, and small at the mouth, which is round, and a little turned vp, very sweete in smell, like vnto Muske, whereof it tooke the name *Muscari*; after the flowers are past, there come three square thicke heads, puffed vp as if it were bladders, made of a spongie substance, wherein are here and there placed blacke round seed : the roote is long, round, and very thicke, and white on the outside, with a little woollinesse on them, being broken, and full of a slimie iuice, whereunto are annexed thicke, fat, and long fibres, which perish not as most of the other Iacinths; and therefore desireth not to bee often remoued, as the other sorts may.

2. *Hyacinthus Botroides maior Moschatus, seu Muscari flore cineritio.*
The Ashcoloured Muske Grape-flower, or Muscari.

This Muscari differeth not in rootes, or forme of leaues or flowers from the former, the chiefe differences are these: the leaues hereof do not appeare so red at the first budding out of the ground, nor are so darke when they are fully growne; the stalke also most vsually hath more store of flowers thereon, the colour whereof at the first budding is a little duskie, and when they are full blowne, are of a bleake, yet bright ashcolour, with a little shew of purple in them, and by long standing change a little more gray; being as sweete, or as some thinke, more sweete then the former : the roote (as I said) is like the former, yet yeeldeth more encrease, and will better endure our cold clymate, although it doth more seldome giue ripe seede.

3. *Hyacinthus Botroides maior Moschatus, siue Muscari flore rubro.*
The red Muske Grape-flower.

This kinde (if there be any such, for I am in some doubt thereof) doth chiefly differ in the colour of the flower from the first, in that this should beare flowers when they are blowne, of a red colour tending to yellownesse.

4. *Hyacinthus Botroides maior Moschatus, siue Muscari flore albo.*
The white Muske Grape-flower.

This also is said to haue (if there bee such an one) his leaues like vnto the second kinde, but of a little whiter greene, and the flowers pale, tending to a white : the roots of these two last are said vsually not to grow to be so great as of the former two.

The Place.

The rootes of the two first sorts, haue been often sent from Constantinople,

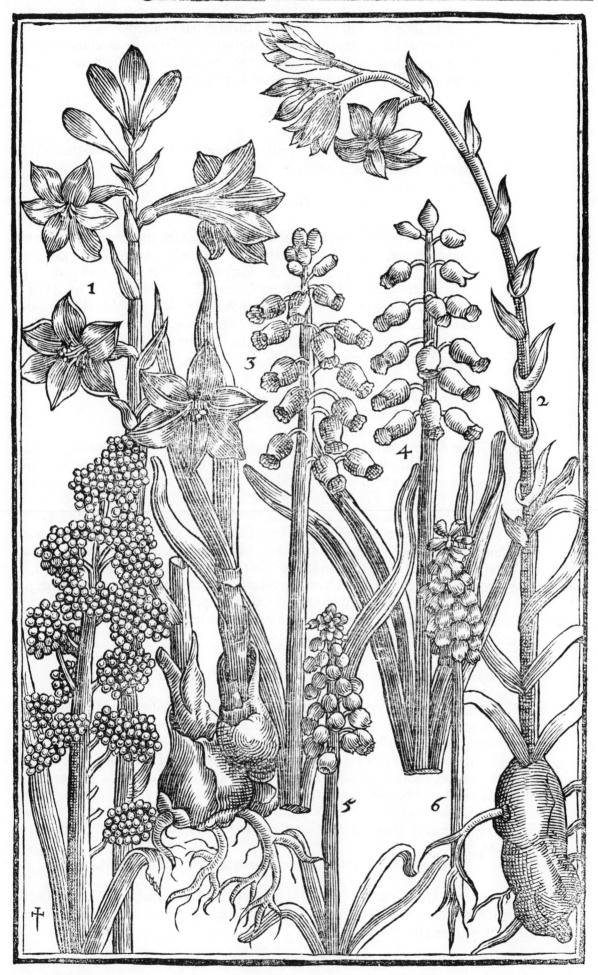

1 *Hyacinthus Indicus maior tuberosa radice.* The greater Indian knobbed Iacinth. 2 *Hyacinthus Indicus minor tuberosa radice.* The lesser Indian knobbed Iacinth. 3 *Muscari flore flauo.* The yellow Muscari. 4 *Muscari flore cineritio.* The ashcoloured Muscari. 5 *Hyacinthus Botroides caruleus amœnus.* The skie coloured Grape-flower. 6 *Hyacinthus Botroides flore albo.* The white Grape flower. 7 *Hyacinthus Botroides ramosus.* The branched Grape-flower.

ple, among many other forts of rootes, and it may be come thither from beyond the Bofphorus in Afia; we haue them in our Gardens.

The other two forts are fprung (it is probable, if they be *in rerum natura*) from the feede of the two former; for we could neuer get fuch from Conftantinople, as if the Turkes had neuer knowledge of any fuch.

The Time.

They flower in March or Aprill, as the yeare is temperate, but the firft is fooneft vp out of the ground.

The Names.

The two former haue beene fent from Turkie by the name of *Mufchoromi* and *Dipcadi*. Matthiolus calleth it *Bulbus vomitorius*, faying that no root doth more prouoke vomit then it. Cafpar Bauhinus doth moft properly call it *Hyacinthus Mofchatus*. It is moft generally called *Mufcari*, by all Herbarifts and Florifts, yet becaufe it doth fo neerely refemble the Grapeflower, I haue named it *Hyacinthus Botroides maior Mufchatus*, to put a difference from the leffer Grape-flowers that follow; in Englifh, The great Muske Grape-flower, or Mufcari.

Hyacinthus Botroides minor caruleus obfcurus.
The darke blew Grape-flower.

This Grape-flower hath many fmall, fat, and weake leaues lying vpon the ground, which are fomewhat brownifh at their firft comming vp, and of a fad greene afterwards, hollow on the vpperfide, and round vnderneath, among which rife vp round, fmooth, weake ftalkes, bearing at the toppe many fmall heauie bottle-like flowers, in fhape like the former Mufcari, but very thicke thruft together, fmaller, and of a very darke or blackifh blew colour, of a very ftrong fmell, like vnto Starch when it is new made, and hot: the root is round, and blackifh without, being compaffed with a number of fmall rootes, or of-fets round about it, fo that it will quickly choke a ground, if it be fuffered long in it. For which caufe, moft men doe caft it into fome by-corner, if they meane to preferue it, or caft it out of the Garden quite.

Alter maior. There is another of this kinde that is greater, both in leafe and flower, and differeth not in colour or any thing elfe.

Hyacinthus Botroides caruleus amanus. The skie coloured Grape-flower.

This Iacinth fpringeth vp with fewer leaues then the firft, and not reddifh, but green at his firft appearing; the leaues, when they are full growne, are long and hollow, like the former, but greener, fhorter, and broader, ftanding vpright, and not lying along vpon the ground as they doe: the flowers grow at the toppe of the ftalke, more fparfedly fet thereon, and not fo thicke together, but like a thinne bunch of grapes, and bottle-like as the former, of a perfect blew or skie-colour, euery flower hauing fome white fpots about the brimmes of them: this hath a very fweet fmell, nothing like the former: this roote is whiter, and doth not fo much encreafe as the former, yet plentifull enough.

Hyacinthus Botroides ramofus. The branched Grape-flower.

Of this kinde, there is another found to grow with many branches of flowers, breaking out from the fides of the greater ftalkes or branches: the leaues as all the reft of the plant is greater then the former.

Hyacinthus Botroides flore albo. The white Grape-flower.

The white Grape-flower hath his greene leaues a little whiter, then the blew or

skie

skie coloured Grape-flower, his flowers are very pure white, alike sparsedly set on the stalkes, but a little lower and smaller then it, in all other things there is no difference.

Hyacinthus Botroides flore albo rubente. The blush Grape-flower.

The roote of this Grape-flower groweth greater, then either the skie coloured, or white Grape-flower, and seldome hath any small rootes or of-sets, as the other haue : his leaues also are larger, and somewhat broader ; the flowers are of a pale, or bleake blush colour out of a white, and are a little larger, and grow a little higher and fuller of flowers then the white.

The Place.

They naturally grow in many places both of Germany and Hungary ; in Spaine likewise, and on Mount Baldus in Italy, and Narbone in France, about the borders of the fields : we haue them in our Gardens for delight.

The Time.

These flower from the beginning of March, or sooner sometimes, vntill the beginning of May.

The Names.

They are most commonly called *Botroides*, but more truely *Botryodes*, of Βότρυς the Greeke word, which signifieth a bunch or cluster of grapes : Lobelius calleth the white one, *Dipcadi flore albo*, transferring the name *Dipcadi*, whereby the *Muscari* is called to this Iacinth, as if they were both one. Their seuerall names, whereby they are knowne and called, are set downe in their titles. The Dutchmen call them *Driuekens*, as I said before. Some English Gentlewomen call the white Grape-flower Pearles of Spaine.

1. *Hyacinthus Comosus albus.* The white haired Iacinth.

This Iacinth doth more neerly resemble the Grape-flowers, then the faire haired Iacinths that follow, whereof it beareth the name, in that it hath no haire or threeds at the toppe of the stalke or sides, as they : and therefore I haue placed it next vnto them, and the other to follow it, as being of another kinde. The root hereof is blackish, a little long and round, from whence rise vp three or foure leaues, being smooth and whitish, long, narrow, and hollow, like a trough or gutter on the vpperside : among which the stalke riseth vp a foote high or more, bearing at the toppe diuers small flowers, somewhat like the former, but not so thicke set together, being a little longer, and larger, and wider at the mouth, and as it were diuided into six edges, of a darke whitish colour, with some blacker spots about the brimmes on the inside : the heads or seede-vessels are three square, and somewhat larger, then the heads of any of the former lesser Grape-flowers, wherein is contained round blacke seede.

2. *Hyacinthus Comosus Byzantinus.* The Turkie faire haired Iacinth.

This other Iacinth which came from Constantinople, is somewhat like the former, but that it is bigger, both in roote, and leafe, and flower, and bearing greater store of flowers on the head of the stalke : the lower flowers, although they haue short stalkes at their first flowring, yet afterwards the stalkes grow longer, and those that are lower, stand out further then those that are highest, whose foot-stalkes are short, and almost close to the stemme, and of a more perfect purple then any below, which are of a duskie greenish purple colour : the whole stalke of flowers seem like a Pyramis, broad belowe, and small aboue, or as other compare it, to a water sprinkle ; yet neither of both these Iacinths haue any threeds at the tops of the stalkes, as the other following haue.

3. *Hyacinthus*

3. *Hyacinthus Comosus maior purpureus.*
The great purple faire haired Iacinth.

This faire haired Iacinth hath his leaues softer, longer, broader, and lesse hollow then the former, lying for the most part vpon the ground : the stalke riseth vp in the midst of the leaues, being stronger, higher, and bearing a greater and longer head of flowers also then they : the flowers of this stand not vpon such long foote-stalkes, but are shorter below, and close almost to the stalke aboue, hauing many bright purplish blew threeds, growing highest aboue the flowers, as it were in a bush together, euery one of these threeds hauing a little head at the end of them, somewhat like vnto one of the flowers, but much smaller : the rest of the flowers below this bush, are of a sadder or deader purple, and not so bright a colour, and the lowest worst of all, rather enclining to a greene, like vnto the last Turkie kinde : the whole stalke with the flowers vpon it, doth somewhat resemble a long Purse tassell, and thereupon diuers Gentlewomen haue so named it : the heads and seede are like vnto the former, but greater : the roote is great and white, with some rednesse on the outside.

4. *Hyacinthus Comosus ramosus purpureus.*
The faire haired branched Iacinth.

The leaues of this Iacinth are broader, shorter, and greener then of the last, not lying so weakly on the ground, but standing somewhat more vpright: the stalke riseth vp as high as the former, but branched out on euery side into many tufts of threeds, with knappes, as it were heads of flowers, at the ends of them, like vnto the head of threeds at the toppe of the former Iacinth, but of a little darker, and not so faire a blewish purple colour : this Iacinth doth somewhat resemble the next Curld haire Iacinth, but that the branches are not so fairely composed altogether of curled threeds, nor of so excellent a faire purple or Doue colour, but more duskie by much : the roote is greater and shorter then of the next, and encreaseth faster.

5. *Hyacinthus Pennatus, siue Comosus ramosus elegantior.*
The faire Curld-haire Iacinth.

This admirable Iacinth riseth vp with three or foure leaues, somewhat like vnto the leaues of the Muske Grape-flower, but lesser ; betweene which riseth vp the stalke about a foote high, or somewhat more, bearing at the toppe a bush or tuft of flowers, which at the first appearing, is like vnto a Cone or Pineapple, and afterwards opening it selfe, spreadeth into many branches, yet still retaining the forme of a Pyramis, being broad spread below, and narrow vp aboue : each of these branches is againe diuided into many tufts of threeds or strings, twisted or curled at the ends, and of an excellent purple or Doue colour, both stalkes and haires. This abideth a great while in his beauty, but afterwards all these flowers (if you will so call them) do fall away without any seede at all, spending it selfe as it should seeme in the aboundance of the flowers : the roote is not so great as the last, but white on the outside.

The Place.

The two first haue been sent diuers times from Constantinople, the third is found wilde in many places of Europe, and as well in Germany, as in Italy. The two last are onely with vs in Gardens, and their naturall places are not knowne vnto vs.

The Time.

The three former kindes doe flower in Aprill, the two last in May.

The Names.

The first and second haue no other names then are expressed in their titles.

1 *Hyacinthus Comosus albus.* The white haired Iacinth. 2 *Hyacinthus Comosus Byzantinus.* The Turkie faire haired Iacinth.
3 *Hyacinthus Comosus maior purpureus.* The purple faire haired Iacinth, or Purse tassels. 4 *Hyacinthus Comosus ramosus, siue Calamistratus.* The faire haired branched Iacinth. 5 *Hyacinthus Pennatus, siue Comosus elegantior.* The faire curld haire Iacinth.

tles. The third is called of some onely *Hyacinthus maior*, and of others *Hyacinthus comosus maior*: We call it in English, The purple faire haired Iacinth, becauſe of his tuft of purple threeds, like haires at the toppe, and (as I ſaid) of diuers Gentlewomen, purple taſſels. The fourth is called by ſome as it is in the title, *Hyacinthus comosus ramosus*, and of others *Hyacinthus Calamiſtratus*. And the laſt or fifth is diuerſly called by diuers, Fabius Columna in his *Phytobaſanos* the ſecond part, calleth it *Hyacinthus Sanneſius*, becauſe hee firſt ſaw it in that Cardinals Garden at Rome. Robin of Paris ſent to vs the former of the two laſt, by the name of *Hyacinthus Pennatus*, and *Hyacinthus Calamiſtratus*, when as others ſent the laſt by the name *Pennatus*, and the other by the name of *Calamiſtratus*; but I thinke the name *Cincinnatus* is more fit and proper for it, in that the curled threeds which ſeeme like haires, are better expreſſed by the word *Cincinnus*, then *Calamiſtrum*, this ſignifying but the bodkin or inſtrument wherewith they vſe to friſle or curle the haire, and that the buſh of haire it ſelfe being curled. Some alſo haue giuen to both theſe laſt the names of *Hyacinthus Comosus Parnaſsi*, the one fairer then the other. Of all theſe names you may vſe which you pleaſe; but for the laſt kinde, the name *Cincinnatus*, as I ſaid, is the more proper, but *Pennatus* is the more common, and *Calamiſtratus* for the former of the two laſt.

1. *Hyacinthus Orientalis Brumalis, ſiue praecox flore albo.*
The white Winter Orientall Iacinth.

This early Iacinth riſeth vp with his greene leaues (which are in all reſpects like to the ordinary Orientall Iacinths, but ſomewhat narrower) before Winter, and ſometimes it is in flower alſo before Winter, and is in forme and colour a plaine white Orientall Iacinth, but ſomewhat leſſer, differing onely in no other thing, then the time of his flowring, which is alwayes certaine to be long before the other ſorts.

2. *Hyacinthus Orientalis Brumalis, ſiue praecox flore purpureo.*
The purple Winter Orientall Iacinth.

The difference of colour in this flower cauſeth it to bee diſtinguiſhed, for elſe it is of the kindred of the Orientall Iacinths, and is, as the former, more early then the reſt that follow: Vnderſtand then, that this is the ſame with the former, but hauing fine blewiſh purple flowers.

3. *Hyacinthus Orientalis maior praecox, dictus Zumbul Indi.*
The greateſt Orientall Iacinth, or Zumbul Indi.

The roote of this Orientall Iacinth, is vſually greater then any other of his kinde, and moſt commonly white on the outſide, from whence riſe vp one or two great round ſtalkes, ſpotted from within the ground, with the lower part of the leaues alſo vpward to the middle of the ſtalkes, or rather higher, like vnto the ſtalkes of Dragons, but darker; being ſet among a number of broad, long, and ſomewhat hollow greene leaues, almoſt as large as the leaues of the white Lilly: at the toppe of the ſtalkes ſtand more ſtore of flowers, then in any other of this kinde, euery flower being as great as the greateſt ſort of Orientall Iacinths, ending in ſix leaues, which turne at the points, of a faire blewiſh purple colour, and all ſtanding many times on one ſide of the ſtalkes, and many times on both ſides.

4. *Hyacinthus Orientalis vulgaris diuerſorum colorum.*
The ordinary Orientall Iacinth.

The common Orientall Iacinth (I call it common, becauſe it is now ſo plentifull in all Gardens, that it is almoſt not eſteemed) hath many greene leaues, long, ſomewhat broad and hollow, among which riſeth vp a long greene round ſtalke, beſet from the middle thereof almoſt, with diuers flowers, ſtanding on both ſides

of

of the stalkes, one aboue another vnto the toppe, each whereof next vnto the foote-stalke is long, hollow, round, and close, ending in six small leaues laid open, and a little turning at the points, of a very sweete smell : the colours of these flowers are diuers, for some are pure white, without any shew of other colour in them : another is almost white, but hauing a shew of blewnesse, especially at the brims and bottomes of the flowers. Others againe are of a very faint blush, tending towards a white : Some are of as deepe a purple as a Violet ; others of a purple tending to red-nesse, and some of a paler purple. Some againe are of a faire blew, others more wat-chet, and some so pale a blew, as if it were more white then blew : after the flowers are past, there rise vp great three square heads, bearing round blacke seede, great and shining : the roote is great, and white on the outside, and oftentimes purplish also, flat at the bottome, and small at the head.

There is a kinde of these Iacinths, whose flowers are of a deepe purplish Violet co-lour, hauing whitish lines downe the backe of euery leafe of the flower, which turne themselues a little backwards at the points. *Flore purpureo violaceo lineis albicantis in dorso.*

There is another, whose flowers stand all opening one way, and not on all sides, but are herein like the great Zumbul Indi, before set out. *Floribus antror-sum respicienti-bus.*

There is againe another kinde which flowreth later then all the rest, and the flow-ers are smaller, standing more vpright, which are either white or blew, or mixt with white and purple. *Serotinus ere-ctis floribus diuersorum co-lorum.*

5. *Hyacinthus Orientalis folioso caule.* The bushy stalked Orientall Iacinth.

This strange Iacinth hath his rootes, leaues, and flowers, like vnto the former Ori-entall Iacinths : the onely difference in this is, that his stalke is not bare or naked, but hath very narrow long leaues, growing dispersedly, and without order, with the flow-ers thereon, which are blew, and hauing for the most part one leafe, and sometimes two at the foote, or setting on of euery flower, yet sometimes it happeneth, some flow-ers to be without any leafe at the bottome, as nature, that is very variable in this plant, listeth to play : the heads and seede are blacke and round, like the other also.

6. *Hyacinthus Orientalis flore duplici.* The bleake Orientall Iacinth once double.

This double Iacinth hath diuers long leaues, like vnto the other Orientall Iacinths, almost standing vpright, among which riseth vp a stalke, brownish at the first, but growing greene afterwards, bearing many flowers at the toppe, made like the flowers of the former Iacinths, and ending in six leaues, greene at the first, and of a blewish white when they are open, yet retaining some shew of greennesse in them, the brims of the leaues being white ; from the middle of each flower standeth forth another small flower, consisting of three leaues, of the same colour with the other flower, but with a greene line on the backe of each of these inner leaues : in the middle of this lit-tle flower, there stand some threeds tipt with blacke : the smell of this flower is not so sweete as of the forme ; the heads, seede, and rootes are like the former.

7. *Hyacinthus Orientalis flore pleno cæruleo, vel purpuro violaceo.* The faire double blew, or purple Orientall Iacinth.

The leaues of these Iacinths are smaller, then the leaues of most of the other for-mer sorts ; the stalkes are shorter, and smaller, bearing but three or foure flowers on the heads of them for the most part, which are not composed like the last, but are more faire, full, and double of leaues, where they shew out their full beauties, and of a faire blew colour in some, and purple in others, smelling pretty sweete ; but these doe sel-dome beare out their flowers faire ; and besides, haue diuers other flowers that will be either single, or very little double vpon the same stalke.

8. *Hyacinthus Orientalis candidissimus flore pleno.* The pure white double Orientall Iacinth.

This double white Iacinth hath his leaues like vnto the single white Orientall Ia-cinth ;

cinth; his ſtalke is likewiſe long, ſlender, and greene, bearing at the toppe two or three flowers at the moſt, very double and full of leaues, of a pure white colour, without any other mixture therein, hanging downe their heads a little, and are reaſonable ſweete. I haue this but by relation, not by ſight, and therefore I can giue no further aſſurance as yet.

The Place.

All theſe Orientall Iacinths, except the laſt, haue beene brought out of Turkie, and from Conſtantinople : but where their true originall place is, is not as yet vnderſtood.

The Time.

The two firſt (as is ſaid) flower the earlieſt, ſometimes before Chriſtmas, but more vſually after, and abide a great while in flower, in great beauty, eſpecially if the weather be milde, when as few or no other flowers at that time are able to match them. The other greateſt kinde flowreth alſo earlier then the reſt that follow, for the moſt part. The ordinary kindes flower ſome in March, and ſome in Aprill, and ſome ſooner alſo ; and ſo doe the double ones likewiſe. The buſhy ſtalked Iacinth flowreth much about the ſame time.

The Names.

The former two ſorts are called *Hyacinthus Orientalis Brumalis*, and *Hyacinthus Orientalis præcox flore albo*, or *cæruleo*. The third is called of many *Zumbul Indicum*, or *Zumbul Indi*, and corruptly *Simboline* ; of others, and that more properly, *Hyacinthus Orientalis maior præcox*. The Turkes doe call all Iacinths *Zumbul*, and by adding the name of *Indi*, or *Arabi*, do ſhew from what place they are receiued. In Engliſh, The greateſt Orientall Iacinth; yet ſome doe call it after the Turkiſh name *Zumbul Indi*, or *Simboline*, as is ſaid before. The reſt haue their names ſet downe in their titles, which are moſt fit for them.

Hyacinthus Hiſpanicus minor Orientalis facie.
The little Summer Orientall Iacinth.

This little Iacinth hath foure or fiue long narrow greene leaues, lying vpon the ground, among which riſeth vp a ſlender ſmooth ſtalke, about a ſpanne high or more, bearing at the toppe many ſlender bleake blew flowers, with ſome white ſtripes and edges to be ſeene in moſt of them, faſhioned very like vnto the flowers of the Orientall Iacinth, but much ſmaller : the flower hath no ſent at all; the ſeede is like the ſeede of the Engliſh Iacinth, or Hareſ bels : the roote is ſmall and white.

Flore cæruleo.

There is another of this kinde, differing in nothing but in the colour of the flower, which is pure white.

Flore albo.

There is alſo another, whoſe flowers are of a fine delayed red colour, with ſome deeper coloured veines, running along the three outer leaues of the flower, differing in no other thing from the former.

Flore rubente.

The Place.

Theſe plants haue been gathered on the Pyrenæan Mountaines, which are next vnto Spaine, from whence, as is often ſaid, many rare plants haue likewiſe been gathered.

The Time.

They flower very late, euen after all or moſt of the Iacinths, in May for the moſt part.

The

1 *Hyacinthus Orientalis brumalis.* The Winter Orientall Iacinth. 2 *Zumbul Indi.* The greatest Orientall Iacinth. 3 *Hyacinthus Orientalis vulgaris.* The ordinary Orientall Iacinth. 4 *Hyacinthus Orientalis folioso caule.* The bushy stalked Orientall Iacinth. 5 *Hyacinthus Orientalis flore duplici.* The Orientall Iacinth once double. 6 *Hyacinthus Orientalis flore pleno cæruleo.* The faire double blew Orientall Iacinth.

The Names.

They are called eyther *Hyacinthus Hispanicus minor Orientalis facie*, as it is in the title, or *Hyacinthus Orientalis facie*, that is to say, The lesser Spanish Iacinth, like vnto the Orientall: yet some haue called them, *Hyacinthus Orientalis serotinus minor*, The lesser late Orientall Iacinth, that thereby they may be knowne from the rest.

Hyacinthus Hispanicus obsoletus. The Spanish dunne coloured Iacinth.

This Spanish Iacinth springeth very late out of the ground, bearing foure or fiue short, hollow, and soft whitish greene leaues, with a white line in the middle of euery one of them, among which rise vp one or more stalkes, bearing diuers flowers at the toppes of them, all looking one way, or standing on the one side, hanging downe their heads, consisting of six leaues, three whereof being the outermost, lay open their leaues, and turne back the ends a little again: the other three which are innermost, do as it were close together in the middle of the flower, without laying themselues open at all, being a little whitish at the edges: the whole flower is of a purplish yellow colour, with some white and green as it were mixed among it, of no sent at all: it beareth blacke and flat seede in three square, great, and bunched out heads: the roote is reasonable great, and white on the outside, with many strong white fibres at it, which perish not yearely, as the fibres of many other Iacinths doe, and as it springeth late, so it holdeth his greene leaues almost vntill Winter.

Mauritanicus. There hath been another hereof brought from about Fez and Marocco in Barbary, which in all respects was greater, but else differed little.

Maximus Æthiopicus. There was another also brought from the Cape of good Hope, whose leaues were stronger and greener then the former, the stalke also thicker, bearing diuers flowers, confusedly standing vpon longer foote-stalkes, yet made after the same fashion, but that the three inner leaues were whitish, and dented about the edges, otherwise the flowers were yellow and greenish on the inside.

The Place.

These plants grow in Spaine, Barbary, and Ethiopia, according as their names and descriptions doe declare.

The Time.

The first flowreth not vntill Iune; for, as I said, it is very late before it springeth vp out of the ground, and holdeth his leaues as is said, vntill September, in the meane time the seede thereof ripeneth.

The Names.

They haue their names according to the place of their growing; for one is called *Hyacinthus Hispanicus obsoletioris coloris.* The other is called also *Hyacinthus Mauritanicus.* And the last, *Hyacinthus Æthiopicus obsoletus.* In English, The Spanish, Barbary, or Ethiopian Iacinth, of a dunne or duskie colour.

Hyacinthus Anglicus Belgicus, vel Hispanicus.
English Hares-bels, or Spanish Iacinth.

Our English Iacinth or Hares-bels is so common euery where, that it scarce needeth any description. It beareth diuers long and narrow greene leaues, not standing vpright, nor yet fully lying vpon the ground, among which springeth vp the stalke, bearing at the toppe many long and hollow flowers, hanging downe their heads all
forwards

forwards for the moſt part, parted at the brimmes into ſix parts, turning vp their points a little againe, of a ſweetiſh, but heady ſent, ſomewhat like vnto the Grape-flower: the heads for ſeede are long and ſquare, wherein is much blacke ſeede: the colour of the flowers are in ſome of a deeper blew, tending to a purple; in others of a paler blew, or of a bleake blew, tending to an aſh colour: Some are pure white, and ſome are party coloured, blew and white; and ſome are of a fine delayed purpliſh red or bluſh colour, which ſome call a peach colour. The rootes of all ſorts agree, and are alike, being white and very ſlimie; ſome whereof will be great and round, others long and ſlender, and thoſe that lye neare the toppe of the earth bare, will be greene.

Hyacinthus Hiſpanicus maior flore campanulæ inſtar.
The greater Spaniſh bell-flowred Iacinth.

This Spaniſh bell-flowred Iacinth, is very like the former Engliſh or Spaniſh Iacinth, but greater in all parts, as well of leaues as flowers, many growing together at the toppe of the ſtalke, with many ſhort greene leaues among them, hanging downe their heads, with larger, greater, and wider open mouths, like vnto bels, of a darke blew colour, and no good ſent.

The Place.

The firſt groweth in many places of England, the Lowe-Countries, as we call them, and Spaine, but the laſt chiefly in Spaine.

The Time.

They flower in Aprill for the moſt part, and ſometimes in May.

The Names.

Becauſe the firſt is more frequent in England, then in Spain, or the Lowe-Countries, it is called with vs *Hyacinthus Anglicus*, The Engliſh Iacinth; but it is alſo called as well *Belgicus*, as *Hiſpanicus*: yet Dodonæus calleth it *Hyacinthus non ſcriptus*, becauſe it was not written of by any Authour before himſelfe. It is generally knowne in England by the name of Hare-bels. The other Spaniſh Iacinth beareth his name in his title.

Hyacinthus Eriophorus. The Woolly Iacinth.

This Woolly Iacinth hath many broad, long, and faire greene leaues, very like vnto ſome of the Iacinths, but ſtiffer, or ſtanding more vpright, which being broken, doe yeeld many threeds, as if a little fine cotton wooll were drawne out: among theſe leaues riſeth vp a long greene round ſtalke, a foote and a halfe high or more, whereon is ſet a great long buſh of flowers, which blowing open by degrees, firſt below, and ſo vpwards, are very long in flowring: the toppe of the ſtalke, with the flowers, and their little footſtalkes, are all blew, euery flower ſtanding outright with his ſtalke, and ſpreading like a ſtarre, diuided into ſix leaues, hauing many ſmall blew threeds, ſtanding about the middle head, which neuer gaue ripe ſeede, as farre as I can heare of: the root is white, ſomewhat like the root of a Muſcari, but as full of wooll or threeds, or rather more, then the leaues, or any other part of it.

The Place.

This hath been ſent diuers times out of Turkie into England, where it continued a long time as well in my Garden as in others, but ſome hard froſty Winters cauſed it to periſh with me, and diuers others, yet I haue had it againe from a friend, and doth abide freſh and greene euery yeare in my Garden.

The

The Time.

This flowred in the Garden of M^r.Richard Barnefley at Lambeth, onely once in the moneth of May, in the yeare 1606. after hee had there preferued it a long time : but neither he, nor any elfe in England that I know, but thofe that faw it at that time, euer faw it beare flower, either before or fince.

The Names.

It is called by diuers *Bulbus Eriophorus*, or *Laniferus*, that is, Woolly Bulbous ; but becaufe it is a Iacinth, both in roote, leafe, and flower, and not a *Narciffus*, or Daffodill, it is called *Hyacinthus Eriophorus*, or *Laniferus*, The Woolly Iacinth. It is very likely, that Theophraftus in his feuenth Book & thirteenth Chapter, did meane this plant, where hee declareth, that garments were made of the woolly fubftance of a bulbous roote, that was taken from between the core or heart of the roote (which, as hee faith, was vfed to be eaten) and the outermoft fhels or peelings ; yet Clufius feemeth to faften this woolly bulbous of Theophraftus, vpon the next Iacinth of Spaine.

Hyacinthus Stellatus Baticus maior, vulgò Perüanus.
The great Spanifh Starry Iacinth, or of Peru.

This Iacinth (the greateft of thofe, whofe flowers are fpread like a ftarre, except the two firft Indians) hath fiue or fix, or more, very broad, and long greene leaues, fpread vpon the ground, round about the roote, which being broken are woolly, or full of threeds, like the former : in the middle of thefe leaues rifeth vp a round fhort ftalke, in comparifon of the greatneffe of the plant (for the ftalke of the Orientall Iacinth is fometimes twice fo high, whofe roote is not fo great) bearing at the toppe a great head or bufh of flowers, fafhioned in the beginning, before they bee blowne or feparated, very like to a Cone or Pineapple, and begin to flower belowe, and fo vpwards by degrees, euery flower ftanding vpon a long blackifh blew foote-ftalke, which when they are blowne open, are of a perfect blew colour, tending to a Violet, and made of fix fmall leaues, laid open like a ftarre ; the threeds likewife are blewifh, tipt with yellow pendents, ftanding about the middle head, which is of a deeper blew, not hauing any good fent to be perceiued in it, but commendable only for the beauty of the flowers : after the flowers are paft, there come three fquare heads, containing round blacke feede : the roote is great, and fomewhat yellowifh on the outfide, with a knobbe or bunch at the lower end of the roote, (which is called the feate of the roote) like vnto the Mufcari, Scylla, and many other bulbous rootes, at which hang diuers white, thicke, and long fibres, whereby it is faftened in the ground, which perifh not euery yeare, but abide continually, and therefore doth not defire much remouing.

Hyacinthus Stellatus Baticus, fiue Perüanus flore albo.
The great white Spanifh ftarry Iacinth.

This other Spanifh Iacinth is in moft parts like vnto the former, but that his leaues are not fo large, nor fo deep a greene : the ftalks of flowers likewife hath not fo thicke a head, or bufh on it, but fewer and thinner fet : the flowers themfelues alfo are whitifh, yet hauing a fmall dafh of blufh in them : the threeds are whitifh, tipt with yellow pendents : the feede and rootes are like vnto the former, and herein confifteth the difference betweene this and the other forts.

Hyacinthus Stellatus Baticus, fiue Perüanus flore carneo.
The great blufh coloured Spanifh Starry Iacinth.

This likewife differeth little from the two former, but onely in the colour of the flowers ;

1. *Hyacinthus Orientalis farie.* The little Summer Orientall Iacinth. 2. *Hyacinthus Mauritanicus.* The Barbary Iacinth. 3. *Hyacinthus obsoletus Hispanicus.* The Spanish duskie Iacinth. 4. *Hyacinthus Hispanicus flore campanula.* The greater Spanish bel-flowred Iacinth. 5. *Hyacinthus Anglicus.* The English Iacinth or Harebels. 6. *Hyacinthus Eriophorus.* The Woolly Iacinth. 7. *Hyacinthus Stellaris Baticus maior, sive Peruanus.* The great Spanish starry Iacinth, or of Peru.

flowers; for this being found growing among both the other, hath h is head of flowers as great and large as the first, but the buds of his flowers, before they are open, are of a deepe blush colour, which being open, are more delayed, and of a pleasant pale purple, or blush colour, standing vpon purplish stalkes : the heads in the middle are whitish, and so are the threeds compassing it, tipt with yellow.

The Place.

These doe naturally grew in Spaine, in the Medowes a little off from the Sea, as well in the Island Gades, vsually called Cales, as likewise in other parts along the Sea side, as one goeth from thence to Porto Santa Maria, which when they be in flower, growing so thicke together, seeme to couer the ground, like vnto a tapistry of diuers colours, as I haue beene credibly enformed by Guillaume Boel, a Freeze-lander borne, often before and hereafter remembred, who being in search of rare plants in Spaine, in the yeare of our Lord 1607. after that most violent frosty Winter, which perished both the rootes of this, and many other fine plants with vs, sent mee ouer some of these rootes for my Garden, and affirmed this for a truth, which is here formerly set downe, and that himselfe gathered those he sent mee, and many others in the places named, with his owne hands; but hee saith, that both that with the white, and with the blush flowers, are farre more rare then the other.

The Time.

They flower in May, the seede is ripe in Iuly.

The Names.

This hath beene formerly named *Eriophorus Peruanus*, and *Hyacinthus Stellatus Peruanus*, The Starry Iacinth of Peru, being thought to haue grown in Peru, a Prouince of the West Indies; but he that gaue that name first vnto it, eyther knew not his naturall place, or willingly imposed that name, to conceale it, or to make it the better esteemed. It is most generally receiued by the name *Hyacinthus Peruanus*, from the first imposer thereof, that is, the Iacinth of Peru : but I had rather giue the name agreeing most fitly vnto it, and call it as it is indeede *Hyacinthus Stellatus Bæticus*, The Spanish Starry Iacinth; and because it is the greatest that I know hath come from thence, I call it, The great Starry Iacinth of Spaine, or Spanish Iacinth.

Hyacinthus Stellatus vulgaris, siue Bifolius Fuchsij.
The common blew Starry Iacinth.

This Starry Iacinth (being longest knowne, and therefore most common) riseth out of the ground, vsually but with two browne leaues, yet sometimes with three, inclosing within them the stalke of flowers, the buds appearing of a darke whitish colour, as soone as the leaues open themselues, which leaues being growne, are long, and hollow, of a whitish greene on the vpper side, and browne on the vnder side, and halfe round, the browne stalke rising vp higher, beareth fiue or sixe small starre-like flowers thereon, consisting of six leaues, of a faire deepe blew, tending to a purple. The seede is yellowish, and round, contained in round pointed heads, which by reason of their heauinesse, and the weaknesse of the stalke, lye vpon the ground, and often perish with wet and frosts, &c. The roote is somewhat long, and couered with a yellowish coate.

Hyacinthus stellatus flore albo. The white Starry Iacinth.

The white Starry Iacinth hath his leaues like the former, but greene and fresh, not browne, and a little narrower also : the buddes for flowers at the first appeare a little blush, which when they are blowne, are white, but yet retaine in them a small shew of that blush colour.

We

We haue another, whofe flowers are pure white, and fmaller then the other, the *Flore niueo*, leaues whereof are of a pale frefh greene, and fomewhat narrower.

Hyacinthus Stellatus flore rubente. The blufh coloured Starry Iacinth.

The difference in this from the former, is onely in the flowers, which are of a faire blufh colour, much more eminent then in the others, in all things elfe alike.

Hyacinthus Stellatus Martius, fiue præcox cæruleus. The early blew Starry Iacinth.

This Iacinth hath his leaues a little broader, of a frefher greene, and not browne at all, as the firft blew Iacinth of Fuchfius laft remembred: the buds of the flowers, while they are enclofed within the leaues, and after, when the ftalke is gowne vp, doe remaine more blew then the buds of the former: the flowers, when they are blowne open, are like the former, but fomewhat larger, and of a more liuely blew colour: the roote alfo is a little whiter on the outfide. This doth more feldome beare feede then the former.

Hyacinthus Stellatus præcox flore albo. The white early Starry Iacinth.

There is alfo one other of this kinde, that beareth pure white flowers, the green leafe thereof being a little narrower then the former, and no other difference.

Hyacinthus Stellatus præcox flore fuaue rubente. The early blufh coloured Starry Iacinth.

This blufh coloured Iacinth is very rare, but very pleafant, his flowers being as large as the firft of this laft kinde, and fomewhat larger then the blufh of the other kinde: the leaues and rootes differ not from the laft recited Iacinth.

The Place.

All thefe Iacinths haue beene found in the Woods and Mountaines of Germany, Bohemia, and Auftria, as Fuchfius and Gefner doe report, and in Naples, as Imperatus and others doe teftifie. Wee cherifh them all with great care in our Gardens, but efpecially the white and the blufh of both kindes, for that they are more tender, and often perifh for want of due regard.

The Time.

The common kindes, which are firft exprefled, flower about the middle of February, if the weather bee milde, and the other kindes fometimes a fortnight after, that is, in March, but ordinarily much about the fame time with the former.

The Names.

The firft is called in Latine *Hyacinthus Stellatus vulgaris*, and *Hyacinthus Stellatus bifolius*, and *Hyacinthus Stellaris Fuchfij*, and of fome *Hyacinthus Stellatus Germanicus*; wee might very well call the other kinde, *Hyacinthus Stellatus vulgaris alter*, but diuers call it *Præcox*, and fome *Martius*, as it is in the title. In Englifh they may bee feuerally called: the firft, The common; and the other, The early Starry Iacinth (notwithftanding the firft flowreth before the other) for diftinction fake.

The *Hyacinthus* feemeth to be called *Vacinium* of Virgil in his Eclogues; for hee alwayes reckoneth it among the flowers that were vfed to decke Garlands, and neuer among fruits, as fome would haue it. But in that hee calleth it *Vacinium nigrum*, in feuerall places, that doth very fitly anfwer the

common

common receiued custome of those times, that called all deepe blew colours, such as are purples, and the like, blacke; for the Violet it selfe is likewise called blacke in the same place, where he calleth the *Vacinium* blacke; so that it seemeth thereby, that he reckoned them to be both of one colour, and we know the colour of the Violet is not blacke, as we doe distinguish of blacke in these dayes. But the colour of this Starry Iacinth, being both of so deepe a purple sometimes, so neare vnto a Violet colour, and also more frequent, then any other Iacinth with them, in those places where Virgil liued, perswadeth me to thinke, that Virgil vnderstood this Starry Iacinth by *Vacinium* : Let others iudge otherwise, if they can shew greater probabilitie.

1. *Hyacinthus Stellatus Byzantinus nigra radice.*
The Starry Iacinth of Turkie with the blacke roote.

This Starry Iacinth of Constantinople hath three or foure fresh greene, thinne, and long leaues, of the bignesse of the English Iacinth, but not so long, betweene which riseth vp a slender lowe stalke, bearing fiue or six small flowers, dispersedly set thereon, spreading open like a starre, of a pale or bleake blew colour : the leaues of the flowers are somewhat long, and stand as it were somewhat loosly, one off from another, and not so compactly together, as the flowers of other kindes : it seldome beareth ripe seede with vs, because the heads are so heauie, that lying vpon the ground, they rotte with the wet, or are bitten with the frosts, or both, so that they seldome come to good: the roote is small in some, and reasonable bigge in others, round, and long, white within, but couered with deepe reddish or purplish peelings, next vnto it, and darker and blacker purple on the outside, with some long and thicke white fibres, like fingers hanging at the bottome of them, as is to be seene in many other Iacinths : the roote it selfe for the most part doth runne downewards, somewhat deep into the ground.

2. *Hyacinthus Stellatus Byzantinus maior.*
The greater Starry Iacinth of Constantinople.

This Iacinth may rightly be referred to the former Iacinth of Constantinople, and called the greater, it is so like thereunto, that any one that knoweth that, will soone say, that this is another of that sort, but greater as it is in all his parts, bearing larger leaues by much, and more store, lying vpon the ground round about the roote : it beareth many lowe stalkes of flowers, as bleake, and standing as loosly as the former : onely the roote of this, is not black on the outside, as the other, but three times bigger.

3. *Hyacinthus Stellatus Byzantinus alter, siue flore boraginis.*
The other Starry Iacinth of Constantinople.

This other Iacinth hath for the most part onely foure leaues, broader and greener then the first, but not so large or long as the second : the stalke hath fiue or six flowers vpon it, bigger and rounder set, like other starry Iacinths, of a more perfect or deeper blew then either of the former, hauing a whitish greene head or vmbone in the middle, beset with six blew chiues or threeds, tipt with blacke, so closly compassing the vmbone, that the threeds seeme so many prickes stucke into a clubbe or head; some therefore haue likened it to the flower of Borage, and so haue called it : after the flowers are past, come vp round white heads, wherein is contained round and white seede : the roote is of a darke whitish colour on the outside, and sometimes a little reddish withall.

The Place.

The first and the last haue beene brought from Constantinople; the first among many other rootes, and the last by the Lord Zouch, as Lobel witnesseth. The second hath been sent vs out of the Lowe-Countries, but from whence they had it, we do not certainly know. They growe with vs in our Gardens sufficiently.   The

The Time.

Thefe flower in Aprill, but the firft is the earlieft of the reft, and is in flower prefently after the early Starry Iacinth, before defcribed.

The Names.

The former haue their names in their titles, and are not knowne vnto vs by any other names that I know; but as I faid before, the laft is called by fome, *Hyacinthus Boraginis flore.* The firft was fent out of Turkie, by the name of *Sufam giul,* by which name likewife diuers other things haue beene fent, fo barren and barbarous is the Turkifh tongue.

Hyacinthus Stellatus Æftivus maior. The greater Summer Starry Iacinth.

This late Iacinth hath diuers narrow greene leaues, lying vpon the ground, fome-what like the leaues of the Englifh Iacinth, but ftiffer and ftronger; among which rifeth vp a round ftiffe ftalke, bearing many flowers at the toppe thereof, and at euery foote-ftalke of the flowers a fmall fhort leafe, of a purplifh colour: the flowers are ftarre-like, of a fine delayed purplifh colour, tending to a pale blew or afh colour, ftriped on the backe of euery leafe, and hauing a pointed vmbone in the middle, with fome whitifh purple threeds about it, tipt with blew: the feede is blacke, round, and fhining, like vnto the feede of the Englifh Iacinth, but not fo bigge: the roote is round and white, hauing fome long thicke rootes vnder it, befides the fibres, as is vfuall in many other Iacinths.

Hyacinthus Stellatus Æftivus minor. The leffer Summer Starry Iacinth.

This leffer Iacinth hath diuers very long, narrow, and fhining greene leaues, fpread vpon the ground round about the roote, among which rifeth vp a very fhort round ftalke, not aboue two inches high, carrying fix or feuen fmall flowers thereon, on each fide of the ftalke, like both in forme and colour vnto the greater before defcribed, but leffer by farre: the feede is blacke, contained in three fquare heads: the roote is fmall and white, couered with a browne coate, and hauing fome fuch thicke rootes among the fibres, as are among the other.

The Place.

Both thefe Iacinths grow naturally in Portugall, and from thence haue been brought, by fuch as feeke out for rare plants, to make a gaine and profit by them.

The Time.

They both flower in May, and not before: and their feed is ripe in Iuly.

The Names.

Some doe call thefe *Hyacinthus Lufitanicus,* The Portugall Iacinth. Clufius, who firft fet out the defcriptions of them, called them as is expreffed in their titles; and therefore we haue after the Latine name giuen their Englifh, according as is fet downe. Or if you pleafe, you may call them, The greater and the leffer Portugall Iacinth.

Hyacinthus Stellaris flore cinereo. The afh coloured Starry Iacinth.

This afh coloured Iacinth, hath his leaues very like vnto the leaues of the Englifh Iacinth, and fpreading vpon the ground in the fame manner, among which rife vp one or two ftalkes, fet at the toppe with a number of fmall ftarre-like flowers, bufhing bigger

ger below then aboue, of a very pale or white blew, tending to an ash colour, and very sweete in smell : the seede is blacke and round, like vnto the seede of the English Iacinth, and so is the roote, being great, round, and white; so like, I say, that it is hard to know the one from the other.

The Place.

The certaine originall place of growing thereof, is not knowne to vs.

The Time.

It flowreth in Aprill.

The Names.

Some doe call this *Hyacinthus Someri*, Somers Iacinth, becaufe as Lobel faith, he brought it firft into the Lowe-Countries, eyther from Conftantinople, or out of Italy.

Hyacinthus Stellatus Lilifolio & radice cæruleo.
The blew Lilly leafed Starre Iacinth.

This Iacinth hath six or feuen broad greene leaues, fomewhat like vnto Lilly leaues, but shorter (whereof it tooke his name as well as from the roote) spread vpon the ground, and lying close and round : before the ftalke rifeth out from the middle of these leaues, there doth appeare a deepe hollow place, like a hole, to bee feene a good while, which at length is filled vp with the ftalke, rifing thence vnto a foote or more high, bearing many ftarre-like flowers at the toppe, of a perfect blew colour, neare vnto a Violet, and fometimes of paler or bleake blew colour, hauing as it were a fmall cuppe in the middle, diuided into six peeces, without any threeds therein : the feede is blacke and round, but not shining : the roote is fomewhat long, bigge belowe, and fmall aboue, like vnto the fmall roote of a Lilly, and compofed of yellow fcales, as a Lilly, but the fcales are greater, and fewer in number.

Hyacinthus Stellatus Lilifolius albus. The white Lilly leafed Starre Iacinth.

The likeneffe of this Iacinth with the former, caufeth me to be briefe, and not to repeate the fame things againe, that haue already been expreffed : You may therefore vnderftand, that except in the colour of the flower, which in this is white, there is no difference betweene them.

Flore carneo. I heare of one that should beare blush coloured flowers, but I haue not yet feene any fuch.

The Place.

These Iacinths haue been gathered on the Pyrenæan Hils, in that part of France that is called Aquitaine, and in fome other places.

The Time.

These flower in Aprill, and fometimes later.

The Names.

Becaufe the roote is fo like vnto a Lilly, as the leafe is alfo, it hath moft properly beene called *Hyacinthus Stellatus Lilifolio & radice*, or for breuity *Lilifolius*, that is, The Starry Lilly leafed Iacinth. It is called *Sarahng* by the Inhabitants where it groweth, as Clufius maketh the report from Venerius, who further faith, that by experience they haue found the cattell to fwell and dye, that haue eaten of the leaues thereof.

Hyacinthus

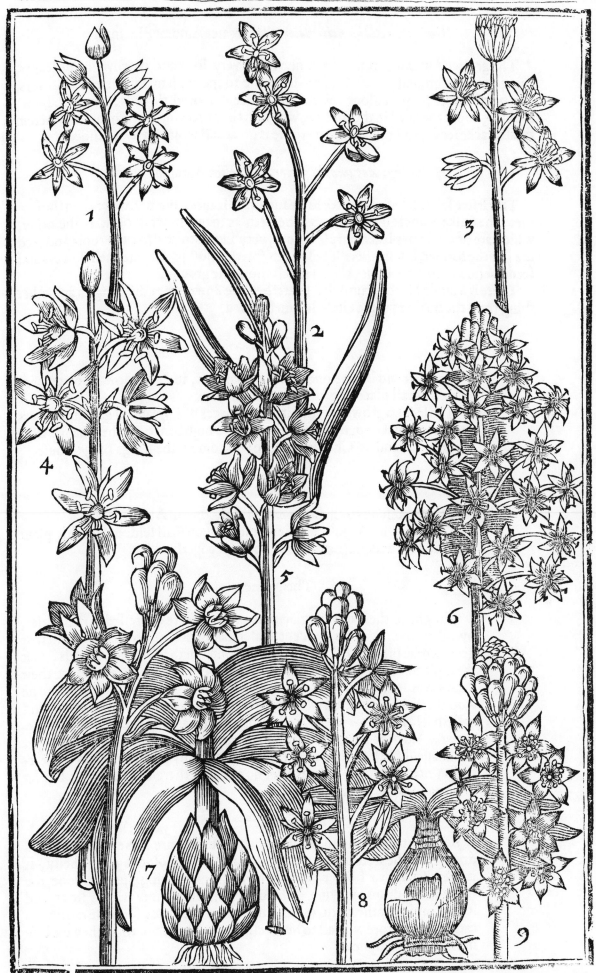

1 *Hyacinthus stellatus pracox caruleus.* The early blew starry Iacinth. 2 *Hyacinthus stellatus pracox albus,* The early white starry Iacinth. 3 *Hyacinthus stellatus Byzantinus nigra radice.* The Turkie starry Iacinth with a blacke roote. 4 *Hyacinthus Byzantinus alter siue flore Boraginis,* The other starry Iacinth f Constantinople. 5 *Hyacinthus astivus maior.* The greater Summer starry Iacinth. 6 *Hyacinthus stellatus flore cinereo,* The ash coloured starry Iacinth. 7 *Hyacinthus stellatus Lilifolius.* The Lilly leafed starre Iacinth. 8 *Hyacinthus Autumnalis.* The Autumne Iacinth. 9 *Scilla alba siue Hyacinthus marinus,* The Sea Onion or Squill.

Hyacinthus Autumnalis maior. The greater Autumne Iacinth.

The greater Autumne Iacinth hath fiue or six very long and narrow greene leaues, lying vpon the ground; the stalkes are set at the toppe with many starre-like flowers, of a pale blewish purple colour, with some pale coloured threeds, tipt with blew, standing about the head in the middle, which in time growing ripe, containeth therein small blacke seede, and roundish : the roote is great and white on the outside.

Hyacinthus Autumnalis minor. The lesser Autumne Iacinth.

This lesser Iacinth hath such like long and small leaues, but narrower then the former : the stalke is not full so high, but beareth as many flowers on it as the other, which are of a pale or bleake purple colour, very like vnto it also : the roote and seed are like the former, but smaller. These both for the most part, beare their flowers and seede before the greene leaues rise vp much aboue the ground.

Flore albo. There is a kinde hereof found that beareth white flowers, not differing in any other thing from the smaller purple kinde last mentioned.

The Place.

The first and last are onely kept in Gardens, and not knowne to vs where their naturall place of growing wilde may be.

The second groweth wilde in many places of England. I gathered diuers rootes for my Garden, from the foote of a high banke by the Thames side, at the hither end of Chelsey, before you come at the Kings Barge-house.

The Time.

The greatest flowreth in the end of Iuly, and in August.

The other in August and September, you shall seldome see this plant with flowers and greene leaues at one time together.

The Names.

They haue their names giuen them, as they are expressed in their titles, by all former Writers, except Daleschampius, or hee that set forth that great worke printed at Lyons; for hee contendeth with many words, that these plants can bee no Iacinths, because their flowers appeare before their leaues in Autumne, contrary to the true Iacinth, as he saith: and therefore he would faine haue it referred to *Theophrastus bulbus in libro primo cap.*12. and calleth it his *Tiphyum* mentioned in that place, as also *Bulbus æstiuus Dalechampy.* Howsoeuer these things may carry some probability in them, yet the likenesse both of rootes, and flowers especially, hath caused very learned Writers to entitle them as is set downe, and therefore I may not but let them passe in the like manner.

The Vertues.

Both the rootes and the leaues of the Iacinths are somewhat cold and drying, but the seede much more. It stayeth the loosnesse of the belly. It is likewise said to hinder young persons from growing ripe too soone, the roote being drunke in wine. It helpeth them also whose vrine is stopt, and is auaileable for the yellow Iaundise; but as you heare some are deadly to cattell, I therefore wish all to bee well aduised which of these they will vse in any inward physicke.

Scilla alba. The Sea Onion or Squill.

As I ended the discourse of both the true and the bastard Daffodils, with the Sea kindes

kindes of both forts ; fo I thinke it not amiffe, to finifh this of the Iacinths with the defcription of a Sea Iacinth, which (as you fee) I take to be the *Scilla*, or Sea Onion , all his parts fo nearely refembling a Iacinth, that I know not where to ranke him better then in this place, or rather not any where but here. You fhall haue the defcription thereof, and then let the iudicious paffe their fentence, as they thinke meeteft.

The Squill or Sea Onion (as many doe call it) hath diuers thicke leaues, broad, long, greene, and hollowifh in the middle, and with an eminent or fwelling ribbe all along the backe of the leafe, (I relate it as I haue feene it, hauing fhot forth his leaues in the fhip by the way, as the Mariners that brought diuers rootes from out of the Straights, did fell them to mee and others for our vfe) lying vpon the ground, fomewhat like vnto the leaues of a Lilly : thefe fpring vp after the flowers are paft, and the feed ripe, they abiding all the Winter, and the next Spring, vntill the heate of the Summer hath fpeat and confumed them, and then about the end of Auguft, or beginning of September, the ftalke with flowers arifeth out of the ground a foote and a halfe high, bearing many ftarre-like flowers on the toppe, in a long fpike one aboue another, flowring by degrees, the loweft firft, and fo vpwards, whereby it is long in flowring, very like, as well in forme as bigneffe, to the flowers of the great Starre of Bethlehem (thefe flowers I haue likewife feene fhooting out of fome of the rootes, that haue been brought in the like manner:) after the flowers are paft, there come vp in their places thicke and three fquare heads, wherin is contained fuch like flat, black, and round feed, as the Spanifh duskie Iacinth before defcribed did beare, but greater : the root is great & white, couered with many peelings or couerings, as is plainly enough feen to any that know them, and that fometimes wee haue had rootes, that haue beene as bigge as a pretty childes head, and fometimes two growing together, each whereof was no leffe then is faid of the other.

Scilla rubra fiue Pancratium verum. The red Sea Onion.

The roote of this Squill, is greater oftentimes then of the former, the outer coates or peelings being reddifh, bearing greater, longer, ftiffer, and more hollow leaues, in a manner vpright : this bringeth fuch a like ftalke and flowers, as the former doth , as Fabianus Ilges, Apothecary to the Duke of Briga, did fignifie by the figure thereof drawne and fent to Clufius.

The Place.

They grow alwayes neare the Sea, and neuer farre off from it, but often on the very baich of the Sea, where it wafheth ouer them all along the coafts of Spaine, Portugal, and Italy, and within the Straights in many places : it will not abide in any Garden farre from the Sea, no not in Italy, as it is related.

The Time.

The time wherein they flower, is expreffed to be in Auguft and September : the feede to be ripe in October and Nouember, and the greene leaues to fpring vp in Nouember and December.

The Names.

Thefe are certainly the true kindes of *Scilla* that fhould bee vfed in medicines, although (as Clufius reporteth) the Spaniards forbade him to tafte of the red Squill, as of a moft ftrong and prefent poifon. Pliny hath made more forts then can be found out yet to this day with vs : that *Scilla* that is called *Epimenidia*, becaufe it might be eaten, is thought to be the great *Ornithogalum*, or Starre of Bethlehem. *Pancratium* is, I know, and as I faid before, referred to that kinde of baftard Sea Daffodill, which is fet forth before in the end of the hiftory of the baftard Daffodils ; and diuers alfo would make the *Narciffus tertius Matthioli*, which I call the true Sea Daffodill, to be a *Pancratium*; but feeing Diofcorides (and no other is againft him)

M   maketh

maketh *Pancratium* to be a kinde of Squill with reddish rootes, I dare not vphold their opinion against such manifest truth.

The Vertues.

The Squill or Sea Onion is wholly vsed physically with vs, because wee can receiue no pleasure from the sight of the flowers. Pliny writeth, that Pithagoras wrote a volume or booke of the properties thereof, for the singular effects it wrought; which booke is lost, yet the diuers vertues it hath is recorded by others, to be effectuall for the spleene, lungs, stomach, liuer, head and heart; and for dropsies, old coughs, Iaundise, and the wormes; that it cleareth the sight, helpeth the tooth-ache, cleanseth the head of scurfe, and running sores; and is an especiall Antidote against poison: and therefore is vsed as a principall ingredient into the *Theriaca Andromachi*, which we vsually call Venice Treakle. The Apothecaries prepare hereof, both Wine, Vinegar, and Oxymel or Syrupe, which is singular to extenuate and expectorate tough flegme, which is the cause of much disquiet in the body, and an hinderer of concoction, or disgestion in the stomach, besides diuers other wayes, wherein the scales of the rootes, being dryed, are vsed. And Galen hath sufficiently explained the qualities and properties thereof, in his eight Booke of Simples.

Chap. XII.

Ornithogalum. Starre of Bethlehem.

AFter the Family of the Iacinths, must needes follow the kindes of Starre-flowers, or Starres of Bethlehem, as they are called, for that they doe so nearely resemble them, that diuers haue named some of them Iacinths, and referred them to that kindred: all of them, both in roote, leafe, and flower, come nearer vnto the Iacinths, then vnto any other plant. They shall therefore bee next described, euery one in their order, the greatest first, and the rest following.

Ornithogalum Arabicum. The great Starre-flower of Arabia.

This Arabian Starre-flower hath many broad, and long greene leaues, very like vnto the leaues of the Orientall Iacinth, but lying for the most part vpon the ground, among which riseth vp a round greene stalke, almost two foote high, bearing at the toppe diuers large flowers, standing vpon long foote-stalkes, and at the bottome of euery one of them a small short pointed greene leafe: these flowers are made of six pure white leaues a peece, laid open as large as an ordinary Daffodill, but of the forme of a Starre Iacinth, or Starre of Bethlehem, which close as they doe euery night, and open themselues in the day time, especially in the Sunne, the smell whereof is pretty sweete, but weake: in the middle of the flower is a blackish head, composed with six white threeds, tipt with yellow pendents: the seede hath not beene obserued with vs: the roote is great and white, with a flat bottome, very impatient of our cold Winters, so that it seldome prospereth or abideth with vs; for although sometimes it doe abide a Winter in the ground, yet it often lyeth without springing blade, or any thing else a whole yeare, and then perisheth: or if it doe spring, yet many doe not beare, and most after their first bearing doe decay and perish. But if any be desirous, to know how to preserue the roote of this plant, or of many other bulbous rootes that are tender, such as the great double white Daffodill of Constantinople, and other fine Daffodils, that come from hot Countries; let them keepe this rule: Let either the roote be planted in a large pot, or tubbe of earth, and housed all the Winter, that so it may bee defended from the frosts; Or else (which is the easier way) keepe the roote out of the ground euery yeare, from September, after the leaues and stalkes are past, vntill February, in
some

some dry, but not hot or windy place, and then plant it in the ground vnder a South wall, or such like defended place, which will spring, and no doubt prosper well there, in regard the greatest and deepest frosts are past after February, so that seldome any great frosts come after, to pierce so deepe as the roote is to be set, or thereby to doe any great harme to it in such a place.

The Place.

This hath been often sent out of Turkie, and likewise out of Italy; I had likewise two rootes sent mee out of Spaine by Guillaume Boel before re-membred, which (as hee said) hee gathered there, but they prospered not with me, for want of the knowledge of the former rule. It may be likely that Arabia is the place, from whence they of Constantinople receiue it.

The Time.

It flowreth in May, if it be of the first yeares bringing; or in Iune, if it haue been ordered after the manner before set downe.

The Names.

It hath been sent out of Italy by the name of *Lilium Alexandrinum*, The Lilly of Alexandria, but it hath no affinity with any Lilly. Others call it *Hyacinthus Arabicus*; and the Italians, *Iacintho del pater nostro*: but it is no Ia-cinth neither, although the flowers be like some of them. Some also would referre it to a *Narcissus* or Daffodill, and it doth as little agree with it, as with a Lilly, although his flowers in largenesse and whitenesse resemble a Daffodill. Clusius hath most fitly referred it to the stocke or kindred of *Ornithogala*, or Starres of Bethlehem, as wee call them in English, and from the Turkish name, *Zumbul Arabi*, entituled it *Ornithogalum Arabicum*, although *Zumbul*, as I haue before declared, is with them, a Iacinth, wee may call it in English, The Arabian Starre-flower, or Starre of Bethlehem, or the great Starre-flower of Arabia.

1. *Ornithogalum maximum album.*
The greatest white Starre-flower, or Starre of Bethlehem.

This great Starre-flower hath many faire, broad, long, and very fresh green leaues, rising vp very early, and are greater, longer, and greener then the leaues of any Ori-entall Iacinth, which doe abide greene, from the beginning or middle of Ianuary, or before sometimes, vntill the end of May, at which time they begin to fade, and the stalke with the head of flowers beginneth to rise, so that it will haue either few or no leaues at all, when the flowers are blowne: the stalke is strong, round, and firme, rising two foote high or more, bearing at the toppe a great bush of flowers, seeming at the first to be a great greene eare of corne, for it is made spike-fashion, which when the flowers are blowne, doth rise to be very high, slender or small at the head aboue, and broad spread and bushing below, so that it is long in flowring; for they flower below first, and so vpwards by degrees: these flowers are snow white, without any line on the backside, and is therein like vnto the former, as also in whitenesse, but nothing so large, with a white vmbone or head in the middle, beset with many white threeds, tipt with yellow: the seede is blacke and round, contained in three square heads: the roote is great, thicke, and short, and somewhat yellowish on the outside, with a flat bottome, both like the former, and the next that followeth.

2. *Ornithogalum maius spicatum album.*
The great white spiked Starre-flower.

This spiked Starre-flower in his growing, is somewhat like vnto the last described,

but

but springeth not vp so early, nor hath his leaues so greene, or large, but hath broad, long, whitish greene hollow leaues, pointed at the end, among which riseth vp the stalke, which is strong and high, as the former, hauing a great bush of flowers at the toppe, standing spike-fashion, somewhat like the former, flowring in the same maner by degrees, first below, and so vpwards; but it is not so thicke set with flowers, nor so farre spread at the bottome as it, the flowers also are not so white, and each of the leaues of them haue a greene line downe the backe, leauing the edges on both sides white: after the flowers are past, the heads for seede grow three square, like the other, bearing such like blacke seede therein: the roote hereof is vsually bigger then the last, and whiter on the outside.

3. *Ornithogalum Pannonicum*. The Hungarian Starre-flower.

This Hungarian Starre-flower shooteth out diuers narrow, long, whitish greene leaues, spread vpon the ground before Winter, which are very like vnto the leaues of Gilloflowers, and so abide aboue ground, hauing a stalke rising in the middle of them the next Spring, about halfe a foote high or thereabouts, bearing many white flowers at the toppe, with greene lines downe the backe of them, very like vnto the ordinary Starres of Bethlehem: the roote is greater, thicker, and longer then the ordinary Starres, and for the most part, two ioyned together, somewhat grayish on the out side.

4. *Ornithogalum vulgare.* The Starre of Bethlehem.

The ordinary Starre of Bethlehem is so common, and well knowne in all countries and places, that it is almost needlesse to describe it, hauing many greene leaues with white lines therein, and a few white flowers set about the toppe of the stalke, with greenish lines downe the backe: the roote is whitish, and encreaseth aboundantly.

5. *Asphodelus bulbosus Galeni, siue Ornithogalum maius flore subnirescente.* The bulbous Asphodill, or greene Starre-flower.

Diuers haue referred this plant vnto the Asphodils, because (as I thinke) the flowers hereof are straked on the backe, and the leaues long and narrow, like vnto the Asphodils; but the roote of this being bulbous, I rather (as some others doe) ioyne it with the *Ornithogala*, for they also haue strakes on the backe of the flowers. It hath many whitish greene leaues, long and narrow, spread vpon the ground, which spring vp in the beginning of the yeare, and abide vntill May, and then they withering, the stalke springeth vp almost as high as the first, hauing many pale yellowish greene flowers, but smaller, and growing more sparsedly about the stalke vpon short foot-stalkes, but in a reasonable long head spike-fashion: the seede is like vnto the second kinde, but smaller: the roote is somewhat yellowish, like the first great white kinde.

The Place.

The first is onely nursed in Gardens, his originall being not well knowne, yet some attribute it vnto *Pannonia* or Hungary. The second hath been found neare vnto Barcinone, and Toledo in Spaine. The third was found in Hungary by Clusius. Our ordinary euery where in the fields of Italy and France, and (as it is said) in England also. And the last groweth likewise by the corne fields in the vpper Hungary.

The Time.

They flower in Aprill and May, and sometimes in Iune.

The Names.

The first is called by Clusius *Ornithogalum maximum album*, because it is greater

1 *Ornithogalum Arabicum*. The great ſtarre-flower of Arabia. 2 *Ornithogalum maximum album* The greateſt white ſtarre flower. 3 *Ornithogalum maius ſpicatum album*. The great white ſpiked ſtarre flower. 4 *Ornithogalum Pannonicum album*. The Hungarian ſtarre-flower. 5 *Aſphodelus bulboſus Galeni, ſiue Ornithogalum maius ſubuireſcente flore* The bulbed Aſphodill, or greene ſtarre flower. 6 *Ornithogalum Hiſpanicum minus*. The little ſtarre-flower of Spaine. 7 *Ornithogalum luteum*. The yellow ſtarre-flower of Bethlehem. 8 *Ornithogalum Neapolitanum*. The ſtarre-flower of Naples.

greater then the next, which hee tooke formerly for the greatest : but it might more fitly, in my iudgement, bee called *Asphodelus bulbosus albus* (if there be any *Asphodelus bulbosus* at all) becaufe this doth fo nearly refemble that, both in the early fpringing, and the decay of the greene leaues, when the ftalkes of flowers doe rife vp. Diuers alfo doe call it *Ornithogalum Pannonicum maximum album.*

The fecond hath his name in his title, as moft authors doe fet it downe, yet in the great Herball referred to Dalechampius, it is called *Ornithogalum magnum Myconi.*

The third hath his name from the place of his birth, and the other from his popularity, yet Dodonæus calleth it *Bulbus Leucanthemos.*

The laft is called by diuers *Asphodelo-hyacinthinus*, and *Hyacintho-asphodelus Galeni.* Dodonæus calleth it *Asphodelus fæmina*, and *Asphodelus bulbofus.* But Lobel, and Gerrard from him, and Dodonæus, doe make this to haue white flowers, whereas all that I haue feene, both in mine owne, and in others Gardens, bore greenifh flowers, as Clufius fetteth it truely downe. Lobel feemeth in the defcription of this, to confound the *Ornithogalum* of Mompelier with it, and calleth it *Asphodelus hyacinthinus forte Galeni*, and faith that fome would call it *Pancratium Monspeliense*, and *Asphodelus Galeni.* But as I haue fhewed, the *Ornithogalum fpicatum* and this, doe plainly differ the one from the other, and are not both to be called by one name, nor to be reckoned one, but two diftinct plants.

Ornithogalum Æthiopicum. The Starre-flower of Æthiopia.

The leaues of this plant are a foote long, and at the leaft an inch broad, which being broken, are no leffe woolly then the woolly Iacinth : the ftalke is a cubit high, ftrong and greene ; from the middle whereof vnto the toppe, ftand large fnow white flowers, vpon long, greene, thicke foot-ftalkes, and yellowifh at the bottome of the flower ; in the middle whereof ftand fix white threeds, tipt with yellow chiues, compaffing the head, which is three fquare, and long containing the feede : the roote is thicke and round, fomewhat like the *Asphodelus Galeni.*

The Place.

This plant was gathered by fome Hollanders, on the Weft fide of the Cape of good Hope.

The Time.

It flowred about the end of Auguft with thofe that had it.

The Names.

Becaufe it came from that part of the continent beyond the line, which is reckoned a part of Æthiopia, it is thereupon fo called as it is fet downe.

Ornithogalum Neopolitanum. The Starre-flower of Naples.

This beautifull plant rifeth out of the ground very early, with foure or fiue hollow pointed leaues, ftanding round together, of a whitifh greene colour, with a white line downe the middle of euery leafe on the infide, fomewhat narrow, but long, (Fabius Columna faith, three foot long in Italy, but it is not fo with vs) in the middle of thefe leaues rifeth vp the ftalke, a foote and a halfe high, bearing diuers flowers at the toppe, euery one ftanding in a little cuppe or huske, which is diuided into three or foure parts, hanging downe very long about the heads for feede : after the flower is paft, thefe flowers doe all hang downe their heads, and open one way, although their little foot-ftalkes come forth on all fides of the greater ftalke, being large, and compofed of fix long leaues, of a pure white on the infide, and of a blewifh or whitifh greene colour

on

on the outfide, leauing the edges of euery leafe white on both fides : in the middle of thefe flowers ftand other fmall flowers, each of them alfo made of fix fmall white leaues a peece, which meeting together, feeme to make the fhew of a cuppe, within which are contained fix white threeds, tipt with yellow, and a long white pointell in the middle of them, being without any fent at all : after the flowers are paft, come vp great round heads, which are too heauie for the ftalke to beare ; and therefore lye downe vpon the leaues or ground, hauing certaine lines or ftripes on the outfide, wherein is contained round, blacke, rough feede : the roote is great and white, and fomewhat flat at the bottome, as diuers of thefe kindes are, and doe multiply as plentifully into fmall bulbes as the common or any other.

The Place.

This Starre-flower groweth in the Medowes in diuers places of Naples, as Fabius Columna, and Ferrantes Imperatus doe teftifie, from whence they haue been fent. And Matthiolus, who fetteth out the figure thereof among his Daffodils, had (it fhould feeme) feene it grow with him.

The Time.

It flowreth in May, although it begin to fpring out of the ground oftentimes in Nouember, but moft vfually in Ianuary : the feede is ripe in Iuly.

The Names.

Matthiolus reckoneth this (as is faid) among the Daffodils, for no other refpect, as i conceiue, then that he accounted the middle flower to bee the cuppe or trunke of a Daffodill, which it doth fomewhat refemble, and fetteth it forth in the fourth place, whereupon many doe call it *Narciffus quartus Matthioli*, The fourth Daffodill of Matthiolus. Fabius Columna calleth it *Hyacinthus aruorum Ornithogali flore*. Clufius (to whom Imperatus fent it, in ftead of the Arabian which hee defired) calleth it of the place from whence he receiued it, *Ornithogalum Neopolitanum*, and we thereafter call it in Englifh, The Starre-flower of Naples.

Ornithogalum Hifpanicum minus. The little Starre-flower of Spaine.

Clufius hath fet forth this plant among his *Ornithogala* or Starre-flowers, and although it doth in my minde come nearer to a *Hyacinthus*, then to *Ornithogalum*, yet pardon it, and let it paffe as he doth. From a little round whitifh roote, fpringeth vp in the beginning of the yeare, fiue or fix fmall long green leaues, without any white line in the middle of them, among which rife vp one or two fmall ftalkes, an hand length high or better, bearing feuen or eight, or more flowers, growing as it were in a tuft or vmbell, with fmall long leaues at the foote of euery ftalke, the lower flowers being equall in length with the vppermoft, of a pale whitifh blew or afh colour, with a ftrake or line downe the backe of euery leafe of them, with fome white threeds ftanding about a blewifh head in the middle : thefe flowers paffe away quickly, and giue no feed, fo that it is not knowne what feede it beareth.

The Place.

This groweth in Spaine, and from thence hath been brought to vs.

The Time.

It flowreth in May.

The Names.

It hath no other name then is fet down in the title, being but lately found out.   1. *Orni-*

1.*Ornithogalum album vnifolium.* The white ſtarre-flower with one blade.

This little ſtarre-flower I bring into this place, as the fitteſt in my opinion where to place it, vntill my minde change to alter it. It hath a very ſmall round white roote, from whence ſpringeth vp one very long and round greene leafe, like vnto a ruſh, but that for about two or three inches aboue the ground, it is a little flat, and from thence ſpringeth forth a ſmall ſtalke not aboue three or foure inches high, bearing at the top thereof three or foure ſmall white flowers, conſiſting of ſix leaues a peece, within which are ſix white chiues, tipt with yellow-pendents, ſtanding about a ſmall three ſquare head, that hath a white pointell ſticking as it were in the middeſt thereof : the flower is pretty and ſweete, but not heady.

Ornithogalum luteum. The yellow Starre of Bethlehem.

This yellow Starre-flower riſeth vp at the firſt, with one long, round, greeniſh leafe, which openeth it ſelfe ſomewhat aboue the ground, and giueth out another ſmall leafe, leſſer and ſhorter then the firſt, and afterward the ſtalke riſeth from thence alſo, being foure or fiue inches high, bearing at the toppe three or foure ſmall green leaues, and among them foure or fiue ſmall yellow ſtarre-like flowers, with a greeniſh line or ſtreake downe the backe of euery leafe, and ſome ſmall reddiſh yellow threeds in the middle : it ſeldome giueth ſeede : the roote is round, whitiſh, and ſomewhat cleare, very apt to periſh, if it bee any little while kept dry out of the ground, as I haue twice tryed to my loſſe.

The Place.

The firſt grew in Portugall, and Cluſius firſt of all others deſciphers it. The other is found in many places both of Germany and Hungary, in the moiſter grounds.

The Time.

The firſt flowreth in May : the other in Aprill, and ſometimes in March.

The Names.

Carolus Cluſius calleth the firſt *Bulbus vnifolius*, or *Bolbine*, but referreth it not to the ſtocke or kindred of any plant ; but (as you ſee) I haue ranked it with the ſmall ſorts of *Ornithogalum*, and giue it the name accordingly.

The other is referred for likeneſſe of forme, and not for colour, vnto the *Ornithogala*, or Starres of Bethlehem. It is called by Tragus and Fuchſius *Bulbus ſilueſtris*, becauſe of the obuiouſneſſe. Cordus taketh it to be *Siſyrinchium*. Lacuna calleth it *Bulbus eſculentus*. Lobel and others in theſe dayes generally, *Ornithogalum luteum*, and wee thereafter in Engliſh, The yellow Starre-flower, or Starre of Bethlehem.

The Vertues.

The firſt kinde being but lately found out, is not knowne to be vſed. The rootes of the common or vulgar, are (as Matthiolus ſaith) much eaten by poore people in Italy, either rawe or roaſted, being ſweeter in taſte then any Cheſnut, and ſeruing as well for a neceſſary food as for delight. It is doubtfull whether any of the reſt may be ſo vſed ; for I know not any in our Land hath made any experience.

There are many other ſorts of Starre-flowers, which are fitter for a generall then this Hiſtory ; and therefore I referre them thereunto.

CHAP.

Chap. XIII.

Moly. Wilde Garlicke.

VNto the former Starre-flowers, muſt needes bee ioyned another tribe or kindred, which carry their ſtraked flowers Starre-faſhion, not ſpikewiſe, but in a tuft or vmbell thicke thruſt or ſet together. And although diuers of them ſmell not as the former, but moſt of their firſt Grandfathers houſe, yet all doe not ſo; for ſome of them are of an excellent ſent. Of the whole Family, there are a great many which I muſt leaue, I will onely ſelect out a few for this our Garden, whoſe flowers for their beauty of ſtatelineſſe, forme, or colour, are fit to bee entertained, and take place therein, euery one according to his worth, and are accepted of with the louers of theſe delights.

1. *Moly Homericum, vel potius Theophraſti.*
The greateſt Moly of Homer.

Homers Moly (for ſo it is moſt vſually called with vs) riſeth vp moſt commonly with two, and ſometimes with three great, thicke, long, and hollow guttured leaues, of a whitiſh greene colour, very neare the colour of the Tulipa leafe, hauing ſometimes at the end of ſome of the leaues, and ſometimes apart by it ſelfe, a whitiſh round ſmall button, like vnto a ſmall bulbe, the like whereof alſo, but greater, doth grow betweene the bottome of the leaues and the ſtalke neare the ground, which being planted when it is ripe, will grow into a roote of the ſame kinde: among theſe leaues riſeth vp a round, ſtrong, and tall ſtalke, a yard high or better, bare or naked vnto the toppe, where it beareth a great tuft or vmbell of pale purpliſh flowers, all of them almoſt ſtanding vpon equall foot-ſtalkes, or not one much higher then another, conſiſting of fiue leaues a peece, ſtriped downe the backe with a ſmall pale line, hauing a round head or vmbone with ſome threeds about it in the midſt: Theſe flowers doe abide a great while blowne before they vade, which ſmell not very ſtrong, like any Onion or Garlicke, but of a faint ſmell: and after they are paſt come the ſeede, which is blacke, wrapped in white cloſe huskes: the roote groweth very great, ſometimes bigger then any mans cloſed fiſt, ſmelling ſtrong like Garlicke, whitiſh on the outſide, and greene at the toppe, if it be but a while bare from the earth about it.

2. *Moly Indicum ſiue Caucaſon.* The Indian Moly.

The Indian Moly hath ſuch like thicke large leaues, as the Homers Moly hath, but ſhorter and broader, in the middle whereof riſeth vp a ſhort weake ſtalke, almoſt flat, not hauing any flowers vpon it, but a head or cluſter of greeniſh ſcaly bulbes, incloſed at the firſt in a large thinne skinne, which being open, euery bulbe ſheweth it ſelfe, ſtanding cloſe one vnto another vpon his foot-ſtalke, of the bigneſſe of an Acorne, which being planted, will grow to bee a plant of his owne kinde: the roote is white and great, couered with a darke coate or skinne, which encreaſeth but little vnder ground; but beſides that head, it beareth ſmall bulbes aboue the ground, at the bottome of the leaues next vnto the ſtalke, like vnto the former.

The Place.

Both theſe doe grow in diuers places of Spaine, Italy, and Greece; for the laſt hath been ſent out of Turkie among other rootes. Ferrantes Imperatus a learned Apothecary of Naples, ſent it to diuers of his friends in theſe parts, and hath deſcribed it in his naturall hiſtory among other plants, printed in the Italian tongue. It grew alſo with Iohn Tradeſcante at Canterbury, who ſent me the head of bulbes to ſee, and afterwards a roote, to plant it in my Garden.

The

The Time.

The firſt flowreth in the end of May, and abideth vnto the midſt of Iuly, and ſometimes longer. The other beareth his head of bulbes in Iune and Iuly.

The Names.

We haue receiued them by their names expreſſed in their titles, yet the laſt hath alſo been ſent by the name of *Ornithogalum Italicum,* but as all may eaſily ſee, it is not of that kindred.

1. *Moly montanum Pannonicum bulbiferum primum.* The firſt bulbed Moly of Hungary.

This firſt Hungarian Moly hath three or foure broad and long greene leaues, folded together at the firſt, which after open themſelues, and are carried vp with the ſtalke, ſtanding thereon one aboue another, which is a foote high; at the toppe whereof doe grow a few ſad reddiſh bulbes, and betweene them long footſtalkes, bearing flowers of a pale purpliſh colour; after which followeth blacke ſeede, incloſed in roundiſh heads: the roote is not great, but white on the outſide, very like vnto the roote of Serpents Moly, hereafter deſcribed, encreaſing much vnder ground, & ſmelling ſtrong.

2. *Moly montanum Pannonicum bulbiferum ſecundum.* The ſecond bulbed Moly of Hungary.

The ſecond Moly hath narrower greene leaues then the former: the ſtalke is about the ſame height, and beareth at the toppe a great cluſter of ſmall greene bulbes, which after turne of a darker colour; from among which come forth long foot-ſtalks, whereon ſtand purpliſh flowers: the roote is couered with a blackiſh purple coate or ſkinne.

3. *Moly Serpentinum.* Serpents Moly.

This Moly muſt alſo be ioyned vnto the bulbous Molyes, as of kindred with them, yet of greater beauty and delight, becauſe the bulbes on the heads of the ſmall ſtalkes are redder, and more pleaſant to behold: the ſtalke is lower, and his graſſie winding leaues, which turne themſelues (whereof it tooke the name) are ſmaller, and of a whiter greene colour: it beareth among the bulbes purpliſh flowers alſo, but more beautifull, the ſent whereof is nothing ſo ſtrong: the roote is ſmall, round, and whitiſh, encreaſing into a number of ſmall rootes, no bigger then peaſe round about the greater roote.

4. *Moly caule & folijs triangularibus.* The three cornered Moly.

This three ſquare Moly hath foure or fiue long, and ſomewhat broad pale greene leaues, flat on the vpper ſide, and with a ridge downe the backe of the leafe, which maketh it ſeeme three ſquare: the ſtalke which riſeth vp a foote and a halfe high or better, is three ſquare or three cornered alſo, bearing at the toppe out of a ſkinnie huſke diuers white flowers, ſomewhat large and long, almoſt bell-faſhion, with ſtripes of greene downe the middle of euery leafe, and a few chiues tipt with yellow in the middle about the head, wherein when it is ripe, is incloſed ſmall blacke ſeede: the roote is white on the outſide, and very like the yellow Moly; both roote, leafe, and flower hath a ſmacke, but not very ſtrong of Garlicke.

5. *Moly Narciſsinis folijs.* Daffodill leafed Moly.

This Moly hath many long, narrow, and flat greene leaues, very like vnto the leaues of a Daffodill, from whence it tooke his name (or rather of the early greater *Leucoium bulboſum,*

1 *Moly Homericum vel potius Theophrasti.* The greatest Moly of Homer. 2 *Moly Indicum siue Caucason.* The Indian Moly. 3 *Moly Pannonicum bulbiferum.* The bulbed Moly of Hungary. 4 *Moly Serpentinum.* Serpents Moly. 5 *Moly purpureum Neapolitanum.* The purplish Moly of Naples. 6 *Moly caule & folijs triangularibus.* The three cornered Moly. 7 *Moly latifolium flore luteo.* The yellow Moly. 8 *Moly Dioscoridum Hispanicum.* The Spanish Moly of Dioscorides. 9 *Moly Zibettinum vel Moschatinum.* The sweete smelling Moly of Mompelier. 10 *Moly serotinum Coniferum.* The late Pine-apple Moly.

bulbosam, or bulbed Violet before described, ioyned next vnto the Daffodils, becaufe it is fo like them) among which rifeth vp two or three ftalkes fometimes, each of a foot and a halfe high, bearing at the toppe, inclofed in a skinny hofe, as all the Molyes haue, a number of fmall purplifh flowers, which doe not long abide, but quickly fade: the feede is blacke as others are; the roote is fometimes knobbed, and more often bulbed, hauing in the knobs fome markes of the old ftalkes to be feene in them, and fmelleth fomewhat like Garlicke, whereby it may be knowne.

6. *Moly montanum latifolium luteo flore.* The yellow Moly.

The yellow Moly hath but one long and broad leafe when it doth not beare flower, but when it will beare flower, it hath two long and broad leaues, yet one alwaies longer and broader then the other, which are both of the fame colour, and neare the bignefle of a reafonable Tulipa leafe: betweene thefe leaues groweth a flender ftalke, bearing at the toppe a tuft or vmbell of yellow flowers out of a skinnie hofe, which parteth three wayes, made of fix leaues a peece, laid open like a Starre, with a greenifh backe or outfide, and with fome yellow threeds in the middle: the feede is blacke, like vnto others: the roote is whitifh, two for the moft part ioyned together, which encreafeth quickly, and fmelleth very ftrong of Garlicke, as both flowers and leaues doe alfo.

7. *Moly Pyrenæum purpureum.* The purple mountaine Moly.

This purple Moly hath two or three leaues, fomewhat like the former yellow Moly, but not fo broad, nor fo white: the ftalke hath not fo many flowers thereon, but more fparingly, and of an vnpleafant purple colour: the roote is whitifh, fmelling fomewhat ftrongly of Garlicke, but quickly perifheth with the extremity of our cold Winters, which it will not abide vnlefle it be defended.

8. *Moly montanum latifolium purpureum Hifpanicum.* The purple Spanifh Moly.

This Moly hath two broad and very long greene leaues, like vnto the yellow Moly, in this, that they doe compafle one another at the bottome of them, between which rifeth vp a ftrong round ftalke, two foote high or more, bearing at the toppe, out of a thinne huske, a number of faire large flowers vpon long foot-ftalkes, confifting of fix leaues a peece, fpread open like a Starre, of a fine delayed purple or blufh colour, with diuers threeds of the fame colour, tipt with yellow, ftanding about the middle head: betweene the ftalke and the bottome of the leaues it hath fome fmall bulbes growing, which being planted, will foone fpring and encreafe: the roote alfo being fmall and round, with many fibres thereat, hath many fmall bulbes fhooting from them; but neither roote, leafe, nor flower, hath any ill fent of Garlicke at all.

9. *Moly purpureum Neapolitanum.* The purple Moly of Naples.

The Neapolitane Moly hath three or foure fmall long greene leaues fet vpon the ftalke after it is rifen vp, which beareth a round head of very fine purple flowers, made of fix leaues a peece, but fo clofing together at the edge, that they feeme like vnto fmall cuppes, neuer laying themfelues open, as the other doe; this hath fome fent of his originall, but the roote more then any part elfe, which is white and round, quickly encreafing as moft of the Molyes doe.

10. *Moly pyxidatum argenteum Hifpanicum.* The Spanifh filuer cupped Moly.

This Spanifh Moly hath two or three very long rufh like leaues, which rife vp with the ftalke, or rather vanifh away when the ftalke is rifen vp to bee three foote high or more, bearing a great head of flowers, ftanding clofe at the firft, but afterwards fprea-ding much one from another, euery flower vpon a long foote-ftalke, being of a white

filuer

filuer colour, with ftripes or lines on euery fide, and fafhioned fmall and hollow, like a cuppe or boxe : the feede I could neuer obferue, becaufe it flowreth fo late, that the Winter hindereth it from bearing feede with vs : the roote is fmall and round, white, and in a manner tranfparent, at leaft fo fhining, as if it were fo, and encreafeth nothing fo much, as many of the other forts : this hath no ill fent at all, but rather a pretty fmell, not to bee mifliked.

11. *Moly ferotinum Coniferum.* The late Pineapple Moly.

This late Moly that was fent me with the laft defcribed, and others alfo from Spain, rifeth vp with one long greene leafe, hollow and round vnto the end, towards this end on the one fide, breaketh out a head of flowers, enclofed in a thinne skinne, which after it hath fo ftood a good while, (the leafe in the meane time rifing higher, and growing harder, becommeth the ftalke) breaketh, and fheweth a great bufh or head of buds for flowers, thicke thruft together, fafhioned very like vnto the forme of a Pineapple (from whence I gaue it the name) of the bignefs of a Walnut : after this head hath ftood in this manner a moneth or thereabouts, the flowers fhew themfelues to bee of a fine delayed or whitifh purple colour, with diuers ftripes in euery of them, of the fame cup-fafhion with the former, but not opening fo plainly, fo that they cannot bee difcerned to bee open, without good heede and obferuation. It flowreth fo late in Autumne, that the early frofts doe quickly fpoile the beauty of it, and foone caufe it to rotte : the roote is fmall and round, and fhining like the laft, very tender alfo, as not able to abide our fharpe Winters, which hath caufed it vtterly to perifh with me.

12. *Moly Diofcorideum.* Diofcorides his Moly.

The roote of this fmall Moly is tranfparent within, but couered with a thicke yellowifh skinne, of the bigneffe of an Hafell Nut, or fomewhat bigger, which fendeth forth three or foure narrow graffie leaues, long and hollow, and a little bending downwards, of a whitifh greene colour, among which rifeth vp a flender weake ftalke, a foot and a halfe high, bearing at the toppe, out of a thinne skinne, a tuft of milke white flowers, very like vnto thofe of Ramfons, which ftand a pretty while in their beauty, and then paffe away for the moft part without giuing any feede : this hath little or no fent of Garlicke.

We haue another of this fort that is leffer, and the flowers rounder pointed.

13. *Moly Diofcorideum Hifpanicum.* The Spanifh Moly of Diofcorides.

This Moly came vnto me among other Molyes from Spaine, and is in all things like vnto the laft defcribed, but fairer, larger, and of much more beauty, as hauing his white flowers twice as great as the former ; but (as it feemeth) very impatient of our Winters, which it could not at any hand endure, but quickly perifhed, as fome others that came with it alfo.

14. *Moly Mofchatinum vel Zibettinum Monfpelienfe.* The fweete fmelling Moly of Mompelier.

This fweete Moly, which I haue kept for the laft, to clofe vp your fenfes, is the fmalleft, and the fineft of all the reft, hauing foure or fiue fmall greene leaues, almoft as fine as haires, or like the leaues of the Feather-graffe : the ftalke is about a foote high, bearing fiue or fix or more fmall white flowers, laid open like Starres, made of fix leaues a peece, of an excellent fweete fent, refembling Muske or Ciuet ; for diuers haue diuerfly cenfured of it. It flowreth late in the yeare, fo that if the precedent Summer bee either ouer moift, or the Autumne ouer early cold, this will not haue that fweete fent, that it will haue in a hot drie time, and befides muft be carefully refpected : for it will hardly abide the extremity of our fharpe Winters.

N   The

The Place.

The places of thefe Molyes, are for the moft part expreffed in their titles, or in their defcriptions.

The Time.

The time is fet downe, for the moft part to bee in Iune and Iuly, the reft
later.

The Names.

To make further relation of names then are expreffed in their tiles, were
needleffe; let thefe therefore fuffice.

The Vertues.

All thefe forts of Molyes are fmall kindes of wilde Garlicke., and are to
be vfed for the fame purpofes that the great Garden Garlicke is, although
much weaker in their effects. For any other efpeciall property is in any of
thefe, more than to furnifh a Garden of variety, I haue not heard at all.

And thus much may fuffice of thefe kindes for our Garden, referuing manie others
that might be fpoken of, to a generall worke, or to my Garden of Simples, which as
God fhall enable me, and time giue leaue, may fhew it felfe to the world, to abide the
iudicious and criticke cenfures of all.

Chap. XIIII.

Aſphodelus. The Aſphodill.

THere remaine fome other flowers, like vnto the laft defcribed, to be fpecified,
which although they haue no bulbous rootes, yet I thinke them fitteft to bee
here mentioned, that fo I may ioyne thofe of neereft fimilitude together, vntill I haue finifhed the reft that are to follow.

1. Aſphodelus maior albus ramoſus. The great white branched Aſphodill.

The great white Aſphodill hath many long, and narrow, hollow three fquare
leaues, fharpe pointed, lying vpon the ground round about the roote: the ftalke is
fmooth, round, and naked without leaues, which rifeth from the midft of them, diuided at the toppe into diuers branches, if the plant bee of any long continuance, or
elfe but into two or three fmall branches, from the fides of the maine great one,
whereon doe ftand many large flowers Starre-fafhion, made of fix leaues a peece,
whitifh on the infide, and ftraked with a purplifh line downe the backfide of euery
leafe, hauing in the middle of the flowers fome fmall yellow threeds: the feede is
blacke, and three fquare, greater then the feede of Bucke wheate, contained in
roundifh heads, which open into three parts: the roote is compofed of many tuberous long clogges, thickeft in the middle, and fmaller at both ends, faftened together
at the head, of a darke grayifh colour on the outfide, and yellow within.

2. Aſphodelus albus non ramoſus. The white vnbranched Aſpodill.

The vnbranched Aſphodill is like vnto the former, both in leaues and flowers, but
that the flowers of this are whiter, and without any line or ftrake on the backe fide,
and

and the ſtalkes are without branches : the rootes likewiſe are ſmaller, and fewer, but made after the ſame faſhion.

3. *Aſphodelus maior flore carneo.* The bluſh coloured Aſphodill.

This Aſphodill is like to the laſt in forme of leaues and branches, and differeth in this, that his leaues are marked with ſome ſpots, and the flowers are of a bluſh or fleſh colour, in all other things alike.

4. *Aſphodelus minimus albus.* The leaſt white Aſphodill.

This leaſt Aſphodill hath foure or fiue very narrow long leaues, yet ſeeming three ſquare like the greateſt, bearing a ſmall ſtalke, of about a foote high among them, without any branches, and at the toppe a few white flowers, ſtraked both within and without, with a purpliſh line in the middle of euery leafe. The rootes are ſuch like tuberous clogges as are in the former, but much leſſer.

5. *Aſphodelus albus minor ſiue Fiſtuloſus.* The little hollow white Aſphodill.

This little white Aſphodill hath a number of leaues growing thicke together, thicker and greener then thoſe of the ſmall yellow Aſphodill, or Kings Speare next following, among which riſeth vp diuers round ſtalkes, bearing flowers from the middle to the toppe, Starre-faſhion, with ſmall greene leaues among them, which are white on the inſide, and ſtriped on the backe with purple lines, like vnto the firſt deſcribed : the ſeede, and heads containing them, are three ſquare, like the ſeede of the little yellow Aſphodill : the rootes of this kinde are not glandulous, as the former, but ſtringie, long and white : the whole plant is very impatient of our cold Winters, and quickly periſheth, if it be not carefully preſerued, both from the cold, and much wet in the Winter, by houſing it ; and then it will abide many yeares : for it is not an annuall plant, as many haue thought.

6. *Aſphodelus luteus minor, ſiue Haſtula regia.* The ſmall yellow Aſphodill, or Kings ſpeare.

This ſmall yellow Aſphodill, which is vſually called the Kings ſpeare, hath many long narrow edged leaues, which make them ſeeme three ſquare, of a blewiſh or whitiſh greene colour : the ſtalke riſeth vp three foote high oftentimes, beſet with ſmall long leaues vp vnto the very flowers, which grow thicke together ſpike-faſhion one aboue another, for a great length, and wholly yellow, laid open like a Starre, ſomewhat greater then the laſt white Aſphodill, and ſmaller then the firſt, which when they are paſt yeeld round heads, containing blacke cornered ſeede, almoſt three ſquare : the rootes are many long yellow ſtrings, which ſpreading in the ground, doe much encreaſe.

The Place.

All theſe Aſphodils doe grow naturally in Spaine and France, and from thence were firſt brought vnto vs, to furniſh our Gardens.

The Time.

All the glandulous rooted Aſphodils doe flower ſome in May, and ſome in Iune ; but the two laſt doe flower, the yellow or laſt of them in Iuly, and the former white one in Auguſt and September, and vntill the cold and winter hinder it.

The Names.

Their ſeuerall names are giuen them in their titles, as much as is fit for

this

this discourse. For to shew you that the Greckes doe call the stalke of the great Asphodill ΑνθεριχΘ, and the Latines *Albucum*, or what else belongeth to them, is fitter for another worke, vnto which I leaue them.

The bastard Asphodils should follow next in place, if this worke were fit for them ; but because I haue tyed my selfe to expresse onely those flowers and plants, that for their beauty, or sent, or both, doe furnish a Garden of Pleasure, and they haue none, I leaue them to a generall History of plants, or that Garden of Simples before spoken of, and will describe the Lilly Asphodils, and the *Phalangia* or Spider-worts, which are remaining of those, that ioyne in name or fashion, and are to be here inserted, before I passe to the rest of the bulbous rootes.

1. *Liliasphodelus phœniceus.* The gold red Day Lilly.

Because the rootes of this and the next, doe so nearely agree with the two last recited Asphodils, I haue set them in this place, although some doe place them next after the Lillies, because their flowers doe come nearest in forme vnto Lillies ; but whether you will call them Asphodils with Lilly flowers, as I thinke it fittest, or Lillies with Asphodill rootes, or Lillies without bulbous rootes, as others doe, I will not contend. The red Day Lilly hath diuers broad and long fresh greene leaues, folded at the first as it were double, which after open, and remaine a little hollow in the middle ; among which riseth vp a naked stalke three foot high, bearing at the toppe many flowers, one not much distant from another, and flowring one after another, not hauing lightly aboue one flower blown open in a day, & that but for a day, not lasting longer, but closing at night, and not opening againe ; whereupon it had his English name, The Lilly for a day : these flowers are almost as large as the flowers of the white Lilly, and made after the same fashion, but of a faire gold red, or Orange tawny colour. I could neuer obserue any seede to follow these flowers ; for they seeme the next day after they haue flowred, (except the time be faire and dry) to bee so rotten, as if they had lyen in wet to rotte them, whereby I thinke no seede can follow : the rootes are many thicke and long yellow knobbed strings, like vnto the small yellow Asphodill rootes, but somewhat greater, running vnder ground in like sort, and shooting young heads round about.

2. *Liliasphodelus luteus.* The yellow Day Lilly.

I shall not neede to make a repetition of the description of this Day Lilly, hauing giuen you one so amply before, because this doth agree thereunto so nearely, as that it might seeme the same ; these differences onely it hath, the leaues are not fully so large, nor the flower so great or spread open, and the colour thereof is of a faire yellow wholly, and very sweet, which abideth blowne many daies before it fade, and hath giuen blacke round seede, growing in round heads, like the heads of the small yellow Asphodill, but not so great.

Clusius hath set downe, that it was reported, that there should be another Liliasphodill with a white flower, but we can heare of none such as yet ; but I rather thinke, that they that gaue that report might be mistaken, in thinking the Sauoye Spider-wort to be a white Liliasphodill, which indeede is so like, that one not well experienced, or not well regarding it, may soone take one for another.

The Place.

Their originall is many moist places in Germany.

The Time.

They flower in May and Iune.

The Names.

They are called by some *Liliago*, and *Lilium non bulbosum*, and *Liliaspho-*
                                                                                    delus,

1 *Asphodelus maior albus ramosus.* The great white branched Asphodill. 2 *Asphodelus minor albus seu fistulosus.* The little hollow white Asphodill. 3 *Asphodelus minor luteus, siue Hastula regia.* The small yellow Asphodill, or Kings speare. 4 *Liliasphodelus luteus.* The yellow Day Lilly. 5 *Liliasphodelus phæniceus.* The gold red Day Lilly.

delus. In Englifh we call them both Day Lillies, but the name doth not fo well agree with the laft, as with the firft, for the caufes aboue fpecified.

The Vertues.

The rootes of Afphodill hath formerly beene had in great account, but now is vtterly neglected; yet by reafon of their fharpeneffe they open and cleanfe, and therefore fome haue of late vfed them for the yellow Iaundife. The Day Lillies haue no phyficall vfe that I know, or haue heard.

CHAP. XV.

Phalangium. Spider-wort.

THefe plants doe fo nearely refemble thofe that are laft fet forth, that I thinke none that knowes them, will doubt, but that they muft follow next vnto them, being fo like vnto them, and therefore of the faireft of this kinde firft.

1. *Phalangium Allobrogicum.* The Sauoye Spider-wort.

The Sauoye Spider-wort fpringeth vp with foure or fiue greene leaues, long and narrow, yet broader at the bottome, narrower pointed at the end, and a little hollow in the middle; among which rifeth vp a round ftiffe ftalke, a foote and a halfe high, bearing at the toppe one aboue another, feuen or eight, or more flowers, euery one as large almoft as the yellow Day Lilly laft defcribed, but much greater then in any other of the Spider-worts, of a pure white colour, with fome threeds in the middle, tipt with yellow, and a fmall forked pointell: after the flowers are paft, the heads or feede veffels grow almoft three fquare, yet fomewhat round, wherein is contained blackifh feede: the rootes are many white, round, thicke, brittle ftrings, ioyned together at the head, but are nothing fo long, as the rootes of the other *Phalangia* or Spider-worts.

2. *Phalangium maius Italicum album.* The great Italian Spider-wort.

This great Spider-wort hath diuers long and narrow leaues fpread vpon the ground, and not rifing vp as the former, and not fo broad alfo as the former, but fomewhat larger then thofe that follow: the ftalke is bigger, but feldome rifeth vp fo high as the next, whereof this is a larger kinde, hauing a long vnbranched ftalke of white flowers, laid open like ftarres as it hath, but fomewhat greater: the rootes are long and white, like the next, but fomewhat larger.

3. *Phalangium non ramofum vulgare.* Vnbranched Spider-wort.

The leaues of this Spider-wort doe feeme to bee little bigger or longer then the leaues of graffe, but of a more grayifh green colour, rifing immediately from the head or tuft of rootes; among which rife vp one or two ftalkes, fometimes two or three foote long, befet toward the toppe with many white Starre-like flowers, which after they are paft turne into fmall round heads, containing blacke feede, like vnto the feed of the little yellow Afphodill, but leffer: the rootes are long white ftrings, running vnder ground.

4. *Phalangium ramofum.* Branched Spider-wort.

The branched Spider-wort hath his leaues fomewhat broader then the former, and of a more yellowifh greene colour: the ftalke hereof is diuerfly branched at the top, bearing many white flowers, like vnto the former, but fmaller: the feedes and rootes are like the former in all things.

The

1 *Phalangium Allobrogicum.* The Sauoye Spider-wort. 2 *Phalangium non ramoſum.* Vnbranched Spider-wort. 3 *Phalangium ramoſum.* branched Spider-wort. 4 *Phalangium
Ephemerum Virginianum.* Iohn Tradeſcante's Spider-wort.

The Place.

The first groweth on the Hils neare vnto Sauoye, from whence diuers, allured with the beauty of the flower, haue brought it into these parts.

The second came vp in my Garden, from the seede receiued out of Italy. The others grow in Spaine, France, &c.

The Time.

The vnbranched Spider-wort most commonly flowreth before all the other, and the branched a moneth after it: the other two about one time, that is, towards the end of May, and not much after the vnbranched kinde.

The Names.

The first (as I said before) hath beene taken to be a white Lilliasphodill, and called *Liliasphodelus flore albo*; but Clusius hath more properly entituled it a *Phalangium*, and from the place of his originall, gaue him his other denomination, and so is called of most, as is set downe in the title.

The other haue no other names then are expressed in their titles, but only that Cordus calleth them *Liliago*; and Dodonæus, *lib.4. hist.plant.* would make the branched kinde to bee *Moly alterum Plinij*, but without any good ground.

The Vertues.

The names *Phalangium* and *Phalangites* were imposed on these plants, because they were found effectuall, to cure the poyson of that kinde of Spider, called *Phalangium*, as also of Scorpions and other Serpents. Wee doe not know, that any Physitian hath vsed them to any such, or any other purpose in our dayes.

5. *Phalangium Ephemerum Virginianum Ioannis Tradescant.*
The soon fading Spider-wort of Virginia, or Tradescant his Spider-wort.

This Spider-wort is of late knowledge, and for it the Christian world is indebted vnto that painfull industrious searcher, and louer of all natures varieties, Iohn Tradescant (sometimes belonging to the right Honourable Lord Robert Earle of Salisbury, Lord Treasurer of England in his time, and then vnto the right Honourable the Lord Wotton at Canterbury in Kent, and lastly vnto the late Duke of Buckingham) who first receiued it of a friend, that brought it out of Virginia, thinking it to bee the Silke Grasse that groweth there, and hath imparted hereof, as of many other things, both to me and others; the description whereof is as followeth:

From a stringie roote, creeping farre vnder ground, and rising vp againe in many places, springeth vp diuers heads of long folded leaues, of a grayish ouer-worne greene colour, two or three for the most part together, and not aboue, compassing one another at the bottome, and abiding greene in many places all the Winter; otherwhere perishing, and rising anew in the Spring, which leaues rise vp with the great round stalke, being set thereon at the ioynts, vsually but one at a ioynt, broad at the bottome where they compasse the stalke, and smaller and smaller to the end: at the vpper ioynt, which is the toppe of the stalke, there stand two or three such like leaues, but smaller, from among which breaketh out a dozen, sixteene, or twenty, or more round green heads, hanging downe their heads by little foot-stalkes, which when the flower beginneth to blow open, groweth longer, and standeth vpright, hauing three small pale greene leaues for a huske, and three other leaues within them for the flower, which lay themselues open flat, of a deepe blew purple colour, hauing an vmbone or small head in the middle, closely set about with six reddish, hairy, or feathered threeds, tipt with yellow pendents: this flower openeth it selfe in the day, & shutteth vsually at night,

night, and neuer openeth againe, but perisheth , and then hangeth downe his head a-
gaine ; the greene huske of three leaues, closing it selfe againe into the forme of a head,
but greater , as it was before, the middle vmbone growing to bee the seede vessell,
wherein is contained small , blackish, long seede : Seldome shall any man see aboue
one , or two at the most of these flowers blowne open at one time vpon the stalke,
whereby it standeth in flowring a long time , before all the heads haue giuen out their
flowers.

The Place.

This plant groweth in some parts of Virginia, and was deliuered to Iohn
Tradescant.

The Time.

It flowreth from the end of May vntill Iuly , if it haue had greene leaues
all the Winter, or otherwise, vntill the Winter checke his luxuriousnesse.

The Names.

Vnto this plant I confesse I first imposed the name, by considering duely
all the parts thereof, which vntill some can finde a more proper, I desire
may still continue, and to call it *Ephemerum Virginianum Tradescanti*, Iohn
Tradescante's Spider-wort of Virginia, or *Phalangium Ephemerum Virginianum*, The soone fading or Day Spider-wort of Virginia.

The Vertues.

There hath not beene any tryall made of the properties since wee had it,
nor doe we know whether the Indians haue any vse thereof.

Chap. XVI.

Colchicum. Medowe Saffron.

TO returne to the rest of the bulbous and tuberous rooted plants , that remaine
to bee entreated of , the *Colchica* or Medowe Saffrons are first to bee handled,
whereof these later dayes haue found out more varieties, then formerly were
knowne ; some flowring in the Spring, but the most in Autumne , and some bearing
double, but the greatest part single flowers : whereof euery one in their order, and
first of our owne Country kindes.

1. *Colchicum Anglicum album.* The white English Medowe Saffron.

It is common to all the Medowe Saffrons, except that of the Spring, and one other,
to beare their flowers alone in Autumne or later, without any green leaues with them,
and afterwards in February, their greene leaues : So that I shall not neede to make manie descriptions, but to shew you the differences that consist in the leaues, and colours
of the flowers ; and briefly to passe (after I haue giuen you a full description of the
first) from one vnto another, touching onely those things that are note worthy. The
white English Medowe Saffron then doth beare in Autumne three or foure flowers at
the most, standing seuerally vpon weake foote-stalkes, a fingers length or more aboue
the ground, made of six white leaues, somewhat long and narrow, and not so large as
most of the other kindes , with some threeds or chiues in the middle, like vnto the
Saffron flowers of the Spring, wherein there is no colour of Saffron , or vertue to that
effect : after the flowers are past and gone, the leaues doe not presently follow, but
the roote remaineth in the ground without shew of leafe aboue ground, most part of
the Winter, and then in February there spring vp three or foure large and long greene
leaues,

leaues, when they are fully growne vp, ftanding on the toppe of a round, weake, green, and fhort foote-ftalke, fomewhat like the leaues of white Lillies, but not fo large, and in the middeft of thefe leaues, after they haue been vp fometime, appeare two or three loofe skinny heads, ftanding in the middle of the leaues vpon fhort, thicke, greene ftalkes, and being ripe, conteine in them round fmall brownifh feede, that lye as it were loofe therein, and when the head is dry, may bee heard to rattle being fhaken: the roote is white within, but coucred with a thicke blackifh skinne or coate, hauing one fide thereof at the bottome longer then the other, with an hollowneffe alfo on the one fide of that long eminence, where the flowers rife from the bottome, and fhooting downe from thence a number of white fibres, whereby it is faftened in the ground: the greene leaues afterwards rifing from the top or head of the roote.

2. *Colchicum Anglicum purpureum.* The purple Englifh Medowe Saffron.

There is no difference at all in this Medowe Saffron from the former, but only in the colour of the flowers, which as they were wholly white in the former, fo in this they are of a delayed purple colour, with a fmall fhew of veines therein.

3. *Colchicum Pannonicum album.* The white Hungary Medowe Saffron.

The greateft difference in this *Colchicum* from the former Englifh white one, is, that it is larger both in roote, leafe, and flower, and befides, hath more ftore of flowers together, and continuing longer in beauty, without fading fo foone as the former, and are alfo fomewhat of a fairer white colour.

4. *Colchicum Pannonicum purpureum.* The purple Hungary Medowe Saffron.

This purple Medowe Saffron is fomewhat like vnto the white of this kinde, but that it beareth not fo plentifully as the white, nor doth the roote grow fo great; but the flowers are in a manner as large as they, and of the like pale delayed purple colour, or fomewhat deeper, as is in the purple Englifh, with fome veines or markes vpon the flowers, making fome fhew of a checker on the out fide, but not fo confpicuous, as in the true checkerd kindes. Wee haue a kinde hereof is party coloured with white ftreakes and edges, which abide conftant, and hath been raifed from the feede of the former.

5. *Colchicum Byzantinum.* Medowe Saffron of Conftantinople.

This Medowe Saffron of Conftantinople hath his leaues fo broad and large, that hardly could any that neuer faw it before, iudge it to be a *Colchicum*; for they are much larger then any Lilly leaues, and of a darke greene colour: the flowers are correfpondent to the leaues, larger and more in number then in any of the former purple kindes, of the fame colour with the laft purple kinde, but of a little deeper purple on the infide, with diuers markes running through the flowers, like vnto it, or vnto checkers, but yet fomewhat more apparantly: the roote is in the middle greater and rounder then the others, with a longer eminence, whereby it may eafily bee knowne from all other forts.

6. *Colchicum Lafitanicum Fritillaricum.* The checkerd Medowe Saffron of Portugall.

The flowers of this Medowe Saffron are larger and longer then the flowers of either the Englifh or Hungarian, and almoft as large as the laft before mentioned, and of the fame colour, but a little deeper, the fpots and markes whereof are fomewhat more eafie to be feene euen a farre off, like vnto the flower of a Fritillaria, from whence it tooke his fignificatiue name: the leaues of this Medowe Saffron doe rife vp fooner then in any other of the Autumne kindes; for they are alwayes vp before Winter, and are foure or fiue in number, fhort rather then long, broad belowe, and pointed at the end, canaled or hollow, and ftanding round aboue the ground, one encompaffing another at the bottome, like the great Spanifh Starre Iacinth, called the Iacinth of Peru,
but

1 *Colchicum Pannonicum.* The Hungarian Medow Saffron. 2 *Colchicum Byzantinum.* Medowe Saffron of Constantinople. 3 *Colchicum Lusitanicum Fritillaricum.* The checkerd Medowe Saffron of Portugall 4 *Colchicum Neapolitanum Fritillaricum.* The checkerd Medowe Saffron of Naples. 5 *Colchicum Fritillaricum Chiense.* The checkerd Medowe Saffron of Chio or Sio. 6 *Colchicum Hermodactylum.* Physicall Medowe Saffron.

but shorter, and of a pale or grayish greene colour, differing from the colour of all the other Medowe Saffrons : the roote is like the roote of the English or Hungarian without any difference, but that it groweth somewhat greater. It is one of the first Medowe Saffrons that flower in the Autumne.

7. *Colchicum Neapolitanum Fritillaricum.*
The checkerd Medowe Saffron of Naples.

This checkerd Medowe Saffron of Naples, is very like vnto the last recited checkerd Saffron of Portugall, but that the flower is somewhat larger, yet sometimes very little, or not at all : the greatest marke to distinguish them is, that the flowers of this are of a deeper colour, and so are the spots on the flowers likewise, which are so conspicuous, that they are discerned a great way off, more like vnto the flowers of a deepe Fritillaria, then the former, and make a goodlier and a more glorious shew : the leaues of this doe rise vp early after the flowers, and are somewhat longer, of a darker greene colour, yet bending to a grayish colour as the other, not lying so neatly or round, but stand vp one by another, being as it were folded together : neither of both these last named checkerd Medowe Saffrons haue giuen any seede in this Countrey, that euer I could learne or heare of, but are encreased by the roote, which in this is like the former, but a little bigger.

8. *Colchicum Fritillaricum Chiense.*
The checkerd Medowe Saffron of Chio or Sio.

This most beautifull Saffron flower riseth vp with his flowers in the Autumne, as the others before specified doe, although not of so large a size, yet farre more pleasant and delightfull in the thicke, deepe blew, or purple coloured beautifull spots therein, which make it excell all others whatsoeuer : the leaues rise vp in the Spring, being smaller then the former, for the most part three in number, and of a paler or fresher greene colour, lying close vpon the ground, broad at the bottome, a little pointed at the end, and twining or folding themselues in and out at the edges, as if they were indented. I haue not seene any seede it hath borne : the roote is like vnto the others of this kinde, but small and long, and not so great : it flowreth later for the most part then any of the other, euen not vntill Nouember, and is very hard to be preserued with vs, in that for the most part the roote waxeth lesse and lesse euery yeare, our cold Country being so contrary vnto his naturall, that it will scarce shew his flower; yet when it flowreth any thing early, that it may haue any comfort of a warme Sunne, it is the glorie of all these kindes.

9. *Colchicum versicolor.* The party coloured Medowe Saffron.

The flowers of this Medowe Saffron most vsually doe not appeare, vntill most of the other Autumne sorts are past, except the last, which are very lowe, scarce rising with their stalkes three fingers breadth aboue the ground, but oftentimes halfe hid within the ground : the leaues whereof are smaller, shorter, and rounder, then in any of the other before specified, some being altogether white, and others wholly of a very pale purple, or flesh colour; and some againe parted, the one halfe of a leafe white, and the other halfe of the same purple, and sometimes striped purple and white, in diuers leaues of one and the same flower : and againe, some will be the most part of the leafe white, and the bottome purple, thus varying as nature list, that many times from one roote may be seene to arise all these varieties before mentioned : these flowers doe stand long before they fade and passe away ; for I haue obserued in my Garden some that haue kept their flower faire vntill the beginning of Ianuary, vntill the extremitie of the Winter frosts and snowes haue made them hide their heads : the leaues therefore accordingly doe rise vp after all other, and are of a brownish or darke greene colour at their first springing vp, which after grow to be of a deepe greene colour : the roote is like the former English or Hungarian kindes, but thicker and greater for the most part, and shorter also.

10. *Colchicum variegatum alterum.* Another party coloured Medowe Saffron.

There is another, whose party coloured flowers rise a little higher, diuersly striped and marked, with a deeper purple colour, and a pale or whitish blush throughout all the leaues of the flower.

11. *Colchicum montanum Hispanicum minus.* The little Spanish Medowe Saffron.

The flowers of this little Medowe Saffron are narrower and smaller then any of the former, and of a deeper reddish purple colour then either the English or Hungarian kindes: the greene leaues also are smaller then any other, lying on the ground, of a deepe or sad greene colour, rising vp within a while after the flowers are past, and doe abide greene all the Winter long: the roote is small and long, according to the rest of the plant, and like in forme to the others.

12. *Colchicum montanum minus versicolore flore.*
The small party coloured Medowe Saffron.

This little kinde differeth not from the Spanish kinde last set forth, but in the varietie of the flower, which is as small as the former; the three inner leaues being almost all white, and the three outer leaues some of them pale or blush, and some party coloured, with a little greene on the backe of some of them.

13. *Colchicum Hermodactilum.* Physicall Medowe Saffron.

This Physicall Medowe Saffron springeth vp with his leaues in Autumne, before his flowers appeare beyond the nature of all the former kindes, yet the flower doth, after they are vp, shew it selfe in the middle of the greene leaues, consisting of six white leaues, with diuers chiues in the middle, and passeth away without giuing any seede that euer I could obserue: the greene leaues abide all the Winter and Spring following, decaying about May, and appeare not vntill September, when (as I said) the flowers shew themselues presently after the leaues are sprung vp.

14. *Colchicum atropurpureum.* The darke purple Medowe Saffron.

The greatest difference in this kinde consisteth in the flower, which at the first appearing is as pale a purple, as the flower of the former Hungarian kinde: but after it hath stood in flower two or three dayes, it beginneth to change, and will after a while become to bee of a very deepe reddish purple colour, as also the little foote-stalke whereon it doth stand: the flower is of the bignesse of the Hungarian purple, and so is the greene leafe: the seede and roote is like the English purple kinde.

15. *Colchicum atropurpureum variegatum.*
The party coloured darke purple Medowe Saffron.

We haue of late gained another sort of this kinde, differing chiefly in the flower, which is diuersly striped thorough euery leafe of the flower, with a paler purple colour, whereby the flower is of great beauty: this might seeme to bee a degeneration from the former, yet it hath abiden constant with me diuers yeares, and giueth seede as plentifully as the former.

16. *Colchicum flore pleno.* Double flowred Medowe Saffron.

The double Medowe Saffron is in roote and leafe very like vnto the English kinde: the flowers are of a fine pale or delayed purple colour, consisting of many leaues set thicke together, which are somewhat smaller, as in the English flower, being narrow and long, and as it were round at the points, which make a very double flower, hauing

O   some

some chiues with their yellow tips, dispersed as it were among the leaues in the middle: it flowreth in September, a little after the first shew of the earlier Medowe Saffrons are past.

17. *Colchicum variegatum pleno flore.*
The party coloured double Medowe Saffron.

We haue another of these double kinds (if it be not the very same with the former, varying in the flower as nature pleaseth oftentimes; for I haue this flower in my garden, as I here set it forth, euery yeare) whose flowers are diuersified in the partition of the colours, as is to be seene in the single party coloured Medowe Saffron before described, hauing some leaues white, and others pale purple, and some leaues halfe white and halfe purple, diuersly set or placed in the double flower, which doth consist of as many leaues as the former, yet sometime this party coloured flower doth not shew it selfe double like the former, but hath two flowers, one rising out of another, making each of them to be almost but single flowers, consisting of eight or ten leaues a peece: but this diuersity is not constant; for the same roote that this yeare appeareth in that manner, the next yeare will returne to his former kinde of double flowers againe.

18. *Colchicum Vernum.* Medowe Saffron of the Spring.

This Medowe Saffron riseth vp very early in the yeare, that is, in the end of Ianuarie sometimes, or beginning, or at the furthest the middle of February, presently after the deepe Frosts and Snowes are past, with his flowers inclosed within three greene leaues, which opening themselues as soone almost as they are out of the ground, shew their buds for flowers within them very white oftentimes, before they open farre, and sometimes also purplish at their first appearing, which neuer shew aboue two at the most vpon one roote, and neuer rise aboue the leaues, nor the leaues much higher then they, while they last: the flower consisteth of six leaues, long and narrow, euery leafe being diuided, both at the bottome and toppe, each from other, and ioyned together onely in the middle, hauing also six chiues, tipt with yellow in the middle, euery chiue being ioyned to a leafe, of a pale red or deepe blush colour, when it hath stood a while blowne, and is a smaller flower then any Medowe Saffron, except the small Spanish kindes onely, but continueth in his beauty a good while, if the extremity of sharpe Frosts and Windes doe not spoile it: the leaues wherein these flowers are enclosed, at their first comming vp, are of a brownish greene colour, which so abide for a while, especially on the outside, but on the inside they are hollow, and of a whitish or grayish greene colour, which after the flowers are past, grow to bee of the length of a mans longest finger, and narrow withall: there riseth vp likewise in the middle of them the head or seede vessell, which is smaller and shorter, and harder then any of the former, wherein is contained small round browne seede: the roote is small, somewhat like vnto the rootes of the former, but shorter, and not hauing so long an eminence on the one side of the bottome.

19. *Colchicum Vernum atropurpureum.* Purple Medowe Saffron of the Spring.

The flower of this Medowe Saffron, is in the rising vp of his leaues and flowers together, and in all things else, like vnto the former, onely the flowers of this sort are at their first appearing of a deeper purple colour, and when they are blowne also are much deeper then the former, diuided in like manner, both at the bottome and toppe as the other, so that they seeme, like as if six loose leaues were ioyned in the middle part, to make one flower, and hath his small chiues tipt with yellow, cleauing in like manner to euery leafe.

The Place.

All these Medowe Saffrons, or the most part of them, haue their places expressed in their titles; for some grow in the fields and medowes of the champion grounds, others on the mountaines and hilly grounds. The English kindes grow in the West parts, as about Bathe, Bristow, Warmister,

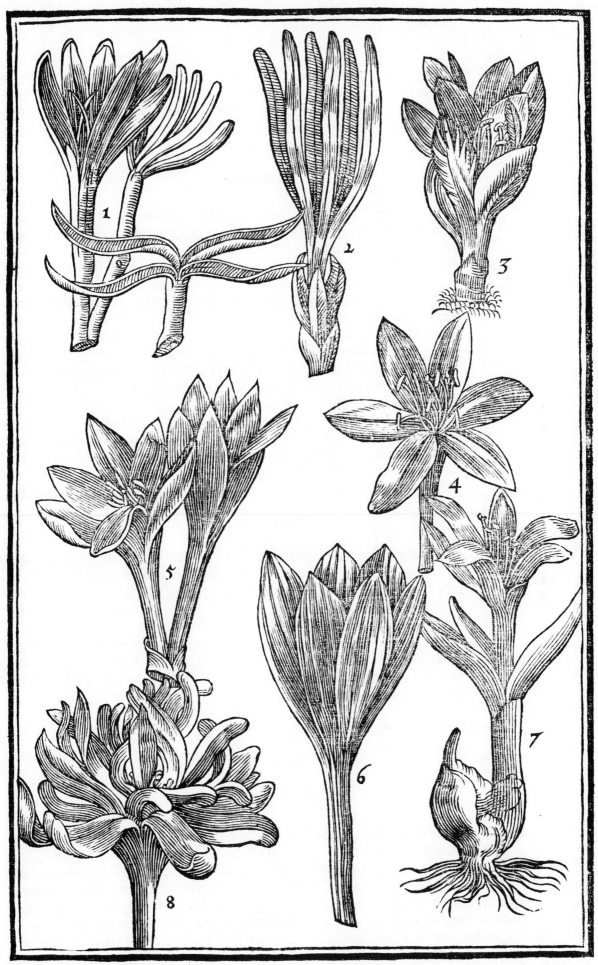

1 *Colchicum montanum Hispanicum.* The little Spanish Medowe Saffron. 2 *Colchicum montanum minus versicolore flore.* The small party coloured Medowe Saffron. 3 *Colchicum versicolor.* The party coloured Medowe Saffron. 4 *Colchicum variegatum alterum.* Another party coloured Medowe Saffron. 5 *Colchicum atropurpureum.* The darke purple Medowe Saffron. 6 *Colchicum atropurpureum variegatum.* The variable darke purple Medowe Saffron. 7 *Colchicum vernum,* Medowe Saffron of the spring. 8 *Colchicum flore pleno,* Double Medowe Saffron.

ster, and other places alfo. The double kindes are thought to come out of Germany.

The Time.

Their times likewife are declared in their feuerall defcriptions: thofe that are earlieft in Autumne, flower in Auguft and September, the later in October, and the lateft in the end of October, and in Nouember. The other are faid to bee of the Spring, in regard they come after the deepe of Winter (which is moft vfually in December and Ianuary) is paft.

The Names.

The generall name to all thefe plants is *Colchicum*, whereunto fome haue added *Ephemerum*, becaufe it killeth within one dayes fpace ; and fome *Strangulatorium*. Some haue called them alfo *Bulbus agreftis*, and *Filius ante Patrem*, The Sonne before the Father, becaufe (as they thinke) it giueth feede before the flower : but that is without due confideration ; for the root of this (as of moft other bulbous plants) after the ftalke of leaues and feede are dry, and paft, may be tranfplanted, and then it beginneth to fpring and giue flowers before leaues, (and therein onely it is differing from other plants) but the leaues and feede follow fucceffiuely after the flowers, before it may be remoued againe ; fo that here is not feede before flowers, but contrarily flowers vpon the firft planting or fpringing, and feede after, as in all other plants, though in a diuers manner.

The *Colchicum Hermodactilum* may feeme very likely to bee the *Colchicum Orientale* of Matthiolus, or the *Colchicum Alexandrinum* of Lobelius : And fome thinke it to be the true *Hermodactilus*, and fo call it, but it is not fo. We doe generally call them all in Englifh Medowe Saffrons, or *Colchicum*, according to the Latine, giuing to euery one his other adiunct to know it by.

The Vertues.

None of thefe are vfed for any Phyficall refpect, being generally held to be deadly, or dangerous at the leaft. Only the true Hermodactile (if it be of this tribe, and not this which is here expreffed) is of great vfe, for paines in the ioynts, and of the hippes, as the *Sciatica*, and the like, to be taken inwardly. Coftæus in his Booke of the nature of plants, faith, that the rootes of our common kindes are very bitter in the Spring of the yeare, and fweet in Autumne, which Camerarius contradicteth, faying, that he found them bitter in Autumne, which were (as he faith) giuen by fome impofters to diuers, as an antidote againft the Plague.

CHAP. XVII.

Crocus. Saffron.

THere are diuers forts of Saffrons, whereof many doe flower in the Spring time, and fome in Autumne, among whom there is but one onely kinde, that is called tame or of the Garden, which yeeldeth thofe blades or chiues that are vfed in meates and medicines, and many wayes profitable for other refpects, none of the reft, which are all wilde kindes, giuing any blade equall vnto thofe of the tame kinde, or for any other vfe, then in regard of their beautifull flowers of feuerall varieties, and as they haue been carefully fought out, and preferued by diuers, to furnifh a Garden of dainty curiofity. To entreate therefore of thefe, I muft, to obferue an orderly declaration, diuide them into two primary families: the former fhall be of thofe that yeeld their pleafant flowers in the Spring of the yeare, and the other that fend out

their

their colours in the Autumne, among whom that *Rex pomary* (as I may so call it) the tame or manured kinde, properly called of the Garden, is to be comprehended, for that it giueth his pleasant flowers at that time among others. I shall againe distribute those of the Spring time into three chiefe colours, that is, into white, purple, and yellow, and vnder euery one of them, comprehend the seuerall varieties that doe belong vnto them; which course I will also hold with those of the Autumne, that thus being rightly ranked, they may the more orderly be described.

1. *Crocus Vernus albus purus minor.*
The smaller pure white Saffron flower of the Spring.

This small Saffron flower springeth vp in the beginning of the yeare, with three or foure small greene leaues, somewhat broader, but much shorter then the true Saffron leaues, with a white line downe the middle of euery leafe: betweene these leaues, out of a white skinne, riseth vp one or two small flowers, made of six leaues a peece, as all the rest in generall are, of a pure white colour, without any mixture in it, which abide not in flower aboue a weeke, or rather lesse, so sodainly is the pleasure of this, and the purple lost: it flowreth not for the most part, vntill a moneth after the yellow Crocus appeareth in flower, and the ordinary stript Crocus is past: the seede is small, round, and reddish, yet not so red as the seede of the yellow, contained in three square heads, yet seldome beareth, but encreaseth by the roote plentifully enough, which is small, round, and flat at the bottome, somewhat white on the outside, but whiter within, shooting out small sprouts on euery side of the roote, which is the best note to know this kinde and the lesser purple, which are both alike, from all other rootes of Saffron flowers.

2. *Crocus albus maior multiflorus.* The great snow white Crocus.

This greater Saffron flower riseth vp vsually with three or foure greene leaues, larger then the former, with a white line in euery one of them: the flowers are greater, and more in number, rising together, but flowring one after another, of a pure snow white colour, and abiding but little longer in flower then the former.

3. *Crocus albus maior alter dictus Masiacus.*
The great white Saffron flower or Crocus of Mesia.

This great white Crocus of Mesia, riseth vp out of the ground, almost as early as the first sort of the yellow, with foure or fiue leaues, being very like vnto the leaues of the yellow Crocus, and as large, with white lines in them: the flowers also are as large as the flowers of the yellow, and many also rising one after another like vnto it, but not of so pure a white colour, as the former or last described, but rather tending to a Milky or Creame colour: the roote is not couered with any reddish, but rather pale skinnes or coates.

4. *Crocus albus Masiacus fundo violaceo.*
The great white Crocus of Mesia with a blew bottome.

There is another of this kinde, like vnto the former in all things, sauing that the bottomes of the flowers of this kinde, with some part of the stalke next the flower, are of a pale shining purple colour, and rising vp a pretty way into the flower; whereas another also of this kind, hath a little shew or marke of blew, and not purple, at the bottome of the flower onely, which maketh a difference.

5. *Crocus albus fundo purpureo.* The white Crocus with a purple bottome.

This Saffron flower is of the same kinde with the first, both in roote, leafe, and flower, in none of them differing from it, but in that the bottome of this flower, with that part of the short foote-stalke next vnto it, is of a violet or purple colour, and sometimes hauing here and there some purple small lines, or spots on the white leaues: it flowreth also with the first white, or somewhat later.

O 3 *6. Crocus*

6. *Crocus vernus albus striatus.* The white stript Crocus.

This stript Saffron flower is likewise neare the same first kind, or first white Crocus, hauing the like leaues and flowers, somewhat larger, but as soone fading almost as it: but herein this flower differeth, that it hath pale blewish lines and spots in all the leaues thereof, and more principally in the three outer leaues: the root is also white on the outside, like the first white, but greater, with young ones growing round about it.

7. *Crocus vernus albus polyanthos versicolor.*
The greater party coloured white Crocus.

The greater party coloured Saffron flower, hath his greene leaues like vnto the second great white Crocus before mentioned, hauing more flowers then any of the former, except the first great white, the leaues whereof haue greater stripes then the last recited Crocus, but of a purple Violet colour, making each leafe seeme oftentimes to haue as much purple as white in them: the roote hereof is somewhat like the second white, but of a little more duskie colour on the outside, and not budding out on the sides at all, or very little.

8. *Crocus vernus albus versicolor.* The lesser party coloured white Crocus.

The leaues and flowers of this other party coloured Crocus, are for bignesse in a manner equall with the last, but hath not so many flowers rising together from the roote: the flower is finely marked with blew strakes on the white flower, but nothing so much as in the former: the roote also is like the last.

9. *Crocus Episcopalis.* The Bishops Crocus.

This party coloured or Bishops Saffron flower, is very like both in leaues and rootes vnto the Neapolitane blew Crocus, but somewhat greater: the flowers doe abide not so long time blowne, and hath all the leaues either wholly white, with blew stripes on both sides of them, or wholly of a fine delayed blew Violet colour, and the three innermost more blew and finely striped, both on the inside and outside of them, and sometimes it hath been seen to haue three leaues white, and three leaues of a pale blew.

10. *Crocus vernus striatus vulgaris.* The ordinary stript Crocus.

There is another sort of stript Saffron flower, which is most common and plentifull in most Gardens, which I must needes bring vnder the ranke of these white kinds, although it differre very notably, both in roote, leafe, and flower, from all of them: the leaues of this rise vp sooner then the yellow or white Crocus, lying spread vpon the ground for the most part, but narrower then any of the former: among these leaues spring vp diuers flowers, almost as large as the former great white Crocus, of a very bleake or pale purple colour, tending to white on the inside, and in many almost white, with some small whitish chiues tipt with yellow in the middle: the three outer leaues are of a yellowish white colour on the backe side of them, stript euery one of them with three broad stripes, of a darke murrey or purple colour, and a little sprinkled with some small purple lines, on both sides of those stripes; but on the inside, of the same pale purple or white colour with the rest: the seede hereof is somewhat darker coloured then of the white, and is more liberall in bearing: the roote is differing from all the former, being rounder and bigger then any of them, except the kindes of Misia, yet somewhat flat withall, not hauing any shootes from the sides, but setting off into rootes plentifully, hauing a round circle compassing the bottome of the roote, which easily falleth away, when it is taken vp out of the ground, and couered with a browne coate, somewhat neare the colour of the yellow Crocus, but not altogether so bright: it flowreth vsually the first of all these sorts, or with the first of the early yellowes.

1 *Crocus vernus albus minor.* The small white Saffron flower of the spring. 2 *Crocus vernus Masiscus albus.* The great white Crocus of Misia. 3 *Crocus vernus albus striatus.* The white stript Crocus. 4 *Crocus vernus albus polyanthos versicolor* The party coloured white Crocus. 5 *Crocus albus fundo purpureo.* The white Crocus with a purple bottome. 6 *Crocus vernus Neapolitanus.* The great blew Crocus of Naples. 7 *Crocus vernus purpureus maximus.* The great purple Crocus. 8 *Crocus vernus purpureus striatus.* The purple stript Crocus. 9 *Crocus vernus purpureus Capillarifolio.* The purple Crocus with small leaues. 10 *Crocus vernus flavus striatus.* The yellow stript Crocus. 11 *Crocus vernus luteus versicolor.* The cloth of gold Crocus.

11. *Crocus vernus striatus Turcicus.* The Turkie stript Crocus.

There is another of this kinde, whose flower is a little larger, and of a deeper purple colour, both on the inside and outside; the greene leafe also is bigger, and of a more whitish colour.

12. *Crocus vernus Capillarifolio albus.* The white Crocus with small leaues.

This white Crocus is in all things like vnto the purple of the same kinde, but that the flower of this is wholly white: the full description therefore hereof, you shall haue in that purple with small leaues, of this kinde hereafter set downe, whereunto I referre you.

13. *Crocus vernus purpureus minor.* The smaller purple Crocus.

The smaller purple Saffron flower of the Spring, hath his greene leaues so like vnto the first white flowred Saffron, that they can hardly be distinguished, onely they seem to bee a little narrower: the flower is also much about the same bignesse, or a little bigger, and seldome beareth aboue one flower from a roote, euen as the first doth, of a deepe purple Violet colour, the bottome of the flower, with the vpper part of the stalke next thereunto, being of a deeper or blacker purple; in the middle of the flower are some pale chiues tipt with yellow pendents, and a longer pointell, diuided or forked at the toppe: the roote of this is in all things so like vnto the first white, that it is impossible for the most cunning and conuersant in them, to know the one from the other. This beareth seede very sparingly, as the white doth, and is reddish like vnto it, but recompenseth that defect with a plentifull encrease by the roote: it likewise flowreth at the very same time with the white, and endureth as small a time.

14. *Crocus vernus purpureus maximus.* The greatest purple Crocus.

This great purple Crocus is of the same kinde with the next described, as well in roote as leafe, but greater; for the greene leaues hereof are the greatest and broadest of all other Crocus, with a large white line in the middle of euery one: it springeth vp much later then the former, and doth not shew his flower vntill the other bee past a good while: the flowers also are the largest of all these Crocus of the Spring time, and equalling, if not surpassing that purple kinde that flowreth in Autumne, hereafter set forth, of a very faire and deepe Violet colour, almost as deepe as the former: the seed vessels are large also and white, wherein is contained pale reddish seede, like vnto the next blew kinde, but somewhat greater: the roote is (as I said before) like vnto the next, that is, flat and round, with a duskie coloured outside, whose head for springing in it is as hardly discerned.

Alter Apicibus albidis. We haue one of this kinde, the toppes onely of whose purple flower are whitish, for the breadth of halfe the naile of a mans hand, which abideth constant euery yeare in that manner, and therefore is a difference fit to be remembred.

15. *Crocus vernus Neapolitanus siue cæruleus maior.* The greater blew Crocus of Naples.

This great blew Crocus riseth vp with diuers greene leaues, broader then any of the former (except the last) with a white line running downe the inside of euery leafe, as in the former, among which riseth vp, out of diuers great long white skinnes, diuers large flowers, but not fully so great as the former, consisting of six leaues, of a paler blew or Violet colour then in the former, hauing in the middle of the flowers a few pale threeds, tipt with yellow, and a longer pointell of a gold yellow colour, forked or diuided at the toppe, smelling sweeter then in the former, and abiding a great while longer, being in flower vsually euen with the stript yellow Crocus, or before the former purple, and yeelding more plenty of seede: the roote hereof is not very great, but a little darke on the outside, being round and flat withall, that one can hardly know which is the vpperside thereof.

This

This kinde differeth very little from the former, either in roote, leafe, or flower, for the bigneffe or colour, but that it feemeth to be a little bleaker or paler blew, becaufe it flowreth a little earlier.

16. *Crocus vernus purpureus ftriatus.* The ftript purple Crocus.

The leaues of this ftript purple Saffron flower, are as large and broad as the laft, or rather a little longer : the flowers alfo are as plentifull, and as large, of a fine delayed purple colour on the outfide, with three broad ftrakes or lines downe the backe of the three outer leaues, and of a little deeper purple on the infide, as the other three leaues are alfo of a deeper purple colour, and are ftriped with the fame deepe purple about the ground, or bottome of the leaues : this fometimes yeeldeth three fquare heads, containing in them brownifh feede : the roote is like vnto the laft, and flowreth much about the time of the former.

17. *Crocus vernus purpureus verficolor.* The filuer ftript purple Crocus.

This ftript Saffron flower, is in leaues and flowers fomewhat like vnto the laft ftript purple, but a little fmaller : the flowers are of a little deeper purple through the whole leaues, ftriped with white lines, both on the leaues, and towards the edges, which maketh a peculiar difference from all the reft : the roote of this is not fo flat, though like it, and couered with a darke afh coloured skinne : it flowreth about the fame time.

18. *Crocus purpureus flammeus maior.* The greater purple flame coloured Crocus.

The greene leaues of this Crocus or Saffron flower, are of a reafonable breadth and length, and of a pleafant frefh greenneffe, with a faire broad white line downe the middle of them, but rifing not out of the ground fo early as the next defcribed Crocus : the flowers are likewife of a meane bigneffe, of a pale purple on the outfide, fomewhat whitifh, efpecially the three outer leaues ; but on the infide of a deeper purple, and ftriped with great ftripes like flames, hauing fome chiues in the middle, and a longer one alfo feathered a little at the toppe : the roote is white on the outfide, fomewhat flat and round, but not fo flat as the Neapolitane Crocus before defcribed.

19. *Crocus purpureus flammeus minor.* The leffer purple flame coloured Crocus.

This Crocus hath almoft as broad and long greene leaues as the former, and of the fame verdure, which rife vp earlier then it, and is in flower likewife fomewhat before it, being fmaller for fize by a little, but of as deepe a purple on the outfide, as on the infide, flamed with faire broad ftripes from the middle of the leaues, or fomewhat lower vnto the edges : each of thefe giue feed that is of a pale reddifh colour: the root is very like vnto the former, but a little leffer.

20. *Crocus vernus purpureus Capillarifolio.* The purple Crocus with fmall leaues.

This fmall kinde of Saffron flower rifeth out of the ground, with two or three long and fmall green leaues, very like vnto the leaues of the fine Fether-Graffe hereafter defcribed, ftanding vpright at the firft, but afterwards lying vpon the ground ; among which come the flowers, fometimes three, but moft vfually two vpon one ftalke, if the roote be not young, which then will beare but one on a ftalke, which is very fhort, fo that the flowers fcarce arife aboue the ground, yet laying themfelues open in the day time, if it be faire, and the Sunne doe fhine, otherwife they keepe clofe, and doe not open at all : and after one flower is paft, which doth not laft aboue three or foure dayes at the moft, the others follow, which are of a bleake blewifh purple in the middle of the flower, and of a deeper purple towards the ends or points of the leaues, but of a more fullen or darke purple on the outfide of them, and yellowifh at the bottome, with fome yellow chiues in the middle : the feede is fmall and darker coloured then any of the former Crocus, contained alfo in fmaller heads, ftanding one by another

vpon

vpon the fame fhort foote-ftalke, which then rifeth vp a little higher, fhewing the maner of the ftanding of the flowers, which in their flowring time could not fo eafily bee difcerned : the roote is very fmall and round, hauing one fide at the bottome lower then the other, very like the roote of a *Colchicum* or Medowe Saffron, and fomewhat neare refembling alfo the hoofe of an horfe foote, couered with a very thicke skinne, of a darke or blackifh browne colour : this flowreth the laft of all the former forts of Saffron flowers, euen when they are all paft.

21. *Crocus vernus purpureus ftriatus Capillarifolio.* The ftript purple Crocus with fmall leaues.

This fmall ftript purple Saffron flower hath fuch like leaues, as the laft defcribed hath, betweene which rifeth the flower vpon as fhort a foote-ftalke, confifting of fix leaues like the former, of a faire purple colour on the outfide of the three outer leaues, with three lines or ftrakes downe euery leafe, of a deeper purple colour, and on the infide of a paler purple, as the other three leaues are alfo, with fome chiues tipt with yellow pendents, and a forked pointell in the middle : the roote of this is fomewhat bigger then the former, and rounder, but couered with as thicke and as browne a skinne : it flowreth about the fame time with the former.

22. *Crocus vernus luteus fiue Mafiacus.* The yellow Crocus.

The yellow Crocus or Saffron flower, rifeth vp with three or foure leaues out of the ground, being fomewhat neare the breadth of the great purple kindes, with a white line in them, as in moft of the reft : the flowers ftand in the middle of thefe leaues, and are very large, of a gold yellow colour, with fome chiues, and a forked point in the middle : the feede hereof is of a brighter colour then in any of the other : the roote is great and round, as great or greater then a Wall Nut fometimes, and couered with reddifh skinnes or coates, yeelding more ftore of flowers then moft of the former, and beginning to blowe with the firft forts, or prefently after, but outlaft many of them, and are of a pleafant good fent.

Flore aureo.　Of this kinde we haue fome, whofe flowers are of a deeper gold yellow colour then others, fo that they appeare reddifh withall.

Flore pallido.　And we haue alfo another fort, whofe flowers are very pale, betweene a white and a yellow, not differing in any thing elfe.

Flore viridante luteo.　And another fmaller, whofe flower hath a fhew of greenneffe in the yellow, and more greene at the bottome.

23. *Crocus vernus flauus ftriatus.* The yellow ftript Crocus.

This kinde of yellow ftript Crocus or Saffron flower, rifeth vp with more ftore of narrower and greener leaues then the former, and after the leaues are fpread, there rife vp many yellow flowers from among them, which are not of fo faire and bright a yellow colour, but more dead and fu len, hauing on the backfide of each of the three outtermoft leaues, three fmall ftripes, of an ouer-worne or dull purple colour, with fome chiues and a pointell in the middle : the roote of this kinde, is very like the roote of the former yellow, but fomewhat fmaller and fhorter, and couered with the like reddifh skinnes, but a little fadder : it flowreth not fo early as the former yellow, but abideth almoft as long as it.

24. *Crocus vernus luteus verficolor primus.* The beft cloth of gold Crocus.

The faireft cloth of gold Crocus or Saffron flower, rifeth vp very early, euen with the firft, or the firft of all other Crocus, with three or foure very narrow and fhort leaues, of a whiter colour then any of the former, which by and by after doe fhew forth the flowers, rifing from among them out of the fame white skinne, which includeth the leaues, but are not fo plentifull as the former yellow, being but two or three at the moft, of a faire gold yellow colour, yet fomewhat paler then the firft, hauing

uing on the backe of euery of the three outer leaues, three faire and great stripes, of a faire deepe purple colour, with some small lines at the sides or edges of those purple stripes; on the inside of these flowers, there is no signe or shew of any line or spot, but wholly of a faire gold yellow, with chiues and a fethertopt pointell in the middle: the seede hereof is like the former, but not so red: the roote of this kinde is easily knowne from the roote of any other Saffron flower, because the outer peelings or shels being hard, are as it were netted on the outside, hauing certaine ribbes, rising vp higher then the rest of the skinnes, diuided in the forme of a net-worke, of a darke browne colour, and is smaller and rounder then the former yellow, and not encreasing so plentifully by the roote.

25. *Crocus vernus luteus versicolor alter.* The second cloth of gold, or Duke Crocus.

There is no difference either in roote, leafe, or colour of flower, or time of flowring in this sort from the last before mentioned; for the flower of this is of the same bignesse and colour, the only note of difference is in the marking of the three outer leaues, which haue not three stripes like the former, but are wholly of the same deepe purple colour on the backe of them, sauing that the edges of them are yellow, which is the forme of a Duke Tulipa, and from thence it tooke the name of a Duke Crocus.

26. *Crocus vernus versicolor pallideluteus.* The pale cloth of gold Crocus.

We haue a third sort of this kinde of cloth of gold Crocus, which hath leaues and flowers like the former, but differeth in this, that the colour of the flower is of a paler yellow by much, but stript in the same manner as the first, but with a fainter purple colour: the roote also is netted like them, to shew that this is but a variation of the same kinde.

27. *Crocus vernus versicolor albidoluteus.* The cloth of siluer Crocus.

The chiefest note of difference in this Saffron flower is, that being as large a flower as any of the former of this kinde, it is of so pale a yellowish white, that it is more white then yellow, which some doe call a butter colour: the three outer leaues are striped on the backe of them, with a paler purple blew shining colour, the bottome of the flower, and the vpper part of the stalke, being of the same purple blew colour: the roote of this is also netted as the other, to shew it is a variety of the same kinde.

And thus much for those Saffron flowers that come in the Spring time; now to those that flower in Autumne onely: and first of the true Saffron.

1. *Crocus verus satiuus Autumnalis.* The true Saffron.

The true Saffron that is vsed in meates and medicines, shooteth out his narrow long greene leaues first, and after a while the flowers in the middle of them appeare about the end of August, in September and October, according to the soile, and climate where they growe; these flowers are as large as any of the other former or later sorts, composed of six leaues a peece, of a murrey or reddish purple colour, hauing a shew of blew in them: in the middle of these flowers there are some small yellow chiues standing vpright, which are as vnprofitable, as the chiues in any other of the wilde Saffrons, before or hereafter specified; but besides these, each flower hath two, three, or foure greater and longer chiues, hanging downe vpon or betweene the leaues, which are of a fierie red colour, and are the true blades of Saffron, which are vsed phy-sically or otherwise, and no other: All these blades being pickt from the seuerall flowers, are laid and pressed together into cakes, and afterwards dryed very warily on a Kill to preserue them; as they are to be seene in the shops where they are sold. I neuer heard that euer it gaue seede with any: the roote groweth often to be as great, or greater then a green Wall Nut, with the outer shell on it, couered with a grayish or ash-coloured skin, which breaketh into long hairie threeds, otherwise then in any other roote of Crocus.

2. *Crocus*

2. *Crocus Byzantinus argenteus.* The filuer coloured Autumne Crocus.

This Saffron flower fpringeth vp in October, and feldome before, with three or foure fhort greene leaues at the firft, but growing longer afterwards, and in the midft of them, prefently after they haue appeared, one flower for the moft part, and feldome two, confifting of fix leaues, the three outermoft whereof are fomewhat larger then the other three within, and are of a pale-bleake blew colour, almoft white, which many call a filuer colour, the three innermoft being of a purer white, with fome yellow chiues in the middle, and a longer pointell ragged or fethered at the toppe : this very feldome beareth feede, but when the yeare falleth out to bee very milde ; it is fmall, round, and of a darke colour : the roote is pretty bigge, and rounder then any other Crocus, without any flat bottome, and couered with a darke ruffet skinne.

3. *Crocus Pyrenaus purpureus.* The purple mountaine Crocus.

This purple Saffron flower of the Autumne, rifeth vp but with one flower vfually, yet fometimes with two one after another, without any leaues at all, in September, or fometimes in Auguft, ftanding vpon a longer foote-ftalke then any kinde of Saffron flower, either of the Spring or Autumne, and is as large as the flower of the greateft purple Saffron flower of the Spring, of a very deepe Violet purple colour, which decayeth after it hath ftood blowne three or foure dayes, and becommeth more pale, hauing in the middle fome yellow chiues, and a long fether topt pointell, branched, and rifing fometimes aboue the edges of the flowers : about a moneth after the flowers are paft, and fometimes not vntill the firft of the Spring, there rifeth vp three or foure long and broad greene leaues, with a white line in euery one of them, like vnto the firft purple Vernall kindes, which abide vntill the end of May or Iune : the roote is fmall and white on the outfide, fo like vnto the roote of the leffer Vernall purple or white Crocus, that it cannot be diftinguifhed, vntill about the end of Auguft, when it doth begin to fhoot, and then by the early fhooting vp a long white fprout for flower, it may be knowne. I neuer could obferue it to giue any feede, the Winter (as I thinke) comming on it fo quickly after the flowring, being the caufe to hinder it.

4. *Crocus montanus Autumnalis.* The Autumne mountaine Crocus.

The mountaine Saffron flower fpringeth vp later then any of the former, and doth not appeare vntill the middle or end of October, when all the flowers of the former are paft, appearing firft with three or foure fhort greene leaues, like vnto the Byzantine Crocus, and afterwards the flowers betweene them, which are of a pale or bleake blew tending to a purple, the foote-ftalkes of them being fo fhort, that they fcarce appeare aboue ground at the firft, but after two or three dayes they grow a little higher : the roote is very great and flat bottomed, couered with a grayifh duskie coate or skinne, and encreafeth very little or feldome.

The Place.

The feuerall places of thefe Saffron flowers, are in part fet downe in their titles ; the others haue beene found out, fome in one Countrey, and fome in another, as the fmall purple and white, and ftript white in Spaine : the yellow in Mefia about Belgrade, the great purple in Italy ; and now by fuch friends helpes as haue fent them, they profper as well in our Gardens, as in their naturall places. Yet I muft giue you this to vnderftand, that fome of thefe formerly expreffed, haue been raifed vp vnto vs by the fowing of their feede.

The Time.

Their feuerall times are likewife expreffed in their defcriptions; for fome fhew forth their pleafant flowers in the Spring, wherein for the three firft moneths,

1 *Crocus vernus luteus vulgaris.* The common yellow spring Crocus. 2 *Crocus vernus sativus Autumnalis.* The true Saffron 3 *Crocus Byzantinus argenteus.* The siluer coloured Autumne Crocus. 4 *Crocus Pyrenæus purpureus.* The purple mountaine Crocus. 5 *Crocus montanus Autumnalis.* The Autumne mountaine Crocus. 6 *Sisyrinchium maius.* The greater Spanish Nut.

moneths, our Gardens are furnished with the varietie of one sort or another: the rest in Autumne, that so they might procure the more delight, in yeelding their beauty both early and late, when scarce any other flowers are found to adorne them.

The Names.

I shall not neede to trouble you with an idle tale of the name of Crocus, which were to little purpose, nor to reiterate the former names imposed vpon them; let it suffice that the fittest names are giuen them, that may distinguish them one from another; onely this I must giue you to vnderstand, that the gold yellow *Crocus* or Saffron flower, is the true *Crocus Masiacus*, as I shewed before; and that neither the yellow stript, or cloth of gold (which wee so call after the Dutch name *Gaud Laken)* is the true *Masiacus*, as some suppose; and that the great white Saffron flower, by reason of his likenesse vnto the gold yellow, is called *Crocus albus Masiaci facie*, or *facie lutei*, that is, The white Saffron flower that is like the *Masiacus* or yellow.

The Vertues.

The true Saffron (for the others are of no vse) which wee call English Saffron, is of very great vse both for inward and outward diseases, and is very cordiall, vsed to expell any hurtfull or venemous vapours from the heart, both in the small Pockes, Measels, Plague, Iaundise, and many other diseases, as also to strengthen and comfort any cold or weake members.

CHAP. XVIII.

Sisyrinchium. The Spanish Nut.

I Can doe no otherwise then make a peculiar Chapter of this plant, because it is neither a *Crocus*, although in the roote it come somewhat neare vnto that kinde that is netted; but in no other part agreeing with any the delineaments of a Saffron flower, and therefore could not be thrust into the Chapter amongst them: neither can I place it in the forefront of the Chapter of the *Iris bulbosa*, or bulbous Flowerdeluces, because it doth not belong to that Family: and although the flower thereof doth most resemble a Flowerdeluce, yet in that no other parts thereof doe fitly agree thereunto, I haue rather chosen to seate it by it selfe betweene them both, as partaking of both natures, and so may serue in stead of a bridge, to passe from the one to the other, that is, from the *Crocus* or Saffron flower, to the *Iris bulbosa* or bulbous rooted Flowerdeluce, which shall follow in the next Chapter by themselues.

The Spanish Nut hath two long and narrow, soft and smooth greene leaues, lying for the most part vpon the ground, and sometimes standing vp, yet bending downewards; betweene these leaues riseth vp a small stalke, halfe a foote high, hauing diuers smooth soft greene leaues vpon it, as if they were skinnes, through which the stalke passeth; at the toppe whereof stand diuers flowers, rising one after another, and not all flowring at once: for seldome shall you haue aboue one flower blowne at a time, each whereof doth so quickly passe and fade away, that one may well say, that it is but one dayes flower, or rather the flower of a few houres: the flower it selfe hath nine leaues, like vnto a Flowerdeluce, whereof the three that fall downe, haue in each of them a yellow spot: the other three, which in the Flowerdeluces are hollow and ridged, couering the other three that fall downe, in this stand vpright, and are parted at the ends: the three that stand vp in the middle are small and short: the whole flower is smaller then any Flowerdeluce, but of sundry colours; for some are of an excellent skie colour blew, others of a Violet purple, others of a darker purple colour, and some white, and many others mixed, either pale blew and deepe purple, or white and blew

mixed

mixed or ftriped together very variably, quickly fading as I faid before : the feede is enclofed in fmall cods, fo thinne and tranfparent, that one may eafily fee, and tell the feeds as they lye, which are of a brownifh red colour: the roote is fmall, blackifh and round, wrapped in a thicke skinne or huske, made like vnto a net, or fomewhat like vnto the roote of the cloth of gold Crocus : when the plant is in flower, it is found to haue two rootes one aboue another, whereof the vppermoft is firme and found, and the vndermoft loofe and fpongie, in like manner as is found in the rootes of diuers Orchides or Satyrions, Bee-flowers and the like, and without any good tafte, or fweetneffe at all, although Clufius faith otherwife.

Sifyrinchium Mauritanicum. The Barbary Nut.

There is another of this kinde, not differing from the former in any other notable part, but in the flower, which in this is of a delayed purplifh red colour, hauing in each of the three lower leaues a white fpot, in ftead of the yellow in the former, but are as foone fading as they.

The Place.

The former doe grow very plentifully in many parts both of Spaine and Portugall, where Guillaume Boel, a Dutch man heretofore remembred often in this Booke, found them ; of the fundry colours fpecified, whereas Clufius maketh mention but of one colour that he found.

The other was found in that part of Barbary, where Fez and Morocco do ftand, and brought firft into the Lowe-Countries : but they are both very tender, and will hardly abide the hard Winters of thefe colder regions.

The Time.

The firft flowreth in May and Iune, the laft not vntill Auguft.

The Names.

The name *Sifyrinchium* is generally impofed vpon this plant, by all authors that haue written thereof, thinking it to bee the right *Sifyrinchium* of Theophraftus : but concerning the Spanifh name *Nozelha*, which Clufius faith it is called by in Spaine, I haue beene credibly enformed by the aforenamed Boel, that this roote is not fo called in thofe parts ; but that the fmall or common ftript Crocus is called *Nozelha*, which is fweete in tafte, and defired very greedily by the Shepheards and Children, and that the roote of this *Sifyrinchium* or Spanifh Nut, is without any tafte, and is not eaten. And againe, that there is not two kindes, although it grow greater, and with more flowers, in thofe places that are neare the Sea, where both the wafhing of the Sea water, and the moifture and ayre of the Sea, caufeth the ground to bee more fertile. This I thought good, from the true relation of a friend, to giue the world to vnderftand, that truth might expell errour.

The Vertues.

Thefe haue not been knowne to bee vfed to any Phyficall purpofe, but wholly neglected, vnleffe fome may eate them, as Clufius reporteth.

Chap. XIX.

Iris bulbosa. The bulbous Flowerdeluce.

THe Flowerdeluces that haue bulbous rootes are of two sorts, the one greater then the other : the greater bearing larger and broader leaues and flowers, and the lesser narrower. But before I giue you the descriptions of the vsuall greater kindes, I must needes place one or two in the fore-front that haue no fellowes ; the one is called of Clusius, his broad leafed Flowerdeluce, and the other a Persian, somewhat like vnto it, which although they differ notably from the rest, yet they haue the nearest resemblance vnto those greater kindes, that come next after them.

Iris bulbosa prima latifolia Clusij.
Clusius his first great bulbous Flowerdeluce.

This Flowerdeluce hath diuers long and broad leaues, not stiffe, like all the other, but soft and greenish on the vpperside, and whitish vnderneath ; among which rise vp sometimes seuerall small, short, slender stalkes, and sometimes but one, not aboue halfe a foote high, bearing at the top one flower a peece, somewhat like vnto a Flowerdeluce, consisting of nine leaues, whereof those three that stand vpright, are shorter and more closed together, then in other sorts of Flowerdeluces ; the other three that fall downe, turne vp their ends a little, and those three, that in other Flowerdeluces doe couer them at the bottome, stand like the vpright leaues of other Flowerdeluces, but are parted into two ends, like vnto two small eares : the whole flower is of a faire blew, or pale skie colour in most, with a long stripe in the middle of each of the three falling leaues, and in some white, but more seldome : the roote is reasonable great, round and white, vnder the blackish coates wherewith it is couered, hauing many long thicke white rootes in stead of fibres, which make them seeme to be Asphodill rootes. The flower is very sweete.

Iris bulbosa Persica. The Persian bulbous Flowerdeluce.

This Persian Flowerdeluce is somewhat like vnto the former, both in roote and in leafe, but that the leaues are shorter and narrower, and the flower being much about the same fashion, is of a pale blew russetish colour, each of the three lower falling leaues are almost wholly of a browne purple colour, with a yellow spot in the middle of them : this as it is very rare, so it seldome beareth flowers with vs.

The Place.

The first groweth in many places of Spaine and Portugall, from whence I and others haue often had it for our Gardens, but by reason of the tendernesse thereof, it doth hardly endure the sharpnesse of our cold Winters, vnlesse it be carefully preserued.

The other is said to come from Persia, and therefore it is so entituled, and is as tender to be kept as the other.

The Time.

The first flowreth most vsually not vntill May with vs, yet many times sooner : but in Ianuary and February, as Clusius saith, in the naturall places thereof.

The other is as early oftentimes when it doth flower with vs.

The Names.

Because Clusius by good iudgement referreth the first to the greater kindes of Flowerdeluces, and placeth it in the fore ranke, calling it *Iris bulbosa latifolia prima*, that is, The first broad leafed Flowerdeluce, and all others doe the like, I haue (as you see) in the like manner put it before all the other, and keepe the same name. The Spaniards, as he saith, called it *Lario espadanal*, and they of Corduba, *Lirios azules*.

The other hath no other name then as it is in the title.

1. *Iris bulbosa maior siue Anglica cærulea.*
The blew English bulbous Flowerdeluce.

This bulbous Flowerdeluce riseth vp early, euen in Ianuary oftentimes, with fiue or six long and (narrow, in comparison of any great breadth, but in regard of the other kinde) broad whitish green leaues, crested or straked on the backside, and halfe round, the inside being hollow like a trough or gutter, white all along the inside of the leafe, and blunt at the end ; among which riseth vp a stiffe round stalke, a cubit or two foot high, at the toppe whereof, out of a skinnie huske, commeth forth one or two flowers, consisting of nine leaues a peece, three whereof that are turned downewards, are larger and broader then the other, hauing in each of them a yellow spot, about the middle of the leafe, other three are small, hollow, ridged or arched, couering the lower part next the stalke of those falling leaues, turning vp their ends, which are diuided into two parts, other three stand vpright, and are very small at the bottome of them, and broader toward the toppe : the whole flower is of a faire blew colour ; after the flowers are past, come vp three square heads, somewhat long, and lanke, or loose, containing in them round yellowish seede, which when it is ripe, will rattle by the shaking of the winde in the dry huskes : the roote of this kinde is greater and longer then any of the smaller kindes with narrow leaues, couered with diuers browne skinnes, which seeme to be fraught with long threeds like haires, especially at the small or vpper end of the roote, which thing you shall not finde in any of the smaller kindes.

2. *Iris bulbosa maior purpurea & purpuro violacea.*
The paler or deeper purple great bulbous Flowerdeluce.

These purple Flowerdeluces differ not from the last described, either in roote or leafe : the chiefest difference consisteth in the flowers, which in these are somewhat larger then in the former, and in the one of a deepe blew or Violet purple colour, and in the other of a deepe purple colour, in all other things alike.

There is also another, in all other things like vnto the former, but only in the flower, *Flore cinerea.* which is of a pale or bleake blew, which we call an ash-colour.

3. *Iris bulbosa maior purpurea variegata siue striata.*
The great purple stript bulbous Flowerdeluce.

There is another of the purple kinde, whose flower is purple, but with some veines or stripes of a deeper Violet colour, diuersly running through the whole leaues of the flower.

And another of that bleake blew or ash-colour, with lines and veines of purple in *Flore cinereo* the leaues of the flowers, some more or lesse then other. *striata purpureo.*

And againe another, whose flower is of a purple colour like vnto the second, but *Flore purpureo* that round about that yellow spot, in the middle of each of the three falling leaues (as *orbe cinereo.* is vsuall in all the bulbous Flowerdeluces)there is a circle of a pale blew or ash-colour, the rest of the leafe remaining purple, as the other parts of the flower is.

4. *Iris bulbosa maior flore rubente.*
The great peach coloured bulbous Flowerdeluce.

There is another of these greater kindes, more rare then any of the former, not differing in roote, leafe, or flower, from the former, but onely that the flower in this is of a pale reddish purple colour, comming somewhat neare vnto the colour of a peach blossome.

5. *Iris*

5. *Iris bulbosa maior siue latifolia alba.*
The great white bulbous Flowerdeluce.

The great white bulbous Flowerdeluce, riseth not vp so early out of the ground as the blew or purple doth, but about a moneth or more after, whose leaues are somewhat larger, and broader then of the others: the stalke is thicker and shorter, bearing vsually two very large and great flowers, one flowring a little before the other, yet oftentimes both in flower together in the end, of a bleake blewish white colour, which wee call a siluer colour, while they are in the budde, and before they be blowne open, but then of a purer white, yet with an eye or shew of that siluer colour remaining in them, the three falling leaues being very large, and hauing that yellow spot in the middle of each of them: the seedes are likewise inclosed in heads, like vnto the blew or purple kindes, but larger, and are of a reddish yellow colour like them: the roote likewise is not differing, but greater.

6. *Iris bulbosa maior alba variegata.*
The great white stript bulbous Flowerdeluce.

This white stript Flowerdeluce, is in roote, leafe, and flower, and in manner of growing, like vnto the former white Flowerdeluce; the onely difference is in the marking of the flower, being diuers from it: for this hath in the white flower great veines, stripes, or markes, of a Violet blew colour, dispersed through the leaues of the flower very variably, which addeth a superexcellent beauty to the flower.

7. *Iris bulbosa maior siue latifolia versicolor.*
The great party coloured bulbous Flowerdeluce.

There is no difference in this from the former, but in the flower, which is of a whitish colour in the three falling leaues, hauing a circle of ash-colour about the yellow spot, the three rigged leaues being likewise whitish, but ridged and edged with that ash-colour, and the three vpright leaues of a pale blewish white colour, with some veines therein of a blewish purple.

Parietas.   There hath beene brought vnto vs diuers rootes of these kindes, with the dryed flowers remaining on them, wherein there hath beene seene more varieties, then I can well remember to expresse, which variety it is very probable, hath risen by the sowing of the seeds, as is truely obserued in the narrower leafed kinde of Flowerdeluce, in the Tulipa, and in some other plants.

Flore luteo.   Wee haue heard of one of this kinde of broad leafed Flowerdeluces, that should beare a yellow flower, in the like manner as is to be seene in the narrow leafed ones: but I haue not seene any such, and therefore I dare report no further of it, vntill time hath discouered the truth or falshood of the report.

The Place.

Lobelius is the first reporter, that the blew Flowerdeluce or first kinde of these broad leafed Flowerdeluces, groweth naturally in the West parts of England; but I am in some doubt of the truth of that report: for I rather thinke, that some in their trauels through Spaine, or other parts where it groweth, being delighted with the beauty of the flower, did gather the rootes, and bring them ouer with them, and dwelling in some of the West parts of England, planted them, and there encreasing so plentifully as they doe, they were imparted to many, thereby in time growing common in all Countrey folkes Gardens thereabouts. They grow also, and all the other, and many more varieties, about Tholouse, from whence Plantinianus Gassanus both sent and brought vs them, with many other bulbous rootes, and rare plants gathered thereabouts.

The

1 *Iris bulbosa latifolia prima Clusii.* Clusius his first great bulbous Flowerdeluce. 2 *Iris bulbosa maior cærulea sive Anglica.* The great blew or English bulbous Flowerdeluce. 3 *Iris bulbosa maior purpurea variegata.* The great purple stript bulbous Flowerdeluce. 4 *Iris bulbosa angustifolia maior alba.* The greater white narrow leafed bulbous Flowerdeluce. 5 *Iris bulbosa angustifolia versicolor.* The party coloured narrow leated bulbous Flowerdeluce. 6 *Iris bulbosa angustifolia Africana.* The purple African bulbous narrow leafed Flowerdeluce.

The Time.

Thefe doe flower vfually in the end of May, or beginning of Iune, and their feede is ripe in the end of Iuly or Auguft.

The Names.

Lobel calleth the firft Englifh blew Flowerdeluce, *Hyacinthus Poetarum flore Iridis, & propter Hyacinthinum colorem,id eft violaceum dictus:* but I know not any great good ground for it, more then the very colour; for it is neither of the forme of a Lilly, neither hath it thofe mourning markes imprinted in it, which the Poet faineth to bee in his Hyacinth. It is moft truely called an *Iris*, or Flowerdeluce (and there is great difference betweene a Lilly and a Flowerdeluce, for the formes of their flowers) becaufe it anfwereth thereunto very exactly, for the flower, and is therefore called vfually by moft, either *Iris bulbofa Anglica*, or *Iris bulbofa maior fiue latifolia*, for a difference betweene it, and the leffer with narrow leaues : In Englifh, eyther The great Englifh bulbous Flowerdeluce, or the great broad leafed bulbous Flowerdeluce, which you will, adding the other name, according to the colour.

And thus much for thefe broad leafed bulbous Flowerdeluces, fo much as hath come to our knowledge. Now to the feuerall varieties of the narrow leafed bulbous Flowerdeluces, fo much likewife as we haue been acquainted with.

Iris bulbofa minor fiue anguftifolia alba.
The fmaller white or narrow leafed bulbous Flowerdeluce.

This firft Flowerdeluce, which beareth the fmaller flower of the two white ones, that are here to bee defcribed, fpringeth out of the ground alwaies before Winter, which after breaketh forth into foure or fiue fmall and narrow leaues, a foote long or more, of a whitifh greene on the infide, which is hollow and chanalled, and of a blewifh greene colour on the outfide, and round withall : the ftalke of this kinde is longer and flenderer then the former, with fome fhorter leaues vpon it, at the toppe whereof, out of fhort skinny leaues, ftand one or two flowers, fmaller, fhorter, and rounder then the flowers of the former broad leafed Flowerdeluces, but made after the fame proportion with nine leaues, three falling downewards, with a yellow fpot in the middle, other three are made like a long arch, which couer the lower part next the ftalke of thofe falling leaues, and turne vp at the ends of them, where they are diuided into two parts : the other three ftand vpright, betweene each of the three falling leaues, being fomewhat long and narrow : the flower is wholly (fauing the yellow fpot) of a pure white colour, yet in fome hauing a fhew of fome blew throughout, and in others towards the bottome of the three vpright leaues : after the flowers are paft, there rife vp fo many long cods or feede veffels, as there were flowers, which are longer and fmaller then in the former, and a little bending like a Cornet, with three round fquares, and round pointed alfo, which diuiding it felfe when the feede is ripe into three parts, doe fhew fix feuerall cells or places, wherein is contained fuch like round reddifh yellow feedes, but fmaller then the former : the roote is fmaller and fhorter then the former, and without any haires or threeds, couered with browne thin skinnes, and more plentifull in giuing encreafe.

Iris bulbofa anguftifolia alba flore maiore.
The greater white narrow leafed bulbous Flowerdeluce.

I fhall not neede to make a feuerall defcription to euery one of thefe Flowerdeluces that follow, for that were but to make often repetition of one thing, which being once done, as it is, may well ferue to expreffe all the reft, and but onely to adde the efpe-

ciall

ciall differences, either in leafe or flower, for bigneffe, colour, or forme, as is expedient to expreffe and diftinguifh them feuerally. This greater white bulbous Flowerdeluce is like vnto the laft defcribed in all parts, fauing that it is a little larger and higher, both in leafe, ftalke, and flower, and much whiter then any of thefe mixed forts that follow, yet not fo white as the former : the roote hereof is likewife a little bigger and rounder in the middle.

Albefcent.
Milke white. There is another, whofe falling leaues haue a little fhew of yellowneffe in them, and fo are the middle ridges of the arched leaues, but the vpright leaues are more white, not differing in roote or leafe from the firft white.

Argentea.
Siluer colour. And another, whofe falls are of a yellowifh white, like the laft, the arched leaues are whiter, and the vpright leaues of a blewifh white, which we call a filuer colour.

Albida.
Whitifh. Another nath the fals yellowifh, and fometimes with a little edge of white about them, and fometimes without ; the vpright leaues are whitifh, as the arched leaues are, yet the ridge yellower.

Albida labris luteis.
White with yellow fals. Another hath his fals yellow, and the vpright leaues white, all thefe flowers are about the fame bigneffe with the firft.

Albida anguftior.
The narrow white. But we haue another, whofe flower is fmaller, and almoft as white as the fecond, the lower leaues are fmall, and doe as it were ftand outright, not hauing almoft any fal at all, fo that the yellow fpot feemeth to be the whole leafe, the arched leaues are not halfe fo large as in the former, and the vpright leaues bowe themfelues in the middle, fo that the tops doe as it were meete together.

And another of the fame, whofe falling leaues are a little more eminent and yellow, with a yellower fpot.

Aurea fiue lutea Hifpanica.
The Spanifh yellow. We haue another kinde that is called the Spanifh yellow, which rifeth not vp fo high, as ordinarily moft of the reft doe, and is wholly of a gold yellow colour.

Pallide lutea.
Straw colour. There is another, that vfually rifeth higher then the former yellow, and is wholly of a pale yellow, but deeper at the fpot.

Albida lutea.
Pale Straw colour. There is alfo another like vnto the pale yellow, but that the falling leaues are whiter then all the reft of the flower.

Mauritanica flaua ferotina minor.
The fmall Barbary yellow. There is a fmaller or dwarfe kinde, brought from the backe parts of Barbary, neare the Sea, like vnto the yellow, but fmaller and lower, and in ftead of vpright leaues, hath fmall fhort leaues like haires : it flowreth very late, after all others haue almoft giuen their feede.

Verficolor Hifpanica cærulea labris albis.
The party coloured Spanifh We haue another fort is called the party coloured Spanifh bulbous Flowerdeluce, whofe falling leaues are white, the arched leaues of a whitifh filuer colour, and the vpright leaues of a fine blewifh purple.

Diuerfitas.
The diuerfity or variation of this flower. Yet fometimes this doth vary ; for the falling leaues will haue either an edge of blew, circling the white leaues, the arched leaues being a little blewer, and the vpright leaues more purple.

Or the fals will be almoft wholly blew, edged with a blewer colour, the arched leaues pale blew, and the vpright leaues of a purplifh blew Violet colour.

Or the fals white, the arched leaues pale white, as the vpright leaues are. Or not of fo faire a blewifh purple, as the firft fort is.

Some of them alfo will haue larger flowers then others, and be more liberall in bearing flowers : for the firft fort, which is the moft ordinary, feldome beareth aboue one flower on a ftalke, yet fometimes two. And of the others there are fome that wil beare vfually two and three flowers, yet fome againe will beare but one. All thefe kindes fmell fweeter then many of the

Cærulea fiue purpurea minor Lufitanica præcox.
The fmall early purple Portugall. other, although the moft part be without fent.

There is another kinde, that is fmaller in all the parts thereof then the former, the ftalke is flender, and not fo high, bearing at the toppe one or two fmall flowers, all wholly of a faire blewifh purple, with a yellow fpot

in

in euery one of the three falling leaues, this vſually flowreth early, euen with the firſt bulbous Flowerdeluces.

Purpurea ma-ior.
The greater purple.

We haue another purple, whoſe flower is larger, and ſtalke higher, and is of a very reddiſh purple colour, a little aboue the ground, at the foote or bottome of the leaues and ſtalke : this flowreth with the later ſort of Flowerdeluces.

Purpurea ſerotina
The late purple.

There is another, whoſe flower is wholly purple, except the yellow ſpot, and flowreth later then any of the other purples.

Purpura rubeſcens labris cæruleis.
A reddiſh purple with blew fals.
Purpura rubeſcens labris albido cæruleis.
A reddiſh purple with whitiſh blew fals.
Purpurea labris luteis.
Party coloured purple & yellow
Purpurea labris ex albido cæruleo & luteo mixtis.
Party coloured purple with ſtript yellow fals.
Subpurpurea labris luteis.
Pale purple with yellow fals.
A paler purple-Subcærulea labris luteis.
Party coloured blew and yellow
Crinis coloris elegantioris.
A faire haire colour.
Altera obſoletior.
A dull haire colour.

There is yet another purple, whoſe vpright leaues are of a reddiſh purple, and the falling leaues of a blew colour.

And another of a reddiſh purple, whoſe falling leaues are of a whitiſh blew colour, in nothing elſe differing from the laſt.

Another hath his falling leaues of a faire gold yellow, without any ſtripe, yet in ſome there are veines running through the yellow leaues, and ſome haue an edge of a ſullen darke colour about them : the vpright leaues in euery of theſe, are of a Violet purple.

Another is altogether like this laſt, but that the falling leaues are of a pale blew and yellow, trauerſing one the other, and the arched leaues of a pale purpliſh colour.

Another hath his vpright leaues of a paler purple, and the falling leaues yellow.

And another little differing from it, but that the arched leaues are whitiſh.

Another whoſe vpright leaues are of a pale blew, and the falling leaues yellow.

And another of the ſame ſort, but of a little paler blew.

We haue another ſort, whoſe vpright leaues are of a faire browniſh yellow colour, which ſome call a *Fuille mort*, and others an haire colour ; the falling leaues yellow.

And another of the ſame colour, but ſomewhat deader.

Iris bulboſa Africana ſerpentaria caule.
The purple or murrey bulbous Barbary Flowerdeluce.

This Flowerdeluce as it is more ſtrange (that is, but lately knowne and poſſeſſed by a few) ſo it is both more deſired, and of more beauty then others. It is in all reſpects, of roote, leafe, and flower, for the forme like vnto the middle ſort of theſe Flowerdeluces, onely the loweſt part of the leaues and ſtalke, for an inch or thereabouts, next vnto the ground, are of a reddiſh colour, ſpotted with many ſpots, and the flower, being of a meane ſize, is of a deepe purpliſh red or murrey colour the whole flower throughout, except the yellow ſpot in the middle of the three lower or falling leaues, as is in all others.

Purpura cærulea obſoleta labris fuſcis.
The duskie party coloured purple.

And laſtly, there is another ſort, which is the greateſt of all theſe narrow leafed Flowerdeluces, in all the parts of it ; for the roote is greater then any of the other, being thicke and ſhort : the leaues are broader and longer, but of the ſame colour: the ſtalke is ſtronger and higher then any of them, bearing two or three flowers, larger alſo then any of the reſt, whoſe falling leaues are of a duskie yellow, and ſometimes with veines and borders about the brimmes, of another dunne colour, yet hauing that yellow ſpot that is in all : the arched leaues are of a ſullen pale purpliſh yellow, and the vpright leaues of a dull or duskie blewiſh purple colour : the heads or hornes for ſeede are likewiſe greater, and ſo is the ſeede alſo a little.

The Place.

Theſe Flowerdeluces haue had their originall out of Spaine and Portugall, as it is thought, except thoſe that haue riſen by the ſowing, and thoſe which are named of Africa.

The

The Time.

Thefe flower in Iune, and fometimes abide vnto Iuly, but vfually not fo early as the former broad leafed kindes, and are foone fpoiled with wet in their flowring.

The Names.

The feuerall names, both in Latine and Englifh, are fufficient for them as they are fet downe; for we know no better.

The Vertues.

There is not any thing extant or to be heard, that any of thefe kindes of Flowerdeluces hath been vfed to any Phyficall purpofes, and ferue onely to decke vp the Gardens of the curious.

And thus much for thefe forts of bulbous Flowerdeluces, and yet I doubt not, but that there are many differences, which haue rifen by the fowing of the feede, as many may obferue from their owne labours, for that euery yeare doth fhew forth fome variety that is not feene before. And now I will conuert my difcourfe a while likewife, to paffe through the feuerall rankes of the other kindes of tuberous rooted Flowerdeluces, called Flagges.

Chap. XX.

Iris latifolia tuberofa. The Flagge or Flowerdeluce.

THere are two principall kindes of tuberous or knobby rooted Flowerdeluces, that is, the tall and the dwarfe, or the greater and the leffer; the former called *Iris maior* or *latifolia*, and the other *Iris minor*, or rather *Chamæiris*; and each of thefe haue their leffer or narrow leafed kindes to bee comprehended vnder them: Of all which in their order. And firft of that Flowerdeluce, which for his excellent beautie and raritie, deferueth the firft place.

Iris Chalcedonica fiue Sufiana maior. The great Turkie Flowerdeluce.

The great Turkie Flowerdeluce, hath diuers heads of long and broad frefh greene leaues, yet not fo broad as many other of thofe that follow, one folded within another at the bottome, as all other of thefe Flowerdeluces are: from the middle of fome one of thofe heads (for euery head of leaues beareth not a flower) rifeth vp a round ftiffe ftalke, two foote high, at the toppe whereof ftandeth one flower (for I neuer obferued it to beare two) the largeft almoft, but rareft of all the reft, confifting of nine leaues, like the others that follow, but of the colour almoft of a Snakes skinne, it is fo diuerfly fpotted; for the three lower falling leaues are very large, of a deepe or darke purple colour, almoft blacke, full of grayifh fpots, ftrakes, and lines through the whole leaues, with a blacke thrume or freeze in the middle of each of them: the three arched leaues that couer them, are of the fame darke purple colour, yet a little paler at the fides, the three vpper leaues are very large alfo, and of the fame colour with the lower leaues, but a little more liuely and frefh, being fpeckled and ftraked with whiter fpots and lines; which leaues being laid in water, will colour the water into a Violet colour, but if a little Allome be put therein, and then wrung or preffed, and the iuice of thefe leaues dryed in the fhadow, will giue a colour almoft as deepe as Indico, and may ferue for fhadowes in limming excellent well: the flower hath no fent that can be perceiued, but is onely commendable for the beauty and rarity thereof: it feldome beareth feedes in thefe cold Countries, but when it doth, it is contained in great heads,

being

being brownish and round, but not so flat as in other sorts, the roots are more browne on the outside, and growing tuberous thicke, as all other that are kept in Gardens.

Iris Chalcedonica siue Susiana minor. The lesser Turkie Flowerdeluce.

There is another hereof little differing, but that the leafe is of a more yellowish greene colour, and the flower neither so large or faire, nor of so perspicuous markes and spots, nor the colour of that liuely (though darke) lustre.

The Place.

These haue been sent out of Turkie diuers times among other things, and it should seeme, that they haue had their originall from about Susis, a chiefe Citie of Persia.

The Time.

They flower in May most vsually, before any of the other kindes.

The Names.

They haue been sent vnto vs, and vnto diuers other in other parts, from Constantinople vnder the name of *Alaia Susiana*, and thereupon it hath been called, both of them and vs, either *Iris Chalcedonica*, or *Susiana*, and for distinction *maior* or *minor* : In English, The Turkie Flowerdeluce, or the Ginnie Hen Flowerdeluce, the greater or the lesser.

Iris alba Florentina. The white Flowerdeluce.

The great white Flowerdeluce, hath many heads of very broad and flat long leaues, enclosing or folding one within another at the bottome, and after a little diuided one from another toward the top, thin edged, like a sword on both sides, and thicker in the middle: from the middle of some of these heads of leaues, riseth vp a round stiffe stalk, two or three foot high, bearing at the top one, two, or three large flowers, out of seuerall huskes or skins, consisting of nine leaues, as all the other do, of a faire white colour, hauing in the middle of each of the three falling leaues, a small long yellow frize or thrume, as is most vsuall in all the sorts of the following Flowerdeluces, both of the greater and smaller kindes : after the flowers are past, come the seed, inclosed in thicke short pods, full fraught or stored with red roundish and flat seede, lying close one vpon another : the roote is tuberous or knobby, shooting out from euery side such like tuberous heads, lying for the most part vpon or aboue the ground, and fastened within the ground with long white strings or fibres, which hold them strongly, and encreaseth *Flore pallido.* fast. There is another like vnto this last in all things, sauing that the colour of the flower is of a more yellowish white, which we vsually call a Straw colour.

Iris alba maior Versicolor. The white party coloured Flowerdeluce.

This variable Flowerdeluce is like vnto the former, but that the leaues are not so large and broad, the flower hereof is as large almost, and as white as the former, but it hath a faire list or line of a blewish purple downe the backe of euery one of the three vpright leaues, and likewise round about the edges, both of the vpper and lower leaues, and also a little more purplish vpon the ridge of the arched leaues, that couer the falling leaues : the roote hereof is not so great as of the former white, but a little slenderer and browner.

Iris Dalmatica maior. The great Dalmatian Flowerdeluce.

This greater Flowerdeluce of Dalmatia, hath his leaues as large and broad as any of the Flowerdeluces whatsoeuer, his stalke and flower doe equall his other proportion,

tion, onely the colour of the flower is differing, being of a faire watchet or bleake blew colour wholly, with the yellow frize or thrum downe the middle of the lower or falling leaues, as before is said to be common to all these sorts of Flowerdeluces; in all other parts it little differeth, sauing onely this is obserued to haue a small shew of a purplish red about the bottome of the greene leaues.

Iris purpurea siue vulgaris. The common purple Flowerdeluce.

This Flowerdeluce, which is most common in Gardens, differeth nothing at all from those that are formerly described, either in roote, leafe, or flower for the forme of them, but onely that the leaues of this are not so large as the last, and the flower it selfe is of a deep purple or Violet colour, and sometimes a little declining to rednesse, especially in some places.

Sometimes this kinde of Flowerdeluce will haue flowers of a paler purple colour, *Purpurea pallidior versicolor.* comming neare vnto a blew, and sometimes it will haue veines or stripes of a deeper blew, or purple, or ash colour, running through all the vpper and lower leaues.

There is another like vnto this, but more purple in the fals, and more pale in the *Cerulea labris purpureis.* vpright leaues.

Iris Asiatica cerulea. The blew Flowerdeluce of Asia.

This Flowerdeluce of Asia, is in largenesse of leaues like vnto the Dalmatian, but beareth more store of flowers on seuerall branches, which are of a deeper blew colour, and the arched leaues whitish on the side, and purplish on the ridges, but in other things like vnto it.

There is another neare vnto this, but that his leaues are a little narrower, and his *Purpurea* flowers a little more purple, especially the vpper leaues.

Iris Damascena. The Flowerdeluce of Damasco.

This is likewise altogether like the Flowerdeluce of Asia, but that it hath some white veines in the vpright leaues.

Iris Lusitanica biflora. The Portugall Flowerdeluce.

This Portugall Flowerdeluce is very like the common purple Flowerdeluce, but that this is not so large in leaues, or flowers, and that it doth often flower twice in a yeare, that is, both in the Spring, and in the Autumne againe, and besides, the flowers haue a better or sweeter sent, but of the like purple or Violet colour as it is, and comming forth out of purplish skins or huskes.

Iris Camerarij siue purpurea versicolor maior. The greater variable coloured purple Flowerdeluce.

The greater of the variable purple Flowerdeluces, hath very broad leaues, like vnto the leaues of the common purple Flowerdeluce, and so is the flower also, but differing in colour, for the three lower leaues are of a deepe purple colour tending to rednesse, the three arched leaues are of the colour with the vpper leaues, which are of a pale or bleake colour tending to yellownesse, shadowed ouer with a smoakie purplish colour, except the ridges of the arched leaues, which are of a more liuely purple colour.

Iris purpurea versicolor minor. The lesser variable purple Flowerdeluce.

This Flowerdeluce differeth not in any thing from the last, but onely that it hath narrower greene leaues, and smaller and narrower flowers, else if they be both conferred together, the colours will not seeme to varie the one from the other any whit at all.

There is another somewhat neare vnto these two last kindes, whose huskes from *Altera minus fuliginea.* whence

Q

whence the flowers doe ſhoote forth, haue purple veines in them, and ſo haue the falling purpliſh leaues, and the three vpright leaues are not ſo ſmoakie, yet of a dun purple colour.

Iris cærulea verſicolor. The blew party coloured Flowerdeluce.

This party coloured Flowerdeluce hath his leaues of the ſame largeneſſe, with the
leſſer variable purple Flowerdeluce laſt deſcribed, and his flowers diuerſly marked :
for ſome haue the fals blew at the edges, and whitiſh at the bottome, the arched leaues
of a yellowiſh white, and the vpright leaues of a whitiſh blew, with yellowiſh edges.
Some againe are of a darker blew, with browniſh ſpots in them. And ſome are ſo pale
a blew, that we may well call it an aſh-colour: And laſtly, there is another of this ſort,
whoſe vpright leaues are of a faire pale blew, with yellowiſh edges, and the falling
leaues parted into two colours, ſometimes equally in the halfe, each ſide ſutable to the
other in colour : And ſometimes hauing the one leafe in that manner : And ſometimes but with a diuers coloured liſt in them ; in the other parts both of flower and
leafe, like vnto the other.

Iris lutea variegata. The yellow variable Flowerdeluce.

This yellow variable Flowerdeluce loſeth his leaues in Winter, contrary to all the
former Flowerdeluces, ſo that his roote remaineth vnder ground without any ſhew of
leafe vpon it : but in the beginning of the Spring it ſhooteth out faire broad leaues,
falling downwards at the points or ends, but ſhorter many times then any of the former, and ſo is the ſtalke likewiſe, not riſing much aboue a foote high, whereon are ſet
two or three large flowers, whoſe falling leaues are of a reddiſh purple colour, the three
that ſtand vpright of a ſmoakie yellow, the arched leaues hauing their ridges of a
bleake colour tending to purple, the ſides being of the former ſmoakie yellow colour,
with ſome purpliſh veines at the foote or bottome of all the leaues : the roote groweth
ſomewhat more ſlender and long vnder ground, and of a darker colour then manie of
the other.

Varietas.　　Another ſort hath the vpright leaues of a reaſonable faire yellow, and ſtand more
vpright, not bowing downe as moſt of the other, and the purple fals haue pale edges.
Some haue their greene leaues party coloured, white and greene, more or leſſe, and ſo
are the huskes of the flowers, the arched leaues yellow, as the vpright leaues are, with
purpliſh veines at the bottome. And ſome haue both the arched and vpright leaues of
ſo pale a yellow, that we may almoſt call it a ſtraw colour, but yellower at the bottome, with purple veines, and the falling leaues purple, with two purple ſpots in them.

And theſe are the ſorts of the greater tuberous or Flagge Flowerdeluces that haue
come to our knowledge : the next hereunto are the leſſer or narrow leafed kindes to be
deſcribed ; and firſt of the greateſt of them.

1. *Iris anguſtifolia Tripolitana aurea.* The yellow Flowerdeluce of Tripoly.

This Flowerdeluce I place in the forefront of the narrow leafed Flowerdeluces,
for the length of the leaues, compared with the breadth of them ; it may fitly bee called a narrow leafed Flowerdeluce, although they be an inch broad, which is broader
then any of them that follow, or ſome of thoſe are ſet downe before, but as I ſaid, the
length make them ſeem narrow, and therfore let it take vp his roome in this place, with
the deſcription that followeth. It beareth leaues a yard long, or not much leſſe, and
an inch broad, as is ſaid before, or more, of a ſad greene colour, but not ſhining : the
ſtalke riſeth vp to be foure or fiue foote high, being ſtrong and round, but not very
great, bearing at the toppe two or three long and narrow gold yellow flowers, of the
faſhion of the bulbous Flowerdeluces, as the next to bee deſcribed is, without any
mixture or variation therein : the heads for ſeede are three ſquare, containing within
them many flat cornered ſeedes : the roote is long and blackiſh, like vnto the reſt that
follow, but greater and fuller.

1 *Iris Chalcedonica siue Susiana maior.* The great Turkie Flowerdeluce. 2 *Iris alba Florentina.* The white Flowerdeluce. 3 *Iris lati folia variegata.* The variable Flowerdeluce. 4 *Chamæiris latifolia maior.* The greater dwarfe Flowerdeluce.

2. *Iris angustifolia maior carulea.*
The greater blew Flowerdeluce with narrow leaues.

This kinde of Flowerdeluce hath his leaues very long and narrow, of a whitish greene colour, but neither so long or broad as the last, yet broader, thicker and stiffer then any of the rest with narrow leaues that follow : the stalke riseth sometimes no higher then the leaues, and sometimes a little higher, bearing diuers flowers at the top, successiuely flowring one after another, and are like vnto the flowers of the bulbous Flowerdeluces, but of a light blew colour, and sometimes deeper : after the flowers are past, rise vp six cornered heads, which open into three parts, wherein is contained browne seede, almost round : the roote is small, blackish and hard, spreading into many long heads, and more closely growing or matting together.

3. *Iris angustifolia purpurea marina.* The purple narrow leafed Sea Flowerdeluce.

This Sea Flowerdeluce hath many narrow hard leaues as long as the former, and of a darke greene colour, which doe smell a little strong : the stalke beareth two or three flowers like the former, but somewhat lesse, and of a darke purple or Violet colour : in seede and roote it is like the former.

4. *Iris angustifolia purpurea versicolor.*
The variable purple narrow leafed Flowerdeluce.

The leaues of this Flowerdeluce are very like the former Sea Flowerdeluce, and do a little stinke like them ; the flowers are differing, in that the vpper leaues are wholly purple or violet, and the lower leaues haue white veines, and purple running one among another : the seede and rootes differ not from the former purple Sea kinde.

5. *Iris angustifolia minor Pannonica siue versicolor Clusij.*
The small variable Hungarian Flowerdeluce of Clusius.

This Hungarian Flowerdeluce (first found out by Clusius, by him described, and of him tooke the name) riseth vp with diuers small tufts of leaues, very long, narrow, and greene, growing thicke together, especially if it abide any time in a place ; among which riseth vp many long round stalkes, higher then the leaues, bearing two or three, or foure small flowers, one aboue another, like the former, but smaller and of greater beauty : for the lower leaues are variably striped with white and purple, without any thrume or fringe at all; the vpper leaues are of a blewish fine purple or Violet colour, & so are the arched leaues, yet hauing the edges a little paler : the heads for seede are smaller, and not so cornered as the other, containing seedes much like the former, but smaller : the roote is blacke and small, growing thicker and closer together then any other, and strongly fastened in the ground, with a number of hard stringie rootes : the flowers are of a reasonable good sent.

6. *Iris angustifolia maior flore duplici.* The greater double blew Flowerdeluce.

This Flowerdeluce, differeth not either in roote or leafe from the first great blew Flowerdeluce of Clusius, but onely in that the leaues grow thicker together, and that the flowers of this kinde are as it were double with many leaues confusedly set together, without any distinct parts of a Flowerdeluce, and of a faire blew colour with many white veines and lines running in the leaues ; yet oftentimes the stalke of flowers hath but two or three small flowers distinctly set together, rising as it were out of one huske.

7. *Iris angustifolia minor alba Clusij.*
The small white Flowerdeluce of Hungary.

This likewise differeth little from the former Hungarian Flowerdeluce of Clusius,
but

1 *Iris angustifolia Tripolitana.* The yellow Flowerdeluce of Tripoli. 2 *Iris angustifolia maior carulea.* The greater blew Flowerdeluce with narrow
leaues. 3 *Iris angustifolia minor Pannonica siue versicolor Clusii.* The small variable Hungarian Flowerdeluce of Clusius. 4 *Iris angustifolia maior flore duplici.*
The greater double blew Flowerdeluce. 5 *Chamæiris angustifolia minor.* The lesser Grasse Flowerdeluce. 6 *Iris tuberosa.* The veluet Flowerdeluce.

but that the leafe is of a little paler greene colour, and the flower is of a faire whitifh colour, with fome purple at the bottome of the leaues.

Next after thefe narrow leafed Flowerdeluces, are the greater and fmaller forts of dwarfe kindes to follow ; and laftly, the narrow or graffe leafed dwarfe kindes, which will finifh this Chapter of Flowerdeluces.

1. *Chamæiris latifolia maior alba.* The greater white dwarfe Flowerdeluce.

This dwarfe Flowerdeluce hath his leaues as broad as fome of the leffer kindes laft mentioned, but not fhorter ; the ftalke is very fhort, not aboue halfe a foote high or thereabouts, bearing moft commonly but one flower, feldome two, which are in fome of a pure white, in others paler, or fomewhat yellowifh through the whole flower, except the yellow frize or thrume in the middle of euery one of the falling leaues : after the flowers are paft, come forth great heads, containing within them round pale feed: the roote is fmall, according to the proportion of the plant aboue ground, but made after the fafhion of the greater kindes, with tuberous peeces fpreading from the fides, and ftrong fibres or ftrings, whereby they are faftened in the ground.

2. *Chamæiris latifolia maior purpurea.* The greater purple dwarfe Flowerdeluce.

There is no difference either in roote, leafe, or forme of flower in this from the former dwarfe kinde, but onely in the colour of the flower, which in fome is of a very deepe or blacke Violet purple, both the toppes and the fals : in others the Violet purple is more liuely, and in fome the vpper leaues are blew, and the lower leaues purple, yet all of them haue that yellow frize or thrume in the middle of the falling leaues, that the other kindes haue.

There is another that beareth purple flowers, that might be reckoned, for the fmalneffe and fhortneffe of his ftalke, to the next kinde, but that the flowers and leaues of this are as large as any of the former kindes of the fmaller Flowerdeluces.

3. *Chamæiris latifolia minor alba.* The leffer white dwarfe Flowerdeluce.

There is alfo another fort of thefe Flowerdeluces, whofe leaues and flowers are leffe, and wherein there is much variety. The leaues of this kinde, are all for the moft part fomewhat fmaller, narrower, and fhorter then the former : the ftalke with the flower vpon it fcarce rifeth aboue the leaues, fo that in moft of them it may be rather called a foote-ftalke, fuch as the Saffron flowers haue, and are therefore called of manie ἀκαυλοι, without ftalkes ; the flowers are like vnto the firft defcribed of the dwarfe kindes, and of a whitifh colour, with a few purplifh lines at the bottome of the vpper leaues, and a lift of greene in the falling leaues.

Another hath the flowers of a pale yellow, called a Straw colour, with whitifh ftripes and veines in the fals, and purplifh lines at the bottome of the vpper leaues.

4. *Chamæiris latifolia minor purpurea.* The leffer purple dwarfe Flowerdeluce.

The difference of this from the former, confifteth more in the colour then forme of the flower, which is of a deep Violet purple, fometimes paler, and fometimes fo deep, that it almoft feemeth blacke : And fometimes the fals purplifh, and the vpper leaues blew. Some of thefe haue a fweete fent, and fome none.

There is another of a fine pale or delayed blew colour throughout the whole flower.

5. *Chamæiris latifolia minor fuauerubens.* The leffer blufh coloured dwarfe Flowerdeluce.

This Flowerdeluce hath the falling leaues of the flower of a reddifh colour, and the thrumes blew : the vpper and arched leaues of a fine pale red or flefh colour, called a blufh colour ; in all other things it differeth not, and fmelleth little or nothing at all.

6. *Chamæiris*

6. *Chamæiris latifolia minor lutea verficolor.*
The leffer yellow variable dwarfe Flowerdeluce.

The falling leaues of this Flowerdeluce are yellowifh, with purple lines from the middle downewards, fometimes of a deeper, and fometimes of a paler colour, and white thrumes in the middle, the vpper leaues are likewife of a yellowifh colour, with purple lines in them : And fometimes the yellow colour is paler, and the lines both in the vpper and lower leaues of a dull or dead purple colour.

3. *Chamæiris latifolia minor cærulea verficolor.*
The leffer blew variable dwarfe Flowerdeluce.

The vpper leaues of this flower are of a blewifh yellow colour, fpotted with purple in the broad part, and at the bottome very narrow : the falling leaues are fpread ouer with pale purplifh lines, and a fmall fhew of blew about the brimmes : the thrume is yellow at the bottome, and blewifh aboue: the arched leaues are of a blewifh white, being a little deeper on the ridge.
And fometimes the vpper leaues are of a paler blew rather whitifh, with the yellow: both thefe haue no fent at all.

8. *Chamæiris marina purpurea.* The purple dwarfe Sea Flowerdeluce.

This fmall Flowerdeluce is like vnto the narrow leafed Sea Flowerdeluce before defcribed, both in roote, leafe, and flower, hauing no other difference, but in the fmalneffe and lowneffe of the growing, being of the fame purple colour with it.

9. *Chamæiris anguftifolia maior.* The greater Graffe Flowerdeluce.

This Graffe Flowerdeluce hath many long and narrow darke greene leaues, not fo ftiffe as the former, but lither, and bending their ends downe againe, among which rife vp diuers ftalkes, bearing at the toppe two or three fweete flowers, as fmall as any of them fet downe before, of a reddifh purple colour, with whitifh yellow and purple ftrakes downe the middle of the falling leaues : the arched leaues are of a horfe flefh colour all along the edges, and purple vpon the ridges and tips that turne vp againe : vnder thefe appeare three browne aglets, like vnto birds tongues : the three vpper leaues are fmall and narrow, of a perfect purple or Violet colour : the heads for feede haue fharper and harder cornered edges then the former : the feedes are fomewhat grayifh like the former, and fo are the rootes, being fmall, blacke, and hard, growing thicke together, faftened in the ground with fmall blackifh hard ftrings, which hardly fhoote againe if the roote be remoued.

10. *Chamæiris anguftifolia minor.* The leffer Graffe Flowerdeluce.

This Flowerdeluce is in leaues, flowers, and rootes fo like the laft defcribed, that but onely it is fmaller and lower, it is not to be diftinguifhed from the other. And this may fuffice for thefe forts of Flowerdeluces, that furnifh the Gardens of the curious louers of thefe varieties of nature, fo farre forth as hath paffed vnder our knowledge. There are fome other that may be referred hereunto, but they belong to another hiftory; and therefore I make no mention of them in this place.

The Place.

The places of moft of thefe are fet downe in their feuerall titles; for fome are out of Turkie, others out of Hungaria, Dalmatia, Illyria, &c. as their names doe import. Thofe that grow by the Sea, are found in Spaine and France.

The

The Time.

Some of these do flower in Aprill, some in May, and some not vntill Iune.

The Names.

The names expressed are the fittest agreeing vnto them, and therefore it is needlesse againe to repeate them. Many of the rootes of the former or greater kindes, being dryed are sweete, yet some more then other, and some haue no sent at all: but aboue all the rest, that with the white flower, called of Florence, is accounted of all to be the sweetest root, fit to be vsed to make sweete powders, &c. calling it by the name of *Orris* rootes.

Iris tuberosa. The Veluet Flowerdeluce.

Vnto the Family of Flowerdeluces, I must needes ioyne this peculiar kinde, because of the neare resemblance of the flower, although it differ both in roote and leafe; lest therefore it should haue no place, let it take vp a roome here in the end of the Flower-deluces, with this description following. It hath many small and foure square leaues, two foote long and aboue sometimes, of a grayish greene colour, stiffe at the first, but afterwards growing to their full length, they are weak and bend downe to the ground: out of the middle, as it were of one of these leaues, breaketh out the stalke, a foot high and better, with some leaues thereon, at the toppe whereof, out of a huske riseth one flower, (I neuer saw more on a stalke) consisting of nine leaues, whereof the three that fall downe are of a yellowish greene colour round about the edges, and in the middle of so deepe a purple, that it seemeth to be blacke, resembling blacke Veluet: the three arched leaues, that couer the lower leaues to the halfe, are of the same greenish colour that the edges and backside of the lower leaues are: the three vppermost leaues, if they may be called leaues, or rather short peeces like eares, are green also, but wherein a glimpse of purple may be seene in them: after the flower is past, there followeth a round knob or whitish seede vessell, hanging downe by a small foote-stalke, from be-tweene the huske, which is diuided as it were into two leaues, wherein is contained round white seede. The roote is bunched or knobbed out into long round rootes, like vnto fingers, two or three from one peece, one distant from another, and one longer then another, for the most part of a darkish gray colour, and reddish withall on the outside, and somewhat yellowish within.

The Place.

It hath beene sent out of Turkie oftentimes (as growing naturally there-abouts) and not knowne to grow naturally any where else.

The Time.

It flowreth in Aprill or May, sometimes earlier or later, as the Spring falleth out to be milde or sharpe.

The Names.

Matthiolus contendeth to make it the true *Hermodactylus*, rather from the shew of the rootes, which (as is said) are like vnto fingers, then from any other good reason: for the rootes hereof eyther dry or greene, do nothing resemble the true *Hermodactyli* that are vsed in Physicke, as any that know-eth them may easily perceiue, either in forme or vertue. It is more truely referred to the Flowerdeluces, and because of the tuberous rootes, called *Iris tuberosa*, although all the Flowerdeluces in this Chapter haue tuberous rootes,

rootes, yet this much differing from them all. In Engliſh it is vſually called, The Veluet Flowerdeluce, becauſe the three falling leaues ſeeme to be like ſmooth blacke Veluet.

The Vertues.

Both the rootes and the flowers of the great Flowerdeluces, are of great vſe for the purging and cleanſing of many inward, as well as outward diſeaſes, as all Authors in Phyſicke doe record. Some haue vſed alſo the greene rootes to cleanſe the skinne, but they had neede to be carefull that vſe them, leſt they take more harme then good by the vſe of them. The dryed rootes called *Orris* (as is ſaid) is of much vſe to make ſweete powders, or other things to perfume apparrell or linnen. The iuice or decoction of the green roots doth procure both neezing to be ſnuft vp into the noſtrils, and vomiting very ſtrongly being taken inwardly.

Chap. XXI.

Gladiolus. Corne Flagge.

NExt vnto the Flagges or Flowerdeluces, come the *Gladioli* or Corne Flagges to bee entreated of, for ſome reſemblance of the leaues with them. There are hereof diuers ſorts, ſome bigger and ſome leſſer, but the chiefeſt difference is in the colour of the flowers, and one in the order of the flowers. Of them all in their ſeuerall orders.

Gladiolus Narbonenſis. The French Corne Flagge.

The French Corne Flagge riſeth vp with three or foure broad, long, and ſtiffe greene leaues, one as it were out of the ſide of another, being ioyned together at the bottome, ſomewhat like vnto the leaues of Flowerdeluces, but ſtiffer, more full of ribbes, and longer then many of them, and ſharper pointed: the ſtalke riſeth vp from among the leaues, bearing them on it as it riſeth, hauing at the toppe diuers huskes, out of which come the flowers one aboue another, all of them turning and opening themſelues one way, which are long and gaping, like vnto the flowers of Foxegloue, a little arched or bunching vp in the middle, of a faire reddiſh purple colour, with two white ſpots within the mouth thereof, one on each ſide, made like vnto a Lozenge that is ſquare and long pointed: after the flowers are paſt, come vp round heads or ſeede veſſels, wherein is contained reddiſh flat ſeede, like vnto the ſeede of the Fritillaria, but thicker and fuller: the roote is ſomewhat great, round, flat, and hard, with a ſhew as if it were netted, hauing another ſhort ſpongie one vnder it, which when it hath done bearing, and the ſtalke dry, that the roote may be taken vp, ſticketh cloſe to the bottome, but may be eaſily taken away, hauing vſually a number of ſmall rootes encreaſed about it, the leaſt whereof will quickly grow, ſo that if it be ſuffered any long time in a Garden, it will rather choake and peſter it, then be an ornament vnto it.

Gladiolus Italicus binis floribus ordinibus. The Italian Corne Flagge.

The Italian Corne Flagge is like vnto the French in roote, leafe, and flower, without any other difference, then that the roote is ſmaller and browner, the leafe and ſtalke of a darker colour, and the flowers (being of a little darker colour like the former, and ſomewhat ſmaller) ſtand out on both ſides of the ſtalke.

Gladiolus Byzantinus. Corne Flagge of Conſtantinople.

This Corne Flagge that came firſt from Conſtantinople, is in all things like vnto the French Corne Flagge laſt deſcribed, but that it is larger, both in rootes, leaues, and flowers,

flowers, and likewiſe that the Flowers of this, which ſtand not on both ſides, are of a deeper red colour, and flower later, after all the reſt are paſt : the roote hereof being netted as plainly as any of the former, is as plentifull alſo to giue encreaſe, but is more tender and leſſe able to abide our ſharpe cold Winters.

Gladiolus flore rubente. Bluſh Corne Flagge.

This bluſh kinde is like vnto the French Corne Flagge in all reſpects , ſauing onely that the flowers are of a pale red colour, tending to whiteneſſe, which wee vſually call a bluſh colour.

Gladiolus flore albo. White Corne Flagge.

This white Corne Flagge alſo differeth not from the laſt, but onely that the rootes are whiter on the outſide, the leaues are greener, without any brownneſſe or darkneſſe as in the former, and the flowers are ſnow white.

Gladiolus purpureus minor. The ſmall purple Corne Flagge.

This alſo differeth not from any of the former, but onely in the ſmalneſſe both of leafe, ſtalke, and flowers, which ſtand all on the one ſide, like vnto the French kinde, and of the ſame colour : the roote of this kinde is netted more then any other.

The Place.

They grow in France and Italy, the leaſt in Spaine, and the Byzantine, as it is thought, about Conſtantinople, being (as is ſaid) firſt ſent from thence. Iohn Tradeſcante aſſured mee, that hee ſaw many acres of ground in Barbary ſpread ouer with them.

The Time.

They all flower in Iune and Iuly, and the Byzantine lateſt, as is ſaid before.

The Names.

It hath diuers names; for the Latines call it *Gladiolus*, of the forme of a ſword, which the leafe doth reſemble. The Romanes *Segetalis*, becauſe it groweth in the Corne fields. Some call it *Victorialis rotunda*, to put a difference between it, and the *longa*, which is a kinde of Garlicke. Plinie ſaith, that *Gladiolus* is *Cypirus*, but to decide that controuerſie, and many others, belongeth to another diſcourſe, this being intended only for pleaſure. Gerrard miſtaketh the French kinde for the Italian.

The Vertues.

The roote being bruiſed, and applyed with Frankinſenſe (and often of it ſelfe without it) in the manner of a pultis or plaiſter, is held of diuers to be ſingular good to draw out ſplinters , thornes, and broken bones out of the fleſh. Some take it to be effectuall to ſtirre vp Venerie, but I ſomewhat doubt thereof : For Galen in his eighth Booke of Simples, giueth vnto it a drawing, digeſting, and drying faculty.

CHAP.

1 *Gladiolus Narbonensis* The French Corne Flagge. 2 *Gladiolus Italicus.* The Italian Corne Flagge. 3 *Gladiolus Byzantinus.* Corne Flagge of Constantinople. 4 *Palma Christi mas* The great male handed Satyrion. 5 *Orchis Hermaphroditica candida.* The white Butterflie Orchis. 6 *Orchis Melittias siue apifera.* The Bee flower or Bee Orchis. 7 *Dens Caninus flore purpurante.* Dogges tooth Violet with a pale purplish flower. 8 *Dens Caninus flore albo.* Dogges tooth Violet with a white flower.

Chap. XXII.

Orchis siue Satyrium. Bee flowers.

ALthough it is not my purpose in this place, to giue a generall history of all the sorrs of Orchides, Satyrions, and the rest of that kinde; yet becaufe many of them are very pleasant to behold, and, if they be planted in a conuenient place, will abide some time in Gardens, so that there is much pleasure taken in them: I shall intrude some of them for curiosities sake, to make vp the prospect of natures beautifull variety, and only entreate of a few, leauing the rest to a more ample declaration.

1. *Satyrium Basilicum siue Palma Christi mas.*
The greater male handed Satyrion.

This handed Satyrion hath for the most part but three faire large greene leaues, neare vnto the ground, spotted with small blackish markes: from among which riseth vp a stalke, with some smaller leaues thereon, bearing at the toppe a bush or spike of flowers, thicke set together, euery one whereof is made like a body, with the belly broader belowe then aboue, where it hath small peeces adioyned vnto it: the flower is of a faire purple colour, spotted with deeper purple spots, and hauing small peeces like hornes hanging at the backes of the flowers, and a small leafe at the bottome of the foote-stalke of euery flower: the rootes are not round, like the other Orchides, but somewhat long and flat, like a hand, with small diuisions belowe, hanging downe like the fingers of a hand, cut short off by the knockles, two alwayes growing together, with some small fibres or strings aboue the heads of these rootes, at the bottome of the stalke.

2. *Satyrium Basilicum siue Palma Christi femina.*
The female handed Satyrion.

This female Satyrion hath longer and narrower leaues then the former, and spotted with more and greater spots, compassing the stalke at the bottome like the other: this beareth likewise a bush of flowers, like vnto the other, but that each of these haue heads like hoods, whereas the former haue none: in some they are white with purple spots, and in others of a reddish purple, with deep or darke coloured spots: the roots are alike.

3. *Orchis Hermaphroditica candida.* The white Butterflie Orchis.

The rootes of this kinde take part with both the sorts of *Orchis* and *Satyrium*, being neither altogether round, nor fully handed, and thereupon it tooke the name, to signifie both kindes: the leaues are two in number, seldome more, being faire and broad, like vnto the leaues of Lillies, without any spot at all in them: at the toppe of the stalke stand many white flowers, not so thicke set as the first or second, euery one being fashioned like vnto a white Butterflie, with the wings spread abroad.

4. *Orchis Melitias siue apifera.* The Bee flower or Bee Orchis.

This is a small and lowe plant for the most part, with three or foure small narrow leaues at the bottome: the stalke is seldome aboue halfe a foote high, with foure or fiue flowers thereon one aboue another, hauing round bodies, and somewhat flat, of a kind of yellowish colour, with purple wings aboue them, so like vnto an honey Bee, that it might soone deceiue one that neuer had seene such a flower before: the roots are two together, round and white, hauing a certaine *muccilaginesse* or clamminesse within them, without any taste almost at all, as all or the most part of these kindes haue.

5. *Orchis Sphegodes.* Gnats Satyrion.

The leaues of this Orchis are somewhat larger then of the Bee flower, the stalke also
somewhat

somewhat higher : the flowers are fewer on the toppe, but somewhat larger then of the Bee flowers, made to the resemblance of a Gnat or great long Flie : the rootes are two round bulbes, as the other are.

6. *Orchis Myodes.* Flie Orchis.

The Flie Orchis is like vnto the last described, both in leafe and roote, the diffe-rence is in the flower, which is neither so long as the Gnat Satyrion, nor so great as the Bee Orchis, but the neather part of the Flie is blacke, with a list of ash-colour crossing the backe, with a shew of legges hanging at it : the naturall Flie seemeth so to bee in loue with it, that you shall seldome come in the heate of the day, but you shall finde one sitting close thereon.

The Place.

These grow in many places of England, some in the Woods, as the But-terflie, and the two former handed Satyrions : others on dry bankes and barren balkes in Kent, and many other places.

The Time.

They flower for the most part in the beginning or middle of May, or thereabouts.

The Names.

Their seuerall names are expressed in their titles, so much as may suffice for this discourse.

The Vertues.

All the kindes of Orchis are accounted to procure bodily lust, as well the flowers distilled, as the rootes prepared.

The rootes boyled in red Wine, and afterwards dryed, are held to bee a singular good remedie against the bloody Flixe.

Chap. XXIII.

Dens Caninus. Dogs tooth Violet.

VNto the kindes of Orchides, may fitly be ioyned another plant, which by many is reckoned to be a *Satyrium*, both from the forme of roote and leafe, and from the efficacy or vertue correspondent thereunto. And although it cannot be the *Satyrium Erythronium* of Dioscorides, as some would entitle it, for that as I haue shewed before, his *Satyrium tryphillum* is the Tulipa without all doubt ; yet becauſe it differeth very notably, and carrieth more beauty and respect in his flower then they, I shall entreate thereof in a Chapter by it selfe, and set it next vnto them.

Dens Caninus flore albo. Dogs tooth Violet with a white flower.

The white Dogs tooth hath for his roote a white bulbe, long and small, yet vsually greater then either of the other that follow, bigger belowe then aboue, with a small peece adioyning to the bottome of it, from whence rise vp in the beginning of the Spring, after the Winter frosts are past, two leaues for the most part (when it will flower, or else but one, and neuer three together that euer I saw) closed together when they first come vp out of the ground, which inclose the flower betweene them : the leaues when they are opened do lay themselues flat on the ground, or not much aboue it, one opposite vnto the other, with the stalke and the flower on it standing betweene them, which leaues are of a whitish greene colour, long and narrow, yet broader in the

middle

middle then at both ends, growing leſſe by degrees each way, ſpotted and ſtriped all ouer the leaues with white lines and ſpots : the ſtalke riſeth vp halfe a foote high or more, bearing at the toppe one flower and no more, hanging downe the head, larger then any of the other of this kinde that follow, made or conſiſting of ſix white long and narrow leaues, turning themſelues vp againe, after it hath felt the comfort of the Sunne, that they doe almoſt touch the ſtalke againe, very like vnto the flowers of *Cyclamen* or Sowebread : it hath in the middle of the flower ſix white chiues, tipt with darke purple pendents, and a white three forked ſtile in the middle of them: the flower hath no ſent at all, but commendable onely for the beauty and forme thereof: after the flower is paſt, commeth in the place a round head ſeeming three ſquare, containing therein ſmall and yellowiſh ſeede.

Dens Caninus flore purpuraſcente. Dogs tooth with a pale purple flower.

This other Dogs tooth is like vnto the former, but leſſer in all parts, the leafe whereof is not ſo long, but broad and ſhort, ſpotted with darker lines and ſpots : the flower is like the other, but ſmaller, and of a delayed purple colour, very pale ſometimes, and ſometimes a little deeper, turning it ſelfe as the other, with a circle round about the vmbone or middle, the chiues hereof are not white, but declining to purple: the roote is white, and like vnto the former, but leſſer, as is ſaid before.

Dens Caninus flore rubro. Dogs tooth with a red flower.

This is in all things like vnto the laſt, both for forme and bigneſſe of flower and leafe : the chiefe difference conſiſteth in this, that the leaues hereof are of a yellowiſh mealy greene colour, ſpotted and ſtreaked with redder ſpots and ſtripes, and the flower of a deeper reddiſh purple colour, and the chiues alſo more purpliſh then the laſt, in all other things it is alike.

The Place.

The ſorts of *Dens Caninus* doe growe in diuers places ; ſome in Italy on the Euganean Hils, others on the Apenine, and ſome about Gratz, the chiefe Citie of Stiria, and alſo about Bayonne, and in other places.

The Time.

They flower in March moſt vſually, and many times in Aprill, according to the ſeaſonableneſſe of the yeare.

The Names.

Cluſius did call it firſt *Dentali,* and Lobel, and from him ſome others *Satyrium,* and *Erythronium,* but I haue ſaid enough hereof in the beginning of the Chapter. It is moſt commonly called *Dens Caninus,* and we in Engliſh, either Dogs tooth, or Dogs tooth Violet. Geſner called it *Hermodactylus,* and Matthiolus *Pſeudohermodactylus.*

The Vertues.

The roote hereof is held to bee of more efficacy for venereous effects, then any of the Orchides and Satyrions.
They of Stiria vſe the rootes for the falling ſickneſſe.
Wee haue had from Virginia a roote ſent vnto vs, that wee might well iudge, by the forme and colour thereof being dry, to be either the roote of this, or of an Orchis, which the naturall people hold not onely to be ſingular to procure luſt, but hold it as a ſecret, loth to reueale it.

Chap.

Chap. XXIIII.

Cyclamen. Sowebread.

THe likenesse of the flowers, and the spotting of the leaues of the *Dens Caninus*, with these of the *Cyclamen* or Sowebread, maketh mee ioyne it next thereunto : as also that after the bulbous rooted plants I might begin with the tuberous that remaine, and make this plant the beginning of them. Of this kinde there are diuers forts, differing both in forme of leaues and time of flowring : for some doe flower in the Spring of the yeare, others afterwards in the beginning of Summer : but the most number in the end of Summer, or beginning of Autumne or Harueft, whereof some haue round leaues, others cornered like vnto Iuie, longer or shorter, greater or smaller. Of them all in order, and first of those that come in the Spring.

1. *Cyclamen Vernum flore purpureo.* Purple flowred Sowebread of the Spring.

This Sowebread hath a smaller roote then most of the others, yet round and blackish on the outside, as all or most of the rest are (I speake of them that I haue seene ; for Clusius and others doe report to haue had very great ones) from whence rise vp diuers round, yet pointed leaues, and somewhat cornered withall, greene aboue, and spotted with white spots circlewise about the leafe, and reddish vnderneath, which at their first comming vp are folded together ; among which come the flowers, of a reddish purple colour and very sweete, euery one vpon a small, long, and slender reddish foote-ftalke, which hanging downe their heads, turne vp their leaues againe : after the flowers are past, the head or seede veffell shrinketh downe, winding his footeftalke, and coyling it selfe like a cable, which when it toucheth the ground, there abideth hid among the leaues, till it be growne great and ripe, wherein are contained a few small round seedes, which being presently sowne, will growe first into round rootes, and afterwards from them shoote forth leaues.

2. *Cyclamen Vernum flore albo.* White flowred Sowebread of the Spring.

The white flowring Sowebread hath his leaues like the rormer, but not fully so much cornered, bearing small snow white flowers, as sweete as the other : and herein consifteth the chiefeft difference, in all other things it is alike.

3. *Cyclamen Vernum Creticum flore albo.* White Candy Sowebread of the Spring.

This Sowebread is somewhat like the former white kinde, but that the leaues grow much larger and longer, with more corners at the edges, and more eminent spots on them : the flowers also somewhat longer and larger, and herein consifteth the whole difference.

4. *Cyclamen Æftivum.* Summer Sowebread.

Summer Sowebread hath round leaues like vnto the Romane Sowebread, but somewhat cornered, yet with shorter corners then the Iuie leafed Sowebread, full of white spots on the vpperside of the leaues, and very purple vnderneath, sometimes they haue fewer spots, and little or no purple vnderneath : the flowers hereof are as small, as purple, and as sweete, as the purple Sowebread of the Spring time : the roote hereof is likewife small, blacke, and round.

5. *Cyclamen Romanum rotundifolium.* Romane Sowebread with round leaues.

The Romane Sowebread hath round leaues, somewhat like vnto the common Sowebread, but not fully so round pointed at the ends, a little cornered sometimes also, or as it were indented, with white spots round about the middle of the leaues,

R 2 and

and very conspicuous, which make it seeme the more beautifull : the flowers appeare in Autumne, and are shorter, and of a deeper purplish red colour then the Iuie Sowebread, rising vp before the leaues for the most part, or at least with them, and little or nothing sweete : the roote is round and blacke, vsually not so flat as it, but growing sometimes to bee greater then any other kinde of Sowebread. There is sometimes some variety to be seene, both in the leaues and flowers of this kinde ; for that sometime the leaues haue more corners, and either more or lesse spotted with white : the flowers likewise of some are larger or lesser, longer or rounder, paler or deeper coloured one then another. This happeneth most likely from the sowing of the seede, causing the like variety as is seene in the Iuie leafed Sowebread. It doth also many times happen from the diuersity of soyles and countries where they grow : the seed of this, as of all the rest, is small and round, contained in such like heads as the former, standing almost like the head of a Snake that is twined or folded within the body thereof. This and the other Autumnall kindes, presently after their sowing in Autumne, shoote forth leaues, and so abide all the Winter, according to their kinde.

Varietas.

6. *Cyclamen folio hederæ autumnale.* Iuie leafed Sowebread.

The Iuie leafed Sowebread groweth in the same manner that the former doth, that is, bringeth forth flowers with the leaues sometimes, or most commonly before them, whose flowers are greater then the common round leafed Sowebread, somewhat longer then the former Romane or Italian Sowebreads, and of a paler purple colour, almost blush, without that sweete sent as is in the first kinde of the Spring : the greene leaues hereof are more long then round, pointed at the ends, and hauing also one or two corners on each side, sometimes much spotted on the vpperside with white spots and marks, and sometimes but a little or not at all ; and so likewise sometimes more or lesse purple vnderneath : all the leaues and flowers doe stand vsually euery one seuerally by themselues, vpon their owne slender foote-stalkes, as most of all the other kindes doe : but sometimes it happeneth, that both leaues and flowers are found growing from one and the same stalke, which I rather take to be accidentall, then naturall so to continue : the seede hereof is like the former kindes, which being sowne produceth variety, both in the forme of the leaues, and colour and smell of the flowers : some being paler or deeper, and some more or lesse sweete then others : the leaues also, some more or lesse cornered then others : the root groweth to be great, being round and flat, and of a blackish browne colour on the outside.

Varietas.

7. *Cyclamen autumnale hederæfolio flore albo.*
Iuie leafed Sowebread with white flowers.

There is one of this kinde, whose leaues are rounder, and not so much cornered as the former, flowring in Autumne as the last doth, and whose flowers are wholly white, not hauing any other notable difference therein.

8. *Cyclamen autumnale angustifolium.* Long leafed Sowebread.

This kinde of Sowebread may easily be knowne from all the other kindes, because his leafe is longer and narrower then others, fashioned at the bottome thereof with points, somewhat like vnto *Arum* or Wake Robin leaues : the flowers are like the former sorts for forme, but of a purple colour. There is also another of this kinde in all things like the former, but that the flowers are white.

9. *Cyclamen Antiochenum Autumnale flore purpureo duplici.*
Double flowred Sowebread of Antioch.

This Sowebread of Antioch with double flowers, hath his leaues somewhat round, like vnto the leaues of the Summer Sowbread, but with lesse notches or corners, & full of white spots on them : it beareth flowers on stalks, like vnto others, & likewise some stalks that haue two or three flowers on them, which are very large, with ten or twelue

leaues

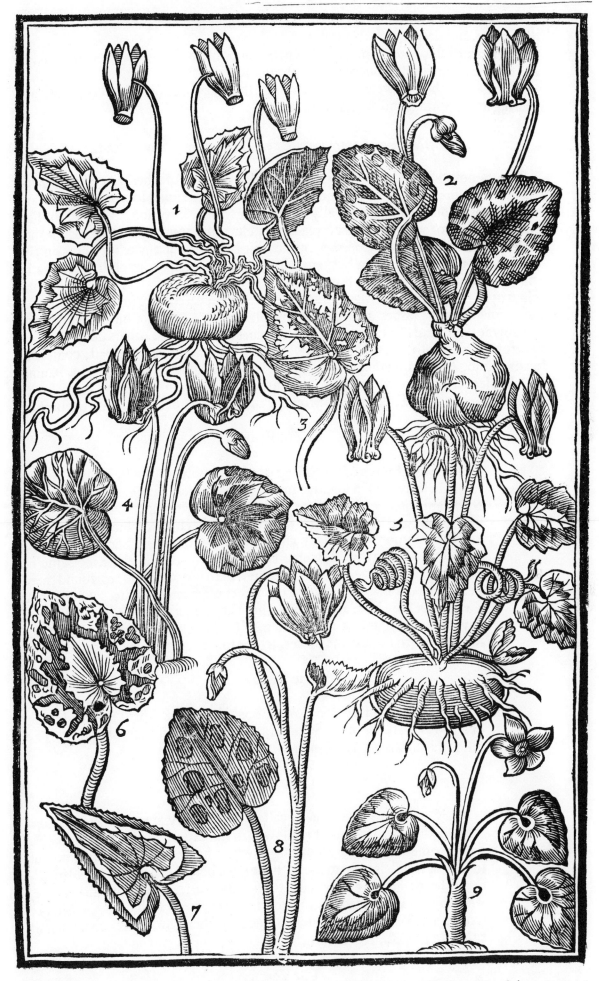

1 *Cyclamen Vernum flore purpureo* Purple flowred Sowebread of the Spring. 2 *Cyclamen aftivum.* Summer Sowebread. 3 *Folium Cyclaminis Cretici vernalis flore candido.* A leafe of Candie Sowebread. 4 *Cyclamen Romanum Autumnale,* Romane Sowebread of the Autumne. 5 *Cyclamen hederæfolio Autumnale.* Iuie leafed Autumne Sowebread. 6 *Folium Cyclaminis Autumnalis flore albo.* A leafe of the Autumne Sowebread with a white flower. 7 *Folium Cyclaminis angustifolij Autumnalis.* A leafe of the long leafed Sowebread. 8 *Cyclamen Antiochenum Autumnale flore amplo purpureo duplici.* The double flowred Sowebread of Antioch. 9 *Cyclamen vulgare folijs retundo.* The common round leafed Sowebread.

leaues a peece, of a faire Peach colour, like vnto the flowers of purple Sowebread of the Spring, and deeper at the bottome.

There are of this kinde some, whose flowers appeare in the Spring, and are as large and double as the former, but of a pure white colour.

There are of these Sowebreads of Antioch, that haue but single flowers, some appearing in the Spring, and others in Autumne.

10. *Cyclamen vulgare folio rotundo.* The common Sowebread.

The common Sowebread (which is most vsed in the Apothecaries Shops) hath many leaues spread vpon the ground, rising from certaine small long heads, that are on the greater round rootes, as vsually most of the former sorts doe, being in the like manner folded together, and after spread themselues into round greene leaues, somewhat like vnto the leaues of *Asarum,* but not shining, without any white spots on the vpperside for the most part, or but very seldome, and reddish or purplish vnderneath, and very seldome greener : the flowers stand vpon small foot-stalkes, and shew themselues open for the most part, before any leaues doe appeare, being smaller and shorter then those with Iuie leaues, and of a pale purple colour, yet sometimes deeper, hanging downe their heads, and turning vp their leaues againe, as all others doe, but more sweete then many other of the Autumne flowers : after the flowers are past, come the heads turning or winding themselues downe in like manner as the other do, hauing such like seede, but somewhat larger, and more vneuen, or not so round at the least : the roote is round, and not flat, of a browner colour, and not so blacke on the outside as many of the others.

The Place.

The Sowebreads of the Spring doe both grow on the Pyrenæan Mountaines in Italy, and in Candy, and about Moinpelier in France ; Antioch in Syria also hath yeelded some both of the Spring and Autumne. Those with round and Iuie leaues grow in diuers places both of France and Italy : and the common in Germany, and the Lowe-Countries. But that Autumne Sowebread with white flowers, is reported to grow in the Kingdome of Naples. I haue very curiously enquired of many, if euer they found them in any parts of England, neare or farther off from the places where they dwell : but they haue all affirmed, that they neuer found, or euer heard of any that haue found of any of them. This onely they haue assured, that there groweth none in the places, where some haue reported them to grow.

The Time.

Those of the Spring doe flower about the end of Aprill, or beginning of May. The other of the Summer, about the end of Iune or in Iuly. The rest some in August, and September, others in October.

The Names.

The Common Sowebread is called by most Writers in Latine, *Panis Porcinus,* and by that name it is knowne in the Apothecaries shops, as also by the name *Arthanita,* according to which name, they haue an ointment so called, which is to be made with the iuice hereof. It is also called by diuers other names, not pertinent for this discourse. The most vsuall name, whereby it is knowne to most Herbarists, is *Cyclamen* (which is the Greeke word) or as some call it *Cyclaminus,* adding thereunto their other seuerall titles. In English, Sowebread.

The Vertues.

The leaues and rootes are very effectuall for the spleene, as the Ointment before remembred plainly proueth, being vsed for the same purpose,

and

and that to good effect. It is vfed alfo for women in long and hard trauels, where there is danger, to accelerate the birth, either the roote or the leafe being applyed. But for any amorous effects, I hold it meere fabulous.

<hr/>

Chap. XXV.

Anemone. Windeflower and his kindes.

THe next tuberous rooted plants that are to follow (cf right in my opinion) are the *Anemones* or Windeflowers, and although fome tuberous rooted plants, that is, the Afphodils, Spiderworts, and Flowerdeluces haue beene before inferted, it was, both becaufe they were in name or forme of flowers futable to them whom they were ioyned vnto, and alfo that they fhould not be feuered and entreated of in two feuerall places: the reft are now to follow, at the leaft fo many of them as be beautifull flowers, fit to furnifh a Florifts Garden, for natures delightfome varieties and excellencies. To diftinguifh the Family of *Anemones* I may, that is, into the wilde kindes, and into the tame or mannured, as they are called, and both of them nourfed vp in Gardens; and of them into thofe that haue broader leaues, and into thofe that haue thinner or more iagged leaues: and of each of them, into thofe that beare fingle flowers, and thofe that beare double flowers. But to defcribe the infinite (as I may fo fay) variety of the colours of the flowers, and to giue to each his true diftinction and denomination, *Hic labor, hoc opus eft*, it farre paffeth my ability I confeffe, and I thinke would grauell the beft experienced this day in Europe (and the like I faid concerning Tulipas, it being as contingent to this plant, as is before faid of the Tulipa, to be without end in yeelding varieties:) for who can fee all the varieties that haue fprung from the fowing of the feede in all places, feeing the variety of colours rifen from thence, is according to the variety of ayres & grounds wherein they are fowne, skill alfo helping nature in ordering them aright. For the feede of one and the fame plant fowne in diuers ayres and grounds, doe produce that variety of colours that is much differing one from another; who then can difplay all the mixtures of colours in them, to fet them downe in fo fmall a roome as this Booke? Yet as I haue done (in the former part of this Treatife) my good will, to expreffe as many of each kinde haue come to my knowledge, fo if I endeauour the like in this, I hope the courteous wil accept it, and hold me excufed for the reft: otherwife, if I were or could be abfolute, I fhould take from my felf and others the hope of future augmentation, or addition of any new, which neuer will be wanting. To begin therefore with the wilde kinds (as they are fo accounted) I fhall firft entreate of the *Pulfatillas* or Pafque flowers, which are certainly kindes of wilde *Anemones*, both in leafe and flower, as may well be difcerned by them that are iudicious (although fome learned men haue not fo thought, as appeareth by their writings) the rootes of them making one fpeciall note of difference, from the other forts of wilde *Anemones*.

1. *Pulfatilla Anglica purpurea.* The purple Pafque flower.

The Pafque or Paffe flower which is of our owne Country, hath many leaues lying on the ground, fomewhat rough or hairie, hard in feeling, and finely cut into many fmall leaues, of a darke greene colour, almoft like the leaues of Carrets, but finer and fmaller, from among which rife vp naked ftalkes, rough or hairie alfo, fet about the middle thereof with fome fmall diuided leaues compaffing them, and rifing aboue thefe leaues about a fpanne, bearing euery one of them one pendulous flower, made of fix leaues, of a fine Violet purple colour, but fomewhat deepe withall, in the middle whereof ftand many yellow threeds, fet about a middle purple pointell: after the flower is paft, there commeth vp in the ftead thereof a bufhie head of long feedes, which are fmall and hoarie, hauing at the end of euery one a fmall haire, which is gray likewife: the roote is fmall and long, growing downewards into the ground, with a tuft of haire at the head thereof, and not lying or running vnder the vpper cruft thereof, as the other wilde *Anemones* doe.

2. *Pulfa-*

2. *Pulsatilla Danica.* The Passe flower of Denmarke.

There is another that was brought out of Denmarke, very like vnto the former, but that it is larger both in roote and leafe, and flower also, which is of a fairer purple colour, not so deepe, and besides, will better abide to bee mannured then our English kinde will, as my selfe haue often proued.

Vtriusque flore albo & flore duplici. Of both these sorts it is said, that some plants haue bin found, that haue borne white flowers. And likewise one that bore double flowers, that is, with two rowes of leaues.

3. *Pulsatilla flore rubro.* The red Passe flower.

Lobel, as I take it, did first set forth this kinde, being brought him from Syria, the leaues whereof are finer cut, the flower smaller, and with longer leaues, and of a red colour.

4. *Pulsatilla flore luteo.* The yellow Passe flower.

The yellow Passe flower hath his leaues cut and diuided, very like vnto the leaues of the first kinde, but somewhat more hairie, greene on the vpperside, and hairie vnderneath : the stalke is round and hoary, the middle whereof is beset with some small leaues, as in the other, from among which riseth vp the stalke of the flower, consisting of six leaues of a very faire yellow colour on the inside, and of a hoary pale yellow on the outside ; after which followeth such an head of hairie thrummes as in the former : the roote is of the bignesse of a mans finger.

5. *Pulsatilla flore albo.* The white Passe flower.

The white Passe flower (which Clusius maketh a kinde of *Anemone*, and yet as hee saith himselfe, doth more nearely resemble the *Pulsatilla*) hath, from amongst a tuft or head of haires, which grow at the toppe of a long blacke roote, many leaues standing vpon long stalkes, which are diuided as it were into three wings or parts, and each part finely cut and diuided, like vnto the Passe flower of Denmarke, but somewhat harder in handling, greenish on the vpperside, and somewhat gray vnderneath, and very hairie all ouer : among these leaues rise vp the stalkes, beset at the middle of them with three leaues, as finely cut and diuided as those belowe, from aboue which standeth the flower, being smaller, and not so pendulous as the former, but in the like manner consisting of six leaues, of a snow white colour on the inside, and a little browner on the outside, with many yellow thrums in the middle : after the flower is past, riseth vp such a like hoary head, composed as it were of many haires, each whereof hath a small seede fastened vnto it, like as the former Passe flowers haue.

The Place.

The first is found in many places of England, vpon dry bankes that lye open to the Sunne.

The second was first brought, as I take it, by Doctor Lobel from Denmarke, & is one of the two kinds, that Clusius saith are common in Germanie, this bearing a paler purple flower, and more early then the other, which is the same with our English, whose flower is so darke, that it almost seemeth blacke.

The red kinde, as Lobel saith, came from Syria.

The yellow Passe flower, which Clusius maketh his third wilde *Anemone*, was found very plentifully growing at the foote of St. Bernards Hill, neare vnto the Cantons of the Switzers.

The white one groweth on the Alpes neare Austria, in France likewise, and other places.

The

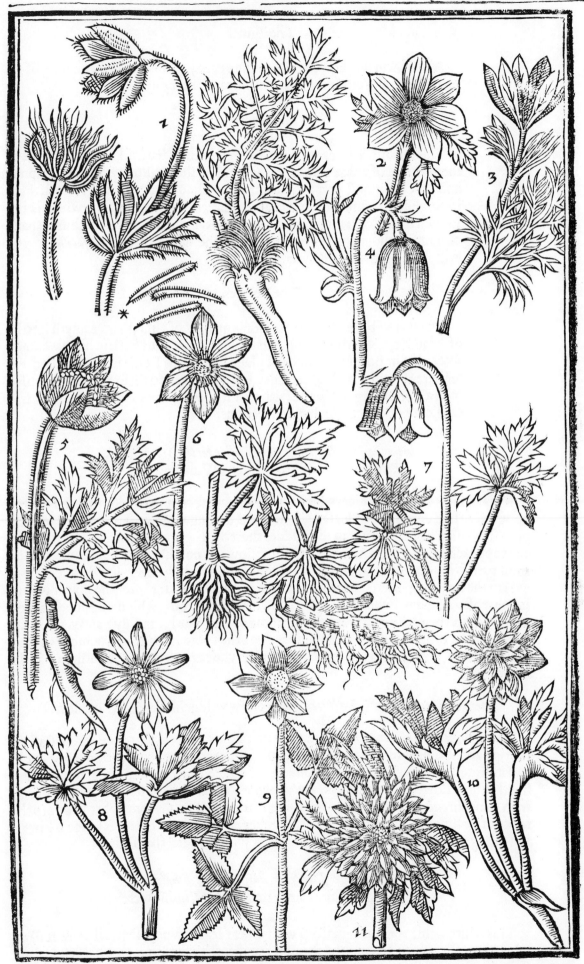

1 *Pulfatilla purpurea cum folio, femine, & radice.* The purple Pafque flower with leafe, feed, and root. 2 *Pulfatilla luteo flore.* The yellow Pafque flower. 3 *Pulfatilla rubra Syriaca Lobelij* Red Pafque flower of Lobel. 4 *Pulfatilla rubra Swertij* Swertz his red Pafque flower. 5 *Pulfatilla flore albo.* White Pafque flower 6 *Anemone filueftris alba Matthioli.* The wilde white broad leafed Windflower. 7 *Anemone filueftris tenuifolia alba.* The wilde fingle white Windflow. 8 *Anemone filueftris tenuifolia lutea.* The yellow wilde thin leafed Windflower. 9 *Anemone filueftris trifolia Dodonei.* The three-leafed wilte Windflower. 10 *Anemone filueftris flore pleno albo.* The double white wilde Windflower. 11 *Anemone filueftris flore pleno purpuree.* The double purple wilde Windflower. ✻ *Semen feparatim diuulfum.* The feed feparated. † *Radix cum folio inferiore.* The roote with a lower leafe.

The Time.

All of them doe flower early in the yeare, that is, in the beginning of A-prill, about which time moſt commonly Eaſter doth fall.

The Names.

Their proper names are giuen to each in their ſeuerall titles, being all of them kindes of wilde *Anemones*, as I ſaid in the beginning of the Chapter, and ſo for the moſt part all Authors doe acknowledge them. We call them in Engliſh, becauſe they flower about Eaſter, Paſque Flower, which is the French name for Eaſter, or *Euphoniægratia*, Paſſe Flower, which may paſſe currant, without any further deſcant on the name, or elſe *Pulſatilla*, if you will, being growne old by cuſtome.

The Vertues.

The ſharpe biting and exulcerating quality of this plant, cauſeth it to be of little vſe, notwithſtanding Ioachimus Camerarius ſaith in his *Hortus Me-dicus*, that in Boruſſia, which is a place in Italy, as I take it, the diſtilled wa-ter hereof is vſed with good ſucceſſe, to be giuen to them that are troubled with a Tertian Ague; for he ſaith that it is *medicamentum* ἐκφρακτικὸν, that is, a medicine of force to helpe obſtructions.

Anemone ſilueſtris latifolia alba ſiue tertia Matthioli.
The white wilde broad leafed Windflower.

This Windflower hath diuers broad greene leaues, cut into diuiſions, and dented about, very like vnto a broad leafed Crowfoote, among which riſeth vp a ſtalke, ha-uing ſome ſuch like cut leaues in the middle thereof, as growe below, but ſmaller; on the toppe whereof ſtandeth one large white flower, conſiſting of fiue leaues for the moſt part, with ſome yellow threads in the middle, ſtanding about ſuch a greene head as is in the tame or garden *Anemones*, which growing greater after the flower is paſt, is compoſed of many ſmall ſeedes, wrapped in white wooll, which as ſoone as they are ripe, raiſe themſelues vp from the bottome of the head, and flye away with the winde, as the other tame or garden kindes doe: the roote is made of a number of long blacke ſtrings, encreaſing very much by running vnderground, and ſhooting vp in di-uers places.

Anemone ſilueſtris tenuifolia lutea. The yellow wilde thin leafed Windflower.

The yellow wilde *Anemone* riſeth vp with one or two ſmall round naked ſtalkes, bearing about the middle of them, ſmall, ſoft, and tender iagged leaues, deeply cut in and indented on the edges about, from aboue which doth grow the ſtalke, bearing ſmall yellow flowers, ſtanding vpon weake foote-ſtalkes, like vnto a ſmall Crowfoot, with ſome threads in the middle: the roote is long and ſmall, ſomewhat like vnto the roote of Pollipodie, creeping vnder the vpper cruſt of the earth: this kinde is lower, and ſpringeth ſomewhat earlier then the other wilde kindes that follow.

Anemone ſilueſtris tennifolia alba ſimplex.
The ſingle white thin leafed wilde Windflower.

This white wilde *Anemone* riſeth vp with diuers leaues vpon ſeuerall long ſtalkes, which are ſomewhat like vnto the former, but that they are ſomewhat harder, and not ſo long, nor the diuiſions of the leaues ſo finely ſnipt about the edges, but a little broader, and deeper cut in on euery ſide: the flowers hereof are larger and broader then the former, white on the inſide, and a little purpliſh on the outſide, eſpecially at
the

the bottome of the flower next vnto the ſtalke : the roote of this is very like vnto the laſt.

There is another of this kinde, whoſe flowers are purple, in all other things it is like *Purpurea.* vnto the white.

And likewiſe another, with a bluſh or carnation coloured flower. *Coccinea ſiue ſuaue rubens.*

There is one that is onely nurſed vp with vs in Gardens, that is ſomewhat like vnto theſe former wilde *Anemones* in roote and leafe, but that the flower of this, being pure white within, and a little purpliſh without, conſiſting of eight or nine ſmall round *Peregrina alba.* pointed leaues, hath ſometimes ſome leaues vnder the flower, party coloured white and greene : the flower hath likewiſe a greene head, like a Strawberry, compaſſed a-bout with white threads, tipt with yellow pendents.

And another of the ſame kinde with the laſt, whoſe flower conſiſting of eight or *Peregrina viridis.* nine leaues, is of a greeniſh colour, except the foure outermoſt leaues, which are a lit-tle purpliſh, and diuided at the points into three parts ; the middle part is of a greeniſh white colour, with a greene head in the middle as the other.

Anemone ſilueſtris trifolia Dodonæi. The three leafed wilde Windflower.

This wilde *Anemone* hath his rootes very like vnto the former kindes ; the leaues are alwaies three ſet together at the toppe of ſlender ſtalkes, being ſmall and indented a-bout, very like vnto a three leafed Graſſe, but ſmaller : the flower conſiſteth of eight ſmall leaues, ſomewhat like vnto a Crowfoote, but of a whitiſh purple or bluſh colour, with ſome white threads, and a greene rough head in the middle.

Anemone ſilueſtris flore pleno albo. The double white wilde Windflower.

This double kinde is very like vnto the ſingle white kinde before deſcribed, both in his long running rootes, and thin leaues, but ſomewhat larger : the flowers hereof are very thicke and double, although they be ſmall, and of a faint ſweete ſent, very white after it is full blowne for fiue or ſix dayes, but afterwards it becommeth a little purpliſh on the inſide, but more on the outſide : this neuer giueth ſeede (although it haue a ſmall head in the middle) like as many other double flowers doe.

Anemone ſilueſtris flore pleno purpureo. The double purple wilde Windflower.

This double purple kinde hath ſuch like iagged leaues as the laſt deſcribed hath, but more hoarie vnderneath : the flower is of a fine light purple toward the points of the leaues, the bottomes being of a deeper purple, but as thicke, and full of leaues as the former, with a greene head in the middle, like vnto the former : this kinde hath ſmall greene leaues on the ſtalkes vnder the flowers, cut and diuided like the lower leaues.

The Place.

The firſt broad leafed *Anemone* groweth in diuers places of Auſtria and Hungary. The yellow in diuers woods in Germany, but not in this Coun-trey that euer I could learne. The other ſingle wilde kindes, ſome of them are very frequent throughout the moſt places of England, in Woods, Groues, and Orchards. The double kindes were found, as Cluſius ſaith, in the Lowe-Countries, in a Wood neare Louaine.

The Time.

They flower from the end of March (that is the earlieſt) and the begin-ning of Aprill, vntill May, and the double kindes begin within a while after the ſingle kinds are paſt.

The Names.

They are called *Ranunculi ſiluarum,* and *Ranunculi nemorum,* and as Cluſius

ſius would haue them, *Leimonia* of Theophraſtus; they are generally called of moſt Herbariſts *Anemones ſilueſtres*, Wilde *Anemones* or Windflowers. The Italians call them *Gengeuo ſaluatico*, that is, Wilde Ginger, becauſe the rootes are, beſides the forme, being ſomewhat like ſmall Ginger, of a biting hot and ſharpe taſte.

Anemone Luſitanica ſiue hortenſis latifolia flore ſimplici luteo. The ſingle Garden yellow Windflower or Anemone.

This ſingle yellow Anemone or Windflower hath diuers broad round leaues, ſome-what diuided and endented withall on the edges, browniſh at the firſt riſing vp out of the ground, and almoſt folded together, and after of a ſad greene on the vpperſide, and reddiſh vnderneath; among which riſe vp ſmall ſlender ſtalkes, beſet at the middle of them with two or three leaues, more cut and diuided then thoſe belowe, with ſmall yellow flowers at the toppe of them, conſiſting of ten or twelue leaues a peece, hauing a few yellow threads in the middle of them, ſtanding about a ſmall greene head, which in time growing ripe hath ſmall flat ſeede, incloſed within a ſoft wooll or downe, which is eaſily blowne away with the winde: the roote groweth downe-ward into the ground, diuerſly ſpread with branches here and there, of a browniſh yellow on the outſide, and whitiſh within, ſo brittle, that it can hardly bee touched without breaking.

Anemone latifolia flore luteo duplici. The double yellow Anemone or Windflower.

This double yellow Anemone hath ſuch broad round leaues as the ſingle kinde hath, but ſomewhat larger or ranker: the ſtalkes are beſet with larger leaues, more deeply cut in on the edges: the flowers are of a more pale yellow, with ſome purpliſh veines on the outſide, and a little round pointed; but they are all on the inſide of a faire yellow colour, conſiſting of two rowes of leaues, whereof the innermoſt is the narrower, with a ſmall greene head in the middle, compaſſed with yellow threads as in the former: the roote is like the roote of the ſingle; neither of theſe haue any good ſent, and this ſpringeth vp and flowreth later then the ſingle kinde.

Anemone latifolia purpurea ſtellata ſiue papaueracea. The purple Starre Anemone or Windflower.

The firſt leaues of this purple Anemone, which alwayes ſpring vp before Winter, (if the roote be not kept too long out of the ground,) are ſomewhat like the leaues of *Sanicle* or Selfe-heale, but the reſt that follow are more deeply cut in and iagged; among which riſe vp diuers round ſtalkes, beſet with iagged leaues as all other Anemones are, aboue which leaues, the ſtalkes riſing two or three inches high, beare one flower a peece, compoſed of twelue leaues or more, narrow and pointed, of a bleake purple or whitiſh aſh-colour, ſomewhat ſhining on the outſide, and of a fine purple co-lour tending to a murrey on the inſide, with many blackiſh blew threads or thrummes in the middle of the flower, ſet about a head, whereon groweth the ſeede, which is ſmall and blacke, incloſed in ſoft wooll or downe, which flieth away with the winde, carrying the ſeede with it, if it be not carefully gathered: the roote is blackiſh on the outſide, and white within, tuberous or knobby, with many fibres growing at it.

Anemone purpurea Stellata altera. Another purple Starre Anemone.

There is ſo great diuerſity in the colours of the flowers of theſe broad leafed kinds of Anemones or Windflowers, that they can very hardly be expreſſed, although in their leaues there is but little or no difference. I ſhall not neede therefore to make ſeuerall deſcriptions of euery one that ſhall be ſet downe; but it will be ſufficient, I thinke, to giue you the diſtinctions of the flowers: for as I ſaid, therein is the greateſt and chiefeſt difference. This other Starre Anemone differeth not from the former in leafe or flower, but onely that this is of a more pale ſullen colour on the outſide, and of a paler purple colour on the inſide.

There

Viola purpurea There is another, whole flower hath eight leaues, as many of them that follow haue (although diuers forts haue but fix leaues in a flower) and is of a Violet purple, and therefore is called, The Violet purple Anemone.

Varietas. Of all thefe three forts laft defcribed, there be other that differ only in hauing white bottomes, fome fmaller and fome larger.

Purpurea striata. There is alfo another of the fame Violet purple colour with the former, but a little paler, tending more to redneffe, whofe flowers haue many white lines and ftripes through the leaues, and is called, The purple ftript Anemone.

Carnea vivacissima simplex. There is another, whofe greene leaues are fomewhat larger, and fo is the flower likewife, confifting of eight leaues, and fometimes of more, of the colour of Carnation filke, fometimes pale and fometimes deeper, with a whitifh circle about the bottome of the leaues, which circle in fome is larger, and more to be feene then in others, when the flower layeth it felfe open with the heate of the Sunne, hauing blewifh threads in the middle. This may be called, the Carnation Anemone.

Persiciviolacea. We haue another, whofe flower is betweene a Peach colour and a Violet, which is vfually called a Gredeline colour.

Cochenille. And another of a fine reddifh Violet or purple, which we call, The Cochenille Anemone.

Cardinalis. And another of a rich crimfon red colour, and may be called, The Cardinall Anemone.

Sanguinea. Another of a deeper, but not fo liuely a red, called, The bloud red Anemone.

Cramesina. Another of an ordinary crimfon colour, called, The crimfon Anemone.

Coccinea. Another of a Stamell colour, neere vnto a Scarlet.

Incarnata. Another of a fine delayed red or flefh colour, and may bee called, The Incarnadine Anemone.

Incarnata Hispanica. Another whofe flower is of a liuely flefh colour, fhadowed with yellow, and may be called, The Spanifh Incarnate Anemone.

Rubescens. Another of a faire whitifh red, which we call, The Blufh Anemone.

Moschutella. Another whofe flower confifteth of eight leaues, of a darke whitifh colour, ftript all ouer with veines of a fine blufh colour, the bottomes being white, this may be called, The Nutmegge Anemone.

Enfumata. Another whofe flower is of a pale whitifh colour, tending to a gray, fuch as the Monkes and Friers were wont to weare with vs, and is called, A Monkes gray.

Pauo maior simplici flore. There is another, whofe leafe is fomewhat broader then many or moft of the Anemones, comming neare vnto the leafe of the great double Orenge coloured Anemone; the flower whereof is fingle, confifting of eight large or broad leaues, very neare vnto the fame Orenge colour, that is in the double flower hereafter defcribed, but fomewhat deeper. This is vfually called in Latine, *Pauo maior simplici flore*, and we in Englifh, The great fingle Orenge tawnie Anemone.

Pauo minor. There is likewife of this kinde another, whofe flower is leffer, and called, The leffer Orenge tawnie Anemone.

Varietas magna ex seminio. There is befides thefe expreffed, fo great a variety of mixt colours in the flowers of this kinde of Anemone with broad leaues, arifing euery yeare from the fowing of the feede of fome of the choifeft and fitteft for that purpofe, that it is wonderfull to obferue, not onely the variety of fingle colours, but the mixture of two or three colours in one flower, befides the diuerfity of the bottomes of the flowers, fome hauing white or yellowifh bottomes, and fome none, and yet both of the fame colour; and likewife in the thrums or threads in the middle: But the greateft wonder of beauty is in variety of double flowers, that arife from among the other fingle ones, fome hauing two or three rowes of leaues in the flowers, and fome fo thicke of leaues as a double Marigold, or double Crowfoote, and of the fame feuerall colours that are in the fingle flowers, that it is almoft impoffible to expreffe them feuerally, and (as is faid before) fome falling out to bee double in one yeare, which will proue fingle or leffe double in another,

other, yet very many abiding conftant double as at the firft ; and therefore let this briefe recitall be fufficient in ftead of a particular of all the colours.

Anemone Chalcedonica maxima verficolor.
The great double Windflower of Conftantinople.

This great Anemone of Conftantinople hath broader and greener leaues then any of the former kindes, and not fo much diuided or cut in at the edges, among which rife vp one or two ftalkes, (feldome more from one roote) hauing fome leaues about the middle of the ftalke, as other Anemones haue, and bearing at the toppes of the ftalkes one large flower a peece, very double, whofe outermoft leaues being broadeft, are greenifh at the firft, but afterwards red, hauing fometimes fome greene abiding ftill in the leaues, and the red ftriped through it : the other leaues which are within thefe are fmaller, and of a perfect red colour ; the innermoft being fmalleft, are of the fame red colour, but turned fomewhat inward, hauing no thrummes or threads in the middle, as the former haue, and bearing no feede : the roote is blackifh on the outfide, and white within, thicke and tuberous as the other kindes, but thicker fet and clofe together, not fhooting any long flender rootes as others doe. Some Gentlewomen call this Anemone, The Spanifh Marigold.

Anemone Chalcedonica altera fiue Pauo maior flore duplici.
The great double Orenge tawney Anemone.

This other great Anemone of Conftantinople hath his large leaues fo like vnto the laft, that one can hardly diftinguifh them afunder ; the ftalke hath alfo fuch like leaues fet vpon it, bearing at the toppe a faire large flower, confifting of many leaues fet in two or three rowes at the moft, but not fo thicke or double as the laft, yet feeming to be but one thicke rowe of many fmall and long leaues, of an excellent red or crimfon colour, wherein fome yellow is mixed, which maketh that colour is called an Orenge tawney ; the bottomes of the leaues are red, compaffed with a whitifh circle, the thrummie head in the middle being befet with many darke blackifh threads : the roote is like the former.

Anemone Superitica fiue Cyparifsia. The double Anemone of Cyprus.

This Anemone (which the Dutchmen call Superitz, and as I haue beene enformed, came from the Ifle of Cyprus) hath leaues very like the laft double Anemone, but not altogether fo large : the flower confifteth of fmaller leaues, of colour very neare vnto the laft double Orenge coloured Anemone, but more thicke of leaues, and as double as the firft, although not fo great a flower, without any head in the middle, or thrums about it as is in the laft, and differeth not in the roote from either of them both.

Somewhat like vnto this kinde, or as it were betweene this and the firft kinde of thefe great double Anemones, we haue diuers other forts, bearing flowers very thicke and double ; fome of them being white, or whitifh, or purple, deeper or paler, and fome of a reddifh colour tending to Scarlet or a Carnation colour, and fome alfo of a blufh or flefh colour, and diuers other colours, and all of them continue conftant in their colours.

Anemone Cacumeni Maringi fiue Perfica. The double Perfian Anemone.

This rare Anemone, which is faid to come out of Perfia to Conftantinople, and from thence to vs, is in leafe and roote very like vnto the former double Anemones before defcribed ; onely the flower hereof is rather like vnto the fecond great double Orenge coloured Anemone, vfually called *Pauo maior flore pleno*, being compofed of three rowes of leaues, the outtermoft rowe confifting of ten or twelue larger leaues, and thofe more inward leffer and more in number, but all of them variably mixed with white, red, and yellow, hauing the bottomes of the leaues white : but inftead of a middle head with thrums about it, as the other hath, this hath a few narrow leaues, of a deepe yellow colour in the middle of the flower, ftanding vpright.

Hauing

Hauing thus farre proceeded in the two parts of the kindes of Anemones or Wind-flowers, it remaineth to entreate of the reft, which is thofe Anemones which haue thin cut leaues, whereof fome haue reckoned vp thirty forts with fingle flowers, which I confeffe I haue not feene; but fo many as haue come to my knowledge, I fhall here fet downe.

Anemone tenuifolia fiue Geranifolia cærulea.
The Watchet Anemone or Storkes bill leafed Windflower.

This firft Windflower with thin cut leaues, rifeth not out of the ground vntill the great Winter frofts be paft, that is, about the middle or end of February, and are fomewhat brownifh at their firft appearing, but afterwards fpread into wings of greene leaues, fomewhat broader then the reft that follow, diuided into three parts, & each part into three leaues, euery one cut in about the edges, one ftanding againft another vpon a long flender foote-ftalke, and the end leafe by it felfe : among thefe rifeth vp two or three greene ftalkes, garnifhed with fuch like thin leaues as are at the bottome, from aboue which rife the flowers, but one vpon a ftalke, confifting of fourteene or fifteene fmall pale blew or watchet leaues, leffer then any of the fingle kindes that follow, compaffing many whitifh threads, and a fmall greene head in the middle, fomewhat like the head of the wilde Crowfoote, wherein is contained fuch like feede : the roote is blackifh without, thrufting out into long tuberous peeces, fomewhat like vnto fome of the broad leafed Anemones.

Alba. Of this kinde there is another, whofe leaues are not browne at their firft rifing, but greene, and the flowers are white, in other things not differing.

Anemone tenuifolia purpurea vulgaris.
The ordinary purple Anemone with thin leaues.

This purple Anemone which is moft common, and therefore the leffe regarded, hath many winged leaues ftanding vpon feuerall ftalkes, cut and diuided into diuers leaues, much like vnto the leaues of a Carrot ; among which rife vp ftalkes with fome leaues thereon (as is vfuall to the whole Family of Anemones, both wilde and tame, as is before faid;) at the toppes whereof ftand the flowers, made of fix leaues moft vfu-ally, but fometimes they will haue feuen or eight, being very large, and of a perfect purple Violet colour, very faire and liuely : the middle head hath many blackifh thrums or threads about it, which I could neuer obferue in my Gardens to beare feed : the roote is fmaller, and more fpreading euery way into fmall long flat tuberous parts, then any other kindes of fingle or double Anemones.

Carnea pallida. There is another very like in leafe and roote vnto the former, but the flower is nothing fo large, and is whitifh, tending to a blufh colour, and of a deeper blufh colour toward the bottome of the flower, with blackifh blew thrums in the middle, and giueth no feede that I could euer obferue.

Carnea viuida vnguibus albis. There is likewifewife another like vnto the laft in leafe and flower, but that the flower is larger then it, and is of a liuely blufh colour, the leaues hauing white bottomes.

Alba venis pur-pureis. And another, whofe flower is white, with purple coloured veines and ftripes through euery leafe, and is a leffer flower then the other.

Anemone tenuifolia coccinea fimplex. The fingle Scarlet Anemone with thin leaues.

The leaues of this Scarlet Windflower are fomewhat like vnto the former, but a lit-tle broader, and not fo finely cut and diuided : the flower confifteth of fix reafonable large leaues, of an excellent red colour, which we call a Scarlet ; the bottomes of the leaues are large and white, and the thrums or threads in the middle of a blackifh pur-ple colour : the roote is tuberous, but confifting of thicker peeces, fomewhat like vnto the rootes of the broad leafed Anemones, but fomewhat browne, and not fo blacke, and moft like vnto the roote of the double Scarlet Anemone.

Coccinea abfq; vnguibus. There is another of this kinde, whofe flower is neare vnto the fame co-lour, but this hath no white bottomes at all in his leaues.

Wc

Flore holoſe-riceo. We haue another which hath as large a flower as any ſingle, and is of an Orient deepe red crimſon Veluet colour.

Sanguinea. There is another of a deeper red colour, and is called, The bloud red ſingle Anemone.

Rubra fundo luteo. And another, whoſe flower is red with the bottomes yellow.

Coccinea dilu-tior. Another of a perfect crimſon colour, whereof ſome haue round pointed leaues, and others ſharpe pointed, and ſome a little lighter or deeper then others.

Alba ſtamini-bus purpureis. There is alſo one, whoſe flower is pure white with blewiſh purple thrums in the middle.

Carnea Hiſpa-nica. And another, whoſe flower is very great, of a kinde of ſullen bluſh co-lour, but yet pleaſant, with blewiſh threads in the middle.

Alba carneis venis. And another with bluſh veines in euery leafe of the white flower.

Alba purpureis vnguibus. And another, the flower whereof is white, the bottomes of the leaues being purple.

Purpuraſcens. Another whoſe flower conſiſteth of many ſmall narrow leaues, of a pale purple or bluſh colour on the outſide, and ſomewhat deeper within.

Facie florum pomi ſimplex. There is another like in leafe and roote vnto the firſt Scarlet Anemone, but the flower hereof conſiſteth of ſeuen large leaues without any bot-tomes, of a white colour, hauing edges, and ſome large ſtripes alſo of a car-nation or fleſh colour to bee ſeene in them, marked ſomewhat like an Ap-ple bloſſome, and thereupon it is called in Latine, *Anemone tenuifolia ſim-plex alba inſtar florum pomi*, or *facie florum pomi*, that is to ſay in Engliſh, The ſingle thin leafed Anemone with Apple bloſſome flowers.

Multiplex. I haue heard that there is one of this kinde with double flowers.

1. *Anemone tenuifolia flore coccineo pleno vulgaris.* The common double red or Scarlet Anemone.

The leaues of this double Anemone are very like vnto the leaues of the ſingle Scar-let Anemone, but not ſo thin cut and diuided as that with the purple flower: the flower hereof when it firſt openeth it ſelfe, conſiſteth of ſix and ſometimes of ſeuen or eight broad leaues, of a deepe red, or excellent Scarlet colour, the middle head being thick cloſed, and of a greeniſh colour, which after the flower hath ſtood blowne ſome time, doth gather colour, and openeth it ſelfe into many ſmall leaues, very thicke, of a more pale red colour, and more Stamell like then the outer leaues : the root of this is thicke and tuberous, very like vnto the root of the ſingle Scarlet Anemone.

2. *Anemone tenuifolia flore coccineo pleno variegata.* The party coloured double Crimſon Anemone.

We haue a kinde hereof, varying neither in roote, leafe, or forme of flower from the former, but in the colour, in that this will haue ſometimes the outer broad leaues party coloured, with whitiſh or bluſh coloured great ſtreakes in the red leaues both in-ſide and outſide ; as alſo diuers of the middle or inner leaues ſtriped in the ſame man-ner : the roote hereof giueth fairer flowers in ſome yeares then in others, and ſome-times giue flowers all red againe.

3. *Anemone tenuifolia flore coccineo ſaturo pleno.* The double crimſon Veluet Anemone.

Wee haue another alſo, whoſe flower is of a deepe Orenge tawny crimſon colour, neare vnto the colour of the outer leaues, of the leſſer French Marigold, and not diffe-ring from the former in any thing elſe.

4. *Anemone tenuifolia flore pleno ſuauerubente.* The greater double bluſh Anemone.

There is ſmall difference to be diſcerned, either in the roote or leaues of this from the

the former double Scarlet Anemone, sauing that the leaues hereof are a little broader, and seeme to bee of a little fresher greene colour : the flower of this is as large almost, and as double as the former, and the inner leaues likewise almost as large as they, being of a whitish or flesh colour at the first opening of them, but afterwards become of a most liuely blush colour ; the bottomes of the leaues abiding of a deeper blush, and with long standing, the tops of the leaues will turne almost wholly white againe.

5. *Anemone tenuifolia flore albo pleno.* The double white Anemone.

This double white Anemone differeth little from the former blush Anemone, but in that it is smaller in all the parts thereof, and also that the flower hereof being wholly of a pure white colour, without any shew of blush therein, hath the middle thrummes much smaller and shorter then it, and not rising vp so high, but seeme as if they were chipped off euen at the toppes.

6. *Anemone tenuifolia flore pleno albicante.* The lesser double blush Anemone.

This small double blush Anemone differeth very little from the double white last recited, but onely in the colour of the flower : for they are both much about the bignesse one of another, the middle thrums likewise being as small and short, and as euen aboue, onely the flower at the first opening is almost white, but afterwards the outer leaues haue a more shew of blush in them, and the middle part a little deeper then they.

7. *Anemone tenuifolia flore pleno purpureo violaceo.* The double purple Anemone.

This double purple Anemone is also of the same kindred with the first double red or Scarlet Anemone for the form or doublenesse of the flower, consisting but of six or seuen leaues at the most in this our Country, although in the hotter it hath ten or twelue, or more as large leaues for the outer border, and as large small leaues for the inner middle also, and almost is double, but of a deepe purple tending toward a Violet colour, the outer leaues being not so deepe as the inner : the roote and leafe commeth neare vnto the single purple Anemone before described, but that the roote spreadeth not so small and so much.

8. *Anemone tenuifolia flore pleno purpureo cæruleo.*
The double blew Anemone.

This Anemone differeth not in any thing from the former double purple, but onely that the flower is paler, and more tending to a blew colour.

9. *Anemone tenuifolia flore pleno roseo.* The double Rose coloured Anemone.

The double Rose coloured Anemone differeth also in nothing from the former double purple, but onely in the flower, which is somewhat smaller, and not so thicke and double, and that it is of a reddish colour, neare vnto the colour of a pale red Rose, or of a deepe coloured Damaske.

10. *Anemone tenuifolia flore pleno carneo viuacissimo.*
The double Carnation Anemone.

This Anemone, both in roote, leafe, and flower, commeth nearest vnto the former double white Anemone, for the largenesse and doublenesse of the flower, and in the smalnesse of the middle thrums, and euennesse at the toppes of them, being not so large and great a flower as the double purple, either in the inner or outter leaues, but yet is very faire, thicke and double, and of a most liuely Carnation silke colour, very deepe, both the outer leaues and middle thrums also so bright, that it doth as it were amaze, and yet delight the minde of the beholder, but by long standing in the Sun, waxe a little paler, and so passe away as all the most beautifull flowers doe.

Anemone tenuifolia simplex purpurea. The single purple Anemone with thin cut leaues. 2 *Anemone tenuifolia simplex alba pura.* The single pure white Anemone. 3 *Anemone tenuifolia simplex chermesina.* The single bright Crimson Anemone. 4 *Anemone tenuifolia simplex sanguinea.* The single bloud red Anemone. 5 *Anemone tenuifolia simplex facie florum pomi.* The single Apple bloome Anemone. 6 *Anemone tenuifolia simplex purpurascens.* The single purplish blush Anemone. 7 *Anemone tenuifolia simplex alba unguibus carneis.* The single white Anemone with blush bottomes. 8 *Anemone tenuifolia flore pleno coccinea.* The double red or ordinary Scarlet Anemone. 9 *Anemone tenuifolia flore pleno rubrofusca sen e Amarantina.* The double purple Veluet Anemone. 10 *Anemone tenuifolia flore pleno purpuro violaceo.* The double blewish purple Anemone. 11 *Anemone tenuifolia flore pleno incarnedini coloris sericei viuacissimi.* The double Carnation Anemone, or of a liuely Carnation silke colour.

11. *Anemone tenuifolia flore rubro fusco pleno coma Amarantina.*
The double purple Veluet Anemone.

This double Veluet Anemone is in all things like the laſt deſcribed Carnation A-
nemone, but ſomewhat larger, the difference conſiſteth in the colour of the flower,
which in this is of a deep or ſad crimſon red colour for the outer leaues, and of a deep
purple Veluet colour in the middle thrums, reſembling the colour of the leſſer *Amaranthus purpureus*, or Purple flower gentle hereafter deſcribed, whereof it tooke the
name, which middle thrums are as fine and ſmall, and as euen at the toppes as the
white or laſt Carnation Anemones.

12. *Anemone tenuifolia flore pleno tricolor.*
The double purple Veluet Anemone of three colours.

This double Anemone alſo is very like the laſt deſcribed Anemone, but that in the
middle of the purple thrums, there thruſteth forth a tuft of threads or leaues of a more
light crimſon colour.

And thus much for the kindes of Anemones or Windflowers, ſo farre forth as haue
hitherto come to our knowledge; yet I doubt not, but that more varieties haue beene
elſewhere collected, and will be alſo in our Countrey daily and yearly obſerued by
diuers, that raiſe them vp from ſowing the ſeede, wherein lyeth a pretty art, not yet familiarly knowne to our Nation, although it be very frequent in the Lowe-Countries,
where their induſtry hath bred and nouriſhed vp ſuch diuerſities and varieties, that
they haue valued ſome Anemones at ſuch high rates, as moſt would wonder at,
and none of our Nation would purchaſe, as I thinke. And I doubt not, if wee would
be as curious as they, but that both our ayre and ſoyle would produce as great variety,
as euer hath been ſeene in the Lowe-Countries; which to procure, if any of our Nation will take ſo much paines in ſowing the ſeedes of Anemones, as diuers haue done of
Tulipas: I will ſet them downe the beſt directions for that purpoſe that I haue learned,
or could by much ſearch and tryall attaine vnto; yet I muſt let them vnderſtand thus
much alſo, that there is not ſo great variety of double flowers raiſed from the ſeede of
the thin leafed Anemones, as from the broad leafed ones.

Firſt therefore (as I ſaid before) concerning Tulipas, there is ſome ſpeciall choice to
be made of ſuch flowers, whoſe ſeed is fitteſt to be taken. Of the *Latifolias*, the double
Orenge tawney ſeede being ſowne, yeeldeth pretty varieties, but the purples, and
reds, or crimſons, either *Latifolias* or *Tenuifolias*, yeeld ſmall variety, but ſuch as draw
neareſt to their originall, although ſome be a little deeper or lighter then others. But
the light colours be they which are the chiefe for choice, as white, aſh-colour, bluſh
or carnation, light orenge, ſimple or party coloured, ſingle or double, if they beare
ſeede, which muſt bee carefully gathered, and that not before it bee thorough ripe,
which you ſhall know by the head; for when the ſeede with the wollineſſe beginneth
to riſe a little of it ſelfe at the lower end, it muſt bee then quickly gathered, leſt the
winde carry it all away. After it is thus carefully gathered, it muſt be laid to dry for a
weeke or more, which then being gently rubbed with a little dry ſand or earth, will
cauſe the ſeede to be ſomewhat better ſeparated, although not thoroughly from the
woollineſſe or downe that compaſſeth it.

Within a moneth at the moſt after the ſeede is thus gathered and prepared, it muſt
be ſowne; for by that meanes you ſhall gaine a yeare in the growing, ouer that you
ſhould doe if you ſowed it in the next Spring.

If there remaine any woollineſſe in the ſeede, pull it in ſunder as well as you can,
and then ſowe your ſeede reaſonable thin, and not too thicke, vpon a plaine ſmooth
bed of fine earth, or rather in pots or tubbes, and after the ſowing, ſift or gently ſtraw
ouer them ſome fine good freſh mould, about one fingers thickneſſe at the moſt for the
firſt time: And about a moneth after their firſt ſpringing vp, ſift or ſtraw ouer them
in like manner another fingers thickneſſe of fine earth, and in the meane time if the
weather proue dry, you muſt water them gently and often, but not to ouerglut them
with moiſture; and thus doing, you ſhall haue them ſpring vp before Winter, and
grow

grow pretty ftrong, able to abide the fharpe Winter in their nonage, in vfing fome little care to couer them loofely with fome fearne, or furfe, or beane hame, or ftraw, or any fuch, which yet muft not lye clofe vpon them, nor too farre from them neither.

The next Spring after the fowing, if you will, but it is better if you ftay vntill Auguft, you may then remoue them, and fet them in order by rowes, with fufficient diftance one from another, where they may abide, vntill you fee what manner of flower each plant will beare, which you may difpofe of according to your minde.

Many of them being thus ordered (if your mould be fine, loofe, and frefh, not ftonie, clayifh, or from a middin) will beare flowers the fecond yeare after the fowing, and moft or all of them the third yeare, if the place where you fowe them, be not annoyed with the fmoake of Brewers, Dyers, or Maultkils, which if it be, then will they neuer thriue well.

Thus much haue I thought good to fet downe, to incite fome of our owne Nation to be induftrious; and to helpe them forward, haue giuen fuch rules of directions, that I doubt not, but they will vpon the tryall and view of the variety, proceede as well in the fowing of Anemones as of Tulipas.

I cannot (Gentlewomen) withhold one other fecret from you, which is to informe you how you may fo order Anemones, that after all others ordinarily are paft, you may haue them in flower for two or three moneths longer then are to be feene with any other, that vfeth not this courfe I direct you.

The ordinary time to plant Anemones, is moft commonly in Auguft, which will beare flower fome peraduenture before Winter, but moft vfually in February, March, and Aprill, few or none of them abiding vntill May; but if you will keepe fome roots out of the ground vnplanted, vntill February, March, and Aprill, and plant fome at one time, and fome at another, you fhall haue them beare flower according to their planting, thofe that fhall be planted in February, will flower about the middle or end of May, and fo the reft accordingly after that manner: And thus may you haue the pleafure of thefe plants out of their naturall feafons, which is not permitted to be enioyed in any other that I know, Nature being not fo prone to bee furthered by art in other things as in this. Yet regard, that in keeping your Anemone rootes out of the ground for this purpofe, you neither keep them too dry, nor yet too moift, for fprouting or rotting; and in planting them, that you fet them not in too open a funny place, but where they may be fomewhat fhadowed.

The Place.

I fhall not need to fpend much time in relating the feuerall places of thefe Anemones, but onely to declare that the moft of them that haue not beene raifed from feed, haue come from Conftantinople to vs; yet the firft broad leafed or yellow Anemone, was firft found in Portugall, and from thence brought into thefe parts. And the firft purple Starre Anemone in Germanie, yet was the fame fent among others from Conftantinople alfo. And the firft thin cut leafed Anemone came firft out of Italy, although many of that fort haue come likewife from Conftantinople. And fo haue the double red or Scarlet Anemones, and the great double blufh, which I firft had by the gift of Mr. Humfrey Packington of Worcefterfhire Efquire, at Haruington.

The Time.

The times of their flowring are fufficiently expreffed in the defcriptions, or in the rules for planting.

The Names.

The Turkifh names whereby the great double broad leafed kindes haue beene fent vnto vs, were *Giul Catamer*, and *Giul Catamer lale*; And *Binizade, Binizante*, and *Galipoli lale* for the thinne cut leafed Anemones. All Authors haue called them *Anemones*, and are the true *Herba venti*. We

Wee call them in English eyther Anemones, after the Greeke name, or Windflowers, after the Latine.

The Vertues.

There is little vse of these in Physicke in our dayes, eyther for inward or outward diseases; onely the leaues are vsed in the Ointment called *Marciatum*, which is composed of many other hot herbes, and is vsed in cold griefes, to warme and comfort the parts. The roote, by reason of the sharpenesse, is apt to drawe downe rheume, if it be tasted or chewed in the mouth.

Chap. XXVI.

Aconitum. Wolfebane.

THere be diuers sorts of Wolfebanes which are not fit for this booke, but are reserued for a generall History or Garden of Simples, yet among them there are some, that notwithstanding their euill quality, may for the beauty of their flowers take vp a roome in this Garden, of whom I meane to entreate in this place: And first of the Winter Wolfesbane, which for the beauty, as well as the earlinesse of his flowers, being the first of all other, that shew themselues after Christmas, deserueth a prime place; and therefore for the likenesse of the rootes vnto the Anemones, I ioyne it next vnto them.

1. *Aconitum Hyemale.* The Winters Wolfesbane.

This little plant thrusteth vp diuers leaues out of the ground, in the deepe of Winter oftentimes, if there be any milde weather in Ianuary, but most commonly after the deepe frosts, bearing vp many times the snow vpon the heads of the leaues, which like vnto the Anemone, doe euery leafe rise from the roote vpon seuerall short foote-stalkes, not aboue foure fingers high, some hauing flowers in the middle of them, (which come vp first most vsually) and some none, which leaues stand as it were round, the stalke rising vp vnder the middle of the leafe, deeply cut in and gashed to the middle stalke almost, of a very faire deepe greene colour, in the middle whereof, close vnto the leafe, standeth a small yellow flower, made of six leaues, very like a Crowfoote, with yellow threads in the middle : after the flower is fallen, there rise vp diuers small hornes or cods set together, wherein are contained whitish yellow round seede. The roote is tuberous, so like both for shape and colour vnto the rootes of Anemones, that they will easily deceiue one not well experienced, but that it is browner and smoother without, and yellow within, if it be broken.

2. *Aconitum flore albido, siue Aconitum luteum Ponticum.* The whitish yellow Wolfesbane.

This Wolfesbane shooteth not out of the ground vntill the Spring be well begun, and then it sendeth forth great broad greene leaues, deeply cut in about the edges, not much vnlike the leaues of the great wilde Crowfoote, but much greater; from among which leaues riseth vp a strong stiffe stalke, three foote high, hauing here and there leaues set vpon it, like vnto the lowest, but smaller; the toppe of the stalke is diuided into three or foure branches, whereon are set diuers pale yellow flowers, which turne at the last to be almost white, in fashion like almost vnto the flowers of the Helmet flower, but much smaller, and not gaping so wide open : after the flowers are past come vp diuers short poddes, wherein is contained blacke seede : the roote is made of a number of darke browne strings, which spread and fasten themselues strongly in the ground.

3. *Napellus*

3. *Napellus verus flore cæruleo*. Blew Helmet flower or Monkes hood.

The Helmet flower hath diuers leaues of a fresh greene colour on the vpperside, and grayish vnderneath, much spread abroad and cut into many slits and notches, more then any of the Wolfebanes; the stalke riseth vp two or three foot high, beset to the top with the like leaues, but smaller : the toppe is sometimes diuided into two or three branches, but more vsually without, whereon stand many large flowers one aboue another, in forme very like vnto a Hood or open Helmet, being composed of fiue leaues, the vppermost of which and the greatest, is hollow, like vnto an Helmet or Headpeece, two other small leaues are at the sides of the Helmet, closing it like cheekes, and come somewhat vnder, and two other which are the smallest hang down like labels, or as if a close Helmet were opened, and some peeces hung by, of a perfect or faire blew colour, (but grow darker, hauing stood long) which causeth it be so nourished vp in Gardens, that their flowers, as was vsuall in former times, and yet is in many Countrey places, may be laid among greene herbes in windowes and roomes for the Summer time: but although their beauty may be entertained for the vses aforesaid, yet beware they come not neare your tongue or lippes, lest they tell you to your cost, they are not so good as they seeme to be : in the middest of the flower, when it is open and gapeth wide, are seene certaine small threads like beards, standing about a middle head, which when the flower is past, groweth into three or foure, or more small blackish pods, containing in them blacke seede : the rootes are brownish on the outside, and white within, somewhat bigge and round aboue, and small downewards, somewhat like vnto a small short Carrot roote, sometimes two being ioyned at the head together. But the name *Napellus* anciently giuen vnto it, doth shew they referred the forme of the roote vnto a small Turnep.

Anthora. The wholsome Helmet flower, or counterpoison Monkes hood.

This wholsome plant I thought good to insert, not onely for the forme of the flower, but also for the excellent properties thereof, as you shall haue them related hereafter. The rootes hereof are small and tuberous, round and somewhat long, ending for the most part in a long fibre, and with some other small threads from the head downeward : from the head whereof riseth vp diuers greene leaues, euery one seuerally vpon a stalke, very much diuided, as finely almost as the leaues of Larkes heeles or spurres : among which riseth vp a hard round stalke, a foote high and better, with some such leaues thereon as grow belowe, at the toppe whereof stand many small yellowish flowers, formed very like vnto the former whitish Wolfesbane, bearing many blacke seedes in pods afterwards in the like manner.

Many more sorts of varieties of these kindes there are, but these onely, as the most specious, are noursed vp in Florists Gardens for pleasure; the other are kept by such as are Catholicke obseruers of all natures store.

The Place.

All these grow naturally on Mountaines, in many shadowie places of the Alpes, in Germany, and elsewhere.

The Time.

The first flowreth (as is said) in Ianuary, and February, and sometimes vntill March be well spent, and the seede is soone ripe after.
The other three flower not vntill Iune and Iuly.

The Names.

The first is vsually called *Aconitum hyemale Belgarum*. Lobelius calleth it
Bulbosus

Bulbofus vnifolius Batrachoides, Aconitum Elleboraceum, and *Ranunculus Monophyllos*, and fome by other names. Moft Herbarifts call it *Aconitum hyemale*, and we in Englifh thereafter, Winters Wolfesbane; and of fome, Yellow Aconite.

The fecond is called by moft Writers, *Aconitum luteum Ponticum*: Some alfo *Lupicida*, *Luparia*, and *Canicida*, of the effect in killing Wolues and Dogs: And fome, becaufe the flower is more white then yellow, doe call it *Aconitum flore albido*, we call it in Englifh, The whitifh yellow Aconite, or Wolfesbane, but fome after the Latine name, The yellow Wolfesbane.

The third is called generally *Napellus*, and *Verus*, becaufe it is the true *Napellus* of the ancient Writers, which they fo termed from the forme of a Turnep, called *Napus* in Latine.

The fourth is called *Aconitum Salutiferum, Napellus Moyfis, Antora* and *Anthora, quafi Antithora*, that is, the remedy againft the poifonfull herbe *Thora*, in Englifh according to the title, eyther wholfome Helmet flower, or counterpoifon Monkes hood.

The Vertues.

Although the firft three forts of plants be very poifonfull and deadly, yet there may bee very good vfe made of them for fore eyes (being carefully applyed, yet not to all forts of fore eyes neither without difcretion) if the diftilled water be dropped therein.

The rootes of the counterpoifon Monkes hood are effectuall not onely againft the poifon of the poifonfull Helmet flower, and all others of that kinde, but alfo againft the poifon of all venemous beafts, the plague or peftilence, and other infectious difeafes, which raife fpots, pockes, or markes in the outward skinne, by expelling the poifon from within, and defending the heart as a moft foueraigne Cordiall. It is vfed alfo with good fuccefse againft the wormes of the belly, and againft the paines of the Wind collick.

Chap. XXVII.
Ranunculus. The Crowfoote.

NExt vnto the Aconites, of right are to follow the *Ranunculi*, or Crowfeete, for the nearenefse both of forme, of leaues, and nature of the plants, although lefse hurtfull, yet all of them for the moft part being fharpe and exulcerating, and not without fome danger, if any would be too bold with them. The whole Family of the *Ranunculi* is of a very large extent, and I am conftrained within the limits of a Garden of Pleafure; I muft therefore felect out onely fuch as are fit for this purpofe, and fet them here downe for your knowledge, leauing the reft for that other generall worke, which time may perfect and bring to light, if the couetous mindes of fome that fhould be moft affected towards it, doe not hinder it: or if the helpe of generous fpirits would forward it.

1. *Ranunculus montanus albus humilior*. The lowe white mountaine Crowfoot.

This lowe Crowfoote hath three or foure broad and thicke leaues, almoft round, yet a little cut in and notched about the edges, of a fine greene and fhining colour on the vpperfide, and not fo green vnderneath, among which rifeth a fmall fhort ftalke, bearing one fnow white flower on the toppe, made of fiue round pointed leaues, with diuers yellow threads in the middle, ftanding about a greene head, which in time groweth to be full of feede, in forme like vnto a fmall greene Strawberry: the roote is compofed of many white ftrings.

Duplici flore.   There is another of this lowe kinde, whofe leaues are fomewhat more deeply cut in on the edges, and the flower larger, and fometimes a little double, as it were with two rowes of leaues, in other things not differing from the former.

2. *Ranunculus*

2. *Ranunculus montanus albus maior vel elatior.*
The great single white mountaine Crowfoote.

The leaues of this Crowfoote are large and greene, cut into three, and sometimes into fiue speciall diuisions, and each of them besides cut or notched about the edges, somewhat resembling the leaues of the Globe Crowfoote, but larger : the stalke is two foote and a halfe high, hauing three small leaues set at the ioynt of the stalke, where it brancheth out into flowers, which stand foure or fiue together vpon long foote-stalkes, made of fiue white leaues a peece, very sweete, and somewhat larger then the next white Crowfoote, with some yellow threads in the middle compassing a greene head, which bringeth seede like vnto other wilde Crowfeete : the roote hath many long thicke whitish strings, comming from a thicke head.

3. *Ranunculus montanus albus minor.* The lesser single white Crowfoote.

This Crowfoote hath faire large spread leaues, cut into fiue diuisions, and somewhat notched about the edges, greene on the vpperside, and paler vnderneath, hauing many veines running through the leaues : the stalke of this riseth not so high as the former, although this be reasonable tall, as being neare two foote high, spread into many branches, bearing such like white flowers, as in the former, but smaller : the seede of this is like the former, and so are the rootes likewise.

4. *Ranunculus albus flore pleno.* The double white Crowfoot.

The double white Crowfoote is of the same kinde with the last single white Crowfoote, hauing such like leaues in all respects : the onely difference is in the flowers, which in this are very thicke and double. Some doe make mention of two sorts of double white Crowfeete, one somewhat lower then another, and the lower likewise bearing more store of flowers, and more double then the higher : but I confesse, I haue neuer seene but one sort of double, which is the same here expressed, not growing very high, and reasonably well stored with flowers.

5. *Ranunculus præcox Rutæfolio siue Coriandrifolio.*
The early Coriander leafed Crowfoote.

This Crowfoote hath three or foure very greene leaues, cut and diuided into many small peeces, like vnto the wing of leaues of Rue, or rather like the lower leaues of the Coriander(for they well resemble either of them)euery of them standing vpon a long purplish stalke, at the toppe whereof groweth the flower alone, being composed or made of twelue small white leaues, broad pointed, and a little endented at the ends, somewhat purplish on the outside, and white on the inside, sustained by diuers small greene leaues, which are in stead of a cup or huske : in the middle of the flower are many small white threads, tipt with yellow pendents, standing about a small greene head, which after groweth to bee full of seedes like a Strawberry, which knobs giue small blackish seede : the roote is white and fibrous.

6. *Ranunculus Thalictrifolio maior.* The great colombine leafed Crowfoot.

The lower leaues of this Crowfoote haue long stalkes, and are very like vnto the smaller leaues of Colombines, or the great Spanish *Thalictrum,* which hath his leaues very like vnto a Colombine, foure or fiue rising from the roote : the stalke riseth about a foote and a halfe high, somewhat reddish, beset here and there with the like leaues, at the toppe whereof stand diuers small white flowers, made of fiue leaues a peece, with some pale white threads in the middle : the seede is round and reddish, contained in small huskes or hornes : the roote is made of a bush or tuft of white strings.

T   7. *Ranunculus*

7. Ranunculus Thalictrifolio minor Asphodeli radice.
The small white Colombine leafed Crowfoote.

This small Crowfoote hath three or foure winged leaues spread vpon the ground, standing vpon long stalkes, and consisting of many small leaues set together, spreading from the middle ribbe, euery leafe somewhat resembling both in shape and colour the smallest and youngest leaues of Colombines: the flowers are white, standing at the toppe of the stalkes, made of fiue round leaues: the root hath three or foure thick, short, and round yellowish clogs hanging at the head, like vnto the Asphodill roote. The great Herball of Lyons, that goeth vnder the name of *Dalischampius*, saith, that Dr. Myconus found it in Spaine, and sent it vnder the name of Oenanthe; and therefore Ioannes Molineus who is thought to haue composed that booke, set it among the vmbelliferous plants, because the Oenanthes beare vmbels of flowers and seede, and haue tuberous or cloggy rootes; but with what iudgement, let others say, when they haue compared the vmbels of flowers and seede of the Oenanthes, with the flowers and seede of this plant, and whether I haue not more properly placed it among the *Ranunculi* or Crowfeete, and giuen it a denomination agreeable to his forme.

8. Ranunculus Globosus. The Globe Crowfoot.

This Crowfoote (which in the Northerne countries of England where it groweth plentifully, is called Locker goulous) hath many faire, broad, darke greene leaues next the ground, cut into fiue, sixe, or seuen diuisions, and iagged besides at the edges; among which riseth vp a stalke, whereon are set such like leaues as are belowe, but smaller, diuided toward the toppe into some branches, on the which stand seuerall large yellow flowers, alwayes folded inward, or as a close flower neuer blowing open, as other flowers doe, consisting of eleuen leaues for the most part, set or placed in three rowes, with many yellow threads in the middle, standing about a greene rough head, which in time groweth to be small knops, wherein are contained blacke seede: the roote is composed of many blackish strings.

9. Ranunculus pratensis flore multiplici. The double yellow field Crowfoot.

There is little or no difference in the leaues of this double Crowfoot, from those of the single kindes that growe in euery medowe, being large and diuided into foure or fiue parts, and indented about the edges, but they are somewhat smaller, and of a fresher greene: the flowers stand on many branches, much diuided or separated, being not very great, but very thicke and double: the roote runneth and creepeth vnder ground like as the single doth.

10. Ranunculus Anglicus maximus multiplex.
The Garden double yellow Crowfoot or Batchelours buttons.

This great double Crowfoote, which is common in euery Garden through England, hath many great blackish greene leaues, iagged and cut into three diuisions, each to the middle ribbe: the stalkes haue some smaller leaues on them, and those next vnder the branches long and narrow: the flowers are of a greenish yellow colour, very thicke and double of leaues, in the middle whereof riseth vp a small stalke, bearing another double flower, like to the other, but smaller: the roote is round, like vnto a small white Turnep, with diuers other fibres annexed vnto it.

11. Ranunculus Gramineus. Grasse leafed Crowfoot.

The leaues of this Crowfoote are long and narrow, somewhat like vnto Grasse, or rather like the leaues of single Gilloflowers or Pinckes, being small and sharpe pointed, a little hollow, and of a whitish greene colour: among these leaues rise vp diuers slender stalkes, bearing one small flower at the toppe of each, consisting of fiue yellow
leaues,

1 *Aconitum hyemale.* Winter Wolfesbane. 2 *Aconitum flore albido siue luteum Ponticum.* The whitish yellow Wolfesbane. 3 *Napellus verus.* Blew Helmets or Monkes hood. 4 *Anthora.* The counterpoison Monkes hood. 5 *Ranunculus humilis albus simplex.* The single white low Crow-foot. 6 *Ranunculus humilis albus duplici flore.* The double lowe white Crowfoot. 7 *Ranunculus Coriandrifolio* The early Corianderleafed Crow foot. 8 *Ranunculus montanus elatior albus.* The great single white mountain Crowfoot. 9 *Ranunculus montanus albus flore pleno* The double white mountain Crowfoot. 10 *Ranunculus Thalictrifolio mouter.* The lesser Colombine leafed Crowfoot. 11 *Ranunculus globosus.* The globe Crowfoot.

leaues, with some threads in the middle : the roote is composed of many thicke, long, round white strings.

There is another of this kinde that beareth flowers with two rowes of leaues, as if it were double, differing in nothing else.

12. *Ranunculus Lusitanicus autumnalis.* The Portugall Autumne Crowfoot.

This Autumne Crowfoote hath diuers broad round leaues lying on the ground, set vpon short foote-stalkes, of a faire greene colour aboue, and grayish vnderneath, snipt all about the edges, hauing many veines in them, and sometimes swelling as with bli-sters or bladders on them; from among which rise vp two or three slender and hairy stalkes, bearing but one small yellow flower a peece, consisting of fiue and sometimes of six leaues, and sometimes of seuen or eight, hauing a few threads in the middle, set about a small greene head, like vnto many of the former Crowfeete, which bringeth small blacke seede : the roote is made of many thicke short white strings, which seeme to be grumous or kernelly rootes, but that they are somewhat smaller, and longer then any other of that kinde.

13. *Ranunculus Creticus latifolius.* The broad leafed Candy Crowfoot.

This Crowfoote of Candy, hath the greatest and broadest leaues of all the sorts of Crowfeete, being almost round, and without any great diuisions, but onely a few notches about the edges here and there, as large or larger sometimes then the palme of a mans hand; among which riseth vp the stalke, not very high when it doth first flower, but afterwards, as the other flowers doe open themselues, the stalke groweth to be a foote and a halfe high, or thereabouts, hauing some leaues on it, deeply cut in or diuided, and bearing many faire yellow flowers, consisting of fiue leaues a peece, being somewhat whitish in the middle, when the flower hath stood blowne a little time : the roote is composed of a number of small kernelly knobs, or long graines, set thicke together. This flowreth very early, being vsually in flower before the end of March, and oftentimes about the middle thereof.

14. *Ranunculus Creticus albus.* The white Candy Crowfoote.

The leaues of this Crowfoote are very like vnto the leaues of the red Crowfoote of Tripoli or Asia, hereafter set downe, being somewhat broad and indented about the edges, some of the leaues being also cut in or gashed, thereby making it as it were three diuisions, of a pale greene colour, with many white spots in them : the stalke ri-seth vp a foote high, with some leaues on it, more diuided then the lower, and diuided at the toppe into two and sometimes into three branches, each of them bearing a faire snow white flower, somewhat large, included at the first in a brownish huske or cup of leaues, which afterwards stand vnder the flowers, consisting of fiue white large round pointed leaues, in the middle whereof is set many blackish purple thrums, compassing a small long greene head, composed of many scales or chaffie whitish huskes, when they are ripe, which are the seede, but vnprofitable in all that euer I could obserue : the rootes are many small graines or kernels, set together as in the former, and much about the same colour, that is, of a darke or duskie grayish colour, but much smaller.

Alba purpureis oris & venis. There is another of this kinde, whose flowers haue purple edges, and sometimes some veines of the same purple in the leaues of the flowers, not differing in any other thing from the former.

Alba oris rubris. And another, whose edges of the flowers are of a bright red colour.

15. *Ranunculus Creticus flore argenteo.* The Argentine, or cloth of siluer Crowfoot.

The greene leaues of this Crowfoote are as small and thinne, cut in or diuided on the edges, as the last two sorts; the stalke riseth vp somewhat higher, and diuided into some branches, bearing at the toppe of euery of them one flower, somewhat smaller then the former, composed of six, seuen, and sometimes of eight small round pointed

leaues,

1 *Ranunculus gramineus flore simplici & duplici.* The single and the double grasse Crowfoot. 2 *Ranunculus Lusitanicus Autumnalis.* The Portugall Autumne Crowfoot. 3 *Ranunculus Creticus latifolius.* The broad leafed Candy Crowfoot. 4 *Ranunculus Anglicus maximus multiplex.* The double English Crowfoot. 5 *Ranunculus pratensis flore multiplici.* The double yellow field Crowfoot. 6 *Ranunculus Creticus albus.* The white Candy Crowfoot. 7 *Ranunculus Asiaticus flore albo vel pallido varis.* The white or the straw coloured Crowfoot with red tops or edges. 8 *Ranunculus Tripolitanus flore rubro simplici.* The single red Crowfoot of Tripoli. 9 *Ranunculus Asiaticus flore rubro ample.* The large single red Crowfoot of Asia. 10 *Ranunculus Asiaticus flore rubro pleno.* The double red Crowfoot of Asia. 11 *Caltha palustris flore pleno.* Double Marsh Marigold or Batchelours buttons.

leaues, of a whitiſh yellow bluſh colour on the inſide wholly, except ſometimes a lit-tle ſtript about the edges : but the outſide of euery leafe is finely ſtript with crimſon ſtripes, very thicke, ſomewhat like vnto a Gilloflower : in the middle riſeth vp a ſmall blacke head, compaſſed about with blackiſh blew threads or thrums, which head is as vnfruitfull for ſeede in our Countrey as the former. This flower hath no ſuch greene leaues vnder it, or to encloſe it before it be blowne open as the former : the rootes are in all things like the former.

16. *Ranunculus Aſiaticus ſiue Tripolitanus flore rubro.*
The ſingle red Crowfoote of Aſia or Tripoli.

The lower leaues of this red Crowfoote are alwayes whole without diuiſions, be-ing onely ſomewhat deeply indented about the edges, but the other that riſe after them are more cut in, ſometimes into three, and ſometimes into fiue diuiſions, and notched alſo about the edges : the ſtalke riſeth higher then any of the former, and hath on it two or three ſmaller leaues, more cut in and diuided then thoſe belowe : at the toppe whereof ſtandeth one large flower, made of fiue leaues, euery one being nar-rower at the bottome then at the toppe, and not ſtanding cloſe and round one to an-other, but with a certaine diſtance betweene, of a duskie yellowiſh red colour on the outſide, and of a deepe red on the inſide, the middle being ſet with many thrums of a darke purple colour : the head for ſeede is long, and ſcaly or chaffie, and idle in like manner as the reſt : the roote is made of many graines or ſmall kernels ſet together, and cloſing at the head, but ſpreading it ſelfe, if it like the ground, vnder the vpper cruſt of the earth into many rootes, encreaſing from long ſtrings, that runne from the middle of the ſmall head of graines, as well as at the head it ſelfe.

17. *Ranunculus Aſiaticus flore amplo rubro.* The large ſingle red Crowfoot of Aſia.

There hath come to vs out of Turkie, together with the former, among many other rootes, vnder the ſame title, a differing ſort of this Crowfoote, whoſe leaues weare broader, and much goaler; the flower alſo larger, and the leaues thereof broader, ſometimes eight in a flower, ſtanding round and cloſe one to another, which maketh the fairer ſhew : in all other things it is like the former.

18. *Ranunculus Aſiaticus flore rubro vario ſimplici.*
The red ſtript ſingle Crowfoote of Aſia.

This party coloured Crowfoote differeth not eyther in roote or leafe from the for-mer, the chiefeſt difference is in the flower, which being red, ſomewhat like the for-mer, hath yet ſome yellow ſtripes or veines through euery leafe, ſometimes but little, and ſometimes ſo much, that it ſeemeth to bee party coloured red and yellow : this ſort is very tender ; for we haue twice had it, and yet periſhed with vs.

19. *Ranunculus Aſiaticus flore luteo vario ſimplici.*
The yellow ſtript ſingle Crowfoote of Aſia.

There is little difference in the roote of this Crowfoote from the laſt deſcribed, but the leaues are much different, being very much diuided, and the flower is large, of a fine pale greeniſh yellow colour, conſiſting of ſix and ſeuen, and ſometimes of eight or nine round leaues ; the toppes whereof haue reddiſh ſpots, and the edges ſometimes alſo, with ſuch purpliſh thrums in the middle that the other haue. None of theſe for-mer Crowfeete with kernelly rootes, haue euer beene found to haue giuen ſo good ſeed in England, as that being ſowne, any of them would ſpring vp ; for hereof tryall hath been often made, but all they haue loſt their labour, that haue beſtowed their paines therein, as farre as I know.

20. *Ranunculus Asiaticus flore rubro pleno.*
The double red Crowfoote of Asia.

The double red Crowfoote hath his rootes and leaues so like vnto the single red kinde, that none can perceiue any difference, or know the one from the other, vntill the budde of the flower doe appeare, which after it is any thing forward, may be perceiued to be greater and fuller then the budde of the single kinde. This kinde beareth most vsually but one faire large double flower on the toppe of the stalke, composed of many leaues, set close together in three or foure rowes, of an excellent crimson colour, declining to Scarlet, the outter leaues being larger then the inner ; and in stead of thrummes, hath many small leaues set together : it hath likewise six small narrow greene leaues on the backside of the flower, where the stalke is fastened to the flower.

There is of this double kinde another sort, whose flower is of the same colour with *Polifero flore.* the former, but out of the middle of the flower ariseth another double flower, but smaller.

The Place.

These plants grow naturally in diuers Countries ; some in France, and Germany, and some in England, some in Spaine, Portugall, and Italy, and some haue been sent out of Turkie from Constantinople, and some from other parts, their titles for the most part descrying their Countries.

The Time.

Some of them flower early, as is set downe in their descriptions, or titles. The others in Aprill and May. The white Candy Crowfoote, and the other single and double sorts of Asia, about the same time, or somewhat later, and one in Autumne, as it is set downe.

The Names.

The names that are giuen seuerally to them may well serue this worke, that thereby they may bee distinguished one from another : For to set downe any further controuersie of names, how fitly or vnfitly they haue beene called, and how variably by diuers former Writers, is fitter for a generall History, vnto which I leaue what may be said, both concerning these and the rest : Onely this I would giue you to vnderstand, that the Turkie kindes haue been sent to vs vnder the names of *Teroboles* for the single, and *Teroboles Catamer lale* for the double, and yet oftentimes, those that haue been sent for double, haue proued single, so little fidelity is to bee found among them.

The Vertues.

All or most of these plants are very sharpe and exulcerating, yet the care and industry of diuers learned men haue found many good effects in many of them. For the rootes and leaues both of the wilde kindes, and of some of these of the Garden, stamped and applyed to the wrifts, haue driuen away the fits in Feuers. The roote likewise of the double English kinde is applyed for pestilent sores, to helpe to breake them, by drawing the venome to the place. They helpe likewise to take away scarres and markes in diuers places of the body.

CHAP.

Chap. XXVIII.

Caltha palustris flore pleno. Double Marsh Marigold.

AS an appendix to the Crowfeete, I muſt needes adde this plant, yet ſeuerally by it ſelfe, becauſe both it and his ſingle kinde are by moſt adioyned there-unto, for the neare reſemblance both in ſhape and ſharpeneſſe of quality. The ſingle kinde I leaue to the Ditch ſides, and moiſt grounds about them, as the fitteſt places for it, and onely bring the double kinde into my Garden, as fitteſt for his goodly proportion and beauty to be entertained, and haue place therein.

The double Marſh Marigold hath many broad and round greene leaues, a little endented about the edges, like vnto the ſingle kinde, but not altogether ſo large, eſpecially in a Garden where it ſtandeth not very moiſt: the ſtalkes are weake, round, hollow, and greene, diuided into three or foure branches at the toppe, with leaues at the ſeuerall ioynts, whereon ſtand very double flowers, of a gold yellow colour: the fiue outer leaues being larger then any of the reſt that are encompaſſed by them, which fall away after they haue ſtood blowne a great while (for it endureth in flower a moneth or more, eſpecially if it ſtand in a ſhadowie place) without bearing any ſeed: the rootes are compoſed of many thicke, long, and round whitiſh ſtrings, which runne downe deep into the ground, and there are faſtened very ſtrongly.

The Place.

This plant groweth naturally in diuers Marſhes, and moiſt grounds in Germany, yet in ſome more double then in others; it hath long agoe beene cheriſhed in our Gardens.

The Time.

It flowreth in Aprill or May, as the yeare proueth earlier or later: all his leaues doe in a manner quite periſh in Winter, and ſpring anew in the end of February, or thereabouts.

The Names.

There is great controuerſie among the learned about the ſingle kinde, but thereof I ſhall not neede to ſpeake in this place; if God permit I may in a fitter. This is called generally in Latine, *Caltha palustris multiplex*, or *flore pleno*. And wee in Engliſh (after the Latine, which take *Caltha* to be that which wee vſually call *Calendula*, a Marigold) The double Marſh Marigold.

The Vertues.

The roote hereof is ſharpe, comming neare vnto the quality of the Crowfeete, but for any ſpeciall property, I haue not heard or found any.

Chap. XXIX.

Hepatica nobilis siue trifolia. Noble Liuerwort.

NExt vnto the Crowfeete are to follow the Hepaticas, becaufe of the likeneffe with them, feeming to be fmall Crowfeete in all their parts, but of another and more wholfome kinde. Their diuerfity among themfelues confifteth chiefly in the colour of the flowers, all of them being fingle, except one which is very thicke and double.

1. *Hepatica flore cæruleo fimplici maior.*
The great fingle blew Hepatica or noble Liuerwort.

The flowers of this Hepatica doe fpring vp, blow open, and fometimes fhed and fall away, before any leaues appeare or fpread open. The rootes are compofed of a bufh of blackifh ftrings, from the feuerall heads or buttons whereof, after the flowers are rifen and blowne, arife many frefh greene leaues, each feuerally ftanding vpon his foot-ftalke, folded together, and fomewhat browne and hairy at their firft comming, which after are broad, and diuided at the edges into three parts : the flowers likewife ftand euery one vpon his owne feuerall foote-ftalke, of the fame height with the leaues for the moft part, which is about foure or fiue fingers breadth high, made of fix leaues moft vfually, but fometimes it will haue feuen or eight, of a faire blew colour, with many white chiues or threads in the middle, ftanding about a middle green head or vmbone, which after the flower is fallen groweth greater, and fheweth many fmall graines or feede fet clofe together (with three fmall greene leaues compaffing them vnderneath, as they did the flower at the bottome) very like the head of feed of manie Crowfeete.

2. *Hepatica minor flore pallido cæruleo.* The fmall blew Hepatica.

The leaues of this Hepatica are fmaller by the halfe then the former, and grow more aboundantly, or bufhing thicke together : the flowers (when it fheweth them, for I haue had the plant halfe a fcore yeares, and yet neuer faw it beare flower aboue once or twice) are of a pale or bleake blew colour, not fo large as the flowers of the former.

3. *Hepatica flore purpureo.* Purple Hepatica or noble Liuerwort.

This Hepatica is in all things like vnto the firft, but onely the flowers are of a deeper blew tending to a Violet purple : and therefore I fhall not neede to reiterate the former defcription.

4. *Hepatica flore albo minor.* The leffer white Hepatica.

The flowers of this Hepatica are wholly white, of the bigneffe of the red or purple, and the leaues fomewhat fmaller, and of a little whiter or paler greene colour, elfe in all other things agreeing with the former.

5. *Hepatica alba magno flore.* The great white Hepatica.

There is no other difference herein from the laft, but that the flower being as white, is as large as the next.

6. *Hepatica albida siue argentea.* Afh-coloured or Argentine Hepatica.

Both the leaues and the flowers of this Hepatica are larger then any of the former, except the laft : the flowers hereof at the firft opening feeme to bee a of blufh afh-colour, which doe fo abide three or foure dayes, decaying ftill vntill it turne almoft
white,

white, hauing yet still a shew of that blush ash-colour in them, till the very last.

7. *Hepatica alba straminibus rubris.* White Hepatica with red threads.

There is no difference between this Hepatica and the first white one, sauing that the threads in the middle of the flower, being white, as in the former, are tipt at the ends with a pale reddish colour, which adde a great beauty to the flowers.

8. *Hepatica flore rubro.* Red Hepatica or noble Liuerwort.

The leaues of this Hepatica are of a little browner red colour, both at their first comming vp, and afterwards, especially in the middle of the leafe more then any of the former: the flowers are in forme like vnto the rest, but of a bright blush, or pale red colour, very pleasant to behold, with white threads or chiues in the middle of them.

9. *Hepatica flore purpureo multiplici siue pleno.*
The double purple Hepatica.

The double Hepatica is in all things like vnto the single purple kinde, sauing onely that the leaues are larger, and stand vpon longer foote-stalkes, and that the flowers are small buttons, but very thicke of leaues, and as double as a flower can be, like vnto the double white Crowfoote before described, but not so bigge, of a deepe blew or purple colour, without any threads or head in the middle, which fall away without giuing any seede.

10. *Hepatica flore caruleo pleno.* The double blew Hepatica.

In the colour of this flower, consisteth the chiefest difference from the last, except one may say it is a little lesse in the bignesse of the flower, but not in doublenesse of leaues.

The Place.

All these plants with single flowers grow naturally in the Woods, and shadowie Mountaines of Germany in many places, and some of them in Italy also. The double kinde likewise hath been sent from Alphonsus Pantius out of Italy, as Clusius reporteth, and was also found in the Woods, neare the Castle of Starnbeg in Austria, the Lady Heusenstains possession, as the same Clusius reporteth also.

The Time.

These plants doe flower very early, and are of the first flowers that shew themselues presently after the deepe frosts in Ianuary, so that next vnto the Winter Wolfesbane, these making their pride appeare in Winter, are the more welcome early guests. The double kinde flowreth not altogether so early, but sheweth his flower, and abideth when the others are past.

The Names.

They haue obtained diuers names; some calling them *Hepatica, Hepatica nobilis, Hepaticum trifolium, Trifolium nobile, Trifolium aureum,* and some *Trinitas,* and *Herba Trinitatis.* In English you may call them either Hepatica, after the Latine name, as most doe, or Noble Liuerwort, which you please.

The Vertues.

These are thought to coole and strengthen the liuer, the name importing as much; but I neuer saw any great vse of them by any the Physitians of our London Colledge, or effect by them that haue vsed them in Physicke in our Country.
CHAP.

1 *Hepatica flore albo amplo simplici.* The large white Hepatica. 2 *Hepatica flore rubro simplici.* The red Hepatica. 3 *Hepatica flore purpureo pleno.* The double purple Hepatica. 4 *Geranium tuberosum.* Knobbed Cranes bill. 5 *Geranium Batrachoides flore albo vel cæruleo.* The blew or white Crowfoote Cranes bill. 6 *Geranium Hematodes.* The red Rose Cranes bill. 7 *Geranium Romanum striatum.* The variable stript Cranes bill. 8 *Geranium Creticum.* Candy Cranes bill.

CHAP. XXX.

Geranium. Storkes bill or Cranes bill.

AS was said before concerning the Crowfeet, of their large extent and restraint, the like may be said of the Storkes bils or Cranes bils ; for euen of these as of them, I must for this worke set forth the descriptions but of a few, and leaue the rest to a generall worke.

1. *Geranium tuberosum vel bulbosum.* Bulbous or knobbed Cranes bill.

The knobbed Cranes hath three or foure large leaues spread vpon the ground, of a grayish or rather dusty greene colour, euery one of them being as it were of a round forme, but diuided or cut into six or seuen long parts or diuisions, euen vnto the middle, which maketh it seeme to be so many leaues, each of the cuts or diuisions being deeply notched or indented on both sides; among which riseth vp a stalke a foote high or better, bearing thereon diuers pale but bright purple flowers, made of fiue leaues a peece, after which come small heads with long pointed beakes, resembling the long bill of a Storke or Crane, or such like bird, which after it is ripe, parteth at the bottome where it is biggest, into foure or fiue seedes, euery one whereof hath a peece of the beake head fastened vnto it, and falleth away if it bee not gathered : the roote is tuberous and round, like vnto the roote of the *Cyclamen* or ordinary Sowbread almost, but smaller, and of a darke russet colour on the outside, and white within, which doth encrease vnder ground, by certaine strings running from the mother root into small round bulbes, like vnto the rootes of the earth Chesnut, and will presently shoote leaues, and quickly grow to beare flowers, but will not abide to be kept long dry out of the ground, without danger to be vtterly spoiled.

Geranium Batrachoides flore cæruleo. The blew Crowfoote Cranes bill.

This Crowfoote Cranes bill hath many large leaues, cut into fiue or six parts or diuisions, euen to the bottome, and iagged besides on the edges, set vpon very long slender foote-stalkes, very like the leaues of the wilde Crowfoot ; from among which rise vp diuers stalkes with great ioynts, somewhat reddish, set with leaues like the former : the toppes of the stalkes are spread into many branches, whereon stand diuers flowers, made of fiue leaues a peece, as large as any of the wilde or field Crowfeete, round pointed, of a faire blew or watchet colour, which being past, there doe arise such heads or bils, as other of the Cranes bils haue : the roote is composed of many reddish strings, spreading in the ground, from a head made of diuers red heads, which lye oftentimes eminent aboue the ground.

Geranium Batrachoides flore albo. The white Crowfoote Cranes bill.

This Cranes bill is in leafe and flower altogether like the former, the onely difference betweene them consisteth in the colour of the flower, which in this is wholly white, and as large as the former : but the roote of this hath not such red heads as the other hath.

Geranium Batrachoides flore albo & cæruleo vario.
The party coloured Crowfoote Cranes bill.

The flowers of this Cranes bill are variably striped and spotted, and sometimes diuided, the one halfe of euery leafe being white, and the other halfe blew, sometimes with lesser or greater spots of blew in the white leafe, very variably, and more in some years then in others, that it is very hard to expresse all the varieties that may be obserued in the flowers, that blow at one time. In all other parts of the plant, it is so like vnto the former, that vntill it be in flower, the one cannot be knowne from the other.

Geranium

5. *Geranium Batrachoides alterum flore purpureo.*
Purple Crowfoote Cranes bill.

This purple Cranes bill hath many leaues rifing from the roote, fet vpon long foot-ftalkes, fomewhat like vnto the other, yet not fo broad, but more diuided or cut, that is, into feuen or more flits, euen to the middle, each whereof is likewife cut in on the edges more deeply then the former; the ftalkes are fomewhat knobbed at the ioynts, fet with leaues like vnto the lower, and bearing a great tuft of buds at the toppes of the branches, which breake out into faire large flowers, made of fiue purple leaues, which doe fomewhat refemble the flower of a Mallow, before it be too full blowne, each whereof hath a reddifh pointell in the middle, and many fmall threads compaffing it, this vmbell or tuft of buds doe flower by degrees, and not all at once, and euery flower abideth open little more then one day, and then fheddeth, fo that euery day yeeldeth frefh flowers, which becaufe they are fo many, are a long while before they are all paft or fpent: after the flowers are paft, there arife fmall beake heads or bils, like vnto the other Cranes bils, with fmall turning feede: the roote is compofed of a great tuft of ftrings, faftened to a knobby head.

6. *Geranium Romanum verficolor fiue ftriatum.* The variable ftript Cranes bill.

This beautifull Cranes bill hath many broad yellowifh greene leaues arifing from the roote, diuided into fiue or fix parts, but not vnto the middle as the firft kindes are: each of thefe leaues hath a blackifh fpot at the bottome corners of the diuifions, the whole leafe as well in forme as colour and fpots, is very like vnto the leafe of the *Geranium fufcum*, or fpotted Cranes bill, next following to be defcribed, but that the leaues of this are not fo large as the other: from among thefe leaues fpring vp fundry ftalkes a foote high and better, ioynted and knobbed here and there, bearing at the tops two or three fmall white flowers, confifting of fiue leaues a peece, fo thickly & variably ftriped with fine fmall reddifh veines, that no green leafe that is of that bigneffe can fhew fo many veines in it, nor fo thick running as euery leafe of this flower doth: in the middle of the flower ftandeth a fmall pointell, which when the flower is paft doth grow to be the feed veffell, wheron is fet diuers fmall feeds, like vnto the fmall feedes of other Cranes bils: the root is made of many fmall yellow threads or ftrings.

7. *Geranium fufcum fiue maculatum.* Swart tawny or fpotted Cranes bill.

The leaues of this Cranes bill are in all points like the laft defcribed, as well in the forme and diuifions as colour of the leaues, being of a yellowifh greene colour, but larger and ftronger by much: the ftalkes of this rife much higher, and are ioynted or knobbed with reddifh knees or ioynts, on the tops whereof ftand not many although large flowers, confifting of fiue leaues a peece, each whereof is round at the end, and a little fnipt round about, and doe bend or turne themfelues backe to the ftalkewards, making the middle to be higheft or moft eminent; the colour of the flower is of a darke or deepe blackifh purple, the bottome of euery leafe being whiter then the reft; it hath alfo a middle pointell ftanding out, which afterwards bring forth feede like vnto others of his kinde: the roote confifteth of diuers great ftrings, ioyned to a knobby head.

8. *Geranium Hematodes.* The red Rofe Cranes bill.

This Cranes bill hath diuers leaues fpread vpon the ground, very much cut in or diuided into many parts, and each of them againe flit or cut into two or three peeces, ftanding vpon flender long foote-ftalkes, of a faire greene colour all the Spring and Summer, but reddifh in Autumne: among thefe leaues fpring vp flender and weake ftalkes, befet at euery ioynt (which is fomewhat reddifh) with two leaues for the moft part, like vnto the lower: the flowers grow feuerally on the toppe of the ftalkes, and not many together in bunches or branches, as in all other of the Cranes bils, euery flower being as large as a fingle Rofe Campion flower, confifting of fiue large leaues,

V of

of a deeper red colour then in any other Cranes bill at the first opening, and will change more blewish afterwards : when the flower is past, there doth arise such like beakes as are in others of the same kinde, but small : the roote is hard, long, and thicke, with diuers branches spreading from it, of a reddish yellow colour on the outside, and whitish within, which abideth and perisheth not, but shooteth forth some new greene leaues, which abide all the Winter, although those that turne red doe fall away.

Geranium Creticum. Candy Cranesbill.

Candy Cranesbill beareth long and tender stalkes, whereon growe diuers broad and long leaues, cut in or iagged on the edges : the toppes of the stalkes are branched into many flowers, made of fiue leaues of a reasonable bignesse, and of a faire blew or watchet colour, with a purplish pointell in the middle, which being past, there follow beake heads like other Cranes bils, but greater, containing larger, greater, and sharper pointed seede, able to pierce the skinne, if one be not warie of it : the roote is white and long, with some fibres at it, and perisheth when it hath perfected his seede, and will spring of it owne sowing many times, if the Winter be not too sharpe, otherwise (being annuall) it must be sowne in the Spring of the yeare.

The Place.

Most of these Cranes bils are strangers vnto vs by nature, but endenizond in our English Gardens. It hath beene reported vnto mee by some of good credit, that the second or Crowfoot Cranes bill hath been found naturally growing in England, but yet I neuer saw it, although I haue seen many sorts of wilde kindes in many places. Matthiolus saith that the first groweth in Dalmatia and Illyria very plentifully. Camerarius, Clusius, and others, that most of the rest grow in Germany, Bohemia, Austria, &c. The last hath his place recorded in his title.

The Time.

All these Cranes bils doe for the most part flower in Aprill, and May, and vntill the middle of Iune. The variable or stript Cranes bill is vsually the latest of all the rest.

The Names.

The first is vsually called *Geranium tuberosum*, of some *Geranium bulbosum*, of the likenesse of the roote vnto a bulbe : It is without controuersie *Geranium primum* of Dioscorides. The second is called *Geranium Gratia Dei*, of others, *Geranium cæruleum*. The blew Cranes bill Lobel calleth it *Batrachoides*, because both leafe and flower are like vnto a Crowfoote, and the affinity with the Cranes bils in the seede causeth it rather to be referred to them then to the Crowfeete. The stript Cranes bill is called by some *Geranium Romanum*. The last sauing one is called *Geranium Hæmatodes*, or *Sanguineum*, of Lobel *Geranium Gruinale Hæmatodes supinum radice repente*. In English it may be called after the Greek and Latine, The bloudy Cranes bill, but I rather call it, The Rose Cranes bill, because the flowers are as large as single Roses, or as the Rose Campion. Some of them are called in many places of England Bassinets.

The Vertues.

All the kindes of Cranes bils are accounted great wound herbes, and effectuall to stay bleedings, yet some more then others. The Emperickes of Germanie, as Camerarius saith, extoll it wonderfully, for a singular remedie against the Stone, both in the reines and bladder.

CHAP.

Chap. XXXI.

Sanicula guttata maior. Spotted Sanicle.

Hauing long debated with my selfe, where to place this & the other plants that follow in the two next Chapters, I haue thought it not amisse for this worke to set them downe here, both before the Beares eares, which are kindes of Sanicle, as the best Authors doe hold, and after the Cranes bils, both for some qualities somewhat resembling them, and for some affinity of the flowers with the former.

The spotted Sanicle hath many small round leaues, bluntly endented about the edges, somewhat like vnto the leaues of our white Saxifrage, of a full greene colour aboue, and whitish hairy, and somewhat reddish withall vnderneath: the stalkes are set here and there with the like leaues, rising a foote and a halfe high or more, very much diuided at the toppe into sundry small branches, bearing many very small white flowers, consisting of fiue small leaues, wherein are many small red spots to be seene, as small as pins points, of a pretty sweete sent, almost like Hawthorne flowers, in the middle whereof are many small threads compassing a head, which when it is ripe containeth small blacke seede: the roote is scaly, or couered with a chaffie matter, hauing many small white fibres vnderneath, whereby it is fastened in the ground.

There is another of this kinde, like both in roote, leafe, and flower to the former, *Minor non guttata.* the onely difference is, that this is lesser then the former, and hath no spots in the flower, as the other hath.

We haue also another smaller kinde then the last, both in leafe and flower, the leaues *Minus guttata.* whereof are smaller, but rounder, and more finely snipt or indented about the edges, like the teeth of a fine sawe: the stalke is little aboue a span high, hauing many small white flowers spotted as the first, but with fewer spots.

The Place.

These growe in the shadowie Woods of the Alpes, in diuers places, and with vs they more delight in the shade then the sunne.

The Time.

All these Sanicles doe flower in May, and continue flowring vntill Iune, and the seede soone ripeneth after: the rootes abide all the Winter, with some leaues on them, springing a fresh in the beginning of the yeare.

The Names.

The former two are called by Clusius *Sanicula montana*, and by others *Sanicula guttata*: by Lobel *Geum Alpinum*. The third or last hath been sent vs vnder the name of *Sanicula montana altera minor*.

The Vertues.

The name imposed on these plants doe certainly assure vs of their vertues, from the first founders, that they are great healers, and from their taste, that they are great binders.

V 2   Chap.

Chap. XXXII.

Cotyledon altera Matthioli. Spotted Nauelwort.

THis spotted Nauelwort, as many doe call it, hath many thicke small leaues, not so broad as long, of a whitish greene colour, lying on the ground in circles, after the manner of the heads of Houseleeke, and dented about the edges, from the middle whereof sometimes (for it doth not flower euery yeare in many places) ariseth vp a stalke, scarce a foote high, beset with such like leaues as are belowe, but somewhat longer : from the middle of the stalke vp to the top it brancheth forth diuersly, with a leafe at euery ioynt, bearing three or foure flowers on euery branch, consisting of fiue white leaues, spotted with small red spots, like vnto the spotted Sanicle, but with fewer and greater spots, hauing a yellowish circle or eye in the bottome of euery flower, and many whitish threads with yellowish tips in it : the seede is small and blacke, contained in small round heads : the roote is small, long, and threadie, shooting out such heads of leaues, which abide all the Winter, those that beare flower perishing.

Cotyledon altera minor. Small dented Nauelwort.

There is another like vnto that before described in most things, the differences be these : It hath shorter leaues then the former, and dented about the edges in the like manner : the flowers hereof are white, but greater, made of six leaues, and most vsually without any spots at all in them, some are seene to haue spots also : the heads or seede vessels are more cornered then the former.

Cotyledon altera flore rubro stellato. Small red flowred Nauelwort.

This hath also many heads of leaues, but more open, which are longer, greener, and sharper pointed then eyther of the former, somewhat reddish also, and not dented about the edges, but yet a little rough in handling : the stalke ariseth from among the leaues, being somewhat reddish, and the leaues thereon are reddish pointed, diuided at the toppe into many branches, with diuers flowers thereon, made of twelue small long leaues, standing like a starre, of a reddish purple colour, with many threads therein, set about the middle head, which is diuided at the toppe into many small ends, like pods or hornes, containing therein very small seede : the root is small like the former.

Sedum serratum flore rubente maculato. The Princes Feather.

This kinde of Sengreene is composed of heads of larger, broader, and thinner leaues then any of the former, of a sadder greene colour, somewhat vneuenly endented about the edges, and not so close set together, but spreading forth into seuerall heads like as the former sorts doe, although not so plentifully ; from the middle of diuers of which heads rise vp brownish or reddish stalkes, set with smaller leaues thereon to the middle thereof, and then brancheth forth into seuerall sprigs, set with diuers small reddish flowers consisting of fiue leaues a peece, the innerside of which are of a pale red, somewhat whitish, spotted with many small bloud red spots, as small almost as pins points, with some small threads in the middle, standing about a small greene head, which turneth into the seede vessell, parted foure wayes at the head, wherein is contained small blackish seede : the rootes are small threads, which spread vnder the ground, and shoote vp seuerall heads round about it.

The Place.

All these growe in Germany, Hungarie, Austria, the Alpes, and other such like places, where they cleaue to the rocke it selfe, that hath but a crust of earth on it to nourish them. They will abide in Gardens reasonable well, if they be planted in shadowie places, and not in the sun.

The

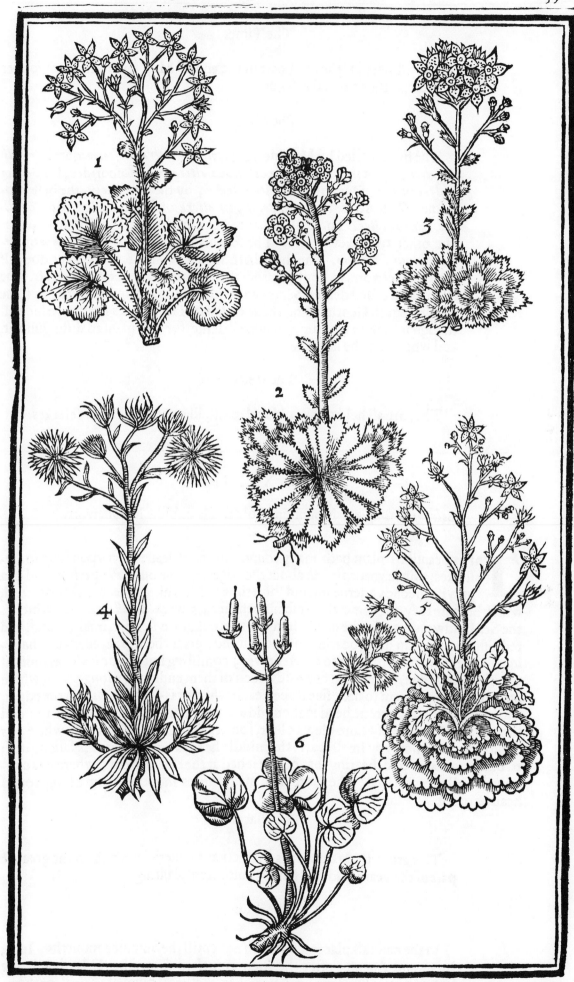

1 *Sanicula guttata.* Spotted Sanicle. 2 *Cotyledon altera Matthioli.* Spotted Nauelwort. 3 *Cotyledon altera minor.* Small dented Nauelwort. 4 *Cotyledon altera flore rubro stellato.* Small red flowred Nauelwort. 5 *Sedum serratum flore rubente maculato.* The Princes Feather. 6 *Soldanella Alpina.* Blew Moonwort.

V 3

The Time.

They flower for the moſt part in the end of May, and ſometimes ſooner or later, as the yeare falleth out.

The Names.

The firſt is called by Matthiolus, *Cotyledon altera Dioſcoridis*, and *Vmbilicus alter*, but it is not the true *Cotyledon altera* of Dioſcorides ; for *Sedum vulgare maius*, Our common Houſeleeke, by the conſent of the beſt moderne Writers, is the true *Cotyledon altera* of Dioſcorides, or *Vmbilicus Veneris alter*. I hold it rather to bee a kinde of ſmall Houſeleeke, as the other two likewiſe are. The ſecond is called by ſome *Aizoum* or *Sedum minus ſerratum*. The third hath his name in his title. Wee doe call them Nauelworts in Engliſh rather then Houſeleekes, *Euphoniæ gratia*. The laſt may be called dented Sengreene with reddiſh ſpotted flowers, but ſome of our Engliſh Gentlewomen haue called it, The Princes Feather, which although it be but a by-name, may well ſerue for this plant to diſtinguiſh it, and whereby to be knowne.

The Vertues.

They are all held to be cold and moiſt, like vnto other Houſeleekes.

Chap. XXXIII.

Soldanella Alpina. Mountaine Soldanella or blew Moonewort.

THis beautifull plant hath many round and hard leaues, ſet vpon long footeſtalkes, a little vneuenly cut about the edges, greene on the vpperſide, and of a grayiſh greene vnderneath, and ſometime reddiſh like the leaues of Sowbread, which becauſe they doe ſomewhat reſemble the leaues of *Soldanella marina*, which is the Sea Bindweede, tooke the name thereof : the ſtalkes are ſlender, ſmall, round, and reddiſh, about a ſpan high, bearing foure or fiue flowers at the toppe, euery one hanging downe their heads, like vnto a Bell flower, conſiſting but of one leafe (as moſt of the Bindweeds doe) plated into fiue folds, each of them ending in a long point, which maketh the flower ſeem to haue fiue leaues, each whereof is deeply cut in on the edges, and hauing a round greene head in the middle, with a pricke or pointell at the end thereof : the flower is of a faire blew colour, ſometimes deeper or paler, or white, as nature liſteth without any ſmell at all : the middle head, after the flower is fallen, riſeth to be a long round pod, bearing that pricke it had at the end thereof, wherein is contained ſmall greeniſh ſeede : the roote hath many fibres ſhooting from a long round head or roote.

The Place.

This groweth on the Alpes, which are couered with ſnow the greateſt part of the yeare, and will hardly abide tranſplanting.

The Time.

In the naturall places it flowreth not vntill the Summer moneths, Iune, Iuly, and Auguſt, after the ſnow is melted from the Hils, but being brought into Gardens, it flowreth in the beginning of Aprill, or thereabouts.

The

The Names.

This plant, by reason of the likenesse of leaues with *Soldanella*, as was before said, is called by many *Soldanella*, but yet is no Bindweede ; and therefore I rather call it in English a Mountaine Soldanella, then as Gerrard doth, Mountaine Bindweede. It is likewise called by some, *Lunaria minor cerulea*, The lesser blew Lunary or Moonwort, and so I would rather haue it called.

The Vertues.

They that imposed the name of *Lunaria* vpon this plant, seeme to referre it to the wound or consolidating herbes, but because I haue no further relation or experience, I can say no more thereof vntill tryall hath taught it. Some also from the name *Soldanella*, which is giuen it, because of the likenesse of the leaues, haue vsed it to help the Dropsie, for which the Sea plant is thought to be effectuall.

Chap. XXXIIII.

Auricula Vrsi. Beares eares.

THere are so many sundry and seuerall sorts of Beares eares, the variety consisting as well in the differing colours of the flowers, as the forme and colour of the leaues, that I shall not comprehend and set downe vnto you all the diuersities by many, that are risen vp to those that haue beene industrious in the sowing of the seedes of the seuerall sorts of them ; yet if you accept of these that I doe here offer vnto you, I shall giue you the knowledge of others, as time, occasion, and the view of them shall enable me. And because they are without all question kinds of Cowslips, I haue set them downe before them in the first place, as being of more beautie and greater respect, or at the least of more rarity various. To dispose them therefore into order, I shall ranke them vnder three principall colours, that is to say, Red or Purple, White, and Yellow, and shew you the varieties of each of them (for so many as are come to my knowledge) apart by themselues, and not promiscuously as many others haue done.

1. *Auricula Vrsi flore purpureo.* Purple Beares eare, or The Murrey Cowslip.

This purple Beares eare or Cowslip hath many greene leaues, somewhat long and smooth, narrow from the bottome of the leafe to the middle, and broad from thence to the end, being round pointed, and somewhat snipt or endented about the edges ; in the middle of these leaues, and sometimes at the sides also, doe spring round greene stalkes foure or fiue fingers high, bearing at the top many flowers, the buds whereof, before they are blowne, are of a very deepe purple colour, and being open, are of a bright, but deepe purple, vsually called a Murrey colour, consisting of fiue leaues a peece, cut in at the end as it were into two, with a whitish ring or circle at the bottome of each flower, standing in small greene cups, wherein after the flowers are fallen, are contained very small heads, not rising to the height of the cups, bearing a small pricke or pointell at the toppe of them, wherein is little blackish seede : the roote hath many whitish strings fastened to the maine long roote, which is very like vnto a Primrose or Cowslip roote, as it is in all other parts besides.

2. *Auricula Vrsi purpurea absq́; orbe.* The murrey Cowslip without eyes.

There is another of this kinde, whose leafe is somewhat lesse, as the flower is also,

but

but of the fame colour, and fometimes fomewhat redder, tending to a Scarlet, without any circle at the bottome of the flower, in no other things differing from it.

3. *Auricula Vrſi minor flore tannetto.* Tawney Beares eares.

The leaues of this kinde haue a greater fhew of mealineſſe to be feene in them, and not much fmaller then the former, yet fnipt or endented about the ends like vnto them : the flowers are many, of the fame fafhion with the former, but fmaller, each whereof is of as deepe a murrey or tawnie colour when it is blowne, as the buds of the former are before they are blowne, hauing a white circle at the bottome of the flower, and yellowifh in the middle belowe the circle.

4. *Auricula Vrſi flore rubro ſaturo orbe luteo.*
Deepe or bloud red Beares eares with eyes.

This kinde hath fmall and long greene leaues, nothing mealy, but fnipt about the edges, from the middle of the leaues forwards to the ends : the flowers hereof are of a deepe red colour, tending to a bloud red, with a deepe yellow circle, or rather bottome in the middle.

Auricula Vrſi flore rubro ſaturo abſque orbe.

There is another of this kinde, whofe leaues are fomewhat mealy, and fmaller then any (that I haue feene) that haue mealy leaues : the flowers are of the fame deepe red colour with the laft defcribed, yet hath no circle or bottome of any other colour at all.

5. *Auricula Vrſi flore purpuro caruleo.* The Violet coloured Beares eare.

We haue another, whofe leaues are fomewhat mealy and large ; the flowers whereof are of a paler purple then the firft, fomewhat tending to a blew.

6. *Auricula Vrſi flore obſoleto magno.* The Spaniards blufh Beares eare.

This great Beares eare hath as large leaues as any other of this kindred whatfoeuer, and whitifh or mealy withall, fomewhat fnipt about the edges, as many other of them are : the flowers ftand at the toppe of a ftrong and tall ftalke, larger then any of the other that I haue feene, being of a duskie blufh colour, refembling the blufh of a Spaniard, whofe tawney skinne cannot declare fo pure a blufh as the Englifh can ; and therefore I haue called it the Spaniards blufh.

7. *Auricula Vrſi flore rubello.* Scarlet or light red Beares eares.

The leaues of this kinde are very like the leaues of the firft purple kinde, but that they are not fo thicke; of a little paler greene colour, and little or nothing fnipt about the edges : the flowers are of a bright, but pale reddifh colour, not halfe fo deepe as the two laft with white circles in the bottomes of them, in other things this differeth not from others.

8. *Auricula Vrſi Roſeo colore.* The Rofe coloured Beares eare.

We haue another, whofe leafe is a little mealy, almoft as large as any of the former, whofe flowers are of a light red colour, very neare the colour of an ordinary Damaske Rofe, with a white eye at the bottome.

9. *Auricula Vrſi flore caruleo folio Boraginis.*
Blew Beares eares with Borage leaues.

This plant is referred to the kindred or family of the Beares eares, onely for the forme of the flower fake, which euen therein it doth not affimilate to the halfe ; but becaufe it hath paffed others with that title, I am content to infert it here, to giue you
the

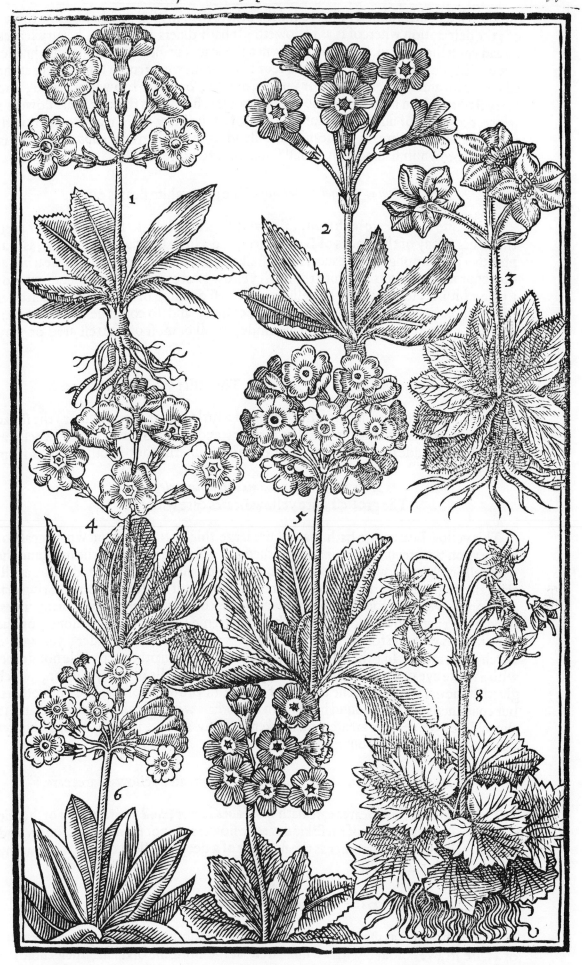

1 *Auricula Vrsi flore purpureo.* Purple Cowſlips or Beares eares. 2 *Auricula Vrsi flore tannetto.* Tawney Beares eare. 3 *Auricula Vrsi flore & folio Boraginis.* Blew Beares eares with Borage leaues. 4 *Auricula Vrsi flore carneo* Bluſh Beares eare. 5 *Auricula Vrsi maxima lutea flore eleganti.* The greateſt faire yellow Beares eares with eyes. 6 *Auricula Vrsi altera flore luteo.* The yellow Beares eare. 7 *Auricula Vrsi crinis coloris ſiue flore fuſco.* The haire coloured Beares eare. 8 *Cortuſa Matthioli.* Beares eare Sanicle.

the knowledge thereof, and rather to fatisfie others then my felfe with the place thereof : the defcription whereof is as followeth : It hath diuers broad rough hairy leaues fpread vpon the ground, fomewhat like vnto the leaues of Borage for the roughnefle, but not for the largeneffe; the leaues hereof being fomewhat rent in fome places at the edges : from among thefe leaues rife vp one, or two, or more brownifh, round, and hairy ftalkes, a fpan high or thereabouts, bearing at the toppes three or foure flowers a peece, confifting of fiue large pointed leaues, of a faire blew or light azur colour, with fome fmall yellow threads in the middle, ftanding in fmall greene cups : the roote is long and brownifh, hauing many fmall fibres annexed vnto it.

10. *Auricula Vrfi maior flore albo.* The great white Beares eare.

This white Beares eare hath many faire whitifh greene leaues, fomewhat paler then the leaues of any of the kindes of Beares eares, and a little fnipt about the ends, as manie other are : among thefe leaues rife vp ftalkes foure or fiue inches high, bearing at the toppe many flowers like vnto the fmall yellow Beares eare hereafter fet downe, of a pale whitifh colour, tending to yellow at the firft opening of the flower, which after two or three dayes change into a faire white colour, and fo continue all the while it flowreth : the roote is like the purple kinde, as all or moft of the reft are, or very little differing.

11. *Auricula Vrfi minor flore albo.* The leffer white Beares eare.

The leffer Beares eare hath fmaller leaues, of a little darker green colour : the ftalke and flowers are likewife leffer then the former, and haue no fhew of yellowneffe at all, eyther in budde or flower, but is pure white, differing not in other things from the reft.

12. *Auricula Vrfi maxima lutea flore eleganti.* The greateft faire yellow Beares eare with eyes.

This yellow Beares eare hath many faire large thicke leaues, fomewhat mealy or hoary vpon the greenneffe, being larger then any other kinde, except the fixth, and the next yellow that followeth, fmooth about the edges, and without any endenting at all : the ftalke is great, round, and not higher then in other of the former, but bearing manie more flowers thereon then in any other kinde, to the number of thirty many times, ftanding fo round and clofe together, that they feeme to be a Nofegay alone, of the fame fafhion with the former, but that the leaues are fhorter and rounder, yet with a notch in the middle like the reft, of a faire yellow colour, neither very pale nor deepe, with a white eye or circle in the bottome, about the middle of euery flower, which giueth it the greater grace : the feede is of a blackifh browne colour, like vnto others, but contained in greater round heads then any other, with a fmall pointell fticking in the middle : the roote is greater and thicker then any other, with long ftrings or fibres like vnto the other forts, but greater.

13. *Auricula Vrfi maior lutea folio in cauo.* The greater yellow Beares eare.

This greater yellow Beares eare hath his leaues larger, and more mealy or hoarie then the laft, or any other of thefe kindes : the flowers are not fo many, but longer, and not fo thicke thrufting together as the firft, but of a deeper yellow colour, without any eye or circle in the middle.

14. *Auricula Vrfi maior flore pallido.* The great Straw coloured Beares eare.

This hath almoft as mealy leaues as the laft, but nothing fo large ; the flowers are of a faire ftrawe colour, with a white circle at the bottome of them, thefe three laft haue no fhew or fhadow of any other colour in any part of the edge, as fome others that follow haue.

15. *Auricula*

15. *Auricula Vrsi minor flore pallente.* The lesser straw coloured Beares eare.

We haue another, whose leafe is lesse mealy, or rather pale green, and a little mealy withall; the flowers whereof are of a paler yellow colour then the last, and beareth almost as many vpon a stalke as the first g eat yellow.

16. *Auricula Vrsi minor lutea.* The lesser yellow Beares eares.

The leaues of this Beares eare are nothing so large as either of the three former yellow kindes, but rather of the bignesse of the first white kinde, but yet a little larger, thicker, and longer then it, hauing vnder the greennesse a small shew of mealinesse, and somewhat snipt about the edges: the flowers are of a pale yellow colour, with a little white bottome in them: the seed and rootes are like vnto the other kindes.

17. *Auricula Vrsi flore flauo.* The deepe yellow or Cowslip Beares eare.

This kinde hath somewhat larger leaues then the last, of a yellowish greene colour, without any mealinesse on them, or endenting about the edges, but smooth and whole: the flowers are not larger but longer, and not laide open so fully as the former, but of as deepe a yellow colour as any Cowslip almost, without any circle in the bottome: neither of these two last haue any shew of other colour then yellow in them, sauing the white in the eye.

18. *Auricula Vrsi versicolor prima siue flore rubescente.* The blush Beares eare.

The blush Beares eare hath his leaues as large, and as hoary or mealy as the third greater yellow, or straw coloured Beares eare; among which riseth vp a stalke about foure inches high, bearing from six to twelue, or more faire flowers, somewhat larger then the smaller yellow Beares eare before described, hauing the ground of the flower of a darke or dunne yellow colour, shadowed ouer a little with a shew of light purple, which therefore we call a blush colour, the edges of the flower being tipt with a little deeper shew of that purple colour, the bottome of the flower abiding wholly yellow, without any circle, and is of very great beauty, which hath caused me to place it in the forefront of the variable coloured Beares eares. And although some might thinke it should be placed among the first ranke of Beares eares, because it is of a blush colour, yet seeing it is assuredly gained from some of the yellow kindes by sowing the seede, as many other sorts are, as may be seene plainly in the ground of the flower, which is yellow, and but shadowed ouer with purple, yet more then any of the rest that follow; I thinke I haue giuen it his right place: let others of skill & experience be iudges herein.

19. *Auricula Vrsi crinis coloris.* Haire coloured Beares eares.

The leaues of this kinde are more mealy like then the last blush kinde, but somewhat longer and larger, and snipt about the edges in the same manner, from the middle of the leafe forwards: the flower is vsually of a fine light browne yellow colour, which wee doe vsually call an Haire colour, and sometimes browner, the edges of the flower haue a shew or shadow of a light purple or blush about them, but more on the outside then on the inside.

20. *Auricula Vrsi versicolor lutea.* The yellow variable Beares eare.

This variable Beares eare hath his greene leaues somewhat like vnto the deepe yellow, or Cowslip Beares eare before described, but somewhat of a fresher greene, more shining and smaller, and snipt about the edges towards the ends, as many of those before are: the flowers are of a faire yellow colour, much laid open when it is full blowne, that it seemeth almost flat, dasht about the edges onely with purple, being more yellow in the bottome of the flower, then in any other part.

21. *Auricula*

21. *Auricula Vrsi versicolor lutescente viridi flore.* The variable green Beares eare.

This kinde of Beares eare hath greene leaues, very like vnto the last described, and snipt in the like manner about the edges, but in this it differeth, that his leaues do turne or fold themselues a little backwards : the flowers are of a yellowish greene colour, more closed then the former, hauing purplish edges, especially after they haue stood blowne some time, and haue little or none at the first opening : these haue no circles at all in them.

Many other varieties are to be found, with those that are curious conseruers of these delights of nature, either naturally growing on the mountaines in seuerall places, from whence they (being searched out by diuers) haue been taken and brought, or else raised from the seede of some of them, as it is more probable : for seuerall varieties haue beene obserued (and no doubt many of these before specified) to bee gotten by sowing of the seedes, euery yeare lightly shewing a diuersity, not obserued before, either in the leafe, diuers from that from whence it was taken, or in the flowers. I haue onely set downe those that haue come vnder mine owne view and not any by relation, euen as I doe with all or most of the things contained in this worke.

The Place.

Many of these goodly plants growe naturally on mountaines, especially the Alpes, in diuers places ; for some kindes that growe in some places, doe not in others, but farre distant one from the other. There hath likewise some beene found on the Pyrenæan mountaines, but that kinde with the blew flower and Borage leafe, hath beene gathered on the mountaines in Spaine, and on the Pyrenæans next vnto Spaine.

The Time.

They all flower in Aprill and May, and the seede is ripe in the end of Iune, or beginning of Iuly, and sometimes they will flower againe in the end of Summer, or in Autumne, if the yeare proue temperate, moist, and rainie.

The Names.

It is very probable, that none of these plants were euer knowne vnto the ancient Writers, because we cannot be assured, that they may be truely referred vnto any plant that they name, vnlesse we beleeue Fabius Columna, that it should be *Alisma* of Dioscorides, for thereunto hee doth referre it. Diuers of the later Writers haue giuen vnto them diuers names, euery one according to his owne conceit. For Gesner calleth it *Lunaria arthritica,* and *Paralytica Alpina.* Matthiolus accounteth it to bee of the kindred of the Sanicles, and saith, that in his time it was called by diuers Herbarists, *Auricula Vrsi,* which name hath since bin receiued as most vsuall. We in English call them Beares eares, according to the Latine, or as they are called by diuers women, French Cowslips ; they may be called Mountaine Cowslips, if you will, for to distinguish betweene them and other Cowslips, whereof these are seuerall kindes.

Sanicula Alpina siue Cortusa Matthioli. Beares eare Sanicle.

I cannot chuse but insert this delicate plant in the end of the Beares eares, for that it is of so neare affinity, although it differ much in the forme of the leaues, the description whereof is in this manner: The leaues that spring vp first are much crumpled, and as it were folded together, which afterwards open themselues into faire, broad, and roundish leaues, somewhat rough or hairy, not onely cut into fine diuisions, but somewhat notched also about the edges, of a darke greene colour on the vpperside, and

more

more whitish greene vnderneath; amongst these leaues riseth vp one or two naked round stalkes, fiue or six inches high, bearing at the toppes diuers small flowers, somewhat sweete, like vnto the first purple Beares eare, hanging downe their heads, consisting of fiue small pointed leaues a peece, of a darke reddish purple colour, with a white circle or bottome in the middle, and some small threads therein : after the flowers are past, there come small round heads, somewhat longer then any of the Beares eares, standing vpright vpon their small foot-stalkes, wherein is contained small round and blackish seede : the roote consisteth of a thicke tuft of small whitish threads, rather then rootes, much enterlaced one among another : the leaues of this plant dye downe euery yeare, and spring vp a new in the beginning of the yeare, whereas all the Beares eares doe hold their leaues greene all the Winter, especially the middlemost, which stand like a close head, the outermost for the most part perishing after seed time.

The Place.

This groweth in many shadowie Woods both of Italy and Germany; for both Clusius hath described it, finding it in the Woods of Austria and Stiria; and Matthiolus setteth it downe, hauing receiued it from Anthonius Cortusus, who was President of the Garden at Padua, and found it in the woody mountaines of Vicenza, neare vnto Villestagna, whereon (as Matthiolus saith) there is found both with white flowers as well as with blew, but such with white flowers or blew we neuer could see or heare further of.

The Time.

It flowreth much about the time of the Beares eares, or rather a little later, and the seede is ripe with them.

The Names.

Clusius calleth it *Sanicula montana*, and *Sanicula Alpina*, and referreth it to the *Auricula Vrsi*, or Beares eare, which it doth most nearly resemble : but Matthiolus referreth it to the *Cariophyllata* or *Auens*, making it to be of that tribe or family, and calleth it *Cortusa* of him that first sent it him. Wee may call it eyther *Cortusa*, as for the most part all Herbarists doe, or Beares eare Sanicle as Gerrard doth.

The Vertues.

All the sorts of Beares eares are Cephalicall, that is, conducing helpe for the paines in the head, and for the giddinesse thereof, which may happen, eyther by the sight of steepe places subiect to danger, or otherwise. They are accounted also to be helping for the Palsey, and shaking of the ioynts; and also as a Sanicle or wound-herbe. The leaues of the *Cortusa* taste a little hot, and if one of them bee laide whole, without bruising, on the cheeke of any tender skind woman, it will raise an orient red colour, as if some *fucus* had beene laide thereon, which will passe away without any manner of harme, or marke where it lay : This is Cortusus his obseruation. Camerarius in his *Hortus Medicus* saith, that an oyle is made thereof, that is admirable for to cure wounds.

X CHAP.

Chap. XXXV.

Primula veris & Paralysis. Primroses and Cowslips.

WE haue so great variety of Primroses and Cowslips of our owne Country breeding, that strangers being much delighted with them, haue beene often furnished into diuers Countries, to their good content: And that I may set them downe in some methodicall manner, as I haue done other things, I will first set downe all the sorts of those we call Primroses, both single and double, and afterwards the Cowslips with their diuersities, in as ample manner as my knowledge can direct me. And yet I know, that the name of *Primula veris* or Primrose, is indifferently conferred vpon those that I distinguish for *Paralyses* or Cowslips. I doe therefore for your better vnderstanding of my distinction betweene Primroses and Cowslips, call those onely Primroses that carry but one flower vpon a stalke, be they single or double, except that of Master Hesket, and that with double flowers many vpon a stalke, set out in Gerards Herball, which is his onely, not found (as I thinke) *in rerum natura*, I am sure, such a one I could neuer heare of: And those Cowslips, that beare many flowers vpon a stalke together constantly, be they single or double also. I might otherwise distinguish them also by the leafe; that all the Primroses beare their long and large broad yellowish greene leaues, without stalkes most vsually; and all the Cowslips haue small stalkes vnder the leaues, which are smaller, and of a darker greene, as vsually, but that this distinction is neither so certaine and generall, nor so well knowne.

1. *Primula veris flore albo.* The single white Primrose.

The Primrose that groweth vnder euery bush or hedge, in all or most of the Woods, Groues, and Orchards of this Kingdome, I may well leaue to his wilde habitation, being not so fit for a Garden, and so well knowne, that I meane not to giue you any further relation thereof: But we haue a kinde hereof which is somewhat smaller, and beareth milke white flowers, without any shew of yellownesse in them, and is more vsually brought into Gardens for the rarity, and differeth not from the wilde or ordinary kinde, either in roote or leafe, or any thing else, yet hauing those yellow spots, but smaller, and not so deepe, as are in the other wilde kinde.

2. *Primula veris flore viridi simplici.* The single greene Primrose.

The single greene Primrose hath his leaues very like vnto the greater double Primrose, but smaller, and of a sadder greene colour: the flowers stand seuerally vpon long foot stalkes, as the first single kinde doth, but larger then they, and more laide open, of the same, or very neare the same yellowish greene colour that the huske is of, so that at the first opening, the huske and the flower seeme to make one double greene flower, which afterwards separating themselues, the single flower groweth aboue the huske, and spreadeth it selfe open much more then any other single Primrose doth, growing in the end to be of a paler greene colour.

3. *Primula veris flore viridante & albo simplici.*
The single greene and white Primrose.

The leaues of this differ in a manner nothing from the former, neither doth the flower but only in this, that out of the large yellowish green huskes, which contain the flowers of the former, there commeth forth out of the middle of each of them either a small peece of a whitish flower, or else a larger, sometimes making vp a whole flower, like an ordinary Primrose.

4. *Primula veris flore viridi duplici.* The double greene Primrose.

This double Primrose is in his leaues so like the former single greene kindes, that the

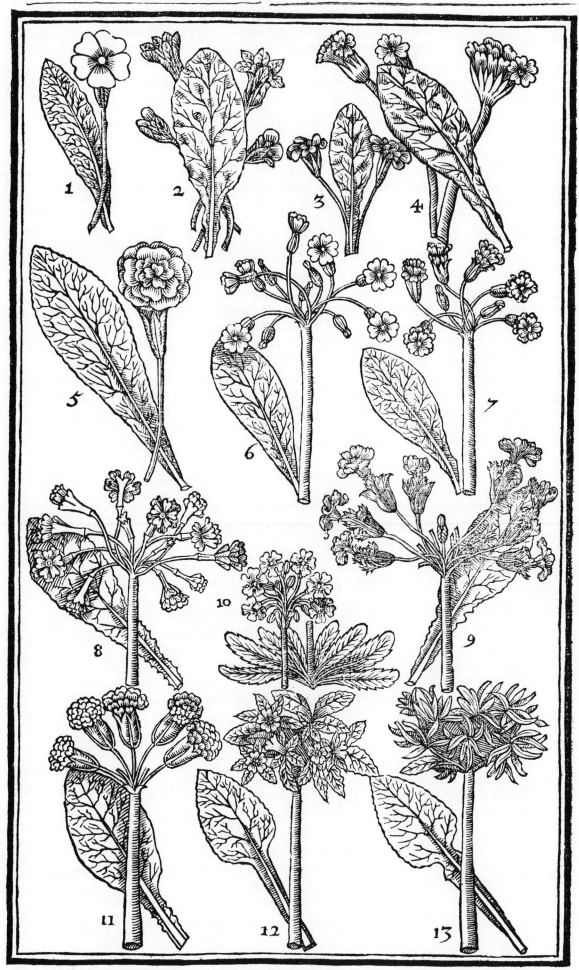

1 *Primula veris flore albo.* The white Primrose. 2 *Primula veris flore viridi & albo simplici,* The green and white Primrose. 3 *Primula veris flore viridi duplici.* The double green Primrose. 4 *Primula veris Hesketi.* Master Heskets double Primrose. 5 *Primula veris flore pleno vulgaris.* The ordinary double Primrose 6 *Paralysis veris flore viridante simplici* The single green Cowslip. 7 *Paralysis flore geminato odorato* Double Cowslips or hose in hose. 8 *Paralysis inodora flore geminato.* Double Oxelips hose in hose. 9 *Paralysis flore & calice crispo.* Curld Cowslips or Gaskins. 10 *Paralysis minor angustifolia flore rubro.* Red Birds eyen. 11 *Paralysis hortensis flore pleno vulgaris.* Double Paigles. 12 *Paralysis fatua.* The foolish Cowslip, or Iacke an Apes on horse backe. 13 *Paralysis flore viridi roseo calamistrato.* The double greene feathered Cowslip.

the one cannot be knowne from the other vntill it come to flower, and then it beareth vpon euery ftalke a double green flower, of a little deeper green colour then the flower of the former fingle kinde confifting but of two rowes of fhort leaues moft vfually, and both of an equall height aboue the huske, abiding a pretty time in flower, efpecially if it ftand in any fhadowed place, or where the Sun may come but a while vnto it.

5. *Primula veris Hesketi flore multiplici feparatim diuifo.* Mafter Heskets double Primrofe.

Mafter Heskets double Primrofe is very like vnto the fmall double Primrofe, both in leafe, roote, and heigth of growing, the ftalke not rifing much higher then it, but bearing flowers in a farre different manner ; for this beareth not only fingle flowers vpon feuerall ftalkes, but fomtimes two or three fingle flowers vpon one ftalk, and alfo at the fame time a bigger ftalke, and fomewhat higher, hauing one greene huske at the toppe thereof, fometimes broken on the one fide, and fometimes whole, in the middle whereof ftandeth fometimes diuers fingle flowers, thruft together, euery flower to be feene in his proper forme, and fometimes there appeare with fome whole flowers others that are but parts of flowers, as if the flowers were broken in peeces, and thruft into one huske, the leaues of the flowers (being of a white or pale Primrofe colour, but a little deeper) feldome rifing aboue the height of the very huske it felfe ; and fometimes as I haue obferued in this plant, it will haue vpon the fame ftalke, that beareth fuch flowers as I haue here defcribed vnto you, a fmall flower or two, making the ftalke feeme branched into many flowers, whereby you may perceiue, that it will vary into many formes, not abiding conftant in any yeare, as all the other forts doe.

6. *Primula hortenfis flore pleno vulgaris.* The ordinary double Primrofe.

The leaues of this Primrofe are very large, and like vnto the fingle kind, but fomewhat larger, becaufe it groweth in gardens : the flowers doe ftand euery one feuerally vpon flender long footeftalkes, as the fingle kinde doth, in greenifh huskes of a pale yellow colour, like vnto the field Primrofe, but very thicke and double, and of the fame fweete fent with them.

7. *Primula veris flore duplici.* The fmall double Primrofe.

This Primrofe is both in leafe, roote, and flower, altogether like vnto the laft double Primrofe, but that it is fmaller in all things ; for the flower rifeth not aboue two or three fingers high, and but twice double, that is, with two rowes of leaues, yet of the very fame Primrofe colour that the former is of.

8. *Paralyfis vulgaris pratenfis flore flauo fimplici odorato.* The Common field Cowflip.

The common fielde Cowflip I might well forbeare to fet downe, being fo plentifull in the fields : but becaufe many take delight in it, and plant it in their gardens, I will giue you the defcription of it here. It hath diuers green leaues, very like vnto the wilde Primrofe, but fhorter, rounder, ftiffer, rougher, more crumpled about the edges, and of a fadder greene colour, euery one ftanding vpon his ftalke, which is an inch or two long : among the leaues rife vp diuers round ftalkes, a foote or more high, bearing at the toppe many faire yellow fingle flowers, with fpots of a deeper yellow, at the buttome of each leafe, fmelling very fweete. The rootes are like to the other Primrofes, hauing many fibres annexed to the great roote.

9. *Paralyfis altera odorata flore pallido polyanthos.* The Primrofe Cowflip.

The leaues of this Cowflip are larger then the ordinary fielde Cowflip, and of a darke yellowifh greene colour : the flowers are many ftanding together, vpon the toppes of the ftalkes, to the number of thirty fometimes vpon one ftalke, as I haue counted them in mine owne Garden, and fometimes more, euery one hauing a longer

foote

foote ſtalke then the former, and of as pale a yellowiſh colour almoſt as the fielde Primroſe, with yellow ſpots at the bottome of the leaues, as the ordinary hath, and of as ſweet a ſent.

10. *Paralyſis flore viridante ſimplici.* The ſingle greene Cowſlip.

There is little difference in leafe or roote of this from the firſt Cowſlip, the chiefeſt varietie in this kinde is this, that the leaues are ſomewhat greener, and the flowers being in all reſpects like in forme vnto the firſt kinde, but ſomewhat larger, are of the ſame colour with the greene huskes, or rather a little yellower, and of a very ſmall ſent ; in all other things I finde no diuerſitie, but that it ſtandeth much longer in flower before it fadeth, eſpecially if it ſtand out of the Sunne.

11. *Paralyſis flore & calice criſpo.* Curl'd Cowſlips or Gallegaskins.

There is another kinde, whoſe flowers are folded or crumpled at the edges, and the huskes of the flowers bigger than any of the former, more ſwelling out in the middle, as it were ribbes, and crumpled on the ſides of the huskes, which doe ſomewhat reſemble mens hoſe that they did weare, and tooke the name of Gallegaskins from thence.

12. *Paralyſis flore geminato odorato.*
Double Cowſlips one within another, or Hoſe in Hoſe.

The only difference of this kinde from the ordinary field Cowſlip is, that it beareth one ſingle flower out of another, which is as a greene huske, of the like ſent that the firſt hath, or ſomewhat weaker

13. *Paralyſis flore flauo ſimplici inodoro abſque calicibus.* Single Oxe lippes.

This kinde of Cowſlip hath leaues much like the ordinary kinde, but ſomewhat ſmaller : the flowers are yellow like the Cowſlip, but ſmaller, ſtanding many vpon a ſtalke, but bare or naked, that is, without any huske to containe them, hauing but little or no ſent at all ; not differing in any thing elſe from the ordinary Cowſlip.

14. *Paralyſis flore geminato inodora.* Double Oxelips Hoſe in Hoſe.

As the former double Cowſlip had his flowers one within another, in the very like manner hath this kinde of Cowſlip or Oxelippe, ſauing that this hath no huske to containe them, no more then the former ſingle Oxelippe hath, ſtanding bare or naked, of the very ſame bigneſſe each of them, and of the ſame deepe yellow colour with it, hauing as ſmall a ſent as the former likewiſe.

Wee haue another of this kinde, whoſe leaues are ſomewhat larger, and ſo are the *Flore pallidiore* flowers alſo, but of a paler yellow colour.

15. *Paralyſis inodora calicibus diſſectis.* Oxelips with iagged huskes.

This kinde differeth not from the firſt Oxelip in the ſmalneſſe of the greene leaues, but in the flower, which ſtanding many together on a reaſonable high ſtalke, and being very ſmall and yellow, ſcarce opening themſelues or layde abroade as it, hath a greene huske vnder each flower, but diuided into ſixe ſeuerall ſmall long peeces.

16. *Paralyſis flore fatuo.* The Franticke, or Fooliſh Cowſlip :
Or Iacke an apes on horſe backe.

Wee haue in our gardens another kinde, not much differing in leaues from the former Cowſlip, and is called Fantaſticke or Fooliſh, becauſe it beareth at the toppe of the ſtalke a buſh or tuft of ſmall long greene leaues, with ſome yellow leaues, as it were peeces of flowers broken, and ſtanding among the greene leaues. And ſometimes

ſome

some stalkes among those greene leaues at the toppe (which are a little larger then when it hath but broken peeces of flowers) doe carry whole flowers in huskes like the single kinde.

17. *Paralysis minor flore rubro.* Red Birds eyes.

This little Cowslippe(which will hardly endure in our gardens, for all the care and industrie we can vse to keepe it) hath all the Winter long, and vntill the Spring begin to come on, his leaues so closed together, that it seemeth a small white head of leaues, which afterwards opening it selfe, spreadeth round vpon the ground, and hath small long and narrow leaues, snipt about the edges, of a pale greene colour on the vpperside,& very white or mealy vnderneath, among these leaues rise vp one or two stalks, small & hoary,halfe a foot high,bearing at the top a bush or tuft of much smaller flowers,standing vpon short foot stalkes, somewhat like vnto Cowslips,but more like vnto the Beares eares,of a fine reddish purple colour, in some deeper,in others paler, with a yellowish circle in the bottomes of the flowers, like vnto many of the Beares eares, of a faint or small sent : the seede is smaller than in any of the former kindes, and so are the rootes likewise, being small, white and threddy.

18. *Paralysis minor flore albo.* White Birds eyes.

This kinde differeth very little or nothing from the former, sauing that it seemeth a little larger both in leafe and flower, and that the flowers hereof are wholly white, without any great appearance of any circle in the bottome of them, vnlesse it be well obserued, or at least being nothing so conspicuous, as in the former.

Flore geminato. These two kindes haue sometimes, but very seldome, from among the middle of the flowers on the stalke,sent out another small stalke,bearing flowers theron likewise.

19. *Paralysis hortensis flore pleno.* Double Paigles or Cowslips.

The double Paigle or Cowslip hath smaller and darker greene leaues then the single kinde hath, and longer stalkes also whereon the leaues doe stand : it beareth diuers flowers vpon a stalke, but not so many as the single kinde, euery one whereof is of a deeper and fairer yellow colour then any of the former, standing not much aboue the brimmes of the huskes that hold them, consisting of two or three rowes of leaues set round together, which maketh it shew very thicke and double, of a prettie small sent, but not heady.

20. *Paralysis flore viridante pleno.* Double greene Cowslips.

This double greene Cowslip is so like vnto the single greene kinde formerly expressed, that vntill they be neare flowring, they can hardly be distinguished: but when it is in flower, it hath large double flowers, of the same yellowish greene colour with the single, and more laid open then the former double Paigle.

21. *Paralysis flore viridante siue calamistrato.*
The greene Rose Cowslip, or double greene feathered Cowslip.

There is small difference in the leaues of this double kinde from the last, but that they are not of so darke a greene: the chiefest difference consisteth in the flowers, which are many, standing together at the toppes of the stalkes, but farre differing from all other of these kindes : for euery flower standing vpon his owne stalke, is composed of many very small and narrow leaues, without any huske to containe them, but spreading open like a little Rose, of a pale yellowish greene colour, and without any sent at all, abiding in flower, especially if it stand in a shadowie place out of the sunne, aboue two moneths, almost in as perfect beauty, as in the first weeke.

The Place.

All these kindes as they haue been found wilde, growing in diuers places

in

in England, ſo they haue been tranſplanted into Gardens, to be there nouriſhed for the delight of their louers, where they all abide, and grow fairer
then in their naturall places, except the ſmall Birds eyes, which will (as I
ſaid) hardly abide any culture, but groweth plentifully in all the North
Countries, in their ſqually or wet grounds.

The Time.

Theſe doe all flower in the Spring of the yeare, ſome earlier and ſome
later, and ſome in the midſt of Winter, as they are defended from the colds
and froſts, and the mildneſſe of the time will permit: yet the Cowſlips doe
alwayes flower later then the Primroſes, and both the ſingle and double
greene Cowſlips lateſt, as I ſaid in their deſcriptions, and abide much after
all the reſt.

The Names.

All theſe plants are called moſt vſually in Latine, *Primulæ veris*, *Primulæ
pratenſes*, and *Primulæ ſiluarum*, becauſe they ſhew by their flowring the new
Spring to bee comming on, they being as it were the firſt Embaſſadours
thereof. They haue alſo diuers other names, as *Herba Paralyſis*, *Arthritica*,
Herba Sancti Petri, *Claues Sancti Petri*, *Verbaſculum odoratum*, *Lunaria arthritica*, *Phlomis*, *Aliſma ſiluarum*, and *Aliſmatis alterum genus*, as Fabius Columna calleth them. The Birds eyes are called of Lobel in Latine, *Paralytica Alpina*, *Sanicula anguſtifolia*, making a greater and a leſſer. Others call
them *Sanicula anguſtifolia*, but generally they are called *Primula veris minor*.
I haue (as you ſee) placed them with the Cowſlips, putting a difference betweene Primroſes and Cowſlips. And ſome haue diſtinguiſhed them, by
calling the Cowſlips, *Primula veris Elatior*, that is, the Taller Primroſe, and
the other *Humilis*, Lowe or Dwarfe Primroſes. In Engliſh they haue in like
manner diuers names, according to ſeuerall Countries, as Primroſes, Cowſlips, Oxelips, Palſieworts, and Petty Mulleins. The firſt kindes, which are
lower then the reſt, are generally called by the name of Primroſes (as I
thinke) throughout England. The other are diuerſly named; for in ſome
Countries they call them Paigles, or Palſieworts, or Petty Mulleins, which
are called Cowſlips in others. Thoſe are vſually called Oxelips, whoſe
flowers are naked, or bare without huskes to containe them, being not ſo
ſweete as the Cowſlip, yet haue they ſome little ſent, although the Latine
name doth make them to haue none. The Franticke, Fantaſticke, or Fooliſh
Cowſlip, in ſome places is called by Country people, Iacke an Apes on
horſe-backe, which is an vſuall name with them, giuen to many other
plants, as Daiſies, Marigolds, &c. if they be ſtrange or fantaſticall, differing in the forme from the ordinary kinde of the ſingle ones. The ſmalleſt
are vſually called through all the North Country, Birds eyen, becauſe of
the ſmall yellow circle in the bottomes of the flowers, reſembling the eye
of a bird.

The Vertues.

Primroſes and Cowſlips are in a manner wholly vſed in Cephalicall diſeaſes, either among other herbes or flowers, or of themſelues alone, to eaſe
paines in the head, and is accounted next vnto Betony, the beſt for that purpoſe. Experience likewiſe hath ſhewed, that they are profitable both for
the Palſie, and paines of the ioynts, euen as the Beares eares are, which
hath cauſed the names of *Arthritica*, *Paralyſis*, and *Paralytica*, to bee giuen
them. The iuice of the flowers is commended to cleanſe the ſpots or marks
of the face, whereof ſome Gentlewomen haue found good experience.

CHAP.

CHAP. XXXVI.

Pulmonaria. Lungwort, or Cowslips of Ierusalem.

ALthough thefe plants are generally more vfed as Pot-herbes for the Kitchen, then as flowers for delight, yet becaufe they are both called Cowflips, and are of like forme, but of much leffe beauty, I haue ioyned them next vnto them, in a diftinct Chapter by themfelues, and fo may paffe at this time.

1. *Pulmonaria maculofa.* Common fpotted Cowflips of Ierufalem.

The Cowflip of Ierufalem hath many rough, large, and round leaues, but pointed at the ends, ftanding vpon long foot ftalkes, fpotted with many round white fpots on the vpperfides of the fad greene or browne leaues, and of a grayer greene vnderneath: among the leaues fpring vp diuers browne ftalkes, a foote high, bearing many flowers at the toppe, very neare refembling the flowers of Cowflips, being of a purple or red-difh colour while they are buds, and of a darke blewifh colour when they are blowne, ftanding in brownifh greene huskes, and fometimes it hath beene found with white flowers: when the flowers are paft, there come vp fmall round heads, containing blacke feed: the roote is compofed of many long and thicke blacke ftrings.

2. *Pulmonaria altera non maculofa.* Vnfpotted Cowflips of Ierufalem.

The leaues of this other kinde are not much vnlike the former, being rough as they are, but fmaller, of a fairer greene colour aboue, and of a whiter greene vnderneath, without any fpots at all vpon the leaues: the flowers alfo are like the former, and of the fame colour, but a little more branched vpon the ftalkes then the former: the rootes alfo are blacke like vnto them.

3. *Pulmonaria anguftifolia.* Narrow leafed Cowflips of Ierufalem.

The leaues hereof are fomewhat longer, but not fo broad, and fpotted with whitifh fpots alfo as the former: the ftalke hereof is fet with the like long hairy leaues, but fmaller, being a foote high or better, bearing at the toppe many flowers, ftanding in huskes like the firft, being fomewhat reddifh in the bud, and of a darke purplifh blew colour when they are blowne open: the feede is like the former, all of them doe well refemble Bugloffe and Comfrey in moft parts, except the roote, which is not like them, but ftringie, like vnto Cowflips, yet blacke.

The Place.

The Cowflips of Ierufalem grow naturally in the Woods of Germany, in diuers places, and the firft kinde in England alfo, found out by Iohn Goodier, a great fearcher and louer of plants, dwelling at Maple-durham in Hampfhire.

The Time.

They flower for the moft part very early, that is, in the beginning of Aprill.

The Names.

They are generally called in Latine, *Pulmonaria,* and *maculofa,* or *non ma-culofa,* is added for diftinctions fake. Of fome it is called *Symphitum macu-lofum,* that is, fpotted Comfrey. In Englifh it is diuerfly called; as fpotted Cowflips of Ierufalem, Sage of Ierufalem, Sage of Bethlehem, Lungwort, and

and spotted Comfrey, and it might bee as fitly called spotted Buglosse, whereunto it is as like as vnto Comfrey, as I said before.

The Vertues.

It is much commended of some, to bee singular good for vlcered lungs, that are full of rotten matter. As also for them that spit bloud, being boyled and drunke. It is of greatest vse for the pot, being generally held to be good, both for the lungs and the heart.

Chap. XXXVII.

1. *Buglossum & Borrago.* Buglosse and Borage.

ALthough Borage and Buglosse might as fitly haue been placed, I confesse, in the Kitchen Garden, in regard they are wholly in a manner spent for Physicall properties, or for the Pot, yet because anciently they haue been entertained into Gardens of pleasure, their flowers hauing been in some respect, in that they haue alwaies been enterposed among the flowers of womens needle-worke, I am more willing to giue them place here, then thrust them into obscurity, and take such of their tribe with them also as may fit for this place, either for beauty or rarity.

The Garden Buglosse and Borage are so well knowne vnto all, that I shall (I doubt) but spend time in waste to describe them ; yet not vsing to passe ouer any thing I name and appropriate to this Garden so sleightly, they are thus to bee knowne : Buglosse hath many long, narrow, hairy, or rough sad greene leaues, among which rise vp two or three very high stalks, branched at the top, whereon stand many blew flowers, consisting of fiue small round pointed leaues, with a small pointell in the middle, which are very smooth, shining, and of a reddish purple while they are buds, and not blowne open, which being fallen, there groweth in the greene huske, wherein the flower stood, three or foure roundish blacke seedes, hauing that thread or pointell standing still in the middle of them : the roote is blacke without, and whitish within, long, thicke, and full of slimie iuice (as the leaues are also) and perisheth not euery yeare, as the roote of Borage doth.

2. *Borrago.* Borage.

Borage hath broader, shorter, greener, and rougher leaues then Buglosse, the stalkes hereof are not so high, but branched into many parts, whereon stand larger flowers, and more pointed at the end then Buglosse, and of a paler blew colour for the most part (yet sometimes the flowers are reddish, and sometimes pure white) each of the flowers consisting of fiue leaues, standing in a round hairy whitish huske, diuided into fiue parts, and haue a small vmbone of fiue blackish threads in the middle, standing out pointed at the end, and broad at the bottome : the seed is like the other : the root is thicker and shorter then the roote of Buglosse, somewhat blackish without also, and whitish within, and perisheth after seede time, but riseth of it owne seede fallen, and springeth in the beginning of the yeare.

3. *Borrago semper virens.* Euerliuing Borage.

Euerliuing Borage hath many broad greene leaues, and somewhat rough, more resembling Comfrey then Borage, yet not so large as either ; the stalkes are not so high as Borage, and haue many small blew flowers on them, very like to the flowers of Buglosse for the forme, and Borage for the colour : the rootes are blacke, thicker then either of them, somewhat more spreading, and not perishing, hauing greene leaues all the Winter long, and thereupon tooke his name.

4. *Anchusa.*

4. *Anchusa*. Sea Buglosse or Alkanet.

The Sea Buglosse or Alkanet hath many long, rough, narrow, and darke greene leaues, spread vpon the ground (yet some that growe by the Sea side are rather hoarie and whitish) among these leaues riseth vp a stalke, spread at the toppe into many branches, whereon stand the flowers in tufts, like vnto the Garden Buglosse, or rather Comfrey, but lesser; in some plants of a reddish blew colour, and in others more red or purplish, and in others of a yellowish colour: after which come the seedes, very like vnto Buglosse, but somewhat longer and paler: the roote of most of them being transplanted, are somewhat blackish on the outside, vntill the later end of Summer, and then become more red: for those that grow wilde, will be then so red, that they will giue a very deepe red colour to those that handle them, which being dryed keepe that red colour, which is vsed to many purposes; the roote within being white, and hauing no red colour at all.

5. *Limonium Rauwolfij*. Marsh Buglosse.

This Limonium (which I referre here to the kindes of Buglosse, as presuming it is the fittest place where to insert it) hath many long, narrow, and somewhat rough leaues lying vpon the ground, waued or cut in on both sides, like an Indenture, somewhat like the leaues of Ceterach or Miltwast, among which rise vp two or three stalkes, somewhat rough also, and with thin skinnes like wings, indented on both sides thereof also, like the leaues, hauing three small, long, rough, and three square leaues at euery ioynt where it brancheth forth; at the toppe whereof stand many flowers vpon their foote stalkes, in such a manner, as is not seene in any other plant, that I know: for although that some of the small winged foot stalkes are shorter, and some longer, standing as it were flatwise, or all on one side, and not round like an vmbell, yet are they euen at the toppe, and not one higher than another; each of which small foote stalkes doe beare foure or fiue greenish heads or huskes, ioyned together, out of each of which doe arise other pale or bleake blew stiffe huskes, as if they were flowers, made as it were of parchment, which hold their colour after they are dry a long time; and out of these huskes likewise, doe come (at seuerall times one after another, and not all at one time or together) white flowers, consisting of fiue small round leaues, with some white threds in the middle: after these flowers are past, there come in their places small long seede, inclosed in many huskes, many of those heads being idle, not yeelding any good seede, but chaffe, especially in our Countrey, for the want of sufficient heate of the Sunne, as I take it: the roote is small, long, and blackish on the outside, and perisheth at the first approach of Winter.

The Place.

Borage and Buglosse grow onely in Gardens with vs, and so doth the *Semper virens*, his originall being vnknowne vnto vs. Alkanet or Sea Buglosse groweth neare the Sea, in many places of France, and Spaine, and some of the kindes also in England. But the Limonium or Marshe Buglosse groweth in Cales, and Malacca in Spaine, and is found also in Syria, as Rauwolfius relateth: and in other places also no doubt; for it hath beene sent vs out of Italie, many yeares before eyther Guillaume Boel found it in Cales, or Clusius in Malacca.

The Time.

Borage and Buglosse doe flower in Iune, and Iuly, and sometimes sooner, and so doth the euer-liuing or neuer dying Borage, but not as Gerrard saith, flowring Winter and Summer, whereupon it should take his name, but leaueth flowring in Autumne, and abideth greene with his leaues all the Winter,

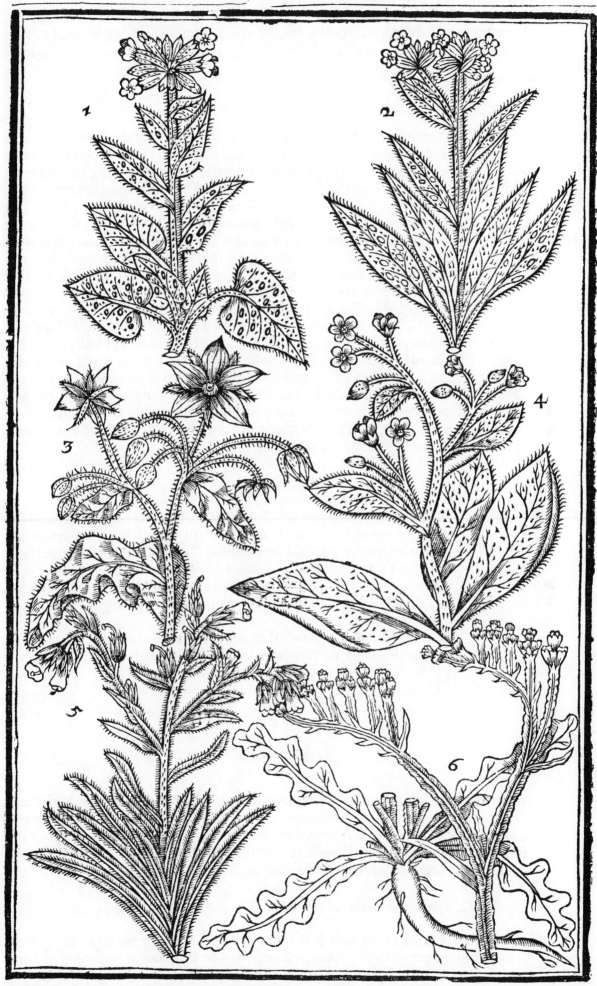

1 *Pulmonaria latifolia maculosa.* Cowslips of Ierusalem.   2 *Pulmonaria angustifolia.* Narrow leafed Cowslips of Ierusalem.   3 *Borrago.* Borage.   4 *Borrago semper virens.* Euerliuing Borage.   5 *Anchusa.* Sea Buglosse or Alkanet.   6 *Limonium Rauwolsij.* Marsh Buglosse.

ter, flowring the next Spring following. The other flower not vntill Iuly, and ſo continue, eſpecially the Marſhe Bugloſſe vntill September bee well ſpent, and then giueth ſeede, if early froſts ouertake it not; for it ſeldome commeth to be ripe.

The Names.

Our ordinary Borage by the conſent of all the beſt moderne Writers, is the true *Bugloſſum* of Dioſcorides, and that our Bugloſſe was vnknowne to the ancients. The *Borago ſemper virens*, Lobel calleth *Bugloſſum ſemper virens*, that is, Euer-liuing, or greene Bugloſſe : but it more reſembleth Borage then Bugloſſe ; yet becauſe Bugloſſe abideth greene, to auoyde that there ſhould not be two *Bugloſſa ſemper virentia*, I had rather call it Borage then Bugloſſe. Anchuſa hath diuers names, as Dioſcorides ſetteth downe. And ſome doe call it *Fucus herba*, from the Greeke word, becauſe the roote giuing ſo deepe a colour, was vſed to dye or paint the skinne. Others call it *Bugloſſum Hiſpanicum*, in Engliſh Alkanet, and of ſome Orchanet, after the French. Limonium was found by Leonhartus Rauwolfius, neere vnto ſoppa, which he ſetteth downe in the ſecond Chapter of the third booke of his trauayles, and from him firſt knowne to theſe parts : I haue, as you ſee, referred it to the kindes of Bugloſſe, for that the flowers haue ſome reſemblance vnto them, although I know that *Limonium genuinum* is referred to the Beetes. Let it therefore here finde a place of reſidence, vntill you or I can finde a fitter; and call it as you thinke beſt, eyther Limonium as Rauwolfius doth, or Marſhe Bugloſſe as I doe, or if you can adde a more proper name, I ſhall not be offended.

The Vertues.

Borage and Bugloſſe are held to bee both temperate herbes, beeing vſed both in the pot and in drinkes that are cordiall, eſpecially the flowers, which of Gentlewomen are candid for comfitts. The Alkanet is drying, and held to be good for wounds, and if a peece of the roote be put into a little oyle of Peter or Petroleum, it giueth as deepe a colour to the oyle, as the Hypericon doth or can to his oyle, and accounted to be ſingular good for a cut or greene wound.

The Limonium hath no vſe that wee know, more then for a Garden; yet as Rauwolfius ſaith, the Syrians vſe the leaues as ſallats at the Table.

Chap. XXXVIII.

Lychnis. Campions.

THere bee diuers ſorts of Campions, as well tame as wilde, and although ſome of them that I ſhall here entreate of, may peraduenture be found wilde in our owne Countrey, yet in regard of their beautifull flowers, they are to bee reſpected, and nourſed vp with the reſt, to furniſh a garden of pleaſure; as for the wilde kindes, I will leaue them for another diſcourſe.

1. *Lychnis Coronaria rubra ſimplex.* The ſingle red Roſe Campion.

The ſingle red Roſe Campion hath diuers thicke, hoary, or woolly long greene leaues, abiding greene all the winter, and in the end of the ſpring or beginning of ſummer, ſhooteth forth two or three hard round woolly ſtalkes, with ſome ioynts thereon, and at euery ioynt two ſuch like hoary greene leaues as thoſe below, but ſmaller, diuerſly branched at the toppe, hauing one flower vpon each ſeuerall long foot ſtalke,

consisting

confifting of fiue leaues, fomewhat broade and round pointed, of a perfect red crimfon colour, ftanding out of a hard long round huske, ridged or crefted in foure or fiue places ; after the flowers are fallen there come vp round hard heads, wherein is contained fmall blackifh feed : the roote is fmall, long and wooddy, with many fibres annexed vnto it, and fhooteth forth anew oftentimes, yet perifheth often alfo.

2. *Lychnis Coronaria alba fimplex.* The white Rofe Campion.

The white Rofe Campion is in all things like the red, but in the colour of the flower, which in this is of a pure white colour.

3. *Lichnis Coronaria albefcens fiue incarnata maculata & non maculata.* The blufh Rofe Campion fpotted and not fpotted.

Like vnto the former alfo are thefe other forts, hauing no other difference to diftinguifh them, but the flowers, which are of a pale or bleake whitifh blufh colour, efpecially about the brims, as if a very little red were mixed with a great deale of white, the middle of the flower being more white ; the one being fpotted all ouer the flower, with fmall fpots and ftreakes, the other not hauing any fpot at all.

4. *Lychnis Coronaria rubra multiplex.* The double red Rofe Campion.

The double red Rofe Campion is in all refpects like vnto the fingle red kinde, but that this beareth double flowers, confifting of two or three rowes of leaues at the moft, which are not fo large as the fingle, and the whole plant is more tender, that is, more apt to perifh, then any of the fingle kindes.

5. *Lychnis Chalcedonica flore fimplici miniato.* Single Nonefuch, or Flower of Briftow, or Conftantinople.

This Campion of Conftantinople hath many broad and long greene leaues, among which rife vp fundry ftiffe round hairy ioynted ftalks three foot high, with two leaues euery ioynt : the flowers ftand at the toppes of them, very many together, in a large tuft or vmbell, confifting of fiue fmall long leaues, broade pointed, and notched-in in the middle, of a bright red orenge colour, which being paft, there come in their places fmall hard whitifh heads or feede veffels, containing blacke feede, like vnto the feede of fweet Williams, and hauing but a fmall fent ; the roote is very ftringie, faftening it felfe very ftrongly in the ground, whereby it is much encreafed.

Of the fingle kinde there is alfo two or three other forts, differing chiefly in the colour of the flowers. The one is pure white. Another is of a blufh colour wholly, without variation. And a third is very variable ; for at the firft it is of a pale red, and after a while groweth paler, vntill in the end it become almoft fully white ; and all thefe diuerfities of the flowers are fometimes to bee feene on one ftalke at one and the fame time. *Flore albo.* *Et carneo.* *Verficolore*

6. *Lychnis Chalcedonica flore miniato pleno.* Double Flower of Briftow, or Nonefuch.

This glorious flower being as rare as it is beautifull, is for rootes beeing ftringie, for leaues and ftalkes being hairy and high, and for the flowers growing in tufts, altogether like the firft fingle kinde : but herein confifteth the chiefeft difference, that this beareth a larger vmbell or tuft of flowers at the toppe of the ftalke, euery flower confifting of three or foure rowes of leaues, of a deeper orenge colour then it, which addeth the more grace vnto it, but paffeth away without bearing feede, as moft other double flowers doe, yet recompenceth that defect with encreafe from the roote.

7. *Lychnis plumaria filueftris fimplex & multiplex.* The featherd wilde Campion fingle and double.

The leaues of this wilde Campion are fomewhat like the ordinary white wilde

Y   Campion

Campion, but not so large, or rather resembling the leaues of sweete Williams, but that they grow not so close, nor so many together : the stalkes haue smaller leaues at the ioynts then those belowe, and branched at the toppe, with many pale, but bright red flowers, iagged or cut in on the edges, like the feathered Pinke, whereof some haue taken it to be a kinde, and some for a kinde of wilde William, but yet is but a wilde Campion, as may be obserued, both by his huske that beareth the flowers, and by the grayish roundish seede, being not of the Family of Pinkes and Gillowers, but (as I said) of the Campions : the roote is full of strings or fibres.

Flore pleno   The double kinde is very like vnto the single kinde, but that it is lower and smaller, and the flowers very double.

8. *Lychnis siluestris flore pleno rubro.* Red Batchelours buttons.

The double wilde Campion (which of our Countrey Gentlewomen is called Batchelours buttons) is very like both in rootes, leaues, stalkes, and flowers vnto the ordinary wilde red Campion, but somewhat lesser, his flowers are not iagged, but smooth, and very thicke and double, so that most commonly it breaketh his short huske, wherein the flower standeth on the one side, seldome hauing a whole huske, and are of a reddish colour.

9. *Lychnis siluestris flore albo pleno.* White Batchelours buttons.

As the leaues of the former double Campion was like vnto the single kinde that had red flowers, so this hath his leaues like vnto the single white kinde, differing in no other thing from it, but in the doublenesse of the flowers, which by reason of the multiplicity of leaues in them thrusting forth together, breaketh his huskes wherein the flowers doe stand, as the other doth, and hath scarce one flower in many that is whole.

10. *Ocymoides arborea semper virens.* Strange Bassil Campion.

This Strange Campion (for thereunto it must bee referred) shooteth forth many round, whitish, wooddy, but brittle stalkes, whereon stand diuers long, and somewhat thicke leaues, set by couples, narrow at the bottome, and broader toward the point, of a very faire greene and shining colour, so that there is more beauty in the greene leaues, which doe so alwaies abide, then in the flowers, which are of a pale red or blush colour, consisting of fiue small long broad pointed leaues, notched in the middle, which doe not lye close, but loosly as it were hanging ouer the huskes : after the flowers are past, there come heads that containe blackish seede : the roote is small, hard, white, and threadie.

11. *Muscipula Lobelij siue Ben rubrum Monspeliensium.* Lobels Catch Flie.

I must needes insert this small plant, to finish this part of the Campions, whereunto it belongeth, being a pretty toye to furnish and decke out a Garden. It springeth vp (if it haue beene once sowne and suffered to shed) in the later end of the yeare most commonly, or else in the Spring with fiue or six small leaues, very like vnto the leaues of Pinkes, and of the same grayish colour, but a little broader and shorter, and when it beginneth to shoote vp for flower, it beareth smaller leaues on the clammy or viscous stalkes (fit to hold any small thing that lighteth on it) being broad at the bottome compassing them, and standing two at a ioynt one against another : the toppes of the stalkes are diuersly branched into seuerall parts, euery branch hauing diuers small red flowers, not notched, but smooth, standing out of small, long, round, stript huskes, which after the flowers are past, containe small grayish seede : the roote is small, and perisheth after it hath giuen seede ; but riseth (as is before said) of its owne seede, if it be suffered to shed.

The Place.

The Rose Campions, Flowers of Bristow, or None such, the Bassil Campion,
pion,

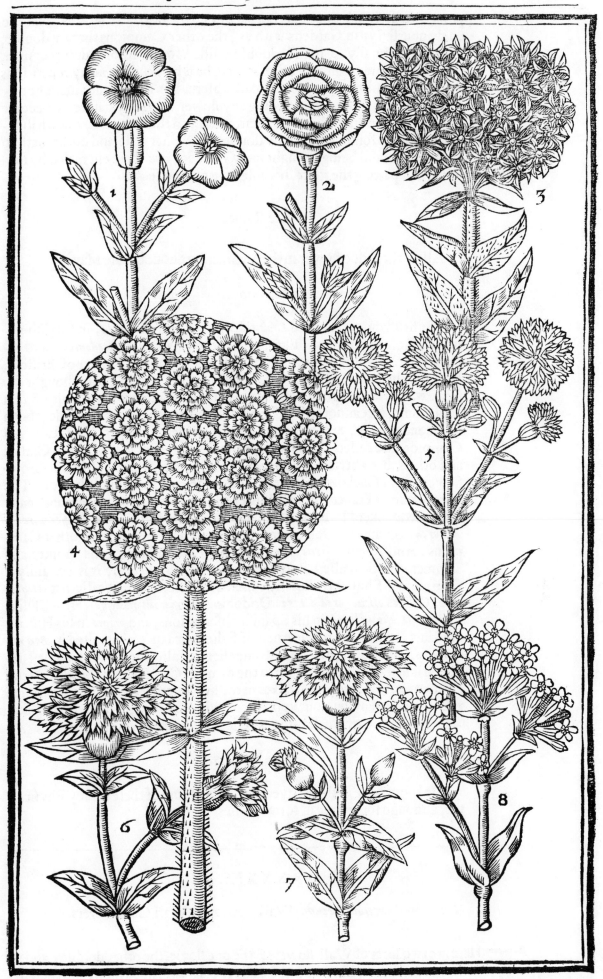

1 *Lychnis Coronaria simplex.* Single Rose Campion. 2 *Lychnis Coronaria rubra multiplex.* The double red Rose Campion. 3 *Lychnis Chalcedonica simplex.* Single None such, or flower of Bristow. 4 *Lychnis Chalcedonica flore pleno* Double None such, or flower of Bristow. 5 *Lychnis plumaria multiplex.* Pleasant in sight. 6 *Lychnis sylvestris flore pleno rubro.* Red Batchelours Buttons. 7 *Lychnis sylvestris flore pleno albo.* White Batchelours Buttons. 8 *Muscipula Lobelij.* Lobels Catch Flie.

Y 2

pion, and the Catch Flie, haue been sent vs from beyond the Seas, and are onely noursed vp in Gardens with vs; the other Campions that are double, haue been naturally so found double wilde (for no art or industry of man, that euer I could be assured of to be true, be it by neuer so many repetitions of transplantations, and planeticall obseruations (as I haue said in he beginning of this worke) could bring any flower, single by nature, to become double, notwithstanding many affirmations to that purpose, but whatsoeuer hath been found wilde to be double, nature her selfe, and not art hath so produced it) and being brought into Gardens, are there encreased by slipping, and parting the roote, because they giue no seede.

The Time.

All of them doe flower in the Summer, yet none before May.

The Names.

The first kindes are called *Lychnides satiuæ*, and *coronariæ*, in English generally Rose Campions. The next is called *Lychnis Chalcedonica*, and *Byzantina*; in English, of some Nonesuch, and of others Flower of Bristow, and after the Latine, Flower of Constantinople, because it is thought the seede was first brought from thence; but from whence the double of this kinde came, we cannot tell. The names of the others of this kinde, both single and double, are set downe with their descriptions. The feathered Campions are called *Armoraria pratensis*, and *Flos Cuculi*, and of Clusius and others thought to be *Odontitis Pliny*. Some call them in English Crowflowers, and Cuckowe-Flowers; and some call the double hereof, The faire Maide of France. The Bassil Campions were sent ouer among many other seedes out of Italy, by the name of *Ocimoides arborea semper virens*. *Arborea*, because the stalke is more wooddy and durable then other Campions: And *semper virens*, because the leaues abide greene Winter and Summer. Clusius calleth it *Lychnis semper virens*, because it is certainly a Campion. The last is diuersly called of Authors; Lobel calleth it *Muscipula*: Others *Armoraria altera*: Dodonæus *Armerius flos quartus*. Clusius *Lychnis siluestris altera*, in his Spanish obseruations, and *prima* in his History of plants, and saith, the learned of Salmantica in Spaine called it, *Ben rubrum*, as Lobel saith, they of Mompelier doe also: and by that name I receiued it first out of Italy. It hath the name of Catch Flie, of *Muscipula* the Latine word, because the stalkes in the hot Summer dayes haue a certaine viscous or clammy humour vpon them, whereby it easily holdeth (as I said before) whatsoeuer small thing, as Flies, &c. lighteth vpon it.

The Vertues.

We know none in these dayes, that putteth any of these to any Physicall vse, although some haue in former times.

CHAP. XXXIX.

Keiri siue Leucoium luteum. Wall-flowers, or Wall Gilloflowers.

THere are two sorts of Wall-flowers, the one single, the other double, and of each of them there is likewise some differences, as shall be shewed in their descriptions.

1. *Keiri siue Leucoium luteum simplex vulgare.* Common single Wall-flowers.

The common single Wall-flower which groweth wilde abroad, and yet is brought into Gardens, hath sundry small, narrow, long, and darke greene leaues, set without order vpon small round whitish wooddy stalkes, which beare at the tops diuers single yellow flowers one aboue another, euery one hauing foure leaues a peece, and of a very sweete sent: after which come long pods, containing reddish seede: the roote is white, hard and thready.

2. *Keiri siue Leucoium luteum simplex maius.* The great single Wall flower.

There is another sort of single Wall flower, whose leaues as well as flowers are much larger then the former: the leaues being of a darker and shining greene colour, and the flowers of a very deepe gold yellow colour, and vsually broader then a twentie shilling peece of gold can couer: the spike or toppe of flowers also much longer, and abiding longer in flower, and much sweeter likewise in sent: the pods for seede are thicker and shorter, with a small point at the end: this is flower to encrease into branches, as also to be encreased by the branches, and more tender to be preserued; for the hard frosts doe cause it to perish, if it be not defended from them

3. *Keiri simplex flore albo.* White Wall-flower.

This Wall-flower hath his leaues as greene as the great kinde, but nothing so large: the flowers stand at the toppe, but not in so long a spike, and consisteth of foure leaues, of a very white colour, not much larger then the common kinde, and of a faint or weaker sent: the pods are nothing so great as the former great one: this is more easie to be propagated and encreased also, but yet will require some care in defending it from the colds of the Winter.

4. *Keiri siue Leucoium luteum vulgare flore pleno.* Common double Wall-flowers.

This ordinary double Wall-flower is in leaues and stalke very like vnto the first single kinde, but that the leaues hereof are not of so deepe a greene colour: the flowers stand at the top of the stalkes one aboue another, as it were a long spike, which flower by degrees, the lowest first, and so vpwards, by which it is a long time in flowring, and is very double, of a gold yellow colour, and very sweete.

5. *Keiri siue Leucoium luteum alterum flore pleno.* Pale double Wall-flowers.

Wee haue another sort of this kinde of double Wall-flower, whose double flowers stand not spike-fashion as the former, but more open spread, and doe all of them blowe open at one time almost, and not by degrees as the other doth, and is of a paler yellow colour, not differing in any thing else, except that the greene leaues hereof are of a little paler greene then it.

6. *Keiri siue Leucoium luteum maius flore pleno ferrugineo.*
Double red Wall-flowers.

We haue also another sort of double Wall-flower, whose leaues are as greene, and almost as large as the great single yellow kinde, or full as bigge as the leaues of the white Wall-flower: the flowers hereof are not much larger then the ordinary, but are of a darker yellow colour then the great single kinde, and of a more brownish or red colour on the vnderside of the leaues, and is as it were striped.

7. *Keiri siue Leucoium maximum luteum flore pleno.*
The greatest double yellow Wall-flower.

This great double Wall-flower is as yet a stranger in England, and therefore what I

here

here write is more vpon relation (which yet I beleeue to be moſt true) then vpon ſight and ſpeculation. The leaues of this Wall flower are as greene and as large, if not larger then the great ſingle kinde : the flowers alſo are of the ſame deepe gold yellow colour with it, but much larger then any of the former double kindes, and of as ſweet a ſent as any, which addeth delight vnto beauty.

The Place.

The firſt ſingle kind is often found growing vpon old wals of Churches, and other houſes in many places of England, and alſo among rubbiſh and ſtones. The ſingle white and great yellow, as well as all the other double kindes, are nourſed vp in Gardens onely with vs.

The Time.

All the ſingle kindes doe flower many times in the end of Autumne, and if the Winter be milde all the Winter long, but eſpecially in the moneths of February, March, and Aprill, and vntill the heate of the Spring doe ſpend them : but the other double kindes doe not continue flowring in that manner the yeare throughout, although very early ſometimes, and very late alſo in ſome places.

The Names.

They are called by diuers names, as *Viola lutea*, *Leucoium luteum*, and *Keiri*, or *Cheiri*, by which name it is chiefly knowne in our Apothecaries ſhops, becauſe there is an oyle made thereof called *Cheirinum :* In Engliſh they are vſually called in theſe parts, Wall-flowers : Others doe call them Bee-flowers; others Wall-Gilloflowers, Winter Gilloflowers, and yellow Srocke-Gilloflowers, but we haue a kinde of Stocke-Gilloflower that more fitly deſerueth that name, as ſhall be ſhewed in the Chapter following

The Vertues.

The ſweetneſſe of the flowers cauſeth them to be generally vſed in Noſegayes, and to decke vp houſes; but phyſically they are vſed in diuers manners : As a Conſerue made of the flowers, is vſed for a remedy both for the Appoplexie and Palſie. The diſtilled water helpeth well in the like manner. The oyle made of the flowers is heating and reſoluing, good to eaſe paines of ſtrained and pained ſinewes.

Chap. XL.

Leucoium. Stocke-Gilloflower.

There are very many ſorts of Stocke-Gilloflowers both ſingle and double, ſome of the fields and mountaines, others of the Sea marſhes and medowes; and ſome nourſed vp in Gardens, and there preſerued by ſeede or ſlippe, as each kinde is apteſt to bee ordered. But becauſe ſome of theſe are fitter for a generall Hiſtory then for this our Garden of Pleaſure, both for that diuers haue no good ſent, others little or no beauty, and to be entreated of onely for the variety, I ſhall ſpare ſo many of them as are not fit for this worke, and onely ſet downe the reſt.

1. *Leucoium ſimplex ſativum diuerſorum colorum.*
Garden Stocke-Gilloflowers ſingle of diuers colours.

Theſe ſingle Stocke-Gilloflowers, although they differ in the colour of their flow-

ers,

1 *Keiri siue Leucoium luteum vulgare.* Common Wall-flowers. 2 *Keiri siue Leucoium luteum maius simplex.* The great single Wall-flower. 3 *Keiri siue Leucoium luteum flore pleno vulgare.* Ordinary double Wall-flowers. 4 *Keiri maius flore pleno ferrugineo* The great double red Wall-flower. 5 *Leucoium sativum simplex.* Single Stocke-Gilloflowers. 6 *Leucoium sativum simplex flore striato.* Single stript Stocke-Gilloflowers.

ers, yet are in leafe and manner of growing, one so like vnto another, that vntill they come to flower, the one cannot be well knowne that beareth red flowers, from another that beareth purple; and therfore one description of the plant shall serue, with a declaration of the sundry colours of the flowers. It riseth vp with round whitish woody stalkes, two, three, or foure foot high, whereon are set many long, and not very broad, soft, and whitish or grayish greene leaues, somewhat round pointed, and parted into diuers branches, at the toppes whereof grow many flowers, one aboue another, smelling very sweet, consisting of foure small, long, and round pointed leaues, standing in small long huskes, which turne into long and flat pods, sometimes halfe a foote long, wherein is contained flat, round, reddish seedes, with grayish ringes or circles about them, lying flat all along the middle rib of the pod on both sides : the roote is long, white, and woody, spreading diuers wayes. There is great variety in the colours of the flowers: for some are wholly of a pure white colour, others of a most excellent crimson red colour, others againe of a faire red colour, but not so bright or liuely as the other, some also of a purplish or violet colour, without any spot, marke, or line in them at all. There are againe of all these colours, mixed very variably, as white mixed with small or great spottes, strakes or lines of pure or bright red, or darke red, and white with purple spots and lines; and of eyther of them whose flowers are almost halfe white, and halfe red, or halfe white, and halfe purple. The red of both sorts, and the purple also, in the like manner spotted, striped, and marked with white, differing neyther in forme, nor substance, in any other point.

2. *Leucoium satiuum albido luteum simplex.*
The single pale yellow Stocke-Gilloflower.

There is very little difference in this kind from the former, for the manner of growing, or forme of leaues or flower. Only this hath greener leaues, and pale yellow almost white flowers, in all other things alike : this is of no great regard, but only for rarity, and diuersity from the rest.

3. *Leucoium Melancholicum.* The Melancholick Gentleman.

This wilde kinde of stocke gilloflower hath larger, longer and greener leaues then any of the former kindes, vneuenly gashed or sinuated on both edges lying on the ground, and a little rough or hairy withall: from among which rise vp the stalks, a yard high or more, and hairy likewise, bearing theron here and there some such like leaues as are below, but smaller, and at the top a great number of flowers, as large or larger then any of the former single kindes, made of 4. large leaues a peece also, standing in such like long huskes, but of a darke or sullen yellowish colour : after which come long roundish pods, wherein lye somewhat long but rounder and greater seede then any stocke gilloflower, and nearer both in pod and seede vnto the *Hesperis* or Dames Violet : this perisheth not vsually after seede bearing, although sometimes it doth.

4. *Leucoium marinum Syriacum.* Leuant stocke gilloflowers.

This kind of stocke gilloflower riseth vp at the first with diuers long and somewhat broad leaues, a little vneuenly dented or waued on the edges, which so continue the first yeare after the sowing : the stalke riseth vp the next yeare to bee two foot high or more, bearing all those leaues on it that it first had, which then do grow lesse sinuated or waued then before : at the top whereof stand many flowers, made of foure leaues a peece, of a delayed purple colour, but of a small sent which turne into very long and narrow flat pods, wherein are contained flat seed like the ordinary stocke gilloflowers, but much larger and of a darke or blackish browne colour : the root is white, and groweth deepe, spreading in the ground, but growing woody when it is in seede, and perisheth afterwards.

5. Leucoÿ alterum genus, flore tam multiplici quam simplici ex seminio oriundum.
Another fort of Stocke gilloflowers bearing as well double
as fingle flowers from feede.

This kinde of Stocke gilloflower differeth neyther in forme of leaues, ftalkes, nor flowers from the former, but that it oftentimes groweth much larger and taller ; fo that whofoeuer fhall fee both thefe growing together, fhall fcarce difcerne the difference, onely it beareth flowers, eyther white, red or purple, wholly or entire, that is, of one colour, without mixture of other colour in them (for fo much as euer I haue obferued, or could vnderftand by others) which are eyther fingle, like vnto the former, or very thicke and double, like vnto the next that followeth ; but larger, and growing with more ftore of flowers on the long ftalke. But this you muft vnderftand withall, that thofe plants that beare double flowers, doe beare no feede at all, and is very feldome encreafed by flipping or cutting, as the next kinde of double is : but the onely way to haue double flowers any yeare, (for this kinde dyeth euery winter, for the moft part, after it hath borne flowers, and feldome is preferued) is to faue the feedes of thofe plants of this kinde that beare fingle flowers, for from that feede will rife, fome that will beare fingle, and fome double flowers, which cannot bee diftinguifhed one from another, I meane which will be fingle and which double, vntill you fee them in flower, or budde at the leaft. And this is the only way to preferue this kinde : but of the feed of the former kinde was neuer known any double flowers to arife, and therefore you muft be carefull to marke this kinde from the former.

6. Leucoium flore pleno diuerforum colorum.
Double Stocke Gillowflowers of diuers colours.

This other kinde of Stock gilloflower that beareth onely double flowers, groweth not fo great, nor fpreadeth his branches fo farre, nor are his leaues fo large, but is in all things fmaller, and lower, and yet is woody, or fhrubby, like the former, bearing his flowers in the like manner, many vpon a long ftalke, one aboue another, and very double, but not fo large as the former double, although it grow in fertile foyle, which are eyther white, or red, or purple wholly, without any mixture, or elfe mixed with fpots and ftripes, as the fingle flowers of the firft kinde, but more variably, and not in all places alike, neuer bearing feede, but muft be encreafed, only by the cutting of the young fproutes or branches, taken in a fit feafon : this kinde perifheth not, as the former double kinde doth, fo as it bee defended in the winter from the extreame frofts, but efpecially from the fnow falling, or at the leaft remaining vpon it.

7. Leucoium fatiuum luteum flore pleno.
The double yellow Stocke Gilloflower.

This double yellow Stock gilloflower is a ftranger in England, as far as I can learne, neyther haue I any further familiaritie with him, then by relation from Germany, where it is affirmed to grow only in fome of their gardens, that are curious louers of thefe delights, bearing long leaues fomewhat hoary or white, (and not greene like vnto the Wallflower, whereunto elfe it might be thought to be referred) like vnto the Stock gilloflowers, as the ftalkes and branches alfo are, and bearing faire double flowers, of a faire, but pale yellow colour. The whole plant is tender, as the double Stock gilloflowers are, and muft be carefully preferued in the winter from the coldes, or rather more then the laft double, left it perifh.

The Place.

The fingle kindes, efpecially fome of them, grow in Italie, and fome in Greece, Candy, and the Ifles adiacent, as may be gathered out of the verfes in Plutarches Booke *De Amore fraterno:*

Inter

Inter Echinopodas velut, asperam & inter Ononim,
Interdum crescunt mollia Leucoia.

Which sheweth, that the soft or gentle stocke gilloflowers doe sometimes grow among rough or prickely Furse and Cammocke. The other sorts are only to be found in gardens.

The Time.

They flower in a manner all the yeare throughout in some places, especially some of the single kindes, if they stand warme, and defended from the windes and cold : the double kindes flower sometimes in Aprill, and more plentifully in May, and Iune ; but the double of seed, flowreth vsually late, and keepeth flowring vnto the winter, that the frostes and colde mistes doe pull it downe.

The Names.

It is called *Leucoium, & Viola alba* : but the name Leucoium (which is in English the white Violet) is referred to diuers plants ; we call it in English generally, Stocke gilloflower, (or as others doe, Stocke gillouer) to put a difference betweene them, and the Gilloflowers and Carnations, which are quite of another kindred, as shall be shewne in place conuenient.

The Vertues.

These haue no great vse in Physick that I know : only some haue vsed the leaues of the single white flowred kinde with salt, to be laid to the wrests of them that haue agues, but with what good successe I cannot say, if it happen well I thinke in one (as many such things else will) it will fayle in a number.

Chap. XLI.

1. *Hesperis, siue Viola Matronalis.* Dames Violets, or Queenes Gilloflowers.

THe ordinary Dames Violets, or Queene Gilloflowers, hath his leaues broader, greener, and sharper pointed, then the Stock gilloflowers, and a little endented about the edges : the stalkes grow two foot high, bearing many greene leaues vpon them, smaller then those at the bottome, and branched at the toppe, bearing many flowers, in fashion much like the flowers of stocke gilloflowers, consisting of foure leaues in like manner, but not so large, of a faint purplish colour in some, and in others white, and of a pretty sweet sent, especially towards night, but in the day time little or none at all : after the flowers are past, there doe come small long and round pods, wherein is contained, in two rowes, small and long blacke seede : the roote is wholly composed of stringes or fibres, which abide many yeares, and springeth fresh stalks euery yeare, the leaues abiding all the Winter.

2. *Hesperis Pannonica.* Dames Violets of Hungary.

The leaues of this Violet are very like the former, but smoother and thicker, and not at all indented, or cut in on the edges : the flowers are like the former, but of a sullen pale colour, turning themselues, and seldome lying plaine open, hauing many purple veines, and streakes running through the leaues of the flowers, of little or no sent in the day time, but of a very sweete sent in the euening and morning ; the seedes are alike also, but a little browner.

3 *Lysimachia*

1 *Leucoium Melancholicum.* Sullen Stocke-Gilloflowers. 2 *Leucoium sativum flore pleno.* Double Stocke-Gilloflowers. 3 *Leucoium sativum flore pleno vario.* Party coloured Stocke-Gilloflowers. 4 *Leucoium marinum Syriacum.* Leuant Stocke-Gilloflowers. 5 *Hesperis vulgaris.* Dames Violets or Winter Gilloflowers. 6 *Lysimachia lutea siliquosa Virginiana.* The tree Primrose of Virginia. 7 *Viola Lunaris sive Bolbonach.* The white S'attin flower.

3. *Lysimachia lutea siliquosa Virgiana.* The tree Primrose of Virginia.

Vnto what tribe or kindred I might referre this plant, I haue stood long in suspence, in regard I make no mention of any other *Lysimachia* in this work: left therfore it should lose all place, let me ranke it here next vnto the Dames Violets, although I confesse it hath little affinity with them. The first yeare of the sowing the seede it abideth without any stalke or flowers lying vpon the ground, with diuers long and narrow pale greene leaues, spread oftentimes round almost like a Rose, the largest leaues being outermost, and very small in the middle: about May the next yeare the stalke riseth, which will be in Summer of the height of a man, and of a strong bigge size almost to a mans thumbe, round from the bottome to the middle, where it groweth crested vp to the toppe, into as many parts as there are branches of flowers, euery one hauing a small leafe at the foote thereof: the flowers stand in order one aboue another, round about the tops of the stalks, euery one vpon a short foot-stalke, consisting of foure pale yellow leaues, smelling somewhat like vnto a Primrose, as the colour is also (which hath caused the name) and standing in a greene huske, which parteth it selfe at the toppe into foure parts or leaues, and turne themselues downewards, lying close to the stalke: the flower hath some chiues in the middle, which being past, there come in their places long and cornered pods, sharpe pointed at the vpper end, and round belowe, opening at the toppe when it is ripe into fiue parts, wherein is contained small brownish seed: the roote is somewhat great at the head, and wooddy, and branched forth diuersly, which perisheth after it hath borne seede.

The Place.

The two first grow for the most part on Hils and in Woods, but with vs in Gardens onely.

The last, as may be well vnderstood by the title, came out of Virginia.

The Time.

They flower in May, Iune, and Iuly.

The Names.

The name of *Hesperis* is imposed by most Herbarists vpon the two first plants, although it is not certainly knowne to be the same that Theophrastus doth make mention of, in his sixth Booke and twenty fiue Chapter *de causis plantarum*: but because this hath the like effects to smell best in the euening, it is (as I said) imposed vpon it. It is also called *Viola Marina Matronalis, Hyemalis, Damascena* and *Muschatella*: In English, Dames Violets, Queens Gilloflowers, and Winter Gilloflowers.

The last hath his Latine name in the title as is best agreeing with it, and for the English, although it be too foolish I confesse, yet it may passe for this time till a fitter be giuen, vnlesse you please to follow the Latine, and call it Virginia Loose-strife.

The Vertues.

I neuer knew any among vs to vse these kindes of Violets in Physicke, although by reason of the sharpe biting taste, Dodonæus accounteth the ordinary sort to be a kinde of Rocket, and saith it prouoketh sweating, and vrine: and others affirme it to cut, digest, and cleanse tough flegme. The Virginian hath not beene vsed by any that I know, either inwardly or outwardly.

CHAP.

CHAP. XLII.

Viola Lunaris siue Bolbonach. The Sattin flower.

VNto the kindes of Stocke-Gilloflowers I thinke fittest to adioyne these kindes of Sattin-flowers, whereof there are two sorts, one frequent enough in all our Countrie, the other is not so common.

1. *Viola Lunaris vulgaris*. The common white Sattin flower.

The first of these Sattin flowers, which is the most common, hath his leaues broad belowe, and pointed at the end, snipt about the edges, and of a darke greene colour: the stalkes are round and hard, two foot high, or higher, diuided into many branches, set with the like leaues, but smaller: the tops of the branches are beset with many purplish flowers, like vnto Dames Violets, or Stocke-Gilloflowers, but larger, being of little sent: after the flowers are past, there come in their places round flat thin cods, of a darke colour on the outside, but hauing a thinne middle skinne, that is white and cleare shining, like vnto very pure white Sattin it selfe, whereon lye flat and round brownish seede, somewhat thicke and great: the rootes perish when they haue giuen their seede, and are somewhat round, long, and thicke, resembling the rootes of *Lilium non bulbosum*, or Day Lilly, which are eaten (as diuers other rootes are) for Sallets, both in our owne Country, and in many places beside.

2. *Viola Lunaris altera seu peregrina*. Long liuing Sattin flower.

This second kinde hath broader and longer leaues then the former, the stalkes also are greener and higher, branching into flowers, of a paler purple colour, almost white, consisting of foure leaues in like manner, and smelling pretty sweete, bearing such like pods, but longer and slenderer then they: the rootes are composed of many long strings, which dye not as the former, but abide, and shoot out new stalkes euery yeare.

The Place.

The first is (as is said) frequent enough in Gardens, and is found wilde in some places of our owne Country, as Master Gerard reporteth, whereof I neuer could be certainly assured, but I haue had it often sent mee among other seedes from Italy, and other places. The other is not so common in Gardens, but found about Watford, as he saith also.

The Time.

They flower in Aprill or May, and sometimes more early.

The Names.

It hath diuers names, as well in English as in Latine; for it is called most vsually *Bolbonach*, and *Viola Lunaris*: Of some *Viola latifolia*, and of others *Viola Peregrina*, and *Lunaria Græca*, *Lunaria maior*, and *Lunaria odorata*, and is thought to be *Thlaspi Crateua*: In English, White Satten, or Satten flower: Of some it is called Honesty, and Penny-flower.

The Vertues.

Some doe vse to eate the young rootes hereof, before they runne vp to flower, as Rampions are eaten with Vinegar and Oyle; but wee know no Physicall vse they haue.

CHAP.

Chap. XLIII.

Linum siluestre & Linaria. Wilde Flaxe and Tode Flaxe.

ALthough neither the manured Line or Flaxe is a plant fit for our Garden, nor many of the wilde forts, yet there are fome, whofe pleafant and delightfull afpeĉt doth entertaine the beholders eyes with good content, and thofe I will fet downe here for varietie, and adioyne vnto them fome of the *Linarias,* or Tode Flaxe, for the neare affinity with them.

1. *Linum siluestre flore albo.* Wilde Flaxe with a white flower.

This kinde of wilde Flaxe rifeth vp with diuers flende branches, a foote high or better, full of leaues, ftanding without order, being broader and longer then the manured Flaxe : the tops of the branches haue diuers faire white flowers on them, compofed of fiue large leaues a peece, with many purple lines or ftrikes in them : the feede veffell as well as the feede, is like vnto the heads and feede of the manured Flaxe: the rootes are white ftrings, and abide diuers yeares, fpringing frefh branches and leaues euery yeare, but not vntill the Spring of the yeare.

2. *Linum siluestre luteum.* Wilde Flaxe with a yellow flower.

This wilde Flaxe doth fo well refemble a kinde of St. Iohns wort, that it will foone deceiue one that doth not aduifedly regard it : For it hath many reddifh ftalkes, and fmall leaues on them, broader then the former wilde Flaxe, but not fo long, which are well ftored with yellow flowers, as large as the former, made of fiue leaues a peece, which being paft, there come fmall flattifh heads, containing blackifh feede, but not fhining like the former : the rootes hereof dye not euery yeare, as many other of the wilde kindes doe, but abide and fhoote out euery yeare.

3. *Linaria purpurea.* Purple Tode Flaxe.

This purple Tode Flaxe hath diuers thicke, fmall, long, and fomewhat narrowifh leaues, fnipt about the edges, of a whitifh greene colour, from among which rife vp diuers ftalkes, replenifhed at the tops with many fmall flowers, ftanding together one aboue another fpike-fafhion, which are fmall and fomewhat fweete, while they are frefh, fafhioned fomewhat like the common Tode flaxe that groweth wilde abroad almoft euery where, but much fmaller, with a gaping mouth, but without any crooked fpurre behinde, like vnto them, fometimes of a fad purple neare vnto a Violet, and fometimes of a paler blew colour, hauing a yellow fpot in the middle or gaping place: after the flowers are paft, there come fmall, hard, round heads, wherein are contained fmall, flat, and grayifh feede : the roote is fmall, and perifheth for the moft part euery yeare, and will fpring againe of it owne fowing, if it be fuffered to fhed it felfe, yet fome hard Winters haue killed the feede it fhould feeme, in that fometimes it faileth to fpring againe, and therefore had neede to be fowne anew in the Spring.

4. *Linaria purpurea odorata.* Sweete purple Tode Flaxe.

The lower leaues of this purple Tode Flaxe are nothing like any of the reft, but are long and broad, endented about the edges, fomewhat refembling the leaues of the greater wilde white Daifie : the ftalke is fet at the bottome with fuch like leaues, but a little more diuided and cut in, and ftill fmaller and fmaller vpward, fo that the vppermoft leaues are very like the common Tode Flaxe, the toppe whereof is branched, hauing diuers fmall flowers growing along vpon them, in fafhion and colour almoft like the laft defcribed Tode Flaxe, but not altogether fo deepe a purple : the heads and feedes are very like the former, but that the feede of this is reddifh : the flowers

in

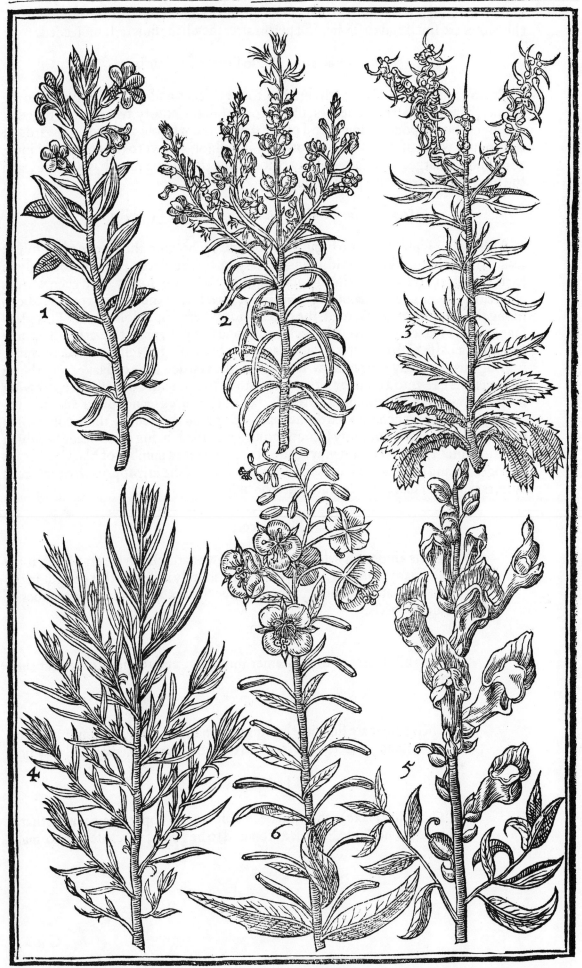

1 *Linum siluestre flore albo.* Wilde Flaxe with a white flower. 2 *Linaria purpurea siue cærulea.* Purple Tode Flaxe. 3 *Linaria purpurea odorata.* Sweete purple Tode Flaxe. 4 *Scoparia siue Beluidere Italorum* Broome Tode Flaxe. 5 *Antirrhinum maius.* The greater Snapdragon. 6 *Chamænerium flore delphinij.* The willowe flower.

in their naturall hot Countries haue a fine fent, but in thefe colder, little or none at all: the rootes are fmall and threadie, and perifh after they haue flowred and feeded.

5. *Linaria Valentina*. Tode Flaxe of Valentia.

This Spanifh Tode Flaxe hath three or fonre thicker and bigger ftalkes then the former, bearing fmall broad leaues, like vnto the fmall Centory, two or three together at a ioynt, round about the lower end of the ftalkes, but without any order vpwards, at the toppes whereof ftand many flowers, in fafhion like vnto the common kinde, and almoft as large, of a faire yellow colour, but the gaping mouth is downie, and the fpurre behinde of a purplifh colour.

6. *Scoparia fiue Beluidere Italorum*. Broome Tode Flaxe.

Although this plant haue no beautifull flowers, yet becaufe the greene plant full of leaues is fo delightfull to behold, being in Italy and other places planted not onely in their Gardens, but fet likewife in pots to furnifh their Windowes, and euen with vs alfo hath growne to be fo dainty a greene bufh, that I haue thought it worthy to be among the delights of my Garden; the defcription whereof is as followeth: This pleafant Broome Flaxe rifeth vp moft vfually with one ftraight vpright fquare ftalke, three foote and a halfe high or better in our Gardens, branching it felfe out diuers waies, bearing thereon many long narrow leaues, like the Garden Line or Flaxe, very thicke fet together, like vnto a bufh, or rather like vnto a faire greene Cypreffe tree, growing broad belowe, and fpire-fafhion vpwards, of a very faire greene colour : at the feuerall ioynts of the branches, towards the tops, and among the leaues, there come forth fmall reddifh flowers, not eafily feene nor much regarded, being of no beauty, which turne into fmall round blackifh gray feede : the rootes are a number of blackifh ftrings fet together, and the whole plant perifheth euery yeare at the firft approach of any cold ayre, as if it neuer had beene fo faire a greene bufh.

The Place.

Thefe kindes of wilde Flaxe doe growe naturally in diuers places, fome in Germany, fome in Spaine, and fome in Italy. Thofe that delight in the beauty of natures variety, doe preferue them, to furnifh vp the number of pleafant afpects.

The Time.

They all flower in the Summer moneths, and foone after perfect their feede.

The Names.

Their names are fufficiently expreffed in their titles, yet I muft giue you to vnderftand, that the laft is called of fome *Linaria magna*, and of others *Ofyris*.

The Vertues.

The wilde Flaxe hath no medecinable vertue appropriate vnto it that is knowne. The Tode Flaxe is accounted to be good, to caufe one to make water.

Chap. XLIIII.

Antirrhinum. Snapdragon.

THere is fome diuerfity in the Snapdragons, fome being of a larger, and others of a leffer ftature and bigneffe; and of the larger, fome of one, and fome of another colour, but becaufe the fmall kindes are of no beautie, I fhall at this time onely entreate of the greater forts.

1. *Antirrhinum album.* White Snapdragon.

The leaues of thefe Snapdragons (for I doe vnder one defcription comprehend the reft) are broader, longer, and greener then the leaues of the Garden Flaxe, or of the wilde Flaxe fet confufedly vpon the tender greene branches, which are fpread on all fides, from the very bottome, bearing at the toppes many flowers, fomewhat refembling the former Tode Flaxe, but much larger, and without any heele or fpurre, of a faire white colour, with a yellow fpot in the mouth or gaping place : after the flowers are paft, there come vp in their places hard round feede veffels, fafhioned fomewhat like vnto a Calues head, the fnout being cut off, wherein is contained fmall blacke feede : the rootes are many white ftrings, which perifh in moft places after they haue giuen feede, notwithftanding any care or paines taken with them to preferue them a-liue, and yet they will abide in fome places where they are defended in the Winter.

2. *Antirrhinum purpureum fiue rofeum.* Purple Snapdragon.

The purple Snapdragon is in ftalkes, leaues, and flowers altogether like the former, and as large and great in euery part, or greater; the only difference is, that this beareth pale Stammell or Rofe coloured flowers, with a yellow fpot in the mouth, and fometimes of a paler colour, almoft blufh.

3. *Antirrhinum variegatum.* Variable Snapdragon.

This variable kinde is fomewhat leffe, and tenderer then the laft defcribed, hauing alfo a reddifh or blufh coloured flower, leffer then the former, but much bigger then the middle kinde of Snapdragon (which is not fet downe in this worke) the yellow fpot in the mouth of it hath fome white about it, and extending to both fides of the fpot : the heads and feede are like the former : the rootes are fmaller, but neuer will abide after they haue giuen flowers and feede.

4. *Antirrhinum luteum.* Yellow Snapdragon.

There is likewife another of thefe kindes, that beareth leaues as large as any of the former, & very faire yellow flowers, as large likewife as they, not differing in any thing elfe from the firft; let not any therefore imagine this to be a *Linaria* or Tode Flaxe : for all parts are anfwerable vnto the Snapdragons.

The Place.

All thefe are nourifhed with vs in our Gardens, although in Spaine and Italy they are found growing wilde.

The Time.

They flower for the moft part the fecond yeare after the fowing, from April vntill Iuly, and the feede is quickly ripe after.

Z 3   The

The Names.

The name *Antirrhinum* is vsually giuen to this plant, although it fully agreeth not eyther with the defcription of Diofcorides, or Theophraftus : It hath alfo diuers other names in Latine, as *Orontium*, *Canis cerebrum Os Leonis, Leo herba, &c.* In Englifh Calues fnout, from the forme of the feede veffels, and Snapdragon, or Lyons mouth, from the forme of the flowers.

The Vertues.

They are feldome or neuer vfed in Phyficke by any in our dayes.

CHAP. XLV.

Chamænerium flore delphinij. The Willowe flower.

THis plant rifeth vp with many ftrong, woddy, round, brownifh great ftalkes, three or foure foote high, befet here and there without order, with one broad and long whitifh greene leafe at a ioynt, fomewhat like vnto a *Lyfimachia*, or Willow herbe, as alfo vnto a Peach leafe, but larger and longer : at the toppe of the branches ftand many flowers one aboue another, of a pale reddifh purple colour, confifting of fiue leaues, fpread open with an heele or fpurre behinde them, with many yellow threads in the middle, much larger then any flower of the Larkes fpurres, and fmelling fomewhat fweete withall ; it beareth a fhew of long pods with feede, but I could neuer obferue the feede : the rootes are like the rootes of *Lyfimachia*, or the ordinary yellow Loofe-ftrife, or Willowe herbe, but greater : running and fpreading vnder ground, and fhooting vp in many places, whereby it filleth a ground that it likes quickly : the ftalkes dye downe euery yeare, and fpring againe in many places farre afunder.

The Place.

Wee haue not knowne where this Willowe flower groweth naturally, but we haue it ftanding in an out corner of our Gardens, to fill vp the number of delightfull flowers.

The Time.

It flowreth not vntill May, and abideth a long while flowring.

The Names.

It may feeme to diuers, that this is that plant that Dodonæus called *Pfeudolyfimachium purpureum minus*, and Lobel feemeth by the name of *Delphinium buccinum* to aime at this plant, but withall calleth it *Chamænerium Gefneri*, and giueth the fame figure that Dodonæus hath for his *Pfeudolyfimachium* : But that is one kinde of plant (which hath fmaller and fhorter ftalkes, and very narrow long leaues, whofe flowers ftand vpon long flender cods, full of downe, with reddifh feede, like vnto the *Lyfimachia filiquofa filueftris*, and rootes that abide many yeares, but creepe not) and this is another, much greater, whofe true figure is not extant in any Author that I know. It is vfually called *Chamænerium flore delphinij*; but the name of *Delphinium buccinum* in my minde may not fo conueniently be applyed vnto it. It is called in Englifh, The Willowe flower, for the likeneffe of the leaues, and the beauty and refpect of the flowers.

The

The Vertues.

There is no vse hereof in Phyficke that euer I could learne, but is onely cherifhed among other forts of flowers, that ferue to decke and fet forth a Garden of varieties.

CHAP. XLVI.

Aquilegia. Colombines.

THere are many forts of Colombines, as well differing in forme as colour of the flowers, and of them both fingle and double carefully nourfed vp in our Gardens, for the delight both of their forme and colours.

1. *Aquilegia vulgaris flore fimplici.* Single Colombines.

Becaufe the whole difference of thefe Colombines ftandeth in the varieties of the forme, and colour of the flowers, and little in the leaues, I fhall not neede to make anie repetitions of the defcription of them, feeing one onely fhall fuffice for each peculiar kinde. The Colombine hath diuers large fpread leaues, ftanding on long ftalkes: euery one diuided in feuerall partitions, and roundly endented about the edges, in colour fomewhat like the leaues of Celondine, that is, of a darke blewifh greene colonr: the ftalkes rife vp fometimes two or three foote high, diuided vfually into many branches, bearing one long diuided leafe at the lower ioynt, aboue which the flowers growe, euery one ftanding on a long ftalke, confifting of fiue hollow leaues, crooked or horned at the ends, turning backward, the open flower fhewing almoft like vnto a Cinquefoile, but more hollow: after the flowers are paft, there arife fmall long cods, foure or fiue together, wherein are contained blacke fhining feede: the rootes are thicke and round, for a little fpace within the ground, and then diuided into branches, ending in many fmall fibres, abiding many yeares, and fhooting a frefh euery Spring from the round heads, that abide all the Winter. The variety of the colours of thefe flowers are very much, for fome are wholly white, fome of a blew or violet colour, others of a blufh or flefh colour, or deepe or pale red, or of a dead purple, or dead murrey colour, as nature lifteth to fhew it felfe.

2. *Aquilegia vulgaris flore pleno.* Double Colombines.

The double Colombines differ not in leafe or manner of growing from the fingle, fo that vntill they come to flower, they cannot bee difcerned one from another; the onely difference is, it beareth very thicke and double flowers, that is, many horned or crooked hollow leaues fet together, and are not fo large as the leaues of the fingle flowers. The variety of colours in this double kinde is as plentifull, or rather more then in the fingle; for of thefe there is party coloured, blew and white, and fpotted very variably, which are not in the fingle kinde, and alfo a very deepe red, very thicke and double, but a fmaller flower, and leffe plentifull in bearing then many of the other double forts. Thefe double kindes doe giue as good feede as the fingle kindes doe, which is not obferued in many other plants.

3. *Aquilegia inuerfis corniculis.* Double inuerted Colombines.

Thefe Colombines are not to be diftinguifhed eyther in roote, leaues, or feed from the former, the flowers onely make the difference, which are as double as the former, but that the heeles or hornes of thefe are turned inward, and ftand out in the middle of the flowers together: there is not that plentifull variety of colours in this kinde, as there is in the former: for I neuer faw aboue three or foure feuerall colours in this
kinde,

kinde, that is, white, purplish, reddish, and a dun or darke ouerworne purplish colour. These double flowers doe likewise turne into pods, bearing seede, continuing his kind, and not varying into the former.

4. *Aquilegia Rosea.* Rose Colombines.

The leaues and other parts of this kinde of Colombine, differ little or nothing from the former, the diuersitie consisteth likewise in the flowers, which although they stand in the same manner seuerally vpon their small stalkes, somewhat more sparingly then the former doe, yet they haue no heeles or hornes, eyther inward or outward, or very seldome, but stand sometimes but with eight or tenne smooth small plaine leaues, set in order one by one in a compasse, in a double rowe, and sometimes with foure or fiue rowes of them, euery one directly before the other, like vnto a small thick double Rose layd open, or a spread Marigold : yet sometimes it happeneth, that some of these flowers will haue two or three of the first rowes of leaues without any heele, and the rest that are inward with each of them a peece of a small horne at them, as the former haue : the colours of these flowers are almost as variable, and as variably mixed as the former double kindes. This likewise giueth seede, preseruing his owne kinde for the most part.

5. *Aquilegia degener.* Degenerate Colombines.

This kinde of Colombine might seeme to some, to bee but a casuall degeneration, and no true naturall kinde, happening by some cause of transplanting, or otherwise by the art of man : but I haue not so found it, in that it keepeth, and holdeth his own proper forme, which is like vnto the double Rose Colombine, but that the outermost row of leaues are larger then any of the rest inwardes, and is of a greenish, or else of a purplish greene colour, and is not altogether so apt to giue good seed like the former.

The Place.

The single kindes haue beene often found in some of the wooddy mountaines of Germany, as Clusius saith, but the double kindes are chiefly cherished in gardens.

The Time.

They flower not vntill May, and abide not for the most part when Iune is past, and in the meane time perfecteth their seede.

The Names.

Costæus doth call this plant *Pothos* of Theophrastus, which Gaza translateth *Desiderium.* Dalechampius vpon Athenæus, calleth it *Diosanthos,* or *Iouis flos* of Theophrastus, who in his sixth Booke and seuenth Chapter reckoneth them both, that is, *Diosanthos* and *Pathos,* to be Summer flowers, but seuerally. Dodonæus *Leoherba,* and Gesner *Leontostomium.* Fabius Columna in his Phytobasanos, vnto whom Clusius giueth the greatest approbation, referreth it to the *Isopyrum* of Dioscorides. All later Writers doe generally call it, eyther *Aquileia, Aquilina,* or *Aquilegia* ; and we in English, generally (I thinke) through the whole Country, Colombines. Some doe call the *Aquilegia rosea, Aquilegia stellata,* The starre Colombine ; because the leaues of the flowers doe stand so directly one by another, besides the doublenesse, that they somewhat represent eyther a Rose or a Starre, and thereupon they giue it the name eyther of a Starre or Rose.

The Vertues.

Some in Spaine, as Camerarius saith, vse to eate a peece of the roote hereof

1 *Aquilegia simplex.* The single Colombine. 2 *Aquilegia flore multiplici.* The double Colombine. 3 *Aquilegia versicolor.* The party coloured Colombine. 4 *Aquilegia inuersis corniculis.* The double inuerted Colombine. 5 *Aquilegia Rosea siue Stellata.* The Rose or the Starre Colombine. 6 *Thalictrum Hispanicum album.* White Spanish tufts.

of fasting, many dayes together, to helpe them that are troubled with the stone in the kidneyes. Others vse the decoction, of both herbe and roote in wine, with a little Ambargrise, against those kinds of swounings, which the Greekes call ἀδυναμία. The seede is vsed for the iaundise, and other obstructions of the liuer. Clusius writeth from the experience of Franciscus Rapard, a chiefe Physician of Bruges in Flanders, that the seede beaten and drunke is effectuall to women in trauell of childe, to procure a speedy deliuerie, and aduiseth a second draught thereof should be taken if the first succeede not sufficiently.

Chap. XLVII.

Thalictrum Hispanicum. Spanish Tufts, or Tufted Colombines.

FRom among the diuersities of this plant, I haue selected out two sorts for this my garden, as hauing more beautie then all the rest; leauing the other to be entreated of, where all in generall may be included. I haue in this place inserted them, for the likenesse of the leaues only, being in no other part correspondent, and in a Chapter by themselues, as it is most fit.

Thalictrum Hispanicum album. White Spanish tufted Colombines.

These plants haue both one forme, in roote, leafe and flower, and therefore neede but one description. The leaues are both for colour and forme so like vnto Colombines leaues (although lesser and darker, yet more spread, and on larger stalkes) that they may easily deceiue one, that doth not marke them aduisedly ; for the leaues are much more diuided, and in smaller parts, and not so round at the ends : the stalkes are round, strong, and three foote high at the least, branching out into two or three parts, with leaues at the seuerall ioynts of them, at the toppes whereof stand many flowers, which are nothing but a number of threads, made like vnto a small round tuft, breaking out of a white skinne, or leafe, which incloseth them, and being vnblowne, shew like vnto little buttons : the colour of these threds or tufts in this are whitish with yellow tips on them, and somewhat purplish at the bottome, hauing a strong but no good sent, and abiding in their beautie (especially if they grow in the shade, and not too hot in the sun) a great while, and then fall away, like short downe or threds : the seed vessels are three square, containing small, long, and round seede ; the rootes are many long yellow stringes, which endure and encrease much.

Thalictrum Montanum purpureum. Purple tufted Colombines.

This purple tufted Colombine differeth onely from the former, in that it is not so high nor so large, and that the colour of the flower or tuft is of a blewish purple colour with yellow tips, and is much more rare then the other.

The Place.

These grow both in Spaine and Italie.

The Time.

They flower in the end of May, or in Iune, and sometime later.

The Names.

Some doe call them *Thalietrum*, and some *Thalictrum*. Others *Ruta pratastris*, and *Ruta pratensis*, and some *Rhabarbarum Monachorum*, or *Pseudo-rhabarbarum*,

rhabarbarum, by reason that the rootes being yellow, haue an opening qualitie, and drying as Rubarbe. In English what other fit Names to giue these then I haue expressed in the titles, I know not.

The Vertues.

The are a little hot and drying withall, good for old Vlcers, as Dioscorides saith, to bring them to cicatrising : in Italy they are vsed against the Plague, and in Saxonye against the Iaundise, as Camerarius saith.

CHAP. XLVIII.

Radix caua. Hollow roote.

THe likenesse of the leaues likewise of this plant with Colombines, hath caused mee to insert it next the other, and although some of this kinde bee of small respect, being accounted but foolish, yet let it fill vp a waste corner, that so no place be vnfurnished.

1. *Radix Caua maior flore albo.* The white Hollow roote.

The leaues of this hollow roote breake not out of the ground, vntill the end of March, or seldome before, and are both for proportion and colour somewhat like vnto the leaues of Colombines, diuided into fiue parts, indented about the edges, standing on small long footestalkes of a whitish greene colour, among which rise vp the stalkes, without any leaues from the bottome to the middle, where the flowers shoote forth one aboue another, with euery one a small short leafe at the foote thereof, which are long and hollow, with a spurre behinde it, somewhat like vnto the flowers of Larckes spurres, but hauing their bellies somwhat bigger, and the mouth not so open, being all of a pure white colour : after the flowers are past, arise small long and round cods, wherein are contained round blackish seede : the roote is round and great, of a yellowish browne colour on the outside, and more yellow within, and hollow vnderneath, so that it seemeth but a shell : yet being broken, euery part will grow : it abideth greene aboue ground but a small time.

2. *Radix Caua maior flore carneo.* Blush colourd Hollow roote.

The blush Hollow roote is in all things like vnto the former, but onely that the flowers hereof are of a delayed red or purple colour, which we call blush : and sometimes of a very deepe red or purple colour ; but very rare to meete with.

3. *Radix Caua minor, seu Capuos fabacea radice.* Small hollow roote.

This small kinde hath his leaues of a blewish greene colour, yet greener and smaller then the former, growing more thicke together : the flowers are like in proportion vnto the former in all respects, but lesser, hauing purplish backes, and white bellyes : standing closer and thicker together vpon the short stalkes : the roote is solid or firme, round and a little long withall, two being vsually ioyned together, yellowish both within and without : but I haue seene the dry roots that came from beyond Sea hither, that haue beene as small as hasell nuts, and somewhat flat with the roundnesse, differing from those that growe with vs, whether the nature thereof is to alter by manuring, I know not.

The Place.

The greater kindes Clusius reporteth he found in many places of Hungarie,

rie, and the other parts neere thereunto : the lesser in the lower Germany, or Low Countries, as we call them.

The Time.

These are most truely to bee reckoned Vernall plants, for that they rise not out of the ground vntill the Spring bee come in, and are gone likewise before it be past, remaining vnder ground all the rest of the yeare, yet the lesser abideth longer aboue ground then the greater.

The Names.

Concerning the former of these, there is a controuersie among diuers, whether it should be *Thesium* of Theophrastus, or *Eriphium* of Galen, but here is no fit place to trauerse those opinions. Some would haue it to bee *Corydalis*, and some referre it to Plinie his *Capnos Chelidonia*, for the likenesse it hath both with Fumeterie and Celandine. It is generally called of all moderne Writers, *Radix Caua*, and we in English thereafter, Hollow roote. The lesser for the firmenesse of his round roote, is vsually called, *Capnos fabacea radice*, and the Dutch men thereafter, 𝔅𝔬𝔬𝔫𝔨𝔢𝔫𝔰 𝔥𝔬𝔩𝔩𝔴𝔬𝔯𝔱𝔢𝔩𝔩: we of the likenesse with the former, doe call it the lesse Hollow roote.

The Vertues.

Some by the bitternesse doe coniecture (for little proofe hath beene had thereof, but in outward cases) that it clenseth, purgeth, and dryeth withall.

Chap. XLIX.

Delphinium. Larkes heeles.

OF Larkes heeles there are two principall kindes, the wilde kinde, and the tame or garden ; the wilde kinde is of two sorts, one which is with vs noursed vp chiefly in gardens, and is the greatest ; the other which is smaller and lower, often found in our plowed landes, and elsewhere : of the former of these wilde sorts, there are double as well as single : and of the tame or more vpright, double also and single : and of each of diuers colours, as shall be set downe.

1. *Delphinium maius siue vulgare.* The ordinary Larkes heeles.

The common Larkes heele spreadeth with many branches much more ground then the other, rather leaning or bending downe to the ground, then standing vpright, whereon are set many small long greene leaues, finely cut, almost like Fennell leaues : the branches end in a long spike of hollow flowers, with a long spurre behinde them, very like vnto the flowers of the Hollow roote last described, and are of diuers seuerall colours, as of a blewish purple colour, or white, or ash colour or red, paler or deeper, as also party coloured of two colours in a flower : after the flowers are past, (which in this kinde abide longer then in the other) there come long round cods, containing very blacke seede : the root is hard after it groweth vp to seede, spreading both abroad and deepe, and perisheth euery yeare, vsually raising it selfe from it own sowing, as well as from the seede sowen in the spring time.

Varietas.

2. *Delphinium vulgare flore pleno.* Double common Larkes heeles.

Of this vulgar kinde there is some difference in the flower, although in nothing else: the flowers stand many vpon a stalke like the former, but euery one of them are as if

three

three or foure small flowers were ioyned together, with euery one his spurre behinde, the greatest flower being outermost, and as it were containing the rest, which are of a pale red, or deepe blush colour : Another of this kinde will beare his flowers with three or foure rowes of leaues in the middle, making a double flower with one spurre behinde onely : and of this kinde there is both with purple, blew, blush, and white flowers, and party coloured also; these doe all beare seed like the single, wherby it is encreased euery yeare.

3. *Delphinium aruense.* Wilde Larkes spurres.

This wilde Larkes spurre hath smaller and shorter leaues, smaller and lower branches, and more thinly or sparsedly growing vpon them, then any of the former : the flowers likewise are neyther so large as any of the former, nor so many growing together, the cods likewise haue smaller seede, and is harder to grow in gardens then any of the former; the most vsuall colour hereof is a pale reddish or blush colour, yet sometimes they are found both white and blew, and sometimes mixt of blew and blush, variably disposed, as nature can when she listeth ; but are much more rare.

4. *Diphinium elatius flore simplici diuersorum colorum.* Single vpright bearing Larkes heeles of many colours.

The difference betweene this and the last is, that the leaues of this are not fully so greene, nor so large; the stalkes grow vpright, to the height of a man, and sometimes higher, hauing some branches thereon, but fewer then the former, and standing likewise vpright, and not leaning downe as the former : the toppes of the stalkes are better stored with flowers then the other, being sometimes two foote long and aboue, of the same fashion, but not altogether so large, but of more diuers and seueral colours, as white, pale, blush, redde deeper or paler, ashcoloured, purple or violet, and of an ouerworne blewish purple, or iron colour : for of all these we haue simple, without any mixture or spot : but we haue other sorts, among the simple colours, that rise from the same seede, and will haue flowers that wil be halfe white, and halfe blush or purple, or one leafe white, and another blush or purple, or else variably mixed and spotted : the seede and seede vessels are like the former but larger and harder.

5. *Delphinium elatius flore pleno diuersorum colorum.* Double vpright Larkes heeles of many colours.

These double Larkes heeles cannot bee knowne from the single of the same kinde, vntill they come towards flowring; for there appeare many flowers vpon the stalkes, in the same manner, and of as many colours almost as of the single, except the party coloured, which stand like little double Roses, layd or spread broade open, as the Rose Colombine without any heeles behinde them, very delightfull to behold, consisting of many small leaues growing together, and after they are fallen there come vp in their places three or foure small cods set together, wherein is contained here and there (for all are not full of seede, as the single kindes) blacke seede, like vnto all the rest, but smaller, which being sowen will bring plants that will beare both single and double flowers againe, and it often happeneth, that it variably altereth in colours from it owne sowing : for none of them hold constantly his owne colour, (so farre as euer I could obserue) but fall into others as nature pleaseth.

6. *Delphinium Hispanicum paruum.* Spanish wilde Larkes spurres.

This small Larkes spurre of Spaine, hath diuers long and broad leaues next the ground, cut-in on both sides, somewhat like vnto the leafe of a Scabious, or rather that kinde of Stœbe, which Lobel calleth *Crupina*, for it doth somewhat neerly resemble the same, but that this is smooth on the edges, and not indented besides the cuts, as the *Crupina* is, being of a whitish greene colour, and somewhat smooth and soft in handling : among the leaues riseth vp a whitish greene stalke, hauing many smaller

A a
leaues

leaues vpon it that grow belowe, but not diuided, branching out into many small stalkes, bearing flowers like vnto the wilde Larkes heeles, but smaller, and of a bleake blewish colour, which being past, there come vp two or three small cods ioyned together, wherein is blacke seede, smaller and rounder then any of the former : the roote is small and thready, quickly perishing with the first cold that ouertaketh the plant.

The Place.

The greatest or first wilde kindes growe among corne in many countries beyond the Seas, and where corne hath beene sowne, and for his beauty brought and nourished in our Gardens: the lesser wilde kinde in some fields of our owne Country. The Spanish kinde likewise in the like places, which I had among many seedes that Guillaume boel brought mee out of Spaine. The first double and single haue been common for many yeares in all countries of this Land, but the tall or vpright single kindes haue been entertained but of late yeares. The double kindes are more rare.

The Time.

These flower in the Summer onely, but the Spanish wilde kinde flowreth very late, so that oftentimes in our Country, the Winter taketh it before it can giue ripe seede : the double kindes, as well the vpright as the ordinary or wilde, are very choise and dainty many times, not yeelding good seede.

The Names.

They are called diuersly by diuers Writers, as *Consolida regalis, Calcaris flos; Flos regius, Buccinum Romanorum,* and of Matthiolus, *Cuminum siluestre alterum Dioscoridis* : but the most vsuall name with vs is *Delphinium* : but whether it be the true *Delphinium* of Dioscorides, or the Poets Hyacinth, or the flower of Aiax, another place is fitter to discusse then this. Wee call them in English Larkes heeles, Larkes spurres, Larkes toes or clawes, and Monkes hoods. The last or Spanish kinde came to mee vnder the name of *Delphinium latifolium trigonum,* so stiled eyther from the diuision of the leaues, or from the pods, which come vsually three together. Bauhinus vpon Matthiolus calleth it, *Consolida regalis peregrina paruo flore.*

The Vertues.

There is no vse of any of these in Physicke in these dayes that I know, but are wholly spent for their flowers sake.

Chap. L.

Balsamina fœmina. The Female Balsam Apple.

IHaue set this plant in this place, for some likenesse of the flower, rather then for any other comparison, euen as I must also with the next that followeth. This plant riseth vp with a thicke round reddish stalke, with great and bunched ioynts, being tender and full of iuice, much like to the stalke of Purslane, but much greater, which brancheth it selfe forth from the very ground, into many stalkes, bearing thereon manie long greene leaues, snipt about the edges, very like vnto the Almond or Peach tree leaues; among which from the middle of the stalkes vpwards round about them, come forth vpon seuerall small short foot-stalkes many faire purplish flowers, of two or three colours in them, fashioned somewhat like the former Larkes heeles, or Monks hoods, but that they are larger open at the mouth, and the spurres behinde crooke or bend downewards : after the flowers are past, there come in their places round rough heads,

1 *Radix Cana maior flore albo.* The white flowred Hollow roote. 2 *Capnos fabacea radice.* The small Hollow roote. 3 *Delphinium flore simplici.* Single Larkes spurs. 4 *Delphinium vulgare flore medio duplici.* Larkes spurs double in the middle. 5 *Delphinium vulgare flore pleno.* Common Larks spurs double. 6 *Delphinium elatius flore pleno.* Double vpright Larkes spurs. 7 *Delphinium Hispanicum parvum.* Small Spanish Larkes spurs. 8 *Balsamina foemina.* The Female Balsam apple. 9 *Nasturtium Indicum.* Indian Cresses, or yellow Larkes spurs.

heads, pointed at the end, greene at the firſt, and a little yellower when they bee ripe, containing within them ſmall round blackiſh ſeede, which will ſoone skippe out of the heads, if they be but a little hardly preſſed betweene the fingers : the rootes ſpread themſelues vnder ground very much from the toppe, with a number of ſmall fibres annexed thereunto : this is a very tender plant, dying euery yeare, and muſt bee ſowne carefully in a pot of earth, and tended and watered in the heate of Summer, and all little enough to bring it to perfection.

The Place.

Wee haue alwaies had the ſeede of this plant ſent vs out of Italy, not knowing his originall place.

The Time.

It floweth from the middle of Iuly, to the end of Auguſt : the ſeed doth ſeldome ripen with vs, eſpecially if the Summer be backward, ſo that wee are oftentimes to ſeeke for new and good ſeede from our friends againe.

The Names.

Some vſe to call it *Charantia fœmina, Balſamina fœmina, Balſamella*, and *Anguillara, Herba Sancta Katharine.* We haue no other Engliſh name to call it by, then the Female Balſame Apple, or *Balſamina.*

The Vertues.

Some by reaſon of the name, would attribute the property of Balme vnto this plant, but it is not ſufficiently knowne to haue any ſuch ; yet I am well perſwaded, there may bee ſome extradinary quality in ſo beautifull a plant, which yet lyeth hid from vs.

Chap. LI.

Naſturtium Indicum. Indian Creſſes, or yellow Larkes heeles.

THe likeneſſe (as I ſaid before) of this flower likewiſe, hauing ſpurres or heeles maketh me ioyne it with the reſt, which is of ſo great beauty and ſweetneſſe withall, that my Garden of delight cannot bee vnfurniſhed of it. This faire plant ſpreadeth it ſelfe into very many long trayling branches, enterlaced one within another very confuſedly (yet doth it not winde it ſelfe with any claſpers about either pole or any other thing, but if you will haue it abide cloſe thereunto, you muſt tye it, or elſe it will lye vpon the ground) foure or fiue foot in length at the leaſt, wherby it taketh vp a great deale of ground : the leaues are ſmooth, greene, and as round as the Penniwort that groweth on the ground, without any cut or inciſure therein at all in any part, the ſtalkes whereof ſtand in the middle of each leafe, and ſtand at euery ioynt of the ſtalke, where they are a little reddiſh, and knobbed or bunched out : the flowers are of an excellent gold yellow colour, and grow all along theſe ſtalkes, almoſt at euery ioynt with the leaues, vpon pretty long foote-ſtalkes, which are compoſed of fiue leaues, not hollow or gaping, but ſtanding open each leafe apart by it ſelfe, two of them, that be larger and longer then the other, ſtand aboue, and the other two that are leſſer belowe, which are a little iagged or bearded on both ſides, and the fift loweſt : in the middle of each of the three lower leaues (yet ſometimes it is but in two of them) there is a little long ſpot or ſtreake, of an excellent crimſon colour, with a long heele or ſpurre behinde hanging downe : the whole flower hath a fine ſmall ſent, very pleaſing, which being placed in the middle of ſome Carnations or Gillo-

flowers

flowers (for they are in flower at the same time) make a delicate Tuffimuffie, as they call it, or Nofegay, both for fight and fent : After the flower is paft, come the feede, which are rough or vneuen, round, greenifh yellow heads, fometimes but one, and fometimes two or three ftanding together vpon one ftalke, bare or naked of them-felues, without any huske, containing a white pulpy kernell : the rootes are fmall, and fpreading vnder ground, which perifh with the firft frofts, and muft be fowne a new euery yeare ; yet there needeth no bed of horfe-dung for the matter : the naturall ground will be fufficient, fo as you defend it a little from thofe frofts, that may fpoile it when it is newly fprung vp, or being yet tender.

The Place.

This goodly plant was firft found in the Weft Indies, and from thence fent into Spaine vnto Monardus and others, from whence all other parts haue receiued it. It is now very familiar in moft Gardens of any curiofity, where it yearly giueth ripe feed, except the yeare be very vnkindly.

The Time.

It flowreth fometimes in Iune, but vfually in Iuly (if it be well defended and in any good ground) and fo continueth flowring, vntill the cold frofts and miftes in the middle or end of October, doe checke the luxurious na-ture thereof, and in the meane time the feede is ripe, which will quickly fall downe on the ground, where for the moft part the beft is gathered.

The Names.

Some doe reckon this plant among the *Clematides* or *Convolvuli*, the Clamberers or Bindweedes ; but (as I faid) it hath no clafpers, neither doth it winde it felfe : but by reafon of the number of his branches, that run one within another, it may feeme to climbe vp by a pole or fticke, which yet doth but onely clofe it, as hauing fomething whereon to leane or reft his branches. Monardus and others call it *Flos fanguineus*, of the red fpots in the flowers, as alfo *Maftnerzo de las Indias*, which is *Nafturtium Indicum*, by which name it is now generally knowne and called, and wee thereafter in Englifh, Indian Crefles, yet it may bee called from the forme of the flow-ers onely, Yellow Larkes heeles.

The Vertues.

The Spaniards and others vfe the leaues hereof in ftead of ordinary Cref-fes, becaufe the tafte is fomewhat fharpe agreeing thereunto, but other Phy-ficall properties I haue heard of none attributed to it.

Chap. LII.

Viola. Violets.

THe Garden Violets (for the Wilde I leaue to their owne place) are fo well knowne vnto all, that either keepe a Garden, or hath but once come into it, that I fhall (I thinke) but lofe labour and time to defcribe that which is fo com-mon. Yet becaufe it is not onely a choife flower of delight, notwitftftanding the po-pularity, and that I let not paffe any thing without his particular defcription, I muft alfo doe fo by this. And hereunto I muft adde that kinde of Violet, which, although it want that fmell of the other, goeth beyond it in variety of dainty colours, called *Viola tricolor & flammea,* or Harts eafes.

1. *Viola simplex Martia.* Single March Violets.

The single Garden Violet hath many round greene leaues, finely fnipt or dented about the edges, ftanding vpon feuerall fmall ftalkes, fet at diuers places of the many creeping branches, which as they runne, doe here and there take roote in the ground, bearing thereon many flowers feuerally at the ioynts of the leaues, which confift of fiue fmall leaues, with a fhort round tayle or fpurre behinde, of a perfect blew purple colour, and of a very fweete fent, it bringeth forth round feede veffels, ftanding likewife vpon their feuerall fmall ftalkes, wherein is contained round white feede: but thefe heads rife not from where the flowers grew, as in all other plants that I know, but apart by themfelues, and being fowne, will produce others like vnto it felfe, whereby there may be made a more fpeedy encreafe to plant a Garden (as I haue done) or any other place, then by flipping, as is the vfuall manner: the rootes fpread both deepe and wide, taking ftrong hold in the ground.

Flore albo. Of this kinde there is another that beareth white flowers, not differing in fmell or any thing elfe from the former.

Flore obfoleto. And alfo another, that beareth flowers of a dead or fad reddifh colour, in all other things alike, fauing that this hath not altogether fo good a fent as the other.

2. *Viola Martia flore multiplici.* Double March Violets.

There is no difference betweene this Violet and the former, in any other thing then in the doublenefe of the flowers, which haue fo many leaues fet and thruft together, that they are like vnto hard buttons. There is of this double kinde both white and purple, as in the fingle; but the white fort is feldome fo thicke and double as the purple: but of the red colour to be double I neuer heard.

3. *Viola flammea fiue tricolor.* Harts eafes or Panfies.

The Harts eafe hath his leaues longer, and more endented or cut in on the edges then the Violet hath, and fomewhat round withall: the ftalkes are vpright, yet weake, and ready to fall downe, and lye vpon the ground, fet here and there with the like leaues, from whence come forth the flowers, of little or no fent at all, made like vnto a Violet, yet more open, and with larger leaues; but fo variably mixed with blew or purple, white and yellow, that it is hard to fet downe all the varieties: For fome flowers will be more white, and but fome fpots of purple or blew in the two vpper leaues, and the lower leaues with fome ftripes of yellow in the middle: others will haue more purple in them then any other colour, both in the vpper and lower leaues, the fide leaues blew, and the middle yellow, and others white and blew with yellow ftripes, as nature lifteth to diftribute their colours: the feede is fmall, whitifh, and round, contained in fmall round heads: the roote perifheth euery yeare, and raifeth it felfe vp plentifully by it owne fowing, if it be fuffered.

4. *Viola tricolor flore duplici.* Double Harts eafe.

We haue in our Gardens another fort, that beareth flowers with more leaues then the former, making it feeme to be twice double, and that onely in Autumne; for the firft flowers are fingle that come in Summer: This is of that fort that beareth purple flowers: And it is to be obferued, that the feed of this kinde will not all bring double flowers, but only fome, if the ground be fit and liking, fo that if you haue once had of this double kinde, you fhall feldome miffe to haue double flowers againe euery yeare of it owne growing or fowing.

5. *Viola flammea lutea maxima.* The great yellow Panfie.

There is one other kinde of Harts eafe, that decketh vp our Gardens not to be forgotten, whofe leaues and flowers are like the former, but more plentifull in ftalkes and branches, and better abideth our Winters: the flowers are larger then any of the

former,

former, of a faire pale yellow colour, with some yellower stripes now and then about the middle; for it is sometimes without any stripes, and also of a little deeper yellow colour: this is to bee encreased by slips, which will soone comprehend in a moist or moistened ground, for that I neuer could obserue that it bore seede.

The Place.

These plants were first wilde, and by manuring brought to be both fairer in colour, and peraduenture of a better sent then when they grew wilde.

The Time.

The Violets flower in March, and sometimes earlier, and if the yeare be temperate and milde, in Autumne againe. The double Violets, as they are later before they flower then the single, so they hold their flowers longer. The Harts ease flowreth seldome vntill May; but then some will abide to flower vntill the end of Autumne almost, especially if the frosts be not early.

The Names.

The Violet is called *Viola nigra, purpurea*, and *Martia*: In English, Violets, March Violets, and purple Violets. The Harts ease is called *Viola flammea, Viola tricolor, Viola multicolor*, and of some, *Iacea, Flos trinitatis*, and *Herba clauellata*: In English, Harts ease, and Pansies, of the French name *Pensees*. Some giue it foolish names, as Loue in idlenesse, Cull mee to you, and Three faces in a hood. The great yellow Harts ease is so called, because it is like in forme, and is the greatest of all other, although it haue not that diuersity of colours in it that the other haue.

The Vertues.

The properties of Violets are sufficiently knowne to all, to coole and moisten: I shall forbeare to recite the many vertues that may be set downe, and onely let you know, that they haue in them an opening or purging quality, being taken either fresh and greene, or dryed, and made into powder, especially the flowers; the dryed leaues will doe the like, but in greater quantity. Costæus in his booke of the nature of all plants saith, that the distilled water of Harts ease, is commended in the French disease, to be profitable, being taken for nine dayes or more, and sweating vpon it, which how true it is, I know not, and wish some better experience were made of it, before we put any great confidence in that assertion.

Chap. LIII.

Epimedium. Barrenwort.

This pretty plant riseth vp out of the ground with vpright, hard, round, small stalkes, a foote and a halfe high, or not two foote high at the highest, diuided into three branches for the most part, each branch whereof is againe diuided for the most part into three other branches, and each of them beare three leaues (seldome either more or lesse) set together, yet each vpon his owne foote-stalke, each leafe being broad, round, and pointed at the end, somewhat hard or dry in feeling, hayrie, or as it were prickly about the edges, but very tenderly, without harme, of a light greene colour on the vpperside, and a little whiter vnderneath: from the middle of the stemme or stalke of leaues doth likewise come forth another long stalke, not much higher then those with the leaues on them, diuided into other branches, each
whereof

whereof hath likewife three flowers, each vpon his owne footeftalke, confifting of eight fmall leaues a peece, yet feeming to be but of foure leaus fpread or layd open flat, for that the foure vppermoft, which are the fmaller and being yellow, doe lye fo clofe on the foure vndermoft, wch are a little broader and red, that they fhew as if they were yellow flowers with red edges, hauing yellow threds tipt with greene, ftanding in the middle of the flowers : the vnderfide of the lower leaues are of a pale yellowifh red, ftriped with white lines : after the flowers are paft, there come fmall long pods, wherin are contained flat reddifh feede : the rootes are fmall, reddifh and hard, fpreading, branching and enterlacing themfelues very much, and is fit to be placed on fome fhady fide of a garden : the whole plant is rather of a ftrong then any good fent, yet is cherifhed for the pleafant varietie of the flowers.

The Place.

Cæfalpinus faith it groweth on the mountaines of Liguria, that is nigh vnto Ligorne, in the Florentine Dominion. Camerarius faith, nigh vnto Vicenzo in Italie. Bauhinus on the Euganian hils, nigh vnto Padoa, and in Romania in fhadowie wet grounds.

The Time.

It flowreth from Iune vntill the end of Iuly, and to the middle of Auguft, if it ftand, as I faid it is fitteft, in a fhadowie place.

The Names.

It is of moft Writers accepted for the true *Epimedium* of Diofcorides, though he faith it is without flower or feede, being therein eyther miftaken, or mif-informed, as he was alfo in *Dictamnus* of Candy, and diuers other plants. From the triple triplicitie of the ftanding of the ftalkes and leaues, and quadriplicitie of the flowers, it might receiue another name in Englifh then is already impofed vpon it : but left I might be thought to be fingular or full of noueltie, let it paffe with the name Barrenwort, as it is in the title.

The Vertues.

It is thought of diuers to agree in the propertie of caufing barrenneffe, as the ancients doe record of *Epimedium.*

Chap. LIIII.

Papauer fatiuum. Garden Poppies.

OF Poppies there are a great many forts, both wilde and tame, but becaufe our Garden doth entertaine none, but thofe of beautie and refpect, I wil onely giue you here a few double ones, and leaue the reft to a general furuey.

1. *Papauer multiplex album.* Double white Poppies.

The double white Poppy hath diuers broade, and long whitifh greene leaues, giuing milke (as all the reft of the plant aboue ground doth, wherefoeuer it is broken) very much rent or torne in on the fides, and notched or indented befides, compaffing at the bottome of them a hard round brittle whitifh greene ftalke, branched towards the toppe, bearing one faire large great flower on the head of euery branch, which before it breaketh out, is contained within a thin skinne, and being blowne open is very thick of leaues, and double, fomewhat iagged at the ends, and of a white colour ; in the

middle

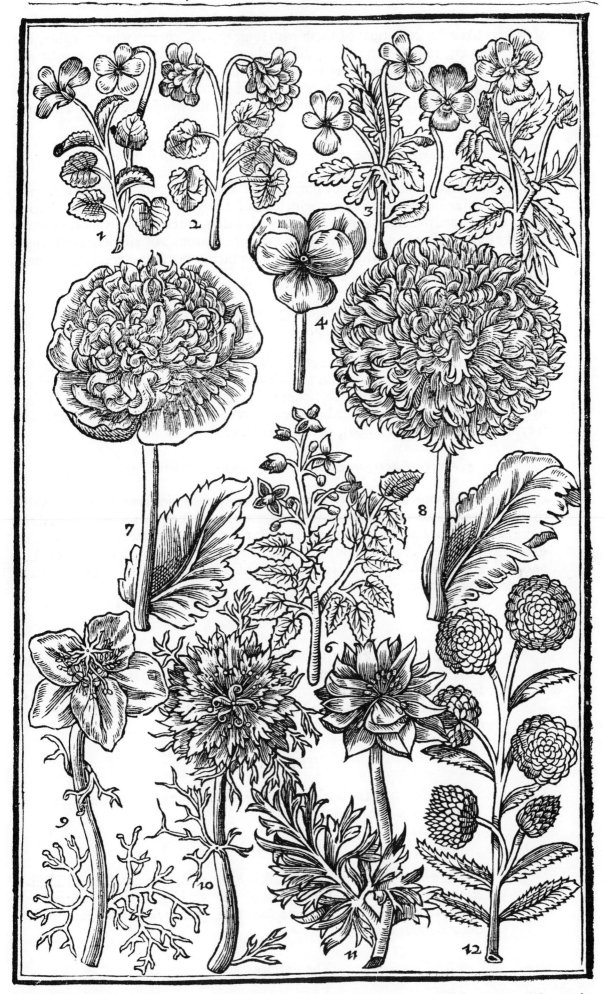

1. *Viola Martia simplex.* Single March Violets. 2 *Viola Martia multiplex.* Double March Violets. 3 *Viola flammea siue tricolor.* Ordinary garden Panfies or Harts eafe. 4 *Viola flammea lutea magna.* Great yellow Panfies. 5 *Viola tricolor duplex* Double Panfies or Harts eafes. 6 *Epimedium.* Barrenwort. 7 *Papauer satiuum flore pleno.* Double garden Poppies. 8 *Papauer satiuum flore pleno laciniato.* Double feathered Poppies. 9 *Nigella Hispanica flore amplo.* Spanish Nigella or Fenell flower. 10 *Nigella multiplex cærulea* Double blew Nigella or Fenell flower. 11 *Nigella duplex flore albo.* Double white Nigella. 12 *Ptarmica flore pleno.* Double wilde Pelletory.

middle whereof ſtandeth a round head or bowle, with a ſtriped crowne on the heade of it, very like a ſtarre, compaſſed about with ſome threds, wherein when it is ripe, is contained ſmall, round, white ſeede, diſpoſed into ſeuerall cels : the roote is hard, wooddy, and long, periſhing euery yeare, and muſt bee new ſowne euery Spring, if they doe not ſpring of their own ſowing, which if it doe, the flowers are ſeldome ſo faire and double as they that are ſowne in the Spring : the whole plant is of a ſtrong heady ſmell.

2. *Papauer multiplex rubeſcens.* Double red or bluſh Poppies.

This other kind of double Poppy differeth not in any other thing from the former, but only in the colour of the flowers, which are of a bright red, tending to a bluſh colour, parted, paned or ſtriped in many places with white, and exceedingly more iagged then the former, almoſt like a feather at the ends, the bottomes of all the leaues being white : the ſeede hereof is white as the former, which is not ſo in any other Poppie, that beareth not a full white flower.

3. *Papauer multiplex nigrum ſiue purpureum.*
Double purple or murry Poppies.

This kinde varyeth both in flowers and ſeede, although neyther in leaues or any other thing from the firſt : the flowers are thicke and double, and ſomewhat iagged at the ends, in ſome more, in ſome leſſe, eyther red or bluſh, or purpliſh red, more or leſſe, or of a ſad murrey or tawney, with browne, or blacke, or tawny bottomes : the ſeede is eyther of a grayiſh blew colour, or in others more blackiſh.

4. *Papauer Rhæas flore multiplici.* The double red field Poppie.

This double Poppie is like the wilde or fielde Poppie, which is well knowne to all to haue longer, narrower, and more iagged greene leaues then the former, the ſtalkes more hairy, and the flower of a deepe yellowiſh red colour, knowne to all. Now this differeth in nothing from it, but in the doubleneſſe of the flower, which is very thicke and double, but not ſo large as the former. This riſeth of ſeede in the like manner as they doe, and ſo to bee preſerued.

The Place.

From what place they haue beene firſt gathered naturally I cannot aſſure you, but we haue had them often and long time in our gardens, being ſent from Italie and other places. The double wilde kindes came from Conſtantinople, which whether it groweth neere vnto it or further off, we cannot tell as yet.

The Time.

They flower in the beginning or middle of Iune at the furtheſt, the ſeede is ripe within a ſmall while after.

The Names.

The generall knowne name to all, is *Papauer*, Poppie : the ſeuerall diſtinctions are according to their colours. Yet our Engliſh Gentlewomen in ſome places, call it by a by-name, Ione ſiluer pinne : *ſubauditur*, Faire without and fowle within.

The Vertues.

It is not vnknowne, I ſuppoſe to any, that Poppie procureth ſleepe, for which cauſe it is wholly and onely vſed, as I thinke : but the water of the
wilde

wilde Poppies, besides that it is of great vse in Pleurisies, and Rheumatick-or thinne Distillations, is found by daily experience, to bee a soueraigne remedy against surfeits ; yet some doe attribute this propertie to the water of the wilde Poppies.

Chap. LV.

Nigella. The Fenell flower, or Nigella.

AMong the many sorts of Nigella, both wilde and tame, both single and double, I will onely set downe three sorts, to be noursed vp in this garden, referring the rest to a Physicke garden, or a generall Historie, which may comprehend all.

1. *Nigella Hispanica flore simplici.* The great Spanish Nigella.

Spanish Nigella riseth vp with diuers greene leaues, so finely cut, and into so many parts, that they are finer then Fenell, and diuided somewhat like the leaues of Larkes heeles, among which rise vp stalkes, with many such like leaues vpon them, branched into three or foure parts, at the toppe of each whereof standeth one faire large flower, like vnto other single Nigella's, consisting of fiue or six leaues sometimes, of a bleake blew, or of a purplish blew colour, with a greene head in the middle, compassed about with seuen or eight small blewish greene flowers, or peeces of flowers rather, made like gaping hoodes, with euery of them a yellowish line thwart or crosse the middle of them, with some threds also standing by them : after the flower is past the head groweth greater, hauing sixe, seuen or eight hornes as it were at the toppe, greater and longer, and standing closer together then any other Nigella, spreading very like a starre, or the crowne of the Poppy head, but larger and longer, each whereof being folded together, openeth a little when the head is ripe, which is greater aboue, and smaller below, and not so round as the others are, containing within them small yellowish greene seede, or not so blacke as the other sorts : the rootes are small and yellow, perishing euery yeare as the others likewise doe.

2. *Nigella Damascena flore multiplici.* Double blew Nigella, or The Fenell flower.

The double Nigella is in leaues, stalkes and rootes, very like vnto the former Nigella, so that the one can very hardly bee discerned from the other before this rise vp to flower, except it be that the leaues hereof are not fully so large as they : the flower consisteth of three or foure rowes of leaues, layde one vpon another, of a pale blew colour, with a greene round head compassed with diuers short threads in the middle, and hauing fiue or sixe such small greene Fenell-like leaues vnder the flower, to beare it vp (as it were) below, which adde a greater grace to the flowers, which at the first sheweth sometimes white, but changeth quickely after : the horned heads hereof are like vnto the heads of the other wilde kinde, which are somewhat rounder and greater, hauing within them blacke vneuen seedes, but without any sent.

3. *Nigella Catrina flore albo multiplici.* Double white Nigella.

This double white Nigella hath such like leaues as the last hath, but somewhat larger, of a yellower greene colour, and not so finely cut and iagged : the flowers are somewhat lesse, and lesser double then the former, and in colour white, hauing no greene leaues vnder the flower, as the former hath, the head whereof in the middle is very like the head of the last double kinde, but not so great, wherein is contained blacke seede for the most part, and sweete like the Romane Nigella, which only is sweet besides this : yet sometimes it is not so blacke, but rather a little more white or yellowish: the roote is yellow, and perisheth as the others euery yeare.

The

The Place.

All thefe, and the reft be found wilde in diuers Countreyes, as France, Spaine, Italie, &c. but wee onely cherifh them in our Gardens for our delight.

The Time.

They flower in the end of Iune, and in Iuly, or thereabouts.

The Names.

They are called *Melanthium, Gith,* and *Nigella,* and of fome *Flos Diuæ Catherinæ.* We may either call them *Nigella* according to the Latine name, or the Fenell flower, as fome doe, becaufe the double blew Nigella hath fmall Fenell-like leaues bearing vp the flower, as I fhewed before in the defcription.

The Vertues.

Thefe Nigella's are nothing fo hot in qualitie as the fingle Romane kind is, as may well be knowne by the fmell of the feede thereof, and therefore are not fit to be vfed in the fteed of it, as many ignorant perfons vfe to doe: for the fingle Romane feede is vfed to helpe paines, and cold diftillations in the head, and to dry vp the rheume. Pena faith, that the preffed oyle of the feede as well taken inwardly as vfed outwardly is an excellent remedy for the hardneffe and fwelling of the fpleene.

Chap. LVI.

Ptarmica filueftris flore pleno. Double wilde Pelletory.

THe double wilde Pelletorie hath ftraight and flender ftalkes, befet with long and narrow leaues, fnipt round about the edges, in all points like vnto the fingle wilde kinde, that groweth common with vs almoft euery where : on the toppes of the ftalkes ftand foure or fiue, or more white flowers, one aboue another, with a greene leafe at the bottome of the footeftalke of euery one of them, beeing fmall, thicke, and very double, with a little yellowifhneffe in the middle of euery flower, like both for forme and colour vnto the flower of the double Featherfew, but fmaller : the rootes are many long ftrings, running here and there in the ground : this hath no fmell at all, but is delightfome only for the double white flowers.

The Place.

It is only cherifhed in fome few Gardens, for it is very rare.

The Time.

It flowreth in the end of Iune or thereabouts.

The Names.

It is called of moft *Ptarmica,* or *Sternutamentoria,* of his qualitie to prouoke neefing ; and of fome *Pyrethrum,* of the hot biting tafte. We vfually call it Double wilde Pelletorie, and fome Sneefewort, but *Elleborus albus* is vfually fo called, and I would not two things fhould be called by one name, for the miftaking and mif-ufing of them.

The

The Vertues.

The properties hereof, no doubt, may well bee referred to the single kinde, beeing of the same qualitie, yet as I take it, a little more milde and temperate.

Chap. LVII.

Parthenium flore pleno. Double Featherfew.

FEatherfew that beareth double flowers is so like vnto the single kinde, that the one cannot be discerned from the other, vntill it come to flower, bearing broad, pale or fresh greene leaues, much cut in on the sides : the stalkes haue such like leaues on them as grow below, from the toppes whereof come forth many double white flowers, like vnto the flowers of the former wilde Pelletory, but larger, and like also vnto the flowers of the double Camomill : the sent whereof is as strong as of the single.

The Place.

We haue this kinde only in Gardens, and as it is thought by others, is peculiar only to our owne Countrey.

The Time.

It flowreth in the end of May, and in Iune and Iuly.

The Names.

It is called diuersly by diuers : Some thinke it to be *Parthenium* of Dioscorides, but not of Galen; for his *Parthenium* is a sweet herbe, and is thought to bee *Amaracus*, that is Marierome : others call it *Matricaria*; and some *Amarella*. Gaza translateth it *Muraleum, Theoph.lib.7.cap.7.* It is generally in these parts of our Country called Double Feauerfew, or Featherfew.

The Vertues.

It is answerable to all the properties of the single kinde which is vsed for womens diseases, to procure their monthly courses chiefly. It is held to bee a speciall remedy to helpe those that haue taken *Opium* too liberally. In Italy some vse to eate the single kinde among other greene herbes, as Camerarius saith, but especially fryed with egges, and so it wholly loseth his strong and bitter taste.

Chap. LVIII.

Chamæmelum. Camomill.

OVr ordinary Camomill is well knowne to all, to haue many smal trayling branches, set with very fine smal leaues, bushing and spreading thicke ouer the ground, taking roote still as it spreadeth : the toppes of the branches haue white flowers, with yellow thrummes in the middle, very like vnto the Featherfew, before described, but somewhat greater, not so hard, but more soft and gentle in handling, and the whole herbe to be of a very sweet sent.

Bb   1.Cha-

1. *Chamæmalum nudum.* Naked Camomill.

We haue another fort of Camomill in fome Gardens, but very rare, like vnto the former, but that it is whiter, finer, and fmaller, and raifeth it felfe vp a little higher, and beareth naked flowers ; that is, without that border of white leaues that is in the former, and confifteth onely of a yellow round thrummie head, fmelling almoft as fweete as the former.

2. *Chamæmalum flore pleno.* Double flowred Camomill.

The double Camomill groweth with his leaues vpon the ground, as the other fin-gle kinde doth, but of a little frefher greene colour, and larger withall : the ftalkes with the flowers on them, doe raife themfelues vp a little higher then the ordinary, and bearing one or two flowers vpon a ftalk, which are compofed of many white leaues fet together in diuers rowes, which make a fine double flower, with a little yellow fpot in the middle for the moft part of euery one, and are much larger then any fingle kinde, fmelling better, and more pleafing then the ordinary : this doth creepe vpon the ground as the other, but is more tender to be kept in the Winter. Yet if you faue the flowers hereof (and fo will the double Featherfew alfo) when they haue ftood long, and ready to fade, and keepe them dry vntill the Spring, and then breaking them or pulling them to peeces, fowe them, there will fpring vp from them Camomill, and alfo Featherfew, that will againe beare double flowers.

The Place.

Our ordinary Camomill groweth wilde in many places of our Country, and as well neare London as in other places. The others are onely found in our Gardens, where they are cherifhed. Bauhinus faith, that the double flowred Camomill is found wilde about Orleance in France.

The Time.

The double kinde is vfually in flower in Iune, before the ordinary kinde, and moft commonly paft before it flowreth, which is not vntill Iuly or Au-guft. The naked Camomill flowreth betweene them both, or later.

The Names.

Camomill is called *Anthemis, Leucanthemis,* and *Leucanthemum,* of the whiteneffe of the flowers ; and *Chamæmalum* of the corrupted Italian name *Camomilla.* Some call the naked Camomill, *Chryfanthemum odoratum.* The double Camomill is called by fome *Chamæmalum Romanum flore multiplici.*

The Vertues.

Camomill is put to diuers and fundry vfes, both for pleafure and profit, both for inward and outward difeafes, both for the ficke and the found, in bathings to comfort and ftrengthen the found, and to eafe paines in the dif-eafed, as alfo in many other formes applyed outwardly. The flowers boy-led in Poffet drinke prouoketh fweat, and helpeth to expell colds, aches, and other griefes. A Syrupe made of the iuice of the double Camomill, with the flowers and white wine, as Bauhinus faith, is vfed by fome againft the Iaundife and Dropfie, caufed by the euill difpofition of the fplene.

CHAP.

1 *Parthenium flore pleno.* Double Featherfew. 2 *Chamæmelum nudum.* Naked Camomill. 3 *Chamæmelum flore pleno* Double Camomill. 4 *Pyrethrum officinarum.* Pelletory of Spaine. 5 *Flos Adonis flore rubro & flore luteo.* Adonis flower both red & yellow. 6 *Helleborus niger feruloceus siue Buphthalmum.* The great Oxe eye or the great yellow Anemone. 7 *Buphthalmum vulgare.* The common yellow Oxe eye.

CHAP. LIX.

Pyrethrum officinarum. Pelletory of Spaine.

I Muft needes adioyne vnto the Camomils this fine and tender plant, for fome neare refemblance it hath with them in face, though not in quality. It is a fmall and lowe plant, bearing many fine greene leaues vpon his flender branches, which leane or lye down vpon the ground, diuided into many parts, yet fomewhat larger and broader then Camomill, the ftalkes whereof are bigger, and more iuicie then it : the flowers that ftand at the toppes of the ftalkes are fingle, but much larger then any Camomill flower, hauing a pale or border of many leaues, white on the vpperfide, and reddifh vnderneath, fet about the yellow middle thrumme ; but not ftanding fo clofe together ioyning at the bottome, as the Camomill flowers doe, but more feuered one from another : it beareth fmall whitifh feede, which is hardly found and difcerned from the chaffe : the roote is long, and growing downe right, of the bigneffe of a mans finger or thumbe in our Countrey, but not halfe fo great where it groweth naturally, with fome fibres and branches from the fides thereof, of a very hot, fharpe, and biting tafte, drawing much water into the mouth, after it hath been chewed a while: the plant with vs is very tender, and will hardly or not at all endure the hardneffe and extremities of our Winters, vnleffe it be very carefully preferued.

The Place.

It groweth in Spaine wilde in many places, and in other hot Countries, where it may feele no frofts to caufe it perifh.

The Time.

It flowreth fo late with vs, that it is not vntill Auguft, that oftentimes we cannot gather ripe feedes from it, before it perifh.

The Names.

The name *Pyrethrum* (taken from πῦρ, that is, *ignis*, fire) is giuen to this plant, becaufe of the heate thereof, and that the roote is fomewhat like in fhew, but fpecially in property vnto the true *Pyrethrum* of Diofcorides, which is an vmbelliferous plant, whofe rootes are greater, and more feruent a great deale, and haue a hayrie bufh or toppe as *Meum*, and many other vmbelliferous plants haue. It is alfo called in Latine, *Saliuaris*, of the effect in drawing much moifture into the mouth, to be fpit out. We doe vfually call it Pelletory of Spaine.

The Vertues.

It is in a manner wholly fpent to draw rheume from the teeth, by chewing it in the mouth, thereby to eafe the tooth-ach, and likewife from the head, in the paines thereof.

Chap. LX.

Flos Adonis flore rubro. Red Adonis flower.

Donis flower may well be accounted a kinde of Camomill, although it hath some especiall differences, hauing many long branches of leaues lying vpon the ground, and some rising vp with the stalke, so finely cut and iagged, that they much resemble the leaues of Mayweed, or of the former *Nigella*: at the top of the stalkes, which rise a foote high or better, stand small red flowers, consisting of six or eight round leaues, hauing a greene head in the middle, set about with many blackish threads, without any smell at all : after the flowers are past, there grow vp heads with many roundish white seedes at the toppes of them, set close together, very like vnto the heads of seede of the great Oxe eye, set downe in the next Chapter, but smaller : the rootes are small and thready, perishing euery yeare, but rising of his owne seede againe, many times before Winter, which will abide vntill the next yeare.

Yellow Adonis flower is like vnto the red, but that the flower is somewhat larger, *Flore luteo.* and of a faire yellow colour.

The Place.

The first groweth wilde in the corn fields in many places of our own country, as well as in others, and is brought into Gardens for the beauties sake of the flower. The yellow is a stranger, but noursed in our Gardens with other rarities.

The Time.

They flower in May or Iune, as the yeare falleth out to be early or late : the seed is soone ripe after, and will quickly fall away, if it be not gathered.

The Names.

Some haue taken the red kinde to be a kinde of Anemone ; other to be *Eranthemum* of Dioscorides : the most vsuall name now with vs is *Flos Adonis*, and *Flos Adonidis* : In English, where it groweth wilde, they call it red Maythes, as they call the Mayweede, white Maythes ; and some of our English Gentlewomen call it Rosarubie : we vsually call it Adonis flower.

The Vertues.

It hath been certainly tryed by experience, that the seed of red Adonis flower drunke in wine, is good to ease the paines of the Collicke and Stone.

Chap. LXI.

Buphthalmum. Oxe eye.

Nder the name *Buphthalmum*, or Oxe eye, are comprehended two or three seuerall plants, each differing from other, both in face and property, yet because they all beare one generall name, I thinke fittest to comprise them all in one Chapter, and first of that which in leafe & seed commeth nearest to the Adonis flower.

1. *Buphthalmum maius siue Helleborus niger ferulaceus.*
Great Oxe eye, or the yellow Anemone.

This great Oxe eye is a beautifull plant, hauing many branches of greene leaues

leaning

leaning or lying vpon the ground for the moſt part, yet ſome ſtanding vpright, which are as fine, but ſhorter then Fenell; ſome of them ending in a ſmall tuft of green leaues, and ſome hauing at the toppes of them one large flower apeece, ſomewhat reddiſh or browniſh on the outſide, while they are in bud, and a while after, and being open, ſhew themſelues to conſiſt of twelue or fourteene long leaues, of a faire ſhining yellow co-lour, ſet in order round about a greene head, with yellow thrums in the middle, laying themſelues open in the ſunne, or a faire day, but elſe remaining cloſe: after the flower is paſt, the head growing greater, ſheweth it ſelfe compact of many round whitiſh ſeede, very like vnto the head of ſeede of the Adonis flower laſt deſcribed, but much greater: the rootes are many long blackiſh fibres or ſtrings, ſet together at the head, very like vnto the rootes of the leſſer blacke Hellebor or Bearefoote, but ſomewhat harder, ſtiffer, or more brittle, and ſeeming without moiſture in them, which abide and encreaſe euery yeare.

2. *Buphthalmum minus, ſeu Anthemis flore luteo.* Small Oxe eye.

This plant might ſeeme to be referred to the Camomils, but that it is not ſweete, or to the Corne-Marigolds, but that the ſtalkes and leaues are not edible: it is therefore put vnder the Oxe eyes, and ſo we will deſcribe it; hauing many weake branches lying vpon the ground, beſet with winged leaues, very finely cut and iagged, ſomewhat like vnto Mayweede, but a little larger: the flowers are like vnto the Corne Marigold, and larger then any Camomill, being wholly yellow, as well the pale or border of leaues, as the middle thrummes: the rootes are ſomewhat tough and long.

3. *Buphthalmum vulgare.* Common Oxe eye.

This Oxe eye riſeth vp with hard round ſtalkes, a foote and a halfe high, hauing many winged leaues vpon them, made of diuers long and ſomething broad leaues, ſnipt about the edges, ſet together ſomewhat like vnto Tanſie, but ſmaller, and not ſo much winged: the flowers ſtand at the toppes of the ſtalkes, of a full yellow colour, both the outer leaues and the middle thrum, and not altogether ſo large as the laſt: the rootes of this kinde periſh euery yeare, and require a new ſowing againe.

The Place.

The firſt groweth in diuers places of Auſtria, Bohemia, and thoſe parts, it hath beene likewiſe brought out of Spaine. The ſecond in Prouence, a country in France. The laſt in diuers places, as well of Auſtria as Morauia, and about Mentz and Norimberg, as Cluſius ſetteth downe. We haue them in our Gardens, but the firſt is of the greateſt reſpect and beauty.

The Time.

The firſt flowreth betimes, oftentimes in March, or at the furtheſt in A-pill; the ſeede is ripe in May, and muſt be quickly gathered, leſt it bee loſt. The other two flower not vntill Iune.

The Names.

The firſt is called *Buphthalmum* of Dodonæus, *Pſeudohelleborus* of Mat-thiolus, *Helleborus niger ferulaceus Theophraſti* by Lobel, of ſome others *Elleborus niger verus*, vſing it for the true blacke Ellebor, but it is much diffe-ring, as well in face as properties. Of others *Seſamoides minus*. Some haue thought it to be a yellow Anemone, that haue looked on it without further iudgement, and by that name is moſt vſually knowne to moſt of our Engliſh Gentlewomen that know it. But it may moſt fitly be called a *Buphthalmum*, as Dodonæus doth, and *Hiſpanicum* or *Auſtriacum*, for diſtinctions ſake. We doe moſt vſually call it *Helleborus niger ferulaceus*, as Lobel doth: Bauhinus calleth

calleth it *Helleborus niger tenuifolius Buphthalmi flore.* The ſecond is called *Buphthalmum Narbonenſe* : In Engliſh, The French, or leſſer Oxe eye, as the firſt is called, The great Oxe eye. The laſt, The common Oxe eye.

The Vertues.

The firſt hath been vſed in diuers places for the true blacke Ellebor, but now is ſufficiently knowne to haue been an errour ; but what Phyſicall property it hath, other then Matthiolus hath expreſſed, to be vſed as Setterwort for cattell, when they rowell them, to put or draw the rootes hereof through the hole they make in the dewe lappe, or other places, for their coughes or other diſeaſes, I know not, or haue heard or read of any. The others likewiſe haue little or no vſe in Phyſicke now a dayes that I know.

Chap. LXII.

Chryſanthemum. Corne Marigold.

ALthough the ſorts of Corne Marigolds, which are many, are fitter for another then this worke, and for a Catholicke Garden of Simples, then this of Pleaſure and Delight for faire Flowers ; yet giue me leaue to bring in a couple : the one for a corner or by-place, the other for your choiſeſt, or vnder a defenced wall, in regard of his ſtatelineſſe.

1. *Chryſanthemum Creticum.* Corne Marigold of Candy.

This faire Corne Marigold hath for the moſt part one vpright ſtalke, two foote high, whereon are ſet many winged leaues, at euery ioynt one, diuided and cut into diuers parts, and they againe parted into ſeuerall peeces or leaues : the flowers growe at the toppes of the ſtalkes, riſing out of a ſcaly head, compoſed of ten or twelue large leaues, of a faire, but pale yellow colour, and more pale almoſt white at the bottome of the leaues, round about the yellow thrumme in the middle, being both larger and ſweeter then any of the other Corne Marigolds : the ſeede is whitiſh and chaffie : the roote periſheth euery yeare.

2. *Chryſanthemum Perüuianum, ſiue Flos Solis.* The golden flower of Peru, or the Flower of the Sunne.

This goodly and ſtately plant, wherewith euery one is now adayes familiar, being of many ſorts, both higher and lower (with one ſtalke, without branches, or with many branches, with a blacke, or with a white ſeede, yet differing not in forme of leaues or flowers one from another, but in the greatneſſe or ſmalneſſe) riſeth vp at the firſt like vnto a Pompion with twe leaues, and after two, or foure more leaues are come forth, it riſeth vp into a great ſtalke, bearing the leaues on it at ſeuerall diſtances on all ſides thereof, one aboue another vnto the very toppe, being ſometimes, and in ſome places, ſeuen, eight, or ten foote high, which leaues ſtanding out from the ſtemme or ſtalke vpon their ſeuerall great ribbed foote-ſtalkes, are very large, broad belowe, and pointed at the end, round, hard, rough, of a ſad greene colour, and bending downewards : at the toppe of the ſtalke ſtandeth one great, large, and broad flower, bowing downe the head vnto the Sunne, and breaking forth from a great head, made of ſcaly greene leaues, like vnto a great ſingle Marigold, hauing a border of manie long yellow leaues, ſet about a great round yellow thrumme, as it were in the middle, which are very like vnto ſhort heads of flowers, vnder euery one whereof there is a ſeede, larger then any ſeede of the Thiſtles, yet ſomewhat like, and leſſer, and rounder then any Gourd ſeede, ſet in ſo cloſe and curious a manner, that when the ſeede is taken out, the head with the hollow places or cels thereof, ſeemeth very like vnto an hony combe ; which ſeede is in ſome plants very blacke, in the hotter countries, or very white,

white, and great, or large, but with vs is neither fo large, blacke, or white; but fome-times blackifh or grayifh. Some fort rifeth not vp halfe the height that others doe, and fome againe beare but one ftemme or ftalke, with a flower at the toppe thereof; and others two or three, or more fmall branches, with euery one his flower at the end; and fome fo full of branches from the very ground almoft, that I haue accounted threefcore branches round about the middle ftalke of one plant, the loweft neare two yards long, others aboue them a yard and a halfe, or a yard long, with euery one his flower thereon; but all fmaller then thofe that beare but one or two flowers, and leffer alfo for the moft part then the flower on the middle ftalke it felfe. The whole plant, and euery part thereof aboue ground hath a ftrong refinous fent of Turpentine, and the heads and middle parts of the flowers doe oftentimes (and fometimes the ioynts of the ftalke where the leaues ftand) fweat out a moft fine thin & cleare Roffin or Turpen-tine, but in fmall quantity, and as it were in drops, in the heate and dry time of the year, fo like both in colour, fmell, and tafte vnto cleare Venice Turpentine, that it cannot be knowne from it : the roote is ftrongly faftened in the ground by fome greater roots branching out, and a number of fmall ftrings, which growe not deepe, but keepe vn-der the vpper cruft of the earth, and defireth much moifture, yet dyeth euery yeare with the firft frofts, and muft be new fowne in the beginning of the Spring.

The Place.

Their places are fet downe in their titles, the one to come out of Candy, the other out of Peru, a Prouince in the Weft Indies.

The Time.

The firft flowreth in Iune, the other later, as not vntill Auguft, and fome-times fo late, that the early frofts taking it, neuer fuffer it to come to ripenefs.

The Names.

The firft hath his name in his title. The fecond, befides the names fet downe, is called of fome *Planta maxima, Flos maximus, Sol Indianus,* but the moft vfuall with vs is, *Flos Solis* : In Englifh, The Sunne Flower, or Flower of the Sunne.

The Vertues.

There is no vfe of either in Phyficke with vs, but that fometimes the heads of the Sunne Flower are dreffed, and eaten as Hartichokes are, and are accounted of fome to be good meate, but they are too ftrong for my tafte.

Chap. LXIII.

Calendula. Marigolds.

SOme haue reckoned vp many forts of Marigolds, I had rather make but two, the fingle and the double; for doubtleffe, thofe that be moft double, rife from the beft feede, which are the middlemoft of the great double, and fome will be leffe double, whofe feede is greater then the reft, according to the ground where it grow-eth; as alfo thofe that be of a paler colour, doe come of the feed of the yellower fort.

1. *Calendula maxima.* The great Garden Marigold.

The Garden Marigold hath round greene ftalkes, branching out from the ground into many parts, whereon are fet long flat greene leaues, broader and rounder at the
point

1 *Chryſanthemum Creticum.* Corne Marigolds of Candy. 2 *Flos Solis.* The Flower of the Sunne. 3 *Calendula.* Marigolds. 4 *Aſter Atticus ſiue Italorum.* The purple Marigold. 5 *Filoſella maior.* Golden Mouſe-eare. 6 *Scorſonera Hiſpanica.* Spanish Vipers graſſe. 7 *Tragopogon.* Goates beard, or goe to bed at noone.

point then any where elfe, and fmaller alfo at the fetting to of the ftalke, where it compaffeth it about : the flowers are fometimes very thicke and double (breaking out of a fcaly clammy greene head) compofed of many rowes of leaues, fet fo clofe together one within another, that no middle thrume can bee feene, and fometimes leffe double; hauing a fmall browne fpot of a thrume in the middle : and fometimes but of two or three rowes of leaues, with a large browne thrume in the middle; euery one whereof is fomewhat broader at the point, and nicked into two or three corners, of an excellent faire deepe gold yellow colour in fome, and paler in others, and of a pretty ftrong and refinous fweete fent : after the flowers are paft, there fucceede heads of crooked feede, turning inward, the outermoft biggeft, and the innermoft leaft : the roote is white, and fpreadeth in the ground, and in fome places will abide after the feeding, but for the moft part perifheth, and rifeth againe of his owne feede. Sometimes this Marigold doth degenerate, and beareth many fmall flowers vpon fhort ftalkes, compaffing the middle flower : but this happeneth but feldome, and therefore accounted but *lufus naturæ*, a play of nature, which fhe worketh in diuers other plants befides.

2. *Calendula fimplex.* The fingle Marigold.

There is no difference betweene this and the former, but that the flowers are fingle, confifting of one rowe of leaues, of the fame colour; eyther paler or deeper yellow, ftanding about a great browne thrumme in the middle : the feed likewife is alike, but for the moft part greater then in the double kindes.

The Place.

Our Gardens are the chiefe places for the double flowers to grow in; for we know not of any other naturall place : but the fingle kinde hath beene found wilde in Spaine, from whence I receiued feede, gathered by Guillaume Boel, in his time a very curious, and cunning fearcher of fimples.

The Time.

They flower all the Summer long, and fometimes euen in winter, if it be milde, and chiefly at the beginning of thofe monethes, as it is thought.

The Names.

They are called *Caltha* of diuers, and taken to be that *Caltha*, wherof both Virgil and Columella haue written. Others doe call them *Calendula*, of the Kalendes, that is the firft day of the monethes, wherein they are thought chiefly to flower; and thereupon the Italians call them, *Fiori di ogni mefe,* that is, The Flowers of euery moneth : We cal them in Englifh generally, eyther Golds, or Marigolds.

The Vertues.

The herbe and flowers are of great vfe with vs among other pot-herbes and the flowers eyther greene or dryed, are often vfed in poffets, broths, and drinkes, as a comforter of the heart and fpirits, and to expel any malignant or peftilential quality, gathered neere thereunto. The Syrupe and Conferue made of the frefh flowers, are vfed for the fame purpofes to good effect.

Chap. LXIIII.

Aster. Starre-wort.

Dioscorides and other of the ancient Writers, haue set forth but one kinde of Starre-wort, which they call *Aster Atticus*, of the place no doubt, where the greatest plentie was found, which was the Countrey of Athens : the later Writers haue found out many other plants which they referre to this kinde, calling them by the same name. It is not my purpose to entreate of them all, neyther doth this garden fitly agree with them : I shall therefore select out one or two from the rest, and giue you the knowledge of them, leauing the rest to their proper place.

1. *Aster Atticus flore luteo.* Yellow Starre-wort.

This Starre-wort riseth vp with two or three rough hairy stalkes, a foote and a halfe high, with long, rough or hairie, brownish, darke greene leaues on them, diuided into two or three branches · at the toppe of euery one whereof standeth a flat scaly head, compassed vnderneath with fiue or sixe long, browne, rough greene leaues, standing like a Starre, the flower it selfe standing in the middle thereof, made as a border of narrow, long, pale yellow leaues, set with a brownish yellow thrume : the roote dyeth euery yeare, hauing giuen his flower.

2. *Aster Atticus Italorum flore purpureo.* Purple Italian Starre-wort.

This Italian Starre-wort hath many wooddy, round brittle stalkes, rising from the roote, somewhat higher then the former, sometimes standing vpright, and otherwhiles leaning downewards, whereon are set many somewhat hard, and rough long leaues, round pointed, without order vp to the toppe, where it is diuided into seuerall branches, whereon stand the flowers, made like vnto a single Marigold, with a border of blewish purple leaues, set about a browne middle thrume, the heads sustaining the flowers, are composed of diuers scaly greene leaues, as is to be seene in the Knapweedes or Matfelons, which after the flowers are past yeelde a certaine downe, wherein lye small blacke and flat seedes, somewhat like vnto Lettice seede, which are carried away with the winde : the roote is composed of many white strings, which perisheth not as the former, but abideth, and springeth afresh euery yeare.

The Place.

The first is found in Spaine, as Clusius, and in France, as Lobel say. The other hath beene found in many places in Germany, and Austria : in Italie also, and other places ; we haue it plentifully in our Gardens.

The Time.

The first flowreth in Summer. And the other not vntill August or September.

The Names.

The first is called *Aster Atticus flore luteo, Bubonium, & Inguinalis*, and of many is taken to be the true *Aster Atticus* of Dioscorides : yet Matthiolus thinketh not so, for diuers good reasons, which hee setteth downe in the Chapter of *Aster Atticus*, as any man may vnderstand, if they will but reade the place, which is too long to bee inserted here. The other is thought by Matthiolus, to bee the truer *Aster Atticus*, (vnto whom I must also consent) and constantly also affirmed to be the *Amellus Virgilij*, as may be seene in the same place : but it is vsually called at this day, *Aster Italorum flore caruleo* or
purpureo,

purpureo. Their Englifh names are fufficiently expreffed in their titles, yet fome call the laft, The purple Marigold, becaufe it is fo like vnto one in form.

The Vertues.

They are held, if they bee the right, to bee good for the biting of a mad dogge, the greene herbe being beaten with old hogs greafe, and applyed; as alfo for fwolne throats : It is likewife vfed for botches that happen in the groine, as the name doth import.

Chap. LXV.

Pilofella maior. Golden Moufe-eare.

SOme refemblance that the flowers of this plant hath with the former Golds, maketh me to infert it in this place, although I know it agreeth not in any other part, yet for the pleafant afpect thereof, it muft bee in this my garden, whofe defcription is as followeth : It hath many broade greene leaues fpread vpon the ground, fpotted with pale fpots, yet more confpicuous at fometimes then at other ; fomewhat hairy both on the vpper and vnderfide, in the middle of thefe leaues rife vp one, two or more blackifh hairy ftalkes, two foote high at the leaft, bare or naked vp to the top, where it beareth an vmbell, or fhort tuft of flowers, fet clofe together vpon fhort ftalkes, of the forme or fafhion of the Haukeweedes, or common Moufe-eare, but fomewhat fmaller, of a deep gold yellow, or orenge tawney colour, with fome yellow threds in the middle, of little or no fent at all: after the flowers are paft, the heads carry fmall, fhort, blacke feede, with a light downie matter on them, ready to bee carried away with the winde, as many other plants are, when they be ripe : the rootes fpread vnder ground, and fhoote vp in diuers other places, whereby it much encreafeth, efpecially if it be fet in any moift or fhadowie place.

The Place.

It groweth in the fhadowie woods of France, by Lions, and Mompelier, as Lobell teftifieth : we keepe it in our gardens, and rather in a fhadowie then funnie place.

The Time.

It flowreth in Somer, and fometimes againe in September.

The Names.

It is called by Lobell, *Pulmonaria Gallorum Hieratij facie* : and the Herbarifts of France take it to be the true *Pulmonaria* of Tragus. Others call it *Hieratium flore aureo.* Pelleterius *Hieratium Indicum.* Some *Pilofella,* or *Auricula muris maior flore aureo.* And fome *Chondrilla flore aureo.* Dalechamptus would haue it to bee *Corchorus,* but farre vnfitly. The fitteft Englifh name we can giue it, is Golden Moufe-eare, which may endure vntill a fitter bee impofed on it : for the name of Grim the Collier, whereby it is called of many, is both idle and foolifh.

The Vertues.

The French according to the name vfe it for the defects of the lunges, but with what good fucceffe I know not.

Chap. LXVI.

Scorsonera. Vipers grasse.

ALthough there be foure or fiue forts of *Scorsonera,* yet I shall here desire you to be content with the knowledge only of a couple.

1. *Scorsonera Hispanica maior.* The greater Spanish Vipers grasse.

This Spanish Vipers grasse hath diuers long, and somewhat broad leaues, hard and crumpled on the edges, and sometimes vneuenly cut in or indented also, of a blewish greene colour: among which riseth vp one stalke, and no more for the most part, two foote high or thereabouts, hauing here and there some narrower long leaues thereon then those below: the toppe of the stalke brancheth it selfe forth into other parts, euery one bearing a long scaly head, from out of the toppe whereof riseth a faire large double flower, of a pale yellow colour, much like vnto the flower of yellow Goates-beard, but a little lesser, which being past, the seede succeedeth, being long, whitish and rough, inclosed with much downe, and among them many other long smooth seedes, which are limber and idle, and are carryed away at the will of the winde: the roote is long, thicke and round, brittle and blacke, with a certaine roughnesse on the outside: but very white within, yeelding a milkie liquor being broken, as euery other part of the plant doth besides, yet the roote more then any other part, and abideth many yeares without perishing.

2. *Scorsonera Pannonica purpurea.* Purple flowred Vipers grasse.

This purple flowred Vipers grasse hath long and narrow leaues, of the same blewish greene colour with the former: the stalke riseth vp a foote and a halfe high, with a few such like leaues, but shorter thereon, breaking at the toppe into two or three parts, bearing on each of them one flower, fashioned like the former, and standing in the like scaly knoppe or head, but of a blewish purple colour, not fully so large, of the sweetest sent of any of this kinde, comming neerest vnto the smell of a delicate perfume.

The Place.

The first is of Spaine. The other of Hungarie and Austrich: which now furnish our gardens.

The Time.

They flower in the beginning of May: the seede is soone ripe after, and then perishing downe to the roote for that yeare, springeth afresh before Winter againe.

The Names.

They are called after the Spanish name *Scorsonera,* which is in Latine *Viperaria,* of some *Viperina,* and *Serpentina:* Wee call them in English Vipers grasse, or *Scorsonera.*

The Vertues.

Manardus as I thinke first wrote hereof, and saith that it hath been found to cure them that are bitten of a Viper, or other such like venemous Creature. The rootes hereof being preserued with sugar, as I haue done often, doe eate almost as delicate as the Eringus roote, and no doubt is good to comfort and strengthen the heart, and vitall spirits. Some that haue vsed the preserued roote haue found it effectuall to expelling winde out of the stomacke, and to helpe swounings and faintnesse of the heart.

Cc

CHAP.

CHAP. LXVII.

Tragopogon. Goates beard.

I Muſt in this place ſet downe but two ſorts of Goates beards; the one blew or aſh-colour, the other red or purple, and leaue the other kindes : ſome to bee ſpoken of in the Kitchin Garden, and others in a Phyſicall Garden.

1. *Tragopogon flore cæruleo.* Blew Goates beard.

All the Goates beards haue long, narrow, and ſomewhat hollow whitiſh greene leaues, with a white line downe the middle of euery one on the vpperſide : the ſtalke riſeth vp greater and ſtronger then the Vipers graſſe, bearing at the toppe a great long head or huske, compoſed of nine or ten long narrow leaues, the ſharpe points or ends whereof riſe vp aboue the flower in the middle, which is thicke and double, ſome-what broad and large ſpread, of a blewiſh aſh-colour, with ſome whitiſh threads a-mong them, ſhutting or cloſing it ſelfe within the greene huske euery day, that it abi-deth blowing, vntill about noone, and opening not it ſelfe againe vntill the next mor-ning : the head or huske, after the flower is paſt, and the ſeede neare ripe, openeth it ſelfe ; the long leaues thereof, which cloſed not before now, falling downe round a-bout the ſtalke, and ſhewing the ſeede, ſtanding at the firſt cloſe together, and the doune at the toppe of them : but after they haue ſtood a while, it ſpreadeth it ſelfe round, and is ready to be carried away with the winde, if it be not gathered : the ſeede it ſelfe is long, round, and rough, like the ſeede of the Vipers graſſe, but greater and blacker : the roote is long, and not very great, but periſheth as ſoone as it hath borne ſeede, and ſpringeth of the fallen ſeede, that yeare remaining greene all Winter, and flowring the next yeare following : the whole yeeldeth milke as the former, but ſome-what more bitter and binding.

2. *Tragopogon purpureum.* Purple Goates beard.

There is little difference in this kind from the former, but that it is a little larger, both in the leafe, and head that beareth the ſeed: the flowers alſo are a little larger, and ſpread more, of a darke reddiſh purple colour, with ſome yellow duſt as it were caſt vpon it, eſpecially about the ends : the roote periſheth in the like manner as the other.

The Place.

Both theſe haue been ſent vs from the parts beyond the Seas, I haue had them from Italy, where no doubt they grow naturally wilde, as the yellow doth with vs : they are kept in our Gardens for their pleaſant flowers.

The Time.

They flower in May and Iune : the ſeede is ripe in Iuly.

The Names.

Their generall name is after the Greeke word *Tragopogon*, which is in La-tine, *Barba hirci* : In Engliſh, Goates beard ; the head of ſeede when it is rea-die to bee carried away with the winde, cauſing that name for the reſem-blance : and becauſe the flower doth euery day cloſe it ſelfe at noone (as I ſaid before) and openeth not againe vntill the next Sunne, ſome haue fitly called it, Goe to bed at noone.

The Vertues.

The rootes of theſe kindes are a little more bitter and more binding alſo

then

then the yellow kinde expreſſed in the Kitchin Garden; and therefore fitter for medicine then for meate, but yet is vſed as the yellow kinde is, which
is more fit for meate then medicine. The diſtilled water is good to waſh
old ſores and wounds.

<hr>

Chap. LXVIII.

Flos Africanus. The French Marigold.

OF the French or African Marigolds there are three kindes as principall, and of
each of them both with ſingle and double flowers : of theſe, ſome diuerſity is obſerued in the colour of the flowers, as well as in the forme or largeneſſe, ſo that as you may here ſee, I haue expreſſed eight differences, and Fabius Columna nine or ten, in regard hee maketh a diuerſity of the paler and deeper yellow colour : and although the leſſer kinde, becauſe of its euill ſent, is held dangerous, yet for
the beauty of the flower it findeth roome in Gardens.

1. *Flos Africanus maior ſiue maximus multiplex.*
The great double French Marigold.

This goodly double flower, which is the grace and glory of a Garden in the time of
his beauty, riſeth vp with a ſtraight and hard round greene ſtalke, hauing ſome creſts
or edges all along the ſtalke, beſet with long winged leaues, euery one whereof is like
vnto the leafe of an Aſh, being compoſed of many long and narrow leaues, ſnipt about
the edges, ſtanding by couples one againſt another, with an odde one at the end, of a
darke or full greene colour : the ſtalke riſeth to be three or foure foote high, and diuideth it ſelfe from the middle thereof into many branches, ſet with ſuch like leaues to
the toppes of them, euery one bearing one great double flower, of a gold yellow colour aboue, and paler vnderneath, yet ſome are of a pale yellow, and ſome betweene
both, and all theſe riſing from one and the ſame ſeede : the flower, before it be blowne
open, hath all the leaues hollow ; but when it is full blowne open, it ſpreadeth it ſelfe
larger then any Prouince Roſe, or equall vnto it at the leaſt, if it be in good earth, and
riſeth out of a long greene huske, ſtriped or furrowed, wherein after the flower is paſt,
(which ſtandeth in his full beauty a moneth, and oftentimes more, and being gathered, may be preſerued in his full beauty for two moneths after, if it be ſet in water)
ſtandeth the ſeede, ſet thicke and cloſe together vpright, which is blacke, ſomewhat flat and long : the roote is full of ſmall ſtrings, whereby it ſtrongly comprehen-
deth in the ground : the flower of this, as well as the ſingle, is of the very ſmell of new
waxe, or of an honie combe, and not of that poiſonfull ſent of the ſmaller kindes.

2. *Flos Africanus maior ſimplex.* The great ſingle French Marigold.

This ſingle Marigold is in all things ſo like vnto the former, that it is hard to diſcerne it from the double, but by the flowers, onely the ſtalke will be browner then
the double ; and to my beſt obſeruation, hath and doth euery yeare riſe from the ſeede
of the double flower : ſo that when they are in flower, you may ſee the difference (or
not much before, when they are in bud) this ſingle flower euer appearing with thrums
in the middle, and the leaues, which are the border or pale ſtanding about them, ſhewing hollow or fiſtulous, which after lay themſelues flat and open (and the double
flower appearing with all his leaues folded cloſe together, without any thrum at all)
and are of a deeper or paler colour, as in the double.

3. *Flos Africanus fiſtuloſo flore ſimplex & multiplex.*
Single and double French Marigolds with hollow leafed flowers.

As the former two greateſt ſorts haue riſen from the ſeede of one and the ſame (I
meane

meane the pod of double flowers) so doe these also, not differing from it in any thing, but that they are lower, and haue smaller greene leaues, and that the flower also being smaller, hath euery leafe abiding hollow, like vnto an hollow pipe, broad open at the mouth, and is of as deepe a yellow colour for the most part as the deepest of the former, yet sometimes pale also.

4. *Flos Africanus minor multiplex.* The lesser double French Marigold.

The lesser double French Marigold hath his leaues in all things like vnto the former, but somewhat lesser, which are set vpon round browne stalkes, not so stiffe or vpright, but bowing and bending diuers wayes, and sometimes leaning or lying vpon the ground : the stalkes are branched out diuersly, whereon are set very faire double flowers like the former, and in the like greene huskes, but smaller, and in some the outermost leaues will be larger then any of the rest, and of a deeper Orenge colour, almost crimson, the innermost being of a deepe gold yellow colour, tending to crimson : the whole flower is smaller, and of a stronger and more vnpleasant sauour, so that but for the beautifull colour, and doublenesse of the flower, pleasant to the eye, and not to any other sense, this kinde would finde roome in few Gardens : the rootes and seedes are like the former, but lesser.

5. *Flos Africanus minor simplex.* The small single French Marigold.

This single kinde doth follow after the last in all manner of proportion, both of stalkes, leaues, seedes, and rootes : the flowers onely of this are single, hauing fiue or six broad leaues, of a deepe yellow crimson colour, with deepe yellow thrummes in the middle, and of as strong a stinking sent, or more then the last.

The Place.

They growe naturally in Africa, and especially in the parts about Tunis, and where old Carthage stood, from whence long agoe they were brought into Europe, where they are onely kept in Gardens, being sowne for the most part euery yeare, vnlesse in some milde Winters. The last single and double kindes (as being more hardy) haue sometimes endured : but that kinde with hollow leafed flowers, as Fabius Columna setteth it downe, is accounted to come from Mexico in America.

The Time.

They flower not vntill the end of Summer, especially the greater kindes: but the lesser, if they abide all the Winter, doe flower more early.

The Names.

They haue been diuersly named by diuers men : Some calling them *Caryophyllus Indicus*, that is, Indian Gilloflowers, and *Tanacetum Peruuianum*, Tansie of Peru, as if it grew in Peru, a Prouince of America ; and *Flos Indicus*, as a flower of the Indies ; but it hath not beene knowne to haue beene brought from thence. Others would haue it to be *Othonna* of Plinie, and others ; some to be *Lycopersicum* of Galen. It is called, and that more truely, *Flos Tunetensis*, *Flos Africanus*, and *Caltha Africana*, that is, the flower of Tunis, the flower of Africa, the Marigold of Africa, and peraduenture *Pedna Panorum*. We in English most vsually call them, French Marigolds, with their seuerall distinctions of greater or smaller, double or single. To that with hollow leafed flowers, Fabius Columna giueth the name of *Fistuloso flore*, and I so continue it.

The

1 *Flos Africanus maximus multiplex.* The greatest double French Marigold.   2 *Flos Africanus maior multiplex.* The greater double French Marigold.
3 *Flos Africanus maximus simplex* The greatest single French Marigold.   4 *Flos Africanus multiplex fistulosus.* The doule hollow French Marigold.
5 *Flos Africanus simplex fistulosus,* The single hollow French Marigold.   6 *Flos Africanus minor multiplex,* Thesmaller double French Marigold.   7 *Flos
Africanus minor multiplex alter.* Another sort of the lesser double French Marigold.   8 *Flos Africanus minor simplex,* The lesser single French Marigold.

The Vertues.

We know no vfe they haue in Phyficke, but are cherifhed in Gardens for their beautifull flowers fake.

CHAP. LXIX.

Caryophyllus hortenfis. Carnations and Gilloflowers.

TO auoide confufion, I muft diuide Gilloflowers from Pinkes, and intreate of them in feuerall Chapters. Of thofe that are called Carnations or Gilloflowers, as of the greater kinde, in this Chapter; and of Pinkes, as well double as fingle, in the next. But the number of them is fo great, that to giue feuerall defcriptions to them all were endleffe, at the leaft needleffe: I will therefore fet downe onely the defcriptions of three (for vnto thefe three may be referred all the other forts) for their fafhion and manner of growing, and giue you the feuerall names (as they are vfually called with vs) of the reft, with their variety and mixture of colours in the flowers, wherein confifteth a chiefe difference. I account thofe that are called Carnations to be the greateft, both for leafe and flower, and Gilloflowers for the moft part to bee leffer in both; and therefore will giue you each defcription apart, and the Orenge tawnie or yellow Gilloflower likewife by it felfe, as differing very notably from all the reft.

1. *Caryophyllus maximus Harwicenfis fiue Anglicus*. The great Harwich or old Englifh Carnation.

I take this goodly great old Englifh Carnation, as a prefident for the defcription of all the reft of the greateft forts, which for his beauty and ftatelineffe is worthy of a prime place, hauing beene alwayes very hardly preferued in the Winter; and therefore not fo frequent as the other Carnations or Gilloflowers. It rifeth vp with a great thicke round ftalke, diuided into feuerall branches, fomewhat thickly fet with ioynts, and at euery ioynt two long greene rather then whitifh leaues, fomewhat broader then Gilloflower leaues, turning or winding two or three times round (in fome other forts of Carnations they are plaine, but bending the points downewards, and in fome alfo of a darke reddifh greene colour, and in others not fo darke, but rather of a whitifh greene colour:) the flowers ftand at the toppes of the ftalkes in long, great, and round greene huskes, which are diuided into fiue points, out of which rife many long and broad pointed leaues, deeply iagged at the ends, fet in order round and comely, making a gallant great double flower, of a deepe Carnation colour, almoft red, fpotted with many blufh fpots and ftrakes, fome greater and fome leffer, of an excellent foft fweete fent, neither too quicke as many others of thefe kinds are, nor yet too dull, and with two whitifh crooked threads like hornes in the middle: this kinde neuer beareth many flowers, but as it is flow in growing, fo in bearing, not to be often handled, which fheweth a kinde of ftatelineffe, fit to preferue the opinion of magnificence: the roote is branched into diuers great, long, wooddy rootes, with many fmall fibres annexed vnto them.

2. *Caryophyllus hortenfis flore pleno rubro*. The red or Cloue Gilloflower.

The red Cloue Gilloflower, which I take as a prefident for the fecond fort, which are Gilloflowers, grow like vnto the Carnations, but not fo thicke fet with ioynts and leaues: the ftalkes are more, the leaues are narrower and whiter for the moft part, and in fome doe as well a little turne: the flowers are fmaller, yet very thicke and double in moft, and the greene huskes wherein they ftand are fmaller likewife then the former: the ends of the leaues in this flower, as in all the reft, are dented or iagged, yet in fome more then in others; fome alfo hauing two fmall white threads, crooked at the ends like hornes, in the middle of the flower, when as diuers other haue none. Thefe

kindes,

1 *Caryophyllus maximus rubro varius.* The great old Carnation or gray Hulo. 2 *Caryophyllus maior rubro & albo varius.* The white Carnation. 3 *Caryophyllus Cantij striatus.* The faire made of Kent. 5 *Caryophyllus Sabaudicus carneus.* The blush Sauadge. 6 *Caryophyllus Xerampelinus.* The Gredeline Carnation. 7 *Caryophyllus distus Grimolo.* The Grimelo or Prince. 8 *Caryophyllus albus maior.* The great white Gilloflower. 6 *Elegans Heroina Bradshawy.* Master Bradshawes dainty Lady.

kindes, and especially this that hath a deepe red crimson coloured flower, doe endure the cold of our winters, and with lesse care is preserued : these sorts as well as the former doe very seldome giue any seede, as far as I could euer obserue or learne.

3. *Caryophyllus Silesiacus flore pleno miniato.*
The yellow or Orenge tawny Gilloflower.

This Gilloflower hath his stalkes next vnto the ground, thicker set, and with smaller or narrower leaues then the former for the most part : the flowers are like vnto the Cloue Gilloflowers, and about the same bignesse and doublenesse most vsually, yet in some much greater then in others; but of a pale yellowish Carnation colour, tending to an Orenge, with two small white threds, crooked at the ends in the middle, yet some haue none, of a weaker sent then the Cloue Gilloflower : this kinde is more apt to beare seede then any other, which is small, black, flat, and long, and being sowen, yeelde wonderfull varieties both of single and double flowers : some being of a lighter or deeper colour then the mother plants : some with stripes in most of the leaues : Others are striped or spotted, like a speckled Carnation or Gilloflower, in diuers sorts, both single and double : Some againe are wholly of the same colour, like the mother plant, and are eyther more or lesse double then it, or else are single with one row of leaues, like vnto a Pinck; and some of these likewise eyther wholly of a crimson red, deeper or lighter, or variably spotted, double or single as a Pinck, or blush eyther single or double, and but very seldome white : yet all of them in their greene leaues little or nothing varying or differing.

Cariophylli maximi.

CARNATIONS.

Caryophyllus maximus dictus Hulo rubro-varius.

THe gray *Hulo* hath as large leaues as the former old Carnation, and as deepely iagged on the edges : it hath a great high stalke, whereon stand the flowers, of a deepe red colour, striped and speckled very close together with a darkish white colour.

Caryophyllus maximus dictus Hulo ruber non variatus.

The red *Hulo* is also a faire great flower, of a stamell colour, deeply iagged as the former, and groweth very comely without any spot at all in it, so that it seemeth to bee but a stamell Gilloflower, saue that it is much greater.

Caryophyllus maximus dictus Hulo caruleo purpureus.

The blew *Hulo* is a goodly faire flower, being of a faire purplish murrey colour, curiously marbled with white, but so smally to be discerned, that it seemeth only purple, it hath so much the Mastrie in it ; it resembleth the Brasill, but that it is much bigger.

Caryophyllus maximus dictus Grimelo siue Princeps.

The *Grimelo* or Prince is a faire flower also, as large as any Chrystall or larger, being of a faire crimson colour, equally for the most part striped with white, or rather more white then red, thorough euery leafe from the bottome, and standeth comely.

Caryophyllus maximus Incarnadinus albus.

The white Carnation or Delicate, is a goodly delightfull fair flower in his pride and perfection, that is, when it is both marbled and flaked, or striped and speckled with white vpon an incarnate crimson colour, beeing a very comely flower, but abideth not constant, changing oftentimes to haue no flakes or strakes of white, but marbled or speckled wholly.

Caryophyllus maximus Incarnadinus Gallicus.

The French Carnation is very like vnto the white Carnation, but that it hath more specks, and fewer stripes or flakes of white in the red, which hath the mastrie of the white.

Caryophyllus maximus Incarnadinus grandis.

The ground Carnation (if it be not the same with the graund or great old Carnation first set downe, as the alteration but of one letter giueth the coniecture) is a thicke flower, but spreadeth not his leaues abroade as others doe, hauing the middle standing higher then the outer leaues, and turning vp their brimmes or edges; it is a sad flower, with few stripes or spots in it : it is very subiect to breake the pod, that the flower seldome commeth faire and right ; the greene leaues are as great as the *Hulo* or Lombard red.

caryophyllus maximus Chrystallinus.

The Chrystall or Chrystalline (for they are both one, howsoeuer some would make them differ) is a very delicate flower when it is well marked, but it is inconstant in the markes, being sometimes more striped with white and crimson red, and sometimes lesse or little or nothing at all, and changing also sometimes to be wholly red, or wholly blush.

caryophyllus maximus flore rubro.

The red Chrystall, which is the red hereof changed, is the most orient flower of all other red Gilloflowers, because it is both the greatest, as comming from the Chrystall, as also that the red hereof is a most excellent crimson.

caryophyllus maximus dictus Fragrans.

The Fragrant is a faire flower, and thought to come from the Chrystall, being as large, but of a blush red colour, spotted with small speckes, no bigger then pinnes points, but not so thicke as in the Pageant.

caryophyllus maximus Sabaudicus varius.

The stript Sauadge is for forme and bignesse equall with the Chrystall or White Carnation, but as inconstant as eyther of them, changing into red or blush ; so that few branches with flowers containe their true mixtures, which are a whitish blush, fairely striped with a crimson red colour, thicke and short, with some spots also among.

caryophyllus maximus Sabaudicus carneus.

The blush Sauadge is the same with the former, the same root of the stript Sauadge, as I said before, yeelding one side or part whose flowers will be eyther wholly blush, or hauing some small spots, or sometimes few or none in them.

caryophyllus maximus Sabaudicus ruber.

The red Sauadge is as the blush, when the colour of the flower is wholly red without any stripes or spots in them, and so abideth long ; yet it is sometimes seene, that the same side, or part, or roote being separate from the first or mother plant, will giue striped and well marked flowers againe.

caryophyllus maximus Oxoniensis.

The Oxeford Carnation is very like vnto the French Carnation, both for forme, largenesse and colour : but that this is of a sadder red colour, so finely marbled with white thereon, that the red hauing the maistry, sheweth a very sad flower, not hauing any flakes or stripes at all in it.

caryophyllus maximus Regius, siue Bristoliensis maior.

The Kings Carnation or ordinary Bristow, is a reasonable great flower, deepely iagged, of a sad red, very smally striped and speckled with white : some of the leaues of the flower on the one side will turne vp their brimmes or edges : the greene leafe is very large.

caryophyllus maximus Granatensis.

The greatest *Granado* is a very faire large flower, bigger then the Chrystall, and almost as bigge as the blew *Hulo* : it is almost equally diuided and stript with purple and white, but the purple is sadder then in the ordinary *Granado* Gilloflower, else it might bee said it were the same, but greater. Diuers haue taken this flower to bee the *Gran Pere*, but you shall haue the difference shewed you in the next ensuing flower.

caryophyllus maximus Gran Pere dictus.

The *Gran Pere* is a fair great flower, and comely for the forme, but of no great beautie for colour, because although it be stript red and white like the Queenes Gilloflower, yet the red is so sad that it taketh away all the delight to the flower.

caryophyllus maximus Camberfine dictus.

The Camberfine is a great flower and a faire, beeing a redde flower, well marked or striped with white, somewhat like vnto a Sauadge

Sauadge, fay fome, but that the red is not crimfon as the Sauadge; others fay the Daintie, but not fo comely : the leaues of the flowers are many, and thruft together, without any due forme of fpreading.

Caryophyllus maximus Longobardicus ruber.

The great Lombard red is a great fad red flower, fo double and thick of leaues, that it moft vfually breaketh the pod, and feldome fheweth one flower among twenty perfect : the blades or greene leaues are as large as the *Hulo.*

Caryophylli majores.

GILLOFLOWERS.

Caryophyllus maior Weftminfterienfis.

THe luftie Gallant or Weftminfter (fome make them to be one flower, and others to bee two, one bigger then the other) at the firft blowing open of the flower fheweth to be of a reafonable fize and comelineffe, but after it hath ftood blowen fome time it fheweth fmaller and thinner : it is of a bright red colour, much ftriped and fpeckled with white.

Caryophyllus maior Briftolienfis purpureus.

The Briftow blew hath greene leaues, fo large, that it would feeme to bring a greater flower then it doth, yet the flower is of a reafonable fize, and very like vnto the ordinary *Granado* Gilloflower, ftriped and fluked in the fame manner, but that the white of this is purer then that, and the purple is more light, and tending to a blew : this doth not abide conftant, but changeth into purple or blufh.

Caryophyllus maior Briftolienfis carneus.

The Briftow blufh is very like the laft both in leafe and flower, the colour only fheweth the difference, which feldome varyeth to be fpotted, or change colour.

Caryophyllus maior Dorobornienfis ruber.

The red Douer is a reafonable great Gilloflower and conftant, being of a faire red thicke poudered with white fpots, and feemeth fomewhat like vnto the ground Carnation.

Caryophyllus maior Dorobornienfis dilutus fiue albus.

The light or white Douer is for forme and all other things more comely then the former, the colour of the flower is blufh, thicke fpotted with very fmall fpots, that it feemeth all gray, and is very delightfull.

Caryophyllus maior Cantii.

The faire maide of Kent, or Ruffling Robin is a very beautiful flower, and as large as the white Carnation almoft: the flower is white, thicke poudered with purple, wherein the white hath the maftrie by much, which maketh it the more pleafant.

Caryophyllus maior Regineus.

The Queenes Gilloflower is a reafonable faire Gilloflower although very common, ftriped red and white, fome great and fome fmall with long ftripes.

Caryophyllus maior elegans.

The Daintie is a comely fine flower, although it be not great, and for the fmallneffe and thinneffe of the flower being red fo finely marked, ftriped and fpeckled, that for the liuelineffe of the colours it is much defired, beeing inferiour to very few Gilloflowers.

Caryophyllus maior Brafilienfis.

The Braffill Gilloflower is but of a meane fize, being of a fad purple colour, thicke poudered and fpeckled with white, the purple herein hath the maftrie, which maketh it fhew the fadder, it is vnconftant, varying much and often to bee all purple : the greene leaues lye matting on the ground.

Caryophyllus maior Granatenfis.

The *Granado* Gilloflower is purple and white, fluked and ftriped very much: this is alfo much fubiect to change purple. There is a greater and a leffer of this kinde, befides the greateft that is formerly defcribed.

The

Caryophyllus Turcicus.

The Turkie Gilloflower is but a small flower, but of great delight, by reason of the well marking of the flower, being most vsually equally striped with red and white.

Caryophyllus Cambrensis Poole.

The Poole flower, growing naturally vpon the rockes neare Cogshot Castle in the Isle of Wight, is a small flower, but very pleasant to the eye, by reason of the comely proportion thereof; it is of a bright pale red, thicke speckled, and very small with white, that it seemeth to bee but one colour, the leaues of the flower are but smally iagged about : it is constant.

Caryophyllus Pegma dilutior.

The light or pale Pageant is a flower of a middle size, very pleasant to behold, and is both constant and comely, and but that it is so common, would be of much more respect then it is : the flower is of a pale bright purple, thicke poudered, and very euenly with white, which hath the mastery, and maketh it the more gracefull.

Caryophyllus Pegma saturatior.

The sad Pageant is the same with the former in forme and bignesse, the difference in colour is, that the purple hath the mastery, which maketh it so sad, that it doth resemble the Brassill for colour, but is not so bigge by halfe.

Caryophyllus Heroina dictus elegans Magistri Bradshawy.

Master Bradshawe his dainty Lady may bee well reckoned among these sorts of Gilloflowers, and compare for neatenesse with most of them : the flower is very neate, though small, with a fine small iagge, and of a fine white colour on the vnderside of all the leaues, as also all the whole iagge for a pretty compasse, and the bottome or middle part of the flower on the vpperside also : but each leafe is of a fine bright pale red colour on the vpperside, from the edge to the middle, which mixture is of wonderfull great delight.

Caryophyllus albus optimus maior Londinensis & alius.

The best white Gilloflower groweth vpright, and very double, the blades growe vpright also, and crawle not on the ground.

The London white is greater and whiter then the other ordinary white, being wholly of one colour.

Caryophyllus maior rubens & minor.

The stamell Gilloflower is well knowne to all, not to differ from the ordinary red or cloue Gilloflower, but only in being of a brighter or light red colour : there is both a greater and a lesser of this kinde.

Caryophyllus purpureus maior & minor.

The purple Gilloflower a greater and a lesse : the stalke is so slender, and the leaues vpon them so many and thicke, that they lye and traile on the ground : the greatest is almost as bigge as a Chrystall, but not so double : the lesse hath a smaller flower.

Caryophyllus Persico violaceus.

The Gredeline Gilloflower is a very neate and handsome flower, of the bignesse of the Cloue red Gilloflower, of a fine pale reddish purple or peach colour, enclining to a blew or violet, which is that colour is vsually called a gredeline colour : it hath no affinity with eyther Purple, Granado, or Pageant.

Caryophyllus purpuro caruleus.

The blew Gilloflower is neither very double nor great, yet round and handsome, with a deepe iagge at the edge, and is of an exceeding deepe purple colour, tending to a tawnie : this differeth from all other sorts, in that the leafe is as greene as grasse, and the stalkes many times red or purple : by the greene leaues it may be knowne in the Winter, as well as in the Summer.

Caryophyllus carneus.

The blush Gilloflower differeth not from the red or stamell, but only in the colour of the flower, which is blush.

Caryophyllus Silesiacus maximus Wittie.

Iohn Wittie his great tawny Gilloflower is for forme of growing, in leafe and flower altogether like vnto the ordinary tawny, the flower onely, because it is the fairest and greatest that any other

ther hath nourſed vp, maketh the difference, as alſo that it is of a faire deepe ſcarlet colour.

There are alſo diuers other Tawnies, either lighter or ſadder, either leſſe or more double, that they cannot be numbered, and all riſing (as I ſaid before) from ſowing the ſeede of ſome of them : beſides the diuerſities of other colours both ſimple and mixed, euery yeare and place yeelding ſome variety was not ſeen with them before : I ſhall neede but onely to giue you the names of ſome of them we haue abiding with vs, I meane ſuch as haue receiued names, and leaue the reſt to euery ones particular denomination.

Of Bluſhes there are many ſorts, as the deepe bluſh, the pale bluſh, the Infanta bluſh, a bluſh enclining to a red, a great bluſh, the faireſt and moſt double of all the other bluſhes, and many others both ſingle and double.

Of Reds likewiſe there are ſome varieties, but not ſo many as of the other colours; for they are moſt dead or deepe reds, and few of a bright red or ſtamell colour; and they are ſingle like Pinkes, either ſtriped or ſpeckled, or more double ſtriped and ſpeckled variably, or elſe

There are neither purple nor white that riſe from this ſeede that I haue obſerued, except one white in one place.

Caryophyllus Sileſiacus ſtriatus.

The ſtriped Tawny are either greater or leſſer, deeper or lighter flowers twenty ſorts and aboue, and all ſtriped with ſmaller or larger ſtripes, or equally diuided, of a deeper or lighter colour : and ſome alſo for the very ſhape or forme will bee more neate, cloſe, and round; others more looſe, vnequall, and ſparſed.

Caryophyllus Sileſiacus marmor-amulus.

The marbled Tawny hath not ſo many varieties as the ſtriped, but is of as great beanty and delight as it, or more : the flowers are greater or ſmaller, deeper or lighter coloured one then another, and the veines or markes more conſpicuous, or more frequent in ſome then in others : but the moſt beautifull that euer I did ſee was with Maſter Ralph Truggie, which I muſt needes therefore call

Heroina Rodolphi florum Imperatoris.

Maſter Tuggies Princeſſe, which is the greateſt and faireſt of all theſe ſorts of variable tawnies, or ſeed flowers, being as large fully as the Prince or Chryſtall, or ſomething greater, ſtanding comely and round, not looſe or ſhaken, or breaking the pod as ſome other ſorts will; the marking of the flower is in this manner : It is of a ſtamell colour, ſtriped and marbled with white ſtripes and veines quite through euery leafe, which are as deeply iagged as the Hulo : ſometimes it hath more red then white, and ſometimes more white then red, and ſometimes ſo equally marked, that you cannot diſcerne which hath the maſtery; yet which of theſe hath the predominance, ſtill the flower is very beautifull, and exceeding delightſome.

Caryophyllus Sileſiacus aſſuloſus

The Flaked Tawny is another diuerſity of theſe variable or mixt coloured flowers, being of a pale reddiſh colour, flaked with white, not alwaies downeright, but often thwart the leaues, ſome more or leſſe then others; the marking of them is much like vnto the Chryſtall : theſe alſo as well as others will be greater or ſmaller, and of greater or leſſe beauty then others.

Caryophyllus Sileſiacus plumatus.

The Feathered Tawny is more rare to meete with then many of the other; for moſt vſually it is a faire large flower and double, equalling the Lumbard red in his perfection : the colour hereof is vſually a ſcarlet, little deeper or paler, moſt curiouſly feathered and ſtreamed with white through the whole leafe.

Caryophyllus Sileſiacus punctatus.

The Speckled Tawny is of diuers ſorts, ſome bigger, ſome leſſe,

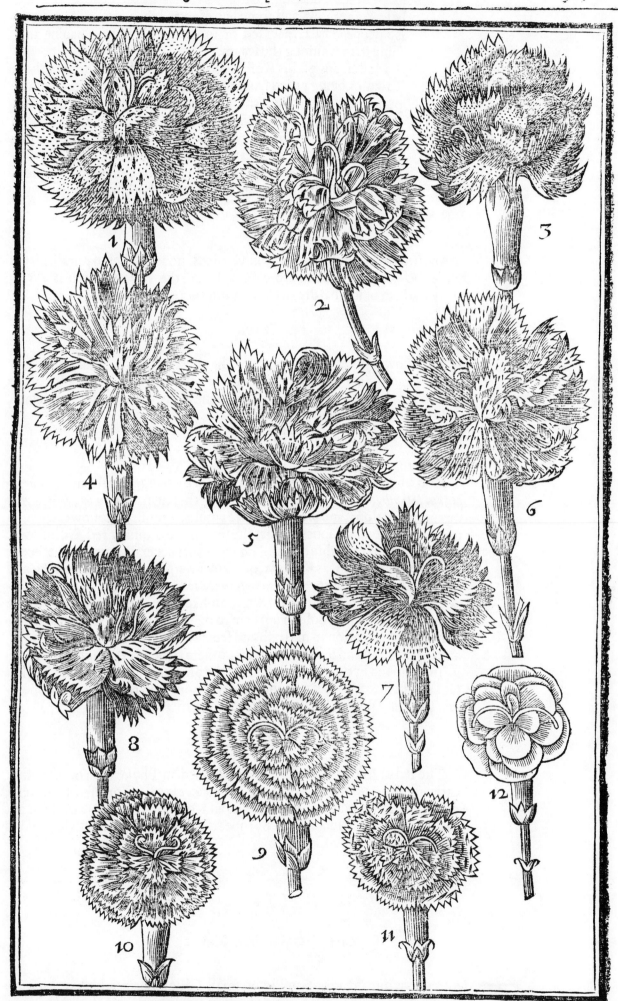

1 *Heroina Radolphi florum Imperatoris Princessa dictus*. Master Tuggie his Princesse. 2 *Caryophyllus Oxoniensis*. The French or Oxford Carnation. 3 *Caryophyllus Westmonasteriensis*. The Gallant or Westminster Gilloflower. 4 *Caryophyllus Bristoliensis*. The Bristow. 5 *Caryophyllus Chrystallinus*. The Chrystall or Chrystalline. 6 *Caryophyllus Sabaudicus striatus*. The stript Sauadge. 7 *Caryophyllus Granatensis maximus*. The Granpere or greatest Granado. 8 *Caryophyllus poramenus*. The Dainty. 9 *Caryophyllus Silesiacus maximus Iagonij Ioannis*. Iohn Witty his great tawny Gilloflower. 10 *Caryophyllus Silesiacus striatus*. The stript Tawny. 11 *Caryophyllus marmor æmulus*. The marbled Tawny. 12 *Caryophyllus roseus rotundus magistri Tuggie*. Master Tuggie his Rose Gilloflower.

lesse, some more, and some lesse spotted then others : Vsually it is a deepe scarlet, speckled or spotted with white, hauing also some stripes among the leaues.

Caryophyllus roseus rotundus Magistri Tuggie.

Master Tuggie his Rose Gilloflower is of the kindred of these Tawnies, being raised from the seede of some of them, and onely possessed by him that is the most industrious preseruer of all natures beauties, being a different sort from all other, in that it hath round leaues, without any iagge at all on the edges, of a fine stamell full colour, without any spot or strake therin, very like vnto a small Rose, or rather much like vnto the red Rose Campion, both for forme, colour, and roundnesse, but larger for size.

The Place.

All these are nourished with vs in Gardens, none of their naturall places being knowne, except one before recited, and the yellow which is *Silesia*; many of them being hardly preserued and encreased.

The Time.

They flower not vntill the heate of the yeare, which is in Iuly (vnlesse it be an extraordinary occasion) and continue flowring, vntill the colds of the Autumne checke them, or vntill they haue wholly out spent themselues, and are vsually encreased by the slips.

The Names.

Most of our later Writers doe call them by one generall name, *Caryophyllus satiuus*, and *flos Caryophylleus*, adding thereunto *maximus*, when wee meane Carnations, and *maior* when we would expresse Gilloflowers, which name is taken from Cloues, in that the sent of the ordinary red Gilloflower especially doth resemble them. Diuers other seuerall names haue beene formerly giuen them, as *Vetonica*, or *Betonica altera*, or *Vetonica altilis*, and *coronaria. Herba Tunica, Viola Damascena, Ocellus Damascenus*, and *Barbaricus*. Of some *Cantabrica Pliny*. Some thinke they were vnknowne to the Ancients, and some would haue them to be *Iphium* of Theophrastus, wherof he maketh mention in his sixth and seuenth Chapters of his sixth booke, among Garland and Summer flowers; others to be his *Dios anthos*, or *Iouis flos*, mentioned in the former, and in other places. We call them in English (as I said before) the greatest kindes, Carnations, and the others Gilloflowers (*quasi* Iuly flowers) as they are seuerally expressed.

The Vertues.

The red or Cloue Gilloflower is most vsed in Physicke in our Apothecaries shops, none of the other being accepted of or vsed (and yet I doubt not, but all of them might serue, and to good purpose, although not to giue so gallant a tincture to a Syrupe as the ordinary red will doe) and is accounted to be very Cordiall.

CHAP. LXX.

Caryophylli siluestres. Pinkes.

THere remaine diuers sorts of wilde or small Gilloflowers (which wee vsually call Pinkes) to be entreated of, some bearing single, and some double flowers, some smooth, almost without any deepe dents on the edges, and some iagged, or as it were feathered. Some growing vpright like vnto Gilloflowers, others creeping

ping

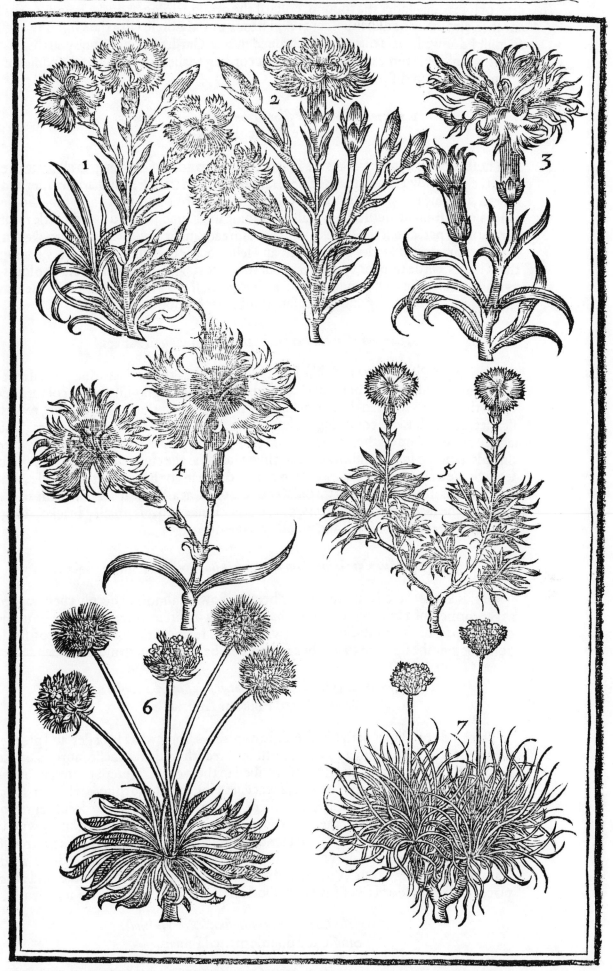

1 *Caryophyllus siluestris simplex.* The vsuall single Pinke. 2 *Caryophyllus multiplex siluestris.* Double Pinkes. 3 *Cariophyllus siluestris pluma-marius.* Feathered or iagged Pinkes. 4 *Caryophyllus Stellatus.* Starre Pinkes. 5 *Caryophyllus retens.* Matted Pinkes. 6 *Caryophyllus me-diterraneus.* The great Thistle or Sea Gilloflower. 7 *Caryophyllus marinus.* The ordinary Thistle or Sea Cushion.

Dd 2

ping or spreading vnder the toppe or cruft of the ground, some of one colour, some of another, and many of diuers colours : As I haue formerly done with the Gilloflowers, so muft I doe with these that are entertained in our Gardens, onely giue you the defcriptions of some three or foure of them, according to their variety, and the names of the reft, with their diftinctions.

1. *Caryophyllus minor filueftris multiplex & fimplex.*
Double and fingle Pinkes.

The fingle and double Pinkes are for forme and manner of growing, in all parts like vnto the Gilloflowers before defcribed, fauing onely that their leaues are fmaller and fhorter, in fome more or leffe then in others, and fo are the flowers alfo : the fingle kindes confifting of fiue leaues vfually (feldome fix) round pointed, and a little fnipt for the moft part about the edges, with fome threads in the middle, either crooked or ftraight : the double kindes being leffer, and leffe double then the Gilloflowers, hauing their leaues a little fnipt or endented about the edges, and of diuers feuerall colours, as fhall hereafter be fet downe, and of as fragrant a fent, efpecially fome of them as they : the rootes are long and fpreading, fomewhat hard and wooddy.

2. *Caryophyllus plumarius.* Feathered or iagged Pinkes.

The iagged Pinkes haue fuch like ftalkes and leaues as the former haue, but fomewhat fhorter and fmaller, or graffe-like, and of a whitifh or grayifh greene colour likewife : the flowers ftand in the like manner at the toppes of the ftalkes, in long, round, flender, greene huskes, confifting of fiue leaues, very much cut in on the edges, and iagged almoft like a feather, of a light red, or bright purple colour, with two white threads ftanding in the middle, crooked like a horne at the end, and are of a very good fent. Some of thefe haue not thofe two crooked threads or hornes in the middle, but haue in their ftead many fmall threads, not crooked at all : the feedes of them all are like vnto the feedes of Gilloflowers, or the other Pinkes, that is, fmall, blacke, long, and flat : the rootes are fmall and wooddy likewife.

3. *Caryophyllus plumarius albus orbe rubro fiue Stellatus.* Starre Pinkes.

Of this kinde there is another fort, bearing flowers almoft as deeply cut or iagged as the former, of a faire white colour, hauing a ring or circle of red about the bottome or lower part of the leaues, and are as fweete as the former : this being fowne of feede doth not giue the ftarre of fo bright a red colour, but becommeth more dunne.

4. *Caryophyllus plumarius Auftriacus fiue Superba Auftriaca.*
The feathered Pinke of Auftria.

This kinde of Pinke hath his firft or lower leaues, fomewhat broader and greener then any of the former Pinkes, being both for breadth and greenneffe more like vnto the Sweete Iohns, which fhall bee defcribed in the next Chapter : the leaues on the ftalkes are fmaller, ftanding by couples at euery ioynt, at the toppes whereof ftand fuch like iagged flowers as the laft defcribed, and as large, but more deeply cut in or iagged round about, fome of them of a purplifh colour, but the moft ordinary with vs are pure white, and of a moft fragrant fent, comforting the fpirits and fenfes a farre off : the feedes and rootes are like vnto the former. Some haue miftaken a kinde of wilde Campion, growing in our Woods, and by the paths fides in Hornfie Parke, and other places, to be this feathered Pinke : but the flowers declare the difference fufficiently.

5. *Caryophyllus minor repens fimplex & multiplex.*
Single and double matted Pinkes.

The matted Pinke is the fmalleft, both for leafe and flower of all other Pinkes that are nourifhed in Gardens, hauing many fhort and fmall graffie greene leaues vpon the

ftalkes,

ftalkes, which as they grow and lye vpon the ground (and not ftanding fo vpright as the former) doe take roote againe, whereby it quickly fpreadeth, and couereth a great deale of ground in a little fpace: the flowers are fmall and round, a little fnipt about the edges, whereof fome are white, and fome red, and fome are white fpotted with red, and fome red fpotted with white, all of them being fingle flowers. But there is another of this kinde, not differing in leafe, but in flower: for that the firft flowers are but once double, or of two rowes of leaues, of a fine reddifh colour, fpotted with filuer fpots: but thofe that follow, are fo thicke and double, that they oftentimes doe breake the pod or huske; being not altogether of fo deepe a red colour, but more pale.

6. *Caryophyllus Mediterraneus fiue Marinus maior.*
Great Sea Gilloflower or Great Thrift.

Vnto thefe kindes of Pinkes I muft needs adde, not only our ordinary Thrift (which is more frequent in gardens, to empale or border a knot, becaufe it abideth greene Winter and Summer, and that by cutting, it may grow thicke, and be kept in what form one lift, rather then for any beautie or the flowers) but another greater kinde, which is of as great beautie and delight almoft as any of the former Pinkes, as well for that the leaues are like vnto Gilloflowers, being longer and larger then any Pinkes, and of a whitifh greene colour like vnto them, not growing long or by couples vpon the ftalkes as Pinkes and Gilloflowers doe, but tufting clofe vpon the ground, like vnto the common Thrift: as alfo that the ftalkes, rifing from among the leaues (being fometimes two foote high (as I haue obferued in my garden) are yet fo flender and weake, that they are fcarce able to beare the heads of flowers, naked or bare, both of leaues and ioynts, fauing only in one place, where at the ioynt each ftalke hath two fmall and very fhort leaues, not rifing vpwards as in all other Gilloflowers, Pinkes, and other herbes, but growing downewards) and doe beare each of them a tuft or vmbell of fmall purplifh, or blufh coloured flowers, at the toppes of them ftanding fomewhat like vnto fweete Williams, but more roundly together, each flower confifting of fiue fmall, round, ftiffe or hardifh leaues, as if they were made of paper, the bottome or middle being hollow, not blowing all at once as the ordinary Thrift, but for the moft part one after another, not fhewing vfually aboue foure or fiue flowers open at one time (fo farre as I could obferue in the plants that I kept) fo that it was long before the whole tuft of flowers were paft; but yet the hoter and dryer the time was, the fooner it would be gone: the feede I haue not perfectly obferued, but as I remember, it was fomewhat like vnto the feede of Scabious; I am fure nothing like vnto Gilloflowers or Pinkes: the roote is fomewhat great, long and hard, and not fo much fpreading in the ground as Gilloflowers or Pinkes.

Caryophyllus Marinus. Thrift, or Sea Cufhion.

Our common Thrift is well knowne vnto all, to haue many fhort and hard greene leaues, fmaller then many of the graffes, growing thicke together, and fpreading vpon the ground: the ftalkes are naked of leaues a fpanne high, bearing a fmall tuft of light purple, or blufh coloured flowers, ftanding round and clofe thrufting together.

Double Pinkes.	Single Pinkes.
THe double white Pinke is onely with more leaues in it then the fingle, which maketh the difference.	THe fingle white ordinary Pinke hath a fingle white flower of fiue leaues, finely iagged about the edges.
The double red Pinke is in the fame manner double, differing from the fingle of the fame colour.	The fingle red Pink is like the white, but that the leaues are not fo much iagged, and the flower is of a pale purplifh red colour.
The double purple Pinke differeth not	

from

from the fingle purple for colour, but only in the doubleneffe of the flower.

The *Granado* Pinke differeth not from
the Gilloflower of the fame name, but in
the fmalneffe both of leaues and flower.

The double Matted Pinke is before defcribed.

The double blufh Pinke is almoft as
great as the ordinary blufh Gilloflower,
and fome haue taken it for one, but the
greene leaues are almoft as fmall as Pinks,
and therefore I referre it to them.

The fingle purple Pinke is of a faire
purple colour, like almoft vnto the purple
Gilloflower.

The great blufh Pinke hath broader and
larger leaues in the flower then any other
Pinke, and of a faire blufh colour.

The white Featherd Pinke hath the
edges of the flower more finely and deeply cut in then the former.

The red or light purple Featherd Pinke
is like the former featherd Pinke, but only
differeth in colour.

The Starre Pinke is a faire flower, finely
iagged on the edges, with a faire red circle at the lower end of the leaues on the infide.
The white featherd Pinke of Auftria is defcribed before. The purple featherd
Pinke of Auftria is fo likewife. The fingle matted Pinke is before defcribed. The
fpeckled Pinke is a fmall flower hauing fmall fpots of red here and there difperfed
ouer the white flower.

Thofe fingle flowers being like vnto Pinkes that rife from the fowing of the orenge
tawney, I bring not into this *claffis*, hauing already fpoken of them in the precedent
Chapter.

The Place.

Thefe are all like as the former, nourifhed in Gardens with vs, although
many of them are found wilde in many places of Auftria, Hungarie, and
Germany, on the mountaines, and in many other places, as Clufius recordeth. The ordinary Thrift groweth in the falt Marfhes at Chattam by Ro-
chefter, and in many other places in England : but the great kinde was gathered in Spaine, by Guillaume Boel that painefull fearcher of fimples,
and the feede thereof imparted to me, from whence I had diuers plants,
but one yeare after another they all perifhed,

The Time.

Many of thefe Pinkes both fingle and double, doe flower before any
Gilloflower, and fo continue vntill Auguft, and fome, moft of the Summer and Autumne.

The Names.

The feuerall titles that are giuen to thefe Pinkes, may fuffice for their particular names : and for their generall they haue beene expreffed in the former Chapter, beeing of the fame kindred, but that they are fmaller, and
more frequently found wilde. The two forts of Thrift are called *Caryophyllus Marinus*. The greater, *Maior & Mediterraneus*; In Englifh, The greater or Leuant Thrift, or Sea Gilloflower. The leffer *Minimus*, and is accounted of fome to be a graffe, and therefore called *Gramen Marinum &
Polyanthemum*; In Englifh, Thrift, Sea graffe, and our Ladies Cufhion,
or Sea Cufhion.

The Vertues.

It is thought by diuers, that their vertues are anfwerable to the Gilloflowers, yet as they are of little vfe with vs, fo I thinke of as fmall effect.

CHAP.

CHAP. LXXI.

Armerius. Sweet Iohns, and sweet Williams.

THese kindes of flowers as they come neerest vnto Pinkes and Gilloflowers, though manifestly differing, so it is fittest to place them next vnto them in a peculiar Chapter.

1. *Armerius angustifolius rubens simplex.* Single red sweete Iohns.

The sweete Iohn hath his leaues broader, shorter and greener then any of the former Gilloflowers, but narrower then sweete Williams, set by couples, at the ioynts of the stalkes, which are shorter then most of the former, and not aboue a foote and a halfe high, at the tops whereof stand many small flowers, like vnto small Pinkes, but standing closer together, and in shorter huskes, made of fiue leaues, smaller then most of them, and more deeeply iagged then the Williams, of a red colour in the middle, and white at the edges, but of a small or soft sent, and not all flowring at once, but by degrees : the seede is blacke, somewhat like vnto the seede of Pinkes, the roote is dispersed diuersly, with many small fibres annexed vnto it.

2. *Armerius angustifolius albus simplex.* Single white sweet Iohns.

This white Iohn differeth not in any thing from the former, but onely that the leafe doth neuer change brownish, and that the flower is of a faire white colour, without any mixture.

3. *Armerius angustifolius duplex.* Double sweet Iohns.

There is of both those former kindes, some whose flowers are once double, that is, consisting of two or three rowes of leaues, and the edges not so deepely iagged ; not differing in any thing else.

4. *Armerius latifolius simplex flore rubro.*
Single red sweet Williams.

The sweet Williams doe all of them spread into many very long trayling branches, with leaues lying on the ground, in the very like manner that the sweete Iohns doe : the chiefe differences betweene them are, that these haue broader, and darker greene leaues, somewhat brownish, especially towards the points, and that the flowers stand thicker and closer, and more in number together, in the head or tuft, hauing many small pointed leaues among them, but harmlesse, as all men know; the colour of the flower is of a deep red, without any mixture or spot at all.

5. *Armerius latifolius flore rubro multiplici.*
Double red sweete Williams.

The double kinde differeth not from the single kinde of the same colour, but only in the doublenesse of the flowers, which are with two rowes of leaues in euery flower.

6. *Armerius latifolius variegatus siue versicolor.*
Speckled sweete Williams, or London pride.

These spotted Williams are very like the first red Williams, in the forme or maner of growing, hauing leaues as broade, and browne sometimes as they, the flowers stand as thicke or thicker, clustring together, but of very variable colours : for some flowers will be of a fine delayed red, with few markes or spots vpon them, and others

will

will bee full peckled or sprinkled with white or siluer spots, circlewise about the middle of the flowers, and some will haue many specks or spots vpon them dispersed: all these flowers are not blowne at one time, but some are flowring, when others are decaying, so that abiding long in their pride, they become of the more respect: The seede is blacke, as all the rest, and not to be distinguished one from another: the roots are some long, and some small and threddy, running vnder the vpper crust of the earth.

7. *Armerius latifolius flore rubro saturo holosericeo.*
Sweet Williams of a deepe red or murrey colour.

The leaues of this kinde seeme to be a little larger, and the ioints a little redder then the former, but in the flower consisteth the chiefest difference, which is of a deepe red, or murrey purple colour, like vnto veluet of that colour, without any spots, but smooth, and as it were soft in handling, hauing an eye or circle in the middle, at the bottome of the leaues.

8. *Armerius latifolius simplex flore albo.*
Single white sweete Williams.

The white kinde differeth not in forme, but in colour from the former, the leaues are not browne at all, but of a fresh greene colour, and the flowers are wholly white, or else they are all one.

The Place.

These for the most part grow wilde in Italie, and other places: we haue them in our Gardens, where they are cherished for their beautifull varietie.

The Time.

They all generally doe flower before the Gilloflowers or Pinkes, or with the first of them: their seede is ripe in Iune and Iuly, and doe all well abide the extremitie of our coldest winters.

The Names.

They all generally are called *Armerius*, or *Armeria*, as some doe write, and distinguished as they are in their titles: Yet some haue called them *Vetonica agrestis*, and others *Herba Tunica, Scarlatea, & Caryophyllus siluestris*: Wee doe in English in most places, call the first or narrower leafed kindes, Sweet Iohns, and all the rest Sweete Williams; yet in some places they call the broader leafed kindes that are not spotted, Tolmeiners, and London tufts: but the speckled kinde is termed by our English Gentlewomen, for the most part, London pride.

The Vertues.

We haue not knowne any of these vsed in Physicke.

CHAP. LXXII.

Bellis. Daisie.

THere be diuers sorts of Daisies, both great and small, both single and double, both wilde growing abroade in the fieldes, and elsewhere, and manured growing only in Gardens: of all which I intend not to entreate, but of those that are of most beautie and respect, and leaue the rest to their proper place.

1. *Bellis*

1 *Armeriꝰ angustifolius simplex.* Single sweete Iohns. 2 *Armerius angustifolius multiplex* Double sweet Iohns. 3 *Armerius latifolius simplex.* Single sweete Williams. 4 *Armerius latifolius versicolor.* Spotted sweet Williams or pride of London. 5 *Armerius latifolius multiplex* Double sweet Williars. 6 *Bellis hortensis minor multiplex.* Double Garden Daisies. 7 *Bellis minor hortensis flore varie.* Double red Daisies stript 8 *Bellis minor hortensis prolifera.* Double fruitfull Daisies or Iacke an Apes on horsebacke. 9 *Bellis cærulea sive Globularia.* Double blew Daisies or blew Globeflower. 10 *Bellis lutea montana sive Globularia lutea montana,* Double yellow Daisies or yellow Globeflower.

1. *Bellis maior flore albo pleno.* The great double white Daisie.

The great Daisie with the double white flower, is in all things so like vnto the great single kinde, that groweth by the high wayes, and in diuers medowes and fields, that there is no difference but in the flower, which is double. It hath many long, and somewhat broad leaues lying vpon the ground, deepely cut in on both sides, somewhat like vnto an oaken leafe; but those that are on the stalkes are shorter, narrower, and not so deeply cut in, but onely notched on the edges: the flowers at the toppe are (as I said) white and double, consisting of diuers rowes of leaues, being greater in compasse then any of the double Daisies that follow, but nothing so double of leaues.

2. *Bellis minor flore rubro simplici.* Single red Daisies.

This single Daisie (like as all the rest of the small Daisies) hath many smooth, greene, round pointed leaues lying on the ground, a little snipt about the edges; from among which rise many slender round foote-stalkes, rather then stalkes or stems, about an hand breadth high at the most, and oftentimes not halfe so high, bearing one flower a peece, consisting of many small leaues, as a pale or border set about a middle thrumme: the leaues of this kinde are almost wholly red, whereas in the wilde they are white or whitish, enclining to red on the edges, the middle being yellow in both sorts: the rootes are many small white threads or strings.

3. *Bellis minor hortensis flore pleno variorum colorum.* Double Garden Daisies of diuers colours.

The leaues of all the double Daisies are in forme like vnto the single ones, but that they are smaller, and little or nothing snipt or notched about the edges: the small stalkes likewise are smaller and lower, but bearing as double flowers as any that growe on the ground, being composed of many small leaues, thicke thrust together, of diuers colours; for some are wholly of a pure white, others haue a little red, either disperled vpon the white leaues, or on the edges, and sometimes on the backes of the leaues: some againe seeme to be of a whitish red, or more red then white, when as indeede they are white leaues disperled among the red; others of a deepe or darke red colour, and some are speckled or striped with white and red through the whole flower: and some the leaues will bee red on the vpperside, and white vnderneath; and some also (but those are very rare) are of a greenish colour.

4. *Bellis minor hortensis prolifera.* Double double Daisies or childing Daisies.

There is no difference either in leafe or roote in this kinde from the former double Daisies: the chiefest variety consisteth in this, that it beareth many small double flowers, standing vpon very short stalkes round about the middle flower, which is vsually as great and double as any of the other double kindes, and is either wholly of a deepe red colour, or speckled white and red as in some of the former kindes, or else greenish, all the small flowers about it being of the same colour with the middlemost.

5. *Bellis cærulea siue Globularia.* Blew Daisies.

The likenesse and affinity that this plant hath with the former, both in the forme of leafe and flower, as also in the name, hath caused me to insert it, and another rare plant of the same kinde, in this place, although they be very rare to be met with in our English Gardens. This beareth many narrower, shorter, and blacker greene leaues then the former, lying round about vpon the ground; among which rise vp slender but stiffe and hard stalks, halfe a foot high or more, set here and there with small leaues, and at the top a small round head, composed of many small blew leaues, somewhat like vnto the head of a Scabious: It hath bin found likewise with a white head of flowers: the roote is hard and stringie: the whole plant is of a bitter taste.

6. *Globularia*

6. Globularia lutea montana. Yellow Daisies.

This mountaine yellow Daisie or Globe-flower hath many thicke, smooth, round pointed leaues, spread vpon the gronnd like the former; among which spring diuers small round rushie stalkes, a foote high, bearing about the middle of them two small leaues at the ioynts, and at the toppes round heads of flowers thrust thicke together, standing in purplish huskes, euery of which flowers do blow or spread into fiue leaues, starre-fashion, and of a faire yellow colour, smelling like vnto broome flowers, with many small threads in the middle compassing a flat pointell, horned or bended two wayes: after the flowers are past rise vp the seede veffels, which are round, swelling out in the middle, and diuided into foure parts at the toppes, containing within them round, flat, blacke seede, with a small cut or notch in them : the roote is a finger long, round and hard, with a thicke barke, and a woddy pith in the middle, of a sharpe drying taste and strong sent : the leaues are also sharpe, but bitter.

The Place.

The small Daisies are all planted, and found onely in Gardens, and will require to be replanted often, left they degenerate into single flowers, or at least into lesse double. The blew Daisie is naturall of Mompelier in France, and on the mountaines in many places of Italy, as also the yellow kinde in the Kingdome of Naples.

The Time.

The Daisies flower betimes in the Spring, and last vntill May, but the last two flower not vntill August or September.

The Names.

They are vsually called in Latine *Bellides*, and in English Daisies. Some call them *Herba Margarita*, and *Primula veris*, as it is likely after the Italian names, of *Marguerite*, and *Fior di prima vera gentile*. The French call them *Pasquettes*, and *Marguerites*, and the Fruitfull sort, or those that beare small flowers about the middle one, *Margueritons* : our English women call them Iacke an Apes on horse-backe, as they doe Marigolds before recited, or childing Daisies : but the Physitians and Apothecaries doe in generall call them, especially the single or Field kindes, *Consolida minor*. The blew Daisie is called *Bellis cærulea*, and *Globularia*, of some *Scabiosæ pumilum genus*. The Italians call it *Botanaria*, because the heads are found like buttons. The yellow, *Globularia montana*, is onely described by Fabius Columna, in his last part of *Phytobasanos*, and by him referred vnto the former *Globularia*, although it differ in some notable points from it.

The Vertues.

The properties of Daisies are certainly to binde, and the roote especially being dryed, they are vsed in medicines to that purpose. They are also of speciall account among those herbes, that are vsed for wounds in the head.

CHAP,

Chap. LXXIII.

Scabiosa. Scabious.

THe forts of Scabious being many, yeeld not flowers of beauty or refpect, fit to bee cherifhed in this our Garden of delight; and therefore I leaue them to the Fields and Woods, there to abide. I haue onely two or three ftrangers to bring to your acquaintance, which are worthy this place.

1. *Scabiofa flore albo.* White flowred Scabious.

This white Scabious hath many long leaues, very much iagged or gafht in on both fides, of a meane bignefle, being neither fo large as many of the field, nor fo fmall as any of the fmall kindes : the ftalkes rife about a foote and a halfe high, or fomewhat higher, at the tops whereof grow round heads, thicke fet with flowers, like in all points vnto the field Scabious, but of a milke white colour.

2. *Scabiofa rubra Auftriaca.* Red Scabious of Auftria.

This red Scabious hath many leaues lying vpon the ground, very like vnto Deuils bit, but not fo large, being fhorter and fnipt, not gafhed about the edges, of a light greene colour; yet (there is another of a darker greene colour, whofe flower is of a deeper red) the ftalkes haue diuers fuch leaues on them, fet by couples at the ioynts as grow belowe, and at the tops fmall heads of flowers, each confifting of fiue leaues, the biggeft flowers ftanding round about in the outer compaffe, as is vfuall almoft in all the kinds of Scabious, of a fine light purple or red colour : after the flowers are paft, come the feede, which is fomewhat long and round, fet with certaine haires at the head thereof, like vnto a Starre : the roote is compofed of a number of flender ftrings, faftened at the head.

3. *Scabiofa rubra Indica.* Red flowred Indian Scabious.

This (reputed Indian) Scabious hath many large faire greene leaues lying on the ground, iagged or cut in on both fides to the middle ribbe, euery peece whereof is narrower then that at the end, which is the broadeft : among thefe leaues rife vp fundry flender and weake ftalkes, yet ftanding vpright for the moft part, fet with fmaller and more iagged leaues at certaine diftances, two or three at euery ioynt, branching forth at the toppe into other fmaller branches, bearing euery one head of flowers, like in forme vnto other Scabioufes, but of an excellent deepe red crimfon colour (and fometimes more pale or delayed) of no fent at all : after which doe come fmall roundifh feede, like vnto the field Scabious : the roote is long and round, compaffed with a great many fmall ftrings, and perifheth vfually as foone as it hath borne out his flowers and feede : otherwife if it doe not flower the firft yeare of the fowing, if it be carefully defended from the extremity of Winter, it will flower the fooner the next yeare, as I my felfe haue often found by experience.

The Place.

The firft is fometimes found wilde in our owne Countrey, but it is very geafon, and hath been fent among other rare feedes from Italy.

The fecond was firft found and written of by Clufius, in Pannonia and Auftria, where it is very plentifull.

The third hath been fent both from Spaine & Italy, and is verily thought to grow naturally in both thofe parts.

The

1 *Scabiosa flore albo.* White flowred Scabious. 2 *Scabiosa rubra Austriaca.* Red Scabious of Austria. 3 *Scabiosa rubra Indica.* Red flowred Indian Scabious. 4 *Cyanus vulgaris minor.* Corn-flower of diuers colours. 5 *Cyanus Baticus.* Spanish Corn-flower. 6 *Cyanus floridus Turcicus.* The braue Sultans flower. 7 *Carthamus satiuus.* Spanish Saffron.

Ee

The Time.

The firſt and ſecond flower earlier then the laſt, for that it flowreth not vntill September or October, (vnleſſe it be not apt to beare the firſt yeare as I before ſaid) ſo that many times (if none be more forward) it periſheth without bearing ripe ſeede, whereby we are oftentimes to ſeeke new ſeede from our friends in other parts.

The Names.

They haue all one generall name of Scabious, diſtinguiſhed eyther by their flower, or place of growing, as in their titles : yet the laſt is called of diuers *Scabioſa exotica,* becauſe they thinke the name *Indica,* is not truely impoſed vpon it.

The Vertues.

Whether theſe kindes haue any of the vertues of the other wilde kinds, I know none haue made any experience, and therefore I can ſay no more of them.

CHAP. LXXIIII.

Cyanus. Corne flower, or blew Bottles.

VNder the name of *Cyanus* are comprehended, not onely thoſe plants which from the excellent blew colour of their flowers (furniſhing or rather peſtering the Corne fieldes) haue peculiarly obtained that name, and which doth much vary alſo, in the colour of the flowers, as ſhallbe ſhewed; but ſome other plants alſo for their neere reſemblance, but with ſeuerall diſtinctions. The *Cyanus maior, Ptarmica Auſtriaca, Ptarmica Imperati,* and many others which may be adioyned vnto them, do more fitly belong to the Garden of Simples, whereunto I leaue them, and will here only entreate of thoſe that may moſt pleaſe the delight of our Gentle Floriſts, in that I labour and ſtriue, to furniſh this our garden, with the chiefeſt choyſe of natures beauties and delights.

1. *Cyanus vulgaris diuerſorum colorum.* Corne flower of diuers colours.

All theſe ſorts of Corne flowers are for the moſt part alike, both in leaues and flowers one vnto another for the forme : the difference betweene them conſiſteth in the varying colour of the flowers : For the leaues are long, and of a whitiſh greene colour, deeply cut in on the edges in ſome places, ſomewhat like vnto the leaues of a Scabious : the ſtalkes are two foote high or better, beſet with ſuch like leaues but ſmaller, and little or nothing ſlit on the edges: the toppes are branched, bearing many ſmall greene ſcaly heads, out of which riſe flowers, conſiſting of fiue or ſixe, or more long and hollow leaues, ſmall at the bottome, and opening wider and greater at the brims, notched or cut in on the edges, and ſtanding round about many ſmall threds in the middle : the colours of theſe flowers are diuers, and very variable ; for ſome are wholly blew, or white, or bluſh, or of a ſad, or light purple, or of a light or dead red, or of an ouerworne purple colour, or elſe mixed of theſe colours, as ſome, the edges white, and the reſt blew or purple, or the edges blew or purple, and the reſt of the flower white, or ſtriped, ſpotted, or halfed, the one part of one colour, and the other of another, the threds likewiſe in the middle varying in many of them ; for ſome will haue the middle thrume of a deeper purple then the outer leaues, and ſome haue white or bluſh leaues, the middle thrume being reddiſh, deeper or paler : After the flowers are paſt, there come ſmall, hard, white and ſhining ſeede in thoſe heads, wrapped

ped or set among a deale of flockie matter, as is most vsuall, in all plants that beare scaly heads : the rootes are long and hard, perishing euery yeare when it hath giuen seede.

2. *Cyanus floridus Turcicus.* The Sultans flower.

As a kinde of these Corne flowers, I must needes adioyne another stranger, of much beautie, and but lately obtained from Constantinople, where, because (as it is said) the great Turke, as we call him, saw it abroade, liked it, and wore it himselfe; all his vassals haue had it in great regard, and hath been obtained from them, by some that haue sent it into these parts. The leaues whereof are greener, and not only gashed, but finely snipt on the edges : the stalkes are three foote high, garnished with the like leaues as are below, and branched as the former, bearing large scaly heads, and such like flowers but larger, hauing eight or nine of those hollow gaping leaues in euery flower, standing about the middle threds (if it be planted in good and fertile ground and be well watered, for it soone starueth and perisheth with drought) the circling leaues are of a fine delayed purple or blush colour, very beautifull to behold ; the seede of this is smaller and blacker, and not enclosed in so much dounie substance, as the former (yet in our Countrey the seede is not so blacke, as it came vnto vs, but more gray) the roote perisheth likewise euery yeere.

3. *Cyanus Bæticus supinus.* The Spanish Corne-flower.

This Spanish kinde hath many square low bending or creeping stalkes, not standing so vpright as the former, but branching out more diuersly ; so that one plant will take vp a great deale of ground : the leaues are broader then any of the rest, softer also, of a pale or whitish greene colour, and not much gashed on the edges : the flowers stand in bigger heads, with foure or fiue leaues vnder euery head, and are of a light pale purple or blush colour ; after which come seede, but not so plentifully, yet wrapped in a great deale of flockie matter, more then any : the roote groweth downe deepe into the ground, but perisheth euery yeare as they doe.

The Place.

The first or former kindes, grow many times in the Corne fields of our own Countrey, as well as of others, especially that sort with a blew flower: but the other sorts or colours are not so frequent, but are nourished in gardens, where they will vary wonderfully.

The second as is before set downe, groweth in Turkie : and the last in Spaine, found out and first sent to vs by that industrious searcher of simples, Guillaume Boel before remembred.

The Time.

The first doe flower in the end of Iune, and in Iuly, and somtimes sooner. The other two later, and not vntill August most commonly, and the seede is soone ripe after.

The Names.

The first is generally called *Cyanus,* and some following the Ditch name, call it *Flos frumenti.* The olde Writers gaue it the name of *Bapti secuba,* which is almost worne out. We doe call them in English, Blew Bottles, and in some places, Corne flowers, after the Ditch names. The second hath beene sent by the name of *Ambreboi,* which whether it be a Turkie or Arabian name, I know not. I haue called it from the place, from whence we had it, *Turcicus,* and for his beauty, *Floridus.* The Turkes themselues as I vnderstand, doe call it The Sultans flower, and I haue done so likewise, that it may bee distinguished from all the other kindes, or else you may call it,

Ee 2 The

The Turkey blush Corne flower, which you pleafe. The laft was fent by the name of *Iacea Bætica*, but I had rather to referre it to the *Cyanus*, or Corne flowers, becaufe the flowers are like vnto the Corne flowers, and not vnto the Iaceas or Knapweedes.

The Vertues.

Thefe had no vfe in Phyficke in Galen and Diofcorides time, in that (as it is thought) they haue made no mention of them : We in thefe dayes doe chiefly vfe the firft kindes (as alfo the greater fort) as a cooling Cordiall, and commended by fome to be a remedy, not onely againft the plague and peftilentiall difeafes, but againft the poifon of Scorpions and Spiders.

Chap. LXXV.

Iacea Marina Bætica. Spanifh Sea Knapweede.

THere are a great many forts of Knapweedes, yet none of them all fit for this our Garden, but this only ftranger, which I haue beene bold to thruft in here, for that it hath fuch like gaping or open flowers, as the former Corne flowers haue, but notably differing, and therefore deferueth a peculiar Chapter, as partaking both with *Cyanus* and *Iacea*. It hath many long and narrow leaues vneuenly dented or waued on both edges (and not notched, gafhed or indented, as many other herbes are) being thicke, flefhie and brittle, a little hairy, and of an ouerworne darke greene colour, among which rife lowe weake ftalkes, with fuch like leaues as grow at the bottome, but fmaller, bearing but here and there a flower, of a bright reddifh purple colour, like in forme vnto the Corne flowers, but much larger, with many threds or thrumes in the middle of the fame colour, ftanding vp higher then any of the former : this flower rifeth out of a large fcaly head, all fet ouer with fmall fharpe (but harmeleffe) white prickles : the feedes are blackifh, like vnto the Knapweedes, and larger then any of the former Corne flowers : the roote is great and thicke, growing deepe into the ground, flefhie and full of a flimie or clammy iuice, and eafie to bee broken, blackifh on the outfide, and whitifh within, enduring many yeares, like as the other Knapweedes, or Matfelons doe, growing in time to be very thicke and great.

The Place.

It groweth naturally by the Sea fide in Spaine, from whence I receiued the feedes of Guillaume Boel, and did abide well in my garden a long time, but is now perifhed.

The Time.

It flowreth in the beginning of Iuly, or thereabouts, and continueth not long in flower : but the head abideth a great while, and is of fome beauty after the flower is paft ; yet feldome giueth good feed with vs.

The Names.

It hath no other name then is fet down in the title, being altogether a Neuelift, and not now to be feene with any fauing my felfe.

The Vertues.

We haue not yet kuown any vfe hereof in Phyfick.

Chap. LXXVI.

Cnicus siue Carthamus sativus. Baſtard or Spaniſh Saffron.

THere are two or three ſorts of *Cnicus* or baſtard Saffrons which I paſſe ouer, as not fit for this Garden, and onely ſet downe this kinde, whoſe flowers are of a fairer and more liuely colour in our Country, then any hath come ouer from Spaine, where they manure it for the profit they make thereof, ſeruing for the dying of Silke eſpecially, and tranſporting great quantities to diuers Countries. It hath large broad leaues, without any prickes at all vpon them in our Country, growing vpon the ſtalke, which is ſtrong, hard, and round, with ſhorter leaues thereon vp to the toppe, where they are a little ſharpe pointed, and prickly about the edges ſometimes, which ſtalke riſeth three or foure foote high, and brancheth it ſelfe toward the toppe, bearing at the end of euery branch one great open ſcaly head, out of which thruſteth out many gold yellow threads, of a moſt orient ſhining colour, which being gathered in a dry time, and kept dry, will abide in the ſame delicate colour that it bare when it was freſh, for a very long time after : when the flowers are paſt, the ſeede when it is come to maturity, which is very ſeldome with vs, is white and hard, ſomewhat long, round, and a little cornered : the roote is long, great, and wooddy, and periſheth quickly with the firſt froſts.

The Place.

It groweth in Spaine, and other hot Countries, but not wilde, for that it is accounted of the old Writers, Theophraſtus and Dioſcorides, to be a manured plant.

The Time.

It flowreth with vs not vntill Auguſt, or September ſometimes, ſo that it hardly giueth ripe ſeede (as I ſaid) neither is it of that force to purge, which groweth in theſe colder Countries, as that which commeth from Spaine, and other places.

The Names.

The name *Cnicus* is deriued from the Greekes, and *Carthamus* from the Arabians, yet ſtill *ſatiuus* is added vnto it, to ſhew it is no wilde, but a manured plant, and ſowne euery where that wee know. Of ſome it is called *Crocus hortenſis,* and *Saraſenicus,* from the Italians which ſo call it. We call it in Engliſh Baſtard Saffron, Spaniſh Saffron, and Catalonia Saffron.

The Vertues.

The flowers are vſed in colouring meates, where it groweth beyond Sea, and alſo for the dying of Silkes : the kernels of the ſeede are onely vſed in Phyſicke with vs, and ſerueth well to purge melancholicke humours.

Chap. LXXVII.

Carduus. Thiſtles.

YOu may ſomewhat maruaile, to ſee mee curious to plant Thiſtles in my Garden, when as you might well ſay, they are rather plagues then pleaſures, and more trouble to weede them out, then to cheriſh them vp, if I made therein no diſtinction or choiſe; but when you haue viewed them well which I bring in, I will

Ee 3   then

then abide your cenfure, if they be not worthy of fome place, although it be but a corner of the Garden, where fomething muft needes be to fill vp roome. Some of them are fmooth, and without prickes at all, fome at the heads onely, and fome all ouer; but yet not without fome efpeciall note or marke worthy of refpect: Out of this difcourfe I leaue the Artichoke, with all his kindes, and referue them for our Kitchin Garden, becaufe (as all know) they are for the pleafure of the tafte, and not of the fmell or fight.

1. *Acanthus fativus.* Garden Beares breech.

The leaues of this kinde of fmooth thiftle (as it is accounted) are almoft as large as the leaues of the Artichoke, but not fo fharp pointed, very deeply cut in and gafhed on both edges, of a fad green & fhining colour on the vpperfide, and of a yellowifh green vnderneath, with a great thicke rib in the middle, which fpread themfelues about the root, taking vp a great deale of ground. After this plant hath ftood long in one place, and well defended from the iniury of the cold, it fendeth forth from among the leaues one or more great and ftrong ftalkes, three or foure foote high, without any branch at all, bearing from the middle to the top many flowers one aboue another, fpike-fafhion round about the ftalke, with fmaller but not diuided greene leaues at euery flower, which is white, and fafhioned fomewhat like vnto a gaping mouth; after which come broad, flat, thicke, round, brownifh yellow feede (as I haue well obferued by them haue beene fent me out of Spaine, and which haue fprung vp, and doe grow with me; for in our Countrey I could neuer obferue any feede to haue growne ripe) the rootes are compofed of many great and thicke long ftrings, which fpread farre in and vnder the ground, fomewhat darkifh on the outfide, and whitifh within, full of a clammy moifture (whereby it fheweth to haue much life) and doe endure our Winters, if they be not too much expofed to the fharpe violence thereof, which then it will not endure, as I haue often found by experience.

2. *Acanthus filueftris.* Wilde or prickly Beares breech.

This prickly Thiftle hath diuers long greenifh leaues lying on the ground, much narrower then the former, but cut in on both fides, thicke fet with many white prickes and thornes on the edges: the ftalke rifeth not vp fo high, bearing diuers fuch like thornie leaues on them, with fuch a like head of flowers on it as the former hath: but the feede hereof (as it hath come to vs from Italy and other places, for I neuer faw it beare feed here in this Country) is blacke and round, of the bigneffe of a fmall peafe: the roote abideth reafonable well, if it be defended fomewhat from the extremity of our Winters, or elfe it will perifh.

3. *Eringium Pannonicum fiue Montanum.* Hungary Sea Holly.

The lower leaues of this Thiftle that lye on the ground, are fomewhat large, round, and broad, hard in handling, and a little fnipt about the edges, euery one ftanding vpon a long foote-ftalke: but thofe that growe vpon the ftalke, which is ftiffe, two or three foote high, haue no foote-ftalke, but encompaffe it, two being fet at euery ioynt, the toppe whereof is diuided into diuers branches, bearing fmall round rough heads, with fmaller and more prickly leaues vnder them, and more cut in on the fides then thofe belowe: out of thefe heads rife many blew flowers, the foote-ftalkes of the flowers, together with the toppes of the branches, are likewife blew and tranfparent, or fhining.

Flore albo. We haue another of this kinde, the whole toppes of the ftalkes, with the heads and branches, are more white then blew: the feede contained in thefe heads are white, flat, and as it were chaffie: the roote is great and whitifh, fpreading farre into many branches, and fomewhat fweete in tafte, like the ordinary Sea Holly rootes.

4. *Carduus mollis.* The gentle Thiftle.

The leaues of this foft and gentle Thiftle that are next vnto the ground, are greene

1 *Acanthus sativus*, Garden Beares breech. 2 *Acanthus siluestris*, Wilde Beares breech. 3 *Eringium Pannonicum* Mountaine Sea
Holly. 4 *Carlina humilis*, The lowe Carline Thistle. 5 *Carduus sphærocephalus maior*. The greater Globe-Thistle 6 *Carduus sphærocephalus minor*. The lesser Globe-Thistle. 7 *Carduus Eriocephalus*. The Friers crowne. 8 *Fraxinella*, Bastard Dittanie.

on the vpperside, and hoary vnderneath, broad at the bottome, somewhat long pointed, and vneuenly notched about the edges, with some soft hairie prickles, not hurting
the handler, euery one standing vpon a short foote-stalke; those that growe about the
middle stalke are like the former, but smaller and narrower, and those next the toppe
smallest, where it diuideth it selfe into small branches, bearing long and scaly heads,
out of which breake many reddish purple threads: the seede is whitish and hard, almost as great as the seede of the greater Centory: the roote is blackish, spreading vnder the ground, with many small fibres fastened vnto it, and abideth a great while.

5. *Carlina humilis.* The lowe Carline Thistle.

This lowe Thistle hath many iagged leaues, of a whitish greene colour, armed with
small sharp white prickles round about the edges, lying round about the root vpon the
ground, in the middle whereof riseth vp a large head, without any stalke vnder it, compassed about with many small and long prickly leaues, from among which the flower
sheweth it selfe, composed of many thin, long, whitish, hard shining leaues, standing
about the middle, which is flat and yellow, made of many thrums or threads like small
flowers, wherein lye small long seede, of a whitish or siluer colour: the roote is somewhat aromaticall, blackish on the outside, small and long, growing downewards into
the ground. There is another of this kinde that beareth a higher stalke, and a redder
flower, but there is a manifest difference betweene them.

6. *Carduus Sphærocephalus siue Globosus maior.* The greater Globe Thistle.

The greatest of these beautifull Thistles, hath at the first many large and long leaues
lying on the ground, very much cut in and diuided in many places, euen to the middle
ribbe, set with small sharpe (but not very strong) thornes or prickles at euery corner of
the edges, greene on the vpperside, and whitish vnderneath: from the middle of these
leaues riseth vp a round stiffe stalke, three foote and a halfe high, or more, set without
order with such like leaues, bearing at the toppe of euery branch a round hard great
head, consisting of a number of sharpe bearded huskes, compact or set close together,
of a blewish greene colour, out of euery one of which huskes start small whitish blew
flowers, with white threads in the middle of them, and rising aboue them, so that the
heads when they are in full flower, make a fine shew, much delighting the spectators:
after the flowers are past, the seede encreaseth in euery one, or the most part of the
bearded huskes, which doe still hold their round forme, vntill that being ripe it openeth it selfe, and the huskes easily fall away one from another, containing within
them a long whitish kernell: the roote is great and long, blackish on the outside, and
dyeth euery yeare when it hath borne seede.

7. *Carduus Globosus minor.* The lesser Globe Thistle.

The lesser kinde hath long narrow leaues, whiter then the former, but cut in and
gashed on the edges very much with some small prickes on them; the stalke is not
halfe so long, nor the heads halfe so great, but as round, and with as blew flowers as the
greater: this seldome giueth ripe seede, but recompenseth that fault, in that the roote
perisheth not as the former, but abideth many yeares.

8. *Carduus Eriocephalus siue Tomentosus.* The Friers Crowne.

This woolly Thistle hath many large and long leaues lying on the ground, cut in on
both sides into many diuisions, which are likewise somewhat vnequally cut in or diuided againe, hauing sharpe white prickles at euery corner of the diuisions, of a dead
or sad greene colour on the vpperside, and somewhat woolly withall, and grayish vnderneath: the stalke is strong and tall, foure or fiue foote high at the least, branching
out into diuers parts, euery where beset with such like leaues as growe below; at the
toppe of euery branch there breaketh out a great whitish round prickly head, flattish at
the toppe, so thicke set with wooll, that the prickles seeme but small spots or haires,
and

and doth so well resemble the bald crowne of a Frier, not onely before it be in flower, but especially after it hath done flowring, that thereupon it deseruedly receiued the name of the Friers Crowne Thistle : out of these heads riseth forth a purple thrumme, such as is to be seene in many other wilde Thistles, which when they are ripe, are full of a flockie or woolly substance, which breake at the toppe shedding it, and the seede which is blackish, flat, and smooth : the roote is great and thicke, enduring for some yeares, yet sometimes perishing, if it be too much exposed to the violence of the frosts in Winter.

The Place.

The first groweth naturally in Spaine, Italy, and France, and in many other hot Countries, and growe onely in Gardens in these colder climates, and there cherished for the beautifull aspect both of the greene plants, and of the stalkes when they are in flower. The Carline Thistle is found both in Germany and Italy in many places, and as it is reported, in some places of the West parts in England. The others are found some in France, some in Hungary, and on the Alpes, and the last in Spaine.

The Time.

They doe all flower in the Summer moneths, some a little earlier or later then others.

The Names.

The first is called *Acanthus satiuus* (because the other that is prickly, is called *siluestris* or *spinosus*) and *Branca vrsina* ; In English, Branck vrsine, and Beares breech. The third is called *Eryngium montanum*, *Alpinum*, and *Pannonicum latifolium* : In English, Mountaine or Hungary Sea Holly. The fourth is called *Carduus mollis*, The gentle Thistle, because it hath no harmfull prickles, although it seeme at the first shew to be a Thistle. The fifth is called of diuers *Chamæleo albus*, and *Carlina*, as if they were both but one plant ; but Fabius Columna hath in my iudgement very learnedly descided that controuersie, making *Carlina* to be *Ixine* of Theophrastus, and *Chamæleo* another differing Thistle, which Gaza translateth *Vernilago*. We call it in English, The Carline Thistle. The other haue their names in their titles, as much as is conuenient for this discourse.

The Vertues.

The first hath alwaies been vsed Physically, as a mollifying herbe among others of the like slimie matter in Glisters, to open the body ; yet Lobel seemeth to make no difference in the vse of them both (that is, the prickly as well as the smooth.) The Carline Thistle is thought to bee good against poysons and infection. The rest are not vsed by any that I know.

CHAP. LXXVIII.

Fraxinella. Bastard Dittany.

Hauing finished those pleasing Thistles, I come to other plants of more gentle handling, and first bring to your consideration this bastard Dittany, whereof there are found out two especiall kindes, the one with a reddish, the other with a whitish flower, and each of these hath his diuersity, as shall be presently declared.

1. *Fraxinella flore rubente.* Bastard Dittany with a reddish flower.

This goodly plant riseth vp with diuers round, hard, brownish stalkes, neare two
foote

foote high, the lower parts whereof are furnished with many winged leaues, somewhat like vnto Liquerice, or a small young Ashe tree, consisting of seuen, nine, or eleuen leaues set together, which are somewhat large and long, hard and rough in handling, of a darkish greene colour, and of an vnpleasant strong resinous sent: the vpper parts of the stalkes are furnished with many flowers, growing spike fashion, at certaine distances one aboue another, consisting of fiue long leaues a peece, whereof foure that stand on the two sides, are somewhat bending vpwards, and the fift hanging downe, but turning vp the end of the leafe a little againe, of a faint or pale red colour, striped through euery leafe with a deeper red colour, and hauing in the middle a tassell of fiue or six long purplish threds, that bowe downe with the lower leafe, and turne vp also the ends againe, with a little freese or thrume at the ends of euery one: after the flowers are past, arise hard, stiffe, rough, clammy huskes, horned or pointed at the end, foure or fiue standing together, somewhat like the seede vessels of the Wolfes-banes, or Colombines, but greater, thicker and harder, wherein is contained round shining blacke seede, greater then any Colombine seede by much, and smaller then Peony seede: the roote is white, large, and spreading many wayes vnder ground, if it stand long: the whole plant, as well roots as leaues and flowers, are of a strong sent, not so pleasing for the smell, as the flowers are beautifull to the sight.

2. *Fraxinella flore rubro.* Bastard Dittaine with a red flower.

This differeth not from the former eyther in roote, leafe or flower for the forme, but that the stalkes and leaues are of a darker greene colour, and that the flowers are of a deeper red colour, (and growing in a little longer spike) wherein the difference chiefly consisteth, which is sufficient to distinguish them.

3. *Fraxinella flore albo.* Bastard Dittanie with a white flower.

The white flowred *Fraxinella* hath his leaues and stalkes of a fresher greene colour then any of the former; and the flowers are of a pure white colour, in forme differing nothing at all from the other.

4. *Fraxinella flore albo cæruleo.* Bastard Dittanie with an ash coloured flower.

The colour of the flower of this *Fraxinella* onely putteth the difference betweene this, and the last recited with a white flower: for this beareth a very pale, or whitish blew flower, tending to an ash colour.

The Place.

All these kindes are found growing naturally, in many places both of Germany, and Italie: and that with the white flower, about Franckford, which being sent me, perished by the way by long and euill carriage.

The Time.

They flower in Iune and Iuly, and the seede is ripe in August.

The Names.

The name *Fraxinella* is most generally imposed on those plants, because of the resemblance of them vnto young Ashes, in their winged leaues. Yet some doe call them *Dictamus albus*, or *Dictamnus albus*, and *Diptamus albus*, as a difference from the *Dictamnus Creticus*, which is a farre differing plant. Some would haue it to be *Tragium* of Dioscorides, but beside other things wherein this differeth from *Tragium*, this yeeldeth no milkie iuice, as Dioscorides saith *Tragium* doth: We in English doe eyther call it *Fraxinella*, or after the other corrupted name of *Dictamus*, Bastard Dittanie.

The

The Vertues.

It is held to be profitable against the stingings of Serpents, against contagious and pestilent diseases, to bring downe the feminine couries, for the paines of the belly and the stone, and in Epilepticall diseases, and other cold paines of the braines : the roote is the most effectuall for all these, yet the seede is sometimes vsed.

Chap. LXXIX.

Legumina. Pulse.

IF I should describe vnto you all the kindes of Pulse, I should vnfold a little world of varieties therein, more knowne and found out in these dayes, then at any time before, but that must bee a part of a greater worke, which will abide a longer time before it see the light. I shall only select those that are fit for this garden, and set them downe for your confideration. All sorts of Pulse may be reduced vnder two generall heads, that is, of Beanes and Peafe, of each whereof there is both tame and wilde : Of Beanes, befides the tame or vsuall garden Beane, and the French or Kidney Beane, (whereof I meane to entreate in my Kitchen garden, as pertinent thereto) there is the Lupine or flat Beane, whereof I meane to entreate here, and the blacke Beane and others which must bee reserued for the Physicke Garden. And of the kindes of Peafe some are fit for this Garden, (whereunto I will adioyne two or three other plants as neereft of affinitie, the flowers of some, and the fruit of others being delightfull to many, and therefore fit for this garden) some for the Kitchen, the rest for the Physicke garden. And first of Lupines or flat Beanes, accepted as delightfull to many, and therefore fit for this garden.

1. *Lupinus satiuus albus.* The white garden Lupine.

The garden Lupine riseth vp with a great round stalke, hollow and somewhat woolly, with diuers branches, whereon grow vpon long footestalkes many broade leaues, diuided into feuen or nine parts, or smaller leaues, equally standing round about, as it were in a circle, of a whitish greene colour on the vpperside, and more woolly vnderneath : the flowers stand many together at feuerall ioynts, both of the greater stalke, and the branches, like vnto beanes, and of a white colour in some places, and in others of a very bleake blew tending to white : after the flowers are past, there come in their places, long, broade, and flat rough cods, wherein are contained round and flat feede, yellowish on the inside, and couered with a tough white skin, and very bitter in taste : the rootes are not very great, but full of small fibres, whereby it fasteneth it selfe strongly in the ground, yet perisheth euery yeare, as all the rest of these kindes doe.

2. *Lupinus cæruleus maximus.* The greater blew Lupine.

The Stemme or stalke of this Lupine is greater then the last before recited, as also the leaues more soft and woolly, and the flowers are of a most perfect blew colour, with some white spots in the middle : the long rough greenish cods are very great and large, wherein are contained hard, flat and round feede, not so white on the outside as the former, but somewhat yellower, greater also, and more rough or hard in handling.

3. *Lupinus cæruleus minor.* The leſſer blew Lupine.

This kinde of wilde Lupine differeth not in the forme of leafe or flower from the former, but only that it is much smaller, the leaues are greener, and haue fewer diuisions in them : the flower is of as deepe a blew colour as the last ; the cods likewise are small and long, containing small round feede, not so flat as the former, but more

discoloured

Minimus. difcoloured or fpotted on the outfide, then the greater kinde is. There is a leffer kind then this, not differing in any thing from this, but that it is leffer.

4. *Lupinus flore luteo.* The yellow Lupine.

The yellow Lupine groweth not vfually fo high, but with larger leaues then the fmall blew Lupine ; the flowers grow in two or three rundles or tufts, round about the ftalke and the branches at the ioynts, of a delicate fine yellow colour, like in fafhion vnto the other kindes, being larger then the laft, but nothing fo large as the greater kindes, and of a fine fmall fent : the feede is round, and not flat, but much about the forme and bigneffe of the fmall blew, or fomewhat bigger, of a whitifh colour on the outfide, fpotted with many fpots.

The Place.

The firft groweth in many places of Greece, and the Eafterne Countries beyond it, where it hath beene anciently cherifhed for their foode, being often watered to take away the bitterneffe. It groweth alfo in thefe Weftern parts, but ftill where it is planted. The great blew Lupine is thought to come from beyond the parts of Perfia, in Caramania. The leffer blew is found very plentifully wilde, in many places both of Spaine and Italy. The laft hath beene brought vs likewife out of Spaine, whereas it is thought it groweth naturally. They all grow now in the gardens of thofe, that are curious louers of thefe delights.

The Time.

They flower in Summer, and their feede is ripe quickly after.

The Names.

They are generally called *Lupini.* Plautus in his time faith, they were vfed in Comedies in ftead of money, when in any Scene thereof there was any fhew of payment, and therefore he calleth them *Aurum Comicum.* And Horace hath this Verfe,

Nec tamen ignorant, quid diftent æra Lupinis,

to fhew that counterfeit money (fuch as counters are with vs, or as thefe Lupines were vfed in thofe times) was eafily knowne from true and currant coine. In Englifh wee vfually call them after the Latine name, Lupines ; and fome after the Dutch name, Figge-beanes, becaufe they are flat and round as a Figge that is preffed ; and fome Flat-beanes for the fame reafon. Some haue called the yellow Lupine, Spanifh Violets : but other foolifh names haue beene giuen it, as Virginia Rofes, and the like, by knauifh Gardiners and others, to deceiue men, and make them beleeue they were the finders out, or great preferuers of rarities, of no other purpofe, but to cheate men of their money : as you would therefore auoyde knaues and deceiuers, beware of thefe manner of people, whereof the skirts of our towne are too pitifully peftered.

The Vertues.

The firft or ordinary Lupine doth fcoure and cleanfe the skin from fpots, morphew, blew markes, and other difcolourings thereof, beeing vfed eyther in a decoction or ponther. Wee feldome vfe it in inward medicines, not that it is dangerous, but of neglect, for formerly it hath beene much vfed for the wormes, &c.

5. *Lathyrus*

1 *Lupinus maior*. The great Lupine. 2 *Lupinus luteus*. The yellow Lupine. 3 *Lathyrus latifolius seu Pisum perenne*. Pease euerlasting. 4 *Pisum quadratum*. The crimson blossomd or square Pease. 5 *Medica cochleata vulgaris* Snailes or Barbary buttons. 6 *Medica spinosa*. Prickly Snailes. 7 *Medica spinosa altera*. Another sort of prickly Snailes. 8 *Medica folliculo lato*. Broad buttons or Snailes. 9 *Medica Lunata*. Halfe Moons. 10 *Hedysarum clypeatum*. The red Sattin flower, or French Honysuckle. 11 *Scorpioides minus*. The lesser Caterpiller. 12 *Scorpioides maius*. The greater Caterpiller. 13 *Orobus Venetus*. Blew vpright Pease euerlasting.

1. *Lathyrus latifolius, siue Pisum perenne.* Pease euerlasting.

This kinde of wilde Pease that abideth long, and groweth euery yeare greater then other, springeth vp with many broade trayling branches, winged as it were on both the sides, diuersly diuided into other smaller branches, at the seuerall ioynts whereof stand two hard, not broad, but somewhat long greene leaues, and diuers twining claspers, in sundry places with the leaues, from betweene the branches and the leaues, at the ioynts towards the toppes, come forth diuers purplish pease like blossomes, standing on a long stemme or stalke, very beautifull to behold, and of a pretty sent or smell : after which come small, long, thin, flat, hard skind cods, containing small round blackish seede : the roote is great and thicke, growing downe deepe into the ground, of the thicknesse sometimes of a mans arme, blackish on the outside, and whitish within, with some branches and a few fibres annexed thereunto.

2. *Orobus Venetus.* Blew vpright euerlasting Pease.

This pretty kinde of Pease blossome beareth diuers slender, but vpright greene branches somewhat cornered, two foote high or thereabouts, hauing at seuerall distances on both sides of them certaine winged leaues, set together vpon long footestalkes one against another, consisting of fix or eight leaues, somewhat broade and pointed, and without any odde one at the end : at the ioynts toward the toppes, betweene the leaues and the stalkes, come forth many flowers set together at the end of a pretty long footestalke, of the fashion of the former Pease blossome, but somewhat smaller, and of a purplish violet colour : after which come slender and long pointed pods rounder then they, wherein is contained small round grayish pease : the roote is blacke, hard or woody, abiding after seede bearing as the former doth, and shooting afresh euery yeare.

3. *Lathyrus annuus siliquis orobi.* Partie coloured Cichelings.

This small Pulse or wild Pease, hath two or three long slender winged branches, with smaller leaues theron then the former, and without any claspers at all on them : the flowers stand single, euery one by it selfe, or two at the most together, the middle leaues whereof that close together are white, and the vpper leaues of a reddish purple colour : after which come long round flattish cods, bunched out in the seuerall places where the seedes lye, like vnto the cods of *Orobus* or the bitter Vetch, but greater: the roote is small and dyeth euery yeare.

4. *Pisum quadratum.* The crimson blossomd or square codded Pease.

This pretty kinde of Pulse might very well for the forme of the leaues, be referred to the kindes of *Lotus* or Trefoiles : but because I haue none of that kindred to entreate of in this Worke, I haue thought fittest to place it here before the Medica's, because both pods and seedes are like also. It hath three or foure small weake stalkes, diuided into many branches, hauing two stalkes of leaues at euery ioynt, and three small soft leaues standing on a very small stalke, comming from the ioynts : the flowers stand for the most part two together, of a perfect red or crimson colour, like in forme almost vnto a Pease blossome; after which come long thicke and round cods, with two skinnes or filmes, running all along the cod at the backe or vpperside, and two other such like filmes, all along the belly or vnder side, which make it seeme foure square, wherein there lye round discoloured Pease, somewhat smaller and harder then ordinary Pease : the roote is small and perisheth euery yeare.

5. *Medica Cochleata vulgaris.* Snailes or Barbary buttons.

The plant that beareth these pretty toyes for Gentlewomen, is somewhat like vnto a Threeleafed grasse or Trefoile, hauing many long trayling branches lying vpon the ground, whereon at diuers places are three small greene leaues, set together at the end of a little footestalke, each of them a little snipt about the edges : at seuerall distances,

from

from the middle of these branches to the ends of them, come forth the flowers, two for the most part standing together vpon a little footstalke, which are of a pale yellow colour, very small, and of the forme of a Peafe bloſſome : after which come ſmooth heads, which are turned or writhen round, almoſt like a Snaile, hard and greene at the firſt, ſomewhat like a greene button (from the formes of both which came their names) but afterwards growing whiter, more ſoft and open, wherein lyeth yellowiſh round and flat ſeede, ſomewhat like vnto the Kidney beane : the roote is ſmall and ſtringie, dying downe euery yeare, and muſt be new ſowne in the ſpring, if you deſire to haue it.

6. *Medica ſpinoſa maior.* Prickly or thorny Snailes, or Buttons.

This kinde of *Medica* is in all things very like vnto the former, both in the long trayling branches, & three leaues alwaies growing together, but a little greater pale yellow flowers, and crooked or winding heads : but herein chiefly conſiſteth the difference, that this kinde hath his heads or buttons harder, a little greater, more cloſed together, and ſet with ſhort and ſomewhat hard prickles, all the head ouer, which being pulled open, haue thoſe prickles ſtanding on each ſide of the filme or ſkinne, whereof the head conſiſteth, ſomewhat like vnto a fiſh bone, and in this kinde goeth all one way ; in which are contained ſuch like ſeedes for the forme, as are in the former, but great and blacke, and ſhining withall.

7. *Medica ſpinoſa altera.* Small thorney Buttons, or Snailes.

This other kinde is alſo like vnto the laſt deſcribed in all other things, except in the heads or buttons, which are a little ſmaller, but ſet with longer and ſofter prickes vpon the filmes, and may eaſily bee diſcerned to goe both forwards and backewards, one enterlacing within another, wherein are contained ſuch like flat and blacke ſhining ſeede, made after the faſhion of a kidney, as are in the former, but ſomewhat ſmaller : the roote periſheth in like manner euery yeare.

8. *Medica lata.* Broade Buttons.

This kinde differeth not from the firſt in leafe or flower, the fruite onely hereof is broade and flat, and not ſo much twined as it.

9. *Medica Lunata.* Halfe Moones.

This is alſo a kinde of theſe Medicke fodders, hauing a trefoyle leafe and yellow flowers like the former ſorts, but both ſomewhat larger, the chiefeſt difference conſiſteth in the head or fruite, which is broade and flat, and not twined like the reſt, but abideth halfe cloſed, reſembling a halfe Moone (and thereupon hath aſſumed both the Latine and Engliſh name) wherein is contained flat ſeede, kidney faſhion like the former.

10. *Hedyſarum clypeatum.* The red Sattin flower.

This red flowred Fitchling, hath many ſtalkes of winged faire greene leaues, that is, of many ſet on both ſides a middle ribbe, whereof that at the end is the greateſt of the reſt : from the ioynts where the leaues ſtand, come forth pretty long ſmall ſtalkes, bearing on them very many flowers, vp to the toppe one aboue another, of an excellent ſhining red or crimſon colour, very like vnto Sattin of that colour, and ſometimes of a white colour, (as Maſter William Coys, a Gentleman of good reſpect in Eſſex, a great and ancient louer and cheriſher of theſe delights, and of all other rare plants, in his life time aſſured me, he had growing in his garden at Stubbers by North Okenden) which are ſomewhat large, and more cloſed together, almoſt flat and not open, as in moſt of the other ſorts : after the flowers are paſt, there come rough, flat, round huskes, ſomewhat like vnto the old faſhioned round bucklers without pikes, three or foure ſtanding one vpon or aboue another, wherein are contained small brownish seede : the roote perisheth the same yeare it beareth seede, for oftentimes it floweth not the first yeare it is sowne.

11. *Scorpioides maius & minus.*
Great and small Caterpillers.

Vnder one description I comprehend both these sorts of Scorpions grasse, or Caterpillers, or Wormes, as they are called by many, whereof the greater hath been known but of late yeares; and ioyne them to these pulses, not hauing a fitter place where to insert them. It is but a small low plant, with branches lying vpon the ground, and somewhat long, broad, and hard leaues theron, among which come forth small stalkes, bearing at the end for the most part, two small pale yellowish flowers, like vnto Tares or Vetches, but smaller, which turne into writhed or crooked tough cods; in the greater sort they are much thicker, rounder and whiter, and lesser wound or turned together then in the smaller, which are slenderer, more winding, yet not closing like vnto the Snailes, and blacker more like vnto a Caterpiller then the other, wherein are contained brownish yellow seede, much like vnto a *Medica :* the rootes of both are small and fibrous, perishing euery yeare.

The Place.

These are found seuerally in diuers and seuerall places, but wee sow and plant them vsually to furnish our gardens.

The Time.

They doe all flower about the moneths of Iune and Iuly, and their seede is ripe soone after : but the second is earlier then the rest.

The Names.

The first is called *Clymenum* of Matthiolus, and *Lathyris* of Lobel and others : but *Lathyris* in Greeke is *Cataputia* in Latine, which is our Spurge, farre differing from this Pulse; and therefore *Lathyrus* is more proper to distinguish them asunder, that two plants so farre vnlike should not bee called by one name : this is also called *Lathyrus latifolius,* becausethere is another called *augustifolius,* that differeth from it also : It is most vsually called with vs, *Pisum perenne,* and in English Pease blossome, or Pease euerlasting. The second is called by Clusius, *Orobus venetus,* becausethe it was sent him from Venice, with another of the same kinde that bore white flowers; yet differeth but little or nothing from that kinde he found in Hungary, that I thinke the seuerall places of their growing only cause them to beare seuerall names, and to be the same in deede. Although I yeeld vnto Clusius the Latine name which doth not sufficiently content mee ; yet I haue thought good to giue it a differing English name, according as it is in the title. The third, because I first receiued it among other seeds from Spaine, I haue giuen it the name, as it is entituled. The fourth is called of some *Sandalida Cretica, & Lotus siliquosus flore rubello, Lotus tetragonolobus, Pisum rubrum, & Pisum quadratum :* We vsually call it in English, Crimson Pease, or square Pease. The *Medica Cochleata* is called of Dodonæus *Trifolium Cochleatum,* but not iudged to be the true *Medica.* Wee call it in English, Medick fodder, Snailes Clauer, or as it is in the title, and so the rest of the Medica's accordingly. The *Hedysarum clypeatum* or *Securidaca* is called of Dodonæus *Onobrichis altera,* and we in English for the likenesse, The red Sattin flower, although some foolishly call it, the red or French Honysuckle. The last is called by Lobel, *Scorpioides bupleurifolio,* I haue called it *minus,* because the greatest sort which came to me out of Spaine was not knowne vnto him : in English they are generally called Caterpillers.

The

The Vertues.

The Medica's are generally thought to feede cattell fat much more then the Medow Trefoile, or Clauergraffe, and therefore I haue known diuers Gentlemen that haue plowed vp fome of their pafture grounds, and fowen them with the feedes of fome Medica's to make the experience. All the other forts are pleafures to delight the curious, and not any way profitable in Phyficke that I know.

CHAP. LXXX.

Pæonia. Peonie.

THere are two principall kindes of Peonie, that is to fay, the Male and the Female. Of the male kinde, I haue onely known one fort, but of the Female a great many; which are thus to be diftinguifhed. The Male his leafe is whole, without any particular diuifion, notch or dent on the edge, & his rootes long & round, diuided into many branches, fomewhat like to the rootes of Gentian or Elecampane, and not tuberous at all. The Female of all forts hath the leaues diuided or cut in on the edges, more or leffe, and hath alwaies tuberous rootes, that is, like clogs or Afphodill rootes, with many great thick round peeces hanging, or growing at the end of fmaller ftrings, and all ioyned to the toppe of the maine roote.

1. *Pæonia mas.* The Male Peonie.

The Male Peonie rifeth vp with many brownifh ftalkes, whereon doe grow winged leaues, that is, many faire greene, and fometimes reddifh leaues, one fet againft another vpon a ftalke, without any particular diuifion in the leafe at all : the flowers ftand at the toppes of the ftalkes, confifting of fiue or fix broade leaues, of a faire purplifh red colour, with many yellow threds in the middle, ftanding about the head, which after rifeth to be the feede veffels, diuided into two, three or foure rough crooked pods like hornes, which when they are ful ripe, open and turn themfelues down one edge to another backeward, fhewing within them diuers round black fhining feede, which are the true feede, being full and good, and hauing alfo many red or crimfon graines, which are lancke and idle, intermixed among the blacke, as if they were good feede, whereby it maketh a very pretty fhew: the roots are great, thick and long, fpreading in the ground, and running downe reafonable deepe.

2. *Pæonia fæmina vulgaris flore fimplici.*
The ordinary fingle Female Peonie.

This ordinary Female Peonie hath many ftalkes, with more ftore of leaues on them then the Male kinde hath, the leaues alfo are not fo large, but diuided or nicked diuerfly on the edges, fome with great and deepe, and others with fmaller cuts or diuifions, and of a darke or dead greene colour : the flowers are of a ftrong heady fent, moft vfually fmaller then the male, and of a more purple tending to a murrey colour, with yellow thrumes about the head in the middle, as the male kinde hath : the heads or hornes with feed are like alfo but fmaller, the feede alfo is blacke, but leffe fhining: the rootes confift, as I faid, of many thicke and fhort tuberous clogs, faftened at the ends of long ftrings, and all from the head of the roote, which is thicke and fhort, and tuberous alfo, of the fame or the like fent with the male.

3. *Pæonia fæmina vulgaris flore pleno rubro.*
The double red Peonie.

This double Peonie as well as the former fingle, is fo frequent in euerie Garden of note, through euery Countrey, that it is almoft labour in vaine to deſcribe it : but yet becauſe I vſe not to paſſe ouer any plant ſo ſlightly, I will ſet down the deſcription briefly, in regard it is ſo common. It is very like vnto the former ſingle female Peony, both in ſtalkes and leaues, but that it groweth ſomewhat higher, and the leaues are of a freſher greene colour : the flowers at the tops of the ſtalkes are very large, thicke, and double (no flower that I know ſo faire, great, and double ; but not abiding blowne aboue eight or ten daies) of a more reddiſh purple colour then the former female kinde, and of a ſweeter ſent : after theſe flowers are paſt, ſometimes come good ſeed, which being ſowne, bring forth ſome ſingle flowers, and ſome double : the rootes are tuberous, like vnto the former female.

4. *Paonia fæmina flore carneo ſimplici.* The ſingle bluſh Peony.

The ſingle bluſh Peony hath his ſtalkes higher, and his leaues of a paler or whiter greene colour then the double bluſh, and more white vnderneath (ſo that it is very probable it is of another kinde, and not riſen from the ſeede of the double bluſh, as ſome might thinke) with many veines, that are ſomewhat diſcoloured from the colour of the leafe running through them : the flowers are very large and ſingle, conſiſting of fiue leaues for the moſt part, of a pale fleſh or bluſh colour, with an eye of yellow diſperſed or mixed therewith, hauing many whitiſh threads, tipt with yellow pendents ſtanding about the middle head : the rootes are like the other female Peonies.

5. *Paonia fæmina flore pleno albicante.* The double bluſh Peony.

The double bluſh Peony hath not his ſtalkes ſo high as the double red, but ſomewhat lower and ſtiffer, bearing ſuch like winged leaues, cut in or diuided here and there in the edges, as all theſe female kindes are, but not ſo large as the laſt : the flowers are ſmaller, and leſſe double by a good deale then the former double red, of a faint ſhining crimſon colour at the firſt opening, but decaying or waxing paler euery day : ſo that after it hath ſtood long (for this flower ſheddeth not his leaues in a great while) it will change ſomewhat whitiſh ; and therefore diuers haue ignorantly called it, the double white Peony : the ſeedes, which ſometimes it beareth, and rootes, are like vnto the former female kindes, but ſomewhat longer, and of a brighter colour on the outſide.

6. *Paonia fæmina Byzantina.* The ſingle red Peony of Conſtantinople.

This red Peony of Conſtantinople is very like in all things vnto the double red Peonie, but that the flowers hereof are ſingle, and as large as the laſt, and that is larger then either the ſingle female, or the male kinde, conſiſting of eight leaues, of a deeper red colour then either the ſingle or double Peonies, and not purpliſh at all, but rather of the colour of an ordinary red Tulipa, ſtanding cloſe and round together : the roots of this kinde haue longer clogs, and not ſo ſhort as of the ordinary female kinde, and of a paler colour on the outſide.

The Place.

All theſe Peonies haue beene ſent or brought from diuers parts beyond the Seas ; they are endenized in our Gardens, where wee cheriſh them for the beauty and delight of their goodly flowers, as well as for their Phyſicall vertues.

The Time.

They all flower in May, but ſome (as I ſaid) abide a ſmall time, and others many weekes.

The Names.

The name *Paonia* is of all the later Writers generally giuen to theſe plants, although they haue had diuers other names giuen by the elder Writers, as *Roſa fatuina, Idæus dactylus, Aglaophotis,* and others, whereof to ſet

downe

1 *Pæonia mas cum ſemine.* The male Peony & the ſeed. 2 *Pæonia fæmina Byzantina.* The female red Peony of Conſtantinople. 3 *Pæonia fæmina flore pleno vulgaris.* The ordinary double Peony. 4 *Pæonia flore pleno albicante.* The double white Peony. 5 *Helleborus vernus atrorubente flore.* The early white Ellebor with a darke red flower. 6 *Helleborus niger verus.* The Chriſtmas flower. 7 *Calceolus Mariæ.* Our Ladies Slipper.

downe the caules, realons, and errours, were to lpend more time then I intend for this worke. Wee call them in Englilh, Peonie, and diltinguilh
them according to their titles.

The Vertues.

The male Peony roote is farre aboue all the relt a molt lingular approued remedy for all Epilepticall dileales, in Englilh, The falling licknelle
(and more elpecially the greene roote then the dry) if the dileale be not too
inueterate, to be boyled and drunke, as allo to hang about the neckes of the
younger lort that are troubled herewith, as I haue found it lufficiently expe-
rimented on many by diuers. The leede likewile is of elpeciall vle for women, for the riling of the mother. The leede of the female kinde, as well
as the rootes, are molt vlually lold, and may in want of the other be (and
lo are generally) vled.

CHAP. LXXXI.

Helleborus niger. Beares foote.

THere are three lorts of blacke Hellebor or Beares foote, one that is the true and
right kinde, whole flowers haue the molt beautifull alpect, and the time of his
flowring molt rare, that is, in the deepe of Winter about Chriltmas, when no
other can bee leene vpon the ground : and two other that are wilde or baltard kindes,
brought into many Gardens for their Phyficall properties ; but I will only ioyne one
of them with the true kinde in this worke, and leaue the other for another.

1. *Helleborus niger verus.* The true blacke Hellebor, or Chriltmas flower.

The true blacke Hellebor (or Beare foote as lome would call it, but that name doth
more fitly agree with the other two baltard kindes) hath many faire greene leaues riling from the roote, each of them ltanding on a thicke round flelhly ltiffe green ltalke,
about an hand breadth high from the ground, diuided into leuen, eight, or nine parts
or leaues, and each of them nicked or dented, from the middle of the leafe to the point-
ward on both lides, abiding all the Winter, at which time the flowers rile vp on luch
lhort thicke ltalkes as the leaues ltand on, euery one by it felfe, without any leafe
thereon for the molt part, or very leldome hauing one lmall lhort leafe not much vnder the flower, and very little higher then the leaues themlelues, conlilting of fiue
broad white leaues, like vnto a great white lingle Role (which lometimes change to
be either lelle or more purple about the edges, as the weather or time of continuance
doth effect) with many pale yellow thrummes in the middle, ltanding about a greene
head, which after groweth to haue diuers cods let together, pointed at the ends like
hornes, lomewhat like the leede vellels of the *Aconitum hyemale*, but greater & thicker,
wherein is contained long, round, and blackilh leede, like the leede of the baltard
kindes : the rootes are a number of brownilh ltrings running downe deepe into the
ground, and faltened to a thicke head, of the bignelle of a finger at the toppe manie
times, and lmaller ltill downewards.

2. *Helleboraster minor.* The leller baltard blacke Hellebor, or Beare foote.

The lmaller Beare foote is in molt things like vnto the former true blacke Hellebor;
for it beareth allo many leaues vpon lhort ltalkes, diuided into many leaues allo, but
each of them are long and narrow, of a blacker greene colour, lnipt or dented on both
edges, which feele lomewhat hard or lharpe like prickes, and perilh euery yeare, but
rile againe the next Spring : the flowers hereof ltand on higher ltalkes, with lome
leaues on them allo, although but very few, and are of a pale greene colour, like in
forme

forme vnto the flowers of the former, but smaller, hauing also many greenish yellow threads or thrums in the middle, and such like heads or seede vessels, and blackish seed: the rootes are stringie and blackish like the former.

The Place.

The first groweth onely in the Gardens of those that are curious, and delight in all sorts of beautifull flowers in our Countrey, but wilde in many places of Germany, Italy, Greece, &c.

The other groweth wilde in many places of England, as well as the other greater sort, which is not here described; for besides diuers places within eight or ten miles from London, I haue seen it in the Woods of Northamptonshire, and in other places.

The Time.

The first of these plants doth flower in the end of December, and beginning of Ianuary most vsually, and the other a moneth or two after, and sometime more.

The Names.

The first is called *Helleborus*, or *Elleborus niger verus*, and is the same that both Theophrastus and Dioscorides haue written of, and which was called *Melampodion*, of Melampus the Goateheard, that purged and cured the mad or melancholicke daughters of Prætus with the rootes thereof. Dodonæus calleth it *Veratrum nigrum primum*, and the other *secundum*: Wee call it in English, The true blacke Hellebor, or the Christmas flower, because (as I said) it is most commonly in flower at or before Christmas. The second is a bastard or wilde kinde thereof, it so nearely resembleth the true, and is called of most of the later Writers, *Pseudoelleborus niger minor*, or *Helleboraster minor*, for a distinction betweene it and the greater, which is not here described: and is called in English, The smaller or lesser Beare foote, and most vsed in Physicke, because it is more plentifull, yet is more churlish and strong in operation then the true or former kinde.

The Vertues.

The rootes of both these kindes are safe medecines, being rightly prepared, to be vsed for all Melancholicke diseases, whatsoeuer others may feare or write, and may be without danger applied, so as care and skill, and not temerary rashnesse doe order and dispose of them.

The powder of the dryed leaues, especially of the bastard kinde, is a sure remedy to kill the wormes in children, moderately taken.

Chap. LXXXII.

Elleborus albus. White Ellebor or Neesewort.

There are two sorts of great white Ellebors or Neeseworts, whereas there was but one kinde knowne to the Ancients; the other being found out of later dayes: And although neither of both these haue any beauty in their flowers, yet because their leaues, being faire and large, haue a goodly prospect, I haue inserted them in this place, that this Garden should not be vnfurnished of them, and you not vnacquainted with them.

1. *Elleborus*

1. *Elleborus albus vulgaris.* White Ellebor or Neesing roote.

The first great white Ellebor riseth at the first out of the ground, with a whitish greene great round head, which growing vp, openeth it selfe into many goodly faire large greene leaues, plaited or ribbed with eminent ribbes all along the leaues, compassing one another at the bottome, in the middle whereof riseth vp a stalke three foot high or better, with diuers such like leaues thereon, but smaller to the middle thereof; from whence to the toppe it is diuided into many branches, hauing many small yellowish, or whitish greene starre-like flowers all along vpon them, which after turne into small, long, three square whitish seede, standing naked, without any huske to containe them, although some haue written otherwise: the roote is thicke and reasonable great at the head, hauing a number of great white strings running downe deepe into the ground, whereby it is strongly fastened.

2. *Elleborus albus præcox siue atrorubente flore.*
The early white Ellebor with reddish flowers.

This other Ellebor is very like the former, but that it springeth vp a moneth at the least before it, and that the leaues are not fully so thicke or so much plaited, but as large or larger, and doe sooner perish and fall away from the plant: the stalke hereof is as high as the former, bearing such like starry flowers, but of a darke or blackish red colour: the seede is like the other: the roote hath no such head as the other (so farre as I haue obserued, both by mine own and others plants) but hath many long white strings fastened to the top, which is as it were a long bulbous scaly head, out of which spring the leaues.

The Place.

The first groweth in many places of Germany, as also in some parts of Russia, in that aboundance, by the relation of that worthy, curious, and diligent searcher and preseruer of all natures rarities and varieties, my very good friend, Iohn Tradescante, often heretofore remembred, that, as hee said, a good ship might be loaden with the rootes hereof, which hee saw in an Island there.

The other likewise groweth in the vpland wooddy grounds of Germanie, and other the parts thereabouts.

The Time.

The first springeth vp in the end or middle of March, and flowreth in Iune. The second springeth in February, but flowreth not vntill Iune.

The Names.

The first is called *Elleborus albus,* or *Helleborus albus,* the letter *H,* as all Schollers know, being but *aspirationis nota:* and *Veratrum album flore viridante,* of some *Sanguis Herculis.* The other is called *Elleborus albus præcox,* and *flore atrorubente,* or *atropurpurante.* We call the first in English, White Ellebor, Neesewort, or Neesing roote, because the powder of the roote is vsed to procure neesing; and I call it the greater, in regard of those in the next Chapter. The other hath his name according to the Latine title, most proper for it.

The Vertues.

The force of purging is farre greater in the roote of this Ellebor, then in the former; and therefore is not carelesly to bee vsed, without extreame danger; yet in contumatious and stubborne diseases it may bee vsed with good

good caution and aduice. There is a Syrupe or Oxymel made hereof in the Apothecaries shops, which as it is dangerous for gentle and tender bodies, so it may be very effectuall in stronger constitutions. Pausanias *in Phocicis*, recordeth a notable stratagem that Solon vsed in besieging the Citie of Cirrheus, *viz.* That hauing cut off the riuer Plistus from running into the Citie, he caused a great many of these rootes to be put into a quantity thereof, which after they had steeped long enough therein, and was sufficiently infected therewith, he let passe into the Citie againe : whereof when they had greedily drunke, they grew so weake and feeble by the superpurgation thereof, that they were forced to leaue their wals vnmand, and not guarded, whereby the Amphyctions their enemies became masters of their Citie. The like stratagems are set downe by diuers other Authors, performed by the helpe of other herbes.

Chap. LXXXIII.

Elleborine. Small or wilde white Ellebor.

THe likenesse of the leaues of these plants, rather then any other faculty with the former white Ellebor, hath caused them to be called *Elleborine*, as if they were smaller white Ellebors. And I for the same cause haue ioyned them next, whereof there are found many sorts : One which is the greater kinde, is of greatest beauty ; the other which are lesser differ not much one from another, more then in the colour of the flowers, whereof I will onely take three, being of the most beautie, and leaue the rest to another worke.

1. *Helleborine vel Elleborine maior, siue Calceolus Mariæ.*
Our Ladies Slipper.

This most beautifull plant of all these kindes, riseth vp with diuers stalkes, a foote and a halfe high at the most, bearing on each side of them broad greene leaues, somewhat like in forme vnto the leaues of the white Ellebor, but smaller and not so ribbed, compassing the stalke at the lower end ; at the tops of the stalkes come forth one, or two, or three flowers at the most, one aboue another, vpon small short foote-stalkes, with a small leafe at the foote of euery stalke : each of these flowers are of a long ouall forme, that is, more long then round, and hollow withall, especially at the vpper part, the lower being round and swelling like a belly : at the hollow part there are two small peeces like eares or flippets, that at the first doe couer the hollow part, and after stand apart one from another, all which are of a fine pale yellow colour, in all that I haue seene (yet it is said there are some found, that are more browne or tending to purple) there are likewise foure long, narrow, darke coloured leaues at the setting on of the flower vnto the stalke, wherein as it were the flower at the first standeth : the whole flower is of a pretty small sent : the seede is very small, very like vnto the seede of the *Orchides* or Satyrions, and contained in such like long pods, but bigger : the roots are composed of a number of strings enterlacing themselues one within another, lying within the vpper crust of the earth, & not spreading deep, of a darke brownish colour.

2. *Elleborine minor flore albo.*
The small or wilde white Ellebor with a white flower.

This smaller wilde white Ellebor riseth vp in the like manner vnto the former, and not much lower, bearing such like leaues, but smaller, and of a whiter greene colour, almost of the colour and fashion of the leaues of Lilly Conually ; the top of the stalke hath many more flowers, but lesser, growing together, spike-fashion, with small short leaues at the stalke of euery flower, which consisteth of fiue small white leaues, with a small close hood in the middle, without any sent at all : the seede and seede vessels are
like

like vnto the former, but ſmaller : the rootes are many ſmall ſtrings, diſperſing themſelues in the ground.

3. Elleborine minor flore purpurante.
The ſmall or wilde white Ellebor with bluſh flowers.

The leaues of this kinde are like vnto the laſt deſcribed, but ſomewhat narrower :
the ſtalkes and flowers are alike, but ſmaller alſo, and of a pale purpliſh or bluſh colour, which cauſeth the difference.

The Place.

The firſt groweth in very many places of Germany, and in other Countries alſo. It groweth likewiſe in Lancaſhire, neare vpon the border of
Yorkeſhire, in a wood or place called the Helkes, which is three miles from
Ingleborough, the higheſt Hill in England, and not farre from Ingleton, as
I am enformed by a courteous Gentlewoman, a great louer of theſe delights, called Miſtris Thomaſin Tunſtall, who dwelleth at Bull-banke,
neare Hornby Caſtle in thoſe parts, and who hath often ſent mee vp the
rootes to London, which haue borne faire flowers in my Garden. The ſecond groweth in many places of England, and with the ſame Gentlewoman
alſo before remembred, who ſent me one plant of this kinde with the other.
The laſt I haue not yet knowne to growe in England; but no doubt many
things doe lye hid, and not obſerued, which in time may bee diſcouered, if
our Country Gentlemen and women, and others, in their ſeuerall places
where they dwell, would be more carefull and diligent, and be aduertiſed
either by themſelues, or by others capable and fit to be imployed, as occaſion and time might ſerue, to finde out ſuch plants as growe in any the cir-
cuits or limits of their habitations, or in their trauels, as their pleaſures or
affaires leade them. And becauſe ignorance is the chiefe cauſe of neglect
of many rare things, which happen to their view at ſometimes, which are
not to be ſeene againe peraduenture, or not in many yeares after, I would
heartily aduiſe all men of meanes, to be ſtirred vp to bend their mindes, and
ſpend a little more time and trauell in theſe delights of herbes and flowers,
then they haue formerly done, which are not onely harmleſſe, but pleaſu-
rable in their time, and profitable in their vſe. And if any would be better
enformed, and certified of ſuch things they know not, I would be willing
and ready to my beſt skill to aduertiſe them, that ſhall ſend any thing vp to
me where I dwell in London. Thus farre I haue digreſſed from the matter
in hand, and yet not without ſome good vſe I hope, that others may make
of it.

The Time.

The two firſt flower earlier then the laſt, and both the firſt about one
time, that is, in the end of Aprill, or beginning of May. The laſt in the
end of May, or in Iune.

The Names.

The firſt is called *Elleborine recentiorum maior*, and *Calceolus Mariæ* : Of
ſome thought to be *Coſmoſandalos*, becauſe it is *Sandali forma*. In Engliſh we
call it our Ladies Slipper, after the Dutch name. The other two leſſer kinds
haue their names in their titles : I haue thought it fit to adde the title of ſmall
white Ellebors vnto theſe, for the forme ſake, as is before ſaid.

The Vertues.

There is no vſe of theſe in Phyſicke in our dayes that I know.

CHAP. LXXXIIII.

Lilium Conuallium. Lilly Conually.

THe remembrance of the Conuall Lilly, spoken of in the precedent Chapter, hath caused me to insert these plants among the rest, although differing both in face and properties; but lest it should lose all place, let it keepe this. It is of two sorts, differing chiefly in the colour of the flowers, the one being white, and the other reddish, as shall be shewed in their descriptions following.

1. *Lilium Conuallium flore albo.* The white Lilly Conually.

The white Conuall or May Lilly, hath three or foure leaues rising together from the roote, one enclosed within another, each whereof when it is open is long and broad, of a grayish shining greene colour, somewhat resembling the leaues of the former wilde Neesewort, at the side whereof, and sometime from the middle of them, riseth vp a small short naked foote-stalke, an hand breadth high or somewhat more, bearing at the toppe one aboue another many small white flowers, like little hollow bottles with open mouths, nicked or cut into fiue or six notches, turning all downewards one way, or on one side of the stalke, of a very strong sweete sent, and comfortable for the memory and senses, which turne into small red berries, like vnto Asparagus, wherein is contained hard white seede: the rootes runne vnder ground, creeping euery way, consisting of many small white strings.

2. *Lilium Conuallium flore rubente.* May Lillies with red flowers.

This other May Lilly differeth neither in roote, leafe, nor forme of flower from that before, but onely in the colour of the flower, which is of a fine pale red colour, being in my iudgement not altogether so sweet as the former.

The Place.

The first groweth aboundantly in many places of England. The other is a stranger, and groweth only in the Gardens of those that are curious louers of rarities.

The Time.

They both flower in May, and the berries are ripe in August.

The Names.

The Latines haue no other name for this plant but *Lilium Conuallium*, although some would haue it to be *Lilium vernum* of Theophrastus, and others *Oenanthe* of the same Author. Gesner thinketh it to be *Callionymus.* Lonicerus to be *Cacalia*, and Fuchsius to be *Ephemerum non lethale:* but they are all for the most part mistaken. We call it in English Lilly Conually, May Lilly, and of some Liriconfancie.

The Vertues.

The flowers of the white kinde are often vsed with those things that help to strengthen the memory, and to procure ease to Apoplecticke persons. Camerarius setteth downe the manner of making an oyle of the flowers hereof, which he saith is very effectuall to ease the paines of the Goute, and such like diseases, to be vsed outwardly, which is thus: Hauing filled a glasse with the flowers, and being well stopped, set it for a moneths space in an Ants hill, and after being drayned cleare, set it by to vse.

Gg CHAP.

CHAP. LXXXV.

Gentiana. Gentian or Fell-wort.

THere are diuers forts of Gentians or Fell-wortes, fome greater, others leffer, and fome very fmall ; many of them haue very beautifull flowers, but becaufe fome are very fuddenly paft, before one would thinke they were blowne open, and others will abide no culture and manuring, I will onely fet forth vnto you two of the greater forts, and three of the leffer kindes, as fitteft, and more familiarly furnifhing our gardens, leauing the reft to their wilde habitations, and to bee comprehended in a generall Worke.

1. *Gentiana maior flore flauo.* The great Gentian.

The great Gentian rifeth vp at the firft, with a long, round and pointed head of leaues, clofing one another, which after opening themfelues, lye vpon the ground, and are faire, long and broad, fomewhat plaited or ribbed like vnto the leaues of white Ellebor or Neefeworte, but not fo fairely or eminently plaited, neyther fo ftiffe, but rather refembling the leaues of a great Plantane : from among which rifeth vp a ftiffe round ftalke, three foote high or better, full of ioynts, hauing two fuch leaues, but narrower and fmaller at euery ioynt, fo compaffing about the ftalke at the lower end of them, that they will almoft hold water that falleth into them : from the middle of the ftalke to the toppe, it is garnifhed with many coronets or rundles of flowers, with two fuch greene leaues likewife at euery ioynt, and wherein the flowers doe ftand, which are yellow, layd open like ftarres, and rifing out of fmall greenifh huskes, with fome threds in the middle of them, but of no fent at all, yet ftately to behold, both for the order, height and proportion of the plant : the feede is browne and flat, contained in round heads, fomewhat like vnto the feede of the *Fritillaria,* or checkerd Daffodill, but browner : the rootes are great, thicke and long, yellow, and exceeding bitter.

2. *Gentiana maior folio Afclepiadis.* Swallow-wort Gentian.

This kinde of Gentian hath many ftalkes rifing from the roote, neere two foote high, whereon grow many faire pale greene leaues, fet by couples, with three ribs in euery one of them, and doe fomewhat refemble the leaues of *Afclepias* or Swallow-wort, that is, broade at the bottome, and fharpe at the point : the flowers grow at the feuerall ioynts of the ftalkes, from the middle vpwards, two or three together, which are long and hollow, like vnto a bell flower, ending in fiue corners, or pointed leaues, and folded before they are open, as the flowers of the Bindeweedes are, of a faire blew colour, fometimes deeper, and fometimes paler : the heads or feede veffels haue two points or hornes at the toppes, and containe within them flat grayifh feed, like vnto the former, but leffe : the rootes hereof are nothing fo great as the former, but are yellow, fmall and long, of the bigneffe of a mans thumbe.

3. *Gentiana minor Cruciata.* Croffe-wort Gentian.

This fmall Gentian hath many branches lying vpon the ground, fcarce lifting themfelues vpright, and full of ioynts, whereat grow vfually foure leaues, one oppofite vnto another, in manner of a Croffe, from whence it tooke his name, in fhape very like vnto *Saponaria* or Sopewort, but fhorter, and of a darker greene colour : at the tops of the ftalkes ftand many flowers, thick thrufting together, and likewife at the next ioynt vnderneath, euery one of them ftanding in a darke blewifh greene huske, and confifting of fiue fmall leaues, the points or ends whereof only appeare aboue the huskes wherein they ftand, and are hardly to be feene, but that they are of a fine pale blew colour, and that many grow together : the feed is fmall and brown, hard, and fomewhat

like

1 *Lilium Conuallium.* Liriconfancy or Lilly Conually. 2 *Gentiana maior.* The great Gentian. 3 *Gentianella verna.* Small Gentian of the Spring. 4 *Gentiana Cruciata.* Croſſewort Gentian. 5 *Pneumonanthe ſeu Gentiana Autumnalis.* Autumne Gentian. 6 *Saponaria flore duplici.* Double flowred Sopewort. 7 *Plantago Roſea.* Roſe Platane.

like vnto the feed of the Marian Violets, or Couentry bels: the roots are fmall and whitifh, difperfing themfelues diuerfly in the ground, of as bitter a tafte almoft as the reft.

4. *Gentianella Verna.* Small Gentian of the Spring.

The fmall Gentian of the Spring hath diuers fmall hard greene leaues, lying vpon the ground, as it were in heads or tufts, fomewhat broade below, and pointed at the end, with fiue ribs or veines therein, as confpicuous as in the former Gentians, among which rifeth vp a fmall fhort ftalke, with fome fmaller leaues thereon, at the toppe whereof ftandeth one faire, large, hollow flower, made bell fafhion, with wide open brimmes, ending in fiue corners or diuifions, of the moft excellent deepe blew colour that can be feene in any flower, with fome white fpots in the bottome on the infide : after the flower is paft, there appeare long and round pods, wherein are contained fmall blackifh feede : the rootes are fmall, long, pale yellow ftrings, which fhoot forth here and there diuers heads of leaues, and thereby encreafe reafonable well, if it finde a fit place, and ground to grow, or elfe will not be nourfed vp, with all the care and diligence can be vfed: the whole plant is bitter, but not fo ftrong as the former.

5. *Gentiana Autumnalis fiue Pneumonanthe.* Calathian Violet or Autumne Gentian.

This Gentian that flowreth in Autumne, hath in fome places higher ftalkes then in others, with many leaues thereon, fet by couples as in other Gentians, but long and narrow, yet fhewing the three ribbes or veines that are in each of them : the toppes of the ftalkes are furnifhed euery one with a flower or two, of an excellent blew purple colour, ending in fiue corners, and ftanding in long huskes : the rootes are fomewhat great at the top, and fpreading into many fmall yellow ftrings, bitter as the reft are.

6. *Saponaria flore duplici.* Double flowred Sopeworte.

Vnto thefe kindes of Gentians, I muft needes adde thefe following plants, for that the former is of fome neere refemblance in leafe with fome of the former. And becaufe the ordinary Sopeworte or Bruifeworte with fingle flowers is often planted in Gardens, and the flowers ferue to decke both the garden and the houfe ; I may vnder the one defcribe them both : for this with double flowers is farre more rare, and of greater beautie. It hath many long and flender round ftalkes, fcarce able to fuftaine themfelues, and ftand vpright, being ful of ioynts and ribbed leaues at them, euery one fomewhat like a fmall Gentian or Plantane leafe : at the toppes of the ftalkes ftand many flowers, confifting of two or three rowes of leaues, of a whitifh or pale purple colour, and of a ftrong fweet fent, fomewhat like the fmell of Iafmin flowers, ftanding in long and thicke pale greene huskes, which fall away without giuing any feede, as moft other double flowers doe that encreafe by the roote, which fpreadeth within the ground, and rifeth vp in fundry diftant places like the fingle.

7 *Plantago Rofea.* Rofe Plantane.

This other plant is in all things like vnto the ordinary Plantane or Ribworte, that groweth wilde abroade in many places, whofe leaues are very large : but in ftead of the long flender fpike, or eare that the ordinary hath, this hath eyther a thicke long fpike of fmall greene leaues vpon fhort ftalkes, or elfe a number of fuch fmall greene leaues layd round-wife like vnto a Rofe, and fometimes both thefe may be feene vpon one and the fame roote, at one and the fame time, which abide a great while frefh vpon the roote, and fometimes alfo giueth feede, efpecially from the more long and flender fpikes.

The Place.

Some of thefe Gentians grow on the toppes of hils, and fome on the fides and foote of them in Germany and other Countreyes : fome of them alfo vpon barren heaths in thofe places, as alfo in our owne Countrey, efpecially the Autumne Gentian, and as it is reported, the Vernall likewise. The single or ordinary Sopeworte is found wilde in many places with vs, but the double came to vs from beyond the Sea, and is scarce known or heard of in England. The Rose Plantaine hath beene long in England, but whether naturall thereof or no, I am not assured.

The Time.

They flower for the most part in Iune and Iuly, but the small Gentian of the Spring flowreth somewhat earlier, and that of the Autumne in August and September.

The Names.

Gentiana is the generall name giuen to the Gentians. We call them in English Gentian, Fellworte, Bitterwort, and Baldmoney. *Saponaria* taketh his name from the scouring qualitie it hath : Wee call it in English Sopewort, and in some places Bruisewort. Some haue thought it to bee *Struthium* of Dioscorides, or at least haue vsed it for the same causes, but therein they are greatly deceiued, as Matthiolus hath very well obserued thereon, and so is Dodonæus, that thought it to be *Alisma.* The Rose Plantaine is so called of the double spikes it carrieth.

The Vertues.

The wonderfull wholsomnesse of Gentian cannot bee easily knowne to vs, by reason our daintie tastes refuse to take thereof, for the bitternesse sake : but otherwise it would vndoubtedly worke admirable cures, both for the liuer, stomacke and lunges. It is also a speciall counterpoison against any infection, as also against the violence of a mad dogges tooth : wilde Sopewort is vsed in many places, to scoure the countrey womens treen, and pewter vessels, and physically some make great boast to performe admirable cures in Hydropicall diseases, because it is diureticall, and in *Lue Veneria,* when other Mercuriall medicines haue failed. The Rose Plantaine no doubt hath the same qualities that the ordinary hath.

Chap. LXXXVI.

Campanula. Bell-flowers.

VNder the title of Bell-flowers are to bee comprehended in this Chapter, not only those that are ordinarily called *Campanula,* but *Viola Mariana,* and *Trachelium* also, whereof the one is called Couentry, the other Canterbury Bells.

1. *Campanula Persicifolio alba, vel cærulea.*
Peach-leafed Bell-flowers white or blew.

The Peach-leafed Bell-flower hath many tufts, or branches of leaues lying vpon the ground, which are long and narrow, somewhat like vnto the leafe of an Almond or Peach tree, being finely nicked about the edges, and of a sad greene colour, from among which rise vp diuers stalkes, two foote high or more, set with leaues to the middle, and from thence vpwards, with many flowers standing on seuerall small footestalkes, one aboue another, with a small leafe at the foote of euery one : the flowers stand in small greene huskes, being small and round at the bottome, but wider open at the brimme, and ending in fiue corners, with a three forked clapper in the middle, set about with some small threds tipt with yellow, which flowers in some plants are pure

white

white, and in others of a pale blew or watchet colour, hauing little or no sent at all : the seede is small, and contained in round flat heads, or seede vessels : the roote is very small, white and threddy, creeping vnder the vpper crust of the ground, so that often-times the heat and drought of the Summer wil goe near to parch and wither it vtterly : it requireth therefore to be planted in some shadowie place.

2. *Campanula maior, siue Pyramidalis*.
The great or steeple Bell-flower.

This great Bell-flower hath diuers stalkes, three foote high or better, whereon grow diuers smooth, darke, greene leaues, broade at the bottome, and small at the point, somewhat vneuenly notched about the edges, and standing vpon longer footestalkes below then those aboue : the flowers are blew, and in some white, not so great or large as the former, but neare of the same fashion, growing thicker and more plenti-fully together, with smaller leaues among them, bushing thicke below, and rising smal-ler and thinner vp to the toppe, in fashion of a *Pyramis* or speere Steeple : the roote is thicke and whitish, yeelding more store of milke being broken (as the leaues and stalks also doe) then any other of the Bell-flowers, euery one whereof doe yeelde milke, some more and some lesse.

3. *Viola Mariana flore albido vel purpureo*.
Couentry Bels white or purple.

The leaues of Couentry Bels are of a pale or fresh greene colour, long, and narrow next vnto the bottome, and broader from the middle to the end, and somewhat round pointed, a little hairy all ouer, and snipt about the edges : the stalkes rise vp the yeare after the sowing, being somewhat hairy also, and branching forth from the roote, into diuers parts, whereon stand diuers leaues, smaller then the former, and of a darker greene colour : at the end of euery branch stand the flowers, in greene huskes, from whence come large, round, hollow Bels, swelling out in the middle, and rising some-what aboue it, like the necke of a pot, and then ending in fiue corners, which are either of a faire or faint white, or of a pale blew purplish colour, and sometimes of a deeper purple or violet : after the flowers are past, there rise vp great square, or cornered seede vessels, wherein is contained in diuers diuisions, small, hard, shining, browne, flat seeds : the roote is white, and being young as in the first yeares sowing, is tender, and often eaten as other Rampions are ; but the next yeare, when it runneth vp to seede, it grow-eth hard, and perisheth : so that it is to be continued by euery other yeares sowing.

4. *Trachelium maius flore albo vel purpureo*.
Great Canterbury Bels white or purple.

The greater Canterbury Bels, or Throateworte, hath many large rough leaues, somewhat like vnto Nettle leaues, being broad and round at the bottome, and pointed at the end, notched or dented on the edges, and euery one standing on a long footstalk : among these leaues rise vp diuers square rough stalkes, diuided at the toppe into diuers branches, whereon grow the like leaues as grow below, but lesser ; toward the ends of the branches stand the flowers, mixed with some longer leaues, euery one in his seuerall huske, which are hollow, long and round, like a bell or cup, wide open at the mouth, and cut at the brimme into fiue corners, or diuisions, somewhat lesser then the Co-uentry Bels, in some of a pure white, and others of a faire deepe purple violet colour, and sometimes paler : after the flowers are past, come smaller and rounder heades then in the former, containing flat seede, but blacker, and not so redde as the last : the roote is hard and white, dispersing it selfe into many branches vnder ground, not perishing euery yeare as the former (although it loseth all the leaues in winter) but abiding many yeares, and encreasing into diuers heades or knobs, from whence spring new leaues and branches.

5. *Trache-*

1 *Campanula persicifolia.* Peach leafed Bell-flower. 2 *Trachelium maius simplex.* Canterbury Bels. * *Trachelium flore duplici.* Double Canterbury Bels. 3 *Viola Mariana.* Couentry Bels. 4 *Trachelium Giganteum.* Giants Throatewort 5 *Trachelium minus.* The lesser Throatewort. 6 *Trachelium Americanum siue Cardinalis planta.* The rich crimson Cardinals flower.

5. *Trachelium maius flore duplici albo & cæruleo.*
Canterbury Bels with double flowers both white and blew.

Of this kinde of Throateworte or Canterbury Bels, there is another sort, not differing in any thing from the former, but in the doublenesse of the flower : For there
is of both the kindes, one that beareth double white flowers, and the other blew : Of
each whereof I receiued plants from friends beyond the Sea, which grow well with
me.

6. *Trachelium Giganteum flore purpurante.*
Pale purple Giants Throateworte.

This Bell-flower, although it hath a Gigantine name, yet did I neuer perceiue it in
my Garden, to rise vp h gher then the former, the epithite beeing in my perswasion,
only giuen for difference sake : the leaues whereof are not so rough, but as large, and
dented about the edges, somewhat larger pointed, and of a fresher greene colour : the
stalkes beare such like leaues on them, but more thinly or dispersedly set, hauing a
flower at the setting on of euery one of the leaues, from the middle vpwards, and are
somewhat like the great Throateworte in forme, but of a pale or bleake reddish purple colour, turning the brims or corners a little backwards, with a forked clapper in
the middle, sufficient eminent and yellow : the seede hereof is white, and plentifull
in the heads, which will abide all the winter vpon the stalkes, vntill all the seede being
shed, the heads remaining seeme like torne rags, or like thin peeces of skin, eaten with
wormes : the roote is great, thicke and white, abiding long without perishing.
Flore albo.   There is another which differeth not any thing but in the flower, which is white.

7. *Trachelium minus flore albo & purpureo.*
Small Throateworte or Canterbury Bells both white and purple.

The lesser Throateworte hath smaller leaues, nothing so broade or hard as the former great kinde, but long, and little or nothing dented about the edges : the stalkes are
square and brownish, if it beare purple flowers, and greene if it beare white flowers,
which in forme are alike, and grow in a bush or tuft, thicke set together, more then any
of the former, and smaller also, being not much bigger then the flowers of the fielde,
or garden Rampions : the roote is lasting, and shooteth afresh euery yeare.

8. *Trachelium Americanum flore ruberrimo, siue Planta Cardinalis.*
The rich crimson Cardinals flower.

This braue plant, from a white roote spreading diuers wayes vnder ground, sendeth
forth many greene leaues, spread round about the head thereof, each whereof is somewhat broade and long, and pointed at the end, finely also sniptabout the edges : from
the middle whereof ariseth vp a round hollow stalke, two foote high at the least, beset
with diuers such leaues as grow below, but longer below then aboue, and branching
out at the toppe aboundantly, euery branch bearing diuers greene leaues on them, and
one at the foote of euery of them also, the toppes whereof doe end in a great large
tuft of flowers, with a small greene leafe at the foote of the stalke of euery flower,
each footestalke being about an inch long, bearing a round greene huske, diuided into
fiue long leaues or points turned downwards, and in the midst of euery of them a most
rich crimson coloured flower, ending in fiue long narrow leaues, standing all of them
foreright, but three of them falling downe, with a long vmbone set as it were at the
backe of them, bigger below, and smaller aboue, and at the toppe a small head, being
of a little paler colour then the flower, but of no sent or smell at all, commendable only for the great bush of so orient red crimson flowers : after the flowers are past, the
seede commeth in small heads, closed within those greene husks that held the flowers,
which is very like vnto the seede vessels of the *Viola Mariana,* or Couentry Bels, and
is small and brownish.

The

The Place.

All these Bell-flowers do grow in our Gardens, where they are cherished for the beautie of their flowers. The Couentry Bels doe not grow wilde in any of the parts about Couentry, as I am credibly informed by a faithfull Apothecary dwelling there, called Master Brian Ball, but are noursed in Gardens with them, as they are in other places. The last groweth neere the riuer of Canada, where the French plantation in America is seated.

The Time.

They flower from May vntill the end of Iuly or August, and in the mean time the seed is ripe : But the Peache-leafed Bell-flowers, for the most part, flower earlier then the other.

The Names.

The first is generally called *Campanula Persicifolia,* in English Peach-leafed Bell-flower. The second is called *Campanula maior, Campanula lactescens Pyramidalis,* and *Pyramidalis Lutetiana* of Lobel, in English, Great or Steeple Bell-flower. The third is vsually called *Viola Mariana,* and of some *Viola Marina.* Lobel putteth a doubt whether it be not *Medium* of Dioscorides, as Matthiolus and others doe thinke ; but in my opinion the thicknesse of the roote, as the text hath it, contradicteth all the rest. We call it generally in English Couentry Bels. Some call it Marian, and some Mercuries Violets. The fourth and fift are called *Trachelium* or *Ceruicaria,* of some *Vvularia,* because many haue vsed it to good purpose, for the paines of the *Vvula,* or Throate : Yet there is another plant, called also by some *Vvularia,* which is *Hippoglossum,* Horse tongue, or Double tongue. The sixt hath his title to descipher it out sufficiently, as is declared. The seuenth is called *Trachelium minus,* and *Ceruiaria minor,* of some *Saponaria altera* ; in English, Small Throateworte, or Small Canterbury Bels. The last hath his name in the title, as it is called in France, from whence I receiued plants for my Garden with the Latine name : but I haue giuen it in English.

The Vertues.

The Peach-Bels as well as the others may safely bee vsed in gargles and lotions for the mouth, throate, or other parts, as occasion serueth. The rootes of many of them, while they are young, are often eaten in sallets by diuers beyond the Seas.

CHAP. LXXXVII.

Campana Carulea siue Convolvulus Caruleus.
Blew Bell flowers, or blew Bindeweede.

THere are two other kindes of Bell-flowers, much differing from the Tribe or Familie of the former, because of their climbing or winding qualitie, which I must needes place next them, for the likenesse of the flowers, although otherwise they might haue beene placed with the other clamberers that follow. Of these there is a greater, and a lesser, and of each likewise some difference, as shall be declared.

I. Con-

1. *Convolvulus cæruleus maior rotundifolius.*
The greater blew Bindweede, or Bell-flower with round leaues.

This goodly plant riseth vp with many long and winding branches, whereby it climbeth and windeth vpon any poles, herbes, or trees, that stand neare it within a great compasse, alwaies winding it selfe contrary to the course of the Sunne: on these branches doe growe many faire great round leaues, and pointed at the end, like vnto a Violet leafe in shape, but much greater, of a sad greene colour: at the ioynts of the branches, where the leaues are set, come forth flowers on pretty long stalkes, two or three together at a place, which are long, and pointed almost like a finger, while they are buds, and not blowne open, and of a pale whitish blew colour, but being blowne open, are great and large bels, with broad open mouths or brims ending in fiue corners, and small at the bottome, standing in small greene huskes of fine leaues: these flowers are of a very deepe azure or blew colour, tending to a purple, very glorious to behold, opening for the most part in the euening, abiding so all the night and the next morning, vntill the Sunne begin to growe somewhat hot vpon them, and then doe close, neuer opening more: the plant carrieth so many flowers, if it stand in a warme place, that it will be replenished plentifully, vntill the cold ayres and euenings stay the luxury thereof: after the flowers are past, the stalkes whereon the flowers did stand, bend downwards, and beare within the huskes three or foure blacke seedes, of the bignesse of a Tare or thereabouts: the rootes are stringy, and perish euery yeare.

2. *Convolvulus trifolius siue hederaceus purpureus.*
The greater purple Bindeweede, or Bell flower with cornered leaues.

The growing and forme of this Bindeweede or Bell flower, is all one with the former, the chiefest differences consisting in the forme of the leafe, which in this is three cornered, like vnto an Iuie leafe with corners; and in the flower, which is of a deeper blew, tending more to a deepe purple Violet, and somewhat more reddish in the fiue plaites of each flower, as also in the bottomes of the flowers.

3. *Convolvulus tennifolius Americanus.* The red Bell-flower of America.

Although this rare plant (because wee seldome haue it, and can as hardly keepe it) be scarce knowne in these cold Countries, yet I could not but make mention of it, to incite those that haue conueniencie to keepe it, to be furnished of it. It springeth vp at the first from the seede with two leaues, with two long forked ends, which abide a long time before they perish, betweene which riseth vp the stalke or stemme, branching forth diuers waies, being of a brownish colour, which windeth it selfe as the former great Bell-flower doth, whereon are set at seuerall ioynts diuers winged leaues, that is to say, many small narrow and long leaues set on both sides of the middle ribbe, and one at the end: from these ioynts arise long stalkes, at the ends whereof stand two or three small, long, hollow flowers, fashioned very like vnto the flowers of a Bindeweede, or the flowers of Tabacco, and ending in the like manner in fiue points, but not so much laide open, being of a bright red colour, plaited as the Bindeweedes or Bell-flowers before they be open, with some few threads in the middle, which turne into long pointed cods, wherein is contained long and blacke seede, tasting hot like Pepper: the roote is small and stringy, perishing euery yeare, and with vs will seldome come to flower, because our cold nights and frosts come so soone, before it cannot haue comfort enough of the Sun to ripen it.

4. *Convolvulus cæruleus minor Hispanicus.*
The Spanish small blew Bindeweede.

This small Bindeweede hath small long leaues, somewhat broader then the next that followeth, and not so broad as the common small Bindeweede (that groweth
euery

euery where wilde on the bankes of fields abroad) set vpon the small trayling branches, which growe aboue two or three foote high : from the middle of these branches, and so vnto the toppes of them, come forth the flowers at the ioynts with the leaues, folded together at the first into fiue plaites, which open into so many corners, of a most excellent faire skie coloured blew (so pleasant to behold, that often it amazeth the spectator) with white bottomes, and yellowish in the middle, which turne into small round white heads, wherein are contained small blackish cornered seede, somewhat like the former, but smaller : the roote is small and threddy, perishing as the former euery yeare : this neuer windeth it selfe about any thing, but leaneth by reason of the weaknesse of the branches, and dyeth euery yeare after seede time, and not to be sowne againe vntill the next Spring.

5. *Convolvulus purpureus Spicafolius*. Lauander leafed Bindeweede.

This small purple Bindeweede, where it naturally groweth, is rather a plague then a pleasure, to whatsoeuer groweth with it in the fields ; yet the beauty of the flower hath caused it to be receiued into Gardens, bearing longer and smaller leaues then the last, and such like small Bell-flowers, but of a sad purple colour : the roote is liuing, as the common kinds are, and springeth againe where it hath been once sowne, without feare of perishing.

The Place.

The first two greater kindes haue beene sent vs out of Italy, but whether they had them from the East Indies, or from some of the Easterne Countries on this side, wee know not : but they thriue reasonable well in our Country, if the yeare be any thing kindly. The next came out of America, as his name testifieth. The lesser blew kinde groweth naturally in many places both of Spaine and Portugall (from whence I first receiued seedes from Guillaume Boel, heretofore remembred.) The last groweth wilde in the fields, about Dunmowe in Essex, and in many other places of our owne Countrey likewise.

The Time.

The three first greater kindes flower not vntill the end of August, or thereabouts, and the seede ripeneth in September, if the colds and frosts come not on too speedily. The lesser kindes flower in Iune and Iuly.

The Names.

The first is called of some *Campana Lazura*, as the Italians doe call it, or *Campana cærulea*, of others *Convolvulus cæruleus maior, siue Indicus*, and *Flos noctis*. Of some *Nil Auicenna*. The second is called *Convolvulus trifolius*, or *hæderaceus*, for the distinction of the leaues. In English wee call them eyther Great blew Bell-flowers, or more vsually, Great blew Bindeweedes. That of America is diuersly called by diuers. It is called *Quamoclit* of the Indians, and by that name it was sent to Ioachinus Camerarius out of Italy, where it is so called still, as Fabius Columna setteth it downe, and as my selfe also can witnesse it, from thence being so sent vnto mee : but Andræas Cæsalpinus calleth it, *Iasminum folio Millefolÿ*, supposing it to be a Iasmine. Camerarius saith, it may not vnfitly be called *Convolvulus tenuifolius*, accounting it a kinde of Bindeweede. Columna entituleth it *Convolvulus pennatus exoticus rarior*, and saith it cannot bee referred to any other kinde of plant then to the Bindeweedes. Hee that published the *Curæ posteriores* of Clusius, giueth it the name of *Iasminum Americanum*, which I would doe also, if I thought it might belong to that Family ; but seeing the face and forme of the plant better agreeing with the Bindeweedes or Bell-flowers,

I haue

I haue (as you fee) inferted it among them, and giuen it that name may bee moft fit for it, efpecially becaufe it is but an annuall plant. The leffer kindes haue their names fufficiently expreffed in their titles.

The Vertues.

We know of no vfe thefe haue in Phyficke with vs, although if the firft be *Nil* of Auicen, both he and Serapio fay it purgeth ftrongly.

Chap. LXXXVIII.

Stramonium. Thorne-Apple.

VNto the Bell-flowers, I muft adioyne three other plants, in the three feuerall Chapters following, for fome affinity of the flowers: and firft of the Thorne-Apples, whereof there are two efpeciall kindes, that is, a greater and a leffer, and of each fome diuerfity, as fhall be fet downe.

1. *Stramonium maius album.* The great white flowred Thorne-Apple.

The greater Thorne-Apple hath a great, ftrong, round greene ftalke, as high as any man, if it be planted in good ground, and of the bigneffe of a mans wreft almoft at the bottome, fpreading out at the toppe into many branches, whereon ftand many very large and broad darke greene leaues, cut in very deeply on the edges, and hauing ma-nie points or corners therein: the flowers come forth at the ioynts, betweene two branches towards the toppe of them, being very large, long, and wide open, ending in fiue points or corners, longer and larger then any other Bell-flowers whatfoeuer: after the flowers are paft, come the fruit, which are thorny long heads, more prickly and greene then the leffer kindes, which being ripe openeth it felfe into three or foure parts, hauing a number of flat blackifh feede within them: the roote is aboundant in fibres, whereby it ftrongly taketh hold in the ground, but perifheth with the firft frofts; yet the feede that is fhed when the fruit is ripe, commeth vp the next yeare.

2. *Stramonium maius purpureum.*
The great purple flowred Thorne-Apple.

This purple Thorne-Apple is in largeneffe of leaues, thickneffe and height of ftalke, greatneffe and forme of flowers and fruit, euery way equall and correfpondent vnto the former, the chiefe differences be thefe: the ftalke is of a darke purple colour; the leaues are of a darker greene, fomewhat purplifh, and the flowers are of light purple or pale Doue colour, enclining to white; and whiter at the bottome.

3. *Stramonium minus feu Nux Metel flore albo.*
The fmaller Thorne-Apple with a white flower.

The fmaller Thorne-Apple rifeth vp with one round ftalke, of the bigneffe of a mans finger, and neuer much aboue two foote high with vs, bearing a few large, broad, fmooth leaues thereon, without any branches at all, which are vneuenly rent or torne about the edges, with many ribs, and fmaller veines running through them, yet leffer by much then the greater kinde: at the ioynts where the leaues ftand, come forth long and large white flowers, with broad or wide open brims, folded together before their opening, as the other former Bell-flowers or Bindeweedes, but hauing their fiue corners more pointed or horned then either they, or the former Thorne-Apples: af-ter the flowers are paft, fucceed fmall fruit, rounder and harder, fet with harder, but blunt prickes then the former, wherein is contained brownifh yellow flat feede,

sticking

1 *Convolvulus maior cæruleus.* The greater blew Bindweed or Bell flower. 2 *Convolvulus trifolius ſeu hederaceus.* The great purple Bindweed. 3 *Convolvulus minor cæruleus Hiſpanicus.* The Spaniſh ſmall blew Bindweed. 4 *Stramonium maius ſeu Pomum ſpinoſum.* The great Thorne Apple. 5 *Datura ſeu Stramonium minus.* The ſmall Thorne Apple. 6 *Stramonium flore duplici.* The double flowred Thorne-Apple. 7 *Stramonium flore geminato.* Double Thorne-Apple one out of another 8 *Tabacco latifolium.* Broad leafed Tabacco. 9 *Mirabilia Peruana.* The Meruaile of the world.

sticking to the inward pulpe : the roote is not very great, but full of strings, and quickly perisheth with the first frosts.

4. *Stramonium minus flore geminato purpurante.*
The small double flowred purple Thorne-Apple.

In the flower of this plant, consisteth the chiefest difference from the former, which is as large as the last, pointed into more hornes or corners, and beareth two flowers, standing in one huske, one of them rising out from the middle of the other, like vnto those kindes of Cowslips and Oxelips, called double, or Hose in hose, before descri-bed, which are of a pale purplish colour on the outside, and almost white within : the fruit is round like the last, and beareth such like seede, so that vntill it bee in flower, their difference can hardly bee discerned : this is more tender then the last, although euen it is so tender, that it seldome beareth ripe seede with vs.

Flore duplici.　Sometimes (for I think it is not another kind) the flower will haue as it were double rowes of leaues, close set together, and not consisting of two, rising so distinctly one aboue another.

The Place.

All these kindes haue been brought or sent vs out of Turkie and Egypt; but Garcias, and Christopherus Acosta, with others, affirme that they grow in the East Indies. The lesser kindes are very rare with vs, because they sel-dome come to maturity ; and therefore we are still to seeke of new seede to sowe. The greater kindes are plentifull enough in our Gardens, and will well abide, and giue ripe fruit.

The Time.

The smaller kindes flower later then the greater ; and therefore their fruit are the sooner spoiled with the cold ayres, dewes, and frosts, that come at the latter end of the yeare : but the greater kinds neuer misse lightly to ripen.

The Names.

Both the greater and smaller kindes are generally called *Stramonium*, *Stramonia*, *Pomum spinosum*, and *Datura*. Bauhinus vpon Matthiolus his Comentaries on Dioscorides, calleth it *Solanum fœtidum spinosum*. Some learned men haue referred it to *Nux Metel*, of the Arabian Authors. Wee call them generally in English, Thorne-Apples, and distinguish them by their titles of greater and lesser, single and double.

The Vertues.

The East Indian lasciuious women performe strange acts with the seed(of the smaller kinde, as I suppose, or it may be of either) giuing it their hus-bands to drinke. The whole plant, but especially the seed, is of a very cold and soporiferous quality, procuring sleep and distraction of senses. A few of the seeds steeped and giuen in drinke, will cause them that take it to seem starke drunke or dead drunke, which fit will within a few houres weare a-way, and they recouer their senses againe, as a drunken man raysed after sleep from his wine. It may therefore (in my opinion) be of safe and good vse to one, that is to haue a legge or an arme cut off, or to be cut for the stone, or some other such like cure to be performed, to take away the sense of paine for the time of doing it ; otherwise I hold it not fit to be vsed without great caution. But the greene leaues of the greater kindes (as also of the lesser, but that with vs they are not so plentifull) are by tryed experience, found to be excellent good for any scalded or burned part, as also to take away any hot inflammations, being made vp into a salue or ointment with suet, waxe, and rossin, &c. or with *Axungia*, that is, Hogs larde.

CHAP.

Chap. LXXXIX.

Tabacco. Indian Henbane, or Tabacco.

THere hath beene formerly but three kindes of Tabacco knowne vnto vs, two of them called Indian, and the third Englifh Tabacco. In thefe later yeares, we haue had in our gardens about London(before the fuppreffing of the planting) three or foure other forts at the leaft, and all of the Indian kinde, hauing fome efpeciall difference, eyther in leafe, or flower, or both : And in regard the flowers of fome of thefe carry a pretty fhew, I fhall only entreate of them, and not of the Englifh kind.

Tabacco latifolium. Broade leafed Tabacco.

The great Indian Tabacco hath many very large, long, thicke, fat and faire greene leaues, ftanding foreright for the moft part, and compaffing the ftalkes at the bottome of them, being fomewhat pointed at the end : the ftalke is greene and round, fixe or feuen foote high at fometimes, and in fome places, in others not paft three or foure foote high, diuided towards the toppe into many branches, with leaues at euery ioynt, and at the toppes of the branches many flowers, the bottomes hereof are long and hollow, and the toppes plaited or folded before they are open, but being open, are diuided fometimes into foure, or more vfually into fiue corners, fomewhat like vnto other of the Bell-flowers, but lying a little flatter open, of a light carnation colour. The feede is very fmall and browne, contained in round heads, that are clammy while they are greene, and pointed at the end : the roote is great, whitifh, and woody at the head, difperfing many long branches, and fmall fibres vnder the ground, whereby it is ftrongly faftened, but perifheth with our violent frofts in the winter, if it be left abroad in the garden, but if it be houfed, or fafely prouided for againft the froftes, the rootes will liue, and fpring afrefh the next yeare.

There is of this kinde another fort, whofe leaues are as large and long as the former, but thicker, and of a more dead greene colour, hanging downe to the ground-ward, and fcarce any ftanding forth-right, as the former, vnleffe they bee very young : the flowers of this kinde are almoft whole, without any great fhew of corners at the brims or edges, in all other things there is no difference.

There is another, whofe large and thicke flat leaues doe compaffe the ftalke at the bottome, and are as it were folded together one fide vnto another: the flowers are of a deeper blufh, or carnation colour, and with longer points and corners then in any of the former ; and in thefe two things confifteth the difference from the others, and is called Verines Tabacco.

Another hath his leaues not fo large and long as the firft, and thefe haue fhort foote-ftalkes, whereon they ftand, and doe not compaffe the ftalke as the other doe : the flower hereof is like the firft, but fmaller, and of a little paler colour.

Tabacco anguftifolium. Narrow leafed Tabacco.

This kinde of Tabacco hath fomewhat lower, and fmaller ftalkes, then any of the former : the leaues hereof are fmaller and narrower, and not altogether fo thicke, but more pointed, and euery one ftanding vpon a footftalke, an inch and a halfe long at the leaft : the flowers hereof ftand thicker together, vpon the fmall branches, fomewhat larger, of a deeper blufh colour, and more eminent corners then in any the former : the feed and roots are alike, and perifh in like manner, vnleffe it be brought into a cellar, or other fuch couert, to defend it from the extremitie of the Winter.

The Place.

America or the Weft Indies is the place where all thefe kindes doe grow naturally, fome in one place, and fome in another, as in Peru, Trinidado,

Hifpani-

Hifpaniola, and almoft in euery Iland and Countrey of the continent there-of : with vs they are cherifhed in gardens, as well for the medicinable qua-lities, as for the beauty of the flowers.

The Time.

It flowreth in Auguft, feldome before, and the feede is ripe quickly after. If it once fowe it felfe in a Garden, it will giue next year after young plants: but for the moft part they will fpring vp late, and therefore they that would haue them more early, haue fowen the feede vpon a bed of dung, and tranfplanted them afterwards.

The Names.

This plant hath gotten many names. The Indians call it in fome places *Petum*, in others *Picielt*, and *Perebecenuc*, as Ouiedus and others doe relate. The Spaniards in the Indies firft called it *Tabacco*, of an Iland where plenty of it grew. It hath in Chriftendome receiued diuers other names, as *Nico-tiana*, of one Nicot a French man, who feeing it in Portugall, fent it to the French Queene, from whom it receiued the name of *Herba Regina*. Lobel calleth it *Sancta herba*, & *Sana fancta Indorum*. Some haue adiudged it to be an *Hiofcyamus*, and therefore call it *Peruvianus*. The moft vfuall name wher-by we call it in Englifh, is Tabacco.

The Vertues.

The herbe is, out of queftion, an excellent helpe and remedy for diuers difeafes, if it were rightly ordered and applyed, but the continuall abufe thereof in fo many, doth almoft abolifh all good vfe in any. Notwithftan-ding if men would apply their wits to the finding out of the vertues, I make no doubt but many ftrange cures would bee performed by it, both inward and outward. For outward application, a Salue made hereof (as is before re-cited of the Thorne apple leaues) cureth vlcers, and wounds of hard cura-tion : And for inward helpes, a Syrupe made of the iuice and fugar, or ho-ney, procureth a gentle vomit (but the dryed leafe infufed in wine much more) and is effectuall in aftmaticall difeafes, if it bee carefully giuen. And likewife cleanfeth cankers and fiftulaes admirably, as hath beene found by late experience. The afhes of Tabacco is often vfed, and with good fuc-ceffe, for cuts in the hands, or other places, and for other fmall greene wounds.

Chap. XC.

Mirabilia Peruviana. The Meruaile of Peru.

THis plant yeeldeth in our Gardens fiue or fixe feuerall varieties of beautifull flowers, as pure white, pure yellow, pure red, white and red fpotted, and red and yellow fpotted. But befides thefe, I haue had fome other forts, among which was one, of a pale purple or peach colour : all which, comming vnto mee out of Spaine with many other, feedes in an vnkindly yeare (an early winter following a cold fummer) perifhed with mee ; yet I plainely might difcerne by their leaues, and manner of growing, to be diuers from them that we now haue and keepe. I fhall need therefore (becaufe the chiefeft difference confifteth almoft in the flowers) to giue only one defcription of the plant, and therein fhew the varieties as is before declared.

Admirabilis. The Meruaile of the World.

The ftalke of this meruellous plant is great and thick, bigger then any mans thumbe, bunched

bunched out or swelling at euery ioynt, in some the stalkes will bee of a faire greene colour, and those will bring white, or white and red flowers : in others they will bee reddish, and more at the ioynts, and those giue red flowers ; and in some of a darker greene colour, which giue yellow flowers ; the stalkes and ioynts of those that will giue red and yellow flowers spotted, are somewhat brownish, but not so red as those that giue wholly red flowers : vpon these stalkes that spread into many branches, doe grow at the ioynts vpon seuerall footestalkes, faire greene leaues, broad at the stalke, and pointed at the end : at the ioynts likewise toward the vpper part of the branches, at the foote of the leaues, come forth seuerall flowers vpon short footestalkes, euery one being small, long and hollow from the bottome to the brimme, which is broade spread open, and round, and consist but of one leafe without diuision, like vnto a Bell flower, but not cornered at all : which flowers, as I said, are of diuers colours, and diuersly marked and spotted, some being wholly white, without any spot in them for the most part, through all the flowers of the plant ; so likewise some being yellow, and some wholly red ; some plants againe being mixed and spotted, so variably either white and red, or purple, (except here and there some may chance to be wholly white, or red or purple among the rest) or red and yellow through the whole plant, (except as before some may chance in this kinde to be eyther wholly red, or wholly yellow) that you shall hardly finde two or three flowers in a hundred, that will bee alike spotted and marked, without some diuersitie, and so likewise euery day, as long as they blow, which is vntill the winters, or rather autumnes cold blastes do stay their willing pronenesse to flower : And I haue often also obserued, that one side of a plant will giue fairer varieties then another, which is most commonly the Easterne, as the more temperate and shadowie side. All these flowers doe open for the most part, in the euening, or in the night time, and so stand blowne open, vntill the next mornings sun beginne to grow warme vpon them, which then close themselues together, all the brims of the flowers shrinking into the middle of the long necke, much like vnto the blew Bindeweede, which in a manner doth so close vp at the sunnes warme heate : or else if the day be temperate and milde, without any sunne shining vpon them, the flowers will not close vp for the most part of that day, or vntill toward night : after the flowers are past, come seuerall seedes, that is, but one at a place as the flowers stood before, of the bignesse (sometimes) of small pease, but not so round, standing within the greene huskes, wherein the flowers stood before, being a little flat at the toppe, like a crowne or head, and round where it is fastened in the cup, of a blacke colour when it is ripe, but else greene all the while it groweth on the stalke, and being ripe is soone shaken downe with the wind, or any other light shaking : the roote is long and round, greater at the head, and smaller downwards to the end, like vnto a Reddish, spreading into two or three, or more branches, blackish on the outside and whitish within. These rootes I haue often preserued by art a winter, two or three (for they will perish if they be left out in the garden, vnlesse it be vnder a house side) because many times, the yeare not falling out kindely, the plants giue not ripe seede, and so we should be to seeke both of seede to sow, and of rootes to set, if this or the like art to keep them, were not vsed ; which is in this manner : Within a while after the first frosts haue taken the plants, that the leaues wither and fall, digge vp the rootes whole, and lay them in a dry place for three or foure dayes, that the superfluous moysture on the outside, may be spent and dryed, which done, wrap them vp seuerally in two or three browne papers, and lay them by in a boxe, chest or tub, in some conuenient place of the house all the winter time, where no winde or moist ayre may come vnto them; and thus you shall haue these rootes to spring a fresh the next yeare, if you plant them in the beginning of March, as I haue sufficiently tryed. But some haue tryed to put them vp into a barrell or firkin of sand, or ashes, which is also good if the sand and ashes be thorough dry, but if it bee any thing moist, or if they giue againe in the winter, as it is vsuall, they haue found the moisture of the rootes, or of the sand, or both, to putrefie the rootes, that they haue beene nothing worth, when they haue taken them forth. Take this note also for the sowing of your seede, that if you would haue variable flowers, and not all of one colour, you must choose out such flowers as be variable while they grow, that you may haue the seede of them : for if the flowers bee of one entire co-lour, you shall haue for the most part from those seedes, plants that will bring flowers all of that colour, whether it be white, red or yellow.

The

The Place.

These plants grow naturally in the West Indies, where there is a perpetuall summer, or at the least no cold frosty winters, from whence the seede hath been sent into these parts of Europe, and are dispersed into euery garden almost of note.

The Time.

These plants flower from the end of Iuly sometimes, or August, vntill the frosts, and cold ayres of the euenings in October, pull them down, and in the meane time the seed is ripe.

The Names.

Wee haue not receiued the seedes of this plant vnder any other name, then *Mirabilia Peruuiana*, or *Admirabilis planta*. In English wee call them, The meruaile of Peru, or the meruaile of the world : yet some Authors haue called it *Gelseminum*, or *Iasminum rubrum*, *& Indicum* : and Bauhinus *Solanum Mexiocanum flore magno*.

The Vertues.

We haue not knowne any vse hereof in Physicke.

CHAP. XCI.

Malua. Mallowes.

OF the kindred of Mallowes there are a great number, some of the gardens, others wilde, some with single flowers, others with double, some with whole leaues, others with cut or diuided : to entreate of them all is not my purpose, nor the scope of this worke, but onely of such whose flowers, hauing beautie and respect, are fit to furnish this garden, as ornaments thereunto. And first of those single kindes, whose flowers come neerest vnto the fashion of the former Bell-flowers, and after to the double ones, which for their brauery, are entertained euery where into euery Countrey womans garden.

1. *Malua Hispanica flore carneo amplo.*
The Spanish blush Mallow.

The Spanish Mallow is in forme and manner of growing, very like vnto our common fielde Mallow, hauing vpright stalkes two or three foote high, spread into diuers branches, and from the bottome to the toppe, beset with round leaues, like vnto our Mallowes, but somewhat smaller, rounder, and lesse diuided, yet larger below then aboue : the flowers are plentifully growing vpon the small branches, folding or writhing their leaues one about another before they bee blowne, and being open consist of fiue leaues, with a long forked clapper therein, of the same colour with the flower : the chiefest difference from the common consisteth in this, that the leaues of these flowers are longer, and more wide open at the brimmes (almost like a Bell-flower) and of a faire blush or light carnation colour, closing at night, and opening all the day : after the flowers are past, there come such like round heads, with small blacke seede, like vnto the common kinde, but somewhat smaller : the roote is small and long, and perisheth euery yeare.

2. *Alcea vulgaris flore carneo.* Vervaine Mallow with blush flowers.

There is a Mallow that hath long stalkes, and flowers like vnto the common wilde

Mal-

Mallow, and of the same deepe colour with it, so that you can hardly know it from the ordinary kinde, which is found growing wilde together with it, but onely by the leafe, which is as round and as large as the former, but cut into many fine diuifions, euen to the ftalke that vpholdeth it, that it feemeth to confift onely of ragges, or peeces of leaues : Of this kinde I take a plante for this garden, growing in all refpects like vnto it, but differing onely in the colour of the flowers, which are of the fame blufh or light carnation colour, or not much differing from the former Spanifh kinde, with fome veines therein of a deeper colour: the root hereof liueth, as the root of the common wilde kinde doth.

3. *Alcea peregrina fiue veficaria.*
Venice Mallow, or Good night at noone.

The Venice Mallow hath long and weake ftalkes, moft vfually lying or leaning vpon the ground, hauing here and there vpon them long leaues and fomewhat broad, cut in or gafhed very deepely on both edges, that it feemeth as if they were diuers leaues fet together, euery one ftanding on a long footeftalke : at the ioynts of thefe ftalkes, where the leaues are fet, come forth feuerall flowers, ftanding vpon long footeftalkes, which are fomewhat larger then any of the former flowers, confifting of fiue leaues, fmall at the bottome, and wide at the brimmes, of a whitifh colour tending to a blufh, and fometimes all white, with fpots at the bottomes of the leaues on the infide, of a very deepe purple or murrey colour, which addeth a great grace to the flower, and hauing alfo a long peftle or clapper in the middle, as yellow as gold : thefe flowers are fo quickly faded and gone, that you fhall hardly fee any of them blowne open, vnleffe it bee betimes in the morning before the Sunne doe grow warme vpon them, for as foone as it feeleth the Sunnes warme heate, it clofeth vp and neuer openeth a-gaine, fo that you fhall very feldome fee a flower blowne open in the day time, after nine a clocke in the morning : after thefe flowers are paft, there rife vp in their places thinne, round, fhining or tranfparent bladders, pointed at the toppe, and ribbed down all along, wherein are contained fmall, round, blackifh feede : the roote is long and fmall, and perifheth euery yeare.

4. *Alcea fruticofa pentaphyllea.* Cinquefoile Mallow.

The ftalkes of this Mallow are very long, hard or wooddy, more then of any of the other Mallowes : at the lower part whereof, and vp to the middle, ftand diuers leaues vpon long footeftalkes, parted or diuided into fiue parts or leaues, and dented about the edges ; but vpwards from the middle to the toppe, the leaues haue but three diuifions : among thefe leaues ftand large wide open flowers, of the colour of the common Mallow : the feede is fmaller then in any other Mallow, but the rootes are great and long, fpreading in the ground like vnto the roots of Marfh Mallowes, fpring-ing vp afrefh euery yeare from the roote.

5. *Sabdarifa feu Alcea Americana.* Thorney Mallowe.

This Thorney Mallowe hath greene leaues next vnto the ground, that are almoft round, but pointed at the end, and dented very much about the edges; the other leaues that growe vpon the ftalke are diuided into three parts, like vnto a trefoile, and fome of them into fiue diuifions, all of them dented about the edges : the ftalke is reddifh, with fome harmeleffe prickles in fundry places thereon, and rifeth vp three or foure foote high in a good ground, a fit place, and a kindly yeare, bearing plenty of flowers vpon the ftalkes, one at the foote of euery leafe, the toppe it felfe ending in a long fpike, as it were of buddes and leaues together : the flowers are of a very pale yellow, tending to a white colour, fpotted in the bottome of each of the fiue leaues, with a deepe purple fpot, broad at the lower part, and ending in a point about the middle of the leafe, which are quickly fading, and not abiding aboue one day, with a long peftle in the middle diuided at the toppe : after the flower is paft, commeth vp a fhort prickly podde, fet within a fmall greene huske or cup that bore the flower. wherein is contai-

ned

ned whitifh, or rather brownifh yellow feede, flat and fomewhat round, like vnto the feedes of Hollyhocke: the roote is ftringie, and quickly perifheth; for it will hardly endure in our cold Country to giue flowers, much leffe feede, vnleffe (as I faid before) it happen in a kindly yeare, and be well planted and tended.

6. *Bamia feu Alcea Ægyptia.* The Mallow of Egypt.

This Mallow is alfo as tender to nourfe vp as the laft, hauing the lower leaues broad like a Marfh Mallow, and of a frefh greene colour; but thofe that growe vpon the ftalke, and vp to the toppe, are diuided into fiue parts or points, but are not cut in to the middle ribbe, like the former Thorney Mallow, yet dented about the edges like vnto them: the flowers growe at the fetting to of the leaues, like vnto a Mallow for forme, but of a whitifh colour; after which come long fiue fquare pointed pods, with hard fhels, wherein are contained round blackifh gray feede, as bigge as a Vetch or bigger: the roote perifheth quickly with vs, euen with the firft frofts.

7. *Althæa frutex flore albo vel purpureo.*
Shrubbe Mallow with a white or purple flower.

There are diuers forts of fhrubbe Mallowes, whereof fome that haue their ftemmes or ftalkes leffe wooddy, dye downe to the ground euery yeare, and others that abide alwayes, are more wooddy: Of the former forts I intend not to fpeake, referring them to a fitter place; and of the other, I will onely giue you the knowledge of one or two in this place, although I doe acknowledge their fitteft place had been to be among the fhrubbes; but becaufe they are Mallowes, I pray let them paffe with the reft of their kindred, and their defcriptions in this manner: Thefe wooddy kindes of fhrub Mallowes haue fomewhat large, long, and diuided leaues, of a whitifh greene colour, foft alfo, and as it were woolly in handling, fet difperfedly on the whitifh hard or wooddy ftalkes: their flowers are large, like vnto a fingle Rofe or Hollyhocke, in the one being white with purple fpots in the bottome; in the other either of a deepe red colour, or elfe of a paler purple, with a deeper bottome, and with veines running in euery leafe: they are fomewhat tender, and would not be fuffered to be vncouered in the Winter time, or yet abroad in the Garden, but kept in a large pot or tubbe, in the houfe or in a warme cellar, if you would haue them to thriue.

8. *Malua hortenfis rofea fimplex & multiplex diuerforum colorum.*
Hollihockes fingle and double of feuerall colours.

I fhall not neede to make many defcriptions of Hollihockes, in regard the greateft difference confifteth in the flowers, which are in fome fingle, in fome double, in fome of one colour, and in others of other colours: for the loweft leaues of Hollihockes are all round, and fomewhat large, with many corners, but not cut in or diuided, foft in handling; but thofe that growe vp higher are much more diuided into many corners: the ftalkes fometimes growe like a tree, at the leaft higher then any man, with diuers fuch diuided leaues on them, and flowers from the middle to the toppe, where they ftand as it were a long fpike of leaues and buds for flowers together: the flowers are of diuers colours, both fingle and double, as pure white, and pale blufh, almoft like a white, and more blufh, frefh and liuely, of a Rofe colour, Scarlet, and a deeper red like a crimfon, and of a darke red like blacke bloud; thefe are the moft efpeciall colours both of fingle and double flowers that I haue feene: the fingle flowers confift of fiue broad and round leaues, ftanding round like vnto fingle Rofes, with a middle long ftile, and fome chiues aboue them: the double flowers are like vnto double Rofes, very thicke, fo that no ftile or vmbone is feene in the middle, and the outermoft rowe of leaues in the flowers are largeft, the innermoft being fmaller and thicke fet together: after the flowers are paft, there come vp as well in the double as fingle, flat round heads, like flat cakes, round about the bottomes whereof growe flat whitifh feede: the roote is long and great at the head, white and tough, like the roote of the common Mallowes, but greater, and will reafonably well abide the Winter.

The

The Place.

The firſt groweth wilde in Spaine. The ſecond in our owne Countrey. The third is thought to growe in Italy and Venice ; but Lobel denieth it, ſaying, that it is there onely in Gardens, and is more plentifull in theſe parts then with them. The fourth Cluſius ſaith he found in many places of Germany. The fifth is ſuppoſed to be firſt brought out of the Weſt Indies, but an Arabicke name being giuen it, maketh me ſomewhat doubtfull how to beleeue it. The ſixth groweth in Egypt, where it is of great vſe, as Proſper Alpinus hath ſet downe in his Booke of Egyptian plants. The ſeuenth groweth in ſome parts both of Spaine and France. The laſt is not found but in Gardens euery where.

The Time.

The firſt, ſecond, third, fourth, and laſt, doe flower from Iune vntill the end of Iuly and Auguſt. The reſt flower very late, many times not vntill September or October.

The Names.

The firſt and ſecond haue their names ſufficiently expreſſed in their titles. The third is diuerſly called, as *Malua horaria, Alcea veſicaria, Alcea Veneta, Alcea Peregrina,* and of Matthiolus, *Hypecoum.* The moſt vſuall Engliſh name is Venice Mallow. The fourth is called *Alcea fruticoſa pentaphyllea,* and *Cannabinifolio,* or *Pentaphyllifolio :* In Engliſh, Cinquefoile Mallow. The fifth hath been ſent vnder the name of *Sabdariſa,* and *Sabdariſſa,* and (as I ſaid) is thought to be brought from America, and therefore it beareth the name of that Country. The ſixth is called in Egypt, *Bamia,* or *Bammia,* and by that name ſent with the addition *del Cayro* vnto it : In Engliſh, Egyptian Mallow, or Mallow of Egypt. The ſeuenth is called *Althæa frutex,* and of ſome *Althæa arborea :* In Engliſh, Shrubbe Mallow, becauſe his ſtemme is wooddie, and abideth as ſhrubbes and trees doe. The eight and laſt is called *Malua hortenſis, Malua Roſea,* and of ſome *Roſa vltra marina :* In Engliſh, of ſome Hockes, and vſually Hollihockes.

The Vertues.

All ſorts of Mallowes, by reaſon of their viſcous or ſlimie quality, doe helpe to make the body ſoluble, being vſed inwardly, and thereby helpe alſo to eaſe the paines of the ſtone and grauell, cauſing them to be the more eaſily voided : being outwardly applyed, they mollifie hard tumors, and helpe to eaſe paines in diuers parts of the body ; yet thoſe that are of moſt vſe, are moſt common. The reſt are but taken vpon credit.

Chap. XCII.

Amaranthus. Flower-gentle.

WE haue foure or fiue ſorts of Flower-gentle to trimme vp this our Garden withall, which doe differ very notably one from another, as ſhall be declared in their ſeuerall deſcriptions ; ſome of which are very tender, and muſt be carefully regarded, and all little enough to cauſe them beare ſeede with vs, or elſe wee ſhall bee to ſeeke euery yeare : others are hardy enough, and will hardly be loſt out of the Garden.

1. Amaranthus

1. *Amaranthus purpureus minor.* The small purple Flower-gentle.

This gallant purple Veluet flower, or Flower-gentle, hath a crested stalke two foote high or more, purplish at the bottome, but greene to the toppe, whereout groweth many small branches, the leaues on the stalkes and branches are somewhat broad at the bottome, and sharpe pointed, of a full greene colour, and often somewhat reddish withall, like in forme vnto the leaues of Blites (whereof this and the rest are accounted *species*, or sorts) or small Beetes : the flowers are long, spikie, soft, and gentle tufts of haires, many as it were growing together, broad at the bottome, and small vp at the toppe, pyramis or steeple-fashion, of so excellent a shining deepe purple colour, tending to a murrey, that in the most excellent coloured Veluet, cannot be seene a more orient colour, (and I thinke from this respect, the French call it *Passe velours*, that is to say, passing Veluet in colour) without any smell at all, which being bruised giueth the same excellent purple colour on paper, and being gathered in his full strength and beauty, will abide a great time (if it be kept out of the winde and sunne in a dry place) in the same grace and colour : among these tufts lye the seede scattered, which is small, very blacke, and shining : the rootes are a few threddy strings, which quickly perish, as the whole plant doth, at the first approach of Winter weather.

2. *Amaranthus Coccineus.* Scarlet Flower-gentle.

The leaues of this Flower-gentle are longer, and somewhat narrower then the former ; the stalke groweth somewhat higher, bearing his long tufts at seuerall leaues, as also at the toppe of the stalkes, many being set together, but separate one from another, and each bowing or bending downe his head, like vnto a Feather, such as is worn in our Gallants and Gentlewomens heads, of an excellent bloudy Scarlet colour : the seede is blacke, like vnto the former : the roote perisheth quicklier, because it is more tender.

3. *Amaranthus tricolor.* Spotted or variable Flower-gentle.

The chiefest beauty of this plant consisteth in the leaues, and not in the flowers ; for they are small tufts growing all along the stalke, which is nothing so high as the former, especially with vs, and at the ioynts with the leaues : the leaues hereof are of the same fashion that the former are, and pointed also ; but euery leafe is to be seene parted into greene, red, and yellow, very orient and fresh (especially if it come to his full perfection, which is in hot and dry weather) diuided not all alike, but in some leaues, where the red or yellow is, there will be greene, and so varying, that it is very pleasant to behold : the seede hereof is blacke and shining, not to bee knowne from the former.

4. *Amaranthus Carnea spica.* Carnation Flower-gentle.

There is another more rare then all the rest, whose leaues are somewhat longer, and narrower then the first, and like vnto the second kinde : the spikes are short, many set together, like branches full of heads or eares of corne, euery one whereof hath some long haires sticking out from them, of a deep blush, tending to a carnation colour.

5. *Amaranthus purpureus maior panniculis sparsis.*
Great Floramour, or purple Flower-gentle.

The great Floramour hath one thicke, tall, crested, browne red stalke, fiue or six foote high, from whence spring many great broad leaues, like vnto the former for the forme, but much larger & redder for the most part, especially the lowest, which brancheth forth into diuers parts, & from between these leaues, & the stalks or branches, as also at the tops of them, stand long, spikie, round, & somewhat flat tufts, of a more reddish purple colour then the first, and diuided also into seuerall parts, wherin when they

are

are full ripe, are to be feen an innumerable company of white feed, ftanding out among the fhort thrums, and do then eafily fall away with a little touching, euery one of thefe white feed hath as it were an hole halfe bored through therin: the root is a great bufh of ftrings, fpreading in the ground, whereby it is ftrongly faftened, yet perifheth euery yeare, after it hath giuen his feede.

The Place.

All thefe plants growe in the Eafterne Countries, as Perfia, Syria, Arabia, &c. except the greateft, which hath been brought out of the Weft Indies, where it is much vfed, efpecially the feeds : they are all, except it, nourfed vp with much care in our Gardens, and yet in a backward or cold yeare they will not thriue, for that they defire much heate : but the greateft doth alwayes giue ripe feede euery yeare.

The Time.

They beare their gallant tufts or fpikes for the moft part in Auguft, and fome not vntill September.

The Names.

The name *Amaranthus* is giuen to all thefe plants, taken from the Greeke word ἀμαράντινος, *non marcefcens*, or *non fenefcens*, that is, neuer waxing old, and is often also impofed on other plants, who haue the fame property, that is, that their flowers being gathered in a fit feafon, will retaine their natiue colour a long time, as fhall be fhewed in the Chapter following. Diuers do thinke the firft to be *Phlox*, or *Flamma* of Theophraftus. The third is called *Gelofia*, or *Celofia* of Tragus. Spigelius in his *Ifagoges* faith, it is generally taken to be *Sophonia*, whereof Plinie maketh mention; and Lobel, to bee the Perfians *Theombroton* of Plinie. The Italians, from whom I had it (by the meanes of Mr. Doctor Iohn More, as I haue had many other rare fimples) call it, *Blito di tre colori*, A three coloured Blite. The fifth, which is the greateft, hath been fent from the Weft Indies by the name of *Quinùa*, as Clufius reporteth. The name Flower-gentle in Englifh, and *Floramour*, which is the French, of *Flos amoris*, and *Paffe velours*, as is before faid, or Veluet flower, according to the Italian, *Fior veluto*, are equally giuen to all thefe plants, with their feuerall diftinctions, as they are expreffed in their titles.

The Vertues.

Diuers fuppofe the flowers of thefe plants doe helpe to ftay the fluxe of bloud in man or woman, becaufe that other things that are red or purple doe performe the fame. But Galen difproueth that opinion very notably, *in lib. 2. & 4. de fimpl. medicament. facultatibus.*

CHAP. XCIII.

Helichryfum, fiue Amaranthus luteus.
Golden Flower-gentle, Goldilockes, or Gold-flower.

THe propinquity of property (as I before faid) hath caufed the affinity in name, and fo in neighbourhood in thefe plants, wherein there are fome diuerfity; and although they differ from them before in many notable points, yet they all agree with themfelues in the golden, or filuer heads or tufts they beare; and therefore I

haue

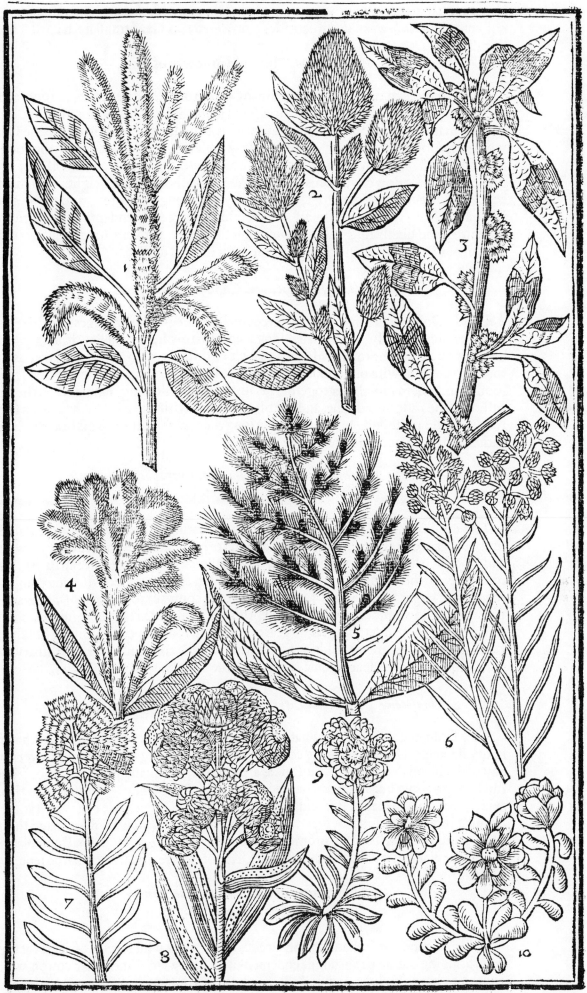

haue comprifed them in one Chapter, and will begin with that which commeth neareft vnto the *Helichryfum* of Diofcorides, or *Aurelia* (as Gaza tranflateth it) of Theophraftus.

1. *Heliochryfum.* The Golden flower of life.

This firft Golden tuft rifeth vp with many hard, round, white ftalkes, a foote and a halfe high, whereon at certaine diftances ftand many fine cut leaues, or rather one leafe cut into many fmall fine parts, almoft as fmall as Fenell, but grayifh, like vnto the Cudweedes or Cotton-weedes (whereof certainly thefe are fpeciall kindes) at the toppes of the ftalkes ftand many round flowers, of a pale gold colour, in an vmbell clofe together, yet euery flower vpon his owne ftalke, and all of an euen height, which will keepe the colour, being gathered, and kept dry for a long time after, and are of a hot and quicke fent : the roote is fmall and wooddy, fpreading vnder the vpper cruft of the earth, and liueth long in his owne naturall place, but very hardly endureth the cold of our Winters, vnleffe they be milde, or it be well defended.

2. *Helichryfum Creticum.* Candy Goldilockes.

Candy Goldilockes hath two or three fmall flender white branches, fet here and there very fcatteringly, with fmall, long, and narrow hoary leaues, hauing yellow heads of flowers at the tops made into vmbels or tufts, not fo round and euen as the former, but longwife one aboue another, the heads being made as it were of fcales, loofly, and not fo clofely fet together, as in the next following, which when they are full ripe, doe paffe into doune, and are blowne away with the winde, hauing a fmall reddifh feede at the end ; but will abide a long time, as the other in his beauty, being gathered in time, as the reft will doe.

3. *Helichryfum Orientale fiue Amaranthus luteus.* Golden Flower-gentle.

This moft beautifull plant is very like vnto the former Candy Goldilockes laft defcribed, but growing vp higher, with many more branches, and more hoary, white, and woolly, hauing alfo long and narrow white leaues, but fomewhat broader, and thicker fet on the branches : the tufts of flowers or vmbels likewife doe confift of longer and larger heads, more fcaly, and clofer compact together, of an excellent pale gold yellow colour, and fhining, with fome yellow threads or thrummes in the middle : the roote dyeth not euery yeare, but liueth long, efpecially in the South and Eaft Countries, where no colds or frofts are felt ; but will require extraordinary care and keeping, and yet fcarce fufficient to preferue it in thefe cold Countries.

4. *Chryfocome fiue Stæchas Citrina.* Golden tufts or Golden Caffidony.

This Golden flower is fomewhat like the former of thefe two laft defcribed, hauing hoary ftalkes and leaues, ftanding confufedly on them, being long, and narrower then any of the former : the tops of the ftalkes are diuided into many parts, each bearing a fmall long yellow head or flower at the toppe, with fome yellow thrummes in them, which heads being many, are diffufedly fet together, like a loofe or fparfed vmbell, keeping their colour long before they wither, and when they are ripe, haue thinne fmall reddifh feede, like Mariorome feede, but fmaller ; the roote is fmall and blacke : the whole plant, as well leaues and flowers, as rootes, are of a ftrong fharpe fent, yet pleafant.

5. *Argyrocome fiue Gnaphalium Americanum.* Liue long or Life euerlafting.

This filuer tuft or Indian Cotton weede, hath many white heads of leafes at their firft fpringing out of the ground, couered with a hoary woollineffe like cotton, which rifing into hard, thicke round ftalkes, containe ftill the fame hoarineffe vpon them, as alfo vpon the long and narrow leaues which are fet thereon, efpecially on the vnder fide,

fide, for the vpper fides are of a darke fhining greene colour : the ftalkes are diuided at the toppe into many fmall branches, each whereof haue many fcaly tufted heads fet together, couered ouer with cotton before their opening, and then diffeuering one from another, abiding very white on the outfide, when they are fully growne, but with a fmall yellow thrume in the midde of euery flower, which in time turne into yellow doune, apt to be blowne away with euery winde : the roots are long and black on the outfide, creeping vnder ground very much.

6. *Gnaphalium montanum flore albo & flore purpureo.* White and purple Cats foote.

This fmall Cudweede or Cottonweede, hath many fmall white woolly leaues growing from the roote, which is compofed of a few fmall blackifh threds, and lying vpon the ground fomewhat like vnto the leaues of a fmall Moufe-eare, but fmaller ; from among which rifeth vp a fmall ftalke of halfe a foote high or thereabouts, befet here and there with fome few leaues, at the top whereof commeth forth a tuft of fmall flowers, fet clofe together, in fome of a pure white, in others of a purple or reddifh colour, in fome of a pale red or blufh, and in others of a white and purple mixt together, which for the beauty is much commended and defired, but will hardly abide to be kept in Gardens, fo vnwilling they are to leaue their naturall abiding.

7. *Gnaphalium Rofeum.* The Cotton Rofe.

This little rofe Cotton weede hath many fuch like woolly leaues, growing as the former from the roote vpon fmall fhort branches, not full an hand breadth high, in fafhion fomewhat like vnto Dayfie leaues, but leffer, and round pointed : at the toppe of euery ftalke or branch, ftandeth one flower, compofed of two rowes of fmall white leaues, layd open like a Starre or a Rofe, as it beareth the name, hauing a round head in the middle made of many yellow threds or thrumes, which falling away, there rifeth vp a fmall round head, full of fmall feedes : the root is fmall, long and threddy.

The Place.

The foure firft plants doe grow naturally in many of the hot Countries of Europe, as Spaine, Italie, and Prouince in France ; as alfo in **Candy**, **Barbary**, and other places, and muft be carefully kept with vs in the winter time. The Liue long was brought out of the Weft Indies, and groweth plentifully in our gardens. The two laft doe grow as well in the colder Countries of Germany, as in France and other places.

The Time.

They all flower in the end of September, if they will fhew out their beauty at all with vs, for fometimes it is fo late, that they haue no faire colour at all, efpecially the foure firft forts.

The Names.

Variable and many are the names that feuerall Writers do call thefe foure firft forts of plants, as *Helichryfum, Heliochryfum,* or *Elichrylum. Eliochryfum, Chryfocome, Coma aurea, Amaranthus luteus, Stoechas Citrina,* and *Aurelia,* with others, needleffe here to be recited : it is fufficient for this worke, to giue you knowledge that their names are fufficient as they are expreffed in their titles : The fift is called *Gnaphalium* by Carolus Clufius, from the likeneffe of the vmbels or tufts of heads, though greater and white : for as I faid before, the Cotton weedes are of kindred with the golden tufts : It hath been called by our Englifh Gentlewomen, Liue long, and Life euerlafting, becaufe of the durabilitie of the flowers in their beautie. The two laft are called

led *Gnaphalium*, according to their titles ; and in Englifh they may paffe vnder thofe names are fet downe with them.

The Vertues.

The foure firft are accounted to bee hot and dry, and the three laft to bee cold and dry : yet all of them may to fome good purpofe bee applyed to rheumaticke heads. The former foure are likewife vfed to caufe vrine, and in baths to comfort and heate cold parts. They are alfo layd in chefts and wardrobes, to keepe garments from moths ; and are worne in the heads and armes of Gentiles and others, for their beautifull afpect.

Chap. XCIIII.

Canna Indica. The Indian flowring Reede.

THere are two kindes or forts of this beautifull plant, the one with a red flower, the other with a yellow, fpotted with reddifh fpots, both which in fome kindly yeares haue borne their braue flowers, but neuer any ripe feede, and doth not abide the extremities of our winters, eyther abroade or vnder couert, vnleffe it meete with a ftoue or hot-houfe, fuch as are vfed in Germany, or fuch other like place: For neyther houfe nor cellar will preferue it, for want of heate.

Canna Indica flore rubro. Red flowred Indian Reede.

This beautifull plant rifeth vp with faire greene, large, broade leaues, euery one rifing out of the middle of the other, and are folded together, or writhed like vnto a paper Coffin (as they call it) fuch as Comfitmakers and Grocers vfe, to put in their Comfits and Spices, and being fpread open, another rifeth from the bottome thereof, folded in the fame manner, which are fet at the ioynts of the ftalke when it is rifen vp, like vnto our water Reede, and growing (if it runne vp for flower) to be three or foure foote high, as I haue obferued in mine owne garden : the flowers grow at the toppe of the ftalke one aboue another, which before their opening are long, fmall, round, and pointed at the end, very like vnto the claw of a Crauife or Sea-Crab, and of the fame red or crimfon colour, but being open, are very like vnto the flower of *Gladiolus* or Corne-flagge, but of a more orient colour then at the firft, and ftanding in a rough huske, wherein afterwards ftandeth a three fquare head, containing therein round blacke feede, of the bigneffe of a peafe : the roote is white and tuberous, growing into many knobs, from whence arife fuch other leaues and ftalkes, whereby it encreafeth very much, if it be righrly kept and defended.

Canna Indica flore flauo punctato.
Yellow fpotted Indian Reede.

This Reede groweth vp with leaues and flowers, in all points fo like vnto the former, that it cannot bee knowne from it, vntill it come to flower, which is of a yellow colour, fpotted with reddifh fpots, without any other difference.

The Place.

Thefe plants grow naturally in the Weft Indies, from whence they were firft fent into Spaine, and Portugall, where Clufius faith he faw them planted by the houfes fides, flowring in winter, which might be in thofe warme Countreyes. We preferue them with great care in our gardens, for the beautifull afpect of their flowers.

The

The Time.

They flower not with vs vntill the end, or middle of Auguſt, at the ſooneſt.

The Names.

They are called of ſome *Canna Indica*, and *Arundo Indica*, of others *Cannacorus*, and of ſome *Flos Cancri*, becauſe the colour of the flowers, as well as the forme of the buds, are ſo like vnto a Sea-Crabs cle, or claw.

The Vertues.

There is not any vſe of theſe in Phyſicke that I know.

Chap. XCV.

Mandragoras. Mandrake.

THe Mandrake is diſtinguiſhed into two kindes, the male and the female ; the male hath two ſorts, the one differing from the other, as ſhall be ſhewed; but of the female I know but one : The male is frequent in many gardens, but the female, in that it is more tender and rare, is nourſed vp but in a few.

Manadrgoras mas. The male Mandrake.

The male Mandrake thruſteth vp many leaues together out of the ground, which being full growne, are faire, large and greene, lying round about the roote, and are larger and longer then the greateſt leaues of any Lettice, whereunto it is likened by Dioſcorides and others : from the middle, among theſe leaues, riſe vp many flowers, euery one vpon a long ſlender ſtalke, ſtanding in a whitiſh greene huske, conſiſting of fiue pretty large round pointed leaues, of a greeniſh white colour, which turne into ſmall round apples, greene at the firſt, and of a pale red colour when they are ripe, very ſmooth and ſhining on the outſide, and of a heady or ſtrong ſtuffing ſmell, wherein is contained round whitiſh flat ſeede : the roote is long and thicke, blackiſh on the outſide, and white within, conſiſting many times but of one long roote, and ſometimes diuided into two branches a little below the head, and ſometimes into three or more, as nature liſteth to beſtow vpon it, as my ſelfe haue often ſeene, by the tranſplanting of many, as alſo by breaking and cutting off of many parts of the rootes, but neuer found harme by ſo doing, as many idle tales haue beene ſet downe in writing, and deliuered alſo by report, of much danger to happen to ſuch, as ſhould digge them vp or breake them ; neyther haue I euer ſeene any forme of man like or woman-like parts, in the rootes of any : but as I ſaid, it hath oftentimes two maine rootes running down-right into the ground, and ſometimes three, and ſometimes but one, as it likewiſe often happeneth to Parſneps, Carrots, or the like. But many cunning counterfeit rootes haue bin ſhaped to ſuch formes, and publickly expoſed to the view of all that would ſee them, and haue been tolerated by the chiefe Magiſtrates of the Citie, notwithſtanding that they haue beene informed that ſuch practices were meere deceit, and vnſufferable; whether this happened through their ouer-credulitie of the thing, or of the perſons, or through an opinion that the information of the truth roſe vpon enuy, I know not, I leaue that to the ſearcher of all hearts : But this you may bee bold to reſt vpon, and aſſure your ſelues, that ſuch formes as haue bin publickly expoſed to be ſeene, were neuer ſo formed by nature, but only by the art and cunning of knaues and deceiuers. and let this be your *Galeatum* againſt all ſuch vaine, idle and ridiculous toyes of mens inuentions.

There

There is likewise another sort of these male Mandrakes, which I first saw at Canterbury, with my very louing and kinde friende Iohn Tradescante, in the garden of the Lord Wotton, whose gardiner he was at that time ; the leaues whereof were of a more grayish greene colour, and somewhat folded together, when as the former kind that grew hard by it, was of the same forme that is before described, and ordinary in all others : but whether the apples were differing from the other, I know not, nor did they remember that euer it had borne any.

Mandragoras fæmina. The female Mandrake.

The female Mandrake doth likewise put vp many leaues together, from the head of the roote, but they are nothing so large, and are of a darker greene colour, narrower also and shining, more crumpled, and of a stronger sent : the flowers are many, rising vp in the middle of the leaues, vpon slender stalkes, as in the male kind, but of a blewish purple colour, which turne into small round fruite or apples, and not long like a peare (as Clusius reporteth that saw them naturally growing in Spaine) greene at the first, and of a pale yellowish colour, when they are full ripe; of a more pleasing, or if you will, of a lesse heady sent then the apples of the male, wherein is contained such like seede, but smaller and blacker : the rootes are like the former, blacke without and white within, and diuided in the same manner as the male is, sometimes with more, and sometimes with fewer parts or branches.

The Place.

They grow in many places of Italie, as Matthiolus reporteth, but especially on Mount Garganus in Apulia. Clusius saith hee found the female in many wet grounds of Spaine, as also in the borders of those medowes that lye neere vnto riuers and water courses. The male is cherished in many Gardens, for pleasure as well as for vse : but the female as is said, is both very rare, and farre more tender.

The Time.

The Male flowreth in March, and the fruit is ripe in Iuly. The Female, if it be well preserued, flowreth not vntill August, or September; so that without extraordinary care, we neuer see the fruite thereof in our gardens.

The Names.

Mandragoras mas is called *albus,* as the *Fæmina* is called *niger,* which titles of blacke and white, are referred vnto the colour of the leaues : the female is called also *Thridacias,* from the likenesse of Lettice, whereunto they say in forme it doth carry some similitude. Dioscorides saith, that in his time the male was called *Morion,* and both of them *Antimelum,* and *Circæa.* Wee call them in English, The male, and the female Mandrake.

The Vertues.

The leaues haue a cooling and drying qualitie, fit for the oyntment *Populeon,* wherein it is put. But the Apples haue a soporiferous propertie, as Leuinus Lemnius maketh mention in his Herball to the Bible, of an experiment of his owne. Besides, as Dioscorides first, and then Serapio, Auicen, Paulus Ægineta, and others also do declare, they conduce much to the cooling and cleansing of an hot *matrix.* And it is probable, that Rachel knowing that they might be profitable for her hot and dry body, was the more earnest with Leah for her Sonne Rubens Apples, as it is set downe *Genesis* 30.*verse* 14. The strong sent of these apples is remembred also, *Cant.* 7.13. although some would diuert the signification of the Hebrew word, דודאים,

vnto

vnto Violets, or some other sweet flowers, in the former place of *Genesis*, and the fruit of *Musa*, or Adams Apples in this place of the *Canticles*. Hamilcar the Carthaginian Captaine is said to haue infected the wine of the Lybians (his enemies against whom he fought) with the apples of Mandrake, whereby they being made exceeding drowsie, he obtained a famous victory ouer them.

Chap. XCVI.

Pomum Amoris. Loue Apples.

ALthough the beautie of this plant consisteth not in the flower, but fruit, yet giue me leaue to insert it here, lest otherwise it haue no place : whereof there are two especiall sorts, which wee comprehend in one Chapter, and distinguish them by *maius* and *minus*, greater and smaller : yet of the greater kinde, we haue nourfed vp in our Gardens two sorts, that differ only in the colour of the fruite, and in nothing else.

Pomum Amoris maius fructu rubro.
Great Apple of Loue the ordinary red sort.

This greater kinde of Loue Apples, which hath beene most frequently cherished with vs, hath diuers long and trayling branches, leaning or spreading vpon the ground, not able to sustaine themselues, whereon doe grow many long winged leaues, that is, many leaues set on both sides, and all along a middle ribbe, some being greater, and others lesse, iagged also and dented about the edges, of a grayish ouer-worne greene colour, somewhat rough or hairy in handling; from among the leaues and the branches come forth long stalkes, with diuers flowers set thereon, vpon seuerall short footstalks, consisting of sixe, and sometimes of eight small long yellow leaues, with a middle pricke or vmbone, which after the flowers are fallen, riseth to be the fruite, which are of the bignesse of a small or meane Pippin, vneuenly bunched out in diuers places, and scarce any full round without bunches, of a faire pale reddish colour, or somewhat deeper, like vnto an Orenge, full of a slimie iuice and watery pulpe, wherein the seede lyeth, which is white, flat and somewhat rough : the roote shooteth with many small strings and bigger branches vnder ground, but perisheth at the first feeling of our winter weather. The fruite hereof by often sowing it in our Land, is become much smaller then I haue here described it : but was at the first, and so for two or three yeares after, as bigge as I haue related it.

Pomum Amoris maius fructu luteo. Yellow Amorous Apples.

Of the same kinde is this other sort of Amorous Apples, differing in nothing but the colour of the fruite, which is of a pale yellow colour, hauing bunches or lobes in the same manner, and seede also like the former.

Pomum Amoris minus, siue Mala Ethiopica parua.
Small Loue Apples.

The small Apples of Loue in the very like manner, haue long weake trayling branches, beset with such like leaues as the greater kinde hath, but smaller in euery part : the flowers also stand many together on a long stalke, and yellow as the former, but much smaller : the fruite are small, round, yellowish red berries, not much bigger then great grapes, wherein are contained white flat seede, like the other, but smaller : the roote perisheth in like manner euery yeare. and therefore must bee new sowen euery spring, if you will haue the pleasure of their sight in the garden; yet some yeares I haue known them rise of their owne sowing in my garden.

The

The Place.

They growe naturally in the hot Countries of Barbary, and Ethiopia; yet some report them to be first brought from Peru, a Prouince of the West Indies. Wee onely haue them for curiosity in our Gardens, and for the amorous aspect or beauty of the fruit.

The Time.

They flower in Iuly and August, and their fruit is ripe in the middle or end of September for the most part.

The Names.

The first is named diuersly by diuers Authors; for Lobel, Camerarius, and others, call them *Poma amoris.* Dodonæus *Aurea Mala.* Gesnerus first, and Bauhinus after him, make it to be a kinde of *Solanum Pomiferum.* Anguillara taketh it to be *Lycopersicum* of Galen. Others thinke it to bee *Glaucium* of Dioscorides. The last is called *Mala Æthiopica parua,* and by that title was first sent vnto vs, as if the former were of the same kinde and country. We call them in English, Apples of Loue, Loue-Apples, Golden Apples, or Amorous Apples, and all as much to one purpose as another, more then for their beautifull aspect.

The Vertues.

In the hot Countries where they naturally growe, they are much eaten of the people, to coole and quench the heate and thirst of their hot stomaches. The Apples also boyled, or infused in oyle in the sunne, is thought to be good to cure the itch, assuredly it will allay the heate thereof.

Chap. XCVII.

Digitalis. Foxegloue.

THere are three principall sorts of Foxegloues, a greater, a middle or meane sort, and a lesser, and of them, three especiall colours, that is, purple, white, and yellow; the common purple kinde that groweth abroad in the fields, I leaue to his wilde habitation: and of the rest as followeth.

1. *Digitalis maxima ferruginea.* Dun coloured Foxegloues.

The leaues of this Foxegloue are long and large, of a grayish green colour, finely cut or dented about the edges, like the teeth of a fine sawe; among which commeth vp a strong tall stalke, which when it was full growne, and with ripe seede thereon, I haue measured to be seuen foot high at the least, wheron grow an innumerable company (as I may so say, in respect of the aboundance) of flowers, nothing so large as the common purple kinde, that groweth wilde euery where in our owne Countrey, and of a kinde of browne or yellowish dunne colour, with a long lippe at euery flower; after them come seede, like the common kinde, but in smaller heads: the rootes are stringie like the ordinary, but doe vsually perish, or seldome abide after it hath giuen seed.

2. *Digitalis maior flore carneo.* Blush coloured Foxegloues.

This kinde of Foxegloues hath reasonable large leaues, yet not altogether so large

as

1 *Canna Indica.* The Indian Reed. 2 *Mandragoras mas.* The male Mandrake. 3 *Pomum amoris maius.* Great Apples of loue. 4 *Digitalis maior flore luteo amplo.* The great yellow Foxegloue. 5 *Digitalis media flore luteo rubente.* Orengé tawny Foxeglouеs. 6 *Digitalis maxima ferruginea.* Dun coloured Foxeglouеs.

as the common field kinde : the flowers are also smaller then the common sort, but of a blush colour.

3. *Digitalis media flore luteo rubente.* Orenge tawnie Foxegloue.

As this Foxegloue is none of the greatest, so also is it none of the smallest; but a sort betweene both, hauing leaues in some proportion correspondent to the lesser yellow Foxegloue, but not so large as the lesser white: the flowers are long and narrow, almost as large as the last white, but nothing so large as the first white, of a faire yellowish browne colour, as if the yellow were ouershadowed with a reddish colour, and is that colour wee vsually call an Orenge tawnie colour : the seede is like the former: the rootes perish euery yeare that they beare seede, which is vsually the second yeare of the springing.

4. *Digitalis maior alba.* The greater white Foxegloue.

This white Foxegloue is in all things so like vnto the purple wilde kinde, that it can hardly be distinguished from it, vnlesse it be in the fresher greennesse and largenesse of the leaues : the flowers are as great in a manner as the purple, but wholly white, without any spot in them : the seed and other things agree in all points.

5. *Digitalis alba altera seu minor.* The lesser white Foxegloue.

We haue in our Gardens another sort of white Foxegloue, whose leaues are like vnto the last described, but not altogether so long or large, and of a darker greene colour : the stalke groweth not so high, as not full three foote : the flowers are pure white, fashioned like vnto the former, but not so great or large, in all other things alike : the rootes hereof did abide sometime in our Gardens, but since perished, and the seede also, since when we neuer could obtaine from any our friends of that kinde againe.

6. *Digitalis maior lutea flore amplo.* The great yellow Foxegloue.

The leaues of this greater yellow Foxegloue, are in forme somewhat like vnto the common purple kinde, but not altogether so large : the stalke groweth to bee three or foure foote high, whereon stand many long hollow pendulous flowers, in shape like the ordinary purple : but somewhat shorter, and more large and open at the brimmes, of a faire yellow colour, wherein are long threads, like as in the others : the roote hereof is greater at the head, and more wooddy then any of the rest, with many smaller fibres, spreading themselues in the ground, and abideth almost as well as our common purple kinde.

7. *Digitalis minor lutea siue pallida.* The small pale yellow Foxegloue.

This small pale yellow Foxegloue hath somewhat short, broad, smooth and darke greene leaues, snipt or dented about the edges very finely : the stalke is two foot high, beset with such like leaues, but lesser: the flowers are moe in number then in any of the rest, except the first and greatest, and growe along the vpper part of the stalke, being long and hollow, like the other, but very small, and of a pale yellow colour almost white : the seede vessels are small like the former, wherein are contained seede like the rest, but smaller : the rootes are stringy, but durable, and seldome perish with any iniury of the extreamest frosts.

The Place.

The great white kinde hath been often, and in many places found wilde in our owne Country, among or hard by the common purple kinde. All the rest are strangers, but cherished in our Gardens.

The

The Time.

They flower in Iune and Iuly, and some in August, their seede becomming ripe quickly after.

The Names.

Onely the name *Digitalis*, is of all Writers giuen vnto these plants ; for it is not knowne to bee remembred of any of the old Authors. Wee call them generally in English, Foxegloue ; but some (as thinking it to bee too foolish a name) doe call them Finger-flowers, because they are like vnto the fingers of a gloue, the ends cut off.

The Vertues.

Foxegloues are not vsed in Physicke by any iudicious man that I know ; yet some Italians of Bononia, as Camerarius saith, in his time vsed it as a wound herbe.

Chap. XCVIII.

Verbascum. Mullein.

There be diuers kindes of Mullein, as white Mullein, blacke Mullein, wooddy Mullein, base Mullein, Moth Mullein, and Ethiopian Mullein, all which to distinguish or to describe, is neither my purpose, nor the intent of this worke, which is to store a Garden with flowers of delight, and sequester others not worthy of that honour. Those that are fit to bee brought to your consideration in this place, are first, the *Blattarias*, or Moth Mulleins, and then the wooddy Mullein, which otherwise is called French Sage, and lastly, the Ethiopian Mullein, whose beauty consisteth not in the flower, but in the whole plant ; yet if it please you not, take it according to his Country for a Moore, an Infidell, a Slaue, and so vse it.

1. *Blattaria lutea odorata.* Sweete yellow Moth Mullein.

The yellow Moth Mullein whose flower is sweete, hath many hard grayish greene leaues lying on the ground, somewhat long and broad, and pointed at the end : the stalks are two or three foot high, with some leaues on them, & branching out from the middle vpwards into many long branches, stored with many small pale yellow flowers, of a pretty sweete sent, somewhat stronger then in the other sorts, which seldome giueth seede, but abideth in the roote, liuing many yeares, which few or none of the others doe.

2. *Blattaria lutea maior siue Hispanica.* The great yellow Moth Mullein.

This Spanish kinde hath larger and greener leaues then the former, and rounder and larger then the next that followeth : the stalke is higher then in any of the Moth Mulleins, being for the most part foure or fiue foote high, whereon toward the toppe growe many goodly yellow flowers, consisting of fiue leaues, as all the rest doe, not so thicke set as the former, but much larger, with some small purplish threads in the middle : the ends whereof are fashioned somewhat like as if a Flie were creeping vp the flower, which turne into round heads, sometimes two or three or more standing together, but vsually one, wherein lye small duskie seed : the roote is not great nor full of threads, and doth perish most vsually hauing giuen seede, except the Winter bee very milde.

3. Blattaria

3. *Blattaria lutea altera vulgatior.* The ordinary yellow Moth Mullein.

This yellow Moth Mullein (which is the moſt frequent in our Gardens) hath longer, and narrower leaues then any of the former, and roundly notched or dented on the edges, of a darke greene colour : the ſtalke is ſometimes branched, but moſt viſually ſingle, whereon ſtand many gold yellow flowers, not fully ſo large as the Spaniſh kinde, but with the like purple threads in the middle : the ſeede is ſmall, and contained in the like round heads, but alwaies euery one ſingle by it ſelfe : the roote periſheth euery yeare that it beareth ſeede.

4. *Blattaria flore luteo purpuraſcente.* Cloth of gold Moth Mullein.

The greateſt point of difference betweene this and the laſt deſcribed, conſiſteth chiefly in the colour of the flower, which in this is of the colour of cloth of gold, that is, the ground yellow, and ouerſhadowed with a bright crimſon colour, which is a fine colour of much delight : the threads in the middle are not ſo purple red as in the former, but much about the colour of the flower : this is not ſo willing to giue ſeede, and will as hardly abide in the roote, and hath out of queſtion riſen from the ſeede of the former.

5. *Blattaria flore albo.* White Moth Mullein.

The leaues of the white Moth Mullein are ſomewhat like vnto the yellow, yet not altogether ſo much roundly notched about the edges, but rather a little dented, with ſharper notches : the ſtalke riſeth as high as the yellow, and hath now and then ſome branches about it: the flowers hereof are pure white, as large and great as the ordinary yellow, or ſomewhat larger, with the like purple threads in the middle, as are in the yellow : the ſeed is like the other; the root periſheth in like maner, and will not endure.

6. *Blattaria flore purpureo.* Purple Moth Mullein.

The Purple Moth Mullein hath his leaues lying on the ground, broader and ſhorter then any of the other, of a more grayiſh greene colour, and without any denting for the moſt part about the edges, ſharpe pointed alſo at the end of the leafe ; among the leaues riſeth vp the ſtalke, not ſo high as either the white or the yellow, and many times branched, bearing many flowers thereon, of the ſame faſhion, and no whit ſmaller, of a faire deepe blewiſh colour tending to redneſſe, the threads in the middle of the flowers being yellow : the ſeede veſſels hereof are ſomewhat ſmaller then any of the former, except the firſt ſweete yellow kinde : the roote hereof is long, thicke, and blackiſh on the outſide, abiding very well from yeare to yeare, and riſeth well alſo from the ſowing of the ſeede.

7. *Blattaria flore cæruleo.* Blew Moth Mullein.

This blew Moth Mullein is in all reſpects like vnto the former purple kinde, ſauing onely in the colour of the flower, which is of a blewiſh violet colour, and is not much inferiour either in greatneſſe of the plant, or in the largeneſſe of the flower, vnto the former purple kinde, and endureth many yeares in the like manner. And theſe be all the ſorts of this kinde of Moth Mullein, that I haue ſeene and nourſed vp for this my Garden, without interpoſing any vnknowne, not ſeene, or vnworthy.

8. *Verbaſcum ſilueſtre ſiue quartum Matthioli.* Wooddy Mullein or French Sage.

Wooddy Mullein or French Sage, hath diuers wooddy branches two or three foot high, very hoary or white, whereon at ſeuerall ioynts ſtand diuers thicke leaues, white alſo and hoary, long, ſomewhat broad, round pointed, and rough, ſomewhat reſembling the leaues of Sage in the forme and roughneſſe, but not in the ſent, whereof our

people

people gaue it the name of Sage, calling it French Sage (when as it is as great a stranger in France as in England, yet they doe with this as with many other things, calling them French, which come from beyond the Seas; as for example, all or most of our bulbous flowers, they call French flowers, &c.) at the toppes of the stalkes and branches, at certaine distances, are placed round about them many gaping flowers, like vnto the flowers of Sage, but yellow: after which now and then come seede, somewhat bigger then the Moth Mulleins, and lesse then the next Mullein of Ethiopia: the roote is wooddy at the toppe, with diuers blackish strings growing from it, and endureth as well aboue ground with his leaues, as vnder it with his rootes.

9. Æthiopis. Ethiopian Mullein.

This Mullein of Ethiopia hath many great, broad, and large leaues lying on the ground, rent or torne in diuers of them very much on the sides, of so hoary a white greene colour, that it farre passeth any of the white Mulleins, that growe wilde abroad in our owne Country; for they are of a yellowish white hoarinesse, nothing so pleasant to looke on as this: in the middle of these leaues riseth vp a square strong stalke, foure or fiue foote high, set full of such like leaues as growe belowe, but much lesser, and lesser still vp to the toppe, all hoary and woolly, as the rest, and diuided into manie branches, spreading farre, and taking vp a great compasse of ground, more then any one roote of Garden Clary, or other such like plant: at each of the stalkes and branches are set two small leaues, and with them, round about the stalkes, stand many small gaping flowers, of a pale bleake blew colour: the seede is almost as large as Garden Clary seede, and of the same forme and colour: the roote is wooddy, and perisheth as soone as it hath borne seede, which is vsually the second yeare after the sowing; for the first yeare it seldome runneth vp to flower.

10. Lamium Pannonicum siue Galeopsis Pannonica. Hungary dead Nettle or the Dragon flower.

Let mee thrust this plant into this place, rather then make a peculiar Chapter, because I haue no other of the same stocke or kindred to be ioyned with it, and is a pretty ornament in a Garden. The leaues whereof are very large, round, and great, rough or full of veines, which make it seeme crumpled, dented or deepely notched about the edges, and of a very darke greene colour, and sometimes brownish, or of a darke reddish colour withall, euery one standing on a long foote-stalke, very like in forme vnto the great white Arch-Angell leaues, but farre larger and blacker: the stalkes are great and foure square, hauing leaues and flowers standing round about them at the ioynts like coronets, which flowers are very great, long, and wide gaping open, of a darke red or purple colour, with some whitenesse or spots in the iawes, and some hairinesse also on the sides, which stand in full flower two or three moneths most vsually, and sometimes longer, after which come brownish seede: the roote is a great tuft or bush of long whitish strings, and encreaseth euery yeare, not fearing the greatest iniuries of our coldest and extreamest Winters.

The Place.

All these plants are strangers in our Countrey, and onely preserued in Gardens, to furnish them with variety; but (as I said) the cloth of gold Moth Mullein hath been raised from seed in our owne Country.

The Time.

The last flowreth first, before all the rest, beginning in Aprill. The Moth Mulleins in May and Iune. The French Sage in Iuly.

The Names.

All the sorts of *Blattaria* may bee comprehended vnder the kindes of

Kk *Verbascum*

Verbascum nigrum, as any one but meanely exercised in the knowledge of plants, may discerne. And although Plinie saith, that Moths doe most frequently haunt where *Blattaria* either groweth, or is laid, yet it is not observed sufficiently in our Country so to doe, notwithstanding the name of Moth Mullein is generally giuen them. The last is generally called with vs *Lamium Pannonicum*, but certainely it is the *Galeosis maxima Pannonica* of Clusius.

The Vertues.

Other qualities I haue not found hath been allotted vnto the *Blattaria* or Moth Mullein, then those of Plinie, to engender Moths. Wee vse none of these plants in Physicke in these daies.

CHAP. XCIX.

Valeriana. Valerian.

THe many sorts of Valerian (or Set-wall as many doe call them) are fitter for a generall worke, or a generall Physicall Garden of Simples, then this of delightfull flowers. I will therefore select out a few, worthy of the place, and offer them to your considerations.

1. *Valeriana rubra Dodonæi.* Red Valerian.

This Valerian hath diuers hard, but brittle whitish greene stalkes, rising from the roote, full of tuberous or swelling ioynts, whereat stand two leaues, on each side one, and now and then some small leaues from betweene them, which are somewhat long and narrow, broadest in the middle, and small at both ends, without either diuision or incisure on the edges, of a pale greene colour : the stalkes are branched at the top into diuers parts, at the ends whereof stand many flowers together, as it were in an vmbell or tuft, somewhat like vnto the flowers of our ordinary Valerian, but with longer neckes, and of a fine red colour, very pleasant to behold, but of no sent of any Valerian : after these flowers haue stood blowne a very great while, they sodainely fall away, and the seede is ripe very quickly after, which is whitish, standing vpon the branches naked, as the Valerians doe, and very like vnto them, with a little white doune at the end of euery one of them, whereby they are soone carried away with the winde : the roote is great, thicke, and white, continuing long, and shooting out new branches euery yeare, and smelling somewhat like a Valerian.

2. *Nardus Montana tuberosa.* Knobbed Mountaine Valerian.

This kinde of Valerian or Spiknard, if you will so call it, hath his first leaues lying on the ground, without any diuision in them at all, being smooth, and of a dark greene colour, which so abide all the winter ; but those that spring vp after, and when it runneth vp to flower, are cut in on the edges, very like vnto the iagged leaues of the great garden Valerian, and so the elder they grow, the more cut and iagged they are : the stalke and flowers are very like the stalke with flowers of the garden Valerian, but of a darke or deepe red colour, and more store of them thrust together, by double the number almost : the seede is like the seede of the great Valerian : the root is tuberous, or knobbed in many parts, round about, aboue and below also, with some fibres shooting from them, whereby it is encreased, and smelleth very like the roote of the garden Setwall, or not altogether so strong.

3. *Valeriana*

1 *Blattaria flore albo.* Moth Mullein with a white flower. 2 *Blattaria flore purpureo.* Moth Mullein with a purple flower. 3 *Verbascum quartum Matthioli.* French Sage. 4 *Æthiopis.* Ethiopian Mullein. 5 *Valeriana rubra Dodonæi.* Red Valerian. 6 *Valeriana Græca.* Greek Valerian. 7 *Lamium Pannoni.um.* Hungary dead Nettle. 8 *Cardamine flore pleno.* Double Cuckowe flower or Ladies smocks.

3. *Valeriana Graca.* Greeke Valerian.

The Greek Valerian hath many winged leaues lying vpon the ground, that is, many small leaues set on both sides of a middle ribbe, very like vnto the wilde Valerian, that groweth by the ditch sides, but much smaller and tenderer, among which rise vp one or two round brittle stalkes, two foote high or thereabouts, whereon are set at the ioynts, such like leaues as grow below, but smaller : the toppes of the stalkes are diuided into many small branches, thicke set together, full with flowers, consisting of fiue small round leaues a peece, layd open like vnto the Cinquefoile flower, with some white threds in the middle, tipt with yellow pendents : the colour of these flowers in some plants, is of a faire bleake blew colour, and in others pure white : And I doe heare of one beyond the Seas (if the report bee true, for I haue not seene such a one) which should beare red flowers : after the flowers are past, there come vp in their places small hard huskes or heads, containing small blackish seedes : the roote is composed of a number of small long blackish threds, fastened together at the head, without any sent at all of a Valerian, eyther in roote or leafe ; and why it should bee called a Valerian I see no great reason, for it agreeth with none of them, in flower or seede, and but onely with the wilde Valerian in leafe, as I said before : but as it is, we so giue it you, and for the flowers sake is receiued into our gardens, to helpe to fill vp the number of natures rarities and varieties.

The Place.

All these Valerians are strangers, but endenizond for their beauties sake in our Gardens. The Mountaine Valerian I had of the liberalitie of my louing friend Iohn Tradescante, who in his trauaile, and search of natures varieties, met with it, and imparted thereof vnto me.

The Time.

They flower in the Summer moneths, and seed quickly after.

The Names.

The first is generally called of most, *Valeriana rubra Dodonæi,* who saith also that some would haue it to be *Behen rubrum.* Some call it *Valerianthon,* others make it a kinde of *Ocimastrum,* and some *Saponaria altera,* with other names, which are to no great purpose to set downe in this place, it beeing fitter for a generall worke to discusse of names, wherein both reading, knowledge and iudgement must bee shewen, to correct errours, and set downe the truth, that one may rest thereon. The others haue their names in their titles sufficient to distinguish them.

The Vertues.

The Mountaine Valerian is of all the the rest here set downe of most vse in Physicke, the rest hauing little or none that I know, although it be much weaker then the great garden kinde, or the Indian Nardus, in whose steed anciently it was vsed, in oyles, oyntments, &c.

Chap. C.

Cardamine. Cuckow flowers, or Ladies smockes.

OF the common sorts of Cuckow flowers that grow by ditch-sides, or in moist medowes, & wet grounds, it is not my purpose here to write, but of one or two other, the most specious or faire of all the tribe, that doe best befit this garden.

1. *Cardamine*

1. *Cardamine flore pleno.* Double Cuckow flowers.

The double *Cardamine* hath a few winged leaues, weake and tender, lying on the ground, very like vnto the single medow kinde; from among which riseth vp a round greene stalke, set here and there, with the like leaues that grow below, the top wherof hath a few branches, whereon stand diuers flowers, euery one vpon a small footestalk, consisting of many small whitish round leaues, a little dasht ouer with a shew of blush, set round together, which make a double flower : the roote creepeth vnder ground, sending forth small white fibres, and shooteth vp in diuers places.

2. *Cardamine trifolia.* Trefoile Ladies smockes.

This small plant hath diuers hard, darke round greene leaues, somewhat vneuen about the edges, alwayes three set together on a blackish small footstalke, among which rise vp small round blackish stalkes, halfe a foote high, with three small leaues at the ioynts, where they branch forth; at the toppes whereof stand many flowers, consisting of foure leaues a peece, of a whitish or blush colour very pale : after which come vp small, thicke and long pods, wherein is contained small round seede: the root is composed of many white threds, from the heads whereof runne out small strings, of a dark purple colour, whereby it encreaseth.

The Place.

The first with the double flower is found in diuers places of our owne Countrey, as neere Micham about eight miles from London; also in Lancashire, from whence I receiued a plant, which perished, but was found by the industrie of a worthy Gentlewoman, dwelling in those parts heretofore remembred, called Mistresse Thomasin Tunstall, a great louer of these delights. The other was sent me by my especial good friend Iohn Tradescante, who brought it among other dainty plants from beyond the Seas, and imparted thereof a roote to me.

The Time.

The last most vsually flowreth before the former, yet not much differing, that is, in the end of Aprill or in May.

The Names.

The first is a double kinde of that plant, that growing wilde abroade, is vsually called *Cardamine altera,* and *Sisymbrium alterum* of Dioscorides, and of some *Flos cuculi,* but not fitly; for that name is more vsually giuen vnto the wilde featherd Campions, both single and double, as is before expressed: yet for want of a fitter name, wee may call it in English, eyther Cuckowe flower, or Ladyes smockes, which you will. The second hath beene sent vnder the name of *Sanicula trifolia,* but the most frequent name now receiued, is *Cardamine trifolia,* and in English Trefoile Ladies spockes.

The Vertues.

The double Ladies smockes are of the same qualitie with the single, and is thought to be as effectuall as Watercresses. The propertie of the other I thinke is not much knowne, although some would make it a wound herbe.

CHAP.

Chap. CI.

Thlaspi Creticum. Candy Tufts.

O F the many forts of *Thlaspi* it is not the fcope of this worke to relate, I will fe-lect but onely two or three, which for their beautie are fit to bee inferted into this garden.

Thlaspi Creticum vmbellatum flore albo & purpureo.
Candy Tufts white and purple.

This fmall plant rifeth feldome aboue a foote and a halfe high, hauing fmall, narrow, long and whitifh greene leaues, notched or dented with three or foure notches on each fide, from the middle to the point-wards ; from among which rife vp the ftalkes, branched from the bottome almoft into diuers fmall branches, at the toppes whereof ftand many fmall flowers, thick thruft together in an vmbell or tuft, making them feeme to be fmall, round, double flowers of many leaues, when as euery flower is fingle, and ftandeth a part by it felfe, of a faire white colour in fome plants, without any fpot, and in others with a purplifh fpot in the centre or middle, as if fome of the middle leaues were purple; in others againe the whole flower is purplifh all ouer, which make a pretty fhew in a garden : the feede is contained in many fmall and flat feed veffels, which ftand together in an vmbell, as the flowers did, in which are contained fomewhat reddifh feede, like vnto fome other forts of *Thlaspi,* called Treakle Muftards : the roote is fmall and hard, and perifheth euery yeare hauing giuen feede.

Thlaspi Mari- We haue another fort, whofe leaues before it fendeth forth any ftalke, are a little
num Baticum. toothed, or finely dented about the edges, and brancheth not fo much out, but carryeth an vmbell of purplifh flowers like vnto the former, and paler yellow feede.

The Place.

Thefe doe grow in Spaine and Candie, not farre from the Sea fide.

The Time.

Thefe *Thlaspi* giue not their flowers vntill the end of Iune, or beginning of Iuly, and the feed is ripe foone after.

The Names.

The firft is named by fome, *Draba,* or *Arabis,* as Dodonæus, but *Draba* is another plant differing much from this. Wee call one fort, *Thlaspi Creti-cum,* and the other *Thlaspi Baticum marinum,* becaufe the one came from Spaine, and the other from Candy ; we giue it in Englifh, the name of Tufts, becaufe it doth fit the forme of the flowers beft, although ordinarily all the *Thlaspi* are Englifhed Wilde Muftardes.

The Vertues.

Candy, or Spanifh Tufts, is not fo fharpe biting in tafte, as fome other of the Thlafpies are, and therefore is not to be vfed in medicines, where *Thlaspi* fhould be in the ftead thereof.

Chap. CII.

Clematis. Clamberers, or Creepers.

HAuing shewed you all my store of herbes bearing fine flowers, let mee now bring to your consideration the rest of those plants, be they Shrubs or Trees, that are cherished in our garden, for the beauty of their flowers chiefly, or for some other beautifull respect: and first I will begin with such as creepe on the ground, without climing, and then such as clime vp by poles, or other things, that are set or grow neere them, fit to make Bowers, and Arbours, or else are like them in forme, in name, or some other such qualitie or propertie.

1. *Clematis Daphnoides, siue Vinca peruinca simplex minor diuersorum colorum.*
Single Perwinkle of diuers colours.

The smaller Perwinkle which not onely groweth wilde in many places, but is most vsuall in our Gardens, hath diuers creeping branches, trayling or running vpon the ground, shooting out small fibres at the ioynts, as it creepeth, taking thereby hold in the ground, and rooteth in diuers places : at the ioynts of these branches stand two small darke greene shining leaues, somewhat like vnto small Baye leaues, but smaller, and at the ioynts likewise with the leaues, come forth the flowers, one at a ioynt, standing vpon a tender footestalke, being somewhat long and hollow, parted at the brims, sometimes into foure leaues, and sometimes into fiue, the most ordinary sort is of a pale or bleake blew colour, but some are pure white, and some of a darke reddish purple colour: the root is in the body of it, little bigger then a rush, bushing in the ground, and creeping with his branches farre about, taking roote in many places, whereby it quickely possesseth a great compasse ; and is therefore most vsually planted vnder hedges, or where it may haue roome to runne.

2. *Vinca peruinca flore duplici purpureo.*
Double purple Perwinkle.

The double Perwinkle is like vnto the former single kinde, in all things except in the flower, which is of that darke reddish purple colour that is in one of the single kindes ; but this hath another row of leaues within the flower, so that the two rowes of leaues causeth it to be called double, but the leaues of these are lesser then the single. I haue heard of one with a double white flower, but I haue not yet seene it.

3. *Clematis Daphnoides siue Peruinca maior.*
The greater Perwinkle.

This greater Perwinkle is somewhat like the former, but greater, yet his branches creepe not in that manner, but stand more vpright, or lesse creeping at the least : the leaues also hereof stand by couples at the ioynts, but they are broader and larger by the halfe : the flowers are larger, consisting of fiue leaues that are blew, a little deeper then the former blew : this plant is farre tenderer to keepe then the other, and therefore would stand warme, as well as in a moist shadowie place.

4. *Clematis altera siue vrens flore albo.*
Burning Clamberer, or Virgins Bower.

This Causticke or burning Climer, hath very long and climing tender branches, yet somewhat woody below, which winde about those things that stand neere it, couered with a brownish greene barke, from the ioynts whereof shoote forth many winged leaues, consisting for the most part of fiue single leaues, that is, two and two together, and one at the end, which are a little cut in or notched on the edges here and
there,

there, but euery part of them is leſſer then the leaues of the next following Climer, without any claſping tendrels to winde about any thing at all : towards the vpper part of the branches, with the ſaid leaues, come forth long ſtalks, wheron ſtand many white flowers cluſtering together, opening the brims into ſixe or eight ſmall leaues, ſpreading like a ſtarre, very ſweet of ſmell, or rather of a ſtrong heady ſent, which after turne into flattiſh and blackiſh ſeede, plumed at the head, which plume or feather flyeth away with the winde after it hath ſtood long, and leaueth the ſeede naked or bare : the roote is white and thicke, fleſhie and tender, or eaſie to be broken, as my ſelfe can well teſtifie, in that deſiring to take a ſucker from the roote, I could not handle it ſo tenderly, but that it broke notwithſtanding all my care. Maſter Gerard in his Herball maketh mention of one of this kinde with double white flowers, which hee ſaith he recouered from the ſeede was ſent him from Argentine, that is Strasborough, whereof hee ſetteth forth the figure with double flowers : but I neuer ſaw any ſuch with him, neither did I euer heare of any of this kinde with double flowers. Cluſius indeed ſaith, that hee receiued from a friend ſome ſeede vnder the name of *Clematis flore albo pleno* : but he doubteth whether there bee any ſuch : the plants that ſprang with him from that ſeede, were like vnto the vpright kinde called *Flammula Matthioli*, or *Iouis creſta*, as he there ſaith : but aſſuredly I haue beene informed from ſome of my eſpeciall friends beyond Sea, that they haue a double white *Clematis*, and haue promiſed to ſend it, but whether it will be of the climing or vpright ſort, I cannot tell vntill I ſee it : but ſurely I doe much doubt whether the double will giue any good ſeede.

5. *Clematis altera ſiue peregrina flore rubro.* Red Ladies Bower.

This Climer hath many limber and weake climing branches like the former, couered with a browne thin outer barke, and greene vnderneath : the leaues ſtand at the ioynts, conſiſting but of three leaues or parts, whereof ſome are notched on one ſide, and ſome on both, without any claſping tendrels alſo, but winding with his branches about any thing ſtandeth next vnto it : the flowers in like manner come from the ſame ioynts with the leaues, but not ſo many together as the former vpon long footſtalkes, conſiſting of foure leaues a peece, ſtanding like a croſſe, of a darke red colour ; the ſeed is flat and round, and pointed at the end, three or foure or more ſtanding cloſe together vpon one ſtalk, without any doune vpon them at all, as in the former : the roots are a bundell of browniſh yellow ſtrong ſtrings, running down deep into the ground, from a bigge head aboue.

6. *Clematis peregrina flore purpureo ſimplici.*
Single purple Ladies Bower.

This Ladies Bower differeth in nothing from the laſt deſcribed, but onely in the colour of the flower, which is of a ſad blewiſh purple colour ; ſo that the one is not poſſible to be known from the other, vntill they be in flower.

7. *Clematis peregrina flore purpureo pleno.*
Double flowred purple Ladies Bower.

This double *Clematis* hath branches and leaues ſo neere reſembling the ſingle kinds, that there can be knowne no difference, vnleſſe it be, that this groweth more goale and great, and yeeldeth both more ſtore of branches from the ground, and more ſpreading aboue : the chiefeſt marke to diſtinguiſh it is the flower, which in this is very thicke and double, conſiſting of a number of ſmaller leaues, ſet cloſe together in order in the middle, the foure outermoſt leaues that encompaſſe them, being much broader and larger then any of the inward, but all of a dull or ſad blewiſh purple colour, the points or ends of the leaues ſeeming a little darker then the middle of them : this beareth no ſeede that euer I could ſee, heare of, or learne by any of credit, that haue nourſed it a great while ; and therefore the tales of falſe deceitfull gardiners, and others, that diliuer ſuch for truth, to deceiue perſons ignorant thereof, muſt not bee creduloufly entertained.

In the great booke of the Garden of the Bishop of Eystot (which place is neere vnto Norimberg) in Germany, I reade of a *Clematis* of this former kinde, whose figure is thereto also annexed, with double flowers of an incarnate, or pale purple tending to a blush colour, whereof I haue not heard from any other place. *clematis peregrina flore caruleo caruleo.*

8. *Flammula Iouis erecta.* Vpright Virgins Bower.

This kinde of *Clematis* hath diuers more vpright stalkes then any of the foure last described, sometimes foure or fiue foote high, or more; yet leaning or bending a little, so that it had some neede of sustaining, couered with a brownish barke; from whence come forth on all sides diuers winged leaues, consisting of fiue or seuen leaues, set on both sides of a middle ribbe, whereof one is at the end: the tops of the stalkes are diuided into many branches, bearing many white sweet smelling flowers on them, like in fashion vnto the white Virgins Bower; after which come such like feather topt seede, which remaine and shew themselues, being flat like the other, when the plumes are blowne abroad: the roote spreadeth in the ground from a thicke head, into many long strings, and fasteneth it selfe strongly in the earth; but all the stalkes dye downe euery yeare, and spring afresh in the beginning of the next.

9. *Clematis cærulea Pannonica.* The Hungarian Climer.

The stalks of this plant stand vpright, & are foure square, bearing at euery ioynt two leaues, which at the first are closed together, and after they are open, are somewhat like vnto the leaues of *Asclepias*, or Swallow-wort: from the tops of the stalks, and sometimes also from the sides by the leaues commeth forth one flower, bending the head downward, consisting of foure leaues, somewhat long & narrow, standing like a crosse, and turning vp their ends a little againe, of a faire blew or skie colour, with a thicke pale yellow short thrumme, made like a head in the middle: after the flower is past, the head turneth into such a like round feather topt ball, as is to be seene in the Trauellers ioy, or *Viorna* (as it is called) that groweth plentifully in Kent, and in other places by the way sides, and in the hedges, wherein is included such like flat seede. These stalkes (like as the last) dye downe to the ground euery yeare, and rise againe in the Spring following, shooting out new branches, and therby encreaseth in the root.

10. *Maracoc siue Clematis Virginiana.* The Virginia Climer.

Because this braue and too much desired plant doth in some things resemble the former Climers, so that vnto what other family or kindred I might better conioyne it I know not; let me I pray insert it in the end of their Chapter, with this description. It riseth out of the ground (very late in the yeare, about the beginning of May, if it be a plant hath risen from the seed of our owne sowing, and if it be an old one, such as hath been brought to vs from Virginia, not till the end thereof) with a round stalke, not aboue a yard and a halfe high (in any that I haue seene) but in hotter Countries, as some Authors haue set it downe, much higher, bearing one leafe at euery ioynt, which from the ground to the middle thereof hath no claspers, but from thence vpwards hath at the same ioynt with the leafe both a small twining clasper, like vnto a Vine, and a flower also: euery leafe is broad at the stalke thereof, and diuided about the middle on both sides, making it somewhat resemble a Figge leafe, ending in three points, whereof the middlemost is longest: the bud of the flower, before it doe open, is very like vnto the head or seede vessell of the ordinary single *Nigella*, hauing at the head or top fiue small crooked hornes, which when this bud openeth, are the ends or points of fiue leaues, that are white on the inside, and lay themselues flat, like vnto an Anemone, and are a little hollow like a scoope at the end, with fiue other smaller leaues, and whiter then they lying betweene them, which were hid in the bud before it opened, so that this flower being full blowne open, consisteth of ten white leaues, laide in order round one by another: from the bottome of these leaues on the inside, rise diuers twined threads, which spread and lay themselues all ouer these white leaues, reaching beyond the points of them a little, and are of a reddish peach colour: towards the bottomes comes likewise of these white leaues there are two red circles, about the breadth of an Oten strawe, one distant from another (and in some flowers there is but one circle seene) which adde a great grace vnto the flower; for the white leaues shew their colour through the peach coloured threads, and these red circles or rings vpon them being also perspicuous, make a tripartite shew of colours most delightfull: the middle part of this flower is hollow, and yellowish; in the bottome whereof riseth vp an vmbone, or round stile, somewhat bigge, of a whitish greene colour, spotted with reddish spots like the stalkes of Dragons, with fiue round threads or chiues, spotted in the like manner, and tipt at the ends with yellow pendents, standing about the middle part of the said vmbone, and from thence rising higher, endeth in three long crooked hornes most vsually (but sometimes in foure, as hath beene obserued in Rome by Dr. Aldine, that set forth some principall things of Cardinall Farnesius his Garden) spotted like the rest, hauing three round greene buttons at their ends: these flowers are of a comfortable sweete sent, very acceptable, which perish without yeelding fruit with vs, because it flowreth so late: but in the naturall place, and in hot Countries, it beareth a small round whitish fruit, with a crowne at the toppe thereof, wherein is contained (while it is fresh, and before it be ouer dried) a sweet

The Iesuites Figure of the Maracoc.

GRANADILLVS FRVTEX INDICVS CHRISTI PASSIONIS IMAGO.

liquor, but when it is dry, the seede within it, which is small, flat, somewhat rough and blacke, will make a ratling noise: the rootes are composed of a number of exceeding long and round yellowish browne strings, spreading farre abroad vnder the ground (I haue seene some rootes that haue beene brought ouer, that were as long as any rootes of *Sarsa parilla*, and a great deale bigger, which to be handsomely laid into the ground, were faine to be coyled like a cable) and shooting vp in seuerall places a good distance one from another, whereby it may be well encreased.

The Place.

The first blew Perwinkle groweth in many Woods and Orchards, by the hedge sides in England, and so doth the white here and there, but the other single and double purple are in our Gardens onely. The great Perwinkle groweth in Prouence of France, in Spaine, and Italy, and other hot Countries, where also growe all the twining Clamberers, as well single as double: but both the vpright ones doe growe in Hungary and thereabouts. The surpassing delight of all flowers came from Virginia. Wee preserue them all in our Gardens.

The Time.

The Perwinkles doe flower in March and Aprill. The Climers not vntill the end of Iune, or in Iuly, and sometimes in August. The Virginian somewhat later in August; yet sometimes I haue knowne the flower to shew it selfe in Iuly.

The Names.

The first is out of question the first *Clematis* of Dioscorides, and called of
many

1 *Thalspi creticum.* Candy tufts. 2 *Vinca peruinca flore simplici.* Single Perwinkle. 3 *Vinca peruinca flore duplici.* Double Perwinkle.
4 *Flammula Matthioli.* Vpright Virgins Bower. 5 *Clematis peregrina flore simplici.* The single Ladies Bower. 6 *Clematis peregrina flore pleno purpureo.* Double flowred Ladies Bower. 7 *Maracoc siue Clematis Virginiana.* The Virginian Climer.

many *Clematis Daphnoides* (but not that plant that is simply called *Daphnoides*, for that is *Laureola*) and is vsually called *Vinca peruinca*: but it is not *Chamædaphne*, for that is another plant, as shall be shewed in his place; some call it *Centunculus*: In Englith wee call it Perwinkle. The other is *Clematis altera* of Dioscorides, and is called also *Clematis peregrina*, whose distinctions are set downe in their titles : In English, Ladies Bower, or Virgins Bower, because they are fit to growe by Arbours, to couer them. The first vpright Clamberer is called, and that rightly of some, *Clematis erecta*, or *surrecta*. Of others, *Flammula frutex*, and *Flammula Iouis*, or *surrecta* : In English, Vpright Virgins Bower. The next is called by Clusius, *Clematis Pannonica cærulea*, who thought it to be *Climeni species*, by the relation of others, at the first, but after entituled it, *Clematis*: In English, the Hungarian Climer. The last may be called in Latine, *Clematis Virginiana*: In Englith, The Virgin or Virginian Climer; of the Virginians, *Maratoc*: of the Spaniards in the West Indies *Granadillo*, because the fruit (as is before said) is in some fashion like a small Pomegranate on the outside; yet the seede within is flattish, round, and blackish. Some superstitious Iesu-ite would faine make men beleeue, that in the flower of this plant are to be seene all the markes of our Sauiours Passion ; and therefore call it *Flos Passionis*: and to that end haue caused figures to be drawne, and printed, with all the parts proportioned out, as thornes, nailes, speare, whippe, pillar, &c. in it, and all as true as the Sea burnes, which you may well perceiue by the true figure, taken to the life of the plant, compared with the figures set forth by the Iesuites, which I haue placed here likewise for euery one to see : but these bee their aduantagious lies (which with them are tolerable, or rather pious and meritorious) wherewith they vse to instruct their people; but I dare say, God neuer willed his Priests to instruct his people with lyes; for they come from the Diuell, the author of them. But you may say I am beside my Text, and I am in doubt you will thinke, I am in this besides my selfe, and so nothing to be beleeued herein that I say. For, for the most part, it is an inherent errour in all of that side, to beleeue nothing, be it neuer so true, that any of our side shall affirme, that contrarieth the assertions of any of their Fathers, as they call them : but I must referre them to God, and hee knoweth the truth, and will reforme or deforme them in his time. In regard whereof I could not but speake (the occasion being thus offered) against such an erroneous opinion (which euen D^r. Aldine at Rome, before remembred, disproued, and contraried both the said figures and name) and seek to disproue it, as doth (I say not almost, but I am affraid altogether) leade many to adore the very picture of such things, as are but the fictions of superstitious brains: for the flower it selfe is farre differing from their figure, as both Aldine in the aforesaid booke, and Robinus at Paris in his *Theatrum Floræ*, doe set forth; the flowers and leaues being drawne to the life, and there exhibited, which I hope may satisfie all men, that will not be perpetually obstinate and contentious.

The Vertues.

Costæus saith hee hath often seene, that the leaues of Perwinkle held in the mouth, hath stayed the bleeding at the nose. The French doe vse it to stay the menstruall fluxes. The other are caust{}icke plants, that is, fiery hot, and blistering the skinne; and therefore (as Dioscorides saith) is profitable to take away the scurfe, leprye, or such like deformities of the skin. What property that of Virginia hath, is not knowne to any with vs I thinke, more then that the liquor in the greene fruit is pleasant in taste; but assuredly it cannot be without some speciall properties, if they were knowne.

CHAP.

Chap. CIII.

Chamælæa. Dwarfe Spurge Oliue, or Dwarfe Baye.

I Haue three forts of *Chamælæa* to bring to your confideration, euery one differing notably from other; two of them of great beauty in their flowers, as well as in the whole plant : the third abiding with greene leaues, although it haue no beauty in the flower, yet worthy of the place it holds.   And vnto thefe I muft adioyne another plant, as comming neareft vnto them in the brauery of the flowers.

1. *Chamælæa Germanica fiue Mezereon floribus dilutioris coloris & faturatioris.*
Dwarfe Baye, or flowring Spurge Oliue.

We haue two forts of this Spurge Oliue or Dwafe Baye, differing onely in the co-lour of the flowers. They both rife vp with a thicke wooddy ftemme, fiue or fix foot high fometimes, or more, and of the thickneffe (if they be very old) of a mans wreft at the ground, fpreading into many flexible long branches, couered with a tough grayifh barke, befet with fmall long leaues, fomewhat like vnto Priuet leaues, but fmaller and paler, and in a manner round pointed : the flowers are fmall, confifting of foure leaues, many growing together fometimes, and breaking out of the branches by themfelues : in the one fort of a pale red at the firft blowing, and more white afterwards; the other of a deeper red in the bloffome, and continuing of a deeper red colour all the time of the flowring, both of them very fweete in fmell : after the flowers are paft, come the berries, which are greene at the firft, and very red afterwards, turning blackifh red, if they ftand too long vpon the branches : the rootes fpread into many tough long bran-ches, couered with a yellowifh barke.

2. *Chamælæa Alpina.* Mountaine Spurge Oliue.

This Mountaine Laurell rifeth vp with a fmall wooddy ftemme, three or foure foot high, or more, branching forth towards the vpper parts into many flender and tough branches, couered with a rough hoary greene barke, befet at the ends thereof with flatter, fuller, and fmaller round pointed leaues then the former, of a grayifh greene colour on the vpperfide, and hoary vnderneath, which abide on the branches in Winter, and fall not away as the former : the flowers are many fet together at the ends of the branches, greater then the former, and confifting of foure leaues a peece, of a light blufh colour, ftanding in fmall grayifh huskes, of little or no fent at all : the fruit followeth, which are fmall long graines or berries, of an excellent red colour, which afterwards turne blacke : the roote is long, and fpreadeth about vnder the vpper part of the earth.

3. *Chamælæa tricoccos.* Widowe Wayle.

This three berried Spurge Oliue hath no great ftemme at all, but the whole plant fpreadeth from the ground into many flexible tough greene branches, whereon are fet diuers narrow, long, darke greene leaues all along the branches, which abide greene all the Winter : the flowers are very fmall, fcarce to be feene, and come forth between the leaues and the ftalke, of a pale yellow colour, made of three leaues ; after which come fmall blackifh berries, three vfually fet together : the roote fpreadeth it felfe in the ground not very farre, being hard and wooddy, and often dyeth, if it bee not well defended from the extremity of our fharpe Winters.

4. *Cneorum Matthioli.* Small Rocke Rofes.

I was long in doubt in what place I fhould difpofe of this plant, whether among the Campions, as Bauhinus, or among thefe, as Clufius doth ; but left my Gorden fhould want it wholly, let it take vp roome for this time here. This gallant plant hath diuers

L l   long.

long, weake, slender, but yet tough branches lying vpon the ground, diuided vsually into other smaller branches, whereon growe many, small, long, and somewhat thicke leaues, somewhat like vnto the leaues of the former *Mezereon*, set without any order to the very tops, from whence doe come forth a tuft of many small flowers together, made or consisting of foure leaues a peece, of a bright red or carnation colour, and very sweete withall, which turne into small round whitish berries, wherein is contained small round seede, couered with a grayish coate or skinne : the roote is long and yellowish, spreading diuers wayes vnder the ground, and abideth many yeares shooting forth new branches.

Flore albo.   It hath beene obserued in some of these plants, to bring forth white flowers, not differing in any thing else.

The Place.

The first sorts growe plentifully in many places of Germany. The second in the mountaines by Sauoye. The third in Prouence and Spaine. The last in diuers parts of Germany, Bohemia, and Austria, and about Franckford.

The Time.

The two first sorts are most vsually in flower about Christmas, or in Ianuary, if the weather be not violent, and sometimes not vntill February. The second flowreth not vntill Aprill. The third in May. The berries of them ripen some in Iune and Iuly; some in August and September, as their flowring is earlier or later. The last flowreth as well in the Spring as in Autumne, so apt and plentifull it is in bearing, and the seede at both times doth ripen soone after.

The Names.

The first is called of some *Chamelæa*, with this addition *Germanica*, that it may differ from the third, which is the true *Chamelæa* of Dioscorides, as all the best Authors doe agree, and is also called *Piper montanum* of the Italians. It is generally called *Mezereon*, and is indeede the true *Mezereon* of the Arabians, and so vsed in our Apothecaries shops, wheresoeuer the Arabians *Mezereon* is appointed, although the Arabians are so intricate and vncertaine in the descriptions of their plants, confounding *Chamelæa* and *Thymelæa* together. Matthiolus maketh it to be *Daphnoides* of Dioscorides; but in my opinion he is therein mistaken : for all our best moderne Writers doe account our *Laureola*, which hath blacke berries, to bee the true *Daphnoides* : the errour of his Countrey might peraduenture drawe him thereunto; but if hee had better considered the text of Dioscorides, that giueth black berries to *Daphnoides*, and red to *Chamædaphne*, he would not so haue written; and truly, I should thinke (as Lobel doth) with better reason, that this *Chamelæa* were Dioscorides *Chamædaphne*, then hee to say it were *Daphnoides* : for the description of *Chamædaphne*, may in all parts be very fitly applyed to this *Chamelæa* : and euen these words, *Semen annexum folijs*, wherein may be the greatest doubt in the description, may not vnfitly bee construed, that as is seene in the plant, the berries growe at the foote of the leaues, about the branches : the faculties indeede that Dioscorides giueth to *Chamædaphne*, are (if any repugnancie be) the greatest let or hinderance, that this *Chamelæa* should not be it : but I leaue the discussing of these and others of the like nature, to our learned Physitians; for I deale not so much with vertues as with descriptions. The second is called of Lobel *Chamelæa Alpina incana*, of Clusius *Chamelæa secunda*, and saith hee had it out of Italy. Wee may call it in English, Mountaine Spurge Oliue, as it is in the description, or Mountaine Laurell, which you will. The last hath the name of *Cneorum*, first giuen it by Matthiolus, which since is continued by all others. Bauhinus (as I said) referreth it to the Mountaine Campions, but Clusius

(as

1 *Chamelæa Germanica seu Mezereon.* Mezereon or Dwarfe Bay. 2 *Chamelæa Alpina.* Mountain Spurge Oliue. 3 *Cneorum Matthioli.* Small Rocke Roses. 4 *Laurus Tinus siue siluestris.* The wild Bay tree. 5 *Oleander siue Laurus Rosea.* The Rose Bay tree. 6 *Laurocerasus.* The Bay Cherrie tree.

(as I doe) to the kindes of *Chamelæa* or *Thymelæa*. For want of an Englifh name I haue (as you fee, and that is according to the name the Germane women, as Clufius faith, doe call it) entituled it the Small Rocke Rofe; which may abide vntill a fitter may be conferred vpon it.

The Vertues.

All thefe plants except the laft, as well leaues as berries, are violent purgers, and therefore great caution is to bee had in the vfe of them. The laft hath not beene applyed for any difeafe that I know.

CHAP. CIII.

Laurus. The Bay Tree.

MY meaning is not to make any defcription of our ordinary Bayes in this place (for as all may very well know, they may be for an Orchard or Courtyard, and not for this Garden) but of two or three other kindes, whofe beautifull afpect haue caufed them to be worthy of a place therein : the one is called *Laurus Tinus*, The wilde Baye : the other *Laurus Rofea* or *Oleander*, The Rofe Bay : and a third is *Laurocerafus*, The Cherry Bay ; which may haue not onely fome refpect for his long bufh of fweet fmelling flowers, but efpecially for the comely ftatelineffe of his gallant euer frefh greene leaues ; and the rather, becaufe with vs in moft places, it doth but *frutefcere*, vfe to bee Shrub high, not *arborefcere*, Tree high, which is the more fit for this Garden.

1. *Laurus Tinus fiue filueftris.* The wilde Bay tree.

This wilde Baye groweth feldome to bee a tree of any height, but abideth for the moft part low, fhooting forth diuers flender branches, whereon at euery ioynt ftand two leaues, long, fmooth, and of a darke greene colour, fomewhat like vnto the leaues of the Female Cornell tree, or between that and Baye leaues : at the toppes of the branches ftand many fmall white fweete fmelling flowers, thrufting together, as it were in an vmbell or tuft, confifting of fiue leaues a peece, the edges whereof haue a fhew of a wafh purple, or light blufh in them, which for the moft part fall away without bearing any perfect ripe fruit in our Countrey : Yet fometimes it hath fmall black berries, as if they were good, but are not. In his naturall place it beareth fmall, round, hard and pointed berries, of a fhining blacke colour, for fuch haue come often to my hands (yet Clufius writeth they are blew) ; but I could neuer fee any fpring that I put into the ground. This that I here defcribe, feemeth to me to be neither of both thofe that Clufius faw growing in Spain and Potugall, but that other, that (as he faith) fprang in the low Countreyes of Italian feede.

2. *Laurus Rofea fiue Oleander.* The Rofe Bay.

Of the Rofe Bay there are two forts, one bearing crimfon coloured flowers, which is more frequent, and the other white, which is more rare. They are fo like in all other things, that they neede but one defcription for both. The ftemme or trunke is many times with vs as bigge at the bottome as a good mans thumbe, but growing vp fmaller, it diuideth it felfe into branches, three for the moft part comming from one ioynt or place, and thofe branches againe doe likewife diuide themfelues into three other, and fo by degrees from three to three, as long as it groweth : the loweft of thefe are bare of leaues, hauing fhed or loft them by the cold of winters, keeping onely leaues on the vppermoft branches, which are long, and fomewhat narrow, like in forme vnto Peach leaues, but thicker, harder, and of a darke greene colour on the vpperfide, and

yellowifh

yellowish greene vnderneath : at the tops of the young branches come forth the flowers, which in the one sort before they are open, are of an excellent bright crimson colour, and being blowen, confist of foure long and narrow leaues, round pointed, somewhat twining themselues, of a paler red colour, almost tending to blush, and in the other are white, the greene leaues also being of a little fresher colour : after the flowers are past, in the hot countries, but neuer in ours, there come vp long bending or crooked flat pods, whose outward shell is hard, almost woody, and of a browne colour, wherein is contained small flat brownish seede, wrapped in a great deale of a brownish yellow doune, as fine almost as silke, somewhat like vnto the huskes of *Asclepias*, or *Periploca*, but larger, flatter and harder ; as my selfe can testifie, who had some of the pods of this Rose bay, brought mee out of Spaine, by Master Doctor Iohn More, the seedes whereof I sowed, and had diuers plants that I raised vp vnto a reasonable height, but they require, as well old as young, to bee defended from the colde of our winters.

3. *Laurocerasus.* The Bay Cherry.

This beautifull Bay in his naturall place of growing, groweth to bee a tree of a reasonable bignesse and height, and oftentimes with vs also if it bee pruined from the lower branches ; but more vsually in these colder Countries, it groweth as a shrub or hedge bush, shooting forth many branches, whereof the greater and lower are couered with a darke grayish greene barke, but the young ones are very greene, whereon are set many goodly, faire, large, thicke and long leaues, a little dented about the edges, of a more excellent fresh shining greene colour, and farre larger then any Bay leafe, and compared by many to the leaues of the *Pomeritron* tree (which because wee haue none in our Countrey, cannot be so well known) both for colour and largenesse, which yeeld a most gracefull aspect : it beareth long stalkes of whitish flowers, at the ioynts of the leaues both along the branches and towards the ends of them also, like vnto the Birds Cherry or *Padus Theophrasti*, which the French men call *Putier & Cerisier blanc*, but larger and greater, consisting of fiue leaues with many threds in the middle: after which commeth the fruite or berries, as large or great as Flanders Cherries, many growing together one by another on a long stalke, as the flowers did, which are very blacke and shining on the outside, with a little point at the end, and reasonable sweete in taste, wherein is contained a hard round stone, very like vnto a Cherry stone, as I haue obserued as well by those I receiued out of Italie, as by them I had of Master Iames Cole a Merchant of London lately deceased, which grew at his house in Highgate, where there is a faire tree which hee defended from the bitternesse of the weather in winter by casting a blanket ouer the toppe thereof euery yeare, thereby the better to preserue it.

The Place.

The first is not certainly knowne from whence it came, and is communicated by the suckers it yeeldeth. The second groweth in Spaine, Italie, Grece, and many other places : that with white flowers is recorded by Bellonius, to grow in Candy. The last, as Matthiolus, and after him Clusius report, came first from Constantinople : I had a plant hereof by the friendly gift of Master Iames Cole, the Merchant before remembred, a great louer of all rarities, who had it growing with him at his countrey house in Highgate aforesaid, where it hath flowred diuers times, and borne ripe fruit also.

The Time.

The first flowreth many times in the end of the yeare before Christmas, and often also in Ianuary, but the most kindly time is in March and Aprill, when the flowers are sweetest. The second flowreth not vntill Iuly. The last in May, and the fruit is ripe in August and September.

The

The Names.

The firſt is called *Laurus ſilueſtris*, and *Laurus Tinus* : in Engliſh Wilde Bay, or Sweete flowring Bay. The ſecond is called *Laurus Roſea, Oleander, Nerium,* and *Rhododendros* : in Engliſh The Roſe Bay, and Oleander. The laſt was ſent by the name of *Trebezon Curmaſi,* that is to ſay, *Dactylus Trapezuntina,* but not hauing any affinitie with any kinde of Date, Bellonius as I thinke firſt named it *Lauroceraſus,* and *Ceraſus Trapezuntina.* Dalechampius thinketh it to bee *Lotus Aphricana,* but Cluſius refuteth it. Thoſe ſtones or kernels that were ſent me out of Italie, came by the name of *Laurus Regia,* The Kings Bay. Wee may moſt properly call it according to the Latine name in the title, The Cherry bay, or Bay Cherry, becauſe his leaues are like vnto Bay-leaues, and both flowers and fruit like vnto the Birdes Cherry or Cluſter Cherry, for the manner of the growing ; and therfore I might more fitly I confeſſe haue placed it in my Orchard among the ſorts of Cherries : but the beautifulneſſe of the plant cauſed mee rather to inſert it here.

The Vertues.

The wilde Bay hath no propertie allotted vnto it in Phyſicke, for that it is not to be endured, the berries being chewed declare it to be ſo violent hot and choking. The Roſe Bay is ſaid by Dioſcorides, to be death to all foure footed beaſts, but contrariwiſe to man it is a remedie againſt the poiſon of Serpents, but eſpecially if Rue bee added vnto it. The Cherry Bay is not knowne with vs to what phyſicke vſe it may be applyed.

Chap. CIIII.

Ceraſus flore multiplici. The Roſe or double bloſſomd Cherry.
Malus flore multiplici. The double bloſſomd Apple tree. And
Malus Perſica flore multiplici. The double bloſſomd Peach tree.

THe beautifull ſhew of theſe three ſorts of flowers, hath made me to inſert them into this garden, in that for their worthineſſe I am vnwilling to bee without them, although the reſt of their kindes I haue tranſferred into the Orchard, where among other fruit trees, they ſhall be remembred : for all theſe here ſet downe, ſeldome or neuer beare any fruite, and therefore more fit for a Garden of flowers, then an Orchard of fruite.

Ceraſus flore pleno vel multiplici.
The Roſe Cherry, or double bloſſomd Cherry.

The double bloſſomed Cherry tree is of two ſorts for the flower, but not differing in any other part, from the ordinary Engliſh or Flanders Cherry tree, growing in the very like manner : the difference conſiſteth in this, that the one of theſe two ſorts hath white flowers leſſe double, that is, of two rowes or more of leaues, and the other more double, or with more rowes of leaues, and beſides I haue obſerued in this greater double bloſſomd Cherry, that ſome yeares moſt of the flowers haue had another ſmaller and double flower, riſing vp out of the middle of the other, like as is to bee ſeene in the double Engliſh Crow-foote, and double redde *Ranunculus* or Crowfoote, before deſcribed : this I ſay doth not happen euery yeare, but ſometimes. Sometimes alſo theſe trees will giue a few berries, here and there ſcattered, and that with leſſe double flowers more often, which are like vnto our Engliſh Cherries both for taſte and bigneſſe. Theſe be very fit to be ſet by Arbours.

Malus

1 *Ceresus flore pleno*. The double blossomd Cherry tree. 2 *Malus flore multiplici* The double blossomd Apple tree. 3 *Malus Persica flore pleno*. The double blossomd Peach tree. 4 *Periclymenum perfoliatum*. Double Honisuckle. 5 *Periclymenum rectum*. Vpright Honisuckle.

Malus flore multiplici. The double bloſſomd Apple tree.

This double bloſſomd Apple tree is altogether like vnto our ordinary Pippin tree in body, branch and leafe, the only difference is in the flower, which is altogether whitiſh, ſauing that the inner leaues towards the middle are more reddiſh, but as double and thicke as our double Damaske Roſes, which fall away without bearing fruit.

Malus Perſica flore multiplici. The double bloſſomd Peach tree.

This Peach tree for the manner of growing, is ſo like vnto an ordinary Peach tree, that vntill you ſee it in bloſſome you can perceiue no difference : the flower is of the ſame colour with the bloſſomes of the Peach, but conſiſting of three or foure, or more rowes of leaues, which fall often away likewiſe without bearing any fruite ; but after it hath abiden ſome yeares in a place doth forme into fruite, eſpecially being planted againſt a wall.

The Place.

Both the Cherry trees are frequent in many places of England, nourſed for their pleaſant flowers. The Apple is as yet a ſtranger. And the Peach hath not been ſeen or knowne, long before the writing hereof.

The Time.

They all flower in April & May, which are the times of their other kinds.

The Names.

Their names are alſo ſufficiently expreſſed to know them by.

The Vertues.

Cherries, Peaches and Apples, are recorded in our Orchard, and there you ſhall finde the properties of their fruit : for in that theſe beare none or very few, their bloſſomes are of moſt vſe to grace and decke the perſons of thoſe that will weare or beare them.

Chap. CV.

Periclymenum. Honyſuckles.

THe Honiſuckle that groweth wilde in euery hedge, although it be very ſweete, yet doe I not bring into my garden, but let it reſt in his owne place, to ſerue their ſenſes that trauell by it, or haue no garden. I haue three other that furniſh my Garden, one that is called double, whoſe branches ſpreade far, and being very fit for an arbour will ſoone couer it : the other two ſtand vpright, and ſpreade not any way far, yet their flowers declaring them to be Honiſuckles, but of leſſe delight, I conſort them with the other.

Periclymenum perfoliatum ſiue Italicum. The double Honiſuckle.

The truncke or body of the double Honiſuckle, is oftentimes of the bigneſſe of a good ſtaffe, running out into many long ſpreading branches, couered with a whitiſh barke, which had neede of ſome thing to ſuſtaine them, or elſe they will fall down to the ground (and therefore it is vſually planted at an arbour, that it may run thereon,

or

or againſt a houſe wall, and faſtened thereto in diuers places with nailes) from whence
ſpring forth at ſeuerall diſtances, and at the ioynts, two leaues, being like in forme vnto the wilde Honiſuckles, and round pointed for the moſt part; theſe branches diuiding themſelues diuers wayes, haue at the toppes of them many flowers, ſet at certaine
diſtances one aboue another, with two greene leaues at euery place; where the flowers
doe ſtand, ioyned ſo cloſe at the bottome, and ſo round and hollow in the middle, that
it ſeemeth like a hollow cuppe or ſawcer of flowers : the flowers ſtand round about
the middle of theſe cuppes or ſawcers, being long, hollow, and of a whitiſh yellow
colour, with open mouthes daſht ouer with a light ſhew of purple, and ſome threds
within them, very ſweet in ſmell, like both in forme and colour vnto the common Honiſuckles, but that theſe cuppes with the flowers in them are two or three ſtanding one
aboue another (which make a far better ſhew then the common, which come forth all
at the heade of the branches, without any greene leaues or cuppes vnder them) and
therefore theſe were called double Honiſuckles.

Periclymenum reɑum fruɑu rubro. Red Honiſuckles.

This vpright Woodbinde hath a ſtraight woody ſtemme, diuided into ſeuerall
branches, about three or foure foote high, couered with a very thinne whitiſh barke,
whereon ſtand two leaues together at the ioynts, being leſſer then the former, ſmooth
and pleine, and a little pointed : the flowers come forth vpon ſlender long footſtalks
at the ioynts where the leaues ſtand, alwayes two ſet together, and neuer more, but
ſeldome one alone, which are much ſmaller then the former, but of the ſame faſhion,
with a little button at the foote of the flower; the buds of the flowers before they
are open are very reddiſh, but being open are not ſo red, but tending to a kinde of yellowiſh bluſh colour : after which come in their places two ſmall red berries, the one
withered for the moſt part, or at leaſt ſmaller then the other, but (as Cluſius ſaith) in
their naturall places they are both full and of one bigneſſe.

Periclymenum reɑum fruɑu cæruleo. Blew berried Honiſuckles.

This other vpright Woodbinde groweth vp as high as the former, or rather ſomewhat higher, couered with a blackiſh rugged barke, chapping in diuers places, the
younger branches whereof are ſomewhat reddiſh, and couered with an hoary doune :
the leaues ſtand two together at the ioints, ſomewhat larger then the former, and more
whitiſh vnderneath : the flowers are likewiſe two ſtanding together, at the end of a
ſlender footeſtalke, of a pale yellowiſh colour when they are blowne, but more reddiſh in the bud : the berries ſtand two together as the former, of a darke blewiſh
colour when they are fully ripe, and full of a red liquour or iuice, of a pleaſant taſte,
which doth not only dye the hands of them that gather them, but ſerueth for a dying
colour to the inhabitants where they grow plentifully, wherein are contained many
flat ſeede : The roote is woody as the former is.

The Place.

The firw groweth in Italie, Spaine, and Prouence of France, but not in
the colder countreyes, vnleſſe it be there planted, as is moſt frequent in our
countrey. The others grow in Auſtria, and Stiria, as Cluſius ſaith, and are
entertained into their gardens onely that are curious.

The Time.

The firſt flowreth vſually in Aprill, the reſt in May.

The Names.

The firſt is called *Periclymenum, Caprifolium perfoliatum,* and *Italicum,*
as a difference from the common kinde : In Engliſh Double Woodbinde,
or

or double Honifuckles. The others, as they are rare, and little knowne, so are their names also : yet according to their Latine, I haue giuen them English names.

The Vertues.

The double Honifuckle is as effectuall in all things, as the single wilde kinde, and besides, is an especiall good wound herbe for the head or other parts. I haue not knowne the vpright kindes vsed in Physicke.

Chap. CVI.

Iasminum siue Gelseminum. Iasmine or Gesmine.

WE haue but one sort of true Iasmine ordinarily in our Gardens throughout the whole Land ; but there is another greater sort, which is farre more tender, brought out of Spaine, and will hardly endure any long time with vs, vnlesse it be very carefully preserued. Wee haue a third kinde called a yellow Iasmine, but differeth much from their tribe in many notable points : but because the flowers haue some likenesse with the flowers of the true Iasmine, it hath been vsually called a Iasmine ; and therefore I am content for this Garden to conioyne them in one Chapter.

1. *Iasminum album.* The white Iasmine.

The white Iasmine hath many twiggy flexible greene branches, comming forth of the sundry bigger boughes or stems, that rise from the roote, which are couered with a grayish darke coloured barke, hauing a white pith within it like the Elder, but not so much : the winged leaues stand alwaies two together at the ioynts, being made of manie small and pointed leaues, set on each side of a middle ribbe, six most vsually on both sides, with one at the end, which is larger, more pointed then any of the rest, and of a darke greene colour : at the toppes of the young branches stand diuers flowers together, as it were in an vmbell or tuft, each whereof standeth on a long greene stalke, comming out of a small huske, being small, long, and hollow belowe, opening into fiue white small, pointed leaues, of a very strong sweete smell, which fall away without bearing any fruit at all, that euer I could learne in our Country ; but in the hot Countries where it is naturall, it is said to beare flat fruit, like Lupines : the rootes spread farre and deepe, and are long and hard to growe, vntill they haue taken strong hold in the ground.

2. *Iasminum Catalonicum.* The Spanish Iasmine.

This Catalonia Iasmine groweth lower then the former, neuer rising halfe so high, and hath slender long greene branches, rising from the toppe of the wooddy stemme, with such like leaues set on them as the former, but somewhat shorter and larger : the flowers also are like vnto the former, and stand in the same manner at the end of the branches, but are much larger, being of a blush colour before they are blowne, and white with blush edges when they are open, exceeding sweete of smell, more strong then the former.

3. *Iasminum luteum, siue Trifolium fruticans alijs Polemonium.* The yellow Iasmine.

This that is called the yellow Iasmine, hath many long slender twiggy branches rising from the roote, greene at the first, and couered with a darke grayish barke afterwards, whereon are set at certaine distances, three small darke greene leaues together, the end leafe being alwaies the biggest : at the ioynts where the leaues come forth,

stand

ftand long ftalkes, bearing long hollow flowers, ending in fiue, and fome in fix leaues, very like vnto the flowers of the firft Iafmine, but yellow, whereupon it is vfually called the Yellow Iafmine : after the flowers are paft, there come in their places round blacke fhining berries, of the bigneffe of a great Peafe, or bigger, full of a purplifh iuyce, which will dye ones fingers that bruife them but a little : the roote is tough, and white, creeping farre about vnder the ground, fhooting forth plentifully, whereby it greatly encreafeth.

The Place.

The firft is verily thought to haue been firft brought to Spaine out of Syria, or thereabouts, and from Spaine to vs, and is to be feene very often, and in many of our Country Gardens. The fecond hath his breeding in Spaine alfo, but whether it be his originall place we know not, and is fcarce yet made well acquainted with our Englifh ayre. The third groweth plentifully about Mompelier, and will well abide in our London Gardens, and any where elfe.

The Time.

The firft flowreth not vntill the end of Iuly. The fecond fomewhat earlier. The third in Iuly alfo.

The Names.

The firft is generally called *Iafminum album*, and *Gelfeminum album* : In Englifh, The white Iafmine. The fecond hath his name in his title, as much as may be faid of it. The third hath been taken of fome to be a *Cytifus*, others iudge it to be *Polemonium*, but the trueft name is *Trifolium fruticans*, although many call it *Iafminum luteum* : In Englifh moft vfually, The yellow Iafmine, for the reafons aforefaid ; or elfe after the Latine name, Shrubbie Trefoile, or Make-bate.

The Vertues.

The white Iafmines haue beene in all times accepted into outward medicines, eyther for the pleafure of the fweete fent, or profit of the warming properties. And is in thefe dayes onely vfed as an ornament in Gardens, or for fent of the flowers in the houfe, &c. The yellow Iafmine, although fome haue adiudged it to be the *Polemonium* of Diofcorides, yet it is not vfed to thofe purpofes by any that I know.

Chap. CVII.

Syringa. The Pipe tree.

Vnder the name of *Syringa*, is contained two fpeciall kinds of Shrubs or Trees, differing one from another ; namely, the *Lilac* of Matthiolus, which is called *Syringa cærulea*, and is of two or three forts : And the *Syringa alba*, which alfo is of two forts, as fhall bee declared.

1. *Lilac fiue Syringa cærulea.* The blew Pipe tree.

The blew Pipe tree rifeth fometimes to be a great tree, as high and bigge in the bodie as a reafonable Apple tree (as I haue in fome places feene and obferued) but moft vfually groweth lower, with many twigs or branches rifing from the roote, hauing as much pith in the middle of them as the Elder hath, couered with a grayifh greene

barke,

barke, but darker in the elder branches, with ioynts ſet at a good diſtance one from another, and two leaues at euery ioynt, which are large, broad, and pointed at the ends, many of them turning or folding both the ſides inward, and ſtanding on long foote ſtalkes : at the toppes of the branches come forth many flowers, growing ſpike-faſhion, that is, a long branch of flowers vpon a ſtalke, each of theſe flowers are ſmall, long, and hollow belowe, ending aboue in a pale blewiſh flower, conſiſting of foure ſmall leaues, of a pretty ſmall ſent : after the flowers are paſt, there come ſometimes (but it is not often in our Country, vnleſſe the tree haue ſtood long, and is grown great, the ſuckers being continually taken away, that it may growe the better) long and flat cods, conſiſting as it were of two ſides, a thin ſkinne being in the midſt, wherein are contained two long flattiſh red ſeede : the rootes are ſtrong, and growe deepe in the ground.

2. *Syringa flore lacteo ſiue argenteo.*
The ſiluer coloured Pipe tree.

This Pipe tree differeth not from the former blew Pipe tree, either in ſtemme or branches, either in leaues or flowers, or manner of growing, but onely in the colour of the flower, which in this is of a milke, or ſiluer colour, which is a kinde of white, wherein there is a thinne waſh, or light ſhew of blew ſhed therein, comming ſomewhat neare vnto an aſh-colour.

3. *Lilac lacinatis folijs.* The blew Pipe tree with cut leaues.

This Pipe tree ſhould not differ from the firſt in any other thing then in the leaues, which are ſaid to be cut in on the edges into ſeuerall parts, as the relation is giuen *à viris fide dignis* ; for as yet I neuer ſaw any ſuch; but I here am bold to ſet it downe, to induce and prouoke ſome louer of plants to obtaine it for his pleaſure, and others alſo.

4. *Syringa flore albo ſimplici.* The ſingle white Pipe tree.

The ſingle white Pipe tree or buſh, neuer commeth to that height of the former, but abideth alwaies like a hedge tree or buſh, full of ſhootes or ſuckers from the roote, much more then the former : the young ſhootes hereof are reddiſh on the outſide, and afterward reddiſh at the ioynts, and grayiſh all the reſt ouer : the young as well as the old branches, haue ſome pith in the middle of them, like as the Elder hath : the leaues ſtand two at a ioynt, ſomewhat like the former, but more rugged or crumpled, as alſo a little pointed, and dented about the edges : the flowers growe at the toppes of the branches, diuers ſtanding together, conſiſting of foure white leaues, like vnto ſmall Muske Roſes, and of the ſame creame colour, as I may call it, with many ſmall yellowiſh threads in the middle, and are of a ſtrong, full, or heady ſent, not pleaſing to a great many, by reaſon of the ſtrange quickneſſe of the ſent : the fruit followeth, being flat at the head, with many leafie ſhels or ſcales compaſſing it, wherein is encloſed ſmall long ſeede : the rootes runne not deepe, but ſpread vnder the ground, with many fibres annexed vnto them.

5. *Syringa Arabica flore albo duplici.*
The double white Pipe tree.

This Pipe tree hath diuers long and ſlender branches, whereon growe large leaues, ſomewhat like vnto the leaues of the former ſingle white kinde, but not ſo rough or hard, and not at all dented about the edges, two alwaies ſtanding one againſt another at euery ioynt of the ſtalke, but ſet or diſpoſed on contrary ſides, and not all vpon one ſide; at the ends whereof come forth diuers flowers, euery one ſtanding on his owne foote-ſtalke, the hoſe or huſke being long and hollow, like vnto the white Iaſmine, and the flowers therin conſiſting of a double rowe of white and round pointed leaues, fiue or ſix in a rowe, with ſome yellowneſſe in the middle, which is hollow, of a very ſtrong and heady ſweet ſent, and abiding a long time flowring, eſpecially in the hotter Countries, but is very tender, and not able to abide any the leaſt cold weather with vs;

1 *Iasminum album vulgare.* The ordinary white Iasmine. 2 *Iasminum Americanum siue Convoluulus Americanus.* The Iasmine or Bindweed of America.
3 *Iasminum luteum vulgare.* The yellow Iasmine. 4 *Lilac seu Syringa cærules.* The blew Pipe tree. 5 *Syringa alba vulgaris.* The single white Syringa or
Pipe tree. 6 *Syringa flore albo duplici.* The double white Syringa. 7 *Sambucus rosea.* The Elder or Gelder Rose.

for the cold windes will (as I vnderftand) greatly moleft it : and therefore muft as charily be kept as Orenge trees with vs, if wee will haue it to abide.

The Place.

The firft groweth in Arabia (as Matthiolus thinketh, that had it from Conftantinople.) We haue it plentifully in our Gardens. The fecond and third are ftrangers with vs as yet. The fourth is as frequent as the firft, or rather more, but his originall is not knowne. The laft hath his originall from Arabia, as his name importeth.

The Time.

The firft, fecond, and third flower in Aprill, the other two not vntill May.

The Names.

The firft is called of Matthiolus *Lilac*, and by that name is moft vfually called in all parts. It is alfo called *Syringa cærulea*, becaufe it commeth neareft vnto thofe woods, which for their pithy fubftance, were made hollow into pipes. It is called of all in Englifh, The blew Pipe tree. It feemeth likely, that Petrus Bellonius in his third Booke and fiftieth Chapter of his obferuations (making mention of a fhrubbe that the Turkes haue, with Iuie leaues alwaies greene, bearing blew or violet coloured flowers on a long ftalke, of the bigneffe and fafhion of a Foxe taile, and thereupon called in their language a Foxe taile) doth vnderftand this plant here expreffed. The certainty whereof might eafily be knowne, if any of our Merchants there refiding, would but call for fuch a fhrubbe, by the name of a Foxe taile in the Turkifh tongue, and take care to fend a young roote, in a fmall tubbe or basket with earth by Sea, vnto vs here at London, which would be performed with a very little paines and coft. The fecond and third, as kindes thereof, haue their names in their titles. The fourth is called by Clufius and others, *Frutex Coronarius*; fome doe call it *Lilac flore albo*, but that name is not proper, in that it doth confound both kindes together. Lobel calleth it *Syringa Italica*. It is now generally called of all *Syringa alba*, that is in Englifh, The white Pipe tree. Some would haue it to bee *Oftrys* of Theophraftus, but Clufius hath fufficiently cleared that doubt. Of others *Liguftrum Orientale*, which it cannot be neither ; for the *Cyprus* of Plinie is Diofcorides his *Liguftrum*, which may be called *Orientale*, in that it is moft proper to the Eafterne Countries, and is very fweete, whofe feede is like vnto Coriander feede. The laft is called by diuers *Syringa Arabica flore albo duplici*, as moft fitly agreeing thereunto. Of Bafilius Beflerus that fet forth the great booke of the Bifhop of Eyftot in Germany his Garden, *Syringa Italica flore albo pleno*, becaufe, as it is likely, hee had it from Italy. It is very likely, that Profper Alpinus in his booke of Egyptian plants, doth meane this plant, which hee there calleth *Sambach, fiue Iafminum Arabicum*. Matthæus Caccini of Florence in his letter to Clufius entituleth it *Syringa Arabica, fiue Iafminum Arabicum, fiue Iafminum ex Gine*, whereby hee declareth that it may not vnfitly be referred to either of them both. We may call it in Englifh as it is in the title, The double white Pipe tree.

The Vertues.

We haue no vfe of thefe in Phyficke that I know, although Profper Alpinus faith, the double white Pipe tree is much vfed in Egypt, to help women in their trauailes of childbirth.

CHAP. CVIII.

Sambucus Rosea. The Elder or Gelder Rose.

ALthough there be diuers kindes of Elders, yet there is but one kinde of Elder Rose, whereof I meane to intreate in this Chapter, being of neare affinity in some things vnto the former Pipe trees, and which for the beauty of it deserueth to be remembred among the delights of a Garden.

Sambucus Rosea. The Gelder Rose.

The Gelder Rose (as it is called) groweth to a reasonable height, standing like a tree, with a trunke as bigge as any mans arme, couered with a darke grayish barke, somewhat rugged and very knotty : the younger branches are smooth and white, with a pithy substance in the middle, as the Elders haue, to shew that it is a kind thereof, whereon are set broad leaues, diuided into three parts or diuisions, somewhat like vnto a Vine leafe, but smaller, and more rugged or crumpled, iagged or cut also about the edges : at the toppes of euery one of the young branches, most vsually commeth forth a great tuft, or ball as it were, of many white flowers, set so close together, that there can be no distinction of any seuerall flower seene, nor doth it seeme like the double flower of any other plant, that hath many rowes of leaues set together, but is a cluster of white leaued flowers set together vpon the stalke that vpholdeth them, of a small sent, which fall away without bearing any fruit in our Country, that euer I could obserue or learne : The roote spreadeth neither farre nor deepe, but shooteth many small rootes and fibres, whereby it is fastened in the ground, and draweth nourishment to it, and sometimes yeeldeth suckers from it.

The Place.

It should seeme, that the naturall place of this Elder is wet and moist grounds, because it is so like vnto the Marsh Elder, which is the single kind hereof. It is onely noursed vp in Gardens in all our Country.

The Time.

It flowreth in May, much about the time of the double Peony flower, both which being set together, make a pleasant variety, to decke vp the windowes of a house.

The Names.

It is generally called *Sambucus Rosea* : In English, The Elder Rose, and more commonly after the Dutch name, the Gelder Rose. Dalechampius seemeth to make it *Thraupalus* of Theophrastus, or rather the single Marsh Elder ; for I thinke this double kinde was not knowne in Theophrastus his time.

The Vertues.

It is not applyed to any Physicall vse that I know.

CHAP. CIX.

Rosa. The Rose tree or bush.

THe great varietie of Roses is much to be admired, beeing more then is to bee seene in any other shrubby plant that I know, both for colour, forme and smell. I haue to furnish this garden thirty sorts at the least, euery one notably diffe-ring from the other, and all fit to be here entertained : for there are some other, that being wilde and of no beautie or smell, we forbeare, and leaue to their wilde habita-tions. To distinguish them by their colours, as white, red, incarnate, and yellow, were a way that many might take, but I hold it not so conuenient for diuers respects : for so I should confound those of diuers sorts one among another, and I should not keepe that methode which to me seemeth most conuenient, which is, to place and ranke e-uery kinde, whether single or double, one next vnto the other, that so you may the bet-ter vnderstand their varieties and differences : I will therefore beginne with the most ancient, and knowne Roses to our Countrey, whether naturall or no I know not, but assumed by our precedent Kings of all others, to bee cognisances of their dignitie, the white Rose and the red, whom shall follow the damaske, of the finest sent, and most vse of all the other sorts, and the rest in their order.

1. *Rosa Anglica alba.* The English white Rose.

The white Rose is of two kindes, the one more thicke and double then the other : The one riseth vp in some shadowie places, vnto eight or ten foote high, with a stocke of a great bignesse for a Rose. The other growing seldome higher then a Damaske Rose. Some doe iudge both these to be but one kinde, the diuersitie happening by the ayre, or ground, or both. Both these Roses haue somewhat smaller and whiter greene leaues then in many other Roses, fiue most visually set on a stalke, and more white vnder-neath, as also a whiter greene barke, armed with sharpe thornes or prickles, whereby they are soone known from other Roses, although the one not so easily from the other : the flowers in the one are whitish, with an eye or shew of a blush, especially towards the ground or bottome of the flower, very thicke double, and close set together, and for the most part not opening it selfe so largely and fully as eyther the Red or Damaske Rose. The other more white, lesse thicke and double, and opening it selfe more, and some so little double as but of two or three rowes, that they might be held to be single, yet all of little or no smell at all. To describe you all the seuerall parts of the Rose, as the bud, the beards, the threds &c. were needlesse, they are so conuersant in euery ones hand, that I shall not neede but to touch the most speciall parts of the varieties of them, and leaue a more exact relation of all things incident vnto them, vnto a generall worke.

2. *Rosa Incarnata.* The Carnation Rose.

The Carnation Rose is in most things like vnto the lesser white rose, both for the growing of the stocke, and bignesse of the flower, but that it is more spreade abroade when it is blown then the white is, and is of a pale blush colour all the flower through-out, of as small a sent as the white one is almost.

Rosa Belgica siue Vitrea.

This kinde of Rose is not very great, but very thicke and double, and is very variable in the flowers, in that they will be so different one from another : some being paler then others, and some as it were blasted, which commeth not casually, but naturally to this rose : but the best flowers (whereof there will bee still some) will be of a bright pale murrey colour, neere vnto the Veluet rose, but nothing so darke a colour.

3. *Rosa Anglica rubra.* The English red Rose.

The red Rose (which I call English, not only for the reason before expressed, but be-cause

cause (as I take it) this Rose is more frequent and vsed in England, then in other places) neuer groweth so high as the damaske Rose bush, but most vsually abideth low, and shooteth forth many branches from the roote (and is but seldome suffered to grow vp as the damaske Rose into standards) with a greene barke, thinner set with prickles, and larger and greener leaues on the vpperside then in the white, yet with an eye of white vpon them, fiue likewise most vsually set vpon a stalke, and grayish or whitish vnderneath. The Roses or Flowers doe very much vary, according to their site and abiding; for some are of an orient, red or deepe crimson colour, and very double (although neuer so double as the white) which when it is full blowne hath the largest leaues of any other Rose; some of them againe are paler, tending somewhat to a damaske; and some are of so pale a red, as that it is rather of the colour of the canker Rose, yet all for the most part with larger leaues then the damaske, and with many more yellow threds in the middle : the sent hereof is much better then in the white, but not comparable to the excellencie of the damaske Rose, yet this Rose being well dryed and well kept, will hold both colour and sent longer then the damaske, bee it neuer so well kept.

4. *Rosa Damascena.* The Damaske Rose.

The Damaske Rose bush is more vsually nourced vp to a competent height to stand alone, (which we call Standards) then any other Rose : the barke both of the stocke and branches, is not fully so greene as the red or white Rose : the leaues are greene with an eye of white vpon them, so like vnto the red Rose, that there is no great difference betweene them, but that the leaues of the red Rose seeme to bee of a darker greene. The flowers are of a fine deepe blush colour, as all know, with some pale yellow threds in the middle, and are not so thicke and double as the white, nor being blowne, with so large and great leaues as the red, but of the most excellent sweet pleasant sent, far surpassing all other Roses or Flowers, being neyther heady nor too strong, nor stuffing or vnpleasant sweet, as many other flowers.

5. *Rosa Prouincialis siue Hollandica Damascena.* The great double Damaske Prouince or Holland Rose.

This Rose (that some call *Centifolia Batauica incarnata*) hath his barke of a reddish or browne colour, whereby it is soone discerned from other Roses. The leaues are likewise more reddish then in others, and somewhat larger, it vsually groweth very like the Damaske rose, and much to the same height : the flowers or roses are of the same deepe blush colour that the damaske roses are, or rather somewhat deeper, but much thicker, broader, and more double, or fuller of leaues by three parts almost, the outer leaues turning themselues backe, when the flower hath stood long blowne, the middle part it selfe (which in all other roses almost haue some yellow threds in them to be seene) being folded hard with small leaues, without any yellow almost at all to be seene, the sent whereof commeth neerest vnto the damaske rose, but yet is short of it by much, howsoeuer many doe thinke it as good as the damask, and to that end I haue known some Gentlewomen haue caused all their damaske stockes to bee grafted with prouince Roses, hoping to haue as good water, and more store of them then of damask Roses ; but in my opinion it is not of halfe so good a sent as the water of damaske Roses: let euery one follow their own fancie.

6. *Rosa Prouincialis rubra.* The red Prouince Rose.

As the former was called *incarnata*, so this is called *Batauica centifolia rubra*, the difference being not very great : the stemme or stocke, and the branches also in this, seeming not to be so great but greener, the barke being not so red ; the leaues of the same largenesse with the former damaske Prouince. The flowers are not altogether so large, thicke and double, and of a little deeper damaske or blush colour, turning to a red Rose, but not comming neere the full colour of the best red Rose, of a sent not so sweete as the damaske Prouince, but comming somewhat neere the sent of the ordinary red rofe, yet exceeding it. This rofe is not fo plentifull in bearing as the damaske Prouince.

7. *Rofa Prouincialis alba.* The white Prouince Rofe.

It is faid of diuers, that there is a white Prouince Rofe, whereof I am not *oculatus teſtis*, and therfore I dare not giue it you for a certaintie, and indeed I haue fome doubt, that it is the greater and more double white rofe, whereof I gaue you the knowledge in the beginning: when I am my felfe better fatisfied, I fhall bee ready to fatisfie others.

8. *Rofa verſicolor.* The party coloured Rofe, of fome Yorke and Lancaſter.

This Rofe in the forme and order of the growing, is neereſt vnto the ordinary damaske rofe, both for ſtemme, branch, leafe and flower : the difference confiſting in this, that the flower (being of the fame largeneſſe and doubleneſſe as the damask rofe) hath the one halfe of it, fometimes of a pale whitifh colour, and the other halfe, of a paler damaske colour then the ordinary; this happeneth fo many times, and fometimes alfo the flower hath diuers ſtripes, and markes in it, as one leafe white, or ſtriped with white, and the other halfe blufh, or ſtriped with blufh, fometimes alfo all ſtriped, or fpotted ouer, and other times little or no ſtripes or markes at all, as nature lifteth to play with varieties, in this as in other flowers : yet this I haue obferued, that the longer it abideth blowen open in the fun, the paler and the fewer ſtripes, markes or fpots will be feene in it : the fmell whereof is of a weake damaske rofe fent.

9. *Rofea Chryſtallina.* The Chryſtall Rofe.

This Rofe is very like vnto the laſt defcribed, both for ſtocke, branch and leafe : the flower hereof is not much different from it, being no great large or double Rofe, but of a meane fize, ſtriped and marked with a deeper blufh or red, vpon the pale coloured leafe, that it feemeth in the marking and beauty thereof, to bee of as much delight as the Chryſtall Gilloflower : this, euen like the former, foone fadeth and paffeth away, not yeelding any great ſtore of flowers any yeare.

10. *Rofa rubra humilis ſiue pumilio.* The dwarfe red Rofe, or Gilloflower Rofe.

This Rofe groweth alwayes low and fmall, otherwife in moſt refpects like vnto the ordinary redde Rofe, and with few or no thornes vpon it : the Flowers or Rofes are double, thicke, fmall and clofe, not fo much fpread open as the ordinary red, but fomewhat like vnto the firſt double white Rofe before expreſſed ; yet in fome places I haue feene them more layde open then thefe, as they grew in my garden, being fo euen at the toppes of the leaues, as if they had been clipt off with a paire of fheeres, and are not fully of fo red a colour as the red Prouince Rofe, and of as fmall or weak fent as the ordinary red Rofe, or not fo much.

11. *Rofa Francafurtenſis.* The Franckford Rofe.

The young fhootes of this Rofe are couered with a pale purplifh barke, fet with a number of fmall prickes like haires, and the elder haue but very few thornes : the flower or rofe it felfe hath a very great bud or button vnder it, more then in any other rofe, and is thicke and double as a red rofe, but fo ſtrongly fwelling in the bud, that many of them breake before they can be full blowen, and then they are of a pale red rofe colour, that is, betweene a red and a damaske, with a very thicke broade and hard vmbone of fhort yellow threds or thrumes in the middle, the huske of the flower hauing long ends, which are called the beards of the rofe, which in all other are iagged in fome of them, in this hath no iagge at all : the fmell is neereſt vnto a red Rofe.

1 *Rosa Damascena.* The Damaske Rose. 2 *Rosa Prouincialis siue Hollandica.* The great Prouince Rose. 3 *Rosa Francafurtensis.* The
Franckford Rose. 4 *Rosa rubra humilis.* The dwarfe red Rose. 5 *Rosa Hungarica.* The Hungarian Rose. 6 *Rosa lutea multiplex.*
The great double yellow Rose.

12. *Rosa Hungarica.* The Hungarian Rose.

The Hungarian Rose hath greene shootes slenderly set with prickes, and seldome groweth higher then ordinarily the red Rose doth ; the stemme or stocke being much about that bignesse : the flower or rose is as great, thicke and double, as the ordinary red Rose, and of the same fashion, of a paler red colour, and beeing neerely looked vpon is finely spotted with faint spots, as it were spreade ouer the red ; the smell wherof is somewhat better then the smell of the ordinary red Rose of the best kinde.

13. *Rosa Holoserica simplex & multiplex.* The Veluet Rose single and double.

The old stemme or stock of the veluet Rose is couered with a dark coloured barke, and the young shootes of a sad greene with very few or no thornes at all vpon them : the leaues are of a sadder greene colour then in most sorts of Roses, and very often seuen on a stalke, many of the rest hauing but fiue: the Rose is eyther single or double : the single is a broade spread flower, consisting of fiue or sixe broade leaues with many yellow threds in the middle : the double hath two rowes of leaues, the one large, which are outermost, the other smaller within, of a very deepe red crimson colour like vnto crimson veluet, with many yellow threds also in the middle ; and yet for all the double rowe of leaues, these Roses stand but like single flowers : but there is another double kinde that is more double then this last, consisting oftentimes of sixteene leaues or more in a flower, and most of them of an equall bignesse, of the colour of the first single rose of this kinde, or somewhat fresher ; but all of them of a smaller sent then the ordinary red Rose.

14. *Rosa sine spinis simplex & multiplex.* The Rose without thornes single and double.

The Rose without thornes hath diuers greene smooth shootes, rising from the root, without any pricke or thorne at all vpon them, eyther young or old: the leaues are not fully so large as of the red rose : the flowers or roses are not much bigger then those of the double Cinamon Rose, thicke set together and short, of a pale red Rose colour, with diuers pale coloured veines through euery leafe of the flower, which hath caused some to call it The marbled Rose, and is of a small sent, not fully equall to the red Rose. The single of this kinde differeth not in any other thing from the former, then in the doublenesse or singlenesse of the flowers, which in this are not halfe so double, nor yet fully single, and are of a paler red colour.

Rosa sine spina flore albo. I haue heard likewise of a white Rose of this kinde, but I haue seene none such as yet, and therefore I can say no more thereof.

15. *Rosa Cinamomea simplex & multiplex.* The Cinamon Rose single and double.

The single Cinamon Rose hath his shootes somewhat red, yet not so red as the double kinde, armed with great thornes, like almost vnto the Eglantine bush, thereby showing, as well by the multiplicitie of his shootes, as the quicknesse and height of his shooting, his wilde nature : On the stemme and branches stand winged leaues, sometimes seuen or more together, which are small and greene, yet like vnto other Roses. The Roses are single, of fiue leaues a peece, somewhat large, and of a pale red colour, like vnto the double kinde, which is in shootes redder, and in all other things like vnto the single, but bearing small, short, thicke and double Roses, somewhat like vnto the Rose without thornes, but a little lesser, of a paler red colour at the end of the leaues, and somewhat redder and brighter toward the middle of them, with many yellow short thrumes ; the small sent of Cinamon that is found in the flowers hath caused it to beare the name.

16. *Rosa lutea simplex.* The single yellow Rose.

This single yellow Rose is planted rather for variety then any other good vse. It often groweth to a good height, his stemme being great and wooddy, with few or no prickes vpon the old wood, but with a number of small prickes like haires, thicke set, vpon the younger branches, of a darke colour somewhat reddish, the barke of the young shootes being of a sad greene reddish colour : the leaues of this Rose bush are smaller, rounder pointed, of a paler greene colour, yet finely snipt about the edges, and more in number, that is, seuen or nine on a stalke or ribbe, then in any other Garden kinde, except the double of the same kinde that followeth next : the flower is a small single Rose, consisting of fiue leaues, not so large as the single Spanish Muske Rose, but somewhat bigger then the Eglantine or sweete Briar Rose, of a fine pale yellow colour, without any great sent at all while it is fresh, but a little more, yet small and weake when it is dryed.

17. *Rosa lutea multiplex siue flore pleno.* The double yellow Rose.

The double yellow Rose is of great account, both for the rarity, and doublenesse of the flower, and had it sent to the rest, would of all other be of highest esteeme. The stemme or stocke, the young shoots or branches, the small hairy prickes, and the small winged leaues, are in all parts like vnto the former single kinde ; the chiefest difference consisteth in the doublenesse of the flower or Rose, which is so thicke and double, that very often it breaketh out on one side or another, and but a few of them abiding whole and faire in our Countrey, the cause whereof wee doe imagine to bee the much moisture of our Countrey, and the time of flowring being subiect to much raine and showers ; many therefore doe either plant it against a wall, or other wayes defend it by couering : againe, it is so plentifull in young shootes or branches, as also in flowers at the toppe of euery branch, which are small and weake for the most part, that they are not able to bring all the flowers to ripenesse ; and therefore most of them fall or wither away without comming to perfection (the remedy that many doe vse for this inconuenience last recited is, that they nippe away most of the buds, leauing but some few vpon it, that so the vigour of the plant may be collected into a few flowers, whereby they may the better come to perfection, and yet euen thus it is hardly effected) which are of a yellowish greene colour in the bud, and before they be blowne open, but then are of a faire yellow colour, very full of leaues, with many short haires rather then leaues in the middle, and hauing short, round, greene, smooth buttons, almost flat vnder them : the flower being faire blowne open, doth scarce giue place for largenesse, thicknesse, and doublenesse, vnto the great Prouence or Holland Rose. This Rose bush or plant is very tender with vs here about London, and will require some more care and keeping then the single of this kinde, which is hardy enough ; for I haue lost many my selfe, and I know but a few about this towne that can nourse it vp kindly, to beare or scarce to abide without perishing ; but abideth well in euery free aire of all or the most parts of this Kingdome : but (as I heare) not so well in the North.

18. *Rosa Moschata simplex & multiplex.* The Muske Rose single and double.

The Muske Rose both single and double, rise vp oftentimes to a very great height, that it ouergroweth any arbour in a Garden, or being set by an house side, to bee ten or twelue foote high, or more, but more especially the single kinde, with many green farre spread branches, armed with a few sharpe great thornes, as the wilder sorts of Roses are, whereof these are accounted to be kindes, hauing small darke greene leaues on them, not much bigger then the leaues of Eglantine : the flowers come forth at the toppes of the branches, many together as it were in an vmbell or tuft, which for the most part doe flower all at a time, or not long one after another, euery one standing on a pretty long stalke, and are of a pale whitish or creame colour, both the single and

the

the double; the single being small flowers, consisting of fiue leaues, with many yellow threads in the middle: and the double bearing more double flowers, as if they were once or twice more double then the single, with yellow thrummes also in the middle, both of them of a very sweete and pleasing smell, resembling Muske: some there be that haue auouched, that the chiefest sent of these Roses consisteth not in the leaues, but in the threads of the flowers.

19. *Rosa Moschata multiplex altera: alijs Damascena alba, vel verisimilior Cinamomea flore pleno albo.* The double white Damaske Muske Rose.

This other kinde of Muske Rose (which with some is called the white Damaske Muske, but more truely the double white Cinamon Rose) hath his stemme and branches also shorter then the former, but as greene: the leaues are somewhat larger, and of a whiter greene colour; the flowers also are somewhat larger then the former double kinde, but standing in vmbels after the same manner, or somewhat thicker, and of the same whitish colour, or a little whiter, and somewhat, although but a little, neare the smell of the other, but nothing so strong. This flowreth at the time of other Roses, or somewhat later, yet much before the former two sorts of Muske Roses, which flower not vntill the end of Summer, and in Autumne; both which things, that is, the time of the flowring, and the sent being both different, shew plainly it cannot be of the tribe of Muske Roses.

20. *Rosa Hispanica Moschata simplex.* The Spanish Muske Rose.

This Spanish Rose riseth to the height of the Eglantine, and sometimes higher, with diuers great greene branches, the leaues whereof are larger and greener then of the former kindes: the flowers are single Roses, consisting of fiue whiter leaues then in any of the former Muske Roses, and much larger, hauing sometimes an eye of a blush in the white, of a very sweete smell, comming nearest vnto the last recited Muske Rose, as also for the time of the flowring.

21. *Rosa Pomifera maior.* The great Apple Rose.

The stemme or stocke of this Rose is great, couered with a darke grayish barke, but the younger branches are somewhat reddish, armed here and there with great and sharpe thornes, but nothing so great or plentifull as in the Eglantine, although it be a wilde kinde: the leaues are of a whitish greene colour, almost like vnto the first white Rose, and fiue alwaies set together, but seldome seuen: the flowers are small and single, consisting of fiue leaues, without any sent, or very little, and little bigger then those of the Eglantine bush, and of the very same deepe blush colour, euery one standing vpon a rough or prickly button, bearded in the manner of other Roses, which when the flowers are fallen growe great, somewhat long and round, peare-fashion, bearing the beards on the tops of them; and being full ripe are very red, keeping the small prickles still on them, wherein are many white, hard, and roundish seedes, very like vnto the seede of the Heppes or Eglantine berries, lying in a soft pulpe, like vnto the Hawthorne berries or Hawes: the whole beauty of this plant consisteth more in the gracefull aspect of the red apples or fruit hanging vpon the bushes, then in the flowers, or any other thing. It seemeth to be the same that Clusius calleth *Rosa Pumila,* but that with me it groweth much higher and greater then he saith his doth.

22. *Rosa siluestris odora siue Eglenteria simplex.* The single Eglantine or sweete Briar bush.

The sweete Briar or Eglantine Rose is so well knowne, being not onely planted in Gardens, for the sweetenesse of the leaues, but growing wilde in many woods and hedges, that I thinke it lost time to describe it; for that all know it hath exceeding long greene shootes, armed with the cruellest sharpe and strong thornes, and thicker set

1 *Roſa ſine ſpinis multiplex.* The double Roſe without thorns. 2 *Roſa Cinamomea flore pleno.* The double Cinamon Roſe. 3 *Roſa Ho-*
loſerica ſimplex. The ſingle Veluet Roſe. 4 *Roſa Holoſerica duplex.* The double Veluet Roſe. 5 *Roſa Moſchata multiplex.* The double
Muske Roſe. 6 *Roſa Moſchata Hiſpanica ſimplex.* The ſingle Spaniſh Muske Roſe. 7 *Roſa Pomifera maior.* The great Apple Roſe.
8 *Roſa ſilueſtris ſiue Eglanteria duplex.* The double Eglantine Roſe.

then is in any Rofe either wilde or tame : the leaues are fmaller then in moft of thofe that are nourfed vp in Gardens, feuen or nine moft vfually fet together on a ribbe or ftalke, very greene and fweete in fmell, aboue the leaues of any other kinde of Rofe : the flowers are fmall fingle blufh Rofes, of little or no fent at all, which turne into reddifh berries, ftuffed within with a dounie or flocky matter or fubftance, wherein doth lye white hard feede.

23. *Rofa filueftris odora fiue Eglenteria flore duplici.*
The double Eglantine.

The double Eglantine is in all the places that I haue feene it a grafted Rofe, (but I doubt not, but that his originall was naturall, and that it may be made naturall againe, as diuers other Rofes are.) It groweth and fpreadeth very well, and with a great head of branches, whereon ftand fuch like leaues as are in the fingle kinde, but a little larger, not fmelling fully fo fweete as it : the flowers are fomewhat bigger then the fingle, but not much, hauing but one other rowe of leaues onely more then the former, which are fmaller, and the outer leaues larger, but of the fame pale reddifh purple colour, and fmelleth fomewhat better then the fingle.

24. *Rofa femper virens.* The euer greene Rofe bufh.

This Rofe or bufh is very like vnto a wilde fingle Eglantine bufh in many refpects, hauing many very long greene branches, but more flender and weake, fo that many times they bend downe againe, not able to fuftaine themfelues without fome helpe, and armed with hooked thornes as other Rofes be; the winged leaues confift of feuen for the moft part, whereof thofe two that are loweft and oppofite, are fmalleft, the next two bigger then they, the third couple bigger then any of the reft belowe, and the end leafe biggeft of all : this proportion generally it holdeth in euery winged leafe through the whole plant, which at the firft comming forth are fomewhat reddifh, with the young branch that fhooteth out with them, but being full growne, are of a deepe greene colour, and fomewhat fhining, dented about the edges, and fall not away from the branches as other Rofes doe, but abide thereon for the moft part all the Winter : the flowers ftand foure or fiue together at the tops of the branches, being fingle Rofes, made of fiue leaues a peece, of a pure white colour, much larger then the ordinary Muske Rofe, and of a fine fent, comming neareft thereunto, with many yellow chiues or threads in the middle.

The Place.

Some of thefe Rofes had their originall, as is thought in England, as the firft and fecond ; for thefe dryed red Rofes that come ouer to vs from beyond the Seas, are not of the kinde of our red Rofe, as may well be perceiued by them that will compare our Englifh dryed leaues with thofe. Some in Germany, Spaine, and Italy. Some againe in Turkie, as the double yellow Rofe, which firft was procured to be brought into England, by Mafter Nicholas Lete, a worthy Merchant of London, and a great louer of flowers, from Conftantinople, which (as wee heare) was firft brought thither from Syria ; but perifhed quickly both with him, and with all other to whom hee imparted it : yet afterwards it was fent to Mafter Iohn de Franqueuille, a Merchant alfo of London, and a great louer of all rare plants, as well as flowers, from which isfprung the greateft ftore, that is now flourifhing in this Kingdome.

The Time.

The Cinamon Rofe is the earlieft for the moft part, which flowreth with vs about the middle of May, and fometimes in the beginning. The ordinary Muske Rofes both fingle and double flower lateft, as is faid. All the other flower much about one time, in the beginning of Iune, or thereabouts, and continue flowring all that moneth, and the next throughout for the moft part, and the red vntill Auguft be halfe paft.

The

The Names.

The feuerall names, whereby they are moſt commonly knowne vnto vs in this Countrey, are expreſſed in their titles; but they are much differing from what they are called in other Countries neare vnto vs, which to compare, conferre, and agree together, were a worke of more paines then vſe: But to proportion them vnto the names ſet downe by Theophraſtus, Pliny, and the reſt of the ancient Authors, were a worke, wherein I might be ſure not to eſcape without falling into errour, as I verily beleeue many others haue done, that haue vndertaken to doe it : I will therefore for this worke deſire that you will reſt contented, with ſo much as hath already been deliuered, and expect an exact definition and complete ſatisfaction by ſuch a methodicall courſe as a generall Hiſtory will require, to be performed by them that ſhall publiſh it.

The Vertues.

The Roſe is of exceeding great vſe with vs; for the Damaske Roſe (beſides the ſuperexcellent ſweete water it yeeldeth being diſtilled, or the perfume of the leaues being dryed, ſeruing to fill ſweete bags) ſerueth to cauſe ſolublenſſe of the body, made into a Syrupe, or preſerued with Sugar moiſt or dry candid. The Damaske Prouince Roſe, is not onely for ſent neareſt of all other Roſes vnto the Damaske, but in the operation of ſolubility alſo. The red Roſe hath many Phyſicall vſes much more then any other, ſeruing for many ſorts of compoſitions, both cordiall and cooling, both binding and looſing. The white Roſe is much vſed for the cooling of heate in the eyes: diuers doe make an excellent yellow colour of the iuyce of white Roſes, wherein ſome Allome is diſſolued, to paint or colour flowers or pictures, or any other ſuch things. There is little vſe of any other ſort of Roſes; yet ſome affirme, that the Muske Roſes are as ſtrong in operation to open or looſen the belly as the Damaske Roſe or Prouince.

Chap. CXI.

Ciſtus. The Holly Roſe or Sage Roſe.

THere are three principall kindes of *Ciſtus,* the male, the female, and the gumme or ſweete ſmelling *Ciſtus* bearing *Ladanum,* called *Ledon.* Of each of theſe three there are alſo diuers ſorts : Of them all to intreate in this worke is not my minde, I will onely ſelect out of the multitude ſome few that are fit for this our Garden, and leaue the reſt to a greater.

1. *Ciſtus mas.* The male Holly Roſe or Sage Roſe.

The male *Ciſtus* that is moſt familiar vnto our Countrey, I meane that will beſt abide, is a ſmall ſhrubby plant, growing ſeldome aboue three or foure foote high with vs, hauing many ſlender brittle wooddy branches, couered with a whitiſh barke, whereon are ſet many whitiſh greene leaues, long and ſomewhat narrow, crumpled or wrinckled as it were with veines, and ſomewhat hard in handling, eſpecially the old ones; for the young ones are ſofter, ſomewhat like vnto Sage leaues for the forme and colour, but much ſmaller, two alwaies ſet together at a ioynt : the flowers ſtand at the toppe of the branches, three or foure together vpon ſeuerall ſlender footſtalkes, conſiſting of fiue ſmall round leaues a peece, ſomewhat like vnto a ſmall ſingle Roſe, of a fine reddiſh purple colour, with many yellow threads in the middle, with-

out

out any ſent at all, and quickly fading or falling away, abiding ſeldome one whole day blowne at the moſt : after the flowers are paſt, there come vp round hard hairie heads in their places, containing ſmall browniſh ſeede : the roote is wooddy, and will abide ſome yeares with vs, if there be ſome care had to keepe it from the extreamity of our Winters froſtes, which both this, and many of the other ſorts and kinds, will not abide doe what we can.

2. *Ciſtus fæmina*. The female Holly Roſe.

The female Holly Roſe groweth lower, and ſmaller then the former male kinde, hauing blackiſh branches, leſſe woody, but not leſſe brittle then it : the leaues are ſomewhat rounder and greener, but a little hard or rough withall, growing in the ſame manner vpon the branches by couples : the flowers grow at the toppes of the branches, like vnto the former, conſiſting of fiue leaues, but ſomewhat leſſer, and wholly white, with yellow threds in the middle, as quickly fading, and of as little ſent as the former : the heads and ſeede are ſomewhat bigger then in the former.

3. *Chamæciſtus Friſicus*. The dwarfe Holly Roſe of Friſeland.

This dwarfe Ciſtus is a ſmall low plant, hauing diuers ſhootes from the rootes, full of leaues that are long and narrow, very like vnto the leaues of the French Spikenard or *Spica Celtica* ; from among which leaues ſhoote forth ſhort ſtalkes, not aboue a ſpan high, with a few ſmaller leaues thereon ; and at the toppes diuers ſmall flowers one aboue another, conſiſting of ſix ſmall round leaues, of a yellow colour, hauing two circles of reddiſh ſpots round about the bottome of the leaues, a little diſtant one from another, which adde much grace to the flover : after the flowers are paſt, there come in their places ſmall round heads, being two forked at the end, containing within them ſmall browniſh chaffie ſeede : the roote is ſmall and ſlender, with many fibres thereat creeping vnder ground, and ſhooting forth in diuers places, whereby it much encreaſeth : the whole plant, and euery part of it, ſmelleth ſtrong without any pleaſant ſent.

4. *Ciſtus annuus*. The Holly Roſe of a yeare.

This ſmall Ciſtus that endureth but a year (and will require to be ſowne euery year, if ye will haue it) riſeth vp with ſtraight, but ſlender hard ſtalkes, ſet here and there confuſedly with long and narrow greeniſh leaues, very like vnto the leaues of the Gum Ciſtus or Ledon, being a little clammy withall : at the toppe of the ſtalkes, and at the ioynts with the leaues, ſtand two or three pale yellow flowers, conſiſting of fiue leaues a peece, with a reddiſh ſpot neere the bottome of euery leafe of the flower, as quickely fading as any of the former : after which follow ſmall three ſquare heades, containing ſmall ſeede, like vnto the firſt female kinde, but ſomewhat paler or yellower : the root is ſmall and woody, and periſheth as ſoone as it hath borne ſeede.

5. *Ciſtus Ledon*. The Gum Ciſtus, or Sweete Holly Roſe.

This ſweete Holly Roſe or Gum Ciſtus, riſeth higher, and ſpreadeth larger then the former male kind doth, with many blackiſh woody branches, whereon are ſet diuers long and narrow darke greene leaues, but whitiſh vnderneath, two alwayes ſtanding together at a ioint, both ſtalks and leaues bedeawed as it were continually with a clammy ſweete moiſture (which in the hot Countries is both more plentifull, and more ſweet then in ours) almoſt tranſparent, and which being gathered by the inhabitants, with certaine inſtruments for that purpoſe (which in ſome places are leather thongs, drawne ouer the buſhes, and after ſcraped off from the thongs againe, and put together) is that kind of blacke ſweet gum, which is called *Ladanum* in the Apothecaries ſhops : at the tops of the branches ſtand ſingle white flowers, like vnto ſingle Roſes, being larger then in any of the former kindes, conſiſting of fiue leaues, whereof euery one hath at the bottome a dark purpliſh ſpot, broad below, and ſmall pointed vpwards, with ſome yellow threds in the middle : after which are paſt, there ariſe cornered

heads,

1 *Ciſtus mas* The male Holly Roſe. 2 *Chamæciſtus Friſicus.* The dwarfe Holly Roſe of Friſia. 3 *Ciſtus Ledon.* The ſweet Holly Roſe or gumme Ciſtus. 4 *Ledum Alpinum.* The mountaine Holly Roſe. 5 *Ledum Sileſiacum.* The ſweet Mary Roſe of Sileſia. 6 *Roſmarinum aureum.* Gilded Roſemary.

heads, containing such small brownish seede as is in the former male kinde : the roote is woody, and spreadeth vnder ground, abiding some yeares, if it be placed vnder a wall, where it may bee defended from the windes that often breake it, and from the extremitie of our winters, and especially the snow, if it lye vpon it, which quickly causeth it to perish.

6. *Ledum Alpinum seu Rosa Alpina.* The Mountaine sweet Holly Rose.

The fragrant smell with properties correspondent of two other plants, causeth me to insert them in this Chapter, and to bring them to your knowledge, as well worthy a fit place in our Garden. The first of them hath diuers slender woody branches, two foote high or thereabouts, couered with a grayish coloured barke, and many times leaning downe to the ground, whereby it taketh roote againe : vpon these branches grow many thicke, short, hard greene leaues, thicke set together, confusedly without order, sometimes whitish vnderneath, and sometimes yellowish : the toppes of the branches are loden with many flowers, which cause them to bend downwards, being long, hollow and reddish, opening into fiue corners, spotted on the outside with many white spots, and of a paler red colour on the inside, of a fine sweet sent : after the flowers are past, there follow small heads, containing small brownish seede : the root is long, hard and woody, abiding better if it comprehend in the ground, then some of the former, because his originall is out of a colder country.

7. *Ledum Silesiacum.* The sweete Mary Rose, or Rosemary of Silesia.

This other sweete plante riseth vp with woody ash-coloured branches two foote high or more, which shoote forth other branches, of a reddish or purplish colour, co-uered with a brownish yellow hoarinesse, on which are set many narrow long greene leaues, like vnto Rosemary leaues, but couered with the like hoarinesse as the stalks are (especially in the naturall places, but not so much being transplanted) and folding the sides of the leaues so close together, that they seeme nothing but ribbes, or stalkes, of an excellent sweet and pleasant sent ; at the ends of the branches there grow certaine brownish scaly heads, made of many small leaues set thicke together, out of which breake forth many flowers, standing in a tuft together, yet seuerally euery one vpon his owne footstalke, consisting of fiue white leaues, with certaine white threds in the middle, smelling very sweete : after which rise small greene heads, spotted with brownish spots, wherein is contained very small, long, yellowish seede : the roote is hard and woodie.

The Place.

The first, second, fourth and fifth, grow in the hot Countries, as Italie, Spaine, &c. The third, and the two last in the colder Countries, as Frise-land, Germanie, Bohemia.

The Time.

They do all flower in the Summer moneths of Iune, Iuly and August, and their seede is ripe quickly after.

The Names.

The first, second, fourth and fift, haue their names sufficiently expressed in their desctiptions. The third was sent vnto Clusius, vnder the name of *Herculus Prisicus*, because of the strong sent : but he referreth it to the kinds of *Chamæcistus*, that is, dwarfe or low *Cistus*, both for the low growth, and for the flowers and seede sake. The sixt is diuersly called; for Clusius calleth it *Ledum Alpinum* : others, *Nerium Alpinum*, making it to bee a Rose Bay. Gesner

Geſner according to the Countrey peoples name, *Roſa Alpina,* and *Roſa Montana.* Lobel calleth it *Balſamum Alpinum,* of the fragrant ſmell it hath, and *Chamærhododendros Chamælæſolio.* And ſome haue called it *Euonymus,* without all manner of iudgement. In Engliſh wee may call it, The Mountaine Roſe, vntill a fitter name be giuen it. The laſt is called of Matthiolus, *Roſmarinum ſilueſtre,* but of Cluſius *Ledum,* referring it to their kindred ; and *Sileſiacum,* becauſe he found it in that Countrey ; or for diſtinction ſake, as he ſaith, it may bee called, *Ledum ſolys Roſmarini,* or *Ledum Bohemicum.* Cordus, as it ſeemeth in his Hiſtory of Plants, calleth it *Chamæpeuce,* as though he did account it a kinde of low Pine, or Pitch tree.

The Vertues.

The firſt, ſecond, and fift, are very aſtringent, effectuall for all ſorts of fluxes of humours. The ſweet Gum called *Ladanum,* made artificially into oyle, is of ſingular vſe for *Alopecia,* or falling of the haire. The ſeed of the fourth is much commended againſt the ſtone of the Kidneyes. The ſweete Roſemary of Sileſia is vſed of the inhabitants, where it naturally groweth, againſt the ſhrinking of ſinewes, crampes, or other ſuch like diſeaſes, wherof their daily experience makes it familiar, being vſed in bathing or otherwiſe.

Chap. CXII.

Roſmarinum. Roſemary.

THere hath beene vſually knowne but one ſort of Roſemary, which is frequent through all this Country : but there are ſome other ſorts not ſo well known ; the one is called Gilded Roſemary ; the other broade leafed Roſemary ; a third I will adioyne, as more rare then all the other, called Double flowred Roſmary, becauſe few haue heard thereof, much leſſe ſeene it, and my ſelfe am not well acquainted with it, but am bold to deliuer it vpon credit.

1. *Libanotis Coronaria ſiue Roſmarinum vulgare.*
Our Common Roſmary.

This common Roſemary is ſo well knowne through all our Land, being in euery womans garden, that it were ſufficient but to name it as an ornament among other ſweete herbes and flowers in our Garden, ſeeing euery one can deſcribe it : but that I may ſay ſomething of it, It is well obſerued, as well in this our Land (where it hath been planted in Noblemens, and great mens gardens againſt bricke wals, and there continued long) as beyond the Seas, in the naturall places where it groweth, that it riſeth vp in time vnto a very great height, with a great and woody ſtemme (of that compaſſe, that (being clouen out into thin boards) it hath ſerued to make lutes, or ſuch like inſtruments, and here with vs Carpenters rules, and to diuers other purpoſes) branching out into diuers and ſundry armes that extend a great way, and from them againe into many other ſmaller branches, wheron are ſet at ſeuerall diſtances, at the ioynts, many very narrow long leaues, greene aboue, and whitiſh vnderneath ; among which come forth towards the toppes of the ſtalkes, diuers ſweet gaping flowers, of a pale or bleake blewiſh colour, many ſet together, ſtanding in whitiſh huskes ; the ſeed is ſmall and red, but thereof ſeldome doth any plants ariſe that will abide without extraordinary care ; for although it will ſpring of the ſeede reaſonable well, yet it is ſo ſmall and tender the firſt yeare, that a ſharpe winter killeth it quickly, vnleſſe it be very well defended : the whole plant as well leaues as flowers, ſmelleth exceeding ſweete.

2. *Roſmarinum ſtriatum, ſiue aureum.* Gilded Roſemary.

This Roſemary differeth not from the former, in forme or manner of growing, nor in the forme or colour of the flower, but only in the leaues, which are edged, or striped, or pointed with a faire gold yellow colour, which fo continueth all the yeare throughout, yet fresher and fairer in Summer then in Winter; for then it will looke of a deader colour, yet fo, that it may be difcerned to be of two colours, green & yellow.

3. *Rofmarinum latifolium.* Broade leafed Rofemary.

This broad leafed Rofemary groweth in the fame manner that the former doth, but that we haue not feene it in our Countrey fince we had it to grow fo great, or with fuch woody ftemmes : the leaues ftand together vpon the long branches after the fame fafhion, but larger, broader and greener then the other, and little or nothing whitifh vnderneath : the flowers likewife are of the fame forme and colour with the ordinary, but larger, and herein confifteth the difference.

4. *Rofmarinum flore duplici.* Double flowred Rofmary.

The double flowred Rofmary thus far differeth from the former, that it hath ftronger ftalkes, not fo eafie to breake, fairer, bigger and larger leaues, of a faire greene colour, and the flowers are double, as the Larkes heele or fpurre : This I haue onely by relation, which I pray you accept, vntill I may by fight better enforme you.

The Place.

Our ordinary Rofmary groweth in Spaine, and Prouence of France, and in others of thofe hot Countryes, neere the Sea fide. It will not abide (vnleffe kept in ftoues) in many places of Germany, Denmarke, and thofe colder Countries. And in fome extreame hard winters, it hath well neere perifhed here in England with vs, at the leaft in many places: but byflipping it is vfually, and yearly encreafed, to replenifh any garden.

The Time.

It flowreth oftentimes twice in the yeare ; in the Spring firft, from April vntill the end of May or Iune, and in Auguft and September after, if the yeare before haue been temperate.

The Names.

Rofmary is called of the ancient Writers, *Libanotis,* but with this diftinction, *Stephanomatica,* that is, *Coronaria,* becaufe there were other plants called *Libanotis,* that were for other vfes, as this for garlands, where flowers and fweete herbes were put together. The Latines call it *Rofmarinum.* Some would make it to be *Cneorum nigrum* of Theophraftus, as they would make Lauander to bee his *Cneorum album,* but Matthiolus hath fufficiently confuted that errour.

The Vertues.

Rofmary is almoft of as great vfe as Bayes, or any other herbe both for inward and outward remedies, and as well for ciuill as phyficall purpofes. Inwardly for the head and heart ; outwardly for the finewes and ioynts : for ciuill vfes, as all doe know, at weddings, funerals, &c. to beftow among friends : and the phyficall are fo many, that you might bee as well tyred in the reading, as I in the writing, if I fhould fet down all that might be faid of it. I will therefore onely giue you a tafte of fome, defiring you will be content therewith. There is an excellent oyle drawne from the flowers alone by the heate of the Sunne, auaileable for many difeafes both inward and outward, and accounted a foueraigne Balfame: it is alfo good to helpe dimneffe neſſe of ſight, and to take away ſpots, markes and ſcarres from the skin; and is made in this manner. Take a quantitie of the flowers of Roſemary, according to your owne will eyther more or leſſe, put them into a ſtrong glaſſe cloſe ſtopped, ſet them in hot horſe dung to digeſt for fourteene dayes, which then being taken forth of the dung, and vnſtopped, tye a fine linnen cloth ouer the mouth, and turne downe the mouth thereof into the mouth of another ſtrong glaſſe, which being ſet in the hot Sun, an oyle will diſtill downe into the lower glaſſe; which preſerue as precious for the vſes before recited, and many more, as experience by practice may enforme diuers.

There is another oyle Chymically drawne, auaileable in the like manner for many the ſame inward and outward diſeaſes, *viz.* for the heart, rheumaticke braines, and to ſtrengthen the memory, outwardly to warme and comfort cold benummed ſinewes, whereof many of good iudgement haue had much experience.

Chap. CXIII.

Myrtus. The Mirtle tree or buſh.

IN the hot Countreyes, there haue been many ſorts of Mirtles found out, naturally growing there, which will not fructifie in this of ours, nor yet abide without extraordinary care, and conueniencie withall, to preſerue them from the ſharpeneſſe of our winters. I ſhall only bring you to view three ſorts in this my Garden, the one with a greater, the other two with leſſer leaues, as the remainder of others which wee haue had, and which are preſerued from time to time, not without much paine and trouble.

1. *Myrtus latifolia*. The greater leafed Mirtle.

The broader leafed Mirtle riſeth vp to the height of foure or fiue foote at the moſt with vs, full of branches and leaues growing like a ſmall buſh, the ſtemme and elder branches whereof are couered with a dark coloured bark, but the young with a green, and ſome with a red, eſpecially vpon the firſt ſhooting forth, whereon are ſet many freſh greene leaues, very ſweet in ſmell, and very pleaſant to behold, ſo neer reſembling the leaues of the Pomegranate tree that groweth with vs, that they ſoone deceiue many that are not expert therein, being ſomewhat broade and long, and pointed at the ends, abiding alwaies green: at the ioynts of the branches where the leaues ſtand, come forth the flowers vpon ſmall footeſtalkes, euery one by it ſelfe conſiſting of fiue ſmall white leaues, with white threds in the middle, ſmelling alſo very ſweet: after the flowers are paſt, there doe ariſe in the hot Countries, where they are naturall, round blacke berries, when they are ripe, wherein are contained many hard white crooked ſeedes, but neuer in this Countrey, as I ſaid before: the roote diſperſeth it ſelfe into many branches, with many fibres annexed thereto.

2. *Myrtus minor, ſeu minore folio*. The ſmaller leafed Mirtle.

The ſmaller leafed Mirtle is a low ſhrub or buſh, like vnto the former, but ſcarce riſing ſo high, with branches ſpreading about the ſtemme, much thicker ſet with leaues then the former, ſmaller alſo, and pointed at the ends, of a little deeper greene colour, abiding greene alſo winter and ſummer, and very ſweete likewiſe: the flowers are white like vnto the former, and as ſweete, but ſhew not themſelues ſo plentifull on the branches: the fruit is blacke in his naturall places, with ſeedes therein as the former.

3. *Myrtus minor rotundiore folio*. Boxe Mirtle.

Wee haue another ſort of this ſmall kinde of Mirtle, ſo like vnto the former both for ſmalneſſe, deepe greene colour of the leaues, and thicke growing of the branches,

that

that it will be thought of moſt, without good heede, and comparing the one with the other, to be the very ſame with the former: but if it bee well viewed, it will ſhew, by the roundneſſe at the ends of the leaues very like vnto the ſmall Boxe leaues, to be another differing kinde, although in nothing elſe. Wee nourſe them with great care, for the beautifull aſpect, ſweete ſent and raritie, as delights and ornaments for a garden of pleaſure, wherein nothing ſhould be wanting that art, care and coſt might produce and preſerue : as alſo to ſet among other euer greene plants to ſort with them.

The Place.

Theſe and many other ſorts of Mirtles grow in Spaine, Portugall, Italie, and other hot Countries in great aboundance, where they make their hedges of them : wee (as I ſaid) keepe them in this Countrey, with very great care and diligence.

The Time.

The Mirtles doe flower very late with vs, not vntill Auguſt at the ſooneſt, which is the cauſe of their not fructifying.

The Names.

They are called in Latine *Myrtus*, and in Engliſh Mirtle tree, without any other diuerſitie of names, for the generall title. Yet the ſeuerall kindes haue had ſeuerall denominations, in Plinies time, and others, as *Romana, Coningala, Terentina, Egyptia, alba, nigra, &c.* which haue noted the differences, euen then well obſerued.

The Vertues.

The Mirtle is of an aſtringent qualitie, and wholly vſed for ſuch purpoſes.

Chap. CXIIII.

Malus Punica ſiue Granata. The Pomegranet tree.

THere are two kindes of Pomegranet trees, The one tame or manured, bearing fruit, which is diſtinguiſhed of ſome into two ſorts, of others into three, that is, into ſower, and ſweet, and into ſower ſweete. The other wilde, which beareth no fruite, becauſe it beareth double flowers, like as the Cherry, Apple, and Peach tree with double bloſſomes, before deſcribed, and is alſo diſtinguiſhed into two ſorts, the one bearing larger, the other leſſer flowers. Of the manured kinde wee haue onely one ſort (ſo farre as we know) for it neuer beareth ripe fruit in this our Countrey) which for the beautifull aſpect, both of the greene verdure of the leaues, and faire proportion and colour of the flowers, as alſo for the raritie, are nourſed in ſome few of their gardens that delight in ſuch rarities : for in regard of the tenderneſſe, there is neede of diligent care, that is, to plant it againſt a brick wall, and defend it conueniently from the ſharpeneſſe of our winters, to giue his Maſter ſome pleaſure in ſeeing it beare flowers : And of the double kinde we haue as yet obtained but one ſort, although I ſhall giue you the knowledge and deſcription of another.

1. *Malus Punica ſatiua.* The tame Pomegranet tree.

This Pomegranet tree groweth not very high in his naturall places, and with vs ſomtimes it ſhooteth forth from the roote many browniſh twigges or branches, or if it bee pruned from them, and ſuffered to grow vp, it riſeth to bee ſeuen or eight foote high,

ſpreading

1 *Myrtus latifolia maior.* The broad leated Myrtle. **2** *Myrtus angustifolia minor.* The small leafed Myrtle. **3** *Myrtus buxifolii minor.* The Boxe leafet Myrtle. **4** *Malus Granatus simplici flore.* The ordinary Pomegranet tree. **5** *Balaustium Romanum seu minus.* The lesser double flowred Pomegranet tree. **6** *Balaustium maius siue Cyprium.* The greater double flowred Pomegranet. **7** *Pseudocapsicum seu Amomum Plinij.* The Winter Cherry tree. **8** *Ficus Indica cum suo fructu.* The Indian Figgetree and his fruit.

spreading into many small and slender branches, here and there set with thornes, and with many very faire greene shining leaues, like in forme and bignesse vnto the leaues of the larger Myrtle before described, euery one hauing a small reddish foote-stalke vpon these branches : among the leaues come forth here and there, long, hard, and hollow reddish cups, diuided at the brimmes, wherein doe stand large single flowers, euery one consisting of one whole leafe, smaller at the bottome then at the brimme, like bels, diuided as it were at the edges into fiue or six parts, of an orient red or crimson colour in the hotter Countries ; but in this it is much more delayed, and tendeth neare vnto a blush, with diuers threads in the middle. The fruit is great and round, hauing as it were a crowne on the head of it, with a thicke tough hard skinne or rinde, of a brownish red colour on the outside, and yellow within, stuffed or packt full of small graines, euery one encompast with a thin skin, wherein is contained a cleare red iuyce or liquor, either of a sweet (as I said before) or sower taste, or betweene them both of a winie taste : the roote disperseth it selfe very much vnder ground.

2. *Balaustium maius siue Malus Punica siluestris maior.*
The greater wilde or double blossomd Pomegranet tree.

The wilde Pomegranet is like vnto the tame in the number of purplish branches, hauing thornes, and shining faire greene leaues, somewhat larger then the former : from the branches likewise shoote forth flowers, farre more beautifull then those of the tame or manured sort, because they are double, and as large as a double Prouince Rose, or rather more double, of an excellent bright crimson colour, tending to a silken carnation, standing in brownish cups or huskes, diuided at the brims vsually into foure or fiue seuerall points, like vnto the former, but that in this kinde there neuer followeth any fruit, no not in the Country, where it is naturally wilde.

3. *Balaustium minus.* The smaller wilde Pomegranet tree.

This smaller kinde differeth from the former in his leaues, being of a darker greene colour, but not in the height of the stemme, or purplishnesse of his branches, or thorns vpon them ; for this doth shew it selfe more like vnto a wilde kind then it : the flowers hereof are much smaller, and not so thicke and double, of a deeper or sadder red Orenge tawny colour, set also in such like cups or huskes.

The Place.

The tame or manured kinde groweth plentifully in Spaine, Portugall, and Italy, and other in other warme and hot countries. Wee (as I said before) preserue it with great care. The wilde I thinke was neuer seene in England, before Iohn Tradescante my very louing good friend brought it from the parts beyond the Seas, and planted it in his Lords Garden at Canterbury.

The Time.

They flower very late with vs, that is, not vntill the middle or end of August, and the cold euenings or frosts comming so soone vpon it, doth not onely hinder it from bearing, but many times the sharpe winters so pinch it, that it withereth it downe to the ground, so that oftentimes it hardly springeth againe.

The Names.

The name *Malus Punica* for the tree, and *Malum Punicum* for the fruit, or *Malus Granata*, and *Malum Granatum*, is the common name giuen vnto this tree, which is called in English the Pomegarnet or Pomegranet tree. The flowers of the tame kinde are called *Cytini*, as Dioscorides saith, although Plinie seemeth either to make *Cytinus* to be the flower of the wilde kinde, or

Balaustium

Balaustium to be the flower of both tame and wilde kinde : but properly, as I take it, *Cytinus* is the cup wherein the flower as well of the tame as wilde kinde doth stand ; for vnto the similitude of them, both the flowers of *Asarum*, and the seede vessels of *Hyosciamus* are compared and resembled, and not vnto the whole flower : the barke or rinde of the fruit is called of diuers *Sidion*, and in the Apothecaries shops *Psidium*, and *cortex Granatorum*. The wilde kinde is called *Malus Punica siluestris* : In English, The wilde Pomegranet tree ; the flower thereof is properly called *Balaustium*. The lesser kind is vsually called *Balaustium Romanum*, as the greater is called *Creticum* and *Cyprinum*, because they growe in Candy and Cyprus.

The Vertues.

The vse of all these Pomegranets is very much in Physicke, to coole and binde all fluxibility both of body and humours : they are also of singular effect in all vlcers of the mouth, and other parts of the body, both of man and woman. There is no part of them but is applyed for some of these respects. The rinde also of the Pomegranet is vsed of diuers in stead of Gaules, to make the best sort of writing Inke, which is durable to the worlds end.

Chap. CXV.

Amonum Plinij seu Pseudocapsicum.
Tree Night shade or the Winter Cherry tree.

I Haue adioyned this plant, for the pleasurable beauty of the greene leaues, and red berries. It groweth vp to be a yard or foure foote high at the most, hauing a small wooddy stemme or stocke, as bigge as ones finger or thumbe, couered with a whitish greene barke, set full of greene branches, and faire greene leaues, somewhat vneuen sometimes on the edges, narrower then any Night shade leaues, and very neare resembling the leaues of the *Capsicum*, or Ginny pepper, but smaller and narrower, falling away in the Winter, and shooting fresh in the Spring of the yeare : the flowers growe often two or three together, at the ioynts of the branches with the leaues, being white, opening starre-fashion, and sometimes turning themselues backe, with a yellow pointell in the middle, very like vnto the flowers of Night shade : after the flowers are past, come forth in their stead small greene buttons, which after turne to be pleasant round red berries, of the bignesse of small Cherries when they are ripe, which with vs vsually ripen not vntill the Winter, or about Christmas, wherein are contained many small whitish seede that are flat : all the whole plant, as well leaues and flowers as seede, are without either smell or taste : the roote hath many yellowish strings and fibres annexed vnto it.

The Place.

The originall place hereof is not well knowne, but is thought to bee the West Indies. It hath been planted of long time in most of these Countries, where it abideth reasonable well, so that some care bee had thereof in the extreamity of the Winter.

The Time.

It flowreth sometimes in Iune, but vsually in Iuly and August, and the fruit is not ripe (as is said) vntill the Winter.

The

The Names.

This plant hath diuers names ; for it is thought to be that kinde of *Amo-mum* that Plinie setteth downe. Dodonæus calleth it *Pseudocapsicum*, for some likenesse in the leafe and fruit vnto the small *Capsicum* or Ginnie Pepper, although much vnlike in the taste and property. Others doe call it *Strichnodendron*, that is, *Solanum arborescens*, and wee in English according thereunto, Tree Night shade. But some Latine asses corrupting the Latine word *Amomum*, doe call it the Mumme tree. Dalechampius calleth it *Solanum Americum*, *seu Indicum*, and saith the Spaniards call it in their tongue, *Guindas de las Indias*, that is, *Cerasa Indiana*, Indian Cherries, which if any would follow, I would not bee much against it : but many Gentlewomen doe call them Winter Cherries, because the fruit is not throughly ripe vntill Winter.

The Vertues.

I finde no Physicall property allotted vnto it, more then that by reason of the insipidity, it is held to be cooling.

Chap. CXVI.

Ficus Indica minor. The smaller Indian Figge tree.

THis Indian Figge tree, if you will call it a tree (because in our Country it is not so, although it groweth in the naturall hot Countries from a wooddy stemme or body into leaues) is a plant consisting only of leaues, one springing out of another, into many branches of leaues, and all of them growing out of one leafe, put into the ground halfe way, which taking roote, all the rest rise out thereof, those belowe for the most part being larger then those aboue ; yet all of them somewhat long, flat, and round pointed, of the thicknesse of a finger vsually, and smallest at the lower end, where they are ioyned or spring out of the other leaues, hauing at their first breaking out a shew of small, red, or browne prickes, thicke set ouer all the vpper side of the leaues, but with vs falling away quickly, leauing onely the markes where they stood : but they haue besides this shew of great prickes, a few very fine, and small, hard, white, and sharpe, almost infensible prickes, being not so bigge as haires on the vnderside, which will often sticke in their fingers that handle them vnaduisedly, neither are they to be discerned vnlesse one look precisely for them : the leaues on the vnderside hauing none of those other great pricks or marks at all, being of a faire fresh pale green colour : out of the vppermost leaues breake forth certaine greene heads, very like vnto leaues (so that many are deceiued, thinking them to be leaues, vntill they marke them better, and be better experienced in them) but that they growe round and not flat, and are broad at the toppe ; for that out of the tops of euery of them shooteth out a pale yellow flower, consisting of two rowes of leaues, each containing fiue leaues a peece, laid open with certaine yellow threads, tipt with red in the middle : this greene head, vntill the flower be past, is not of halfe that bignesse that it attaineth vnto after, yet seldome or neuer commeth vnto perfection with vs, being long and round, like vnto a Figge, small belowe, and greater aboue, bearing vpon the flat or broad head the marke of the flower ; some holding still on them the dryed leaues, and others hauing lost them, shew the hollownesse which they haue in the toppe or middle of the head, the sides round about being raised or standing vp higher : this head or figge in our Country abideth greene on the outside, and little or nothing reddish within (although it abide all the Winter, and the Summer following, as sometimes it doth) for want of that heate and comfort of the Sunne it hath in his naturall place, where it groweth

reddish

reddifh on the outfide, and containing within it a bloudy red clammy iuyce, making the vrine of them that eate of them as red as bloud, which many feeing, were in doubt of themfelues, left their vrine were not very bloud; of what fweetneffe, like a figge, in the naturall places, I am not well affured, yet affirmed: but thofe that haue beene brought vnto me, whofe colour on the outfide was greenifh, were of a reddifh purple within, and contained within them round, fmall, hard feede, the tafte was flat, wate-rifh, or infipide: the roote is neither great, nor difperfeth it felfe very deepe or farre, but fhooteth many fmall rootes vnder the vpper cruft of the earth.

There is a greater kinde hereof, whofe leaues are twice or thrice as bigge, which ha-uing been often brought vs, will feldome abide more then one Summer with vs, our Winters alwaies rotting the leaues, that it could not be longer kept.

The Place.

This Indian Figge tree groweth difperfedly in many places of America, generally called the Weft Indies: The greater kinde in the more re-mote and hot Countries, as Mexico, Florida, &c. and in the Bermudas or Summer Iflands, from whence wee haue often had it. The leffer in Virginia, and thofe other Countries that are nearer vnto vs, which better endureth with vs.

The Time.

It flowreth with vs fometimes in May, or Iune; but (as I faid) the fruit ne-uer commeth to perfection in this Country.

The Names.

Diuers doe take it to bee *Opuntia Pliny*, whereof hee fpeaketh in the 21. Booke and 17. Chapter of his Naturall Hiftory: but he there faith, *Opun-tia* is an herbe, fweete and pleafant to be eaten, and that it is a wonder that the roote fhould come from the leafe, and fo to growe; which words al-though they defcipher out the manner of the growing of this plant, yet be-caufe this is a kinde of tree, and not an herbe, nor to be eaten, it cannot bee the fame: but efpecially becaufe there is an herbe which groweth in the fame manner, or very neare vnto it, one leafe ftanding on the toppe or fide of another, being a Sea plant, fit to be eaten with vinegar and oyle (as many other herbes are that growe in the falt marfhes, or neare the Sea, whereof Sea Purflane is one) which Clufius calleth *Lychen Marinus*, and (as Clufius faith) Cortufus very fitly called *Opuntia marina*, and out of doubt is the ve-rie fame *Opuntia* that Theophraftus maketh mention of, and Plinie out of him. Our Englifh people in Virginia, and the Bermuda Ifland, where it groweth plentifully, becaufe of the form of the fruit, which is fomewhat like to a Peare, & not being fo familiarly acquainted with the growing of Figs, fent it vnto vs by the name of the prickly Peare, from which name many haue fuppofed it to be a Peare indeede, but were therein deceiued.

The Vertues.

There is no other efpeciall property giuen hereunto, by any that haue written of the Weft Indies, then of the colouring of the vrine, as is be-fore faid.

Oo CHAP.

1 *Yuca ſiue Iucca.* The Indian Iucca. 2 *Arbor vitæ.* The tree of life. 3 *Arbor Iudæ.* Iudas tree. 4 *Laburnum.* Beane Trefoile. 5 *Cytiſus.* Tree Trefoile.

Chap. CXVIII.

Arbor vitæ. The tree of life.

THe tree of life riseth vp in some places where it hath stood long, to be a tree of a reasonable great bignesse and height, couered with a redder barke then any other tree in our Country that I know, the wood whereof is firme and hard, and spreadeth abroad many armes and branches, which againe send forth many smaller twigges, bending downewards; from which twiggy or slender branches, being flat themselues like the leaues, come forth on both sides many flat winged leaues, somewhat like vnto Sauine, being short and small, but not pricking, seeming as if they were brayded or folded like vnto a lace or point, of a darke yellowish greene colour, abiding greene on the branches Winter and Summer, of a strong resinous taste, not pleasing to most, but in some ready to procure casting, yet very cordiall and pectorall also to them that can endure it : at the toppes of the branches stand small yellowish dounie flowers, set in small scaly heads, wherein lye small, long, brownish seede, which ripen well in many places, and being sowne, doe spring and bring forth plants, which with some small care will abide the extreamest Winters we haue.

The Place.

The first or originall place where it naturally groweth, as farre as I can learne or vnderstand, is that part of America which the French doe inhabite, about the riuer of Canada, which is at the backe of Virginia Northward, and as it seemeth, first brought by them from thence into Europe, in the time of Francis the first French King, where it hath so plentifully encreased, and so largely beene distributed, that now few Gardens of respect, either in France, Germany, the Lowe-Countries, or England, are without it.

The Time.

It flowreth in the end of May, and in Iune; the fruit is ripe in the end of August and Sptember.

The Names.

All the Writers that haue written of it, since it was first knowne, haue made it to be *Thuyæ genus*, a kinde of Thuya, which Theophrastus compareth vnto a Cypresse tree, in his fifth Book and fifth Chapter : but *Omne simile non est idem*, and although it haue some likenesse, yet I verily beleeue it is *proprium sui genus*, a proper kinde of it owne, not to bee paralleld with any other. For wee finde but very few trees, herbes, or plants in America, like vnto those that growe in Europe, the hither part of Africa, or in the lesser Asia, as experience testifieth. Some would make it to be *Cedrus Lycia*, but so it cannot be. The French that first brought it, called it *Arbor vitæ*, with what reason or vpon what ground I know not : but euer since it hath continued vnder the title of the Tree of life.

The Vertues.

It hath beene found by often experience, that the leaues hereof chewed in the morning fasting, for some few dayes together, haue done much good to diuers, that haue beene troubled with shortnesse of breath, and to helpe to expectorate thinne purulentous matter stuffing the lungs. Other properties I haue not heard that it hath ; but doubtlesse, the hot resinous smell and taste

taſte it hath, both while it is freſh, and after it hath beene long kept dry, doth euidently declare his tenuity of parts, a digeſting and cleanſing quality it is poſſeſſed with, which if any induſtrious would make tryall, hee ſhould finde the effects.

Chap. CXIX.

Arbor Iudæ. Iudas tree.

IVdas tree riſeth vp in ſome places, where it ſtandeth open from a wall, and alone free from other trees (as in a Garden at Battherſey, which ſometimes agoe belonged to Maſter Morgan, Apothecary to the late Queene Elizabeth of famous memory) to be a very great and tall tree, exceeding any Apple tree in height, and equall in bigneſſe of body thereunto (as my ſelfe can teſtifie, being an eye witneſſe thereof) when as it had many ſtalkes of flowers, being in the bud, breaking out of the body of the tree through the barke in diuers places, when as there was no bough or branch near them by a yard at the leaſt, or yet any leafe vpon the tree, which they gathered to put among other flowers, for Noſegayes) and in other places it groweth to bee but an hedge buſh, or plant, with many ſuckers and ſhootes from belowe, couered with a darke reddiſh barke, the young branches being more red or purpliſh: the flowers on the branches come forth before any ſhew or budding of leaues, three or foure ſtanding together vpon a ſmall foote-ſtalke, which are in faſhion like vnto Peaſe bloſſomes, but of an excellent deepe purpliſh crimſon colour: after which come in their places ſo many long, flat, large, and thinne cods, of a browniſh colour, wherein are contained ſmall, blackiſh browne, flat, and hard ſeede: the roote is great, and runneth both deepe, and farre ſpreading in the earth: the leaues come forth by themſelues, euery one ſtanding on a long ſtalke, being hard & very round, like vnto the leafe of the largeſt *Aſarum*, but not ſo thick, of a whitiſh green on the vpper ſide, and grayiſh vnderneath, which fall away euery yeare, and ſpring afreſh after the Spring is well come in, and the buds of flowers are ſprung.

There is another of this kinde, growing in ſome places very high, ſomewhat like the former, and in other places alſo full of twiggy branches, which are greener then the former, as the leaues are likewiſe: the flowers of this kinde are wholly white, and the cods nothing ſo red or browne, in all other things agreeing together. *Flore albo.*

The Place.

The former groweth plentifully in many places of Spaine, Italy, Prouence in France, and in many other places. The other hath beene ſent vs out of Italy many times, and the ſeede hath ſprung very well with vs, but it is ſomewhat tender to keepe in the Winter.

The Time.

The flowers (as I ſaid) appeare before the leaues, and come forth in Aprill and May, and often ſooner alſo, the leaues following ſhortly after; but neither of them beareth perfect ſeede in our Country, that euer I could learne, or know by mine owne or others experience.

The Names.

Some would referre this to *Cercis*, whereof Theophraſtus maketh mention in his firſt Booke and eighteenth Chapter, among thoſe trees that beare their fruit in cods, like as Pulſe doe: and hee remembreth it againe in the fourteenth Chapter of his third Booke, and maketh it not vnlike the white

Poplar

Poplar tree, both in greatneſſe and whiteneſſe of the branches, with the leafe of an Iuie, without corners on the one part, cornered on the other, and ſharpe pointed, greene on both ſides almoſt alike, hauing ſo ſlender long footeſtalkes that the leaues cannot ſtand forthright, but bend downwards, with a more rugged barke then the white Poplar tree. Cluſius thinketh this large deſcription is but an ample deſcription of the third kinde of Poplar, called *Lybica*, the Aſpen tree, which Gaza tranſlateth *Alpina* : but who ſo will well conſider it, ſhall finde it neyther anſwerable to any Poplar tree, in that it beareth not cods as *Cercis* doth ; nor vnto this *Arbor Iudæ*, becauſe it beareth not white branches. Cluſius ſaith alſo, that the learned of Mompelier in his time, referred it to *Colytea* of Theophraſtus in his third booke and ſeuenteenth chapter, where he doth liken it to the leaues of the broadeſt leafed Bay tree, but larger and rounder, green on the vpperſide, and whitiſh vnderneath, and whereunto (as he ſaith) Theophraſtus giueth cods in the fourteenth chapter of the ſame third booke: and by the contracting of their deſcriptions both together, ſaith, they agree vnto this Iudas tree. But I find ſome doubts and differences in theſe places : for the *Colutea* that Dioſcorides mentioneth in the ſaid fourteenth chapter of his third booke, hath (as he ſaith there) a leafe like vnto the Willow, and therefore cannot bee the ſame *Colutea* mentioned in the ſeuenteenth chapter of the ſame third book, which hath a broade Bay leafe : indeede hee giueth ſeede in cods : but that with broade Bay leaues is (as he ſaith) without eyther flower or fruite ; and beſides all this, he ſaith the rootes are very yellow, which is not to bee found in this *Arbor Iudæ*, or Iudas tree: let others now iudge if theſe things can bee well reconciled together. Some haue for the likeneſſe of the cods vnto Beane cods, called it *Fabago*. And Cluſius called it *Siliqua ſilueſtris.* It is generally in theſe dayes called *Arbor Iudæ*, and in Engliſh after the Latine name, vntill a fitter may be had, Iudas tree.

The Vertues.

There is nothing extant in any Author of any Phyſicall vſe it hath, neyther hath any later experience found out any.

Chap. CXX.

Laburnum. Beane Trefoile.

THere be three ſorts of theſe codded trees or plants, one neere reſembling another, whereof *Anagyris* of Dioſcorides is one. The other two are called *Laburnum* ; the larger whereof Matthiolus calleth *Anagyris altera*, and ſo doe ſome others alſo : the third is of the ſame kinde with the ſecond, but ſmaller. I ſhall not for this our Garden trouble you or my ſelfe with any more of them then one, which is the leſſer of the two *Liburnum*, in that it is more frequent, and that it will far better abide then the *Anagyris*, which is ſo tender, that it will hardly endure the winters of our Countrey : and the greater *Laburnum* is not ſo eaſily to be had.

Laburnum. Beane Trefoile.

This codded tree riſeth vp with vs like vnto a tall tree, with a reaſonable great body, if it abide any long time in a place, couered with a ſmooth greene barke; the branches are very long, greene, pliant, and bending any way, whereon are ſet here and there diuers leaues, three alwaies ſtanding together vpon a long ſtalk, being ſomwhat long, and not very narrow, pointed at the ends, greene on the vpperſide, and of a ſiluer ſhining colour vnderneath, without any ſmell at all : at the ioynts of theſe branches, where the leaues ſtand, come forth many flowers, much like vnto broome flowers, but not ſo

large

large or open, growing about a very long branch or ftalke, fometimes a good fpan or more in length, and of a faire yellow colour, but not very deepe ; after which come flat thin cods, not very long or broade, but as tough and hard as the cods of Broome; wherein are contained blackifh feede, like, but much leffe then the feede of *Anagyris vera* (which are as big as a kidney beane, purplifh and fpotted) : the roote thrufteth down deepe into the ground, fpreading alfo farre, and is of a yellowifh colour.

The Place.

This tree groweth naturally in many of the woods of Italie, and vpon the Alpes alfo, and is therefore ftill accounted to be that *Laburnum* that Plinie calleth *Arbor Alpina*. It groweth in many gardens with vs.

The Time.

It flowreth in May, the fruit or cods, and the feedes therein are ripe in the end of Auguft, or in September.

The Names.

This tree (as I faid before) is called of Matthiolus *Anagyris altera, fiue fecunda,* of Cordus, Gefner and others, efpecially of moft now adayes, *Laburnum.* It is probable in my opinion, that this fhould bee that *Colutaea* of Theophraftus, mentioned in the fourteenth Chapter of his third book with the leafe of a Willow ; for if you take any one leafe by it felfe, it may well refemble a Willow leafe both for forme and colour, and beareth fmall feed in cods like vnto pulfe as that doth. Of fome it hath beene taken for a kinde of *Cytifus,* but not truely. We call it in Englifh, Beane Trefoile, in regard of his cods and feede therein, fomewhat like vnto Kidney Beanes, and of the leaues, three alwayes ftanding together, vntill a more proper name may bee giuen it.

The Vertues.

There is no vfe hereof in Phyficke with vs, nor in the naturall place of the growing, faue only to prouoke a vomit, which it will doe very ftrongly.

Chap. CXXI.

Cytifus. Tree Trefoile.

THere are fo many forts of *Cytifus* or Tree trefoiles, that if I fhould relate them all, I fhould weary the Reader to ouerlooke them, whereof the moft part pertaine rather to a generall worke then to this abftract. I fhall not therefore trouble you with any fuperfluous, but only with two, which we haue nourfed vp to furnifh wafte places in a garden.

Cytifus Marantha. Horned Tree Trefoile.

This Tree Trefoile which is held of moft Herbarifts to bee the true *Cytifus* of Diofcorides, rifeth vp to the height of a man at the moft, with a body of the bigneffe of a mans thumbe, couered with a whitifh bark, breaking forth into many whitifh branches fpreading farre, befet in many places with fmall leaues, three alwayes fet together vpon a fmall fhort footeftalke, which are rounder, and whiter then the leaues of Beane Trefoile : at the ends of the branches for the moft part, come forth the flowers three or foure togethers, of a fine gold colour, and of the fafhion of Broome flowers, but

not

not fo large : after the flowers are paft, there come in their places crooked flat thinne cods, of the fafhion of a halfe moone, or crooked horne, whitifh when they are ripe, wherein are contained blackifh feede : the roote is hard and woody, fpreading diuers wayes vnder the ground : the whole plant hath a pretty fmall hot fent.

Cytifus vulgatior. The common Tree Trefoile.

This *Cytifus* is the moft common in this Land, of any the other forts of tree trefoiles, hauing a blackifh colourd barke, the ftemme or body whereof is larger then the former, both for height and fpreading, bearing alfo three leaues together, but fmaller and greener then the former : the flowers are fmaller, but of the fame fafhion and colour: the cods blackifh and thin, and not very long, or great, but leffer then Broome cods, wherein there lyeth fmall blackifh hard feede : the roote is diuerfly difperfed in the ground.

The Place.

The firft groweth in the kingdome of Naples, and no doubt in many other places of Italie, as Matthiolus faith. The other groweth in diuers places of France.

The Time.

They flower for the moft part in May or Iune : the feede is ripe in Auguft or September.

The Names.

The firft (as I faid) is thought of moft to be the true *Cytifus* of Diofcorides, and as is thought, was in thefe later dayes firft found by Bartholomæus Maranta of Naples, who fent it firft to Matthiolus, and thereupon hath euer fince beene called after his name, *Cytifus Marantha.* Some doe call it *Cytifus Lunatus,* becaufe the cods are made fomewhat like vnto an halfe Moone. We call it in Englifh, Horned Tree Trefoile. The other is called *Cytifus vulgaris* or *vulgatior;* in Englifh, The common Tree Trefoile, becaufe we haue not any other fo common.

The Vertues.

The chiefeft vertues that are appropriate to thefe plants, are to procure milke in womens breafts, to fatten pullen, fheep &c. and to be good for bees.

Chap. CXXII.

Colutea. The Baftard Sena Tree.

WEe haue in our Gardens two or three forts of the Baftard Sena tree ; a greater as I may fo call it, and two leffer : the one with round thin tranfparent skins like bladders, wherein are the feede : the others with long round cods, the one bunched out or fwelling in diuers places, like vnto a Scorpions tale, wherein is the feede, and the other very like vnto it, but fmaller.

1. *Colutea Veficaria.* The greater Baftard Sena with bladders.

This fhrub or tree, or fhrubby tree, which you pleafe to call it rifeth vnto the height of a pretty tree, the ftemme or ftock being fometimes of the bigneffe of a mans arme, couered with a blackifh greene rugged barke, the wood whereof is harder then of an
Elder,

Elder, but with an hollowneſſe like a pith in the heart or middle of the branches, which are diuided many wayes, and whereon are ſet at ſeuerall diſtances, diuers winged leaues, compoſed of many ſmall round pointed, or rather flat pointed leaues, one ſet againſt another, like vnto Licoris, or the Hatchet Fitch; among theſe leaues come forth the flowers, in faſhion like vnto Broome flowers, and as large, of a very yellow colour : after which appeare cleare thinne ſwelling cods like vnto thinne tranſparent bladders, wherein are contained blacke ſeede, ſet vpon a middle ribbe or ſinew in the middle of the bladder, which if it be a little cruſhed betweene the fingers, will giue a cracke, like as a bladder full of winde. The roote groweth branched and woody.

2. *Colutæa Scorpioides maior.* The greater Scorpion podded Baſtard Sena.

This Baſtard Sena groweth nothing ſo great or tall, but ſhooteth out diuerſly, like vnto a ſhrub, with many ſhoots ſpringing from the root : the branches are greener, but more rugged, hauing a white barke on the beſt part of the elder growne branches; for the young are greene, and haue ſuch like winged leaues ſet on them as are to be ſeen in the former, but ſmaller, greener, and more pointed : the flowers are yellow, but much ſmaller, faſhioned ſomewhat like vnto the former, with a reddiſh ſtripe downe the backe of the vppermoſt leafe : the long cods that follow are ſmall, long and round, diſtinguiſhed into many diuiſions or dents, like vnto a Scorpions tayle, from whence hath riſen the name : in theſe ſeuerall diuiſions lye ſeuerall blacke ſeede, like vnto the ſeede of Fenigrecke : the roote is white and long, but not ſo woody as the former.

3. *Colutæa Scorpioides minor.* The leſſer Scorpion Baſtard Sena.

This leſſer Baſtard Sena is in all things like the former, but ſomewhat lower, and ſmaller both in leafe, flower, and cods of ſeede, which haue not ſuch eminent bunches on the cods to be ſeene as the former.

The Place.

They grow as Matthiolus ſaith about Trent in Italie, and in other places : the former is frequent enough through all our Countrey, but the others are more rare.

The Time.

They flower about the middle or end of May, and their ſeede is ripe in Auguſt. The bladders of the firſt will abide a great while on the tree, if they be ſuffered, and vntill the winde cauſe them to rattle, and afterwards the skins opening, the ſeed will fall away.

The Names.

The name *Colutæa* is impoſed on them, and by the iudgement of moſt writers, the firſt is taken to bee that *Colutæa* of *Lipara* that Theophraſtus maketh mention of, in the ſeuenteenth chapter of his third booke. But I ſhould rather thinke that the *Scorpioides* were the truer *Colutæa* of Theophraſtus, becauſe the long pods thereof are more properly to bee accounted *ſiliquæ*, then the former which are *veſicæ tumentes*, windy bladders, and not *ſiliquæ* : and no doubt but Theophraſtus would haue giuen ſome peculiar note of difference if he had meant thoſe bladders, and not theſe cods. Let others of iudgement be vmpeeres in this caſe; although I know the currant of writers ſince Matthiolus, doe all hold the former *Colutæa veſicaria* to be the true *Colutæa Lipara* of Theophraſtus. Wee call it in Engliſh, Baſtard Sena, from Ruellius, who as I thinke firſt called it Sena, from the forme of the leaues. The ſecond and third (as I ſaid before) from the forme of the cods receiued their names, as it is in the titles and deſcriptions; yet they may as properly be called *Siliquoſa*, for that their fruite are long cods.

The

The Vertues.

Theophraſtus ſaith it doth wonderfully helpe to fatten ſheepe : But ſure it is found by experience, that if it be giuen to man it cauſeth ſtrong caſtings both vpwards and downwards ; and therefore let euery one beware that they vſe not this in ſteede of good Sena, leſt they feele to their coſt the force thereof.

Chap. CXII.

Spartum Hiſpanicum frutex. Spaniſh Broome.

ALthough Cluſius and others haue found diuers ſorts of this ſhrubby Spartum or Spaniſh Broome, yet becauſe our Climate will nourſe vp none of them, and euen this very hardly, I ſhall leaue all others, and deſcribe vnto you this one only in this manner : Spaniſh Broome groweth to bee fiue or ſixe foote high, with a woody ſtemme below, couered with a darke gray, or aſh-coloured barke, and hauing aboue many pliant, long and ſlender greene twigs, whereon in the beginning of the yeare are ſet many ſmall long greene leaues, which fall away quickly, not abiding long on ; towards the tops of theſe branches grow the flowers, faſhioned like vnto Broom flowers, but larger, as yellow as they, and ſmelling very well ; after which come ſmall long cods, creſted at the backe, wherein is contained blackiſh flat ſeede, faſhioned very like vnto the Kidney beanes : the roote is woody, diſperſing it ſelfe diuers waies.

The Place.

This groweth naturally in many places of France, Spaine and Italie, wee haue it as an ornament in our Gardens, among other delightfull plants, to pleaſe the ſenſes of ſight and ſmelling.

The Time.

It flowreth in the end of May, or beginning of Iune, and beareth ſeede, which ripeneth not with vs vntill it be late.

The Names.

It is called *Spartium Græcorum*, and *Spartum frutex*, to diſtinguiſh it from the ſedge or ruſh, that is ſo called alſo. Of ſome it is called *Geniſta*, and thought not to differ from the other *Geniſta*, but they are much deceiued ; for euen in Spaine and Italie, the ordinary *Geniſta* or Broome groweth with it, which is not pliant, and fit to binde Vines, or ſuch like things withall as this is.

The Vertues.

There is little vſe hereof in Phyſicke, by reaſon of the dangerous qualitie of vomiting, which it doth procure to them that take it inwardly : but being applyed outwardly, it is found to helpe the *Sciatica*, or paine of the hippes.

1 *Colutea vulgaris.* Ordinary baſtard Sene. 2 *Periploca recta Virginiana.* Virginian Silke. 3 *Colutea Scorpioides.* Scorpion baſtard
Sene. 4 *Spartum Hiſpanicum.* Spaniſh Broome. 5 *Liguſtrum.* Priuet. 6 *Saluia variegata.* Party coloured Sage. 7 *Maiorana aurea.*
Gudded Maierome.

CHAP. CXXIIII.

Periploca recta Virginiana. Virginian Silke.

Est this stranger should finde no hospitality with vs, being so beautifull a plant, or not finde place in this Garden, let him be here receiued, although with the last, rather then not at all. It riseth vp with one or more strong and round stalkes, three or foure foote high, whereon are set at the seuerall ioynts thereof two faire, long, and broad leaues, round pointed, with many veines therein, growing close to the stemme, without any foote-stalke : at the tops of the stalkes, and sometimes at the ioynts of the leaues, groweth forth a great bush of flowers out of a thinne skinne, to the number of twenty, and sometimes thirty or forty, euery one with a long foote-stalke, hanging downe their heads for the most part, especially those that are outermost, euery one standing within a small huske of greene leaues, turned to the stalkeward, like vnto the Lysimachia flower of Virginia before described, and each of them consisting of fiue small leaues a peece, of a pale purplish colour on the vpperside, and of a pale yellowish purple vnderneath, both sides of each leafe being as it were folded together, making them seeme hollow and pointed, with a few short chiues in the middle : after which come long and crooked pointed cods standing vpright, wherein are contained flat brownish seede, dispersedly lying within a great deale of fine, soft, and whitish browne silke, very like vnto the cods, seede, and silke of *Asclepias,* or Swallow-wort, but that the cods are greater and more crooked, and harder also in the outer shell : the roote is long and white, of the bignesse of a mans thumbe, running vnder ground very far, and shooting vp in diuers places, the heads being set full of small white grumes or knots, yeelding forth many branches, if it stand any time in a place : the whole plant, as well leaues as stalkes, being broken, yeeld a pale milke.

The Place.

It came to me from Virginia, where it groweth aboundantly, being raised vp from the seede I receiued.

The Time.

It flowreth in Iuly, and the seede is ripe in August.

The Names.

It may seeme very probable to many, that this plant is the same that Prosper Alpinus in the twenty fift Chapter of his Booke of Egyptian plants, nameth *Beidelsar*; and Honorius Bellus in his third and fourth Epistles vnto Clusius (which are at the end of his History of plants) calleth *Ossar frutex* : And Clusius himselfe in the same Booke calleth *Apocynum Syriacum, Palastinum,* and *Ægyptiacum,* because this agreeth with theirs in very many and notable parts ; yet verily I thinke this plant is not the same, but rather another kinde of it selfe : First, because it is not *frutex,* a shrub or wooddy plant, nor keepeth his leaues all the yeare, but loseth both leaues and stalks, dying downe to the ground euery yeare : Secondly, the milke is not causticke or violent, as Alpinus and Bellus say *Ossar* is : Thirdly, the cods are more crooked then those of Clusius, or of Alpinus, which Honorius Bellus acknowledgeth to be right, although greater then those he had out of Egypt : And lastly, the rootes of these doe runne, whereof none of them make any mention. Gerard in his Herball giueth a rude figure of the plant, but a very true figure of the cods with seede, and saith the Virginians call it *wisanck,* and referreth it to the *Asclepias,* for the likenesse of the cods stuffed with

silken

filken doune. But what reafon Cafpar Bauhinus in his *Pinax Theatri Bota-nici* had, to call it (for it is Clufius his *Apocynum Syriacum*) by the name of *Lapathum Ægyptiacum lactefcens filiqua Afclepiadis*, I know none in the world : for but that he would fhew an extreame fingularity in giuing names to plants, contrary to all others (which is very frequent with him) how could he thinke, that this plant could haue any likeneffe or correfponden-cie, with any of the kindes of Dockes, that euer he had feene, read, or heard of, in face, or fhew of leaues, flowers, or feede ; but efpecially in giuing milke. I haue you fee (and that not without iuft and euident caufe) giuen it a differing Latine name from Gerard, becaufe the *Afclepias* giueth no milke, but the *Periploca* or *Apocynum* doth ; and therefore fitter to be referred to this then to that. And becaufe it fhould not want an Englifh name anfwera-ble to fome peculiar property thereof, I haue from the filken doune called it Virginian Silke : but I know there is another plant growing in Virginia, called Silke Graffe, which is much differing from this.

The Vertues.

I know not of any in our Land hath made any tryall of the properties hereof. Captaine Iohn Smith in his booke of the difcouery and defcrip-tion of Virginia, faith, that the Virginians vfe the rootes hereof (if his be the fame with this)being bruifed and applyed to cure their hurts & difeafes.

CHAP. CXXV.

Liguftrum. Primme or Priuet.

Becaufe the vfe of this plant is fo much, and fo frequent throughout all this Land, although for no other purpofe but to make hedges or arbours in Gardens, &c. whereunto it is fo apt, that no other can be like vnto it, to bee cut, lead, and drawne into what forme one will, either of beafts, birds, or men armed, or otherwife: I could not forget it, although it be fo well knowne vnto all, to be an hedge bufh grow-ing from a wooddy white roote, fpreading much within the ground, and bearing ma-nie long, tough, and plyant fprigs and branches, whereon are fet long, narrow, and pointed fad greene leaues by couples at euery ioynt : at the tops whereof breake forth great tufts of fweete fmelling white flowers, which when they are fallen, turne into fmall blacke berries, hauing a purple iuyce within them, and fmall feede, flat on the one fide, with an hole or dent therein : this is feene in thofe branches that are not cut, but fuffered to beare out their flowers and fruit.

The Place.

This bufh groweth as plentifully in the Woods of our owne Couutrey, as in any other beyond the Seas.

The Time.

It flowreth fometimes in Iune, and in Iuly ; the fruit is ripe in Auguft and September.

The Names.

There is great controuerfie among the moderne Writers concerning this plant, fome taking it to be κύπρος of Diofcorides, other to be *Phillyrea* of Di-ofcorides, which followeth next after *Cyprus*. Plinie maketh mention of *Cyprus* in two places ; in the one he faith, *Cyprus* hath the leafe of *Ziziphus,* or the Iuiube tree: in the other he faith, that certain do affirme, that the *Cyprus* of the Eaft Country, and the *Liguftrum* of Italy is one and the fame plant : whereby you may plainly fee, that our Priuet which is *Liguftrum*, cannot be that *Cyprus* of Plinie with Iuiube leaues: Befides, both Diofcorides & Plinie fay, that *Cyprus* is a tree; but all know that *Liguftrum*, Priuet, is but an hedge bufh : Againe, Diofcorides faith, that the leaues of *Cyprus* giue a red colour, but Priuet giueth none. Bellonius and Profper Alpinus haue both recorded, that the true *Cyprus* of Diofcorides groweth plentifully in Egypt, Syria, and thofe Eafterne Countries, and nourfed vp alfo in Conftantinople, and other parts of Greece, being a merchandife of much worth, in that they tranfport the leaues, and young branches dryed, which laid in water giue a yellow colour, wherewith the Turkifh women colour the nailes of their hands, and fome other parts of their bodies likewife, delighting much therein : and that it is not our *Liguftrum*, or Priuet, becaufe *Cyprus* beareth round white feede, like Coriander feede, and the leaues abide greene alwaies vpon the tree, which groweth (if it bee not cut or pruined) to the height of the Pomegranet tree. I haue (I confeffe) beyond the limits I fet for this worke fpoken concerning our Priuet, becaufe I haue had the feede of the true *Cyprus* of Diofcorides fent mee, which was much differing from our Priuet, and although it fprang vp, yet would not abide any time, whereas if it had beene our Priuet, it would haue beene familiar enough to our Countrey.

The Vertues.

It is of fmall vfe in phyficke, yet fome doe vfe the leaues in Lotions, that ferue to coole and dry fluxes or fores in diuers parts.

Chap. CXXVI.

Saluia variegata. Party coloured Sage. And
Maiorana verficolor fiue aurea. Yellow or golden Marierome.

Vnto all thefe flowers of beauty and rarity, I muft adioyne two other plants, whofe beauty confifteth in their leaues, and not in their flowers : as alfo to feparate them from the others of their tribe, to place them here in one Chapter, before the fweete herbes that fhall follow, as is fitteft to furnifh this our Garden of pleafure. This kinde of Sage groweth with branches and leaues, very like the ordinary Sage, but fomewhat fmaller, the chiefeft difference confifteth in the colour of the leaues, being diuerfly marked and fpotted with white and red among the greene : for vpon one branch you fhall haue the leaues feuerally marked one from another, as the one halfe of the leafe white, and the other halfe greene, with red fhadowed ouer them both, or more white then greene, with fome red in it, either parted or fhadowed, or dafht here and there, or more greene then white, and red therein, eyther in the middle or end of the leafe, or more or leffe parted or ftriped with white and red in the greene, or elfe fometimes wholly greene the whole branch together, as nature lifteth to play with fuch varieties : which manner of growing rifing from one and the fame plant, becaufe it is the more variable, is the more delightfull and much refpected.

There is another fpeckled Sage parted with white and greene, but it is nothing of that beauty to this, becaufe this hath three colours euidently to bee difcerned in euery leafe almoft, the red adding a fuperaboundant grace to the reft.

Maiorana aurea fiue verficolor. Yellow or golden Marierome.

This kinde of Marierome belongeth to that fort is called in Latine *Maiorana latifolia,*

lia, which Lobel fetteth forth for *Hyſſopus Græcorum genuina:* In Engliſh Winter Marierome, or pot Marierome : for it hath broader and greater leaues then the ſweete Marierome, and a different vmbell or tuft of flowers. The difference of this from that ſet forth in the Kitchin Garden, confiſteth chiefly in the leaues, which are in Summer wholly yellow in ſome, or but a little greene, or parted with yellow and greene more or leſſe, as nature liſteth to play : but in Winter they are of a darke or dead greene colour, yet recouering it ſelfe againe : the ſent hereof is all one with the pot Marierome.

Wee haue another parted with white and greene, much after the manner with the former.

The Place, Time, Names, and Vertues of both theſe plants, ſhall be declared where the others of their kindes are ſpecified hereafter, and in the Kitchen Garden ; for they differ not in properties.

Chap. CXXVII.

Lauendula. Lauender Spike.

AFter all theſe faire and ſweete flowers before ſpecified, I muſt needes adde a few ſweete herbes, both to accompliſh this Garden, and to pleaſe your ſenſes, by placing them in your Noſegayes, or elſe where, as you liſt. And although I bring them in the end or laſt place, yet are they not of the leaſt account.

1. *Lauendula maior.* Garden Lauender.

Our ordinary Garden Lauender riſeth vp with a hard wooddy ſtemme aboue the ground, parted into many ſmall branches, whereon are ſet whitiſh, long, and narrow leaues, by couples one againſt another ; from among which riſeth vp naked ſquare ſtalkes, with two leaues at a ioynt, and at the toppe diuers ſmall huſkes ſtanding round about them, formed in long and round heads or ſpikes, with purple gaping flowers ſpringing out of each of them : the roote is wooddy, and ſpreadeth in the ground : The whole plant is of a ſtrong ſweete ſent, but the heads of flowers much more, and more piercing the ſenſes, which are much vſed to bee put among linnen and apparrell.

There is a kinde hereof that beareth white flowers, and ſomewhat broader leaues, *Flore albo.* but it is very rare, and ſeene but in few places with vs, becauſe it is more tender, and will not ſo well endure our cold Winters.

2. *Lauendula minor ſeu Spica.* Small Lauender or Spike.

The Spike or ſmall Lauender is very like vnto the former, but groweth not ſo high, neither is the head or ſpike ſo great and long, but ſhorter and ſmaller, and of a more purpliſh colour in the flower : the leaues alſo are a little harder, whiter, and ſhorter then the former ; the ſent alſo is ſomewhat ſharper and ſtronger. This is not ſo frequent as the firſt, and is nouriſhed but in ſome places that are warme, and where they delight in rare herbes and plants.

The Place.

Lauender groweth in Spaine abondantly, in many places ſo wilde, and little regarded, that many haue gone, and abiden there to diſtill the oyle thereof whereof great quantity now commeth ouer from thence vnto vs : and alſo in Lanquedocke, and Prouence in France.

The Time.

It flowreth early in thoſe hot Countries, but with vs not vntill Iune and Iuly.

The

The Names.

It is called of some *Nardus Italica*, and *Lauendula*, the greater is called *Fæmina*, and the lesser *Mas*. We doe call them generally Lauender, or Lauender Spike, and the lesser Spike, without any other addition.

The Vertues.

Lauender is little vsed in inward physicke', but outwardly; the oyle for cold and benummed parts, and is almost wholly spent with vs, for to perfume linnen, apparrell, gloues, leather, &c. and the dryed flowers to comfort and dry vp the moisture of a cold braine.

Chap. CXXVIII.

Stæchas. Sticadoue, Cassidony, or French Lauender.

Cassidony that groweth in the Gardens of our Countrey, may peraduenture somewhat differ in colour, as well as in strength, from that which groweth in hotter Countries; but as it is with vs, it is more tender a great deale then Lauender, and groweth rather like an herbe then a bush or shrub, not aboue a foote and a halfe high, or thereabouts, hauing many narrow long greene leaues like Lauender, but softer and smaller, set at seuerall distances together about the stalkes, which spread abroad into branches: at the tops whereof stand long and round, and sometimes foure square heads, of a darke greenish purple colour, compact of many scales set together; from among which come forth the flowers, of a blewish purple colour, after which follow seede vessels, which are somewhat whitish when they are ripe, containing blackish browne seede within them: the roote is somewhat wooddy, and will hardly abide the iniuries of our cold Winters, except in some places onely, or before it haue flowred: The whole plant is somewhat sweete, but nothing so much as Lauender.

The Place.

Cassidony groweth in the Ilands Stæchades, which are ouer against Marselles, and in Arabia also: we keep it with great care in our Gardens.

The Time.

It flowreth the next yeare after it is sowne, in the end of May, which is a moneth before any Lauender.

The Names.

It is called of some *Lauendula siluestris*, but most vsually *Stæchas*: in English, of some Stickadoue, or French Lauender; and in many parts of England, Cassidony.

The Vertues.

It is of much more vse in physicke then Flauender, and is much vsed for old paines in the head. It is also held to be good for to open obstructions, to expell melancholy, to cleanse and strengthen the liuer, and other inward parts, and to be a Pectorall also.

Chap. CXXIX.

Abrotanum fæmina siue Santolina. Lauender Cotton.

THis Lauender Cotton hath many wooddy, but brittle branches, hoary or of a whitish colour, whereon are set many leaues, which are little, long, and foure square, dented or notched on all edges, and whitish also : at the tops of these branches stand naked stalkes , bearing on euery one of them a larger yellow head or flower, then eyther Tansie or Maudeline, whereunto they are somewhat like, wherein is contained small darke coloured seede : the roote is hard , and spreadeth abroad with many fibres : the whole plant is of a strong sweete sent, but not vnpleasant, and is in many places planted in Gardens, to border knots with, for which it will abide to be cut into what forme you thinke best ; for it groweth thicke and bushy , very fit for such workes, besides the comely shew the plant it selfe thus wrought doth yeeld, being alwayes greene, and of a sweet sent ; but because it quickly groweth great, and will soon runne out of forme, it must be euery second or third yeare taken vp, and new planted.

The Place.

It is onely planted in Gardens with vs, for the vses aforesaid especially.

The Time.

It flowreth in Iuly , and standeth long in the hot time of the yeare in his colour, and so will doe, if it be gathered before it haue stood ouer long.

The Names.

Diuers doe call it as Matthiolus doth, *Abrotanum fæmina*, and *Santolina* ; and some call it *Chamæcyparissus*, because the leaues thereof, are somewhat like the leaues of the Cypresse tree : Wee call it in English generally Lauen-der Cotton.

The Vertues.

This is vsually put among other hot herbes, eyther into bathes , oint-ments, or other things, that are vsed for cold causes. The seede also is much vsed for the wormes.

Chap. CXXX.

Ocimum. Bassill.

BAssill is of two sorts (besides other kindes) for this our Garden, the one whereof is greater, the other lesse in euery part thereof, as shall be shewed.

1. *Ocimum Citratum.* Common Bassill.

Our ordinary Garden Bassill hath one stalke rising from the root, diuersly branched out, whereon are set two leaues alwayes at a ioynt, which are broad, somewhat round, and pointed, of a pale greene colour, but fresh, a little snipt or dented about the edges, and of a strong or heady sent, somewhat like a Pomecitron , as many haue compared it, and thereof call it *Citratum* : the flowers are small and white, standing at the tops of the branches, with two smal leaues at euery ioynt vnder them in some plants green, in o-

thers

thers browne vnder them : after which commeth blackish seede : the roote perisheth at the first approach of winter weather, and is to be new sowen euery yeare.

2. *Ocimum minimum siue Gariophyllatum.* Bush Basill.

The bush Basill groweth not altogether so high, but is thicker spreade out into branches, whereon grow smaller leaues, and thicker set then the former, but of a more excellent and pleasant smell by much : the flowers are white like the former, and the seede blacke also like it, and perisheth as suddenly, or rather sooner then it, so that it requireth more paines to get it, and more care to nourse it, because we seldome or neuer haue any seede of it.

Ocimum Indicum. Indian Basill.

The Indian Basill hath a square reddish greene stalke, a foote high or better, from the ioynts whereof spreade out many branches, with broade fat leaues set thereon, two alwayes together at the ioynt, one against another, as other Basils haue, but somewhat deepely cut in on the edges, and oftentimes a little crumpled, standing vpon long reddish footestalkes, of a darke purple colour, spotted with deeper purple spots, in some greater, in others lesser: the flowers stand at the tops of the stalkes spike-fashion, which are of a white colour, with reddish stripes and veines running through them, set or placed in darke purple coloured huskes : the seede is greater and rounder then the former, and somewhat long withall : the roote perisheth in like manner as the other former doe. The whole plant smelleth strong, like vnto the other Basils.

The Place.

The two last sorts of Basils are greater strangers in our Country then the first which is frequent, and only sowen and planted in curious gardens. The last came first out of the West Indies.

The Time.

They all flower in August, or Iuly at the soonest, and that but by degrees, and not all at once.

The Names.

The first is vsually called *Ocimum vulgare,* or *vulgatius,* and *Ocimum Citratum.* In English, Common or Garden Basill. The other is called *Ocimum minimum,* or *Gariophyllatum,* Cloue Basill, or Bush Basill. The last eyther of his place, or forme of his leaues, being spotted and curled, or all, is called *Ocimum Indicum maculatum, latifolium & crispum.* In English according to the Latine, Indian Basill, broade leafed Basill, spotted or curled Basill, which you please.

The Vertues.

The ordinary Basill is in a manner wholly spent to make sweet, or washing waters, among other sweet herbes, yet sometimes it is put into nosegayes. The Physicall properties are, to procure a cheerefull and merry heart, whervnto the seede is chiefly vsed in pouder, &c. and is most vsed to that, and to no other purpose.

CAPH.

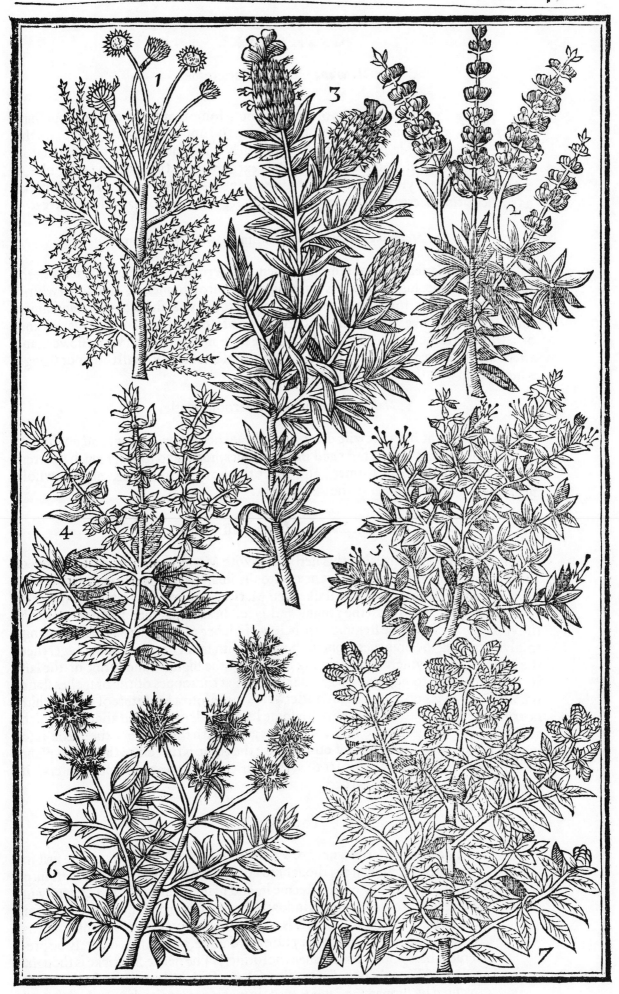

1 *Santolina.* Lauender Cotton. 2 *Lauendula.* Lauender Spike. 3 *Stœchas.* Caſſidony. 4 *Chamædrys.* Germander. 5 *Ocimum minus.* Fine Baſſill. 6 *Marum.* Herbe Maſticke. 7 *Maiorana.* Sweete Marierome.

CHAP. CXXXI.

Maiorana. Sweete Maricrome.

WEe haue many forts of Maricrome ; fome that are fweete, and but Summer plants ; others that are greater and not fo fweet ; and fome alfo that are wilde. Of all thefe I will onely felect fome of the choifeft that are fit for this place, and leaue the other for the next garden, and the garden of fimples, or a generall worke : yet hereunto I will adioyne another fweete plant called Mafticke, as participating neerer with them then with Time, whereunto many doe referre it.

1. *Maiorana maior æstiua.* Common fweet Maricrome.

The fweet Maricrome that is moft ftequently fowen in our Country, is a low herbe little aboue a foote high when it is at the higheft, full of branches, and fmall whitifh foft roundifh leaues, fmelling very fweet : at the toppes of the branches ftand diuers fmall fcaly heads, like vnto knots, (and therefore of fome called knotted Maricrome) of a whitifh greene colour, out of which come here and there fmall white flowers, and afterwards fmall reddifh feede : the roote is compofed of many fmall threds or ftrings, which perifh with the whole plant euery yeare.

2. *Maiorana tenuifolia.* Maricrome gentle.

This Maricrome hath likewife diuers fmall branches, growing low, and not higher then the former, but hauing finer and fmaller leaues, hoary and foft, but much fweeter : the heads are like vnto the former, and fo are the flowers and feede, and the whole plant abiding but a Summer in the like manner.

3. *Marum.* Herbe Mafticke.

The neerer refemblance that this herbe hath with Maricrome then with Tyme (as I faid before) hath made me place it next vnto the fmall fweet Maricrome. It rifeth vp with a greater, and a more woody ftalke then Maricrome, two foote high or better in fome places, where it liketh the ground and ayre, branching out on all fides towards the vpper part, leauing the ftemme bare below, if it bee old, otherwife being young, thinly furnifhing the branches from the bottome with fmall greene leaues, bigger then the leaues of any Tyme, and comming neere vnto the bigneffe and forme of the laft recited finer Maricrome, but of a greener colour : at the toppes of the branches ftand fmall white flowers on a head, which afterwards turne into a loofe tuft of a long white hoary matter, like vnto foft doune, with fome leaues vnderneath and about it, which abide not long on the ftalkes, but are blowne away with the winde : the feede is fo fmall if it haue any, that I haue not obferued it : the roote is threddy : the whole plant is of a fweete refinous fent, ftronger then the Maricrome, and abideth our winters, if it be carefully planted and regarded.

The Place.

The fweete Maricromes grow naturally in hot Countreyes : the firft in Spaine &c. the fecond is thought to come out of Syria, or Perfia firft into Italie, where they much efteeme it, and plant it curioufly and carefully in pots, and fet them in their windowes, beeing much delighted therewith for the fweet fent it hath. The firft is vfually fowen euery yeare in moft gardens with vs : but the fecond is very rare and daintie, and muft as daintely be preferued, being more tender then the former. The herbe Mafticke is thought to be firft brought out of Candie, Clufius faith he found it in Spaine : It is planted by flippes, (and not fowen) in many gardens, and is much replanted

for

for increase, but prospereth onely, or more frequently, in loamie or clay grounds then in any other soyle.

The Time.

The sweete Marieromes beare their knots or scaly heads in the end of Iuly, or in August. Herbe Masticke in Iune many times, or in the beginning of Iuly.

The Names.

The first of the two sweet Marieromes called *Maiorana* in Latine *à maiore cura*, is taken of most writers to be the *Amaracus* or *Sampsuchum* of Dioscorides, Theophrastus and Plinie, although Galen doth seem a little to dissent therefrom. The other sweet Mariicrome hath his name in his title as much as can be said of it. The next is thought by the best of the moderne Writers to be the true *Marum* that Galen preferreth for the excellent sweetnesse, before the former Mariicrome in making the *Oleum*, or *vnguentum Amaricinum*, and seemeth to incline to their opinion that thought *Amaracus* was deriued from *Marum*. It is the same also that Galen and others of the ancient Writers make mention of, to go into the composition of the *Trochisci Hedychroi*, as well as *Amaracus* among the ingredients of the *Theriaca Andromachi*. In English we call it Masticke simply, or Herbe Mastick, both to distinguish it from that Tyme that is called Masticke Tyme, and from the Masticke Tree, or Gum, so called. Some of later times, and Clusius with them, haue thought this to be Dioscorides his *Tragoriganum*, which doth somewhat resemble it : but there is another plant that Matthiolus setteth forth for *Marum*, that in Lobels opinion and mine is the truest *Tragoriganum*, and this the truest *Marum*.

The Vertues.

The sweete Marieromes are not onely much vsed to please the outward senses in nosegayes, and in the windowes of houses, as also in sweete pouders, sweete bags, and sweete washing waters, but are also of much vse in Physicke, both to comfort the outward members, or parts of the body, and the inward also : to prouoke vrine being stopped, and to ease the paines thereof, and to cause the feminine courses. Herbe Masticke is of greater force to helpe the stopping of vrine, then the Mariicrome, and is put into Antidotes, as a remedie against the poyson of venemous Beasts.

Chap. CXXXI.

Thymum. Tyme.

THere are many kindes of Tyme, as they are vsually called with vs, some are called of the garden, and others wilde, which yet for their sweetnesse are brought into gardens, as Muske Tyme, and Lemon Tyme; and some for their beauty, as embroidered or gold yellow Tyme, and white Tyme. But the true Tyme of the ancient Writers, called *Capitatum*, as a speciall note of distinction from all other kindes of Tyme, is very rare to be seene with vs here in England, by reason of the tendernesse, that it will not abide our Winters. And all the other sorts that with vs are called garden Tymes, are indeede but kindes of wilde Tyme, although in the defect or want of the true Tyme, they are vsed in the stead of it. With the Tymes I must doe as I did with the Marieromes in the Chapter before, that is, reserue the most common in vse, for the common vse of the Kitchen, and shew you only those here, that are not put to that vse : and first with the true Tyme, because it is knowne but to a few.

1, Thymum

1. *Thymum legitimum capitatum.* The true Tyme.

The true Tyme is a very tender plant, hauing hard and hoary brittle branches, spreading from a small wooddy stemme, about a foote and a halfe high, whereon are set at seuerall ioynts, and by spaces, many small, long, whitish, or hoary greene leaues, of a quicke sent and taste : at the tops of the branches stand small long whitish greene heads, somewhat like vnto the heads of *Stæchas*, made as it were of many leaues or scales, out of which start forth small purplish flowers (and in some white, as Bellonius saith) after which commeth small seede, that soone falleth out, and if it be not carefully gathered, is soone lost, which made (I thinke) Theophrastus to write, that this Tyme was to be sowne of the flowers, as not hauing any other seede : the root is small and wooddy. This holdeth not his leaues in Winter, no not about Seuill in Spaine, where it groweth aboundantly, as Clusius recordeth, finding it there naked or spoiled of leaues. And will not abide our Winters, but perisheth wholly, roote and all.

2. *Serpillum hortense siue maius.* Garden wilde Tyme.

The wilde Tyme that is cherished in gardens groweth vpright, but yet is lowe, with diuers slender branches, and small round greene leaues, somewhat like vnto small fine Marierome, and smelling somewhat like vnto it : the flowers growe in roundels at the toppes of the branches, of a purplish colour : And in another of this kinde they are of a pure white colour.
There is another also like hereunto, that smelleth somewhat like vnto Muske ; and therefore called Muske Tyme, whose greene leaues are not so small as the former, but larger and longer.

3. *Serpillum Citratum.* Lemon Tyme.

The wilde Tyme that smelleth like vnto a Pomecitron or Lemon, hath many weake branches trayling on the ground, like vnto the first described wilde Tyme, with small darke greene leaues, thinly or sparsedly set on them, and smelling like vnto a Lemon, with whitish flowers at the toppes in roundels or spikes.

4. *Serpillum aureum siue versicolor.* Guilded or embroidered Tyme.

This kinde of wilde Tyme hath small hard branches lying or leaning to the ground, with small party coloured leaues vpon them, diuided into stripes or edges, of a gold yellow colour, the rest of the leafe abiding greene, which for the variable mixture or placing of the yellow, hath caused it to be called embroidered or guilded Tyme.

The Place.

The first groweth as is said before, about Seuill in Spaine, in very great aboundance as Clusius saith; and as Bellonius saith, very plentifully on the mountaines through all Greece. The others growe some in this Country, and some in others : but wee preserue them with all the care wee can in our gardens, for the sweete and pleasant sents and varieties they yeeld.

The Time.

The first flowreth not vntill August; the rest in Iune and Iuly.

The Names.

Their names are seuerally set downe in their titles, as is sufficient to distinguish them ; and therefore I shall not neede to trouble you any further with them.

The

The Vertues.

The true Tyme is a speciall helpe to melancholicke and spleneticke diseases, as also to flatulent humours, either in the vpper or lower parts of the body. The oyle that is Chimically drawne out of ordinary Tyme, is vsed (as the whole herbe is, in the stead of the true) in pils for the head and stomach. It is also much vsed for the toothach, as many other such like hot oyles are.

CHAP. CXXXII.

Hyssopus. Hyssope.

THere are many varieties of Hyssope, beside the common or ordinary, which I reserue for the Kitchen garden, and intend onely in this place to giue you the knowledge of some more rare : *viz.* of such as are noursed vp by those that are curious, and fit for this garden : for there are some other, that must be remembred in the Physicke garden, or garden of Simples, or else in a generall worke.

1. *Hyssopus folijs niueis.* White Hyssope.

This white Hyssope is of the same kinde and smell with the common Hyssope; but differeth, in that this many times hath diuers leaues, that are wholly of a white colour, with part of the stalke also : others are parted, the one halfe white, the other halfe greene, and some are wholly greene, or with some spots or stripes of white within the greene, which makes it delightfull to most Gentlewomen.

2. *Hyssopus folijs cinereis.* Russet Hyssope.

As the last hath party coloured leaues, white and greene, so this hath his leaues of an ash-colour, which of some is called russet; and hath no other difference either in forme or smell.

3. *Hyssopus aureus.* Yellow or golden Hyssope.

All the leaues of this Hyssope are wholly yellow, or but a little greene in them, and are of so pleasant a colour, especially in Summer, that they prouoke many Gentlewomen to weare them in their heads, and on their armes, with as much delight as many fine flowers can giue : but in Winter their beautifull colour is much decayed, being of a whitish greene, yet recouer themselues againe the next Summer.

4. *Hyssopus surculis densis.* Double Hyssope.

As this kinde of Hyssope groweth lower then the former or ordinary kinde, so it hath more branches, slenderer, and not so wooddy, leaning somewhat downe toward the ground, so wonderfully thicke set with leaues, that are like vnto the other, but of a darker greene colour, and somewhat thicker withall, that it is the onely fine sweete herbe, that I know fittest (if any be minded to plant herbes) to set or border a knot of herbes or flowers, because it will well abide, and not growe too wooddy or great, nor be thinne of leaues in one part, when it is thicke in another, so that it may be kept with cutting as smooth and plaine as a table. If it be suffered to growe vp of it selfe alone, it riseth with leaues as before is specified, and flowreth as the common doth, and of the same sent also, not differing in any thing, but in the thicknesse of the leaues on the stalkes and branches, and the aptnesse to be ordered as the keeper pleaseth.

5. *Chamædrys*

Chamædrys. Germander.

Left Germander ſhould be vtterly forgotten, as not worthy of our Garden, ſeeing many (as I ſaid in my treatiſe or introduction to this Garden) doe border knots therewith : let me at the leaſt giue it a place, although the laſt, being more vſed as a ſtrewing herbe for the houſe, then for any other vſe. It is (I thinke) ſufficiently knowne to haue many branches, with ſmall and ſomewhat round endented leaues on them, and purpliſh gaping flowers : the rootes ſpreading far abroad, and riſing vp againe in many places.

The Place.

Theſe Hyſſopes haue beene moſt of them nourſed vp of long time in our Engliſh Gardens, but from whence their firſt originall ſhould be, is not well knowne. The Germander alſo is onely in Gardens, and not wilde.

The Time.

They flower in Iune and Iuly.

The Names.

The ſeuerall names whereby they are knowne to vs, are ſet forth in their titles ; and therefore I neede not here ſay more of them then onely this, that neyther they here ſet downe, nor the common or ordinary ſort, nor any of the reſt not here expreſſed, are any of them the true Hyſſope of the ancient Greeke Writers, but *ſuppoſititia*, vſed in the ſtead thereof. The Germander, from the forme of the leaues like vnto ſmall oaken leaues, had the name *Chamædrys* giuen it, which ſignifieth a dwarfe Oake.

The Vertues.

The common Hyſſope is much vſed in all pectorall medicines, to cut fleagme, and to cauſe it eaſily to be auoided. It is vſed of many people in the Country, to be laid vnto cuts or freſh wounds, being bruiſed, and applyed eyther alone, or with a little Sugar. It is much vſed as a ſweet herbe, to be in the windowes of an houſe. I finde it much commended againſt the Falling Sickneſſe, eſpecially being made into Pils after this manner : Of Hyſſope, Horhound, and Caſtor, of each halfe a dramme, of Peony rootes (the male kinde is onely fit to be vſed for this purpoſe) two drams, of *Aſſa fœtida* one ſcruple : Let them be beaten, and made into pils with the iuyce of Hyſſope ; which being taken for ſeuen dayes together at night going to bed, is held to be effectuall to giue much eaſe, if not thoroughly to cure thoſe that are troubled with that diſeaſe. The vſe of Germander ordinarily is as Tyme, Hyſſope, and other ſuch herbes, to border a knot, whereunto it is often appropriate, and the rather, that it might be cut to ſerue (as I ſaid) for a ſtrewing herbe for the houſe among others. For the phyſicall vſe it ſerueth in diſeaſes of the ſplene, and the ſtopping of vrine, and to procure womens courſes.

Thus haue I led you through all my Garden of Pleaſure, and ſhewed you all the varieties of nature nourſed therein, pointing vnto them, and deſcribing them one after another. And now laſtly (according to the vſe of our old ancient Fathers) I bring you to reſt on the Graſſe, which yet ſhall not be without ſome delight, and that not the leaſt of all the reſt.

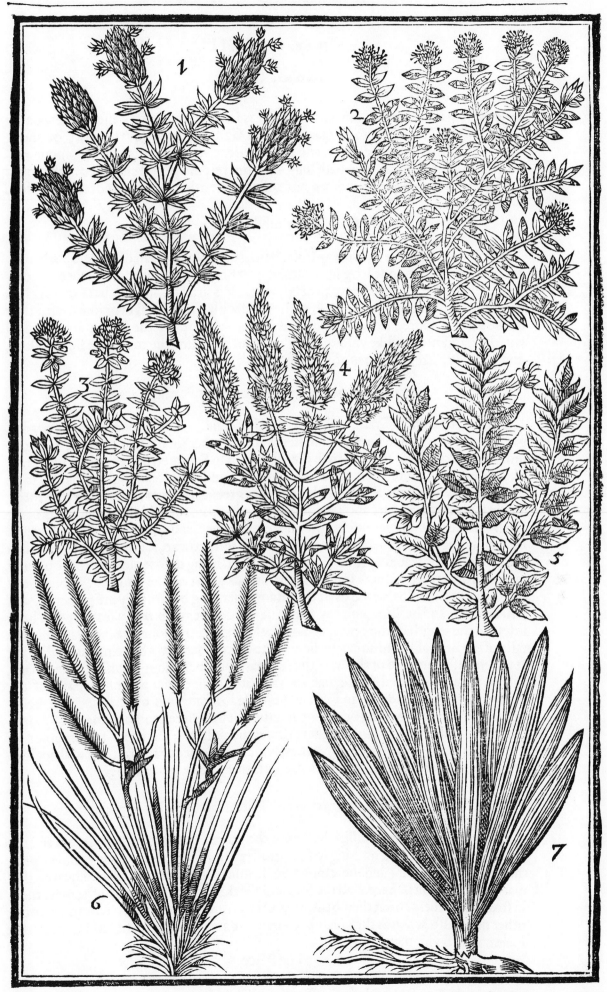

1 *Thymum legitimum.* The true Tyme. 2 *Serpillum maius hortense.* Garden wilde Tyme. 3 *Serpillum Citratum.* Lemon Tyme. 4 *Hyssopus versicolor sue aureus.* Golden Hyssope. 5 *Chamædrys.* Germander. 6 *Spartum Austriacum sue Gramen plumarium minus.* The lesser feather Grasse. 7 *Gramen striatum vel sulcatum.* Painted Grasse or Ladies Laces.

Q q

Chap. CXXXIII.

Gramina. Grasses.

THere are among an infinite number (as I may so say) of Grasses, a few onely which I thinke fit to be planted in this Garden, both for the rarity of them, and also for your delight, and the excellent beauty that is in them aboue many other plants. One of them hath long agoe bin respected, and cherished in the country gardens of many Gentlewomen, and others. The others are knowne but vnto a few.

1. *Gramen striatum.* Painted Grasse or Ladies laces.

This kinde of Grasse hath many stiffe, hard, round stalkes, full of ioynts, whereon are set at euery ioynt one long leafe, somewhat broad at the bottome, where it compasseth the stalke, and smaller to the end, where it is sharpe pointed, hard or rough in handling, and striped all the length of the leafe with white streakes or lines, that they seeme party coloured laces of white and greene : the tops of the stalkes are furnished with long spikie tufts, like vnto the tufts of Couch Grasse : the rootes are small, white, and threddy, like the rootes of other Grasses.

2. *Gramen Plumarium minus.* The lesser Feather-Grasse.

This lesser Feather-Grasse hath many small, round, and very long leaues or blades, growing in tufts, much finer and smaller then any other Grasse that I know, being almost like vnto haires, and of a fresh greene colour in Summer, but changing into gray, like old hay in Winter, being indeede all dead, and neuer reuiuing ; yet hardly to be plucked away vntill the Spring, and then other greene leaues or rushes rise vp by them, and in their stead, and are aboue a foote in length: from the middle of these tufts come forth rounder and bigger rushes, which are the stalkes, and which haue a chaffie round eare about the middle thereof, which when it is full growne, is somewhat higher then the toppes of the leaues or rushes, opening it selfe (being before close) at the top, and shewing forth three or foure long ayles or beards, one aboue another, which bend themselues a little downewards (if they stand ouer long before they are gathered, and will fall off, and be blowne away with the winde) being so finely feathered on both sides, all the length of the beard, and of a pale or grayish colour, that no feather in the taile of the Bird of Paradise can be finer, or to be compared with them, hauing sticking at the end of euery one of them, within the eare, a small, long, whitish, round, hard, and very sharpe pointed graine, like vnto an oaten graine, that part of the stalke of the feather that is next vnder it, and aboue the seede for some two or three inches, being stiffe and hard, and twining or curling it selfe, if it be suffered to stand too long, or to fall away, otherwise being straight as the feather it selfe : the roote is composed of many long, hard, small threddy strings, which runne deepe and far, and will not willingly be remoued, in that it gaineth strength euery yeare by standing.

3. *Gramen Plumarium maius.* The greater Feather-Grasse.

The greater Feather-Grasse is like vnto the lesser, but that both the leaues and the feathers are greater, and nothing so fine, grosser also, and of lesse beauty and respect, though whiter then it ; and therefore is not so much regarded : for I haue knowne, that many Gentlewomen haue vsed the former lesser kinde, being tyed in tufts, to set them in stead of feathers about their beds, where they haue lyen after childe-bearing, and at other times also, when as they haue been much admired of the Ladies and Gentles that haue come to visit them.

The Place.

The first of these Grasses, as Lobel saith, groweth naturally in the woods and hils of Sauoy. It hath long agoe beene receiued into our English gardens.

dens. The fecond, as Clufius faith, in Auftria, from whence alfo (as I take it) the greater came, and are both in the gardens of thofe, that are curious obferuers of thefe delights.

The Time.

The firft is in its pride for the leaues all the Spring and Summer, yeelding his bufh in Iune. The other giue their feather-like fprigs in Iuly and Auguft, and quickly (as I faid) are fhed, if they be not carefully gathered.

The Names.

The firft is called by Lobel *Gramen fulcatum*, or *ftriatum album*, of others *Gramen pictum*. The French call it *Aiguellettes d' armes*, of the fafhion that their Enfignes, Pennons, or Streamers vfed in wars were of, that is, like vnto a party coloured curtaine. In Englifh vfually Ladies laces, and Painted Graffe. The firft of the other two is called *Gramen plumarium* or *plumofum*, and *minus* is added for the diftinction of it. Clufius calleth it *Spartum Auftriacum*, of the likeneffe and place where he found it. The laft is called *Gramen plumarium*, or *plumofum maius*, The greater Feather-Graffe.

The Vertues.

Thefe kindes of Graffes are not in any time or place that I doe heare of applyed to any Phyficall vfe; and therefore of them I will fay no more: but here I will end the prime part of this worke.

THE

THE ORDERING
OF THE KITCHEN
GARDEN.

CHAP. I.

The situation of a Kitchen Garden, or Garden of Herbes, and what sort of manure is fittest to helpe the decaying of the soyle thereof.

Auing giuen you the best rules and instructions that I can for your flower Garden, and all the flowers that are fit to furnish it, I now proceede to your herbe garden, which is not of the least respect belonging to any mans house, nor vtterly to bee neglected for the many vtilities are to be had from it, both for the Masters profit and pleasure, and the meynies content and nourishment : all which if I should here set down, I had a large field to wander in, and matter sufficient to entreat of, but this worke permitteth not that libertie : and I thinke there are but few but eyther know it already, or conceiue it sufficiently in their minds. Passing therefore no further in such discourses, I come to the matter in hand, which is to shew you where the fittest place is for an herbe garden. As before I shewed you that the beautie of any worthy house is much the more commended for the pleasant situation of the garden of flowers, or of pleasure, to be in the sight and full prospect of all the chiefe and choisest roomes of the house; so contrariwise, your herbe garden should bee on the one or other side of the house, and those best and choyse roomes : for the many different sents that arise from the herbes, as Cabbages, Onions, &c. are scarce well pleasing to perfume the lodgings of any house; and the many ouertures and breaches as it were of many of the beds thereof, which must necessarily bee, are also as little pleasant to the sight. But for priuate mens houses, who must like their habitations as they fall vnto them, and cannot haue time or meanes to alter them, they must make a vertue of necessity, and conuert their places to their best aduantage, by making their profit their chiefest pleasure, and making one place serue for all vses. The choyce of ground for this Garden, is (as I said before) where it is fat, fertill and good, there needeth the lesse labour and cost : and contrariwise, where it is cold, wet, dry or barren, there must bee the more helpes still added to keepe it in heart. For this Garden by reason of the much and continuall stirring therein, the herbes and rootes drawing out the substance of the fertilitie thereof more aboundantly then in the former, must be continually holpen with soyle, or else few things of goodnesse or worth will come forward therein. The stable soyle of horses is best and more proper for any colde grounds, for being the hottest, it will cause any the seedes for this Garden to prosper well, and be more forward then in any other ground that is not so holpen. The stable soyle of Cattell is of a colder and moister nature, and is therefore more proper for

Qq 3 the

the hot fandy or grauelly grounds, and although it bee longer before it bee brought to mould then that of horfes, yet it will outlaft it more then twice fo long. Let euery one therefore take according to the nature of the ground fuch helpes as are moft fit and conuenient, as I haue here and before fhewed. But I doe here ingenuoufly confeffe my opinion of thefe forcings and helpings of ground, that howfoeuer it doth much good to fome particular things, which becaufe they delight in heate, and cannot be brought to perfection without it in this our Countrey, which is colder then their naturall from whence they are brought, muft therfore haue artificiall helpes to forward them ; yet for many other things the compoft doth much alter and abate the naturall vigour, and quickeneffe of tafte, that is perceiued in them that grow in a naturall fat or fandy foile that is not fo holpen.

Снар. II.

*The forme of a Garden of herbes for neceffary vfes,
with the ordering thereof.*

AS our former Garden of pleafure is wholly formable in euery part with fquares, trayles, and knots, and to bee ftill maintained in their due forme and beautie : fo on the contrary fide this Garden cannot long conferue any forme, for that euery part thereof is fubiect to mutation and alteration. For although it is conuenient that many herbes doe grow by themfelues on beds, caft out into fome proportion fit for them, as Tyme, Hiffope, Sage, &c. yet many others may bee fowen together on a plot of ground of that largeneffe that may ferue euery mans particular vfe as he fhall haue occafion to employ it, as Reddifh, Lettice and Onions, which after they are growne vp together may be drawne vp and taken away, as there is occafion to fpend them : but Carrots or Parfneps being fowen with others muft bee fuffered to grow laft, becaufe they require a longer time before they be fit to be taken vp. Other herbes require fome large compaffe of ground whereon they may grow of themfelues without any other herbes growing among them, as Artichokes, Cowcumbers, Melons, Pompions. And fome will doe fo with their Cabbages alfo, but the beft and moft frugall way now vfed, is to plant them round about the border of your plot or ground whereon you plant Cowcumbers, Pompions, or other things, in that by this meanes fo much ground will be well faued, and the other things be no whit hindered thereby, which elfe a great deale of ground muft be employed for them apart. So that by this that I haue here faid, you may perceiue the forme of this Garden is for the moft part, to bee ftill out of forme and order, in that the continuall taking vp of the herbes and rootes that are fowen and planted, caufeth the beds or parts of this Garden to lye broken, difmembered, and out of the order that at the firft it was put into. Remember herewithall that (as I faid before) this Garden requireth the continuall helpe of foyle to be brought into it, in that the plenty of thefe manner of herbes and rootes doe fo much wafte the fertilitie and fatneffe of the ground, that without continuall refrefhing it would quickly become fo poore and barren, that it would not yeelde the worth of the feede. The ordinary time to foyle a Garden, is to bring in manure or dung before Chriftmas, and eyther bury it fome fmall depth, not too deepe, or elfe to lay it vpon the ground that the winter froftes may pierce it, and then turne it fhallow into the ground to fow your feeds in the Spring.

Chap. III.

How to order diuers Garden herbes, both for their sowing,
spending, and gathering of the seede.

OVr chiefeſt and greateſt Gardiners now adaies, doe ſo prouide for themſelues euery yeare, that from their owne grounds they gather the ſeede of many herbes that they ſowe againe: for hauing gained the beſt kind of diuers herbes, they will be ſtill furniſhed with the ſame, and be not to ſeeke euery yeare for new that oftentimes will not yeelde them halfe the profit that their choyce ſeede will : I ſay of many herbes, but not of all ; for the beſt of them all hath not ground ſufficient for all ſorts, nor will our climate bring ſome to that perfection that other forraine doth, and therefore the ſeede of ſome things are continually brought from beyond Sea vnto vs. And againe although our chiefe Gardiners doe ſtill prouide their owne ſeede of diuers things from their owne ground, becauſe as I ſaid it is of the beſt kinde, yet you muſt vnderſtand alſo, that good ſtore of the ſame ſortes of ſeeds are brought from beyond the Seas, for that which is gathered in this Land is not ſufficient to ſerue euery mans vſe in the whole Kingdome by many parts ; yet ſtill it is true, that our Engliſh ſeede of many things is better then any that commeth from beyond the Seas : as for example, Reddiſh, Lettice, Carrots, Parſneps, Turneps, Cabbages, and Leekes, of all which I intend to write in this place ; for theſe are by them ſo husbanded, that they doe not ſow their owne grounds with any other ſeede of theſe ſorts but their owne : which that you may know the manner how to doe, I will here ſet it downe, that euery one may haue the beſt directions if they will follow them. Of Reddiſh there are two ſorts, one more early then the other : they vſe therfore to ſow their early Reddiſh firſt, that they may haue the earlieſt profit of them, which is more worth in one fortnight, then in a moneth after. And to effect this they haue ſome artificiall helps alſo, which are theſe: They vſe to digge vp a large plot of ground where they intend to ſow their ſeede a little before or after Chriſtmas, caſting it into high balkes or ridges fiue or ſixe foote aſunder, which they ſuffer to lye and take all the extreame froſts in Ianuary to mellow the earth, and when the froſtes are paſt, they then beginne to bring into it good ſtore of freſh ſtable dung, which they laye neyther too deepe nor too thicke, and couer it with the mould a hand breadth thickneſſe aboue the dung, which doth giue ſuch a warmth and comfort to whatſoeuer is ſowen thereon, that it forceth it forward much ſooner then any other way can doe : And to preuent both the froſtes, and the cold bitter windes which often ſpoyle their ſeede new ſprung vp, they vſe to ſet great high and large mattes made of reedes, tyed together, and faſtened vnto ſtrong ſtakes, thruſt into the ground to keepe them vp from falling, or being blowne down with the winde, which mattes they place on the North and Eaſt ſide to breake the force of theſe winds, and are ſo ſure and ſafe a defence, that a bricke wall cannot better defend any thing vnder it, then this fence will. In this manner they doe euery yeare to bring forward their ſeede to gaine the more by them, and they that will haue Reddiſh early, muſt take the ſame courſe. The other ſort of Reddiſh for the moſt part is ſowen in Februarie, a fortnight after the other at the leaſt, and likewiſe euery moneth after vnto September, that they may haue young continually. For the blacke Reddiſh, although many in many places doe ſowe it in the ſame time, and in the ſame manner that the ordinary is ſowen, yet the nature thereof is to runne vp to ſeede more ſpeedily then the other, if it haue ſo rich ground to grow vpon, and therefore the beſt time to ſow it is in Auguſt, that ſo it may abide all winter, wherein is the chiefeſt time for the ſpending thereof, and to keepe it vntill the beginning of the next yeare from running vp to ſeed: the gathering whereof, as alſo of the other ſort, is all after one manner, that is, to be pulled vp when the pods change whitiſh, and then hanged vpon buſhes, pales, or ſuch other thing, vntill they bee thorough dry, and then beaten or thraſhed out vpon a ſmooth plancher, or vpon clothes, as euery ones ſtore is, and their conueniencie. Lettice is ſowen oftentimes with the early Reddiſh, in the ſame manner before ſaid, that they may haue Lettice likewiſe as early as the time of the year will permit them, which they

they pull vp where they grow too thicke, spending them first, and so taking vp from time to time, vntill they stand two foote in sunder one from another, and beginne to spindle and shoote vp for seede. In this is vsed some arte to make the plants strong to giue the better seede without danger of rotting or spoyling with the wet, which often happeneth to those about whom this caution is not obserued: Before your Lettice is shot vp, marke out the choysest and strongest plantes which are fittest to grow for seede, and from those when they are a foote high, strippe away with your hand the leaues that grow lowest vpon the stalke next the ground, which might rot, spoyle or hinder them from bearing so good seede; which when it is neere to be ripe, the stalkes must be cut off about the middle, and layde vpon mats or clothes in the Sunne, that it may there fully ripen and be gathered ; for it would be blowne away with the winde if it should be suffered to abide on the stalkes long. Parsneps must be sowen on a deep trenched mellow ground, otherwise they may run to seede the first yeare, which then are nothing worth : or else the rootes will be small staruelings and short, and runne into many spires or branches, whereby they will not bee of halfe the worth. Some vse to sow them in August and September, that so they may bee well growne to serue to spend in Lent following, but their best time is in February, that the Summers growth may make them the fairer and greater. When they runne vp to seede, you shall take the principall or middle heades, for those carry the Master seede, which is the best, and will produce the fairest rootes againe. You shall hardly haue all the seede ripe at one instant, for vsually the chiefest heads will be fallen before the other are ripe: you must therefore still looke them ouer, and cut them as they ripen. Carrots are vsually sowen in March and Aprill, and if it chance that some of them doe runne vp for seede the same year, they are to be weeded out, for neyther the seed nor roots of them are good: You must likewise pull them vp when they are too thicke, if you will haue them grow fair, or for seed, that they may grow at the least three or foure foot in sunder: the stalkes of Carrots are limber, and fall downe to the ground ; they must therefore be sustained by poles layde acrosse on stalkes thrust into the ground, and tyed to the poles and stalkes to keepe them vp from rotting or spoyling vpon the ground : the seed hereof is not all ripe at once, but must be tended and gathered as it ripeneth, and layd to dry in some dry chamber or floore, and then beaten out with a stick, and winnowed from the refuse. Turneps are sowne by themselues vpon a good ground in the end of Iuly, and beginning of August, to haue their rootes best to spend in winter ; for it often happeneth that those seedes of Turneps that are sowen in the Spring, runne vp to seede the same yeare, and then it is not accounted good. Many doe vse to sow Turneps on those grounds from whence the same yeare they haue taken off Reddish and Lettice, to make the greater profit of the ground, by hauing two crops of increase in one yeare. The stalkes of Turneps will bend downe to the ground, as Carrots doe, but yet must not be bound or ordered in that manner, but suffered to grow without staking or binding, so as they grow of some good distance in sunder : when the seede beginneth to grow ripe, be very carefull to preserue it from the birds, which will be most busie to deuour them. You shall vnderstand likewise that many doe account the best way to haue the fairest and most principall seede from all these fore-recited herbes, that after they are sowen, and risen to a reasonable growth, they be transplanted into fresh ground. Cabbages also are not only sowen for the vse of their heads to spend for meat, but to gather their seede likewise, which howsoeuer some haue endeauoured to doe, yet few haue gained good seede, because our sharpe hard frostes in winter haue spoyled and rotted their stockes they preserued for the purpose ; but others haue found out a better and a more sure way, which is, to take vp your stocks that are fittest to be preserued, and bring them into the house, and there wrap them eyther in clothes, or other things to defend them from the cold, and hang them vp in a dry place, vntill the beginning of March following, then planting them in the ground, and a little defend them at the first with straw cast ouer them from the cold nights, thereby you may be sure to haue perfect good seede, if your kinde be of the best : Sowe your seed in the moneths of February or March, and transplant them in May where they may stand to grow for your vse, but be carefull to kill the wormes or Caterpillers that else will deuoure all your leaues, and be carefull also that none of the leaues bee broken in the planting, or otherwise rubbed, for that oftentimes hindereth the well closing of them. Leekes are

for

for the moſt part wholly nourſed vp from the ſeede that is here gathered; and becauſe there is not ſo much ſtore of them either ſowne or ſpent, as there is of Onions by the twentieth part, we are ſtill the more carefull to be prouided from our owne labours; yet there be diuers Gardiners in this Kingdome, that doe gather ſome ſmall quantity of Onion ſeede alſo for their owne or their priuate friends ſpending. The ſowing of them both is much about one time and manner, yet moſt vſually Leeks are ſowne later then Onions, and both before the end of March at the furtheſt; yet ſome ſowe Onions from the end of Iuly to the beginning of September, for their Winter prouiſion. Thoſe that are ſowne in the Spring, are to be taken vp and tranſplanted on a freſh bed prepared for the purpoſe, or elſe they will hardly abide a Winter; but hauing taken roote before Winter, they will beare good ſeede in the Summer following : You muſt ſtake both your Leekes and your Onion beds, and with poles laid a croſſe, binde your lopple headed ſtalkes vnto them, on high as well as belowe, or elſe the winde and their owne weight will beare them downe to the ground, and ſpoile your ſeede. You muſt thinne them, that is, pull vp continually after they are firſt ſprung vp thoſe that growe too thicke, as you doe with all the other herbes before ſpoken of, that they may haue the more roome to thriue. Of all theſe herbes and rootes before ſpoken of, you muſt take the likelieſt and faireſt to keepe for your ſeede; for if you ſhould not take the beſt, what hope of good ſeede can you expect? The time for the ſpending of theſe herbes and rootes, not particularly mentioned, is vntill they begin to runne vp for ſeede, or vntill they are to be tranſplanted for ſeede, or elſe vntill Winter, while they are good, as euery one ſhall ſee cauſe.

Chap. IIII.

How to order Artichokes, Melons, Cowcumbers, and Pompions.

THere are certaine other herbes to be ſpoken of, which are wholly nourſed vp for their fruit ſake, of whom I ſhall not need to ſay much, being they are ſo frequent in euery place. Artichokes being planted of faire and large ſlips, taken from the roote in September and October (yet not too late) will moſt of them beare fruit the next yeare, ſo that they be planted in well dunged ground, and the earth raiſed vp like vnto an Anthill round about each roote, to defend them the better from the extreame froſts in Winter. Others plant ſlips in March and Aprill, or ſooner, but although ſome of them will beare fruit the ſame yeare, yet all will not. And indeede many doe rather chooſe to plant in the ſpring then in the fall, for that oftentimes an extreame hard Winter following the new ſetting of ſlips, when they haue not taken ſufficient heart and roote in the ground, doth vtterly pierce and periſh them, when as they that are ſet in the Spring haue the whole Summers growth, to make them ſtrong before they feele any ſharpe froſts, which by that time they are the better able to beare. Muske Melons haue beene begun to bee nourſed vp but of late dayes in this Land, wherein although many haue tryed and endeauoured to bring them to perfection, yet few haue attained vnto it : but thoſe rules and orders which the beſt and skilfulleſt haue vſed, I will here ſet downe, that who ſo will, may haue as good and ripe Melons as any other in this Land. The firſt thing you are to looke vnto, is to prouide you a peece of ground fit for the purpoſe, which is either a ſloping or ſheluing banke, lying open and oppoſite to the South Sunne, or ſome other fit place not ſheluing, and this ground alſo you muſt ſo prepare, that all the art you can vſe about it to make it rich is little enough; and therefore you muſt raiſe it with meere ſtable ſoyle, thorough rotten & well turned vp, that it may be at the leaſt three foote deepe thereof, which you muſt caſt alſo into high beds or balkes, with deepe trenches or furrowes betweene, ſo as the ridges may be at the leaſt a foot and a halfe higher then the furrowes; for otherwiſe it is not poſſible to haue good Melons growe ripe. The choiſe of your ſeede alſo is another thing of eſpeciall regard, and the beſt is held to be Spaniſh, and not French, which hauing once gained, be ſure to haue ſtill of the ſame while they laſt
good,

good, that you may haue the feede of your owne ripe Melons from them that haue eaten them, or faue fome of the beft your felfe for the purpofe. I fay while they laft good; for many are of opinion, that no feede of Muske Melons gathered in England, will endure good to fowe againe here aboue the third yeare, but ftill they muft be renewed from whence you had your choifeft before. Then hauing prepared a hot bed of dung in Aprill, fet your feedes therein to raife them vp, and couer them, and order them with as great care or greater then Cowcumbers, &c. are vfed, that when they are ready, they may be tranfplanted vpon the beds or balkes of that ground you had before prepared for them, and fet them at the leaft two yards in funder, euery one as it were in a hole, with a circle of dung about them, which vpon the fetting being watered with water that hath ftood in the Sunne a day or two, and fo as often as neede is to water, couer them with ftrawe (fome vfe great hollow glaffes like vnto bell heads) or fome fuch other things, to defend them both from the cold euenings or dayes, and the heate of the Sunne, while they are young and new planted. There are fome that take vpon them great skill, that miflike of the raifing vp of Melons, as they doe alfo of Cowcumbers, on a hot bed of horfe dung, but will put two or three feedes in a place in the very ground where they fhall ftand and growe, and thinke without that former manner of forcing them forwards, that this their manner of planting will bring them on faft and fure enough, in that they will plucke away fome of the worft and weakeft, if too many rife vp together in a place; but let them know for certaine, that howfoeuer for Cowcumbers their purpofe and order may doe reafonable well, where the ground is rich and good, and where they ftriue not to haue them fo early, as they that vfe the other way, for Muske Melons, which are a more tender fruit, requiring greater care and trouble in the nourfing, and greater and ftronger heate for the ripening, they muft in our cold climate haue all the art vfed vnto them that may be, to bring them on the more early, and haue the more comfort of the Sunne to ripen them kindly, or elfe they will not bee worth the labour and ground. After you haue planted them as aforefaid, fome of good skill doe aduife, that you be carefull in any dry feafon, to giue them water twice or thrice euery weeke while they are young, but more afterward when they are more growne, and that in the morning efpecially, yea and when the fruit is growne fomewhat great, to water the fruit it felfe with a watering pot in the heate of the day, is of fo good effect, that it ripeneth them much fafter, and will giue them the better tafte and fmell, as they fay. To take likewife the fruit, and gather it at the full time of his ripeneffe is no fmall art; for if it be gathered before his due time to be prefently eaten, it will be hard and greene, and not eate kindly; and likewife if it be fuffered too long, the whole goodneffe will be loft: You fhall therefore know, that it is full time to gather them to fpend prefently, when they begin to looke a little yellowifh on the outfide, and doe fmell full and ftrong; but if you be to fend them farre off, or keepe them long vpon any occafion, you fhall then gather them fo much the earlier, that according to the time of the carriage and fpending, they may ripen in the lying, being kept dry, and couered with woollen clothes: When you cut one to eate, you fhall know it to be ripe and good, if the feede and pulpe about them in the middle be very waterifh, and will eafily be feparated from the meate, and likewife if the meate looke yellow, and be mellow, and not hard or greene, and tafte full and pleafant, and not waterifh: The vfuall manner to eate them is with pepper and falt, being pared and fliced, and to drowne them in wine, for feare of doing more harme. Cowcumbers and Pompions, after they are nourfed vp in the bed of hot dung, are to be feuerally tranfplanted, each of them on a large plot of ground, a good diftance in funder: but the Pompions more, becaufe their branches take vp a great deale more ground, & befides, will require a great deale more watering, becaufe the fruit is greater. And thus haue you the ordering of thofe fruits which are of much efteeme, efpecially the two former, with all the better fort of perfons; and the third kinde is not wholly refufed, of any, although it ferueth moft vually for the meaner and poorer fort of people, after the firft early ripe are fpent.

*The ordering of diuers sorts of herbes for the pot, for meate,
and for the table.*

TYme, Sauory, and Hyssope, are vsually sowne in the Spring on beds by themselues, euerie one a part ; but they that make a gaine by selling to others the
young rootes, to set the knots or borders of Gardens, doe for the most part
sowe them in Iuly and August, that so being sprung vp before Winter, they will be the
fitter to be taken vp in the Spring following, to serue any mans vse that wonld haue
them. Sage, Lauender, and Rosemary, are altogether set in the Spring, by slipping
the old stalkes, and taking the youngest and likeliest of them, thrusting them either
twined or otherwise halfe a foote deepe into the ground, and well watered vpon the
setting ; if any seasonable weather doe follow, there is no doubt of their well thriuing : the hot Sunne and piercing drying Windes are the greatest hinderances to them;
and therefore I doe aduise none to set too soone in the Spring, nor yet in Autumne, as
many doe practise : for I could neuer see such come to good, for the extremity of the
Winter comming vpon them so soone after their setting, will not suffer their young
shootes to abide, not hauing taken sufficient strength in the ground, to maintain themselues against such violence, which doth often pierce the strongest plants. Marierome
and Bassill are sowne in the Spring, yet not too early ; for they are tender plants, and
doe not spring vntill the weather bee somewhat warme : but Bassill would bee sowne
dry, and not haue any water of two or three daies after the sowing, else the seede will
turne to a gelly in the ground. Some vse to sowe the seed of Rosemary, but it seldome
abideth the first Winter, because the young plants being small, and not of sufficient
strength, cannot abide the sharpnesse of some Winters, notwithstanding the couering
of them, which killeth many old plants; but the vsuall way is to slippe and set, and so
they thriue well. Many doe vse to sowe all or the most sorts of Pot-herbes together
on one plot of ground, that they neede not to goe farre to gather all the sorts they
would vse. There are many sorts of them well knowne vnto all, yet few or none doe
vse all sorts, but as euery one liketh; some vse those that others refuse, and some esteem
those not to bee wholesome and of a good rellish, which others make no scruple of.
The names of them are as followeth, and a short relation of their sowing or planting.

Rosemary, Tyme, and Sauorie are spoken of before, and Onions and Leekes.

Mints are to bee set with their rootes in some by-place, for that their rootes doe
creepe so farre vnder ground, that they quickly fill vp the places neare adioyning, if
they be not puld vp.

Clarie is to be sowne, and seedeth and dyeth the next yeare, the herbe is strong, and
therefore a little thereof is sufficient.

Nep is sowne, and dyeth often after seeding, few doe vse it, and that but a little at a
time : both it and Clarie are more vsed in Tansies then in Broths.

Costmarie is to be set of rootes, the leaues are vsed with some in their Broths, but
with more in their Ale.

Pot Marierome is set of rootes, being separated in sunder.

Penniroyall is to be set of the small heads that haue rootes, it creepeth and spreadeth quickly.

Allisanders are to be sowne of seede, the tops of the rootes with the greene leaues
are vsed in Lent especially.

Parsley is a common herbe, and is sowne of seede, it seedeth the next yeare and
dyeth : the rootes are more vsed in broths then the leaues, and the leaues almost with
all sorts of meates.

Fennell is sowne of seede, and abideth many yeares yeelding seede : the rootes also are vsed in broths, and the leaues more seldome, yet serue to trimme vp many
fish meates.

Borage is sowne of seede, and dyeth the next yeare after, yet once being suffered to
seede in a Garden, will still come of it owne shedding.

Buglosse

Buglosse commeth of feede, but abideth many yeares after it hath giuen feede, if it ftand not in the coldeft place of the Garden.

Marigolds are fowne of feede, and may be after tranfplanted, they abide two or three yeares, if they be not fet in too cold a place: the leaues and flowers are both vfed.

Langedebeefe is fowne of feede, which fhedding it felfe will hardly be deftroyed in a Garden.

Arrach is to be fowne of feede, this likewife will rife euery yeare of it owne feed, if it be fuffered to fhed it felfe.

Beetes are fowne of feede, and abideth fome yeares after, ftill giuing feede.

Blites are vfed but in fome places; for there is a generall opinion held of them, that they are naught for the eyes: they are fowne euery yeare of feede.

Bloodwort once fowne abideth many yeares, if the extremity of the frofts kill it not, and feedeth plentifully.

Patience is of the fame nature, and vfed in the fame manner.

French Mallowes are to be fowne of feede, and will come of it owne fowing, if it be fuffered to fhed it felfe.

Ciues are planted onely by parting the rootes; for it neuer giueth any feede at all.

Garlicke is ordered in the fame manner, by parting and planting the rootes euerie yeare.

Thefe be all the forts are vfed with vs for that purpofe, whereas I faid before, none vfeth all, but euery one will vfe thofe they like beft: and fo much fhall fuffice for pot-herbes.

CHAP. VI.

The manner and ordering of many forts of herbes and rootes for Sallets.

IF I fhould fet downe all the forts of herbes that are vfually gathered for Sallets, I fhould not onely fpeake of Garden herbes, but of many herbes, &c. that growe wilde in the fields, or elfe be but weedes in a Garden; for the vfuall manner with many, is to take the young buds and leaues of euery thing almoft that groweth, as well in the Garden as in the Fields, and put them all together, that the tafte of the one may amend the rellifh of the other: But I will only fhew you thofe that are fown or planted in gardens for that purpofe. Afparagus is a principall & deleĉable Sallet herbe, whofe young fhootes when they are a good handfull high aboue the ground, are cut an inch within the ground, which being boyled, are eaten with a little vinegar and butter, as a Sallet of great delight. Their ordering with the beft Gardiners is on this wife: When you haue prouided feede of the beft kinde, you muft fowe it either before Chriftmas, as moft doe, or before the end of February; the later you fowe, the later and the more hardly will they fpring: after they are growne vp, they are to be tranfplanted in Autumne on a bed well trenched in with dung; for elfe they will not bee worth your labour, and fet about a foote diftance in funder, and looke that the more carefull you are in the replanting of them, the better they will thriue, and the fooner growe great: after fiue or fix yeares ftanding they vfually doe decay; and therefore they that ftriue to haue continually faire and great heads, doe from feede raife vp young for their ftore. You muft likewife fee that you cut not your heads or young fhoote too nigh, or too much, that is, to take away too many heads from a roote, but to leaue a fufficient number vncut, otherwife it will kill the heart of your rootes the fooner, caufing them to dye, or to giue very fmall heads or fhootes; for you may well confider with your felfe, that if the roote haue not head enough left it aboue the ground to fhoote greene this yeare, it will not, nor cannot profper vnder ground to giue encreafe the next yeare. The ordering of Lettice I haue fpoken of before, and fhall not neede here to repeate what hath beene already faid, but referre you thereunto for the fowing, planting, &c. onely I will here fhew you the manner of ordering them for Sallets. There are fome forts of Lettice that growe very great, and clofe their heads, which are called Cab-

bage

bage Lettice, both ordinary and extraordinary, and there are other forts of great Lettice that are open, and clofe not, or cabbage not at all, which yet are of an excellent kinde, if they be vfed after that efpeciall manner is fit for them, which is, That when they are planted (for after they are fowne, they muft be tranfplanted) of a reafonable diftance in funder, and growne to be of fome bigneffe, euery one of them muft bee tyed together with baft or thread toward the toppes of the leaues, that by this meanes all the inner leaues may growe whitifh, which then are to be cut vp and vfed : for the keeping of the leaues clofe doth make them tafte delicately, and to bee very tender. And thefe forts of Lettice for the moft part are fpent after Summer is paft, when other Lettice are not to be had. Lambes Lettice or Corne Sallet is an herbe, which abiding all Winter, is the firft Sallet herbe of the yeare that is vfed before any ordinarie Lettice is ready ; it is therefore vfually fowne in Auguft, when the feede thereof is ripe. Purflane is a Summer Sallet herbe, and is to be fowne in the Spring, yet fomewhat late, becaufe it is tender, and ioyeth in warmth ; and therefore diuers haue fowne it vpon thofe beddes of dung, whereon they nourfed vp their Cowcumbers, &c. after they are taken away, which being well and often watered, hath yeelded Sallet vntill the end of the yeare. Spinach is fowne in the Spring, of all for the moft part that vfe it, but yet if it be fowne in Summer it will abide greene all the Winter, and then feedeth quickly : it is a Sallet that hath little or no tafte at all therein, like as Lettice and Purflane ; and therefore Cookes know how to make many a good difh of meate with it, by putting Sugar and Spice thereto. Coleworts are of diuers kinds, and although fome of them are wholly fpent among the poorer fort of people, yet fome kindes of them may be dreffed and ordered as may delight a curious palate, which is, that being boyled tender, the middle ribs are taken cold, and laid in difhes, and vinegar and oyle poured thereon, and fo eaten. Coleflowers are to be had in this Countrey but very feldome, for that it is hard to meete with good feede : it muft bee fowne on beds of dung to force it forward, or elfe it would perifh with the froft before it had giuen his head of flowers, and tranfplanted into verie good and rich ground, left you lofe the benefit of your labours. Endiue is of two forts, the ordinary, and another that hath the edges of the leaues curld or crumpled ; it is to be whited, to make it the more dainty Sallet, which is vfually done in this manner : After they are grown to fome reafonable greatneffe (but in any cafe before they fhoote forth a ftalke in the midft for feede) they are to be taken vp, and the rootes being cut away, lay them to dry or wither for three or foure houres, and then bury them in fand, fo as none of them lye one vpon another, or if you can, one to touch another, which by this meanes will change whitifh, and thereby become verie tender, and is a Sallet both for Autumne and Winter. Succorie is vfed by fome in the fame manner, but becaufe it is more bitter then Endiue, it is not fo generally vfed, or rather vfed but of a verie few : and whereas Endiue will feede the fame yeare it is fowne, and then dye, Succorie abideth manie yeares, the bitterneffe thereof caufing it to be more Phyficall to open obftructions ; and therefore the flowers pickled vp, as diuers other flowers are vfed to be now adaies, make a delicate Sallet at all times when there is occafion to vfe them. Of red Beetes, the rootes are onely vfed both boyled and eaten cold with vinegar and oyle, and is alfo vfed to trimme vp or garnifh forth manie forts of difhes of meate : the feede of the beft kinde will not abide good with vs aboue three yeares, but will degenerate and growe worfe ; and therefore thofe that delight therein muft be curious, to be prouided from beyond Sea, that they may haue fuch as will giue delight. Sorrell is an herbe fo common, and the vfe fo well knowne, both for fawce, and to feafon broths and meates for the found as well as ficke perfons, that I fhall not neede to fay anie more thereof. Cheruill is a Sallet herbe of much vfe, both with French and Dutch, who doe much more delight in herbes of ftronger tafte then the Englifh doe : it is fowne early, and vfed but a while, becaufe it quickly runneth vp to feede. Sweete Cheruill, or as fome call it, Sweete Cis, is fo like in tafte vnto Anife feede, that it much delighteth the tafte among other herbes in a Sallet : the feede is long, thicke, blacke, and cornered, and muft be fowne in the end of Autumne, that it may lye in the ground all the Winter, and then it will fhoote out in the Spring or elfe if it be fowne in the Spring, it will not fpring vp that yeare vntill the next : the leaues (as I faid before) are vfed among other herbes : the rootes likewife are not onely cordiall, but alfo held to be preferuatiue againft the Plague, either greene, dryed, or preferued

R r with

with fugar. Rampion rootes are a kinde of Sallet with a great many, being boyled tender, and eaten cold with vinegar and pepper. Creffes is an herbe of eafie and quick growth, and while it is young eaten eyther alone, or with parfley and other herbes: it is of a ftrong tafte to them that are not accuftomed thereunto, but it is much vfed of ftrangers. Rocket is of the fame nature and qualitie, but fomewhat ftronger in tafte: they are both fowen in the Spring, and rife, feede and dye the fame yeare. Tarragon is an herbe of as ftrong a tafte as eyther Rocket or Creffes, it abideth and dyeth not euery yeare, nor yet giueth ripe feede (as far as euer could bee found with vs) any yeare, but maketh fufficient increafe within the ground, fpreading his roots all abroad a great way off. Muftard is a common fawce both with fifh and flefh, and the feed thereof (and no part of the plant befides) is well knowne how to be vfed being grownded, as euery one I thinke knoweth. The rootes of horfe Radifh likewife beeing grownd like Muftard, is vfed both of ftrangers and our owne nation, as fawce for fifh. Tanfie is of great vfe, almoft with all manner of perfons in the Spring of the yeare: it is more vfually planted of the rootes then otherwife; for in that the rootes fpread far and neere they may be eafily taken away, without any hurt to the reft of the rootes. Burnet, although it be more vfed in wine in the Summer time then any way elfe, yet it is likewife made a fallet herbe with many, to amend the harfh or weak rellifh of fome other herbs. Skirrets are better to be fowen of the feed then planted from the roots, and will come on more fpeedily, and be fairer rootes: they are as often eaten cold as a Sallet, being boyled and the pith taken out, as ftewed with butter and eaten warme. Let not Parfley and Fenell be forgotten among your other Sallet herbes, wherof I haue fpoken before, and therefore need fay no more of them. The flowers of Marigolds pickt cleane from the heads, and pickled vp againft winter, make an excellent Sallet when no flowers are to be had in a garden. Cloue Gilloflowers likewife preferued or pickled vp in the fame manner (which is *ftratum fuper ftratum*, a lay of flowers, and then ftrawed ouer with fine dry and poudered Sugar, and fo lay after lay ftrawed ouer, vntill the pot bee full you meane to keepe them in, and after filled vp or couered ouer with vinegar) make a Sallet now adayes in the higheft efteeme with Gentles and Ladies of the greateft note: the planting and ordering of them both is fpoken of feuerally in their proper places. Goates bearbe that groweth in Gardens only, as well as that which groweth wilde in Medowes, &c. bearing a yellow flower, are vfed as a Sallet, the rootes beeing boyled and pared are eaten cold with vinegar, oyle and pepper; or elfe ftewed with butter and eaten warme as Skirrets, Parfneps &c. And thus haue you here fet downe all thofe moft vfuall Sallets are vfed in this Kingdome: I fay the moft vfuall, or that are nourfed vp in Gardens; for I know there are fome other wilde herbes and rootes, as Dandelion &c. but they are vfed onely of ftrangers, and of thofe whofe curiofitie fearcheth out the whole worke of nature to fatisfie their defires.

Chap. VII.

Of diuers Phyficall herbes fit to be planted in Gardens, to ferue for the efpeciall vfes of a familie.

Hauing thus fhewed you all the herbes that are moft vfually planted in Kitchen Gardens for ordinary vfes, let mee alfo adde a few other that are alfo nourfed vp by many in their Gardens, to preferue health, and helpe to cure fuch fmall difeafes as are often within the compaffe of the Gentlewomens skils, who, to helpe their owne family, and their poore neighbours that are farre remote from Phyfitians and Chirurgions, take much paines both to doe good vnto them, and to plant thofe herbes that are conducing to their defires. And although I doe recite fome that are mentioned in other places, yet I thought it meete to remember them altogether in one place. Angelica, the garden kinde, is fo good an herbe, that there is no part thereof but is of much vfe, and all cordiall and preferuatiue from infectious or contagious difeafes, whether you will diftill the water of the herbe, or preferue or candie the rootes or the greene ftalkes, or vfe the feede in pouder or in diftillations, or decoctions with other things: it is fowen of feede, and will abide vntill

it

it giue feede, and then dyeth. Rue or Herbe grace is a ftrong herbe, yet vfed inwardly againft the plague as an Antidote with Figs and Wall-nuts, and helpeth much againft windy bodies : outwardly it is vfed to bee layde to the wreftes of the hands, to driue away agues : it is more vfually planted of flips then raifed from feede, and abideth long if fharpe froftes kill it not. Dragons being diftilled are held to be good to expell any euill thing from the heart : they are altogether planted of the rootes. Setwall, Valerian, or Capons tayle, the herbe often, but the roote much better, is vfed to prouoke fweating, thereby to expell euill vapours that might annoy the heart : it is only planted of the rootes when they are taken vp, and the young replanted. Afarabacca, the leaues are often vfed to procure vomiting being ftamped, and the ftrained iuice to a little quantitie, put into a draught of ale and drunke, thereby to eafe the ftomacke of many euill and groffe humours that there lye and offend it ; diuers alfo take the leaues and rootes a little boyled in wine, with a little fpice added thereunto, to expell both tertian and quartan agues : the rootes of our Englifh growing is more auaileable for thefe purpofes then any outlandifh : it is planted by the roote ; for I could neuer fee it fpring of feede. Mafterwort commeth fomewhat neere in propertie vnto Angelica, and befides very effectuall to difperfe winde in the bodie, whether of the collicke or otherwife ; as alfo very profitable to comfort in all cold caufes : it yeeldeth feede, but yet is more vfually planted from the rootes being parted. Balme is a cordiall herbe both in fmell and tafte, and is wholly vfed for thofe purpofes, that is, to comfort the heart being diftilled into water either fimple or compound, or the herbe dryed and vfed : it is fet of the rootes being parted, becaufe it giueth no feede that euer I could obferue. Camomill is a common herbe well knowne, and is planted of the rootes in alleyes, in walkes, and on bankes to fit on, for that the more it is troden on, and preffed downe in dry weather, the clofer it groweth, and the better it will thriue : the vfe thereof is very much, both to warme and comfort, and to eafe paines being applyed outwardly after many fafhions : the decoction alfo of the flowers prouoketh fweat, and they are much vfed againft agues. Featherfew is an herbe of greater vfe for women then for men, to diffolue flatulent or windy humours, which caufeth the paines of the mother : fome vfe to take the iuice thereof in drinke for agues : it is as well fowen of the feede as planted of the rootes. Coftmary is vfed among thofe herbes that are put ino ale to caufe it haue a good rellifh, and to be fomewhat phyficall in the moneth of May, and doth helpe to prouoke vrine : it is fet of the rootes being parted. Maudlin is held to be a principall good herbe to open and cleanfe the liuer, and for that purpofe is vfed many wayes, as in ale, in tanfies, and in broths &c. the feed alfo is vfed, and fo is the herbe alfo fometimes, to kill the wormes in children : it is fowen of the feede, and planted alfo of the feparated rootes. Caffidonie is a fmall kinde of Lauender, but differing both in forme and qualitie : it is much vfed for the head to eafe paines thereof, as alfo put among other things to purge melancholicke difeafes : it is fowen of feede, and abideth not a winter vnleffe it bee well defended, and yet hardly giueth ripe feede againe with vs. Smallage is a great opening herbe, and much more then eyther Parfley or Fenell, and the rootes of them all are often vfed together in medicines : it is fowen of feede, and will not bee wanting in a Garden if once you fuffer it to fow it felfe. Cardus Benedictus, or the Bleffed Thiftle, is much vfed in the time of any infection or plague, as alfo to expell any euill fymptome from the heart at all other times. It is vfed likewife to be boyled in poffet drink, & giuen to them that haue an ague, to help to cure it by fweating or otherwife. It is vfually fowen of feed, and dyeth when it hath giuen feed. Winter Cherries are likewife nurfed vp in diuers gardens, for that their propertie is to giue helpe to them that are troubled eyther with the ftopping or heate of their vrine: the herbe and berries are often diftilled, but the berries alone are more often vfed: after it is once planted in a garden it will runne vnder ground, & abide well enough. Celondine is held to bee good for the iaundife, it is much vfed for to cleere dim eyes, eyther the iuice or the water dropped into them : it is fowen of feede, and being once brought into a garden, will hardly be weeded out ; the feede that fheddeth will fo fow it felfe, and therefore fome corner in a garden is the fitteft place for it. Tabacco is of two forts, and both vfed to be planted in Gardens, yet the Englifh kinde (as it is called) is more to be found in our Countrey Gardens then the Indian fort : the leaues of both forts indifferently, that is, of eyther of which is next at hand, being ftamped and boy-

led

led eyther by it felfe, or with other herbes in oyle or hogs fuet, doe make an excellent falue for greene wounds, and alfo to clenfe old vlcers or fores ; the iuice of the greene leaues drunke in ale, or a dryed leafe fteeped in wine or ale for a night, and the wine or ale drunke in the morning, prouoketh to caft, but the dryed leafe much ftronger then the greene : they are fowen of feede, but the Indian kinde is more tender, and will not abide a winter with vs abroade. Spurge that vfually groweth in Gardens, is a violent purger, and therefore it is needfull to be very carefull how it is vfed : the feede is more ordinarily vfed then any other part of the plant, which purgeth by vomiting in fome, and both vpwards and downwards in many ; the iuice of the herbe, but efpecially the milke thereof, is vfed to kill warres : it is fowen of feede, and when it doth once fhed it felfe, it will ftill continue fpringing of the fallen feede. Bearefoote is fowen of feed, and will hardly abide tranfplanting vnleffe it bee while it is young ; yet abideth diuers yeares, if it ftand not in too cold a place. This I fpeake of the greater kinde ; for the lower fmall wilde kind (which is the moft ordinary in this land) will neuer decay : the leaues are fometimes vfed greene, but moft vfually dryed and poudered, and giuen in drinke to them that haue the wormes : it purgeth melancholy, but efpecially the roots. In many Countries of this Land, and elfewhere, they vfe to thruft the ftalk of the great kinde through the eare or dewlap of Kine and Cattell, to cure them of many difeafes. Salomons Seale, or (as fome call it) Ladder to heauen, although it doth grow wilde in many places of this Land, yet is planted in Gardens : it is accounted an excellent wound herbe to confolidate, and binde, infomuch that many vfe it with good fucceffe to cure ruptures, and to ftay both the white and the red fluxe in women : it is planted altogether of the rootes, for I could neuer finde it fpring from the feede, it is fo ftrong. Comfry likewife is found growing wilde in many places by ditch fides, and in moift places, and therefore requireth fome moift places of the garden : it is wholly vfed for knitting, binding, and confolidating fluxes and wounds, to be applyed either inwardly or outwardly : The rootes are ftronger for thofe purpofes then any other parts of the plant. Licoris is much vfed now adaies to bee planted in great quantitie, euen to fill many acres of ground, whereof rifeth a great deale of profit to thofe that know how to order it, and haue fit grounds for it to thriue in ; for euery ground will not be aduantagious : It will require a very rich, deepe and mellow ground, eyther naturall or artificiall ; but for a priuate houfe where a fmall quantitie will ferue, there needeth not fo much curiofitie : it is vfually planted of the top heads, when the lower rootes (which are the Licoris that is vfed) and the runners are cut from them. Some vfe to make an ordinary drinke or beuerage of Licoris, boyled in water as our vfuall ale or beere is with malt, which fermented with barme in the fame manner, and tunned vp, ferueth in ftead thereof, as I am credibly informed: It is otherwife in a manner wholly fpent for colds, coughes and rheumes, to expectorate flegme, but vfed in diuers formes, as in iuice, in decoctions, fyrrups, roules, trochifces, and the greene or dryed roote of it felfe.

And thefe are the moft ordinary Phyficall herbes that are vfed to be planted in gardens for the vfe of any Country familie, that is (as I faid before) farre remote from Phyfitians or Chirurgions abidings, that they may vfe as occafion ferueth for themfelues or their neighbours, and by a little care and paines in the applying may doe a great deale of good, and fometimes to them that haue not wherewith to fpend on themfelues, much leffe on Phyfitians or Chirurgions, or if they haue, may oftentimes receiue leffe good at their hands then at others that are taught by experience in their owne families, to be the more able to giue helpe to others.

The

THE
KITCHEN
GARDEN.

THE SECOND PART,

Ontaining as well all forts of herbes, as rootes and fruits, that are vſually planted in Gardens, to ſerue for the vſe of the Table whether of the poore or rich of our Countrey : but herein I intend not to bring any fruite bearing trees, ſhrubbes, or buſhes ; for I reſerue them for my Orchard, wherin they ſhal be ſet forth. So that in theſe three parts, I ſuppoſe the exquiſite ornament of any worthy houſe is conſummate for the exteriour bounds, the benefit of their riches extending alſo to the furniſhing of the moſt worthy inward parts thereof : but becauſe many take pleaſure in the ſight and knowledge of other herbes that are Phyſicall, and much more in their properties and vertues, if vnto theſe three I ſhould adde a Phyſicke Garden, or Garden of Simples, there would be a quadripartite complement, of whatſoeuer arte or nature, neceſſitie or delight could affect : which to effect (as many my friends haue intreated it at my hands) will require more paines and time then all this worke together : yet to ſatisfie their deſires and all others herein, that would bee enformed in the truth, and reformed of the many errours and ſlips ſet forth and publiſhed heretofore of plants by diuers, I ſhall (God aſſiſting and granting life) labour to performe, that it may ſhew it ſelfe to the light in due conueniencie, if theſe bee well and gratefully accepted. And becauſe I ended with ſome ſweete herbes in the former part, I will in this part beginne with the reſt, which I reſerued for this place, as fitter for the pot and kitchen then for the hand or boſome, and ſo deſcend to other herbes that are for meat or ſallets : and after them to thoſe rootes that are to be eaten, as meate or as ſallets : and laſtly the fruits that grow neere, or vpon the ground, or not much aboue it ; as the Artichoke, &c. in which I make a ſhorter deſcription then I did in the former, rather endeauouring to ſhew what they are, and whereunto they are vſed, then the whole varietie or any exact declaration : which methode, although in ſome ſort it may bee fitting for this purpoſe, yet it is not for an hiſtory or herball : I ſhall therefore require their good acceptance for whoſe ſake I doe it, not doubting, but that I, or others, if they write againe of this ſubiect, may poliſh and amende what formerly hath beene eyther miſ ſet, or not ſo thoroughly expreſſed, beſides ſome additions of new conceits, ſeeing I treade out a new path, and therefore thoſe that follow may the eaſilier ſee the Meanders, and ſo goe on in a direct line.

CHAP.

CHAP. I.

Maiorana latifolia, fiue maior Anglica. Winter, or pot Marierome.

Inter Marierome is a fmall bufhie herbe like vnto fweete Marierome, being parted or diuided into many branches, whereon doe grow broader
and greener leaues, fet by couples, with fome fmall leaues likewife at the
feuerall ioynts all along the branches : at the tops whereof grow a number of fmall
purplifh white flowers fet together in a tuft, which turne into fmall and round feed, bigger then fweet Marierome feede : the whole plant is of a fmall and fine fent, but much
inferiour to the other, and is nothing fo bitter as the fweete Marierome, and thereby
both the fitter and more willingly vfed for meates : the roote is white and threddy,
and perifheth not as the former, but abideth many yeares.

The Vfe of winter Marierome.

The vfe of this Marierome is more frequent in our Land then in others,
being put among other pot-herbes and farfing(or fafeting herbes as they are
called) and may to good profit bee applyed in inward as well as outward
griefes for to comfort the parts, although weaker in effect then fweete
Marieromes.

CHAP. II.

Thymum vulgatius fiue durius. Ordinary Garden Tyme.

He ordinary Garden Tyme is a fmall low wooddy plant with brittle branches,
and fmall hard greene leaues, as euery one knoweth, hauing fmall white purplifh flowers, ftanding round about the tops of the ftalkes : the feed is fmall
and browne, darker then Marierome feed : the root is woody, and abideth well diuers
Winters.

Thymum latifolium. Mafticke Tyme.

This Tyme hath neyther fo wooddy branches, nor fo hard leaues, but groweth
lower, more fpreading, and with fomewhat broader leaues : the flowers are of a purplifh white colour, ftanding in roundles round about the ftalkes, at the ioynts with
leaues at them likewife. This Tyme endureth better and longer then the former, and
by fpreading it felfe more then the former, is the more apt to bee propagated by flipping, becaufe it hath beene feldome feene to giue feede : It is not fo quicke in fent or
tafte as the former, but is fitter to fet any border or knot in a garden, and is for the moft
part wholly employed to fuch vfes.

The Vfe of Tyme.

To fet downe all the particular vfes whereunto Tyme is applyed, were to
weary both the Writer and Reader; I will but only note out a few : for befides the phyficall vfes to many purpofes, for the head, ftomacke, fpleene,
&c. there is no herbe almoft of more vfe, in the houfes both of high and
low, rich and poore, both for inward and outward occafions ; outwardly
for bathings among other hot herbes, and among other fweete herbes for
ftrewings : inwardly in moft forts of broths, with Rofmary, as alfo with
other fafeting (or rather farfing) herbes, and to make fawce for diuers forts
both fifh and flefh, as to ftuffe the belly of a Goofe to bee rofted, and after
put into the fawce, and the pouder with breade to ftrew on meate when it

1 *Maioran maior Anglica*. Pot Marierome. 2 *Thymum vulgatius*. Garden Tyme. 3 *Satureia*. Sauorie. 4 *Hyſſopus*. Hyſſope. 5 *Pulegium*. Penniroyall. 6 *Saluia maior*. Common Sage. 7 *Saluia minor primata*. Sage of vertue.

is rofted, and fo likewife on rofted or fryed fifh. It is held by diuers to bee a fpeedy remedy againft the fting of a Bee, being bruifed and layd thereon.

Chap. III.

Satureia fiue Thymbra. Sauorie.

THere are two forts of of Sauory, the one called Summer, and the other Winter Sauorie : The Summer Sauory is a fmall tender herbe, growing not aboue a foote and a halfe high, or thereabours, rifing vp with diuers brittle branches, flenderly or fparfedly fet with fmall long leaues, foft in handling, at euery ioynt a couple, one againft another, of a pleafant ftrong and quicke fent and tafte : the flowers are fmall and purplifh, growing at the toppes of the ftalkes, with two fmall long leaues at the ioynts vnder them: the feede is fmall, and of a darke colour, bigger then Tyme feede by the halfe : the roote is wooddy, and hath many ftrings, perifhing euery yeare wholly, and muft bee new fowen againe, if any will haue it.

The Winter Sauorie is a fmall low bufhie herbe, very like vnto Hyffope, but not aboue a foote high, with diuers fmall hard branches, and hard darke green leaues thereon, thicker fet together then the former by much, and as thicke as common Hyffope, fometimes with foure leaues or more at a ioynt, of a reafonable ftrong fent, yet not fo ftrong or quicke as the former : the flowers are of a pale purplifh colour, fet at feuerall diftances at the toppes of the ftalkes, with leaues at the ioynts alfo with them, like the former : the roote is woody, with diuers fmall ftrings thereat, and abideth all the winter with his greene leaues : it is more vfually encreafed by flipping or diuiding the roote, and new fetting it feuerally againe in the Spring, then by fowing the feed.

The Vfe of Sauorie.

The Summer Sauorie is vfed in other Countryes much more then with vs in their ordinary diets, as condiment or fawce to their meates, fometimes of it felfe, and fometimes with other herbes, and fometimes ftrewed or layde vpon the difhes as we doe Parfley, as alfo with beanes and peafe, rife and wheate ; and fometimes the dryed herbe boyled among peafe to make pottage.

The Winter Sauorie is one of the (farfing) fafeting herbes as they call them, and fo is the Summer Sauorie alfo fometimes. This is vfed alfo in the fame manner that the Summer Sauorie is, fet downe before, and to the fame purpofes : as alfo to put into puddings, fawfages, and fuch like kindes of meates. Some doe vfe the ponder of the herbe dryed (as I fayd before of Tyme) to mixe with grated bread, to breade their meate, be it fifh or flefh, to giue it the quicker rellifh. They are both effectuall to expell winde.

Chap. IIII.

Hyffopus. Hyffope.

GArden Hyffope is fo well knowne to all that haue beene in a Garden, that I fhall but *actum agere*, to beftow any time thereon, being a fmall bufhie plant, not rifing aboue two foote high, with many branches, woody below, and tender aboue, whereon are fet at certaine diftances, fundry fmall, long and narrow greene leaues : at the toppe of euery ftalke ftand blewifh purple gaping flowers, one aboue another in a long fpike or eare : after which followeth the feede, which is fmall and blackifh : the rootes are compofed of many threddy ftrings ; the whole plant is of a ftrong fweet fent.

The

The Vſe of Hyſſope.

Hyſſope is much vſed in Ptiſans and other drinkes, to help to expectorate flegme. It is many Countrey peoples medicine for a cut or greene wound, being bruiſed with ſugar and applyed. I finde it is alſo much commended againſt the falling ſickeneſſe, eſpecially being made into pils after the manner before rehearſed. It is accounted a ſpeciall remedy againſt the ſting or biting of an Adder, if the place be rubbed with Hyſſope, bruiſed and mixed with honey, ſalt and cummin ſeede. A decoction thereof with oyle, and annointed, taketh away the itching and tingling of the head, and vermine alſo breeding therein. An oyle made of the herbe and flowers, being annointed, doth comfort benummed ſinewes and ioynts.

Chap. V.

Pulegium. Pennyroyall.

Pennyroyall alſo is an herbe ſo well knowne, that I ſhall not neede to ſpend much time in the deſcription of it : hauing many weake round ſtalkes, diuided into ſundry branches, rather leaning or lying vpon the ground then ſtanding vpright, whereon are ſet at ſeuerall ioynts, ſmall roundiſh darke greene leaues : the flowers are purpliſh that grow in gardens, yet ſome that grow wilde are white, or more white then purple, ſet in roundles about the tops of the branches ; the ſtalkes ſhoote forth ſmall fibres or rootes at the ioynts, as it lyeth vpon the ground, thereby faſtening it ſelfe therein, and quickly increaſeth, and ouer-runneth any ground, eſpecially in the ſhade or any moiſt place, and is replanted by breaking the ſprouted ſtalkes, and ſo quickely groweth.

Other ſorts of Pennyroyall are fit for the Phyſicke Garden, or Garden of Simples.

The Vſe of Pennyroyall.

It is very good and wholeſome for the lunges, to expell cold thin flegme, and afterwards to warme and dry it vp : and is alſo of the like propertie as Mintes, to comfort the ſtomacke, and ſtay vomiting. It is alſo vſed in womens baths and waſhings : and in mens alſo to comfort the ſinewes. It is yet to this day, as it hath beene in former times, vſed to bee put into puddings, and ſuch like meates of all ſorts, and therefore in diuers places they know it by no other name then Pudding-graſſe.

The former age of our great Grandfathers, had all theſe hot herbes in much and familiar vſe, both for their meates and medicines, and therewith preſerued themſelues in long life and much health : but this delicate age of ours, which is not pleaſed with any thing almoſt, be it meat or medicine, that is not pleaſant to the palate, doth wholly refuſe theſe almoſt, and therefore cannot be partaker of the benefit of them.

Chap. VI.

Salvia. Sage.

There are two eſpeciall kindes of Sage nourſed vp in our Gardens, for our ordinary vſe, whereof I intend to write in this place, leauing the reſt to his fitter place. Our ordinary Sage is reckoned to bee of two ſorts, white and red,

both

both of them bearing many foure square wooddy ftalkes, in fome whiter, in others redder, as the leaues are alfo, ftanding by couples at the ioynts, being long, rough, and wrinkled, of a ftrong fweete fent : at the tops of the ftalkes come forth the flowers, fet at certaine fpaces one aboue another, which are long and gaping, like vnto the flowers of Clary, or dead Nettles, but of a blewifh purple colour ; after which come fmall round feede in the huske that bore the flower : the roote is wooddy, with diuers ftrings at it : It is more vfually planted of the flips, pricked in the Spring time into the ground, then of the feed.

Saluia minor fiue pinnata. Small Sage or Sage of vertue.

The leffer Sage is in all things like vnto the former white Sage, but that his branches are long and flender, and the leaues much fmaller, hauing for the moft part at the bottome of each fide of the leafe a peece of a leafe, which maketh it fhew like finns or eares : the flowers alfo are of a blewifh purple colour, but leffer. Of this kinde there is one that beareth white flowers.

The Vfe of Sage.

Sage is much vfed of many in the moneth of May fafting, with butter and Parfley, and is held of moft much to conduce to the health of mans body.

It is alfo much vfed among other good herbes to bee tund vp with Ale, which thereupon is termed Sage Ale, whereof many barrels full are made, and drunke in the faid moneth chiefly for the purpofe afore recited : and alfo for teeming women, to helpe them the better forward in their childebearing, if there be feare of abortion or mifcarrying.

It is alfo vfed to be boyled among other herbes, to make Gargles or waters to wafh fore mouths and throates : As alfo among other herbes, that ferue as bathings, to wafh mens legs or bodies in the Summer time, to comfort nature, and warme and ftrengthen aged cold finewes, and lengthen the ftrength of the younger.

The Kitchen vfe is either to boyle it with a Calues head, and being minced, to be put with the braines, vinegar and pepper, to ferue as an ordinary fawce thereunto : Or being beaten and iuyced (rather then minced as manie doe) is put to a rofted Pigges braines, with Currans for fawce thereunto. It is in fmall quantity (in regard of the ftrong tafte thereof) put among other fafting herbes, to ferue as fawce for peeces of Veale, when they are farfed or ftuffed therewith, and rofted, which they call Olliues.

For all the purpofes aforefaid, the fmall Sage is accounted to be of the more force and vertue.

Chap. VII.

Horminum fativum. Garden Clary.

THere is but one fort of Garden Clary, though many wilde, which hath foure fquares ftalks, with broad rough wrinkled whitifh leaues, fomewhat vneuenly cut in on the edges, and of a ftrong fweete fent, growing fome next the ground, & fome by couples vpon the ftalkes : the flowers growe at certaine diftances, with two fmall leaues at the ioynts vnder them, fomewhat like vnto the flowers of Sage, but leffer, and of a very whitifh or bleake blew colour : the feede is of a blackifh browne colour, fomewhat flat, and not fo round as the wilde : the rootes fpread not farre, and perifh euery yeare that they beare flowers and feede. It is altogether to bee fowne of feed in the Spring time, yet fometimes it will rife of it owne fowing.

The

The Vſe of Clary.

The moſt frequent and common vſe of Clary, is for men or women that haue weake backes, to helpe to comfort and ſtrengthen the raines, being made into Tanſies and eaten, or otherwiſe. The ſeede is vſed of ſome to be put into the corner of the eye, if any mote or other thing haue happened into it : but aſſuredly although this may peraduenture doe ſome good, yet the ſeede of the wilde will doe much more. The leaues taken dry, and dipped into a batter made of the yolkes of egges, flower, and a little milke, and then fryed with butter vntill they be criſpe, ſerue for a diſh of meate accepted with manie, vnpleaſant to none.

Chap. VIII.

Nepeta. Nep.

ALthough thoſe that are Herbariſts do know three ſorts of Nep, a greater & two leſſer, yet becauſe the leſſer are not vſuall, but in the Gardens of thoſe that delight in natures varieties, I do not here ſhew you them. That which is vſuall (and called of manie Cat Mint) beareth ſquare ſtalkes, but not ſo great as Clarie, hauing two leaues at euery ioynt, ſomewhat like vnto Balme or Speare Mintes, but whiter, ſofter, and longer, and nicked about the edges, of a ſtrong ſent, but nothing ſo ſtrong as Clary : the flowers growe at the toppes of the ſtalkes, as it were in long ſpikes or heads, ſomewhat cloſe together, yet compaſſing the ſtalkes at certaine ioynts, of a whitiſh colour, for forme and bigneſſe like vnto Balme, or ſomewhat bigger : the rootes are compoſed of a number of ſtrings, which dye not, but keepe greene leaues vpon them all the Winter, and ſhoote anew in the Spring. It is propagated both by the ſeede, and by ſlipping the rootes.

The Vſe of Nep.

Nep is much vſed of women either in baths or drinkes to procure their feminine courſes : as alſo with Clarie, being fryed into Tanſies, to ſtrengthen their backes. It is much commended of ſome, if the iuyce thereof be drunke with wine, to helpe thoſe that are bruiſed by ſome fall, or other accident. A decoction of Nep is auaileable to cure the ſcabbe in the head, or other places of the body.

Chap. IX.

Meliſſa. Baulme.

THe Garden Baulme which is of common knowne vſe, hath diuers ſquare blackiſh greene ſtalkes, and round, hard, darke, greene pointed leaues, growing thereon by couples, a little notched about the edges, of a pleaſant ſweete ſent, drawing neareſt to the ſent of a Lemon or Citron ; and therefore of ſome called *Citrago* : the flowers growe about the toppes of the ſtalkes at certaine diſtances, being ſmall and gaping, of a pale carnation colour, almoſt white : the rootes faſten themſelues ſtrongly in the ground, and endure many yeares, and is encreaſed by diuiding the rootes ; for the leaues dye downe to the ground euery yeare, leauing no ſhew of leafe or ſtalke in the Winter.

The

The Vſe of Baulme.

Baulme is often vſed among other hot and ſweete herbes, to make baths and waſhings for mens bodies or legges, in the Summer time, to warme and comfort the veines and ſinewes, to very good purpoſe and effect, and hath in former ages beene of much more vſe then now adaies. It is alſo vſed by diuers to be ſtilled, being ſteeped in Ale, to make a Baulme water, after the manner they haue beene taught, which they keepe by them, to vſe in the ſtead of *Aqua vitæ*, when they haue any occaſion for their owne or their neighbours Families, in ſuddaine qualmes or paſſions of the heart : but if they had a little better direction (for this is ſomewhat too rude) it would doe them more good that take it : For the herbe without all queſtion is an excellent helpe to comfort the heart, as the very ſmell may induce any ſo to beleeue. It is alſo good to heale greene wounds, being made into ſalues : and I verily thinke, that our forefathers hearing of the healing and comfortable properties of the true naturall Baulme, and finding this herbe to be ſo effectuall, gaue it the name of Baulme, in imitation of his properties and vertues. It is alſo an herbe wherein Bees doe much delight, as hath beene found by experience of thoſe that haue kept great ſtore ; if the Hiues bee rubbed on the inſide with ſome thereof, and as they thinke it draweth others by the ſmell thereof to reſort thither. Plinie ſaith, it is a preſent remedy againſt the ſtinging of Bees.

Chap. X.

Mentha. Mintes.

THere are diuers ſorts of Mints, both of the garden, and wilde, of the woods, mountaines, and ſtanding pooles or waters : but I will onely in this place bring to your remembrance two or three ſorts of the moſt vſuall that are kept in gardens, for the vſes whereunto they are proper.

Red Mint or browne Mint hath ſquare browniſh ſtalkes, with ſomewhat long and round pointed leaues, nicked about the edges, of a darke greene colour, ſet by couples at euery ioynt, and of a reaſonable good ſent : the flowers of this kinde are reddiſh, ſtanding about the toppes of the ſtalkes at diſtances : the rootes runne creeping in the ground, and as the reſt, will hardly be cleared out of a garden, being once therein, in that the ſmalleſt peece thereof will growe and encreaſe apace.

Speare Mint hath a ſquare greene ſtalke, with longer and greener leaues then the former, ſet by couples, of a better and more comfortable ſent, and therefore of much more vſe then any other : the flowers hereof growe in long eares or ſpikes, of a pale red or bluſh colour : the rootes creepe in the ground like the other.

Party coloured or white Mint hath ſquare greene ſtalkes and leaues, ſomewhat larger then Speare Mint, and more nicked in the edges, whereof many are parted, halfe white and halfe greene, and ſome more white then greene, or more green then white, as nature liſteth : the flowers ſtand in long heads cloſe ſet together, of a bluſh colour : the rootes creepe as the reſt doe.

The Vſe of Mintes.

Mintes are oftentimes vſed in baths, with Baulme and other herbes, as a helpe to comfort and ſtrengthen the nerues and ſinewes.

It is much vſed either outwardly applyed, or inwardly drunke, to ſtrengthen and comfort weake ſtomackes, that are much giuen to caſting : as alſo for feminine fluxes. It is boyled in milke for thoſe whoſe ſtomackes are

apt

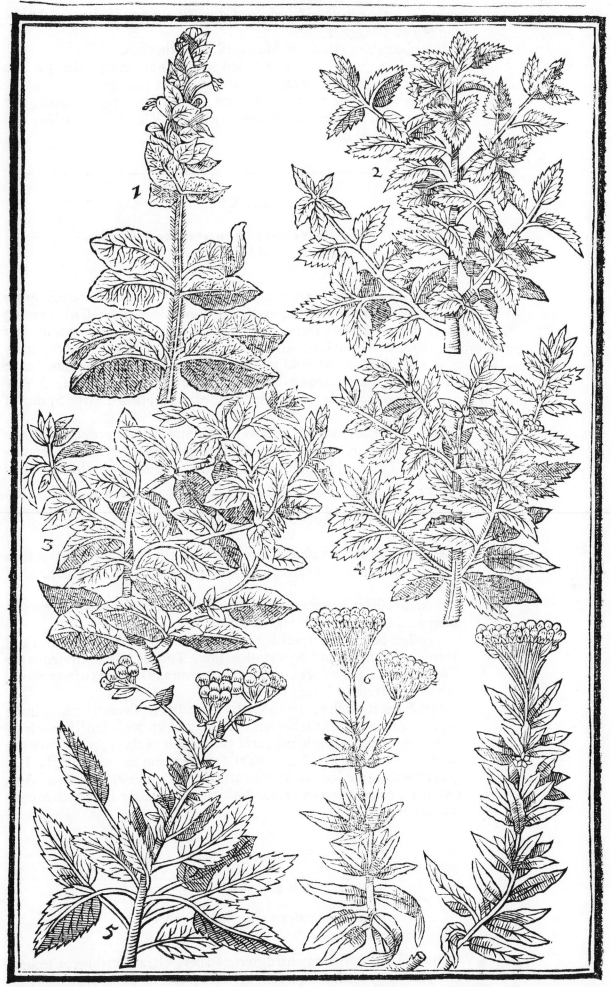

1 *Horminum ſativum.* Garden Clary. 2 *Nepeta.* Nep. 3 *Meliſſa.* Baulme. 4 *Mentha ſatiua.* Garden Mintes.
5 *Balſamita mas, ſeu Ceſtus hortorum.* Coſtmary. 6 *Ageratum.* Maudeline.

apt to caufe it to curdle. And applyed with falt, is a good helpe for the biting of a mad dogge.

It is vfed to be boyled with Mackarell, and other fifh.

Being dryed, is often and much vfed with Penniroyall, to bee put into
puddings : as alfo among peafe that are boyled for pottage.

Where Dockes are not ready at hand, they vfe to bruife Mintes, and
lay them vpon any place that is ftung with Bees, Wafpes, or fuch like, and
that to good purpofe.

Chap. XI.

Balfamita mas & femina, feu Coftus hortorum maior & minor.
Coftmary and Maudeline.

COftmary or Alecoaft is a fweet herbe, bearing many broad and long pale green
leaues, fnipped about the edges, euery one vpon a long foote-ftalke ; among
which rife vp many round greene ftalkes, with fuch like leaues on them, but
leffer vp to the toppe, where it fpreadeth it felfe into three or foure branches, euery
one bearing an vmbell or tuft of gold yellow flowers, fomewhat like vnto Tanfie
flowers, but leffer, which turne into fmall heads, containing fmall flat long feede : the
roote is fomewhat hard and ftringy, and being diuided, is replanted in the Spring of
the yeare for increafe.

Maudeline hath fomewhat long and narrow leaues, fnipt about the edges : the ftalks
are two foot high, bearing many yellow flowers on the tops of the branches, in an vmbell or tuft like vnto Tanfie : the whole herbe is fweete, and fomewhat bitter, and is
replanted by flipping.

The Vfe of Coftmary and Maudeline.

Coftmary is of efpeciall vfe in the Spring of the yeare, among other fuch
like herbes, to make Sage Ale, and thereupon I thinke it tooke the name of
Alecoaft.

It is alfo vfed to be put among other fweete herbes, to make fweete wafhing water, whereof there is great ftore fpent.

The leaues haue an efpeciall vertue to comfort both the ftomack and
heart, and to warme and dry a moift braine. The feede is much vfed in the
Country, to be giuen to children for the wormes, in the ftead of wormfeed,
and fo is the feede of Maudeline alfo.

Maudeline is much vfed with Coftmary and other fweet herbes, to make
fweete wafhing water : the flowers alfo are tyed vp with fmall bundels of
Lauender toppes, thefe being put in the middle of them, to lye vpon the
toppes of beds, preffes, &c. for the fweete fent and fauour it cafteth. It is
generally accounted of our Apothecaries to be the true *Eupatorium* of Auicen, and the true *Ageratum* of Diofcorides ; but Dodonæus feemeth to contradict both.

Chap. XII.

Tanacetum vulgare & crifpum. ### Tanfie.

OVr Garden Tanfie hath many hard greene leaues, or rather wings of leaues ;
for they are many fmall ones, fet one againft another all along a middle ribbe
or ftalke, and fnipt about the edges : in fome the leaues ftand clofer and thicker, and fomewhat crumpled, which hath caufed it to be called double or curld Tan-

                                                                                    fie,

fie, in others thinner and more fparfedly : It rifeth vp with many hard ftalks, whereon growe at the tops vpon the feuerall fmall branches gold yellow flowers like buttons, which being gathered in their prime, will hold the colour frefh a long time : the feede is fmall, and as it were chaffie : the roote creepeth vnder ground, and fhooteth vp againe in diuers places : the whole herbe, both leaues and flowers, are of a fharpe, ftrong, bitter fmell and tafte, but yet pleafant, and well to be endured.

The Vfe of Tanfie.

The leaues of Tanfie are vfed while they are young, either fhred fmall with other herbes, or elfe the iuyce of it and other herbes fit for the purpofe, beaten with egges, and fryed into cakes (in Lent and the Spring of the yeare) which are vfually called Tanfies, and are often eaten, being taken to be very good for the ftomack, to helpe to digeft from thence bad humours that cleaue thereunto: As alfo for weak raines and kidneyes, when the vrine paffeth away by drops : This is thought to be of more vfe for men then for women. The feed is much commended againft all forts of wormes in children.

Chap. XIII.

Pimpinella fiue Sanguiforba. Burnet.

BVrnet hath many winged leaues lying vpon the ground, made of many fmall, round, yet pointed greene leaues, finely nicked on the edges, one fet againft another all along a middle ribbe, and one at the end thereof; from among which rife vp diuers round, and fometimes crefted browne ftalkes, with fome few fuch like leaues on them as growe belowe, but fmaller : at the toppes of the ftalkes growe fmall browne heads or knaps, which fhoote forth fmall purplifh flowers, turning into long and brownifh, but a little cornered feede : the roote groweth downe deepe, being fmall and brownifh : the whole plant is of a ftipticke or binding tafte or quality, but of a fine quicke fent, almoft like Baulme.

The Vfe of Burnet.

The greateft vfe that Burnet is commonly put vnto, is to put a few leaues into a cup with Claret wine, which is prefently to be drunke, and giueth a pleafant quicke tafte thereunto, very delightfull to the palate, and is accounted a helpe to make the heart merrie. It is fometimes alfo while it is young, put among other Sallet herbes, to giue a finer rellifh thereunto. It is alfo vfed in vulnerary drinkes, and to ftay fluxes and bleedings, for which purpofes it is much commended. It hath beene alfo much commended in contagious and peftilentiall agues.

Chap. XIIII.

Hippolapathum fativum, fiue Rhabarbarum Monacherum.
Monkes Rubarbe or Patience.

GArden Patience is a kinde of Docke in all the parts thereof, but that it is larger and taller then many others, with large and long greene leaues, a great, ftrong, and high ftalke, with reddifh or purplifh flowers, and three fquare feede, like as all other Dockes haue : the roote is great and yellow, not hauing any fhew of flefh coloured veines therein, no more then the other kinde with great round thin leaues, commonly called *Hippolapathum rotundifolium*, Baſtard Rubarbe, or Monkes Rubarbe, the properties of both which are of very weake effect : but I haue a kinde of round leafed Dock growing in my Garden, which was ſent me from beyond Sea by a worthy Gentleman, Mr. Dr. Matth. Liſter, one of the Kings Phyſitians, with this title, *Rhaponticum verum*, and firſt grew with me, before it was euer ſeen or known elſewhere in England, wch by proof I haue found to be ſo like vnto the true Rubarbe, or the Rha of Pontus, both for forme and colour, that I dare ſay it is the very true Rubarbe, our climate only making it leſſe ſtrong in working, leſſe heauy, and leſſe bitter in taſte: For this hath great and thicke rootes, as diuerſly diſcoloured with fleſh coloured veines as the true Rubarbe, as I haue to ſhew to any that are deſirous to ſee and know it ; and alſo other ſmaller ſprayes or branches of rootes, ſpreading from the maine great roote, which ſmaller branches may well be compared to the *Rhaponticum* which the Merchants haue brought vs, which we haue ſeene to be longer and ſlenderer then Rubarbe, but of the very ſame colour : this beareth ſo goodly large leaues, that it is a great beauty in a garden to behold them : for I haue meaſured the ſtalke of the leafe at the bottome next the roote to bee of the bigneſſe of any mans thumbe ; and from the roote to the leafe it ſelfe, to bee two foote in length, and ſometimes more ; and likewiſe the leafe it ſelfe, from the lower end where it is ioyned to the ſtalke, to the end or point thereof, to bee alſo two foote in length, and ſometimes more ; and alſo in the broadeſt part of the leafe, to be two foote or more ouer in breadth : it beareth whitiſh flowers, contrary to all other Dockes, and three ſquare browniſh ſeede as other Dockes doe, but bigger, and therefore aſſuredly it is a Docke, and the true Rubarbe of the Arabians, or at the leaſt the true *Rhaponticum* of the Ancients. The figure of the whole plant I haue cauſed to be cut, with a dryed roote as it grew in my garden by it ſelfe, and haue inſerted it here, both becauſe Matthiolus giueth a falſe figure of the true Rubarbe, and that this hath not been expreſſed and ſet forth by any before.

The Vſe of Patience, and of the Rubarbe.

The leaues of Patience are often, and of many vſed for a pot-herbe, and ſeldome to any other purpoſe : the roote is often vſed in Diet-beere, or ale, or in other drinkes made by decoction, to helpe to purge the liuer, and clenſe the blood. The other Rubarbe or *Rhaponticum*, wherof I make mention, and giue you here the figure, I haue tryed, and found by experience to purge gently, without that aſtriction that is in the true Rubarbe is brought vs from the Eaſt Indies, or China, and is alſo leſſe bitter in taſte ; whereby I coniecture it may bee vſed in hot and feaueriſh bodies more effectually, becauſe it doth not binde after the purging, as the Eaſt India Rubarbe doth : but this muſt bee giuen in double quantitie to the other, and then no doubt it will doe as well : The leaues haue a fine acide taſte : A ſyrrupe therefore made with the iuice and ſugar, cannot but be very effectuall in deiected appetites, and hot fits of agues ; as alſo to helpe to open obſtructions of the liuer, as diuers haue often tryed, and found auaileable by experience.

Chap. XV.

Lapathum ſanguineum. Blood-wort.

Among the ſorts of pot-herbes Blood-worte hath alwayes beene accounted a principall one, although I doe not ſee any great reaſon therein, eſpecially ſeeing there is a greater efficacie of binding in this Docke, then in any of the other : but as common vſe hath receiued it, ſo I here ſet it downe. Blood-worte is one of the ſorts of Dockes, and hath long leaues like vnto the ſmaller yellow Docke, but ſtriped with red veines, and ouer-ſhadowed with red vpon the greene leafe, that it ſeemeth almoſt wholly red ſometimes : the ſtalke is reddiſh, bearing ſuch like leaues, but
ſmaller

1 *Tanacetum.* Tanſie. 2 *Pimpinella.* Burnet. 3 *Rhaponticum verum ſeu potius Rhabarbarum verum.* True Raponticke or rather true Rubarbe. 4 *Lapathum ſativum ſeu Patientia.* Monkes Rubarbe or Patience. 5 *Lapathum ſanguineum.* Bloudwort. 6 *Acetoſa.* Sorrell.

ſmaller vp to the toppe, where it is diuided into diuers ſmall branches, whereon grow purpliſh flowers, and three ſquare darke red ſeede, like vnto others : the roots are not great, but ſomewhat long, and very red, abiding many yeares, yet ſometimes ſpoiled with the extremitie of winter.

The Vſe of Blood-worte.

The whole and onely vſe of the herbe almoſt, ſerueth for the pot, among other herbes, and, as I ſaid before, is accounted a moſt eſpeciall one for that purpoſe. The ſeede therof is much commended for any fluxe in man or woman, to be inwardly taken, and ſo no doubt is the roote, being of a ſtipticke qualitie.

Chap. XVI.

Oxalis ſiue Acetoſa. Sorrell.

Sorrell muſt needes bee reckoned with the Dockes, for that it is ſo like vnto them in all things, and is of many called the ſower Docke. Of Sorrels there are many ſorts, but I ſhall not trouble you with any other in this place, then the common Garden Sorrell, which is moſt knowne, and of greateſt vſe with vs ; which hath tender greene long leaues full of iuice, broade, and bicorned as it were, next vnto the ſtalke, like as Arrach, Spinach, and our Engliſh Mercurie haue, of a ſharpe ſower taſte : the ſtalkes are ſlender, bearing purpliſh long heads, wherein lye three ſquare ſhining browne ſeede, like, but leſſer then the other : the root is ſmaller then any of the other Dockes, but browne, and full of ſtrings, and abideth without decaying, hauing greene leaues all the winter, except in the very extremitie thereof, which often taketh away all or moſt of his leaues.

The Vſe of Sorrell.

Sorrell is much vſed in ſawces, both for the whole, and the ſicke, cooling the hot liuers, and ſtomackes of the ſicke, and procuring vnto them an appetite vnto meate, when their ſpirits are almoſt ſpent with the violence of their furious or fierie fits ; and is alſo of a pleaſant relliſh for the whole, in quickning vp a dull ſtomacke that is ouer-loaden with euery daies plenty of diſhes. It is diuers waies dreſſed by Cooks, to pleaſe their Maſters ſtomacks.

Chap. XVII.

Bugloſſum luteum, ſiue Lingua Bouis. Langdebeefe.

Vnto this place may well bee referred our ordinary Borage and Bugloſſe, ſet forth in the former Booke, in regard of the properties whereunto they are much employed, that is, to ſerue the pot among other herbes, as is ſufficiently knowne vnto all. And yet I confeſſe, that this herbe (although it bee called *Bugloſſum luteum,* as if it were a kind of Bugloſſe) hath no correſpondency with Bugloſſe or Borage in any part, ſauing only a little in the leafe; & our Borage or Bugloſſe might more fitly, according to the Greeke name, bee called Oxe tongue or Langdebeefe; and this might in my iudgement more aptly be referred to the kinds of *Hieratium* Hawkeweed, whereunto it neereſt approacheth : but as it is commonly receiued, ſo take it in this place, vntill it come to receiue the place is proper for it. It hath diuers broad and long darke green leaues, lying vpon the ground, very rough in handling, full of ſmall haires or prickes, ready to enter into the hands of any that handle it ; among which riſeth

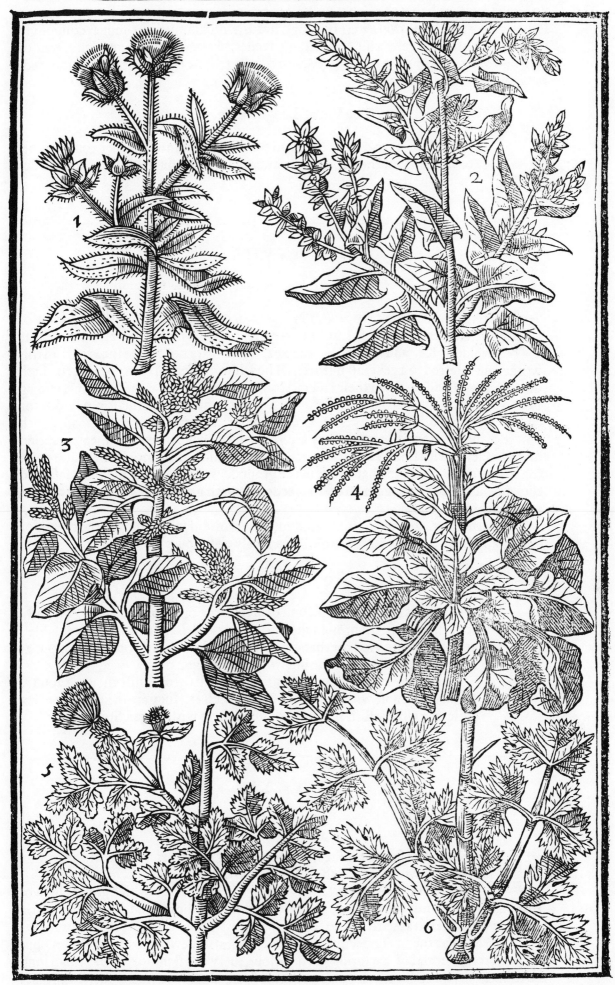

1   *Lingua bouis siue Buglossum luteum.* Langdebeefe.   2   *Atriplex siue Olus aureum.* Arrach.   3   *Blitum.* Blites.
4   *Beta.* Beetes.   5   *Hipposelinum siue Olus atrum.* Allisanders.   6   *Selinum dulce.* Sweete Parsley.

vp a round greene hairy or prickly ſtalk, bearing at the toppe, among a few ſmall greene leaues, diuers ſmall yellow flowers in rough heads, which turne into doune, containing within them browne yellowiſh ſmall long ſeedes, ſomewhat like vnto the ſeede of Hawkeweede : the roote is wooddy, which periſheth quickly after it hath borne ſeed, but is tender while it is young.

The Vſe of Langdebeeſe.

The leaues are onely vſed in all places that I know, or euer could learne, for an herbe for the pot among others, and is thought to bee good to looſen the belly.

Chap. XVIII.

Atriplex ſiue Olus Aureum. Arrach.

THere be diuers kindes of Arrach, or Orach, as ſome doe call them ; ſome of the Garden, whereof I meane to entreate in this place ; others wilde of the Fieldes, &c. and others of the Sea, which are not to bee ſpoken of in this worke, but referred to a generall hiſtorie. The white garden Arrach, or Orach, hath diuers leaues, ſtanding vpon their ſeuerall footeſtalkes, broade at the bottome, ending in two points like an arrow, with two feathers at the head, and ſmall pointed at the end of the leafe, of a whitiſh yellow greene colour, and as it were ſtrewed ouer with flower or meale, eſpecially while they are young : the ſtalke likewiſe is mealy, bearing many branches with ſmall yellow flowers on them, which turne into ſmall leaſie ſeeds : the rooote groweth ſomewhat deepe in the ground, with many ſmall threds faſtened thereto : it quickly ſpringeth vp of the ſeede, groweth great, and fadeth away as ſoon as it hath borne ſeede.

The purple Arrach is in all things like vnto the white, ſauing onely in the colour of the leafe, ſtalke, ſeede, &c. which are all of a mealy duſty purpliſh colour.

The Vſe of Arrach.

Arrach is cold and moiſt, and of a lubricke or ſlippery qualitie, whereby it quickely paſſeth through the ſtomacke and belly, and maketh it ſoluble, and is of many vſed for that purpoſe, being boyled and buttered, or put among other herbes into the pot to make pottage.

There are many diſhes of meate made with them while they are young, for being almoſt without ſauour of themſelues, they are the more conuertible into what relliſh any one will make them with Sugar, Spice &c.

Chap. XIX.

Blitum. Blites.

THere be diuers ſorts of Blites, ſome whereof I haue entreated in the former part of this worke, vnder the title of *Amaranthus,* Flower gentle : others that are nourſed vp in Gardens, I will ſet forth in this place, which are onely two, that haue come to my knowledge, that is, the white and the red, and are of a qualitie as neere vnto Arrach as vnto Beetes, participating of both, and therefore I haue placed them betwixt them. The white Blite hath leaues ſomewhat like vnto Beetes, but ſmaller, rounder, and of a whitiſh greene colour, euery one ſtanding vpon a ſmall long footeſtalke : the ſtalke riſeth vp two or three foote high, with many ſuch like leaues thereon : the flowers grow at the top in long round tufts or cluſters, wherein are contained

tained fmall round feede : the roote is very full of thrcds or ftrings.

The red Blite is in all things like the white, but that his leaues and tufted heades are exceeding red at the firſt, and after turne more purplifh.

The Vſe of Blites.

Blites are vſed as Arrach, eyther boyled of it felfe or ſtewed, which they call Loblolly, or among other herbes to bee put into the pot ; and yet ſome doe vtterly refuſe it, becauſe in diuers it prouoketh caſtings. It is altogether inſipide or without taſte, but yet by reaſon of the moiſt ſlipperie qualitie it hath, it helpeth to looſen the belly. The vnſauorineſſe whereof hath in many Countries growne into a prouerbe, or by-word, to call dull, ſlow, or lazie perſons by that name : They are accounted more hurtfull to the ſtomacke, and ſo to the head and eyes, then other herbes, and therefore they are the leſſe vſed.

Chap. XX.

Beta. Beetes.

THere are many diuerſities of Beetes, ſome growing naturally in our own Coun-try, others brought from beyond Sea ; whereof ſome are white, ſome greene, ſome yellow, ſome red : the leaues of ſome are of vſe only, and the root not vſed : others the roote is only vſed, and not the leaues : and ſome againe, both roote and leafe. The ancient Authors, as by their workes appeare, knew but two ſorts, the white and the blacke Beete, whereof the white is ſufficiently known, and was of them termed *Sicula,* of the later Phyſitians *Sicla,* becauſe it was thought firſt to be brought from Sicilie : the blacke abideth ſome controuerſie ; ſome thinking that our common greene Beete, becauſe it is of a darke greene colour, was that they called the blacke Beete ; others that our fmall red Beete, which is of a darke red colour, was their black Beete, which in my opinion is the more likely : But to come to the matter in hand, and giue you the deſcriptions of them which are in vſe with vs, and leaue controuerſies to ſuch a worke as is fit for them, wherein all ſuch matters may be diſcuſſed at large.

The common white Beete hath many great leaues next the ground (in ſome hot Countries growing to be three foote long, and very broade, in our Countrey they are very large, but nothing neere that proportion) of a whitiſh greene colour ; the ſtalke is great, ſtrong, and ribbed or creſted, bearing great ſtore of leaues vpon it vp to the very toppe almoſt : the flowers grow in very long tufts, ſmall at the ends, and turning down their heads, which are ſmall pale greeniſh yellow burres, giuing cornered prick-ly ſeede : the roote is great, long and hard, when it hath giuen ſeede, of no vſe at all, but abideth a former winter with his leaues vpon it, as all other ſorts following doe.

The common red Beet differeth not from the white Beete, but only that it is not ſo great, and both the leaues and rootes are ſomewhat red : the leaues bee in ſome more red then in others, which haue but red veines or ſtrakes in them, in ſome alſo of a freſh red, in others very darke red : the roote hereof is red, ſpongy, and not vſed to bee eaten.

The common greene Beete is alſo like vnto the white Beete, but of a darke greene colour. This hath beene found neere the ſalt Marſhes by Rocheſter, in the foote-way going from the Lady Leveſons houſe thither, by a worthy, diligent and painefull ob-ſeruer and preſeruer both of plants and all other natures varieties, often remembred before in this worke, called Iohn Tradeſcante, who there finding it, gaue me the know-ledge thereof, and I haue vpon his report ſet it here down in this manner :

The Romane red Beete, called *Beta rapoſa,* is both for leafe and roote the moſt excel-lent Beete of all others : his rootes bee as great as the greateſt Carrot, exceeding red both within and without, very ſweete and good, fit to bee eaten : this Beete groweth higher then the laſt red Beete, whoſe rootes are not vſed to bee eaten : the leaues likewiſe

wife are better of tafte, and of as red a colour as the former red Beete : the roote is fometimes fhort like a Turnep, whereof it took the name of *Rapa* or *rapofa* ; and fometimes as I faid before, like a Carrot and long : the feede is all one with the leffer red Beete.

The Italian Beete is of much refpect, whofe faire greene leaues are very large and great, with great white ribbes and veines therein: the ftalke in the Summer time, when it is growen vp to any height, is fix fquare in fhew, and yellowifh withall, as the heades with feede vpon them feeme likewife.

The great red Beete that Mafter Lete a Merchant of London gaue vnto Mafter Gerrard, as he fetteth it downe in his Herball, feemeth to bee the red kinde of the laft remembred Beete, whofe great ribbes as he faith, are as great as the middle ribbe of the Cabbage leafe, and as good to bee eaten, whofe ftalke rofe with him to the height of eight cubits, and bore plenty of feede.

The Vfe of Beetes.

Beetes, both white, greene and red, are put into the pot among other herbes, to make pottage, as is commonly known vnto all, and are alfo boyled whole, both in France vfually with moft of their boyled meates, and in our Countrey, with diuers that delight in eating of herbes.

The Italian Beete, and fo likewife the laft red Beete with great ribbes, are boyled, and the ribbes eaten in fallets with oyle, vinegar and pepper, and is accounted a rare kinde of fallet, and very delicate.

The roote of the common red Beete with fome, but more efpecially the Romane red Beete, is of much vfe among Cookes to trimme or fet out their difhes of meate, being cut out into diuers formes and fafhions, and is grown of late dayes into a great cuftome of feruice, both for fifh and flefh.

The rootes of the Romane red Beete being boyled, are eaten of diuers while they are hot with a little oyle and vinegar, and is accounted a delicate fallet for the winter ; and being cold they are fo vfed and eaten likewife.

The leaues are much vfed to mollifie and open the belly, being vfed in the decoction of Glifters. The roote of the white kinde fcraped, and made vp with a little honey and falt, rubbed on and layd on the belly, prouoketh to the ftoole. The vfe of eating Beetes is likewife held to bee helpefull to fpleneticke perfons.

CHAP. XXI.

Hippofelinum, fiue Olus atrum. Alifanders.

ALifanders hath beene in former times thought to be the true Macedonian Parfley, and in that errour many doe yet continue : but this place giueth not leaue to difcuffe that doubt : but I muft here only fhew you, what it is, and to what vfe it is put ordinarily for the Kitchen. The leaues of Alifanders are winged or cut into many parts, fomewhat refembling Smallage, but greater, broader, and more cut in about the edges : the ftalkes are round and great, two foote high or better, bearing diuers leaues on them, and at the toppe fpokie roundles of white flowers on feuerall fmall branches, which turne into blacke feede, fomewhat cornered or crefted, of an aromaticall bitter tafte : the roote is blacke without, and white within, and abideth well the firft year of the fowing, perifhing after it hath borne feed.

The Vfe of Alifanders.

The tops of the rootes, with the lower part of the ftalkes of Alifanders, are vfed in Lent efpecially, and Spring of the yeare, to make broth, which although it be a little bitter, yet it is both wholfome, and pleafing to a great

many,

many, by reason of the aromaticall or spicie taste, warming and comforting the stomack, and helping it digest the many waterish and flegmaticke meates are in those times much eaten. The rootes also either rawe or boyled are often eaten with oyle and vinegar. The seede is more vsed physically then the roote, or any other part, and is effectuall to prouoke plenty of vrine in them that pisse by drops, or haue the Strangury : It helpeth womens courses, and warmeth their benummed bodies or members, that haue endured fierce cold daies and nights, being boyled and drunke.

Chap. XXII.

Selinum dulce. Sweete Parsley or sweete Smallage.

THis kinde of sweete Parsley or Smallage, which soeuer you please to call it; for it resembleth Smallage as well in the largenesse of the leaues, as in the taste, yet sweeter and pleasanter, is (as I take it) in this like vnto sweete Fennell (that hath his sweetnesse from his naturall soyle and clymate ; for howsoeuer it bee reasonable sweete the first yeare it is sowne with vs, yet it quickly doth degenerate, and becommeth no better then our ordinarie Fennell afterwards). The first yeare it is sowne and planted with vs (and the first that euer I saw, was in a Venetian Ambassadours Garden in the Spittle yard, neare Bishops gate streete) is so sweete and pleasant, especially while it is young, as if Sugar had beene mingled with it : but after it is growne vp high and large, it hath a stronger taste of Smalladge, and so likewise much more the next yeare ; that it groweth from the seed was gathered here : the leaues are many, spreading farre about the roote, broader and of a fresher greene colour then our ordinary Smalladge, and vpon longer stalkes : the seed is as plentifull as Parsly, being small and very like vnto it, but darker of colour.

The Vse of sweete Parsley.

The Venetians vse to prepare it for meate many waies, both the herbe and the roote eaten rawe, as many other herbes and rootes are, or boyled or fryed to be eaten with meate, or the dryed herbe poudered and strewed vpon meate ; but most vsually either whited, and so eaten rawe with pepper and oyle, as a dainty Sallet of it selfe, or a little boyled or stewed : the taste of the herbe being a little warming, but the seede much more, helpeth cold windy stomackes to digest their meate, and to expell winde.

Chap. XXIII.

Petroselinum & Apium. Parsley and Smalledge.

WE haue three sorts of Parsley in our Gardens, and but one of Smalladge : Our common Parsley, Curld Parsley, and Virginia Parsley ; which last, although it be but of late knowne, yet it is now almost growne common, and of as good vse as the other with diuers. Our common Parsley is so well knowne, that it is almost needlesse to describe it, hauing diuers fresh greene leaues, three alwaies placed together on a stalke, and snipt about the edges, and three stalkes of leaues for the most part growing together : the stalkes growe three or foure foote high or better, bearing spikie heads of white flowers, which turne into small seede, somewhat sharpe and hot in taste : the roote is long and white.

Curld Parsley hath his leaues curled or crumpled on the edges, and therein is the onely difference from the former.

Virginia

Virginia Parſley is in his leafe altogether like vnto common Parſley for the forme, conſiſting of three leaues ſet together, but that the leaues are as large as Smallage leaues, but of a pale or whitiſh greene colour, and of the ſame taſte of our common Parſley: the ſeede hereof is as the leaues, twice if not thrice as bigge as the ordinary Parſley, and periſheth when it hath giuen ſeede, abiding vſually the firſt yeare of the ſowing.

Smallage is in forme ſomewhat like vnto Parſley, but greater and greener, and leſſe pleaſant, or rather more bitter in taſte: the ſeede is ſmaller, and the root more ſtringy.

The Vſe of Parſley.

Parſley is much vſed in all ſorts of meates, both boyled, roaſted, fryed, ſtewed, &c. and being greene it ſerueth to lay vpon ſundry meates, as alſo to draw meate withall. It is alſo ſhred and ſtopped into poudered beefe, as alſo into legges of Mutton, with a little beefe ſuet among it, &c.

The rootes are often vſed to be put into broth, to helpe to open obſtructions of the liuer, reines, and other parts, helping much to procure vrine.

The rootes likewiſe boyled or ſtewed with a legge of Mutton, ſtopped with Parſley as aforeſaid, is very good meate, and of very good relliſh, as I haue proued by the taſte; but the rootes muſt bee young, and of the firſt yeares growth, and they will haue their operation to cauſe vrine.

The ſeed alſo is vſed for the ſame cauſe, when any are troubled with the ſtone, or grauell, to open the paſſages of vrine.

Although Smallage groweth in many places wilde in moiſt grounds, yet it is alſo much planted in Gardens, and although his euill taſte and ſauour doth cauſe it not to be accepted into meates as Parſley, yet it is not without many ſpeciall good properties, both for outward and inward diſeaſes, to helpe to open obſtructions, and prouoke vrine. The iuyce cleanſeth vlcers; and the leaues boyled with Hogs greaſe, healeth felons on the ioynts of the fingers.

Chap. XXIIII.

Fœniculum. Fenell.

THere are three ſorts of Fenell, whereof two are ſweete. The one of them is the ordinary ſweete Fenell, whoſe ſeedes are larger and yellower then the common, and which (as I ſaid before in the Chapter of ſweete Parſley) doth ſoone degenerate in this our Country into the common. The other ſweete Fenell is not much knowne, and called Cardus Fenell by thoſe that ſent it out of Italy, whoſe leaues are more thicke and buſhie then any of the other. Our common Fenell, whereof there is greene and red, hath many faire and large ſpread leaues, finely cut and diuided into many ſmall, long, greene, or reddiſh leaues, yet the thicker tufted the branches be, the ſhorter are the leaues: the ſtalkes are round, with diuers ioynts and leaues at them, growing fiue or ſix foot high, bearing at the top many ſpoakie rundels of yellow flowers: the Common, I meane, doth turne into a darke grayiſh flat ſeede, and the Sweete into larger and yellower: the roote is great, long, and white, and endureth diuers yeares.

The Vſe of Fenell.

Fenell is of great vſe to trimme vp, and ſtrowe vpon fiſh, as alſo to boyle or put among fiſh of diuers ſorts, Cowcumbers pickled, and other fruits, &c. The rootes are vſed with Parſley rootes, to be boyled in broths and drinkes to open obſtructions. The ſeed is of much vſe with other things to expell winde. The ſeede alſo is much vſed to be put into Pippin pies, and diuers

other

1 *Petroſelinum.* Parſley. 2 *Aſium.* Smallage. 3 *Fœniculum.* Fenell. 4 *Anethum.* Dill. 5 *Myrrhis ſiue Cerefolium magnum.* Sweete Cheruill. 6 *Cerefolium vulgare.* Common Cheruill.

other such baked fruits, as also into bread, to giue it the better rellish.

The sweete Cardus Fenell being sent by Sir Henry Wotton to Iohn Tradescante, had likewise a large direction with it how to dresse it; for they vse to white it after it hath been transplanted for their vses, which by reason of the sweetnesse by natnre, and the tendernesse by art, causeth it to be the more delightfull to the taste, especially with them that are accustomed to feede on greene herbes.

Chap. XXV.

Anethum. Dill.

DIll doth much growe wilde, but because in many places it cannot be had, it is therefore sowne in Gardens for the vses whereunto it serueth. It is a smaller herbe then Fenell, but very like, hauing fine cut leaues, not so large, but shorter, smaller, and of a stronger and quicker taste: the stalke is smaller also, and with few ioynts and leaues on them, bearing spoakie tufts of yellow flowers, which turne into thinne, small, and flat seedes: the roote perisheth euery yeare, and riseth againe for the most part of it owne sowing.

The Vse of Dill.

The leaues of Dill are much vsed in some places with Fish, as they doe Fenell; but because it is so strong many doe refuse it.

It is also put among pickled Cowcumbers, wherewith it doth very well agree, giuing vnto the cold fruit a pretty spicie taste or rellish.

It being stronger then Fenell, is of the more force to expell winde in the body. Some vse to eate the seed to stay the Hickocke.

Chap. XXVI.

Myrrhis siue Cerefolium maius & vulgare.
Sweet Cheruill and ordinary Cheruill.

THe great or sweete Cheruill (which of some is called Sweete Cicely) hath diuers great and faire spread winged leaues, consisting of many leaues set together, deeply cut in the edges, and euery one also dented about, very like, and resembling tne leaues of Hemlockes, but of so pleasant a taste, that one would verily thinke, he chewed the leaues or seedes of Aniseedes in his mouth: The stalke is reasonable great, and somewhat cornered or crested about three or foure foote high, at the toppe whereof stand many white spoakie tufts of flowers, which change into browne long cornered great seede, two alwaies ioyned together: the roote is great, blackish on the outside, and white within, with diuers fibres annexed vnto it, and perisheth not, but abideth many yeares, and is of a sweete, pleasant, and spicie hot taste, delightfull vnto many.

The common Cheruill is a small herbe, with slender leaues, finely cut into long peeces, at the first of a pale yellowish greene colour, but when the stalke is growne vp to seede, both stalkes and leaues become of a darke red colour: the flowers are white, standing vpon scattered or thin spread tufts, which turne into small, long, round, and sharpe pointed seedes, of a brownish blacke colour: the roote is small, with diuers long slender white strings, and perisheth euery yeare.

The

The Vſes of theſe Cheruils.

The common Cheruill is much vſed of the French and Dutch people, to bee boyled or ſtewed in a pipkin, eyther by it ſelfe, or with other herbes, whereof they make a Loblolly, and ſo eate it. It is vſed as a pot-herbe with vs.

Sweete Cheruill, gathered while it is young,and put among other herbes for a ſallet,addeth a meruellous good relliſh to all the reſt. Some commend the greene ſeedes ſliced and put in a ſallet of herbes,and eaten with vinegar and oyle, to comfort the cold ſtomacke of the aged. The roots are vſed by diuers, being boyled, and after eaten with oyle and vinegar, as an excellent ſallet for the ſame purpoſe. The preſerued or candid rootes are of ſingular good vſe to warme and comfort a cold flegmaticke ſtomack, and is thought to be a good preſeruatiue in the time of the plague.

Chap. XXVII.

Malua Criſpa. French Mallowes.

THe curld or French Mallow groweth vp with an vpright greene round ſtalke, as high vſually as any man, whereon from all ſides grow forth round whitiſh greene leaues, curld or crumpled about the edges, like a ruffe,elſe very like vnto an ordinary great Mallow leafe : the flowers grow both vpon the ſtalke,and on the other branches that ſpring from them, being ſmall and white ; after which come ſmall caſes with blacke ſeede like the other Mallowes : the roote periſheth when it hath borne ſeede, but abideth vſually the firſt yeare, and the ſecond runneth vp to flower and ſeede.

The Vſe of French Mallowes.

It is much vſed as a pot-herbe,eſpecially when there is cauſe to moue the belly downward, which by his ſlippery qualitie it doth helpe forward. It hath beene in times paſt, and ſo is to this day in ſome places,vſed to be boyled or ſtewed, eyther by it ſelfe with butter, or with other herbes, and ſo eaten.

Chap. XXVIII.

Intubum. Succorie and Endiue.

I Put both Succorie and Endiue into one chapter and deſcription, becauſe they are both of one kindred ; and although they differ a little the one from the other, yet they agree both in this, that they are eaten eyther greene or whited, of many.

Endiue, the ſmooth as well as the curld, beareth a longer and a larger leafe then Succorie,and abideth but one yeare, quickely running vp to ſtalke and ſeede, and then periſheth: whereas Succorie abideth many years,and hath long and narrower leaues, ſomewhat more cut in, or torne on the edges : both of them haue blew flowers, and the ſeede of the ſmooth or ordinary Endiue is ſo like vnto the Succorie, that it is very hard to diſtinguiſh them aſunder by ſight; but the curld Endiue giueth blackiſh and flat ſeede, very like vnto blacke Lettice ſeede : the rootes of the Endiue periſh, but the Succorie abideth.

The Vſe of Succory and Endiue.

Although Succorie bee ſomewhat more bitter in taſte then the Endiues,

yet

yet it is oftentimes, and of many eaten greene, but more vſually being buried a while in ſand, that it may grow white, which cauſeth it to loſe both ſome part of the bitterneſſe, as alſo to bee the more tender in the eating ; and Horace ſheweth it to be vſed in his time, in the 3 2. Ode of his firſt Book, where he ſaith,

Me paſcunt Oliuæ, me Cithorea leueſq̃, Maluæ.

Endiue being whited in the ſame, or any other manner, is much vſed in winter, as a ſallet herbe with great delight ; but the curld Endiue is both farre the fairer, and the tenderer for that purpoſe.

Chap. XXIX.

Spinachia, ſiue Olus Hiſpanicum. Spinach.

SPinach or Spinage is of three ſorts (yet ſome doe reckon of foure, accounting that herbe that beareth no ſeede to be a ſort of it ſelfe, when it is but an accident of nature, as it falleth out in Hempe, Mercury, and diuers other herbes) two that bear prickly ſeed, the one much greater then the other: the third that beareth a ſmooth ſeede, which is more daintie, and nourſed vp but in few Gardens : The common Spinach which is the leſſer of the two prickly ſorts, hath long greene leaues, broad at the ſtalke, and rent, or torne as it were into foure corners, and ſharpe pointed at the ends : it quickly runneth vp to ſtalke, if it be ſowen in the Spring time ; but elſe, if at the end of Summer, it will abide all the winter green, and then ſuddenly in the very beginning of the Spring, runne vp to ſtalke, bearing many leaues both below and at the toppe, where there doth appeare many ſmal greeniſh flowers in cluſters, and after them prickly ſeede : The other greater ſort that hath prickly ſeede, is in all things like the former, but larger both in ſtalke, leafe and ſeede. The ſmooth Spinach hath broader, and a little rounder pointed leaues then the firſt, eſpecially the lower leaues ; for thoſe that grow vpwards vpon the ſtalke, are more pointed, and as it were three ſquare, of as darke a greene colour as the former : at the ſeuerall ioynts of the ſtalkes and branches, ſtand cluſtering many ſmall greeniſh flowers, which turne into cluſters of round whitiſh ſeede, without any prickles at all vpon them : the roote is long, white and ſmall, like vnto the other, with many fibres at it : If it be often cut, it will grow the thicker, or elſe ſpindle vp very thinly, and with but few leaues vpon the ſtalke.

The Vſe of Spinage.

Spinage is an herbe fit for ſallets, and for diuers other purpoſes for the table only ; for it is not knowne to bee vſed Phyſically at all. Many Engliſh that haue learned it of the Dutch people, doe ſtew the herbe in a pot or pipkin, without any other moiſture then it owne, and after the moiſture is a little preſſed from it, they put butter, and a little ſpice vnto it, and make therewith a diſh that many delight to eate of. It is vſed likewiſe to be made into Tartes, and many other varieties of diſhes, as Gentlewomen and their Cookes can better tell then my ſelfe ; vnto whom I leaue the further ordering of theſe herbes, and all other fruits and rootes of this Garden : For I intend only to giue you the knowledge of them, with ſome briefe notes for their vſe, and no more.

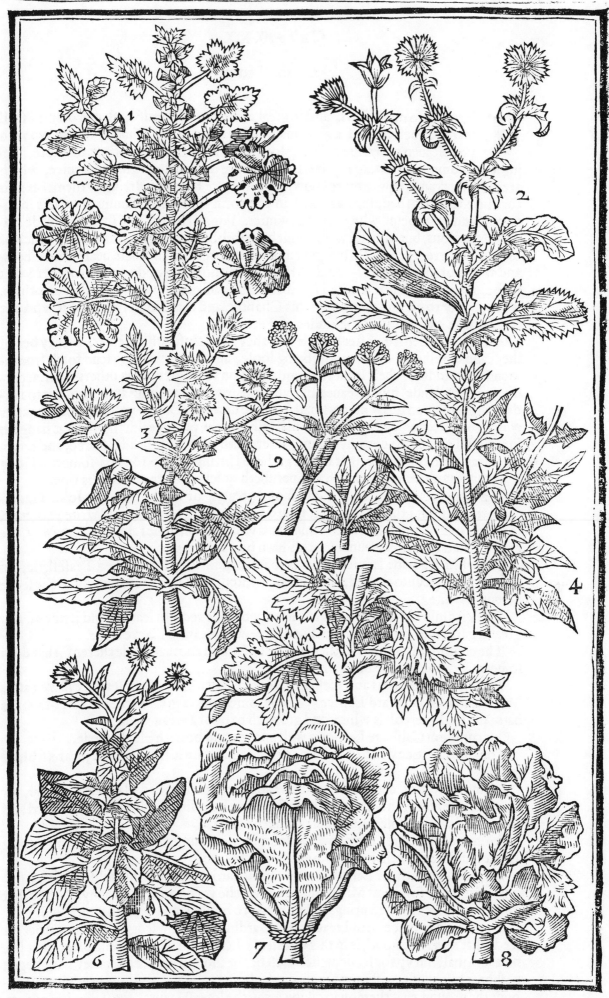

1 *Malua crispa.* French Mallowes. 2 *Endiuia.* Endiue. 3 *Cichorium.* Succory. 4 *Spinachia.* Spinach. 5 *Lactuca crispa.* Curld Lettice. 6 *Lactuca patula.* An open Lettice. 7 *Lactuca capitata vulgaris.* Ordinary cabbage Lettice. 8 *Lactuca capitata Romana.* The great Romane cabbage Lettice. 9 *Lactuca agnina.* Corne Sallet or Lambes Lettice.

CHAP. XXX.

Lactuca. Lettice.

THere are fo many forts, and fo great diuerfitie of Lettice, that I doubt I fhall fcarce be beleeued of a great many. For I doe in this Chapter reckon vp vnto you eleauen or twelue differing forts ; fome of little vfe, others of more, being more common and vulgar ; and fome that are of excellent vfe and feruice, which are more rare, and require more knowledge and care for the ordering of them, as alfo for their time of fpending, as fome in the fpring, fome in fummer, others in autumne, and fome being whited for the winter. For all thefe forts I fhall not neede many defcriptions, but only fhew you which doe cabbage, and which are loofe, which of of them are great or fmall, white, greene or red, and which of them beare white feeds, and which of them blacke. And laftly I haue thought good to adde another Sallet herbe, which becaufe it is called Lambes Lettice of many, or Corne Sallet of others, is put in only to fill vp a number in this Chapter, and that I muft fpeake of it, and not that I thinke it to be any of the kindes of Lettice.

All forts of Lettice, after a while that they haue clofed themfelues, if they bee of the Cabbage kindes, or otherwife being loofe, and neuer clofing, fend forth from among the middle of their leaues a round ftalke (in fome greater, in others leffer, according to their kinde) full of leaues like vnto the lower, branching at the toppe into fundry parts, whereon grow diuers fmall ftar-like flowers, of a pale yellowifh colour; after which come feede, eyther white or blackifh, as the plant yeeldeth, whereat hangeth fome fmall peece of a cottony doune, wherewith the whole head is ftored, and is carried away with the winde, if it be not gathered in time : the roote is fomewhat long and white, with fome fibres at it, and perifheth quickely after the feede is ripe.

The Romane red Lettice is the beft and greateft of all the reft. For Iohn Tradefcante that firft, as I thinke, brought it into England, and fowed it, did write vnto mee, that after one of them had been bound and whited, when the refufe was cut away, the reft weighed feuenteene ounces : this hath blacke feede.

The white Romane Lettice is like vnto it, hauing long leaues like a Teafell, it is in goodneffe next vnto the red, but muft be whited, that it may eate kindly : the feede hereof is white.

The Virginia Lettice hath fingle and very broade reddifh leaues, and is not of any great regard, and therefore is kept but of a few : it beareth blacke feede.

The common Lumbard Lettice that is loofe, and another kinde thereof that doth fomewhat cabbage, haue both white feedes.

The Venice Lettice is an excellent Cabbage Lettice, and is beft to bee fowen after Midfummer for lateward Lettice ; they be fometimes as great as the crowne of a mans hatt : the feede hereof is white, and groweth to be of a meane height.

Our common Cabbage Lettice is well known, and beareth blacke feede.

The curld Lettice which is open, and differeth but little from Endiue, beareth black feede.

Another fort of curld Lettice doth cabbage, and is called Flanders Cropers, or Cropers of Bruges ; this groweth loweft, and hath the fmalleft head, but very hard and round, and white while it groweth : the feed is blacke.

A kinde of Romane Lettice is of a darke green colour, growing as low as the Venice Lettice, and is an excellent kinde, bearing blacke feede.

And laftly our winter Lettice is wonderfull hardy to endure our cold : It is but fingle, and muft be fowen at Michaelmas, but will be very good, before any of the other good forts fowen in the Spring, will be ready to be vfed, and beareth white feed.

To inftruct a nouice (for I teach not a Gardiner of knowledge) how to gather his feede that it may be good, is in this manner : Let him marke out thofe plants that hee meaneth fhall run vp for feede, which muft be the moft likely ; & after they haue begun to fhoote forth ftalkes, ftrip away the lower leaues, for two or three hands breadth aboue the ground, that thereby in taking away the loweft leaues, the ftalke doe not rot, nor the feed be hindered in the ripening.

There

There are two manner of wayes to whiten Lettice to make them eate the more tender : the one is by rayfing vp earth like moale hils, round about the plants while they are growing, which will make them grow white : the other is by tying vp all the loofe leaues round together while it groweth, that fo the clofe tying may make it grow white, and thereby be the more tender.

Lambes Lettice or Corne Sallet is a fmall plant while it is young, growing clofe vpon the ground, with many whitifh greene, long and narrow, round pointed leaues, all the winter, and in the beginning of the fpring (if it bee fowen in autumne, as it is vfuall to ferue for an early fallet) rifeth vp with fmall round ftalkes, with two leaues at euery ioynt, branching forth at the toppe, and bearing tufts of fmall bleake blew flowers, which turne into fmall round whitifh feede : the roote is fmall and long, with fome fmall threds hanging thereat : the whole plant is of a waterifh tafte, almoft infipide.

The Vfe of Lettice.

All forts of Lettice are fpent in fallets, with oyle and vinegar, or as euery one pleafe, for the moft part, while they are frefh and greene, or whited, as is declared of fome of the forts before, to caufe them to eate the more delicate and tender. They are alfo boyled, to ferue for many forts of difhes of meate, as the Cookes know beft.

They all coole a hot and fainting ftomacke.

The iuice of Lettice applyed with oyle of Rofes to the foreheads of the ficke and weake wanting fleepe, procureth reft, and taketh away paines in the head : bound likewife to the cods, it helpeth thofe that are troubled with the Colts euill. If a little camphire be added, it reftraineth immoderate luft : but it is hurtfull to fuch as are troubled with the fhortneffe of breath.

Lambes Lettice is wholly fpent for fallets, in the beginning of the yeare, as I faid, before any almoft of the other forts of Lettice are to be had.

Chap. XXXI.

Portulaca. Purflane.

PVrflane hath many thicke round fhining red ftalkes, full of iuice, lying vpon the ground for the moft part ; whereon are fet diuers long, thicke, pale green leaues, fometimes alone by themfelues, and fometimes many fmall ones together with them ; among which grow fmall yellow flowers, which ftand in little greene huskes, containing blacke feede: the roote is fmall, and perifheth euery yeare, and muft be new fowen in Aprill, in the alleyes of the Garden betweene the beds, as fome haue heretofore vfed, where it may haue the more moifture, or, as I haue feene in fome Gardens, vpon thofe beds of dung that Gardiners haue vfed to nourfe vp their Cowcumbers, Melons, and Pompions, whereon after they haue beene taken away, they haue fowen Purflane, whereif it be much watered, the warmth of the dung, and the water giuen it, the Purflane hath grown great and large, and continued vntill winter.

The Vfe of Purflane.

It is vfed as Lettice in fallets, to coole hot and faint ftomackes in the hot time of the yeare, but afterwards if only for delight, it is not good to bee too prodigall in the vfe thereof.

The feede of Purflane doth coole much any inflammation inward or outward, and doth a little binde withall.

CHAP.

Chap. XXXII.

Dracoherba sine Tarchon & Dracunculus hortensis. Tarragon.

Tarragon hath long and narrow darke greene leaues, growing on slender and brittle round stalkes, two or three foote high, at the tops whereof grow forth long slender spikes of small yellowish flowers, which seldome giue any good seede, but a dustie or chaffie matter, which flieth away with the winde: the roote is white, and creepeth about vnder ground, whereby it much encreaseth: the whole herbe is of a hot and biting taste.

The Vse of Tarragon.

It is altogether vsed among other cold herbes, to temper their coldnesse, and they to temper its heate, so to giue the better rellish vnto the Sallet; but many doe not like the taste thereof, and so refuse it.

There are some Authors that haue held Tarragon not to be an herbe of it owne kinde, but that it was first produced, by putting the seede of Lin or Flaxe into the roote of an Onion, being opened and so set into the ground, which when it hath sprung, hath brought forth this herbe Tarragon, which absurd and idle opinion, Matthiolus by certaine experience saith, hath been found false.

Chap. XXXIII.

Nasturtium hortense. Garden Cresses.

Garden Cresses growe vp to the height of two foote or thereabouts, hauing many small, whitish, broad, endented, torne leaues, set together vpon a middle ribbe next the ground, but those that growe higher vpon the stalkes are smaller and longer: the tops of the stalkes are stored with white flowers, which turne into flat pods or pouches, like vnto Shepheard purse, wherein is contained flat reddish seede: the roote perisheth euery yeare: the taste both of leaues and seedes are somewhat strong, hot, and bitter.

The Vse of Cresses.

The Dutchmen and others vse to eate Cresses familiarly with their butter and bread, as also stewed or boyled, either alone or with other herbes, whereof they make a Hotch potch, and so eate it. Wee doe eate it mixed among Lettice or Purslane, and sometimes with Tarragon or Rocket, with oyle and vinegar and a little salt, and in that manner it is very sauoury to some mens stomackes.

The vse of Cresses physically is, it helpeth to expectorate tough flegme, as also for the paines of the breast; and as it is thought taketh away spots, being laid to with vinegar. The seede is giuen of many to children for the wormes.

1 *Portulaca.* Purſlane. 2 *Dracho herba ſeu Tarchon.* Tarragon. 3 *Eruca ſatiua.* Garden Rocket. 4 *Naſturtium ſatiuum.* Garden Creſſes. 5 *Sinapi.* Muſtard. 6 *Aſparagus.* Aſparagus or Sperage.

CHAP. XXXIIII.

Eruca satiua. Garden Rocket.

OVr Garden Rocket is but a wilde kinde brought into Gardens ; for the true Romane Rocket hath larger leaues ; this hath many long leaues , much torne or rent on the edges, finaller and narrower then the Romane kinde : the flowers hereof are of a pale yellowifh colour, whereas the true is whitifh, confifting of foure leaues : the feede of this is reddifh, contained in finaller and longer pods then the true, which are fhorter and thicker, and the feede of a whitifh yellow colour : the rootes of both perifh as foone as they haue giuen feede. Some haue taken one fort of the wilde kinde for Muftard, and haue vfed the feede for the fame purpofe.

The Vfe of Rocket.

It is for the moft part eaten with Lettice, Purflane, or fuch cold herbes, and not alone , becaufe of its heate and ftrength ; but that with the white feede is milder. The feede of Rocket is good to prouoke vrine , and to ftirre vp bodily luft.

The feede bruifed, and mixed with a little vinegar, and of the gall of an Oxe, cleanfeth the face of freckles, fpots, and blew markes, that come by beatings, fals, or otherwaies.

Matthiolus faith, that the leaues boyled, and giuen with fome Sugar to little children, cureth them of the cough.

The feede is held to be helpfull to fpleneticke perfons ; as alfo to kill the wormes of the belly.

CHAP. XXXV.

Sinapi fativum. Garden Muftard.

THe Muftard that is moft vfuall in this Country, howfoeuer diuers doe for their priuate vfes fowe it in their Gardens or Orchards, in fome conuenient corner, yet the fame is found wilde alfo abroad in many places. It hath many rough long diuided leaues, of an ouerworne greene colour : the ftalke is diuided at the toppe into diuers branches, whereon growe diuers pale yellow flowers, in a great length, which turne into fmall long pods , wherein is contained blackifh feede, inclining to rednefſe, of a fiery fharpe tafte : the roote is tough and white, running deepe into the ground, with many fmall fibres at it.

The Vfe of Muftard.

The feede hereof grownd between two ftones, fitted for the purpofe, and called a Querne, with fome good vinegar added vnto it , to make it liquid and running , is that kinde of Muftard that is vfually made of all forts , to ferue as fawce both for fifh and flefh.

The fame liquid Muftard is of good vfe, being frefh, for Epilepticke perfons, to warme and quicken thofe dull fpirits that are fopite and fcarce appeare, if it be applyed both inwardly and outwardly.

It is with good fuccefſe alfo giuen to thofe that haue fhort breathes , and troubled with a cough in the lungs.

CHAP. XXXVI.

Asparagus. Sperage or Asparagus.

ASparagus riseth vp at the first with diuers whitish greene scaly heads, very brittle or easie to breake while they are young, which afterwards rise vp into very long and slender greene stalkes, of the bignesse of an ordinary riding wand at the bottome of most, or bigger or lesser, as the rootes are of growth, on which are set diuers branches of greene leaues, shorter and smaller then Fennell vp to the toppe, at the ioynts whereof come forth small mossie yellowish flowers, which turne into round berries, greene at the first, and of an excellent red colour when they are ripe, shewing as if they were beades of Corrall, wherein are contained exceeding hard and blacke seede : the rootes are dispersed from a spongious head into many long, thicke, and round strings, whereby it sucketh much nourishment out of the ground, and encreaseth plentifully thereby.

We haue another kinde hereof that is of much greater account, because the shootes are larger, whiter, and being dressed taste more sweete and pleasant, without any other difference.

The Vse of Asparagus.

The first shootes or heads of Asparagus are a Sallet of as much esteeme with all sorts of persons, as any other whatsoeuer, being boyled tender, and eaten with butter, vinegar, and pepper, or oyle and vinegar, or as euery ones manner doth please ; and are almost wholly spent for the pleasure of the pallate. It is specially good to prouoke vrine, and for those that are troubled with the stone or grauell in the reines or kidneyes, because it doth a little open and cleanse those parts.

CHAP. XXXVII.

Brassica. Cabbages and Coleworts.

THere is greater diuersity in the forme and colour of the leaues of this plant, then there is in any other that I know groweth vpon the ground. But this place requireth not the knowledge of all sorts which might be shewen, many of them being of no vse with vs for the table, but for delight, to behold the wonderfull variety of the workes of God herein. I will here therefore shew you onely those sorts that are ordinary in most Gardens, and some that are rare, receiued into some especiall Gardens : And first of Cabbages, and then of Coleworts.

Our ordinary Cabbage that closeth hard and round, hath at the first great large thicke leaues, of a grayish greene colour, with thicke great ribbes, and lye open most part of the Summer without closing, but toward the end of Summer, being growne to haue many leaues, it then beginneth to growe close and round in the middle, and as it closeth, the leaues growe white inward ; yet there be some kindes that will neuer be so close as these, but will remaine halfe open, which wee doe not account to be so good as the other : in the middle of this head, the next yeare after the sowing, in other Countries especially, and sometimes in ours, if the Winter be milde, as may be seene in diuers Gardens (but to preuent the danger of our Winter frosts, our Gardiners now doe vse to take vp diuers Cabbages with their rootes, and tying a cloth or some such thing about the rootes, doe hang them vp in their houses, where they may be defended from cold, and then set them againe after the frosts are past) and then there shooteth out a great thicke stalke, diuided at the toppe into many branches, bearing thereon diuers small flowers, sometime white, but most commonly yellow, made of foure leaues, which turne into long, round, and pointed pods, containing therein small
round

round feede, like vnto Turnep feede : the roote fpreadeth not farre nor deepe, and dyeth vfually in any great frofte ; for a fmall froft maketh the Cabbage eate the tenderer.

The red Cabbage is like vnto the white, laft fpoken of, but differing in colour and greatneffe ; for it is feldome found fo great as the white, and the colour of the leaues is very variable, as being in fome ftript with red, in others more red, or very deepe red or purple.

The fugar loafe Cabbage, fo called becaufe it is fmaller at the toppe then it is at the bottome, and is of two forts, the one white, the other greene.

The Sauoy Cabbadge, one is of a deepe greene coloured leafe, and curld when it is to be gathered ; the other is yellowifh : neyther of both thefe doe clofe fo well as the firft, but yet are vfed of fome, and accounted good.

The Cole flower is a kinde of Coleworte, whofe leaues are large, and like the Cabbage leaues, but fomewhat fmaller, and endented about the edges, in the middle wherof, fometimes in the beginning of Autumne, and fometimes much fooner, there appeareth a hard head of whitifh yellow tufts of flowers, clofely thruft together, but neuer open, nor fpreading much with vs, which then is fitteft to be vfed, the green leaues being cut away clofe to the head : this hath a much pleafanter tafte then eyther the Coleworte, or Cabbage of any kinde, and is therefore of the more regard and refpect at good mens tables.

The ordinary Coleworte is fufficiently knowne not to clofe or cabbage, and giueth feede plentifully enough.

The other Colewortes that are nourfed vp with thofe that delight in curiofities, befides the aforefaid ordinary greene, which is much vfed of Dutchmen, and other ftrangers, are thefe : The Curld Coleworte eyther wholly of a greene colour, or of diuers colours in one plant, as white, yellow, red, purple or crimfon, fo variably mixed, the leaues being curld on the edges, like a ruffe band, that it is very beautifull to behold.

There is alfo another curld Colewort of leffe beauty and refpect, being but a little curld on the edges, whofe leaues are white, edged with red, or green edged with white.

Two other there are, the one of a popingaye greene colour : the other of a fine deepe greene, like vnto the Sauoyes.

Then there is the Cole rape, which is alfo a kinde of Coleworte, that beareth a white heade, or headed ftalke aboue the ground, as bigge as a reafonable Turnep, but longer, and from the toppe thereof fpringeth out diuers great leaues, like vnto Colewortes ; among which rife diuers ftalkes that beare yellow flowers, and feede in pods, almoft as fmall as Muftard feede : the roote is fomewhat long, and very bufhie with threds.

The Vfe of Cabbages and Colewortes.

They are moft vfually boyled in poudered beefe broth vntil they be tender, and then eaten with much fat put among them.

The great ribs of the Popingay, and deepe greene Colewortes, beeing boyled and layde into difhes, are ferued to the table with oyle and vinegar in the Lent time for very good fallets.

In the cold Countries of Ruffia and Mufcouia, they pouder vp a number of Cabbages, which ferue them, efpecially the poorer fort, for their moft ordinary foode in winter ; and although they ftinke moft grieuoufly, yet to them they are accounted good meate.

It is thought, that the vfe of them doth hinder the milke in Nurfes breafts, caufing it to dry vp quickely : but many women that haue giuen fucke to my knowledge haue denyed that affertion, affirming that they haue often eaten them, and found no fuch effect. How it might proue in more delicate bodies then theirs that thus faid, I cannot tell : but Matthiolus auerreth it to encreafe milke in Nurfes breaftes ; fo differing are the opinions of many. The feede groffely bruifed and boyled a little in flefh broth, is a prefent remedie for the Collicke ; the feede and the broth being taken together, eafing them that are troubled therewith of all griping paines : as alfo for the ftone in the kidneyes. A Lohoc or licking Electuary made of the pulpe of

the

1 *Brassica capitata.* Close Cabbage. 2 *Brassica patula.* Open Cabbage. 3 *Brassica Sabaudica crispa.* Curld Sauoye Colewort. 4 *Caulis florida.* Cole flower. 5 *Caulis crispa.* Curld Colewort. 6 *Caulis crispa variata.* Changeable curld Colewort. 7 *Rapocaulis.* Cole rape.

the boyled ſtalkes, and a little honey and Almond milke, is very profitable for ſhortneſſe of breath, and thoſe that are entring into a Conſumption of the lunges. It hath beene formerly held to be helpfull in all diſeaſes: for Criſippus, an ancient Phyſitian, wrote a whole Volume of the vertues, applying it to all the parts of the body : which thing neede not ſeeme wonderfull, in that it is recorded by writers, that the old Romanes hauing expelled Phyſitians out of their Common-wealth, did for many hundred of yeares maintaine their health by the vſe of Cabbages, taking them for euery diſeaſe.

CHAP. XXXVIII.

Siſarum. Skirrets.

AFter all the herbes before rehearſed, fit for ſallets, or otherwiſe to bee eaten, there muſt follow ſuch rootes as are vſed to the ſame purpoſe : and firſt, Skirrets haue many leaues next the ground, compoſed of many ſmall ſmooth green leaues, ſet each againſt other vpon a middle ribbe, and euery one ſnipt about the edges: the ſtalke riſeth vp two or three foote high, ſet with the like leaues, hauing at the toppe ſpoakie tufts of white flowers, which turne into ſmall ſeede, ſomewhat bigger and darker then Parſley ſeede : the rootes be many growing together at one head, beeing long, ſlender, & rugged or vneuen, of a whitiſh colour on the outſide, and more white within, hauing in the middle of the roote a long ſmall hard pith or ſtring : theſe heads are vſually taken vp in February and March, or ſooner if any ſo pleaſe, the greater number of them being broken off to bee vſed, the reſt are planted againe after the heads are ſeparated, and hereby they are encreaſed euery yeare by many ; but it is now adayes more ſowen of the ſeed, which come forwards well enough if the ground be fat and good.

The Vſe of Skirrets.

The rootes being boyled, peeled and pithed, are ſtewed with butter, pepper and ſalt, and ſo eaten ; or as others vſe them, to roule them in flower, and fry them with butter, after they haue beene boyled, peeled and pithed: each way, or any way that men pleaſe to vſe them, they may finde their taſte to be very pleaſant, far beyond any Parſnep, as all agree that taſte them.

Some doe vſe alſo to eate them as a ſallet, colde with vinegar, oyle, &c. being firſt boyled and dreſſed as before ſaid. They doe helpe to prouoke vrine, and as is thought, to procure bodily luſt, in that they are a little windy.

CHAP. XXXIX.

Paſtinaca ſatiua latifolia. Parſneps.

THe common garden Parſnep hath diuers large winged leaues lying vpon the ground, that is, many leaues ſet one by another on both ſides of a middle ſtalk, ſomewhat like as the Skirret hath, but much larger, and cloſer ſet: the ſtalke riſeth vp great and tall, fiue or ſix foot high ſomtimes, with many ſuch leaues thereon at ſeuerall ioynts ; the top whereof is ſpread into diuers branches, whereon ſtand ſpoakie rundles of yellow flowers, which turne into browniſh flat ſeede : the root is long, great and white, very pleaſant to bee eaten, and the more pleaſant if it grow in a fat ſandy ſoyle.

There is another ſort of garden Parſnep, called the Pine Parſnep, that is not common in euery Garden, and differeth from the former in three notable parts. The root is not ſo long, but thicker at the head and ſmaller below ; the ſtalke is neither ſo bigge,

nor

1 *Sisarum.* Skirrets. 2 *Pastinaca latisolia.* Parsneps. 3 *Pastinaca tenuisolia.* Carrets. 4 *Rapum.* Turneps. 5 *Napus sativus.* Navewes.
6 *Raphanus niger.* Blacke Raddish. 7 *Raphanus vulgaris.* Common Raddish.

nor so high ; and the seede is smaller : yet as Iohn Tradescante saith (who hath giuen me the relation of this, and many other of these garden plants, to whom euery one is a debtor) the roote hereof is not altogether so pleasant as the other.

Moreouer the wilde kinde, which groweth in many places of England(and wherof in some places there might be gathered a quarter sacke full of the seede) if it be so ven in Gardens, and there well ordered, will proue as good as the former kinde of Garden Parsneps.

The Vse of Parsneps.

The Parsnep root is a great nourisher, and is much more vsed in the time of Lent, being boyled and stewed with butter, then in any other time of the yeare ; yet it is very good all the winter long. The seede helpeth to dissolue winde, and to prouoke vrine.

CHAP. XL.

Pastina satiua tenuifolia. Carrots.

THe Carrot hath many winged leaues, rising from the head of the roote, which are much cut and diuided into many other leaues, and they also cut and diuided into many parts, of a deepe greene colour, some whereof in Autumne will turne to be of a fine red or purple (the beautie whereof allureth many Gentlewomen oftentimes to gather the leaues, and sticke them in their hats or heads, or pin them on their armes in stead of feathers) : the stalke riseth vp among the leaues, bearing many likewise vpon it, but nothing so high as the Parsnep, being about three foote high, bearing many spoakie tufts of white flowers, which turne into small rough seede, as if it were hairy, smelling reasonable well if it bee rubbed : the roote is round and long, thicke aboue and small below, eyther red or yellow, eyther shorter or longer, according to his kinde ; for there is one kinde, whose roote is wholly red quite throughout ; another whose roote is red without for a pretty way inward, but the middle is yellow.

Then there is the yellow, which is of two sorts, both long and short : One of the long yellow sorts, which is of a pale yellow, hath the greatest and longest roote, and likewise the greatest head of greene, and is for the most part the worst, being spongy, and not firme.

The other is of a deepe gold yellow colour, and is the best, hauing a smaller head, or tuft of greene leaues vpon it.

The shorte rootes are likewise distinguished, into pale and deepe yellow colours.

The Vse of Carrots.

All these sorts being boyled in the broth of beefe, eyther fresh or salt, but more vsually of salted beefe, are eaten with great pleasure, because of the sweetenesse of them : but they nourish lesse then Parsneps or Skirrets.

I haue not often knowne the seede of this Garden kinde to bee vsed in Physicke : but the wilde kinde is often and much vsed to expell winde, &c.

CHAP. XLI.

Rapum hortense. Turneps.

THere are diuers sorts of Turneps, as white, yellow, and red : the white are the most common, and they are of two kinds, the one much sweeter then the other. The yellow and the red are more rare, and noursed vp only by those that are curious : as also the Navewe, which is seene but with very few.

The

The ordinary Garden Turnep hath many large, and long rough greene leaues, with deepe and vneuen gaſhes on both ſides of them : the ſtalke riſeth vp among the leaues about two foote high, ſpread at the toppe into many branches, bearing theron yellow flowers, which turne into long pods, with blackiſh round ſeede in them : the roote is round and white, ſome greater, ſome ſmaller; the beſt kinde is knowne to be flat, with a ſmall pigges tale-like roote vnderneath it; the worſer kinde which is more common in many places of this land, both North and Weſt, is round, and not flat, with a greater pigges tayle-like roote vnderneath.

The yellow kinde doth often grow very great, it is hardly diſcerned from the ordinary kinde while it groweth, but by the greatneſſe and ſpreading of the leaues beeing boyled, the roote changeth more yellow, ſomewhat neare the colour of a Carrot.

The red Turnep groweth vſually greater then any of the other, eſpecially in a good ground, being of a faire red colour on the outſide, but being pared, as white as any other on the inſide. This, as Matthiolus ſaith, doth grow in the Countrey of Anania, where hee hath ſeene an infinite number of them that haue waighed fifty pound a peece, and in ſome places hee ſaith, a hundred pound a peece, both which we would thinke to be incredible, but that we ſee the kind is greatly giuen to grow, and in warme Countries they may ſo thriue, that the bulke or bigneſſe of the roote may ſo farre paſſe the growth of our Countrey, as that it may riſe to that quantity aboue ſpecified.

The Navew gentle is of two kindes, a ſmaller and a greater ; the ſmaller is vſually called in France, *Naveau de Cane*, the roote is ſomewhat long with the roundneſſe; this kinde is twice as bigge as a mans thumbe, and many of them leſſe : The other is long and great, almoſt as big as the ſhort Carrot, but for the moſt part of an vneuen length, and roundneſſe vnto the very end, where it ſpreadeth into diuers ſmall long fibres: neyther of them doth differ much from the Turnep, in leafe, flower or ſeed.

The Vſe of Turneps.

Being boyled in ſalt broth, they all of them eate moſt kindly, and by reaſon of their ſweetneſſe are much eſteemed, and often ſeene as a diſh at good mens tables : but the greater quantitie of them are ſpent at poore mens feaſts. They nouriſh much, and engender moiſt and looſe fleſh, and are very windy. The ſeede of the Navew gentle is (as I take it) called of Andromachus in the compoſition of his Treakle, *Bunias dulcis* : for Dioſcorides and Plinie doe both ſay, that the ſeede of the tame Bunias or Napus is put into Antidotes, and not the ſeede of the wilde, which is more ſharpe and bitter; neyther the ſeede of the Turnep, which is called in Greeke γογυλη, in Latine *Rapum*, becauſe the ſeede is not ſweete.

Chap. XLII.

Raphanus. Raddiſh.

THere are two principall kindes of Garden Raddiſh, the one is blackiſh on the outſide, and the other white ; and of both theſe there is ſome diuiſion againe, as ſhall be ſhewed. Dittander and horſe Raddiſh be reckoned kinds thereof.

The ordinary Raddiſh hath long leaues, vneuenly gaſhed on both ſides, the ſtalke riſeth vp to the height of three or foure foote, bearing many purpliſh flowers at the top, made of foure leaues a peece, which turne into thicke and ſhort pods, wherein are contained round ſeede, greater then Turnep or Coleworte ſeede, and of a pale reddiſh colour : the roote is long, white, and of a reddiſh purple colour on the outſide toward the toppe of it, and of a ſharpe biting taſte.

There is a ſmall kind of Raddiſh that commeth earlier then the former, that we haue had out of the low Countries, not differing in any thing elſe.

The blacke Raddiſh I haue had brought me out of the lowe Countries, where they ſell them in ſome places by the pound, and is accounted with them a rare winter ſallet:

the

the roote of the beſt kinde is blackiſh on the outſide (and yet the ſeede gathered from ſuch an one, hath after the ſowing againe, giuen rootes, whereof ſome haue beene blacke, but the moſt part white on the outſide) and white within, great and round at the head, almoſt like a Turnep, but ending ſhorter then a Raddiſh, and longer then a Turnep, almoſt peare-faſhion, of a firmer and harder ſubſtance then the ordinary Raddiſh, but no leſſe ſharpe and biting, and ſomewhat ſtrong withall; the leaues are ſomewhat ſmaller, and with deeper gaſhes, the flower and ſeede are like the former, but ſmaller.

Another ſort of blacke Raddiſh is like in leafe and ſeede to the former, but the flower is of a lighter purple colour : the roote is longer and ſmaller, and changeth alſo to bee white as the former doth, ſo that I thinke they haue both riſen from one kinde.

The Horſe Raddiſh is a kinde of wilde Raddiſh, but brought into Gardens for the vſe of it, and hath great large and long greene leaues, which are not ſo much diuided, but dented about the edges : the roote is long and great, much ſtronger in taſte then the former, and abideth diuers yeares, ſpreading with branches vnder ground.

Dittander is likewiſe a wilde kinde hereof, hauing long pointed blewiſh greene leaues, and a roote that creepeth much vnder ground : I confeſſe this might haue bin placed among the herbes, becauſe the leaues and not the rootes are vſed; but let it paſſe now with the kindes of Raddiſh.

The Vſe of theſe Raddiſhes.

Raddiſhes doe ſerue vſually as a *ſtimulum* before meat, giuing an appetite thereunto; the poore eate them alone with bread and ſalt. Some that are early ſowen, are eaten in Aprill, or ſooner if the ſeaſon permit ; others come later ; and ſome are ſowen late to ſerue for the end of Summer : but (as of all things elſe) the earlier are the more accepted.

The blacke Raddiſhes are moſt vſed in the winter, (yet ſome in their naturall and not forc'd grounds, haue their rootes good moſt part of the Summer) and therefore muſt bee ſowen after Midſomer ; for if they ſhould bee ſowen earlier, they would preſently runne vp to ſtalke and ſeed, and ſo loſe the benefit of the roote. The Phyſicall propertie is, it is often vſed in medicines that helpe to breake the ſtone, and to auoyde grauell.

The Horſe Raddiſh is vſed Phyſically, very much in Melancholicke, Spleneticke and Scorbuticke diſeaſes. And ſome vſe to make a kinde of Muſtard with the rootes, and eate it with fiſh.

Dittander or Pepperworte is vſed of ſome cold churliſh ſtomackes, as a ſawce or ſallet ſometimes to their meate, but it is too hot, bitter and ſtrong for weake and tender ſtomackes.

Our Gardiners about London vſe great fences of reede tyed together, which ſeemeth to bee a mat ſet vpright, and is as good as a wall to defend the cold from thoſe things that would be defended, and to bring them forwards the earlier.

Chap. XLIII.

Cepæ. Onions.

WEe haue diuers ſorts of Onions, both white and red, flat, round and long, as ſhall be preſently ſhewed : but I will doe with theſe as I doe with the reſt, only giue you one deſcription for them all, and afterwards their ſeuerall names and varieties, as they are to be known by.

Our common Garden Onion hath diuers long greene hollow leaues, ſeeming halfe flat ; among which riſeth vp a great round hollow ſtalke, bigger in the middle then any where elſe, at the toppe whereof ſtandeth a cloſe round head, couered at the firſt with a thin ſkinne, which breaketh when the head is growne, and ſheweth forth a great vmbell

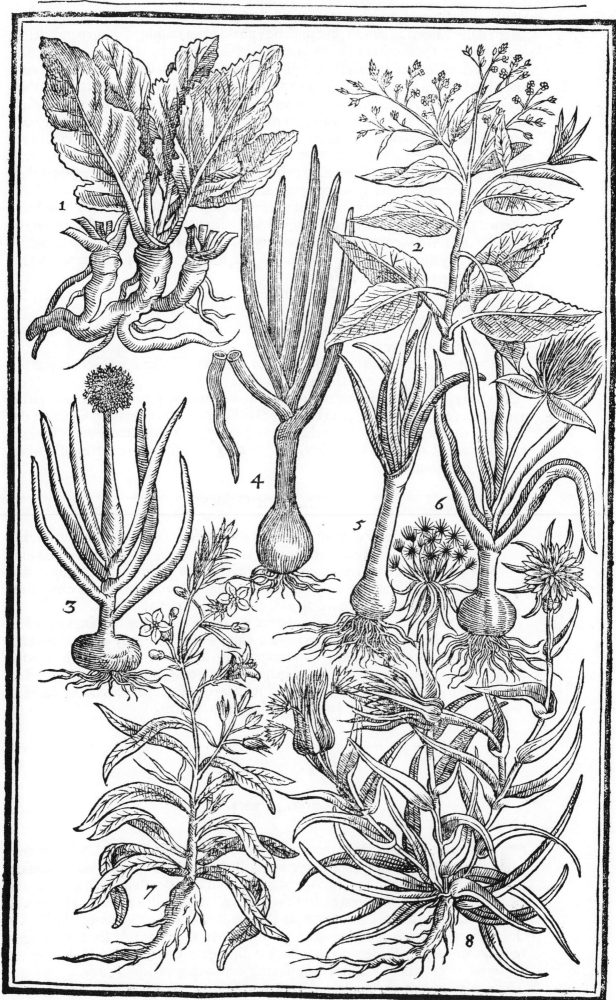

1 *Raphanus rusticanus.*)Horse Raddish. 2 *Lepidium siue Piperitis.* Dittander. 3 *Cepæ rotundæ.* Round Onions. 4 *Cepæ longa.* Long Onions. 5 *Porrum.* Leekes 6 *Allium.* Garlicke. 7 *Rapunculus.* Rampions. 8 *Tragopogon* Goatesbeard.

bell of white flowers, which turne into blacke feede : but then the head is fo heauie that the ftalke cannot fuftaine it, but muft be vpheld from falling to the ground, left it rot and perifh : the roote as all know is round, in fome greater, in others leffer, or flat, in fome red on the outfide only, in others quite thorough out, in fome white, and very fharpe and ftrong, in others milder, and more pleafant, and fome fo pleafant that they may be eaten as an Apple : All thefe kindes of Onions, contrary to the nature of all other bulbous rootes, haue no off-fet, or other roote growing to it, but are euery one alone fingle by themfelues ; and therefore it feemeth, the Latines, as Columella recordeth, haue giuen it the name *Vnio,* and the French it fhould feeme following the Latine, and the Englifh the French, do call it *Oignon* and *Onion,* as an vnite, or as if they were but one and one, and dye euery yeare after feed bearing.

The red flat kinde is moft vfually with vs the ftrongeft of them all, yet I haue had a great red Onion brought mee from beyond Sea, that was as great almoft as two mens fiftes, flat and red quite thoroughout, and very pleafant both to fmell vnto, and to eate, but did quickly degenerate; fo that we plainly fee, that the foyle and climate doth giue great alteration to plants of all forts.

The long kinde wee call St. Omers Onions, and corruptly among the vulgar, St. Thomas Onions.

The other red kinde we call Strasborough Onions, whofe outfide onely is red, and are very fharpe and fierce.

The white Onions both long and flat, are like vnto Chalke-ftones lying vpon the ground, when they are ripe and fit to be gathered.

And laftly, there is the Spanifh Onion, both long and flat, very fweete, and eaten by many like an apple, but as Iohn Tradefcante faith, who hath beene in Spaine, that the Spaniards themfelues doe not eate them fo familiarly, as they doe thofe white Onions that come out of our owne Countrey, which they haue there more plentifully then their fweete Onions.

The Vfe of Onions.

Onions are vfed many wayes, as fliced and put into pottage, or boyled and peeled and layde in difhes for fallets at fupper, or fliced and put into water, for a fawce for mutton or oyfters, or into meate roafted being ftuffed with Parfly, and fo many waies that I cannot recount them, euery one pleafing themfelues, according to their order, manner or delight.

The iuice of Onions is much vfed to be applyed to any burnings with fire, or with Gun-pouder, or to any fcaldings with water or oyle, and is moft familiar for the Country, where vpon fuch fudden occafions they haue not a more fit or fpeedy remedie at hand : The ftrong fmell of Onions, and fo alfo of Garlicke and Leekes, is quite taken away from offending the head or eyes, by the eating of Parfley leaues after them.

Chap. XLIIII.

Porrum. Leekes.

THere be likewife fundry forts of Leekes, both great and fmall. Leekes are very like vnto Onions, hauing long green hollow-like leaues, flattifh on the one fide, and with a ridge or creft on the backe fide : if they bee fuffered to grow vncut, then in the fecond or third yeare after the fowing, they will fend forth a round and flender ftalke, euen quite thoroughout, and not fwollen or bigger in the middle like the Onion, bearing at the toppe a head of purplifh flowers, and blacke feede after them, very like vnto Onion feede, that it is hard to diftinguifh them : the root is long and white, with a great bufh of fibres hanging at it, which they call the beards.

The vnfet Leeke hath longer and flenderer roots then the other, which being tranfplanted, groweth thicker and greater.

The

The French Leeke, which is called the Vine Leeke, is the beſt of all others.
Our common kinde is of two ſorts, one greater then another.
Another ſort encreaſeth altogether by the roote, as Garlicke doth.
And then Ciues, which are the ſmalleſt, and encreaſe aboundantly only by the root.
Some doe account Scalions to be rather a kinde of Onions then Leekes, and call them *Cepa Aſcalonica*, or *Aſcalonitides*, which will quickly ſpend it ſelfe, if it be ſuffered to be vncut; but all Authors affirme, that there is no wilde kinde of Onion, vnleſſe they would haue it to be *Gethyam*, whereof Theophraſtus maketh mention, ſaying, that it hath a long necke (and ſo theſe Scalions haue) and was alſo of ſome called *Gethyllides*, which antiquity accounted to be dedicated to Latona, the mother of Apollo, becauſe when ſhe was bigge with childe of Apollo, ſhe longed for theſe Leekes.

The Vſe of Leekes.

The old World, as wee finde in Scripture, in the time of the children of Iſraels being in Egypt, and no doubt long before, fed much vpon Leekes, Onions, and Garlicke boyled with fleſh; and the antiquity of the Gentiles relate the ſame manner of feeding on them, to be in all Countries the like, which howſoeuer our dainty age now refuſeth wholly, in all ſorts except the pooreſt; yet Muſcouia and Ruſſia vſe them, and the Turkes to this day, (as Bellonius writeth) obſerue to haue them among their diſhes at their tables, yea although they be *Baſhas*, *Cades*, or *Vaiuodas*, that is to ſay, Lords, Iudges, or Gouernours of countries and places. They are vſed with vs alſo ſometimes in Lent to make pottage, and is a great and generall feeding in Wales with the vulgar Gentlemen.
Onions boyled or roſted vnder the embers, and mixed with ſugar and butter, are good for thoſe that are troubled with coughes, ſhortneſſe of breath, and wheeſing. An Onion made hollow at the bottome, and ſome good Treakle put into it, with a little iuyce of Citrons (or Lemons in the ſtead thereof) being well baked together vnder the embers, after the hole is ſtopped againe, and then ſtrained forth, and giuen to one that hath the plague, is very helpefull, ſo as hee be laid to ſweate vpon it.
Ciues are vſed as well to be ſhred among other herbes for the pot, as to be put into a Sallet among other herbs, to giue it a quicker relliſh.
Leekes are held to free the cheſt and lungs from much corruption and rotten flegme, that ſticketh faſt therein, and hard to be auoided, as alſo for them that through hoarſeneſſe haue loſt their voice, if they be eyther taken rawe, or boyled with broth of barley, or ſome ſuch other ſupping, fit and conducing thereunto. And baked vnder hot embers is a remedy againſt a ſurfeit of Muſhromes.
The greene blades of Leekes being boyled and applyed warme to the *Hemorrhoides* or piles, when they are ſwolne and painfull, giue a great deale of eaſe.

CHAP. XLV.

Allium. Garlicke.

I Haue ſpoken of diuers ſorts of Garlicke called Moly, in the former booke: I ſhall neede in this place to ſhew onely thoſe kindes, that this Garden nourſeth vp, and leaue the reſt to his fit time and place.
Garlicke hath many long greene leaues, like vnto Onions, but much larger, and not hollow at all as Onions are: the ſtalke riſeth vp to be about three foote high, bearing ſuch a head at the toppe thereof as Onions and Leekes doe, with purpliſh flowers, and blacke ſeede like Leekes: the roote is white within, couered ouer with many pur-

plish skins, and is diuided into many parts or cloues, which ferue both to fet againe for increafe, and alfo to vfe as neede fhall require, and is of a very ftrong fmell and tafte, as euery one knoweth, paffing either Onions or Leekes, but exceeding wholfome withall for them that can take it.

Allium Vrfinum. Ramfons.

Ramfons are another kinde of Garlicke, and hath two or three faire broad leaues, of a frefh or light greene colour, pointed at the end : the ftalke groweth about an hand length high, bearing many fmall and pure white ftarre-like flowers at the toppe, and afterwards fmall, blacke, and fmooth round feede : the roote is alfo diuided into many parts, whereby it is much encreafed, and is much milder then the former, both in fmell and tafte.

The Vfe of Garlicke.

It being well boyled in falt broth, is often eaten of them that haue ftrong ftomackes, but will not brooke in a weake and tender ftomacke.

It is accounted, and fo called in diuers Countries, The poore mans Treakle, that is, a remedy for all difeafes. It is neuer eaten rawe of any man that I know, as other of the rootes aforefaid, but fodden alwaies and fo taken.

Ramfons are oftentimes eaten with bread and butter, and otherwife alfo, as euery mans affection and courfe of life leadeth him to vfe.

Chap. XLVI.

Rapunculus fiue Rapuntium. Rampions.

Garden Rampions are of two forts, the one greater, the other leffer : the leaues of Rampions are in the one fomewhat broad like a Beete, in the other fomewhat long and narrow, and a little broader at the end, of a light greene colour, lying flat vpon the ground all the firft winter, or yeare of the fpringing, and the next Spring fhooteth forth ftalkes two or three foote high, bearing at the toppe, in the bigger fort, a long flender fpike of fmall horned or crooked flowers, which open their brimmes into foure leaues ; in the leffer many fmall purplifh bels, ftanding vpon feuerall fmall foote-ftalkes, which turne into heads, bearing fmall blackifh feede : the root is white, branched into two or three rootes, of the bigneffe and length of a mans finger or thumbe.

The Vfe of Rampions.

The rootes of both are vfed for Sallets, being boyled, and then eaten with oyle and vinegar, a little falt and pepper.

Chap. XLVII.

Tragopogon. Goates beard.

Goates beard hath many long and narrow leaues, broader at the bottome, and fharper at the end, with a ridge downe the backe of the leafe, and of a pale greene colour ; among which rifeth vp a ftalke of two or three foote high, fmooth and hollow, bearing thereon many fuch like leaues, but fmaller and fhorter, and at the toppe thereof on euery branch a great double yellow flower, like almoft vnto the flower of a Dandelion, which turneth into a head, ftored with doune, and long whitifh feede therein, hauing on the head of euery one fome part of the doune,

and

and is carried away with the winde if it bee neglected : the roote is long and round, somewhat like vnto a Parfnep, but farre fmaller, blackifh on the outfide, and white within, yeelding a milkie iuyce being broken, as all the reft of the plant doth, and of a very good and pleafant tafte. This kinde, as alfo another with narrower leaues, almoft like graffe, growe wilde abroad in many places, but are brought into diuers Gardens. The other two kindes formerly defcribed in the firft part, the one with a purple flower, and the other with an afh-coloured, haue fuch rootes as thefe here defcribed, and may ferue alfo to the fame purpofe, being of equall goodneffe, if any will vfe them in the fame manner ; that is, while they are young, and of the firft yeares fowing, elfe they all growe hard, in running vp to feede.

The Vfe of Goates beard.

If the rootes of any of thefe kindes being young, be boyled and dreffed as a Parfnep, they make a pleafant difh of meate, farre paffing the Parfnep in many mens iudgements, and that with yellow flowers to be the beft.

They are of excellent vfe being in this manner prepared, or after any other fit and conuenient way, to ftrengthen thofe that are macilent, or growing into any confumption.

Chap. XLVIII.

Carum. Carawayes.

CArawayes hath many very fine cut and diuided leaues lying on the ground, being alwaies greene, fomewhat refembling the leaues of Carrots, but thinner, and more finely cut, of a quicke, hot, and fpicie tafte: the ftalke rifeth not much higher then the Carrot ftalke, bearing fome leaues at the ioynts along the ftalke to the toppe, where it brancheth into three or foure parts, bearing fpoakie vmbels of white flowers, which turne into fmall blackifh feede, fmaller then Anifeede, and of a hotter and quicker tafte : the roote is whitifh, like vnto a Parfnep, but much fmaller, more fpreading vnder ground, and a little quicke in tafte, as all the reft of the plant is, and abideth long after it hath giuen feede.

The Vfe of Carawayes.

The rootes of Carawayes being boyled may be eaten as Carrots, and by reafon of the fpicie tafte doth warme and comfort a cold weake ftomacke, helping to diffolue winde (whereas Carrots engender it) and to prouoke vrine, and is a very welcome and delightfull difh to a great many, yet they are fomewhat ftronger in tafte then Parfneps.

The feede is much vfed to bee put among baked fruit, or into bread, cakes, &c. to giue them a rellifh, and to helpe to digeft winde in them are fubiect thereunto.

It is alfo made into Comfits, and put into *Trageas*, or as we call them in Englifh, Dredges, that are taken for the cold and winde in the body, as alfo are ferued to the table with fruit.

CHAP.

Chap. XLIX.

Pappas sine Battatas. Potatoes.

THree sorts of Potatoes are well knowne vnto vs, but the fourth I rest doubtfull of, and dare not affirme it vpon such termes as are giuen vnto it, vntill I may be better informed by mine owne sight.

The Spanish kinde hath (in the Islands where they growe, either naturally, or planted for increase, profit, and vse of the Spaniards that nourse them) many firme and verie sweete rootes, like in shape and forme vnto Asphodill rootes, but much greater and longer, of a pale browne on the outside, and white within, set together at one head; from whence rise vp many long branches, which by reason of their weight and weaknesse, cannot stand of themselues, but traile on the ground a yard and a halfe in length at the least (I relate it, as it hath growne with vs, but in what other forme, for flower or fruit, we know not) whereon are set at seuerall distances, broad and in a manner three square leaues, somewhat like triangled Iuie leaues, of a darke greene colour, the two sides whereof are broad and round, and the middle pointed at the end, standing reasonable close together : thus much we haue seene growe with vs, and no more : the roote rather decaying then increasing in our country.

The Potatoes of Virginia, which some foolishly call the Apples of youth, is another kinde of plant, differing much from the former, sauing in the colour and taste of the roote, hauing many weake and somewhat flexible branches, leaning a little downwards, or easily borne downe with the winde or other thing, beset with many winged leaues, of a darke grayish greene colour, whereof diuers are smaller, and some greater then others : the flowers growe many together vpon a long stalke, comming forth from betweene the leaues and the great stalkes, euery one seuerally vpon a short footstalke, somewhat like the flower of Tabacco for the forme, being one whole leafe six cornered at the brimmes, but somewhat larger, and of a pale blewish purple colour, or pale doue colour, and in some almost white, with some red threads in the middle, standing about a thicke gold yellow pointell, tipped with greene at the end : after the flowers are past, there come vp in their places small round fruit, as bigge as a Damson or Bulleis, greene at the first, and somewhat whitish afterwards, with many white seedes therein, like vnto Nightshade : the rootes are rounder and much smaller then the former, and some much greater then others, dispersed vnder ground by many small threads or strings from the rootes, of the same light browne colour on the outside, and white within, as they, and neare of the same taste, but not altogether so pleasant.

The Potatos of Canada, (which hath diuers names giuen it by diuers men, as Bauhinus vpon Matthiolus calleth it, *Solanum tuberosum esculentum,* Pelleterius of Middleborough in his *Plantarum Synonimia, Heliotropium Indicum tuberosum,* Fabius Columna in the second part of his *Phytobasanos, Flos Solis Farnesianus, siue Aster Peruanus tuberosus:* We in England, from some ignorant and idle head, haue called them Artichokes of Ierusalem, only because the roote, being boyled, is in taste like the bottome of an Artichoke head : but they may most fitly be called, Potatos of Canada, because their rootes are in forme, colour and taste, like vnto the Potatos of Virginia, but greater, and the French brought them first from Canada into these parts) riseth vp with diuers stiffe, round stalkes, eight or tenne foote high in our Country, where they haue scarce shewed their flowers, whereas the very head of flowers in other Countries, as Fabius Columna expresseth it, being of a Pyramis or Sugar loafe fashion, broade spreading below, and smaller pointed vpwards towards the toppe, is neere of the same length, whereon are set large and broade rough greene leaues, very like vnto the leaues of the flower of the Sunne, but smaller, yet growing in the very same manner, round about the stalkes : at the very later end of Summer, or the beginning of Autumne, if the roote bee well planted and defended, it will giue a shew of a few small yellow flowers at the top, like vnto the flowers of *Aster* or Starre-worte, and much smaller then any flower of the Sunne, which come to no perfection with vs : the roote, while the plant

1 *Carum.* Carawayes. 2 *Battatas Hispanorum.* Spaniſh Potatoes. 3 *Papas ſeu Battatas Virginianorum.* Virginia Potatoes. 4 *Battatas de Canada.* Potatoes of Canada, or Artichokes of Ieruſalem.

is growing aboue ground, encreaſeth not to his full growth, but when the Summer is well ſpent, and the ſpringing of the ſtalk is paſt, which is about the end of Auguſt, or in September, then the root is perceiued to be encreaſed in the earth, and will before Autumne be ſpent, that is, in October, ſwell like a mound or hillocke, round about the foote of the ſtalkes, and will not haue his rootes fit to be taken vp, vntill the ſtalkes be halfe withered at the ſooneſt ; but after they be withered, and ſo all the winter long vntill the Spring againe, they are good, and fit to bee taken vp and vſed, which are a number of tuberous round rootes, growing cloſe together ; ſo that it hath beene obſerued, that from one roote, being ſet in the Spring, there hath been forty or more taken vp againe, and to haue ouer-filled a pecke meaſure, and are of a pleaſant good taſte as many haue tryed.

The Vſe of all theſe Potato's.

The Spaniſh Potato's are roaſted vnder the embers, and being pared or peeled and ſliced, are put into ſacke with a little ſugar, or without, and is delicate to be eaten.

They are vſed to be baked with Marrow, Sugar, Spice, and other things in Pyes, which are a daintie and coſtly diſh for the table.

The Comfit-makers preſerue them, and candy them as diuers other things, and ſo ordered, is very delicate, fit to accompany ſuch other banquetting diſhes.

The Virginia Potato's being dreſſed after all theſe waies before ſpecified, maketh almoſt as delicate meate as the former.

The Potato's of Canada are by reaſon of their great increaſing, growne to be ſo common here with vs at London, that euen the moſt vulgar begin to deſpiſe them, whereas when they were firſt receiued among vs, they were dainties for a Queene.

Being put into ſeething water they are ſoone boyled tender, which after they bee peeled, ſliced and ſtewed with butter, and a little wine, was a diſh for a Queene, beeing as pleaſant as the bottome of an Artichoke : but the too frequent vſe, eſpecially being ſo plentifull and cheape, hath rather bred a loathing then a liking of them.

Chap. L.

Cinara. Artichokes.

THe fruits that grow vpon or neere the ground, are next to be entreated of, and firſt of Artichokes, whereof there be diuers kindes, ſome accounted tame and of the Garden, others wilde and of late planted in Gardens, Orchards or Fieldes, of purpoſe to be meate for men.

The Artichoke hath diuers great, large, and long hollowed leaues, much cut in or torne on both edges, without any great ſhew of prickles on them, of a kinde of whitiſh greene, like vnto an aſh colour, whereof it tooke the Latine name *Cinara* : the ſtalke is ſtrong, thicke and round, with ſome skins as it were downe all the length of them, bearing at the toppe one ſcaly head, made at the firſt like a Pine-apple, but after growing greater, the ſcales are more ſeparate, yet in the beſt kindes lying cloſe, and not ſtaring, as ſome other kindes doe, which are eyther of a reddiſh browne, whitiſh, or greeniſh colour, and in ſome broade at the ends, in others ſharpe or prickly : after the head hath ſtood a great while, if it bee ſuffered, and the Summer proue hot and kindly, in ſome there will breake forth at the toppe thereof, a tuft of blewiſh purple thrumes or threds, vnder which grow the ſeede, wrapped in a great deale of dounie ſubſtance : but that roote that yeeldeth flowers will hardly abide the next winter ; but elſe being cut off when it is well growne, that dounie matter abideth cloſe in the middle of the head, hauing the bottome thereof flat and round, which is that matter or ſubſtance that is vſed to be eaten : the roote ſpreadeth it ſelfe in the ground reaſonable

ble

1 *Cinara satiua rubra.* The red Artichoke. 2 *Cinara satiua alba.* The white Artichoke. 3 *Cinara patula.* The French Artichoke. 4 *Cinara siluestris.* The Thistle Artichoke. 5 *Carduus esculentus.* The Chardon.

ble well, yeelding diuers heads of leaues or fuckers, whereby it is increafed.

The white Artichoke is in all things like the red, but that the head is of a whitifh afhe colour, like the leaues, whereas the former is reddifh.

We haue alfo another, whofe head is greene, and very fharpe vpwards, and is common in many places.

Wee haue had alfo another kinde in former times that grew as high as any man, and branched into diuers ftalkes, euery one bearing a head thereon, almoft as bigge as the firft.

There is another kinde, called the Muske Artichoke, which groweth like the French kinde, but is much better in fpending, although it haue a leffer bottome.

The French Artichoke hath a white head, the fcales whereof ftand ftaring far afunder one from another at the ends, which are fharpe : this is well known by this qualitie, that while it is hot after it is boyled, it fwelleth fo ftrong, that one would verily thinke it had bin boyled in ftinking water, which was brought ouer after a great frofte that had well nigh confumed our beft kindes, and are now almoft cleane caft out again, none being willing to haue it take vp the roome of better.

There is a lowe kinde that groweth much about Paris, which the French efteeme more then any other, and is lower then the former French kinde, the head whereof as well as the leaues, is of a frefher greene colour, almoft yellowifh.

Then there is the Thiftle Artichoke, which is almoft a wilde kinde, and groweth fmaller, with a more open and prickly head then any of the former.

And laftly, the Chardon as they call it, becaufe it is almoft of the forme and nature of a Thiftle, or wilde Artichoke. This groweth high, and full of fharpe prickles, of a grayifh colour. Iohn Tradefcante affured mee, hee faw three acres of Land about Bruffels planted with this kinde, which the owner whited like Endiue, and then fold them in the winter : Wee cannot yet finde the true manner of dreffing them, that our Countrey may take delight therein.

All thefe kindes are encreafed by flipping the young fhootes from the root, which being replanted in February, March, or Aprill, haue the fame yeare many times, but the next at the moft, borne good heads.

Wee finde by dayly experience, that our Englifh red Artichoke is in our Countrey the moft delicate meate of any of the other, and therefore diuers thinking it to bee a feuerall kinde, haue fent them into Italie, France, and the Lowe Countries, where they haue not abode in their goodneffe aboue two yeare, but that they haue degenerated ; fo that it feemeth, that our foyle and climate hath the preheminence to nourifh vp this plant to his higheft excellencie.

The Vfe of Artichokes.

The manner of preparing them for the Table is well knowne to the youngeft Houfewife I thinke, to bee boyled in faire water, and a little falt, vntill they bee tender, and afterwardes a little vinegar and pepper, put to the butter, poured vpon them for the fawce, and fo are ferued to the Table.

They vfe likewife to take the boyled bottomes to make Pyes, which is a delicate kinde of baked meate.

The Chardon is eaten rawe of diuers, with vinegar and oyle, pepper and falt, all of them, or fome, as euery one liketh for their delight.

CHAP. LI.

Fabæ & Phaſeoli. Garden and French Beanes.

THe Garden Beane is of two colours, red or blacke, and white, yet both riſe from one ; the ſmall or fielde Beanes I make no mention of in this place ; but the French or Kidney Beane is almoſt of infinite ſorts and colours : we doe not for all that intend to trouble you in this place, with the knowledge or relation of any more then is fit for a Garden of that nature, that I haue propounded it in the beginning.

Our ordinary Beanes, ſeruing for foode for the poorer ſort for the moſt part, are planted as well in fieldes as in gardens, becauſe the quantity of them that are ſpent taketh vp many acres of land to be planted in, and riſe vp with one, two or three ſtalks, according to the fertilitie of the ſoyle, being ſmooth and ſquare, higher then any man oftentimes, whereon are ſet at certaine diſtances, from the very bottome almoſt to the toppe, two long ſmooth fleſhy and thicke leaues almoſt round, one ſtanding by another at the end of a ſmall footeſtalke : betweene theſe leaues and the ſtalke, come forth diuers flowers, all of them looking one way for the moſt part, which are cloſe a little turned vp at the brimmes, white and ſpotted with a blackiſh ſpot in the middle of them, and ſomwhat purpliſh at the foot or bottome, of the forme almoſt of Broome or Peaſe flowers, many of which that grow vpward toward the toppe, doe ſeldome beare fruit, and therefore are gathered to diſtill, and the toppes of the ſtalkes cut off, to cauſe the reſt to thriue the better ; after which grow vp long great ſmooth greene pods, greater then in any other kinde of Pulſe, which grow blacke when they are ripe, and containe within them two, three or foure Beanes, which are ſomewhat flat and round, eyther white or reddiſh, which being full ripe grow blackiſh : the roote hath diuers fibres annexed vnto the maine roote, which dyeth euery yeare.

The French or Kidney Beane riſeth vp at the firſt but with one ſtalke, which afterwards diuideth it ſelfe into many armes or branches, euery one of them being ſo weak, that without they be ſuſtained with ſtickes or poles, whereon with their winding and claſpers they take hold, they would lye fruitleſſe vpon the ground : vpon theſe branches grow forth at ſeuerall places long footeſtalkes, with euery of them three broade, round and pointed greene leaues at the end of them, towards the tops whereof come forth diuers flowers, made like vnto Peaſe bloſſomes, of the ſame colour for the moſt part that the fruit will be or, that is to ſay, eyther white, or yellow, or red, or blackiſh, or of a deepe purple &c. but white is moſt vſuall for our Garden ; after which come long and ſlender flat pods, ſome crooked, and ſome ſtraight, with a ſtring as it were running downe the backe thereof, wherein are contained flattiſh round fruit, made to the faſhion of a kidney : the roote is long, and ſpreadeth with many fibres annexed vnto it, periſhing euery yeare.

The Vſe of theſe Beanes.

The Garden Beanes ſerue (as I ſaid before) more for the vſe of the poore then of the rich : I ſhall therefore only ſhew you the order the poore take with them, and leaue curioſity to them that will beſtow time vpon them. They are only boyled in faire water and a little ſalt, and afterwards ſtewed with ſome butter, a little vinegar and pepper being put vnto them, and ſo eaten : or elſe eaten alone after they are boyled without any other ſawce. The water of the bloſſomes diſtilled, is vſed to take away ſpots, and to cleer the skin. The water of the greene huskes or cods is good for the ſtone.

The Kidney Beanes boyled in water huske and all, onely the ends cut off, and the ſtring taken away, and ſtewed with butter &c. are eſteemed more ſauory meate to many mens pallates, then the former, and are a diſh more oftentimes at rich mens Tables then at the poore.

Chap. LII.

Pisum. Pease.

THere is a very great variety of manured Pease known to vs, and I think more in our Country then in others, whereof some prosper better in one ground and country, and some in others : I shall giue you the description of one alone for all the rest, and recite vnto you the names of the rest.

Garden Pease are for the most part the greatest and sweetest kinds, and are sustained with stakes or bushes. The Field Pease are not so vsed, but growe without any such adoe. They spring vp with long, weake, hollow, and brittle (while they are young and greene) whitish greene stalkes, branched into diuers parts, and at euery ioynt where it parteth one broad round leafe compassing the stalke about, so that it commeth as it were thorough it : the leaues are winged, made of diuers small leaues set to a middle ribbe, of a whitish greene colour, with claspers at the ends of the leaues, whereby it taketh hold of whatsoeuer standeth next vnto it : betweene the leaues and the stalkes come forth the flowers, standing two or three together, euery one by it selfe on his owne seuerall stalke, which are either wholly white, or purple, or mixed white and purple, or purple and blew : the fruit are long, and somewhat round cods, whereof some are greater, others lesser, some thicke and short, some plaine and smooth, others a little crooked at the ends ; wherein also are contained diuers formes of fruit or pease ; some being round, others cornered, some small, some great, some white, others gray, and some spotted : the roote is small, and quickly perisheth.

The kindes of Pease are these :

The Rounciuall.	The gray Pease.
The greene Hasting.	The white Hasting.
The Sugar Pease.	The Pease without skins.
The spotted Pease.	

The Scottish or tufted Pease, which some call the Rose Pease, is a good white Pease fit to be eaten.

The early or French Pease, which some call Fulham Pease, because those grounds thereabouts doe bring them soonest forward for any quantity, although sometimes they miscarry by their haste and earlinesse.

Cicer Arietinum. Rams Ciches.

This is a kinde of Pulse, so much vsed in Spaine, that it is vsually one of their daintie dishes at all their feasts : They are of two sorts, white and red ; the white is onely vsed for meate, the other for medicine. It beareth many vpright branches with winged leaues, many set together, being small, almost round, and dented about the edges : the flowers are either white or purple, according to the colour of the Pease which follow, and are somewhat round at the head, but cornered and pointed at the end, one or two at the most in a small roundish cod.

The Vse of Pease.

Pease of all or the most of these sorts, are either vsed when they are greene, and be a dish of meate for the table of the rich as well as the poore, yet euery one obseruing his time, and the kinde : the fairest, sweetest, youngest, and earliest for the better sort, the later and meaner kindes for the meaner, who doe not giue the deerest price : Or

Being dry, they serue to boyle into a kinde of broth or pottage, wherein many doe put Tyme, Mints, Sauory, or some other such hot herbes, to giue it the better rellish, and is much vsed in Towne and Countrey in the Lent time,

1 *Faba ſatina.* Garden Beanes. 2 *Phaſeoli ſatiui.* French Beanes. 3 *Piſum vulga.e.* Garden Peaſe. 4 *Piſum vmbellatum ſiue Roſeum.* Roſe Peaſe or Scottiſh Peaſe. 5 *Piſum Saccharatum.* Sugar Peaſe. 6 *Piſum maculatum.* Spotted Peaſe. 7 *Cicer Arietinum.* Rams Ciches or Cicers.

time, especially of the poorer sort of people.

It is much vsed likewise at Sea for them that goe long voyages, and is for change, because it is fresh, a welcome diet to most persons therein.

The Rams Ciches the Spaniards call *Grauancos*, and *Garauancillos*, and eate them boyled and stewed as the most dainty kinde of Pease that are, they are of a very good rellish, and doe nourish much; but yet are not without that windy quality that all sorts of Pulse are subiect vnto : they increase bodily lust much more then any other sorts, and as it is thought, doth helpe to encrease seede.

CHAP. LIII.

Cucumer. The Cowcumber.

OF Cowcumbers there are diuers sorts, differing chiefly in the forme and colour of the fruit, and not in the forme of the plant; therefore one description shall serue in stead of all the rest.

The Cowcumber bringeth forth many trailing rough greene branches lying on the ground, all along whereof growe seuerall leaues, which are rough, broad, vneuen at the edges, and pointed at the ends, with long crooked tendrels comming forth at the same ioynt with the leafe, but on the other side therof : between the stalks & the leaues at the ioynts come forth the flowers senerally, euery one standing on a short foot-stalke, opening it selfe into fiue leaues, of a yellowish colour, at the bottome whereof groweth the fruit, long and greene at the first, but when it is thorough ripe, a little yellowish, hauing many furrowes, and vneuen bunches all the length of it, wherein is a white firme substance next vnto the skin, and a cleare pulpe or watery substance, with white flat seede lying dispersed through it : the roote is long and white, with diuers fibres at it.

The kindes.

The first described is called, The long greene Cowcumber.

There is another is called, The short Cowcumber, being short, and of an equall bignesse in the body thereof, and of an vnequall bignesse at both ends.

The long Yellow, which is yellowish from the beginning, and more yellow when it is ripe, and hath beene measured to be thirteene inches long : but this is not that small long Cowcumber, called of the Latines, *Cucumis anguinus*.

Another kinde is early ripe, called The French kinde.

The Dantsicke kinde beareth but small fruit, growing on short branches or runners : the pickled Cowcumbers that are vsually sold are of this kind.

The Muscouie kinde is the smallest of all other, yet knowne, and beareth not aboue foure or fiue at the most on a roote, which are no bigger then small Lemons.

The Vse of Cowcumbers.

Some vse to cast a little salt on their sliced Cowcumbers, and let them stand halfe an houre or more in a dish, and then poure away the water that commeth from them by the salt, and after put vinegar, oyle, &c. thereon, as euery one liketh : this is done, to take away the ouermuch waterishnesse and coldnesse of the Cowcumbers.

In many countries they vse to eate Cowcumbers as wee doe Apples or Peares, paring and giuing slices of them, as we would to our friends of some dainty Apple or Peare.

The pickled Cowcumbers that come from beyond Sea, are much vsed with vs for fawce to meate all the Winter long. Some haue ſtriuen to equall them, by pickling vp our Cowcumbers at the later end of the yeare, when they are cheapeſt, taking the little ones and ſcalding them thoroughly well, which after they put in brine, with ſome Dill or Fenell leaues and ſtalkes : but theſe are nothing comparable to the former, wee either miſſing of the right and orderly pickling of them, or the kinde it ſelfe differing much from ours (as I ſaid of the Dantſicke kinde) for ours are neither ſo tender and firme, nor ſo ſauoury as the other.

The rawe or greene Cowcumbers are fitteſt for the hotter time of the yeare, and for hot ſtomackes, and not to be vſed in colder weather or cold ſtomackes, by reaſon of the coldneſſe, whereby many haue been ouertaken.

The ſeede is vſed phyſically in many medicines that ſerue to coole, and a little to make the paſſages of vrine ſlippery, and to giue eaſe to hot diſeaſes.

Chap. LIIII.

Melo. Milions or Muske Melons.

THere bee diuers ſorts of Melons found out at this day, differing much in the goodneſſe of taſte one from another. This Countrey hath not had vntill of late yeares the skill to nourſe them vp kindly, but now there are many that are ſo well experienced therein, and haue their ground ſo well prepared, as that they will not miſſe any yeare, if it be not too extreme vnkindly, to haue many ripe ones in a reaſonable time : yet ſome will be later then others alwayes.

The Melon is certainly a kinde of Cowcumber, it doth ſo neare reſemble it, both in the manner of his growing, hauing rough trailing branches, rough vneuen leaues, and yellow flowers : after which come the fruit, which is rounder, thicker, bigger, more rugged, and ſpotted on the outſide then the Cowcumber, of a ruſſet colour, and greene vnderneath, which when it groweth full ripe, will change a little yellowiſh, being as deepe furrowed and ribbed as they, and beſides hauing chaps or rifts in diuers places of the rinde : the inward hard ſubſtance is yellow, which onely is eaten : the ſeede which is bigger, and a little yellower then the Cowcumber, lying in the middle onely among the moiſter pulpe : the ſmell and changing of his colour, fore-ſhew their ripeneſſe to them that are experienced : the roote is long, with many fibres at it. The fruit requireth much watering in the hot time of the day, to cauſe them to ripen the ſooner, as I haue obſerued by diuers of the beſt skill therein.

The Vſe of the kindes of Melons.

The beſt Melon ſeede doe come to vs out of Spaine, ſome haue come out of Turkie, but they haue been nothing ſo good and kindly.

Some are called Sugar Melons, others Peare Melons, and others Musk Melons.

They haue beene formerly only eaten by great perſonages, becauſe the fruit was not only delicate but rare ; and therfore diuers were brought from France, and ſince were nourſed vp by the Kings or Noblemens Gardiners onely, to ſerue for their Maſters delight : but now diuers others that haue skill and conueniencie of ground for them, doe plant them and make them more common.

They paire away the outer rinde, and cut out the inward pulpe where the ſeede lyeth, ſlice the yellow firme inward rinde or ſubſtance, & ſo eate it with ſalt and pepper (and good ſtore of wine, or elſe it will hardly diſgeſt) for this is firmer, & hath not that moiſture in it that the Cowcumbers haue. It is alſo more delicate, and of more worth, which recompenſeth the paine.

The ſeed of theſe Melons are vſed as Cowcumbers phyſically, and together with them moſt vſually.

CHAP.

Chap. LV.

Pepo. Pompions.

WE haue but one kinde of Pompion (as I take it) in all our Gardens, not-withstanding the diuersities of bignesse and colour.

The Pompion or great Melon (or as some call it Milion) creepeth vp-on the ground (if nothing bee by it whreeon it may take hold and climbe) with very great, ribbed, rough, and prickly branches, whereon are set very large rough leaues, cut in on the edges with deepe gashes, and dented besides, with many claspers also, which winde about euery thing they meete withall : the flowers are great and large, hollow and yellow, diuided at the brims into fiue parts, at the bottome of which, as it is in the rest, groweth the fruit, which is very great, sometimes of the bignesse of a mans body, and oftentimes lesse, in some ribbed or bunched, in others plaine, and ei-ther long or round, either green or yellow, or gray, as Nature listeth to shew her selfe; for it is but waste time, to recite all the formes and colours may be obserued in them : the inner rinde next vnto the outer is yellowish and firme : the seede is great, flat, and white, lying in the middle of the watery pulpe : the roote is of the bignesse of a mans thumbe or greater, dispersed vnder ground with many small fibres ioyned thereunto.

Gourds are kindes of Melons; but because wee haue no vse of them, wee leaue them vnto their fit place.

The Vse of Pompions.

They are boyled in faire water and salt, or in powdered beefe broth, or sometimes in milke, and so eaten, or else buttered. They vse likewise to take out the inner watery substance with the seedes, and fill vp the place with Pippins, and hauing laid on the couer which they cut off from the toppe, to take out the pulpe, they bake them together, and the poore of the Citie, as well as the Country people, doe eate thereof, as of a dainty dish.

The seede hereof, as well as of Cowcumbers and Melons, are cooling, and serue for emulsions in the like manner for Almond milkes, &c. for those are troubled with the stone.

Chap. LVI.

Fragaria. Strawberries.

THere be diuers sorts of Strawberries, whereof those that are noursed vp in Gar-dens or Orchards I intend to giue you the knowledge in this place, and leaue the other to a fitter; yet I must needs shew you of one of the wilde sorts, which for his strangenesse is worthy of this Garden : And I must also enforme you, that the wilde Strawberry that groweth in the Woods is our Garden Strawberry, but bettered by the soyle and transplanting.

The Strawberry hath his leaues closed together at the first springing vp, which af-terwards spread themselues into three diuided parts or leaues, euery one standing vpon a small long foote-stalke, greene on the vpperside, grayish vnderneath, and snipped or dented about the edges; among which rise vp diuers small stalkes, bearing foure or fiue flowers at the tops, consisting of fiue white round pointed leaues, somewhat yel-lowish in the bottome, with some yellow threads therein; after which come the fruit, made of many small graines set together, like vnto a small Mulberry or Raspis, red-dish when it is ripe, and of a pleasant winy taste, wherein is enclosed diuers small blac-kish seede : the roote is reddish and long, with diuers small threads at it, and sendeth
forth

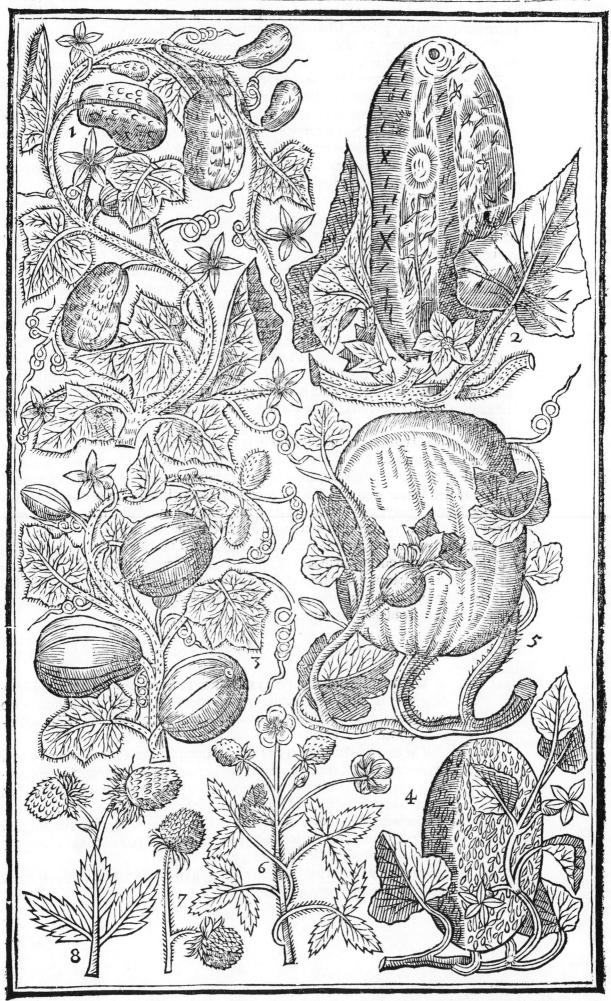

1 *Cucumis vulgaris vulgaris.* The ordinary Cowcumber. 2 *Cucumis Hispanicus.* The long yellow Spanish Cowcumber. 3 *Melo vulgaris.* The ordinary Melon. 4 *Melo maximus optimus.* The greatest Muske Melon. 5 *Pepo.* The Pompion. 6 *Fraga vulgaris.* Common Strawberries. 7 *Fraga Bohemica maxima.* The great Bohemia Strawberries. 8 *Fraga acuseata.* The prickly Strawberry.

forth from the head therof long reddiſh ſtrings running vpon the ground,which ſhoot forth leaues in many places, whereby it is much encreaſed.

The white Strawberry differeth not from the red, but in the colour of the fruite, which is whiter then the former when it is thorough ripe, enclining to redneſſe.

The greene Strawberry likewiſe differeth not, but that the fruit is green on all ſides when it is ripe, ſaue on that ſide the Sun lyeth vpon it,and there it is ſomewhat red.

The Virginia Strawberry carryeth the greateſt leafe of any other, except the Bohemian, but ſcarce can one Strawberry be ſeene ripe among a number of plants; I thinke the reaſon thereof to be the want of ſkill,or induſtry to order it aright. For the Bohemia, and all other Strawberries will not beare kindly, if you ſuffer them to grow with many ſtrings, and therefore they are ſtill cut away.

There is another very like vnto this, that Iohn Tradeſcante brought with him from Bruſſels long agoe, and in ſeuen yeares could neuer ſee one berry ripe on all ſides, but ſtill the better part rotten, although it would euery yeare flower abundantly,and beare very large leaues.

The Bohemia Strawberry hath beene with vs but of late dayes, but is the goodlieſt and greateſt, both for leafe next to the Virginian, and for beauty farre ſurpaſſing all ; for ſome of the berries haue beene meaſured to bee neere fiue inches about. Maſter Queſter the Poſtmaſter firſt brought them ouer into our Country,as I vnderſtand, but I know no man ſo induſtrious in the carefull planting and bringing them to perfection in that plentifull maner,as Maſter Vincent Sion who dwelt on the Banck ſide, neer the old Paris garden ſtaires, who from ſeuen rootes, as hee affirmed to me, in one yeare and a halfe,planted halfe an acree of ground with the increaſe from them,beſides thoſe he gaue away to his friends, and with him I haue ſeene ſuch, and of that bigneſſe before mentioned.

One Strawberry more I promiſed to ſhew you, which although it be a wilde kinde, and of no vſe for meate, yet I would not let this diſcourſe paſſe, without giuing you the knowledge of it. It is in leafe much like vnto the ordinary, but differeth in that the flower, if it haue any, is greene, or rather it beareth a ſmall head of greene leaues, many ſet thicke together like vnto a double ruffe, in the midſt whereof ſtandeth the fruit, which when it is ripe,ſheweth to be ſoft and ſomwhat reddiſh, like vnto a Strawberry,but with many ſmall harmleſſe prickles on them,which may be eaten and chewed in the mouth without any maner of offence, and is ſomewhat pleaſant like a Strawberry : it is no great bearer, but thoſe it doth beare, are ſet at the toppes of the ſtalks cloſe together, pleaſant to behold, and fit for a Gentlewoman to weare on her arme, &c. as a raritie in ſtead of a flower.

The Vſe of Strawberries.

The leaues of Strawberries are alwaies vſed among other herbes in cooling drinkes, as alſo in lotions, and gargles for the mouth and throate : the rootes are ſometimes added to make it the more effectuall,and withall ſomwhat the more binding.

The berries themſelues are often brought to the Table as a reare ſeruice, whereunto claret wine,creame or milke is added with ſugar, as euery one liketh ; as alſo at other times, both with the better and meaner ſort, and are a good cooling and pleaſant diſh in the hot Summer ſeaſon.

The water diſtilled of the berries, is good for the paſſions of the heart, cauſed by the perturbation of the ſpirits, being eyther drunke alone, or in wine ; and maketh the heart merry.

Some doe hold that the water helpeth to clenſe the face from ſpots,and to adde ſome cleereneſſe to the skinne.

CHAP.

Chap. LVII.

Angelica. Garden Angelica.

HAuing thus furnifhed you out a Kitchen Garden with all forts of herbes, roots & fruits fit for it, and for any mans priuate vfe, as I did at the firft appropriate it; let me a little tranfcend, and for the profit & vfe of Country Gentlewomen and others, furnifh them with fome few other herbes, of the moft efpeciall vfe for thofe fhall need them, to be planted at hand in their Gardens, to fpend as occafion fhall ferue, and firft of Angelica.

Angelica hath great and long winged leaues, made of many broade greene ones, diuided one from another vpon the ftalk, which is three foot long or better fomtimes, among which rife vp great thicke and hollow ftalkes with fome few ioynts, whereat doth alwayes ftand two long leaues compaffing the ftalke at the bottome, in fome places at the ioynts fpring out other ftalkes or branches, bearing fuch like leaues but fmaller, and at the tops very large vmbels of white flowers, that turne into whitifh feede fomewhat thicke : the roote groweth great with many branches at it, but quickly perifheth after it hath borne feede : to preferue the roote therefore the better, they vfe to cut it often in the yeare, thereby to hinder the running vp to feede : the whole plant, both leafe, roote and feede, is of an excellent comfortable fent, fauour and tafte.

The Vfe of Angelica.

The diftilled water of Angelica, eyther fimple or compound, is of efpeciall vfe *in deliquinm animi, vel cordis tremores & paffiones,* that is, fwounings, when the fpirits are ouercome and faint, or tremblings and paffions of the heart, to expell any windy or noyfome vapours from it. The green ftalkes or the young rootes being preferued or candied, are very effectuall to comfort and warme a colde and weake ftomacke : and in the time of infection is of excellent good vfe to preferue the fpirits and heart from infection. The dryed roote made into pouder, and taken in wine or other drinke, will abate the rage of luft in young perfons, as I haue it related vnto me vpon credit : A Syrupe made thereof in this manner, is very profitable to expectorate flegme out of the cheft and lunges, and to procure a fweete breath. Into the greene ftalke of Angelica as it ftandeth growing, make a great gafhe or incifion, wherein put a quantitie of fine white Sugar, letting it there abide for three dayes, and after take it forth by cutting a hole at the next ioynt vnder the cut, where the Syrupe refteth, or cut off the ftalke, and turne it downe, that the Syrupe may drayne forth ; which keepe for a moft delicate medicine.

Chap. LVIII.

Dracunculus hortenfis fiue Serpentaria. Dragons.

DRagons rifeth out of the ground with a bare or naked round whitifh ftalke, fpotted very much with purplifh fpots and ftrakes, bearing at the toppe therof a few greene leaues very much diuided on all fides, ftanding vpon long footeftalkes, in the middle whereof (if the roote be old enough) commeth forth a great long huske or hofe, green on the outfide, and of a darke purplifh colour on the infide, with a flender long reddifh peftell or clapper in the middle : the roote is great, round, flat and whitifh on the outfide, and whiter within, very like vnto the rootes of *Arum,* or Wakerobin, and tafting fomewhat fharpe like it.

The

The Vfe of Dragons.

The chiefe vfe whereunto Dragons are applyed, is, that according to an old receiued cuftome and tradition (and not the iudgement of any learned Author) the diftilled water is giuen with Mithridatum or Treakle to expell noyfome and peftilentiall vapours from the heart.

CHAP. LIX.

Ruta. Garden Rue, or Herbe Grace.

GArden Rue or Herbe Grace groweth vp with hard whitifh wooddy ftalkes, whereon are fet diuers branches of leaues, being diuided into many fmall ones, which are fomewhat thicke and round pointed, of a blewifh greene colour: the flowers ftand at the tops of the ftalkes confifting of foure fmall yellow leaues, with a greene button in the middle, and diuers fmall yellow threds about it, which growing ripe, containe within them fmall blacke feede: the roote is white and wooddy, fprea-ding farre in the ground.

The Vfe of Rue.

The many good properties whereunto Rue ferueth, hath I thinke in for-mer times caufed the Englifh name of Herbe Grace to be giuen vnto it. For without doubt it is a moft wholefome herbe, although bitter and ftrong, and could our dainty ftomackes brooke the vfe thereof, it would worke admi-rable effects being carefully and skilfully applyed, as time and occafion did require : but not vndifcreetly or hand ouer head, as many vfe to doe that haue no skill. Some doe rippe vp a beade rowle of the vertues of Rue, as Macer the Poet and others, in whom you fhall finde them fet downe, to bee good for the head, eyes, breaft, liuer, heart, fpleene, &c. In fome places they vfe to boyle the leaues of Rue, and keep them in pickle, to eate them as Sam-pire for the helpe of weake eyes. It is very auaileable in glifters or drinkes againft the winde or the collicke, and to procure vrine that is ftayed by the paines therof. The diftilled water is often vfed for the fame purpofes afore-faid : but beware of the too frequent or ouermuch vfe thereof, becaufe it heateth exceedingly, and wafteth nature mightily.

CHAP. LX.

Carduus Benedictus. The Bleffed Thiftle.

CArduus benedictus or the bleffed Thiftle, hath many weake tender branches ly-ing for the moft part on the ground, whereon are fet long and narrow leaues, much cut in or waued about the edges, hairy or rough in handling, yet without any hard or fharpe thornes or prickles at all, that the tendereft hand may touch them without harme : but thofe that grow toward the toppes of the ftalkes are fomewhat more prickly, and the heads which grow on the tops of the feuerall branches are fome-what fharpe, fet with prickles like a Thiftle : the flower is yellow, and the feede lying within the woolly or flocky doune like to all other thiftles, are blackifh, long and round, with a few haires on the head of them : the roote is white, and perifheth euery yeare after it hath giuen feede.

The Vfe of the bleffed Thiftle.

The diftilled water hereof is much vfed to be drunke againft agues of all fortes, eyther peftilentiall or humorall, of long continuance or of leffe :

but

1 *Angellica.* Angellica. 2 *Dracunculus hortenfis.* Dragons. 3 *Ruta hortenfis.* Garden Rue, or Herbegrace. 4 *Carduus benedictus.* The bleſſed Thiſtle. 5 *Alkakengi five Solanum Halicacabum & Veficarium.* Winter Cherries. 6 *Afarum.* Afarabacca. 7 *Liqueritia.* Licoris.

but the decoction of the herbe giuen in due time, hath the more forcible operation : it helpeth to expell wormes, becaufe of the bitternaffe, and is thereby alfo a friend to the ftomack ouercharged with chollar, and to clenfe the liuer : it prouoketh fweate and vrine, is helpefull to them are troubled with the ftone, and to eafe paines in the fides.

Chap. LXI.

Solanum veficarium, fiue Alkakengi. Winter Cherries.

THe Winter Cherry hath a running or creeping roote in the ground, of the bignefle wany times of ones little finger, fhooting forth at feuerall ioynts in feuerall places, whereby it quickly fpreadeth a great compaffe of ground : the ftalke rifeth not aboue a yard high, whereon are fet many broade and long greene leaues, fomewhat like vnto the leaues of Nightfhade, but larger : at the ioynts whereof come forth whitifh flowers made of fiue leaues a peece, which after turne into green berries, inclofed with thin skins or bladders, which change to bee reddifh when they grow ripe, the berry likewife being reddifh, and as large as a Cherry, wherein are contained many flat and yellowifh feed lying within the pulpe : which being gathered and ftrung vp, are kept all the yeare to be vfed vpon occafion.

The Vfe of Winter Cherries.

The diftilled water of the herbe and fruit together, is often taken of them that are troubled with the fharpneffe or difficultie of vrine, and with the ftone in the kidneyes, or grauel in the bladder : but the berries themfelues either greene or dryed boyled eyther in broth, in wine, or in water, is much more effectuall : It is likewife conducing to open obftructions of the liuer, &c. and thereby to helpe the yellow Iaundife.

Chap. LXII.

Afarum. Afarabacca.

ASarabacca, from a fmall creeping roote fet with many fibres, fhooteth forth diuers heads, and from euery of them fundry leaues, euery one ftanding vpon a long greene ftalke, which are round, thicke, and of a very fad or darke greene colour, and fhining withall : from the rootes likewife fpring vp fhort ftalkes, not fully foure fingers high, at the toppe of euery one of which ftandeth the flower, in fafhion very like the feede veffell of Henbane feede, of a greenifh purple colour, which changeth not his forme, but groweth in time to containe therein fmall cornered feed : the greene leaues abide all the winter many times, but vfually fheddeth them in winter, and recouereth frefh in the fpring.

The Vfe of Afarabacca.

The leaues are much and often vfed to procure vomits, fiue or feuen of them bruifed, and the iuice of them drunke in ale or wine. An extract made of the leaues with wine artificially performed, might bee kept all the yeare thorough, to bee vfed vpon any prefent occafion, the quantitie to bee proportioned according to the conftitution of the patient. The roote worketh not fo ftrongly by vomit, as the leaues, yet is often vfed for the fame purpofe, and befides is held auaileable to prouoke vrine, to open obftructions in the liuer and fpleene, and is put among diuers other fimples, both into Mithridatum and Andromachus Treakle, which is vfually called Venice Treakle. A dram of the dryed roots in pouder giuen in white wine a little before the fit of an ague, taketh away the fhaking fit, & therby caufe the hot fit to be the more remiffe, and in twice taking expell it quite.

Chap.

Chap. LXIII.

Glycyrrhiza siue Liqueritia. Licorice.

ALthough there are two forts of Licorice fet downe by diuers Authors, yet becaufe this Land familiarly is acquainted but with one fort, I fhall not neede for this Garden, to make any further relation of that is vnknowne, but onely of that fort which is fufficiently frequent with vs. It rifeth vp with diuers wooddy ftalks, whereon are fet at feuerall diftances many winged leaues, that is to fay, many narrow long greene leaues fet together on both fides of the ftalke, and an odde one at the end, very well refembling a young Afhe tree fprung vp from the feede : this by many yeares continuance in a place without remouing, and not elfe, will bring forth flowers many ftanding together fpike-fafhion one aboue another vpon the ftalkes, of the forme of Peafe bloffomes, but of a very pale or bleake blew colour, which turne into long fomewhat flat and fmooth cods, wherein is contained fmall round hard feede : the roote runneth downe exceeding deep into the ground, with diuers other fmaller roots and fibres growing with them, and fhoote out fuckers from the maine rootes all about, whereby it is much encreafed, of a brownifh colour on the outfide, and yellow within, of a farre more weake fweete tafte, yet far more pleafing to vs then that Licorice that is brought vs from beyond Sea ; becaufe that, being of a ftronger fweet tafte hath a bitterneffe ioyned with it, which maketh it the leffe pleafing and acceptable to moft.

The Vfe of Licorice.

Our Englifh Licorice is now adaies of more familiar vfe (as I faid before) then the outlandifh, and is wholly fpent and vfed to helpe to digeft and expectorate flegme out of the cheft and lunges, and doth allay the fharpeneffe or faltneffe thereof. It is good alfo for thofe are troubled with fhortneffe of breath, and for all forts of coughes. The iuice of Licorice artificially made with Hyffope water, ferueth very well for all the purpofes aforefaid. It being diffolued with Gum Tragacanth in Rofe water, is an excellent Lohoc or licking medicine to breake flegme, and to expectorate it, as alfo to avoyde thin frothy matter, or thin falt flegme, which often fretteth the lunges. It doth alfo lenifie exulcerated kidneyes, or the bladder, and helpeth to heale them. It is held alfo good for thofe that cannot make their water but by drops, or a fmall deale at a time.

The dryed root finely minced, is a fpeciall ingredient into all Trageas or Dredges, feruing for the purpofes aforefaid, but the vfe of them is almoft wholly left now adaies with all forts.

Thus haue I fhewed you not only the herbes, rootes and fruites, nourfed vp in this Garden, but fuch herbes as are of moft neceffary vfes for the Country Gentlewomens houfes : And now I will fhew you the Orchard alfo.

The

THE
ORDERING OF THE
ORCHARD.

The third part, or ORCHARD.

CHAP. I.

The situation of an Orchard for fruit-bearing trees, and how to amend the defects of many grounds.

S I haue done in the two former parts of this Treatise, so I meane to proceede in this ; first to set downe the situation of an Orchard, and then other things in order : And first, I hold that an Orchard which is, or should bee of some reasonable large extent, should be so placed, that the house should haue the Garden of flowers iust before it open vpon the South, and the Kitchen Garden on the one side thereof, should also haue the Orchard on the other side of the Garden of Pleasure, for many good reasons : First, for that the fruit trees being grown great and tall, will be a great shelter from the North and East windes, which may offend your chiefest Garden, and although that your Orchard stand a little bleake vpon the windes, yet trees rather endure these strong bitter blasts, then other smaller and more tender shrubs and herbes can doe. Secondly, if your Orchard should stand behinde your Garden of flowers more Southward, it would shadow too much of the Garden, and besides, would so binde in the North and East, and North and West windes vpon the Garden, that it would spoile many tender things therein, and so much abate the edge of your pleasure thereof, that you would willingly wish to haue no Orchard, rather then that it should so much anney you by the so ill standing thereof. Thirdly, the falling leaues being still blowne with the winde so aboundantly into the Garden, would either spoile many things, or haue one daily and continuall attending thereon, to cleanse and sweepe them away. Or else to auoide these great inconueniences, appoint out an Orchard the farther off, and set a greater distance of ground betweene. For the ground or soile of the Orchard, what I haue spoken concerning the former Garden for the bettering of the seuerall grounds, may very well serue and be applyed to this purpose. But obserue this, that whereas your Gardens before spoken of may be turned vp, manured, and bettered with soile if they growe out of heart, your Orchard is not so easily done, but must abide many yeares without altering ; and therefore if the ground be barren, or not good, it had the more neede to bee amended, or wholly made good, before you make an Orchard of it ; yet some there be

that

that doe appoint, that where euery tree fhould bee fet, you onely digge that place to make it good : but you muft know, that the rootes of trees runne further after a little times ftanding, then the firft compaffe they are fet in ; and therefore a little compaffe of ground can maintaine them but a little while, and that when the rootes are runne beyond that fmall compaffe wherein they were firft fet, and that they are come to the barren or bad ground, they can thriue no better then if they had beene fet in that ground at the firft, and if you fhould afterwards digge beyond that compaffe, intending to make the ground better further off, you fhould much hurt the fpreading rootes, and put your trees in danger: the fituation of hils in many places is grauelly or chalky, which is not good for trees, becaufe they are both too ftonie, and lacke mellow earth, wherein a tree doth moft ioy and profper, and want moifture alfo (which is the life of all trees) becaufe of the quicke defcent of raine to the lower grounds : and befides all thefe inconueniences there is one more ; your trees planted either on hils or hill fides, are more fubiect to the fury and force of windes to be ouer-turned, then thofe that growe in the lower grounds ; for the ftrongeft and moft forci-ble windes come not vfually out of the North Eaft parts, where you prouide beft de-fence, but from the South and Weft, whence you looke for the beft comfort of the Sunne.   To helpe therefore manie of the inconueniences of the hils fides, it were fit to caufe manie leauels to bee made thereon, by raifing the lower grounds with good earth, and fuftaining them with bricke or ftone wals, which although chargeable, will counteruaile your coft, befide the pleafure of the walkes, and profpect of fo worthy a worke.   The plaine or leuell grounds as they are the moft frequent, fo they are the moft commendable for an Orchard, becaufe the moulds or earths are more rich, or may better and fooner be made fo ; and therefore the profits are the more may be rai-fed from them.   A ftiffe clay doth nourifh trees well, by reafon it containeth moi-fture ; but in regard of the coldneffe thereof, it killeth for the moft part all tender and early things therein: fea-cole afhes therefore, bucke afhes, ftreete foyle, chaulke after it hath lyen abroad and been broken with many yeares frofts and raine, and fheepes dung, are the moft proper and fitteft manure to helpe this kinde of foyle.   The dry fandy foile, and grauelly ground are on the contrary fide as bad, by reafon of too much heate and lacke of moifture : the dung of kine or cattell in good quantity beftowed thereon, will much helpe them.   The amending or bettering of other forts of grounds is fet down toward the end of the firft Chapter of the firft part of this worke, where-vnto I will referre you, not willing to repeate againe the fame things there fet downe. The beft way to auoide and amend the inconueniences of high, boifterous, and cold windes, is to plant Walnut trees, Elmes, Oakes or Afhes, a good diftance without the compaffe of your Orchard, which after they are growne great, will bee a great fafe-guard thereunto, by breaking the violence of the windes from it. And if the foyle of your Orchard want moifture, the conueying of the finke of the houfe, as alfo any o-ther draine of water thereinto, if it may be, will much helpe it.

CHAP. II.

The forme of an Orchard, both ordinary, and of more grace and rarity.

Accordingto the fituation of mens grounds, fo muft the plantation of them of neceffitie be alfo ; and if the ground be in forme, you fhall haue a formall Or-chard : if otherwife, it can haue little grace or forme. And indeed in the elder ages there was fmall care or heede taken for the formality ; for euery tree for the moft part was planted without order, euen where the mafter or keeper found a vacant place to plant them in, fo that oftentimes the ill placing of trees without fufficient fpace be-tweene them, and negligence in not looking to vphold them, procured more wafte and fpoile of fruit, then any accident of winde or weather could doe. Orchards in moft places haue not bricke or ftone wals to fecure them, becaufe the extent thereof being larger then of a Garden, would require more coſt, which euery one cannot vndergoe; and therefore mud wals, or at the beſt a quicke ſet hedge, is the ordinary and moſt vſuall defence it findeth almoſt in all places : but with thoſe that are of ability to compaſſe it with bricke or ſtone wals, the gaining of ground, and profit of the fruit trees planted there againſt, will in ſhort time recompenſe that charge. If you make a doubt how to be ſure that your Orchard wall ſhall haue ſufficient comfort of the Sunne to ripen the fruits, in regard the trees in the Orchard being ſo nigh thereunto, and ſo high withall, will ſo much ſhadow the wall, that nothing will ripen well, becauſe it will want the comfort of the Sunne : you may follow this rule and aduice, to remedy thoſe inconueniences. Hauing an Orchard containing one acre of ground, two, three, or more, or leſſe, walled about, you may ſo order it, by leauing a broad and large walke betweene the wall and it, containing twenty or twenty foure foote (or yards if you will) that the wall ſhall not be hindered of the Sun, but haue ſufficient comfort for your trees, notwithſtanding the height of them, the diſtance betweene them and the wall being a ſufficient ſpace for their ſhadow to fall into : and by compaſſing your Or-

chard on the inſide with a hedge (wherein may bee planted all ſorts of low ſhrubs or buſhes, as Roſes, Cornellian Cherry trees plaſhed lowe, Gooſeberries, Curran trees, or the like) you may encloſe your walke, and keepe both it and your Orchard in better forme and manner, then if it lay open. For the placing of your trees in this Orchard, firſt for the wals: Thoſe ſides that lye open to the South & Southweſt Sunne, are fitteſt to bee planted with your tendereſt and earlieſt fruits, as Apricockes, Peaches, Nectarius, and May or early Cherries : the Eaſt, North and Weſt, for Plums and Quinces, as you ſhall like beſt to place them. And for the Orchard it ſelfe, the ordinary manner is to place them without regard of meaſure or difference, as Peares among Apples, and Plums among Cherries promiſcuouſly ; but ſome keepe both a diſtance and a diuiſion for

euery ſort, without intermingling : yet the moſt gracefull Orchard containeth them all, with ſome others, ſo as they be placed that one doe not hinder or ſpoile another ; and therefore to deſcribe you the modell of an Orchard, both rare for comelineſſe in the proportion, and pleaſing for the profitableneſſe in the vſe, and alſo durable for continuance, regard this figure is here placed for your direction, where you muſt obſerue, that your trees are here ſet in ſuch an equall diſtance one from another euery way, & as is fitteſt for them, that when they are grown great, the greater branches ſhall not gall or rubbe one againſt another; for which purpoſe twenty or ſixteene foot is the leaſt to be allowed for the diſtance euery way of your trees, & being ſet in rowes euery one in the middle diſtance, will be the moſt gracefull for the plantation, and beſides, giue you way ſufficient to paſſe through them, to pruine, loppe, or dreſſe them, as need ſhall require, and may alſo bee brought (if you pleaſe) to that gracefull delight, that euery alley or diſtance may be formed like an arch, the branches of either ſide meeting to be enterlaced together. Now for the ſeuerall ſorts of fruit trees that you ſhall place in this modell, your beſt direction is to ſet Damſons, Bulleis, and your taler growing Plums on the outſide, and your lower Plums, Cherries, and Apples on the inſide, hauing regard, that you place no Peare tree to the Sunward, of any other tree, leſt it ouerſhadow

shadow them : Let your Peare trees therefore be placed behinde, or on the one side of your lower trees, that they may be as it were a shelter or defence on the North & East side. Thus may you also plant Apples among Plums and Cherries, so as you suffer not one to ouer-growe or ouer-toppe another ; for by pruning, lopping, and shredding those that growe too fast for their fellowes, you may still keepe your trees in such a conformity, as may be both most comely for the sight, and most profitable for the yeelding of greater and better store of fruit. Other sorts of fruit trees you may mixe among these, if you please, as Filberds, Cornellian Cherries in standerds, and Medlers : but Seruice trees, Baye trees, and others of that high sort, must be set to guard thereft. Thus haue I giuen you the faireft forme could as yet be deuised ; and from this patterne, if you doe not follow it precisely, yet by it you may proportion your Orchard, be it large or little, be it walled or hedged.

Chap. III.

Of a nourfery for trees, both from fowing the kernels, and planting fit ftockes to graft vpon.

ALthough I know the greater sort (I meane the Nobility and better part of the Gentrie of this Land) doe not intend to keepe a Nursery, to raise vp those trees that they meane to plant their wals or Orchards withall, but to buy them already grafted to their hands of them that make their liuing of it : yet because many Gentlemen and others are much delighted to beftowe their paines in grafting themselues, and efteeme their owne labours and handie worke farre aboue other mens : for their incouragement and fatisfaction, I will here fet downe fome conuenient directions, to enable them to raife an Orchard of all forts of fruits quickly, both by fowing the kernels or ftones of fruit, and by making choife of the beft forts of ftockes to graft on : Firft therefore to begin with Cherries ; If you will make a Nurfery, wherein you may bee ftored with plenty of ftockes in a little fpace, take what quantitie you thinke good of ordinarie wilde blacke Cherrie ftones, cleanfed from the berries, and fowe them, or pricke them in one by one on a peece of ground well turned vp, and large enough for the quantitie of ftones you will beftowe thereon, from the midft of Auguft vnto the end of September, which when they are two or three yeares old, according to their growth, you may remoue them, and fet them anew in fome orderly rowes, hauing pruned their tops and their rootes, which at the next yeares growth after the new planting in any good ground, or at the fecond, will be of fufficient bigneffe to graft vpon in the bud what forts of Cherries you thinke beft : and it is fitteft to graft them thus young, that pruning your ftockes to raife them high, you may graft them at fiue or fix foote high, or higher, or lower, as you fhall fee good, and being thus grafted in the bud, will both more fpeedily and fafely bring forward your grafts, and with leffe danger of lofing your ftockes, then by grafting them in the ftocke : for if the bud take not by inoculating the firft yeare, yet your tree is not loft, nor put in any hazzard of loffe ; but may be grafted anew the yeare following, if you will, in another place thereof, whereas if you graft in the ftocke, and it doe not take, it is a great chance if the ftocke dye not wholly, or at leaft be not fo weakened both in ftrength and height, that it will not bee fit to bee grafted a yeare or two after. In the fame manner as you doe with the blacke, you may deale with the ordinary Englifh red Cherrie ftones, or kernels, but they are not fo apt to growe fo ftraight and high, nor in fo fhort a time as the blacke Cherrie ftones are, and befides are fubiect in time to bring out fuckers from the rootes, to the hinderance of the ftockes and grafts, or at the leaft to the deformitie of your Orchard, and more trouble to the Gardiner, to pull or digge them away. Plumme ftones may bee ordered in this manner likewife, but you muft make choife of your Plums ; for although euery Plumme is not fo fit for this purpofe, as the white Peare Plumme, becaufe it groweth the goaleft and freeft, the barke being fmooth and apteft to be raifed, that they may be grafted vpon ; yet diuers other Plummes may be taken, if they be not at hand, or to be had, as the blacke and red Peare Plumme, the

white

white and red Wheate Plumme, becaufe they are nearest in goodneffe vnto it. Peach ftones will be foone raifed vp to graft other forts of Peaches or Nectorins vpon, but the nature of the Peach roote being fpongie, is not to abide long. As for Almonds, they will be raifed from their ftones to be trees of themfelues; but they will hardly a-bide the remouing, and leffe to bee grafted vpon. Apricocke ftones are the worft to deale withall of any fort of ftone fruit; for although the Apricocke branches are the fitteft ftockes to graft Nectorins of the beft forts vpon, yet thofe that are raifed from the kernels or ftones will neuer thriue to be brought on for this purpofe; but will ftarue and dye, or hardly grow in a long time to be a ftraight and fit ftocke to be grafted, if it be once remoued. Your Cornellian Cherrie trees are wholly, or for the moft part rai-fed from the ftones or kernels; yet I know diuers doe increafe them, by laying in their loweft branches to take roote: and thus much for ftone fruits. Now for Apples and Peares, to be dealt withall in the fame manner as aforefaid. They vfe to take the pref-fing of Crabs whereas Veriuyce is made, as alfo of Cidar and Perry where they are made, and fowing them, doe raife vp great ftore of ftockes; for although the beating of the fruit doth fpoile many kernels, yet there will bee enough left that were neuer toucht, and that will fpring: the Crabbe ftockes fome preferre for the fitteft, but I am fure, that the better Apple and Peare kernels will growe fairer, ftraighter, quicklier, and better to be grafted on. You muft remember, that after two or three yeares you take vp thefe ftockes, and when you haue pruned both toppe and roote, to fet them a-gaine in a thinner and fitter order, to be afterwards grafted in the bud while they are young, as I fhall fhew you by and by, or in the ftocke if you will fuffer them to growe greater. Now likewife to know which are the fitteft ftockes of all forts to choofe, thereon to graft euery of thefe forts of fruits, is a point of fome skill indeede; and therefore obferue them as I doe here fet them downe: for bee you affured, that they are certaine rules, and knowne experiences, whereunto you may truft without being deceiued. Your blacke Cherrie ftockes (as I faid before) are the fitteft and beft for all forts of Cherries long to abide and profper, and euen May or early Cherry will a-bide or liue longer, being grafted thereon, either in the budde or in the ftocke, then on the ordinary red Cherry ftocke; but the red Cherry ftocke is in a manner the onely tree that moft Nurfery men doe take to graft May Cherries on in the ftocke (for it is but a late experience of many, to graft May Cherries in the bud) many alfo doe graft May Cherries on Gafcoigne Cherry ftockes, which doe not onely thriue well, but en-dure longer then vpon any ordinay Cherry ftocke: For indeede the May Cherries that are grafted vpon ordinary red Cherrie ftockes, will hardly hold aboue a dozen yeares bearing well, although they come forwarder at the firft, that is, doe beare foo-ner then thofe that are grafted on Gafcoigne or blacke Cherry ftockes; but as they are earlier in bearing, fo they are fooner fpent, and the Gafcoigne and blacke Cherry ftockes that are longer in comming forward, will laft twice or thrice their time; but many more grafts will miffe in grafting of thefe, then of thofe red Cherry ftockes, and befides, the natures of the Gafcoigne and blacke Cherry ftockes are to rife higher, and make a goodlier tree then the ordinary red ftocke will, which for the moft part fprea-deth wide, but rifeth not very high. The Englifh red Cherry ftocke will ferue very well to graft any other fort of Cherry vpon, and is vfed in moft places of this Land, and I know no other greater inconuenience in it, then that it fhooteth out many fuckers from the roote, which yet by looking vnto may foone bee remoued from doing any harme, and that it will not laft fo long as the Gafcoigne or blacke Cherry ftocke will. May Cherries thus grafted lowe, doe moft vfually ferue to be planted againft a wall, to bring on the fruit the earlier; yet fome graft them high vpon ftandards, although not many, and it is, I thinke, rather curiofity (if they that doe it haue any wals) then anie o-ther matter that caufeth them thus to doe: for the fruit is naturally fmall, though early, and the ftandard Cherries are alwaies later then the wall Cherries, fo that if they can fpare any roome for them at their wals, they will not plant many in ftandards. Now concerning Plummes (as I faid before) for the fowing or fetting of the ftones, fo I fay here for their choife in grafting of them, either in the budde or ftocke. The white Peare Plumme ftocke, and the other there mentioned, but efpecially the white Peare Plumme is the goodlieft, freeft, and fitteft of all the reft, as well to graft all fort of Plummes vpon, as alfo to graft Apricockes, which can be handfomely, and to any

good

good purpofe grafted vpon no other Plum ftocke, to rife to bee worth the labour and paine. All forts of Plums may be grafted in the ftocke, and fo may they alfo in the bud; for I know none of them that will refufe to be grafted in the bud, if a cunning hand performe it well; that is, to take off your bud cleanely and well, when you haue made choice of a fit cyon: for, as I fhal fhew you anon, it is no fmall peece of cunning to chufe your cyon that it may yeeld fit buds to graft withall, for euery plum is not of a like aptnes to yeeld them: But Apricocks cannot be grafted in the ftock for any thing that euer I could heare or learne, but only in the bud, and therefore let your Plum ftocke bee of a reafonable fize for Apricockes efpecially, and not too fmall, that the graft ouergrow not the ftocke, and that the ftocke bee large enough to nourifh the graft. As your Plum ftockes ferue to graft both Apricockes and Plummes, fo doe they ferue alfo very well to graft Peaches of all forts; and although Peach ftockes will ferue to be grafted with Peaches againe, yet the Peach ftocke (as I faid before) will not endure fo long as the Plumme ftocke, and therefore ferueth but for neceffity if Plum ftocks be not ready, or at hand, or for the prefent time, or that they afterwards may graft that fort of Peach on a Plumme ftocke : for many might lofe a good fruit, if when they meete with it, and haue not Plumme ftockes ready to graft it on, they could not be affured that it would take vpon another Peach ftocke or branch, or on the branch of an Apricocke eyther. Plumme ftockes will ferue likewife very well for fome forts of Nectorins; I fay, for fome forts, and not for all : the greene and the yellow Nectorin will beft thriue to be grafted immediately on a Plumme ftocke; but the other two forts of red Nectorins muft not be immediately grafted on the Plumme ftocke, but vpon a branch of an Apricocke that hath beene formerly grafted on a Plumme ftocke, the nature of thefe Nectorins being found by experience to be fo contrary to the Plum ftocke, that it will fterue it, and both dye within a yeare, two or three at the moft : Diuers haue tryed to graft thefe red Nectorins vpon Peach ftockes, and they haue endured well a while; but feeing the Peach ftocke will not laft long it felfe, being ouerweake, how can it hold fo ftrong a nature as thefe red Nectorins, which will (as I faid before) fterue a Plum ftocke that is fufficient durable for any other Plumme?

Apricocke ftockes from the ftones are hardly nurfed vp, and worfe to be remoued, and if a red Nectorin fhould be grafted on an Apricock rayfed from the ftone, and not remoued, I doubt it might happen with it as it doth with many other trees raifed from ftones or kernels, and not remoued, that they would hardly beare fruit : for the nature of moft trees raifed from ftones or kernels, and not remoued, is to fend great downeright rootes, and not to fpread many forwards; fo that if they be not cut away that others may fpreade abroad, I haue feldome feene or known any of them to beare in any reafonable time; and therefore in remouing, thefe great downe-right rootes are alwayes fhred away, and thereby made fit to fhoote others forwards. Hereby you may perceiue, that thefe red Nectorins will not abide to bee grafted vpon any other ftocke well, then vpon an Apricocke branch, although the green and the yellow (as I faid before) will well endure and thriue vpon Plums. The fuckers or fhootes both of Plums and Cherries that rife from their rootes, eyther neare their ftockes, or farther off, fo that they bee taken with fome fmall rootes to them, will ferue to bee ftockes, and will come forward quickly; but if the fuckers haue no fmall roots whereby they may comprehend in the ground, it is almoft impoffible it fhould hold or abide. There is another way to rayfe vp eyther ftockes to graft on, or trees without grafting, which is, by circumcifing a faire and fit branch in this manner : About Midfomer, when the fappe is thoroughly rifen (or before if the yeare be forward) they vfe to binde a good quantity of clay round about a faire and ftraight branch, of a reafonable good fize or bigneffe, with fome conuenient bands, whether it be ropes of hey, or of any other thing, about an handfull aboue the ioynt, where the branch fpreadeth from the tree, and cutting the barke thereof round about vnder the place where the clay is bound, the fap is hereby hindered from rifing, or defcending further then that place fo circumcifed, whereby it will fhoote out fmall knubs and rootes into the clay, which they fuffer fo to abide vntill the beginning of winter, whenas with a fine Sawe they cut off that branch where it was circumcifed, and afterwardes place it in the ground where they would haue it to grow, and ftake it, and binde it faft, which will fhoote forth rootes, and will become eyther a faire tree to beare fruite without grafting, or elfe a fit ftocke to graft on according

ding to the kinde: but oftentimes this kinde of propagation miſſeth, in that it ſendeth not forth rootes ſufficient to cauſe it to abide any long time. Let me yet before I leaue this narration of Plummes, giue you one admonition more, that vpon whatſoeuer Plumme ſtocke you doe graft, yet vpon a Damſon ſtocke that you neuer ſtriue to graft, for it (aboue all other ſorts of Plumme ſtockes) will neuer giue you a tree worth your labour. It remaineth only of ſtone fruit, that I ſpeake of Cornelles, which as yet I neuer ſaw grafted vpon any ſtocke, being as it ſhould ſeeme vtterly repugnant to the nature thereof, to abide grafting, but is wholly rayſed vp (as I ſaid before) eyther from the ſtones, or from the ſuckers or layers. For Peares and Apples your vſuall ſtockes to graft on are (as I ſaid before, ſpeaking of the nurſing vp of trees from the kernels) your Crabbe ſtockes, and they bee accepted in euery Countrey of this Land as they may conueniently be had, yet many doe take the ſtockes of better fruit, whether they bee ſuckers, or ſtockes rayſed from the kernels (and the moſt common and knowne way of grafting, is in the ſtocke for all ſorts of them, although ſome doe vſe whipping, packing on, or inciſing, as euery one liſt to call it: but now we doe in many places begin to deale with Peares and Apples as with other ſtone fruit, that is, graft them all in the bud, which is found the moſt compendious and ſafeſt way both to preſerue your ſtocke from periſhing, and to bring them the ſooner to couer the ſtock, as alſo to make the goodlier and ſtraighter tree, being grafted at what height you pleaſe:) for thoſe ſtockes that are rayſed from the kernels of good fruit (which are for the moſt part eaſily knowne from others, in that they want thoſe thornes or prickles the wilde kindes are armed withall:) I ſay for the moſt part; for I know that the kernels of ſome good fruite hath giuen ſtockes with prickles on them (which, as I thinke, was becauſe that good fruite was taken from a wilde ſtocke that had not beene long enough grafted to alter his wilde nature; for the longer a tree is grafted, the more ſtrength the fruite taketh from the graft, and the leſſe ſtill from the ſtocke) being ſmoother and fairer then the wilde kinds, muſt needes make a goodlier tree, and will not alter any whit the taſte of your fruit that is grafted thereon, but rather adde ſome better relliſh thereunto; for the Crabbe ſtockes yeelding harſh fruite, muſt giue part of their nature to the grafts are ſet thereon, and therefore the taſte or relliſh, as well as ſome other naturall properties of moſt fruits, are ſomewhat altered by the ſtocke. Another thing I would willingly giue you to vnderſtand concerning your fruits and ſtockes, that whereas diuers for curioſity and to try experiments haue grafted Cherries vpon Plumme ſtockes, or Plums on Cherry ſtockes, Apples vpon Peare ſtockes, and Peares vpon Apple ſtockes, ſome of theſe haue held the graft a yeare, two or three peraduenture, but I neuer knew that euer they held long, or to beare fruite, much leſſe to abide or doe well: beſtow not therefore your paines and time on ſuch contrary natures, vnleſſe it be for curioſitie, as others haue done: Yet I know that they that graft peares on a white thorne ſtocke haue had their grafts ſeeme to thriue well, and continue long, but I haue ſeldome ſeene the fruite thereof anſwerable to the naturall wilde Peare ſtocke; yet the Medlar is knowne to thriue beſt on a white thorne. And laſtly, whereas diuers doe affirme that they may haue not only good ſtockes to graft vpon, but alſo faire trees to bear ſtore of fruit from the kernels of Peares or Apples being prickt into the ground, and ſuffered to grow without remouing, and then eyther grafted or ſuffered to grow into great trees vngrafted; and for their bearing of fruite, aſſigne a dozen or twenty yeares from the firſt ſetting of the kernels, and abiding vngrafted, I haue not ſeene or heard that experience to hold certaine, or if it ſhould be ſo, yet it is too long time loſt, and too much fruit alſo, to waite twenty yeares for that profit may be gained in a great deale of leſſe time, and with more certainty. Vnto theſe inſtructions let mee adde alſo one more, which is not much known and vſed, and that is, to haue fruit within foure or fiue years from the firſt ſowing of your ſtones or kernels in this manner: After your ſtones or kernels are two or three yeares old, take the faireſt toppe or branch, and graft it as you would doe any other cyon taken from a bearing tree, and looke what rare fruite, eyther Peare or Apple, the kernell was of that you ſowed, or Peach or Plum &c. the ſtone was ſet, ſuch fruite ſhall you haue within two or three yeares at the moſt after the grafting, if it take, and the ſtocke be good. And thus may you ſee fruit in farre leſſe time then to ſtay vntill the tree from a kernell or ſtone beareth fruit of it ſelfe.

Chap.

CHAP. IIII.

*The divers manners of grafting all sorts of fruits
vsed in our Land.*

THe moſt vſuall manner of grafting in the ſtocke is ſo common and well knovvn
in this Land to euery one that hath any thing to doe with trees or an Orchard,
that I think I ſhall take vpon mee a needleſſe worke to ſet downe that is ſo well
knowne to moſt ; yet how common ſoeuer it is, ſome directions may profit euery one,
without which it is not eaſily learned. And I doe not ſo much ſpend my time and
paines herein for their ſakes that haue knowledge, but for ſuch as not knowing would
faine be taught priuately, I meane, to reade the rules of the arte ſet downe in priuate,
when they would refuſe to learne of a Gardiner, or other by ſight : and yet I diſcom-
mend not that way vnto them to learne by ſight; for one may ſee more in an inſtant by
ſight, then he ſhall learn by his own practice in a great while, eſpecially if he be a little
practiſed before he ſee a cunning hand to doe it. There are many other kindes of grafting, which ſhall be ſpoken of hereafter, and peraduenture euen they that know it well,
may learne ſomething they knew not before.

1. The grafting in the ſtocke, is, to ſet the ſprigge of a good fruit into the body or
ſtocke of another tree, bee it wilde or other, bee it young or old, to cauſe that tree to
bring forth ſuch fruit as the tree bore from whence you took the ſprigge, and not ſuch
as the ſtocke or tree would haue borne, if it had not beene grafted, and is performed in
this manner : Looke what tree or ſtocke you will chuſe to graft on, you muſt with a
ſmall fine ſawe and very ſharpe, whip off, or cut off the head or toppe thereof at what
height you eyther thinke beſt for your purpoſe, or conuenient for the tree : for if you
graft a great tree, you cannot without endangering the whole, cut it downe ſo low to
the ground, as you may without danger doe a ſmall tree, or one that is of a reaſonable
ſize ; and yet the lower or neerer the ground you graft a young tree, the ſafer it is both
for your ſtocke and graft, becauſe the ſappe ſhall not aſcend high, but ſoone giue vigour to the graft to take and ſhoote quickly : After you haue cut off the toppe of your
ſtocke, cut or ſmooth the head thereof with a ſharpe knife, that it may be as plaine and
ſmooth as you can, and then cleaue it with a hammer or mallet, and with a ſtrong knife,
cleauer or cheſſell, either in the middle of it if it be ſmall, or of a reaſonable ſize, or on
the ſides an inch or more within the barke, if it be great : into both ſides of the cleft
put your grafts, or into one if the ſtocke bee ſmaller ; which grafts muſt bee made
fit for the purpoſe on this faſhion : Hauing made choiſe of your grafts from the toppe
branches eſpecially, or from the ſides of that tree wherof you would haue the fruit, and
that they be of a reaſonable good ſize, not too ſmall or too great for your ſtockes, and
of one or the ſame yeares ſhoote ; (and yet many doe cut an inch or more of the olde
wood with the ſprigge of the laſt yeares growth, and ſo graft the old and young together (but both are good, and the old wood no better then the young) cut your graft not
too long, but with two, three or foure eyes or buds at the moſt, which at the lower or
bigger end for an inch long or more (for the greater ſtockes, and an inch or leſſe for the
leſſer ſort) muſt be ſo cut, that it be very thin on the one ſide from the ſhoulders downward, and thicker on the other, and thin alſo at the end, that it may goe downe cloſe into the cleft, and reſt at the ſhoulders on the head of the ſtocke : but take heede that in
cutting your grafts your knife bee very ſharpe that you doe not rayſe any of the barke,
eyther at the ſides or the end, for feare of loſing both your paines and graft, and ſtocke
too peraduenture ; and let not your grafts bee made long before you ſet them, or elſe
put the ends of them in water to keepe them freſh and cleane : when you ſet them you
muſt open the cleft of your ſtocke with a wedge or cheſſell as moſt doe, that the graft
may goe eaſily into it, and that the barke of both graft and ſtocke may ioyne cloſe the
one to the other, which without ſtirring or diſplacing muſt bee ſo left in the cleft, and
the wedge or cheſſell gently pulled forth; but becauſe in the doing hereof conſiſteth
in a manner the whole loſſe or gaine of your paines, graft and ſtocke, to preuent which
inconuenience I doe vſe an iron Inſtrument, the forme whereof is ſhowne in the following

lowing page, marked with the letter A, crooked at both ends, and broade like vnto a
cheſſell, the one bigger, and the other leſſer, to fit all ſorts of ſtockes, and the iron nan-
dle ſomewhat long betweene them both, that being thruſt or knocked downe into the
cleft, you may with your left hand open it as wide as is fit to let in your graft, without
ſtrayning, which being placed, this iron may bee pulled or knocked vp againe without
any moiuing of your graft : when you haue thus done, you muſt lay a good handfull or more (according to the bigneſſe of your ſtocke) of ſoft and well moiſtned clay
or loame, well tempered together with ſhort cut hey or horſe dung, vpon the head of
your ſtocke, as lowe or ſomewhat lower then the cleft, to keepe out all winde, raine or
ayre from your graft vntill Midſomer at the leaſt, that the graft be ſhot forth ſomewhat
ſtrongly, which then if you pleaſe may be remoued, and the cleft at the head only filled
with a little clay to keepe out earewigs, or other things that may hurt your graft.

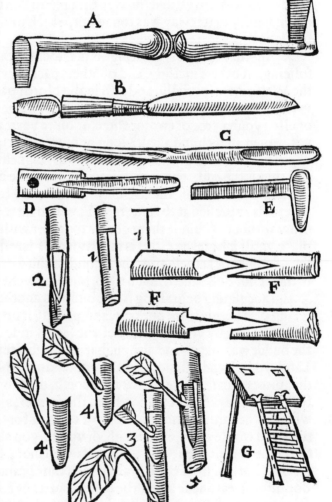

A. The Iron Inſtrument with
cheſſels at each end, the one
bigger and the other leſſer,
to keepe the cleft of the Tree
open vntill the graft bee placed in the ſtocke, which with
a knock vpwards will be eaſily
taken away.

B. The ſmall Penne-knife with a
broad and thinne ended hafte,
to raiſe the ſides both of the
bud and the down-right ſlit in
the body or arme of a Tree to
be grafted in the bud.

C. A pen or quil cut halfe round
to take off a bud from the
branch.

D. An Iuory Inſtrument made to
the ſame faſhion.

E. A ſhielde of braſſe made hollow before to be put into the
ſlit, to keepe it open vntill the
bud be put into its place.

F. The manner of grafting called inciſing or ſplicing.

G. A Ladder made with a ſtoole
at the toppe, to ſerue both to
graft higher or lower, and alſo
to gather fruit without ſpoy-
ling or hurting any buddes or
branches of Trees.

1. The firſt ſlit in the body or
arme of a Tree to be grafted in the bud with the croſſe cut at the head.

2. The ſame ſlit opened on both ſides, ready to receiue the budde ſhould be put therein : theſe ſmall peeces ſerue as well as trees to ſhew the manner and order of the grafting.

3. The branch of a Tree with one budde cut ready to be taken off, and another not yet
touched.

4. The bud cleane taken off from the branch, both the foreſide and backſide.

5. The graft or bud now put into the ſtocke or tree you intend to be grafted : but the
binding thereof is omitted.

2. Inarching is another manner of grafting in the ſtocke, and is more troubleſome,
and more caſuall alſo then the former, and is rather a curioſity then any way of good
ſpeede, certainety or profit, and therefore vſed but of a few. Yet to ſhew you, the

manner thereof, it is thus: Hauing a tree well growne, bee it high or low, yet the lower the better, with young branches well spread, they vse to set stockes round about it, or on the one side as you please ; into which stockes they ingraft the young branches of the well growne tree as they are growing (before they cut them from the tree) by bowing downe the branch they intend to graft, and putting it into the stocke, hauing first cut off the head thereof, and cut a notch in the middle of the head a little slope on both sides, wherein the branch must be fitted : let the branch be cut thinne on the vnderside, only of that length as may suffice to fit the notch in the stocke, leauing about halfe a yarde length of the branch, to rise aboue or beyond the stocke, which beeing bound on, and clayed ouer or couered with red or greene soft waxe, they let so abide, that if it take in the stocke they cut off the branch a little below the grafting place in Nouember following, and remouing the stocke, they haue thus gained a grafted and growne tree the first yeare : but it is vsually seene, that where one branch taketh, three doe misse : yet this manner of grafting was much in vse for May Cherries, when they were first known to vs, and the way thought to be a rare manner of grafting to encrease them, vntill a better way was found out, which now is so common and good also, that this is not now scarce thought vpon.

3. Another kinde of grafting in the stocke is called of some whipping, of some splicing, of others incising, and of others packing on (and as I heare, is much vsed in the West parts especially, and also in the North parts of this Land) and is performed in this manner : Take and slice the branch of a tree (so as the branch be not too bigge) or else a young tree of two, or three, or foure yeares growth at the most, quite off slope wise, about an inch and a halfe long or more, and cut a deep notch in the middle thereof, then fit into it a graft iust of that size or bignesse, cut on both sides with shoulders, and thin at the end, that it may ioyne close in the notch, and neyther bigger or lesser, but that the barke of the one may bee fitted iust to the barke of the other, the figure wherof is expressed at the letters E.F. which shew the one to be with a shoulder & the other without ; binde them gently together with bast, and put clay or waxe ouer the place, vntill it be taken : this is much vsed of late dayes for such young trees as are risen of stones or kernels after the second or third yeares growth, and thriue very well in that it not only saueth much time, but diuers checks by remouing and grafting.

4. Inoculating or grafting in the budde is another manner of grafting, which is the taking of a budde from one tree, and putting it into the barke of another tree, to the end, that thereby you may haue of the same kinde of fruit the tree bare from whence the budde was taken ; and although it bee sufficiently knowne in many places of this Land, yet as I vnderstand, good Gardiners in the North parts, and likewise in some other places, can scarce tell what it meaneth, or at the least how to doe it well. It is performed after a different fashion from the former, although they all tend vnto one end, which is the propagating of trees. You must for this purpose obserue, that for those trees you would graft, either with, or vpon, you choose a fit time in Summer, when the sappe is well risen, and your graft well shot, that the barke will rise easily and cleanly, both of stocke and graft, which time I cannot appoint, becaufe both the years doe differ in earlinesse, and the seuerall parts or countries of this Land likewise one from another, but most vsually in these Southerne parts, from the beginning of Iune vnto the end of it, or to the middle of Iuly, or either somewhat before or after. First (as I said) hauing taken the fittest time of the yeare, you must take especiall care, that your grafts be well growne, and of the same yeares shoote, and also that the buds or eyes haue but single leaues at them, as neare as you can : for I would vtterly refuse those buds that haue aboue two leaues as vnprofitable, either in Peaches or any other fruit ; and therefore see that your grafts or cyons bee taken from the chiefest place of the tree, that is, either from the toppe, or from a sunnie side thereof, and not from the contrarie side if you may otherwise, nor from any vnder-boughes ; for seeing your graft is so small a thing, you had neede take the more care that it be the best and fairest. You must to take off this eye or budde from the sprigge, haue a small sharpe pen-knife, the end of the haft being made flat and thinne, like a chessell or wedge, the figure whereof is set forth at the letter B, and a pen or goose quill cut, to be lesse then halfe round, and to be broad at the end, but not sharpe pointed like a penne, or else such a peece of bone or Iuorie made in that fashion as the quill is, to bee thinne, hollow, or

halfe

halfe round, the figures of both which are marked with the letters C, D. with your knife cut the barke of the bud (hauing firſt cut off the leafe, leauing onely the ſhort foote ſtalke thereof at the bud) about a ſtrawes breadth aboue the eye thereof halfe round, and then from that round or ouerthwart cut, with your knife cut it downe on both ſides of the eye, cloſe to the bud ſlopewiſe about an inch long or thereabouts, that it bee broad at the head aboue the eye, and pointing at the end like a ſheild or ſcutcheon; and then cutting away the reſt of the barke from about it, with the thinne flat end of the haft of your knife raiſe vp both ſides of your bud a little, and with your quill or bone put vnder the barke, raiſe your budde, and thruſt it quite off, beginning at the toppe or head of your eye; but ſee that you thruſt it off cloſe to the wood of the branch or ſprigge, and that you doe not leaue the eye of the budde behinde ſticking vpon the branch; for if that eye be left or loſt, your bud is worth nothing; you muſt caſt it away, and cut another that may haue that eye abiding within the budde on the inſide: you may perceiue if that eye be wanting, if you ſee an emptie hole in the place where the eye ſhould be, to fill it vp on the inſide thereof; thus hauing taken off your bud well and cleanly, which is ſet forth vnto you at the figures 3 and 4. preſently ſet it on the tree you would graft (for your ſmall bud can abide no delay, leſt by taking the ayre too long it become dry, and nothing worth) in this manner: Cut the barke of your tree you would graft in a ſmooth place, at what height you pleaſe, firſt aboue or ouerthwart, and then downe right in the middle thereof, more then an inch long, the figure whereof you ſhall haue at the figure 1. and then raiſe vp both ſides of the barke, firſt one, and then another, with the flat and thinne haft end of your knife, a prettie way inwards (for if the barke will not riſe eaſily, the ſtocke is not then fit to graft vpon) put in your budde into the cleft with the point downewards, holding the ſtalke of the leafe that is with the budde betweene your fingers of the one hand, and opening the cleft with the flat end of your knife with the other hand, that the head of your bud may be put cloſe vnder the ouerthwart cut in the ſtocke or tree (which muſt not be raiſed or ſtirred as the ſides are) & the eye of the bud ſtand iuſt in the middle of the ſlit that is downeright, and then cloſing the barke of the ſtocke or tree ſoftly vnto the bud thus put in with your fingers, let it be bound gently with a ſmall long peece of baſte, or other ſuch like ſoft thing, firſt aboue the eye, & then compaſſing it belowe as cloſe as you can, but not too hard in any caſe, vntil you haue bound it all ouer the ſlit you made, eſpecially the lower end, leſt any winde get in to dry and ſpoile it; and hauing tyed both ends thereof faſt, leaue it ſo for a fornight or ſomewhat more, in which ſpace it will take and hold, if it be well done, which you ſhall perceiue, if the bud abide green, and turne not blacke, when you haue vnlooſed the tying; for if it hold faſt to the tree, and be freſh and good, tye it vp gently againe, and ſo leaue it for a fortnight longer, or a moneth if you will, and then you may take away your binding cleane: this budde will (if no other miſchance happen vnto it) ſpring and ſhoote forth the next yeare, (and ſometimes the ſame yeare, but that is ſeldome) and therefore in the beginning of the yeare, cut off the head of the grafted tree about an handfull aboue the grafted place, vntill the graft be growne ſtrong, and then cut it off cloſe, that the head may be couered with the graft, and doe not ſuffer any buds to ſprout beſides the graft, either aboue or belowe it. If you graft diuers buds vpon one ſtocke (which is the beſt way) let that onely remaine and abide that ſhooteth beſt forth, and rubbe off, or take away the other: the ſeuerall parts of this grafting I haue cauſed to be expreſſed for your further information.

5. Grafting in the ſcutcheon is accounted another kinde of grafting, and differeth verie little from grafting in the budde: the difference chiefly conſiſteth in this, that in ſtead of the downe right ſlit, and that aboue ouerthwart, they take away iuſt ſo much barke of the great tree, as your bud is in bigneſſe, which vſually is a little larger then the former, and placing it therein, they binde it as formerly is ſaid: ſome vſe for this purpoſe a paire of compaſſes, to giue the true meaſure both of bud and ſtocke; this manner of grafting is moſt vſed vpon greater trees, whoſe young branches are too high to graft vpon in the former maner, and whoſe tops they cut off (for the moſt part) at the latter end of the next yeare after the bud is taken: both theſe waies were inuented to ſaue the loſſe of trees, which are more endangered by grafting in the ſtocke,

then

then any of thefe waies ; and befides, by thefe waies you may graft at a farre grea height without loffe.

Of the manner of grafting and propagating all fonts of Rofes.

HAuing now fpoken of the grafting of trees, let mee adioyne the properties of Rofes, which although they better fit a Garden then an Orchard, yet I could not in a fitter place expreffe them then here, both for the name and affinity of grafting, & becaufe I do not expreffe it in the firft part. All forts of Rofes may be grafted (although all forts are not, fome feruing rather for ftockes for others to be grafted on) as eafily as any other tree, & is only performed, by inoculating in the fame maner I haue fet downe in the former Chapter of grafting trees in the bud; for both ftocke and budde muft bee dealt with after the fame fafhion. And although fome haue boafted of grafting Rofes by flicing or whipping, as they call it, or in the ftocke, after the firft manner, fet downe in the former Chapter, yet I thinke it rather a bragge, not hauing feene or heard any true effect proceede from that relation. The fweete Briar or Eglantine, the white and the Damaske Rofes, are the chiefeft ftockes to graft vpon. And if you graft lowe or neare the ground, you may by laying downe that graft within the ground, after it hath bin fhot out well, and of a years growth, by pinning it faft downe with fhort ftickes, athwart or acroffe, caufe that grafted branch, by taking roote, to become a naturall Rofe, fuch as the graft was, which being feparated and tranfplanted after it hath taken root wel, will profper as well as any naturall fucker. And in this maner, by laying downe branchefe at length into the ground, if they be full of fpreading fmall branches, you may increafe all forts of Rofes quickly and plentifully; for they will fhoote forth rootes at the ioynt of euery branch: But as for the manner of grafting white Rofes or Damaske vpon Broome ftalkes or Barbary bufhes, to caufe them to bring forth double yellow Rofes, or vpon a Willowe, to beare greene Rofes, they are all idle conceits, as impoffible to be effected, as other things, whereof I haue fpoken in the ninth Chapter of my firft part, concerning a Garden of flowers, vnto which I referre you to be fatisfied with the reafons therealledged. And it is the more needleffe, becaufe we haue a naturall double yellow Rofe of it owne growing. The fowing of the feedes of Rofes (which are fometimes found vpon moft forts of Rofes, although not euery yeare, and in euerie place) hath bin formerly much vfed; but now the laying downe of the young fhootes is a way for increafe fo much vfed, being fafe and verie fpeedie to take, efpecially for thofe Rofes that are not fo apt to giue fuckers, that it hath almoft taken quite away the vfe of fowing of the feedes of Rofes, which yet if anie one bee difpofed to make the triall, they muft gather the feede out of the round heads, from amongft the doune, wherein they lye verie like vnto the berries of the Eglantine or fweete Briar bufh, and efpecially of thofe Rofes that bee of the more fingle kindes, which are more apt to giue berries for feed then the more double, although fometimes the double Rofes yeeld the like heads or berries. Their time of fowing is in the end of September (yet fome referue them vntill February) and their manner of nourfing is to bee tranfplanted, after the firft or fecond yeares growth, and tended carefully, that while they are young they be not loft for want of moifture in the dry time of Summer.

Chap. VI.

Certaine rules and obseruations in and after grafting, not remembred in the former Chapter.

THe time of some manners of grafting being not mentioned before, must here be spoken of. For the grafting of all sorts of trees in the stocke, the most vsuall time is from the middle of February vntill the middle of March, as the yeare and the countrie is more forward or backward, with vs about London wee neuer passe midde March : but because the May Cherrie is first ripe, and therefore of a very forward nature, it doth require to be grafted somewhat sooner then others. The time of gathering likewise, or cutting your grafts for grafting in the stocke, is to be obserued, that they bee not long gathered before they bee grafted, for feare of being too dry, which I commend, howsoeuer diuers say, if they be long kept they are not the worse; and therefore if you be forced to haue your grafts from farre, or by some other chance to keepe them long, be carefull to keepe them moist, by keeping their ends stucke in moist clay; but if neare hand, neglect no time I say after the cutting of them for their grafting, but either the same, or the next day, or verie speedily after, in the meane time being put into the ground to keepe them fresh. The grafts taken from old trees, because they are stronger, and shoote forth sooner, are to bee sooner grafted then those that are taken from younger trees : of a good branch may bee made two, and sometimes three grafts sufficient for anie reasonable stocke. For whipping, the time is somewhat later then grafting in the stocke, because it is performed on younger trees, which (as I said before) doe not so early bud or shoote forth as the elder. Inarching likewise is performed much about the later end of the grafting time in the stocke; for being both kindes thereof they require the same time of the yeare. The times of the other manners of graftings are before expressed, to bee when they haue shot forth young branches, from whence your buds must be taken; and therefore need not here againe to be repeated. If a graft in the stocke doth happen not to shoote forth when others do (so as it holdeth green) it may perchance shoot out a moneth or two after, & do well, or else after Midsummer, when a second time of shooting, or the after Spring appeareth : but haue an especiall care, that you take not such a graft that shal haue nothing but buds for flowers vpon it, and not an eye or bud for leaues (which you must be carefull to distinguish) for such a graft after it hath shot out the flowers must of necessitie dye, not hauing wherewith to maintaine it selfe. Also if your good graft doe misse, and not take, it doth hazzard your stocke at the first time, yet manie stockes doe recouer to be grafted the second time; but twice to faile is deadly, which is not so in the inoculating of buds in the greene tree : for if you faile therein three, or three times three, yet euerie wound being small, and the tree still growing greene, will quickly recouer it, and not be afterwards seen. Some vse to graft in the stocke the same yeare they remoue the stocke, to saue time, & a second checke by grafting; but I like better both in grafting in the stocke, and in the bud also, that your trees might be planted in the places where you would haue them growe, for a yeare or two at the least before you graft them, that after grafting there should be no remouall, I neede not be tedious, nor yet I hope verie sollicitous to remember many other triuiall, or at the least common knowne things in this matter. First, for the time to remoue trees, young or old, grafted or vngrafted, to be from a fortnight after Michaelmas vntill Candlemas, or if neede be, somewhat after, yet the sooner your remoue is, the better your trees will thriue, except it be in a very moist ground. For the manner or way to set them : *viz.* in the high and dry grounds set them deeper, both to haue the more moisture, and to be the better defended from windes; and in the lower and moister grounds shallower, and that the earth be mellow, well turned vp, and that the finer earth bee put among the small rootes, wherein they may spread, and afterwards gently troden downe, that no hollownesse remaine among the rootes: as also that after setting (if the time be not ouermoist) there may be some water powred to the rootes, to moisten and fasten them the better; and in the dry time of Summer, after the setting, let them not want moisture, if you will

haue

haue them thriue and profper; for the want thereof at that time, hath often killed manie a likely tree.   To ftake and fence them alfo if neede bee after they are new fet,
and fo to continue for two or three yeares after, is verie expedient,left windes or other
cafualties fpoile your paines, and ouerthrow your hopes.   And likewife to defend
your grafts from birds lighting on them, to breake or difplace them, to fticke fome
prickes or fharpe pointed ftickes longer then your graft into your clay, that fo they
may be a fure defence of it: As alfo to tye fome woollen cloathes about the lower end
of your ftockes, or thruft in fome thornes into the ground about the rootes, to defend
them from hauing their barkes eaten by Conies, or hurt by fome other noifome vermine.

Chap. VII.

Obferuations for the dreffing and well keeping of Trees and an Orchard in good order.

THere are two manner of waies to dreffe and keepe trees in good order, that
they may bee both gracefull and fruitfull; the one is for wall trees, the o-
ther is for ftandards: for as their formes are different, fo is their keeping or
ordering.   Wall trees, becaufe they are grafted lowe, and that their branches muft
be plafht or tackt vnto the wall to faften them, are to be fo kept, that all their branches
may be fuffered to growe, that fhoote forth on either fide of the bodie, and led either
along the wall, or vpright, and one to lappe ouer or vnder another as is conuenient,
and ftill with peeces of lifts, parings of felt, peeces of foft leather, or other fuch like
foft thing compaffing the armes or branches, faftened with fmall or great nailes, as
neede requireth, to the wals, onely thofe buds or branches are to be nipped or cut off,
that fhoot forward, and will not fo handfomely be brought into conformity, as is fitting; yet if the branches growe too thicke, to hinder the good of the reft, or too high
for the wall, they may, nay they muft be cut away or lopped off: and if anie dead
branches alfo happen to be on the trees, they muft be cut away, that the reft may haue
the more libertie to thriue. Diuers alfo by carefully nipping away the wafte and fuperfluous buds, doe keepe their trees in conformity, without much cutting. The time
to pruine or plafh, or tye vp wall trees, is vfually from the fall of the leafe, to the beginning of the yeare, when they begin to bloffome, and moft efpecially a little before
or after Chriftmas: but in any cafe not too late, for feare of rubbing off their buds.
Some I know doe plafh and tye vp their wall trees after bearing time, while the leaues
are greene, and their reafon is, the buds are not fo eafie or apt to bee rubbed from the
branches at that time, as at Chriftmas, when they are more growne: but the leaues
muft needes be very cumberfome, to hinder much both the orderly placing, and clofe
faftening of them to the wall.   This labour you muft performe euery yeare in its due
time; for if you fhall neglect and ouerflip it, you fhall haue much more trouble, to
bring them into a fit order againe, then at the firft.   The ftandard trees in an Orchard
muft be kept in another order; for whereas the former are fuffered to fpread at large,
thefe muft be pruined both from fuperfluous branches that ouerload the trees, & make
them leffe fruitfull, as well as leffe fightly, and the vnder or water boughes likewife,
that drawe much nourifhment from the trees, and yet themfelues little the better for
it, I meane to giue fruit. If therefore your Orchard confift of young trees, with a little care and paines it may bee kept in that comely order and proportion it was firft de-
ftined vnto; but if it confift of old growne trees, they will not without a great deale
of care and paines be brought into fuch conformitie, as is befitting good and comely
trees: for the marke of thofe boughes or branches that are cut off from young trees,
will quickly be healed againe, the barke growing quickly ouer them, whereby they
are not worfe for their cutting; but an old tree if you cut off a bough, you muft cut it
clofe and cleanly, and lay a fearcloth of tallow, waxe, and a little pitch melted together vpon the place, to keepe off both the winde, funne, and raine, vntill the barke
haue couered it ouer againe: and in this manner you muft deale with all fuch fhort
ftumps of branches, as are either broken fhort off with the winde, or by carelefneffe or
want

want of skill, or elfe fuch armes or branches as are broken off clofe, or fliued from the body of the tree : for the raine beating and falling into fuch a place, will in fhort time rotte your tree, or put it in danger, befides the deformity. Some vfe to fill vp fuch an hole with well tempered clay, and tacke a cloth or a peece of leather ouer it vntill it be recouered, and this is alfo not amiffe. Your young trees, if they ftand in anie good ground, will bee plentifull enough in fhooting forth branches ; bee carefull therefore if they growe too thicke, that you pruine away fuch as growe too clofe (and will, if they be fuffered, fpoile one another) as they may be beft fpared, that fo the funne, ayre, and raine may haue free acceffe to all your branches, which will make them beare the more plentifully, and ripen them the fooner and the more kindly. If anie boughes growe at the toppe too high, cut them alfo away, that your trees may rather fpread then growe too high. And fo likewife for the vnder boughes, or anie other that by the weight of fruit fall or hang downe, cut them off at the halfe, and they will afterwards rife and fhoote vpwards. You fhall obferue, that at all thofe places where anie branches haue been cut away, the fappe will euer bee readie to put forth : if therefore you would haue no more branches rife from that place, rubbe off or nippe off fuch buddes as are not to your minde, when they are new fhot : and thus you may keep your trees in good order with a little paines, after you haue thus pruined and dreffed them. One other thing I would aduertife you of, and that is how to preferue a fainting or decaying tree which is readie to perifh, if it be not gone too farre or paft cure, take a good quantitie of oxe or horfe bloud, mixe therewith a reafonable quantitie of fheepe or pigeons dung, which being laid to the roote, will by the often raines and much watering recouer it felfe, if there bee anie poffibilitie ; but this muft bee done in Ianuarie or Februarie at the furtheft.

Chap. VIII.

Diuers other obferuations to be remembred in the well keeping of an Orchard.

THere be diuers other things to be mentioned, whereof care muft be had, either to doe or auoide, which I thinke fit in this Chapter promifcuoufly to fet down, that there may be nothing wanting to furnifh you with fufficient knowledge of the care, paines, and cafualties that befall an Orchard : for it hath many enemies, and euery one laboureth as much as in them lye, to fpoile you of your pleafure, or profit, or both, which muft bee both fpeedily and carefully preuented and helped ; and they are thefe : Moffe, Caterpillars, Ants, Earwigs, Snailes, Moales, and Birds. If Moffe begin to ouergrowe your trees, looke to it betimes, left it make your trees barren : Some vfe to hacke, and croffehacke, or cut the barke of the bodies of their trees, to caufe it fall away, but I feare it may endanger your trees. Others do either rubbe it off with a haire cloth, or with a long peece of wood formed like a knife, at the end of a long fticke or pole, which if it bee vfed cauteloufly without hurting the buds, I like better. Caterpillars, fome fmoake them with burning wet ftrawe or hay, or fuch like ftuffe vnder the trees ; but I doe not greatly like of that way : others cut off the boughes whereon they breed, and tread them vnder their feete, but that will fpoile too manie branches ; and fome kill them with their hands : but fome doe vfe a new deuifed way, that is, a pompe made of lattin or tin, fpout-fafhion, which being fet in a tubbe of water vnder or neare your trees, they will caufe the water to rife through it with fuch a force, and through the branches, that it will wafh them off quickly. To deftroy Ants, that eate your fruit before and when it is ripe, fome vfe to annoint the bodies of their trees with tarre, that they may not creepe vp on the branches ; but if that doe not helpe, or you will not vfe it, you muft be carefull to finde out their hill, and turne it vp, pouring in fcalding water, either in Summer, but efpecially if you can in Winter, and that will furely deftroy them. I haue fpoken of Earwigs in the firft part of this worke, entreating of the annoyances of Gilloflowers, and therefore I referre you thereunto : yet one way more I

will

will here relate which some doe vse, and that is with hollow canes of halfe a yard long or more, open at both ends for them to creepe in, and stucke or laid among the branches of your trees, will soone drawe into them many Earwigs, which you may soone kill, by knocking the cane a little vpon the ground, and treading on them with your foote. Snailes must be taken with your hands, and that euerie day, especially in the morning when they will be creeping abroad. Moales by running vnder your trees make them lesse fruitfull, and also put them in danger to be blowne downe, by leauing the ground hollow, that thereby the rootes haue not that strength in the ground, both to shoote and to hold, that otherwise they might haue. Some haue vsed to put Garlicke, and other such like things into their holes, thinking thereby to driue them away, but to no purpose : others haue tryed manie other waies; but no way doth auaile anie thing, but killing them either with a Moale spade, or a trappe made for the purpose as manie doe know : and they must bee watched at their principall hill, and trenched round, and so to be caught. Birds are another enemie both to your trees and fruit; for the Bullfinch will destroy all your stone fruit in the budde, before they flower, if you suffer them, and Crowes, &c. when your Cherries are ripe: for the smaller birds, Lime twigs set either neare your trees, or at the next water where they drinke, will helpe to catch them and destroy them. And for the greater birds, a stone bowe, a birding or fowling peece will helpe to lessen their number, and make the rest more quiet : or a mill with a clacke to scarre them away, vntill your fruit be gathered. Some other annoyances there are, as suckers that rise from the rootes of your trees, which must be taken away euerie yeare, and not suffered to growe anie thing great, for feare of robbing your trees of their liuelihood. Barke bound, is when a tree doth not shoote and encrease, by reason the barke is as it were drie, and will not suffer the sappe to passe vnto the branches : take a knife therefore, and slit the barke downe almost all the length of the tree in two or three places, and it will remedy that euill, and the tree will thriue and come forward the better after. Barke pilled is another euill that happeneth to some trees, as well young as old, either by reason of casuall hurts, or by the gnawing of beasts, howsoeuer it bee, if it bee anie great hurt, lay a plaister thereon made of tallow, tarre, and a little pitch, and binde it thereto, letting it so abide vntill the wound bee healed : yet some doe only apply a little clay or loame bound on with ropes of hay. The Canker is a shrewd disease when it happeneth to a tree; for it will eate the barke round, and so kill the very heart in a little space. It must be looked vnto in time before it hath runne too farre; most men doe wholly cut away as much as is fretted with the Canker, and then dresse it, or wet it with vinegar or Cowes pisse, or Cowes dung and vrine, &c. vntill it be destroyed, and after healed againe with your salue before appointed. There are yet some other enemies to an Orchard : for if your fence be not of bricke or stone, but either a mudde wall, or a quicke set or dead hedge, then looke to it the more carefully, and preuent the comming in of either horse, or kine, sheepe, goates, or deere, hare, or conie; for some of them will breake through or ouer to barke your trees, and the least hole almost in the hedge will giue admittance to hares and conies to doe the like. To preuent all which, your care must be continuall to watch them or auoide them, and to stoppe vp their entrance. A dogge is a good seruant for many such purposes, and so is a stone bowe, and a peece to make vse of as occasion shall serue. But if you will take that medicine for a Canker spoken of before, which is Cowes dung and vrine mixed together, and with a brush wash your trees often to a reasonable height, will keepe hares and conies from eating or barking your trees. Great and cold windes doe often make a great spoile in an Orchard, but great trees planted without the compasse thereof, as Wall-nuts, Oakes, Elmes, Ashes, and the like, will stand it in great stead, to defend it both early and late. Thus haue I shewed you most of the euils that may happen to an Orchard, and the meanes to helpe them, and because the number is great and daily growing, the care and paines must be continuall, the more earnest and diligent, lest you lose that in a moment that hath been growing many yeares, or at the least the profit or beauty of some yeares fruit.

CHAP.

The manner and way how to plant, order, and keepe other trees that beare greene leaues continually.

THe way to order thofe trees that beare their leaues greene continually, is dif-fering from all others that doe not fo : for neyther are they to bee planted or remoued at the time that all other trees are fet, nor doe they require that man-ner of dreffing, pruining and keeping, that others doe. And although many ignorant perfons and Gardiners doe remoue Bay trees, and are fo likewife perfwaded that all other trees of that nature, that is, that carry their greene leaues continually, may bee remoued in Autumne or Winter, as well as all other trees may bee; yet it is certaine it is a great chance if they doe thriue and profper that are fet at that time, or rather it is found by experience, that fcarce one of ten profpereth well that are fo ordered. Now in regard that there be diuers trees and fhrubs mentioned here in this booke that beare euer greene leaues, wherein there is very great beauty, and many take pleafure in them ; as the ordinary Bay, the Rofe Bay, and the Cherry Bay trees, the Indian Figge, the Cypreffe, the Pine tree, the Mirtle and dwarfe Boxe, and many others ; I will here fhew you how to plant and order them, as is fitteft for them. For in that they doe not fhed their greene leaues in winter as other trees doe, you may in reafon be perfwaded that they are of another nature; and fo they are indeede: for fee-ing they all grow naturally in warme Countries, and are from thence brought vnto vs, we muft both plant them in a warmer place, and tranfplant them in a warmer time then other trees be, or elfe it is a great hazzard if they doe not perifh and dye, the cold and frofts in the winter being able to pierce them through, if they fhould bee tranfplanted in winter, before they haue taken roote. You muft obferue and take this therefore for a certaine rule, that you alwaies remoue fuch trees or fhrubbes as are euer greene in the fpring of the yeare, and at no time elfe if you will doe well, that is, from the end of March, or beginning of Aprill, vnto the middle or end of May, efpecially your more dainty and tender plants, fhadowing them alfo for a while from the heate of the Sun, and giuing them a little water vpon their planting or tranfplanting; but fuch water as hath not prefently been drawn from a Well or Pumpe, for that will go neer to kill any plant, but fuch water as hath ftood in the open ayre for a day at the leaft, if not two or three. Yet for dwarfe Boxe I confeffe it may endure one moneth to be earlier planted then the reft, becaufe it is both a more hardy and lowe plant, and thereby not fo much fubiect to the extremitie of the colde : but if you fhould plant it before winter, the frofts would raife it out of the ground, becaufe it cannot fo foone at that time of the yeare take roote, and thereby put it in danger to be loft. Moreouer all of them will not abide the extremitie of our winter frofts, and therefore you muft of neceffity houfe fome of them, as the Rofe Bay, Mirtle, and fome others, but the other forts being fet where they may bee fomewhat defended from the cold windes, froftes, and fnow in winter, with fome couering or fhelter for the time, will reafonably well endure and beare their fruit, or the moft of them. If any be defirous to be furnifhed with ftore of thefe kinds of trees that will be nourfed vp in our Country, he may by fowing the feed of them in fquare or long woodden boxes or chefts made for that purpofe, gaine plenty of them : but hee muft be carefull to couer them in winter with fome ftraw or fearne, or beane hame, or fuch like thing layd vpon croffe fticks to beare it vp from the plants, and after two or three yeares that they are growne fomewhat great and ftrong, they may bee tranfplanted into fuch places you meane they fhall abide : yet it is not amiffe to defend them the firft yeare after they are tranfplanted, for their more fecuritie : the feedes that are moft vfually fowen with vs, are, the Cypreffe tree, the Pine tree, the Baye, the Pyracantha or prickly Corall tree, and the Mirtle : the Rofe Bay I haue had alfo rifen from the feede that was frefh, and brought me from Spaine. But as for Orenge trees, becaufe they are fo hardly preferued in this our cold climate (vnleffe it bee with fome that doe beftow the houfing of them, befides a great deale more of care and re-fpect vnto them) from the bitterneffe of our cold long winter weather (although their

kernels

kernels being put into the ground in the Spring or Summer, and if care bee had of them and conuenient keeping, will abide, and by grafting the good fruite on the crab stocke they may bee in time nurfed vp) I doe not make any other efpeciall account of them, nor giue you any further relation of their ordering. Now for the ordering of thefe trees after they are eyther planted of young fets, or tranfplanted from the feede, it is thus : Firft for Bay trees, the moft vfuall way is to let them grow vp high to bee trees, and many plant them on the North or Eaft fide of their houfes that they may not bee fcorched with the Sunne; but the bitter winters which we often haue, doe pinch them fhrewdly, infomuch that it killeth euen well growne trees fometimes downe to the roote : but fome doe make a hedge of them being planted in order, and keep them low by lopping of them continually, which will make them bufh and fpread. The Cyprefle tree is neuer lopped, but fuffered to grow with all the branches from a foote a-boue the ground, if it may be, ftraight vpright; for that is his natiue grace and greateft beautie, and therefore the more branches doe dye that they muft bee cut away, the more you deforme his propertie. The Pine tree may be vfed in the fame manner, but yet it wil better endure to fuftaine pruining then the Cyprefle, without any fuch defor-mitie. The Laurocerafus or Cherry Bay may be diuerfly formed, that is, it may be ei-ther made to grow into a tall tree by fhredding ftill away the vnder branches, or elfe by fuffering all the branches to grow to be a low or hedge bufh, & both by the fuckers and by laying downe the lower branches into the earth, you may foone haue much increafe ; but this way will caufe it to bee the longer before it beare anie fruit. The Rofe Baye will verie hardlie bee encreafed either by fuckers or by layers, but muft bee fuffered to grow without lopping, topping or cutting. The Pyracantha or Prickly Corall tree may bee made to grow into a reafonable tall tree by fhredding away the lower branches, or it may be fuffered to grow lowe into an hedge bufh, by fuffering all the branches to grow continually, you may alfo propagate it by the fuckers, or by lay-ing downe the lower branches. The Myrtle of all forts abideth a low bufh fpreading his branches full of fweete leaues and flowers, without anie great encreafe of it felfe, yet fometimes it giueth fuckers or fhootes from the rootes : but for the more fpeedie propagating of them, fome doe put the cuttings of them into the earth, and thereby in-creafe them. There are fome other trees that are not of any great refpeft, as the Yew tree, and the Savine bufh, both which may be encreafed by the cuttings, and therefore I need not make any further relation or amplification of them, and to fay thus much of them all, is (I thinke) fufficient for this Worke.

Chap. X.

The ordering, curing, and propagating Vines of all forts.

IN moft places of this countrie there is fmall care or paines taken about the orde-ring of Vines : it fufficeth for the moft part with them that haue anie, to make a frame for it to fpread vpon aboue a mans height, or to tacke it to a wall or win-dow, &c. and fo to let it hang downe with the branches and fruit, vntill the weight thereof, and the force of windes doe teare it downe oftentimes, and fpoile the grapes : and this way doth fomewhat refemble that courfe that the Vineyard keepers obferue in the hot countries of Syria, Spaine, and Italy, and in the furtheft parts of France as I hear likewife : for in moft of thefe hot countries they vfe to plant an Oliue betweene two Vines, and let them runne thereupon. But manie of the other parts of France, &c. doe not fuffer anie trees to growe among their Vines ; and therefore they plant them thicke, and pruine them much and often, and keepe them lowe in comparifon of the other way, faftening them to pearches or poles to hold them vp. And according to that fafhion many haue aduentured to make Vineyards in England, not onely in thefe later daies, but in ancient times, as may wel witneffe the fundrie places in this Land, en-tituled by the name of Vineyards; and I haue read, that manie Monafteries in this King-dome hauing Vineyards, had as much wine made therefrom, as fufficed their couents yeare by yeare : but long fince they haue been deftroyed, and the knowledge how to order a Vineyard is alfo vtterly perifhed with them. For although diuers, both No-

bles

bles and Gentlemen, haue in thefe later times endeauoured to plant and make Vine-yards, and to that purpofe haue caufed Frenchmen, being skilfull in keeping and dref-fing of Vines, to be brought ouer to performe it, yet either their skill failed them, or their Vines were not good, or (the moft likely) the foile was not fitting, for they could neuer make anie wine that was worth the drinking, being fo fmall and heartlefle, that they foone gaue ouer their practice. And indeede the foile is a maine matter to bee chiefly confidered to feate a Vineyard vpon: for euen in France and other hot coun-tries, according to the nature of the foile, fo is the rellifh, ftrength, and durabilitie of the wine. Now although I think it a fruitlefle labour for any man to ftriue in thefe daies to make a good Vineyard in England, in regard not only of the want of knowledge, to make choife of the fitteft ground for fuch Vines as you would plant therupon, but alfo of the true maner of ordering them in our country; but moft chiefly & aboue all others, that our years in thefe times do not fal out to be fo kindly and hot, to ripen the grapes, to make anie good wine as formerly they haue done ; yet I thinke it not amifle, to giue you inftructions how to order fuch Vines as you may nourfe vp for the pleafure of the fruit, to eate the grapes being ripe, or to preferue and keepe them to bee eaten almoft all the winter following : And this may be done without any great or extraordinarie paines. Some doe make a lowe wall, and plant their Vines againft it, and keepe them much about the height thereof, not fuffering them to rife much higher : but if the high bricke or ftone wals of your Garden or Orchard haue buttreffes thereat, or if you caufe fuch to bee made, that they bee fomewhat broade forwards, you may the more conueniently plant Vines of diuers forts at them, and by fticking down a couple of good ftakes at euery buttreffe, of eight or ten foot high aboue ground, tacking a few lathes acrofle vpon thofe ftakes, you may therunto tye your Vines, & carry them ther-on at your pleafure : but you muft be carefull to cut them euery year, but not too late, and fo keepe them downe, and from farre fpreading, that they neuer runne much be-yond the frame which you fet at the buttreffes : as alfo in your cutting you neuer leaue too many ioynts, nor yet too few, but at the third or fourth ioint at the moft cut them off. I doe aduife you to thefe frames made with ftakes and lathes, for the better ripe-ning of your grapes : for in the blooming time, if the branches of your vines bee too neare the wall, the reflection of the Sunne in the day time, and the colde in the night, doe oftentimes fpoile a great deale of fruit, by piercing and withering the tender foot-ftalkes of the grapes, before they are formed, whereas when the bloflomes are paft, and the fruit growing of fome bigneffe, then all the heate and reflection you can giue them is fit, and therefore cut away fome of the branches with the leaues, to admit the more Sunne to ripen the fruit. For the diuers forts of grapes I haue fet them downe in the Booke following, with briefe notes vpon euerie of them, whether white or blacke, fmall or great, early or late ripe; fo that I neede not here make the fame relation again. There doth happen fome difeafes to Vines fometimes, which that you may helpe, I thinke it conuenient to informe you what they are, and how to remedy them when you fhall be troubled with any fuch. The firft is a luxurious fpreading of branches and but little or no fruit : for remedie whereof, cut the branches fomewhat more neere then vfuall, and bare the roote, but take heed of wounding or hurting it, and in the hole put either fome good old rotten ftable dung of Horfes, or elfe fome Oxe blood new taken from the beafts, and that in the middle of Ianuarie or beginning of Febru-arie, which being well tempered and turned in with the earth, let it fo abide, which no doubt, when the comfort of the blood or dung is well foaked to the bottome by the raines that fall thereon, will caufe your Vine to fructifie againe. Another fault is, when a Vine doth not bring the fruit to ripeneffe, but either it withereth before it be growne of any bigneffe, or prefently after the blooming : the place or the earth where fuch a Vine ftandeth, affuredly is too cold, and therefore if the fault bee not in the place, which cannot bee helped without remouing to a better, digge out a good quantity of that earth, and put into the place thereof fome good frefh ground well heartned with dung, and fome fand mixed therewith (but not falt or falt water, as fome doe aduife, nor yet vrine, as others would haue) and this will hearten and ftrengthen your Vine to beare out the frut vnto maturitie. When the leaues of a Vine in the end of Summer or in Autumne, vntimely doe turne either yellow or red, it is a great figne the earth is too hot and drie ; you muſt therefore in ſtead of dung and ſand, as in the former defect is ſaid, put in ſome freſh loame or ſhort clay, well mixed together with ſome of the earth, and ſo let them abide, that the froſts may mellow them. And laſtly, a Vine ſometimes beareth ſome ſtore of grapes, but they are too many for it to bring to ripeneſſe ; you ſhall therefore helpe ſuch a Vine (which no doubt is of ſome excellent kinde, for they are moſt vſually ſubiect to this fault) by nipping away the bloſſomes from the branches, and leauing but one or two bunches at the moſt vpon a branch, vntill the Vine be growne older, and thereby ſtronger, and by this meane inured to beare out all the grapes to ripeneſſe. Theſe be all the diſeaſes I know doe happen to Vines : for the bleeding of a Vine it ſeldome happeneth of it ſelfe, but commeth either by cutting it vntimely, that is, too late in the yeare, (for after Ianuarie, if you will be well aduiſed, cut not any Vine) or by ſome caſuall or wilfull breaking of an arme or a branch. This bleeding in ſome is vnto death, in others it ſtayeth after a certaine ſpace of it ſelfe : To helpe this inconuenience, ſome haue ſeared the place where it bleedeth with an hot iron, which in many haue done but a little good ; others haue bound the barke cloſe with packe-thred to ſtay it ; and ſome haue tied ouer the place, being firſt dried as well as may bee, a plaiſter made with waxe roſſen and turpentine while it is warme. Now for the propagating of them : You muſt take the faireſt and goaleſt ſhot branches of one yeares growth, and cut them off with a peece of the old wood vnto it, and theſe being put into the ground before the end of Ianuarie at the furtheſt, will ſhoote forth, and take roote, and ſo become Vines of the ſame kinde from whence you tooke them. This is the moſt ſpeedy way to haue increaſe : for the laying downe of branches to take roote, doth not yeelde ſuch ſtore ſo plentifully, nor doe ſuckers riſe from the rootes ſo aboundantly ; yet both theſe waies doe yeelde Vines, that being taken from the old ſtockes will become young plants, fit to bee diſpoſed of as any ſhall thinke meete.

Chap. XI.

The way to order and preſerue grapes, fit to be eaten almoſt all the winter long, and ſometimes vnto the Spring.

ALthough it bee common and vſuall in the parts beyond the Sea to dry their grapes in the Sunne, thereby to preſerue them all the year, as the Raiſins of the Sunne are, which cannot bee done in our Countrie for the want of ſufficient heate thereof at that time : or otherwiſe to ſcald them in hot water (as I heare) and afterwards to dry them, and ſo keepe them all the yeare, as our Malaga Raiſins are prepared that are packed vp into Frayles : yet I doe intend to ſhew you ſome other waies to preſerue the grapes of our Countrie freſh, that they may be eaten in the winter both before and after Chriſtmas with as much delight and pleaſure almoſt, as when they were new gathered. One way is, when you haue gathered your grapes you intend to keepe, which muſt be in a dry time, and that all the ſhrunke, dried, or euill grapes in euery bunch be picked away, and hauing prouided a veſſell to hold them, be it of wood or ſtone which you will, and a ſufficient quantitie of faire and cleane drie ſand ; make *ſtratum ſuper ſtratum* of your grapes and the ſand, that is, a lay of ſand in the bottome firſt, and a lay of grapes vpon them, and a lay or ſtrowing againe of ſand vpon thoſe grapes, ſo that the ſand may couer euery lay of grapes a fingers breadth in thickneſſe, which being done one vpon another vntill the veſſell be full, and a lay of ſand vppermoſt, let the veſſell be ſtopped cloſe, and ſet by vntill you pleaſe to ſpend them, being kept in ſome drie place and in no ſellar : let them bee waſhed cleane in faire water to take away the ſand from ſo many you will ſpend at a time. Another way is (which Camerarius ſetteth downe he was informed the Turkes vſe to keepe grapes all the winter vnto the next ſummer) to take ſo much meale of Muſtard ſeede, as will ſerue to ſtrow vpon grapes, vntill they haue filled their veſſels, whereon afterwards they poure new wine before it hath boiled, to fill vp their veſſels therwith, and being ſtopped vp cloſe, they keepe them a certaine time, and ſelling them with their liquour to them that will

vſe

vſe them, they doe waſh the ſeedes or meale from them when they vſe them. Another way is, that hauing gathered the faireſt ripe grapes, they are to be caſt vpon threds or ſtrings that are faſtened at both ends to the ſide walks of a chamber, neere vnto the ſeeling thereof, that no one bunch touch another, which will bee ſo kept a great while, yet the chamber muſt be well defended from the froſts, and cold windes that pierce in at the windowes, leſt they periſh the ſooner : and ſome will dippe the ends of the branches they hang vp firſt in molten pitch, thinking by ſearing vp the ends to keepe the bunches the better ; but I doe not ſee any great likelihood therein. Your chamber or cloſet you appoint out for this purpoſe muſt alſo bee kept ſomewhat warme, but eſpecially in the more cold and froſtie time of the yeare, leſt it ſpoile all your coſt and paines, and fruſtrate you of all your hopes : but although the froſts ſhould pierce and ſpoile ſome of the grapes on a bunch, yet if you be carefull to keepe the place warme, the fewer will be ſpoiled. And thus haue I ſhewed you the beſt directions to order this Orchard rightly, and all the waies I know are vſed in our Countrie to keep grapes good anie long time after the gathering, in regard wee haue not that comfort of a hotter Sun to preſerue them by its heate.

The fruits themſelues ſhall follow euerie one in their order ; the lower ſhrubbes or buſhes firſt, and the greater afterwards.

THE THIRD PART
CALLED
THE ORCHARD,

Ontaining all forts of trees bearing fruit for mans vfe to eate, proper and fit for to plant an Orchard in our climate and countrie : I bound it with this limitation, becaufe both Dates, Oliues, and other fruits, are planted in the Orchards of Spaine, Italy, and other hot countries, which will not abide in ours. Yet herein I will declare whatfoeuer Art, ftriuing with Nature, can caufe to profper with vs, that whofoeuer will, may fee what can bee effected in our countrie. And firft to begin with the lower fhrubbes or bufhes, and after afcend to the higher trees.

Chap. I.

Rubus Idæus. Rafpis.

THe Rafpis berrie is of two forts, white and red, not differing in the forme either of bufh, leafe, or berry, but onely in the colour and tafte of the fruit. The Rafpis bufh hath tender whitifh ftemmes, with reddifh fmall prickes like haires fet round about them, efpecially at the firft when they are young; but when they grow old they become more wooddy and firme, without any fhew of thornes or prickles vpon them, and hath onely a little hairineffe that couereth them : the leaues are fomewhat rough or rugged, and wrinkled, ftanding three or fiue vpon a ftalke, fomewhat like vnto Rofes, but greater, and of a grayer greene colour : the flowers are fmall, made of fine whitifh round leaues, with a dafh as it were of blufh caft ouer them, many ftanding together, yet euery one vpon his owne ftalke, at the tops of the branches; after which come vp fmall berries, fomewhat bigger then Strawberries, and longer, either red or white, made of many graines, more eminent then in the Strawberry, with a kinde of dounineffe caft ouer them, of a pleafant tafte, yet fomewhat fowre, and nothing fo pleafant as the Strawberrie. The white Rafpis is a little more pleafant then the red, wherein there is fmall feede inclofed : the rootes creepe vnder ground verie farre, and fhoote vp againe in many places, much encreafing thereby.

There is another whofe ftemme and branches are wholly without prickles : the fruit is red, and fomewhat longer, and a little more fharpe.

The Vfe of Rafpis.

The leaues of Rafpis may be vfed for want of Bramble leaues in gargles, and other decoctions that are cooling and drying, although not fully to that effect.

A a a 3 The

The Conserue or Syrupe made of the berries, is effectuall to coole an hot ftomacke, helping to refresh and quicken vp thofe that are ouercome with faintnesse.

The berries are eaten in the Summer time, as an afternoones difh, to pleafe the tafte of the ficke as well as the found.

The iuyce and the diftilled water of the berries are verie comfortable and cordiall.

It is generally held of many, but how true I know not, that the red wine that is vfually fold at the Vintners, is made of the berries of Rafpis that grow in colder countries, which giueth it a kinde of harfhnesse: And alfo that of the fame berries growing in hotter climates, which giueth vnto the wine a more pleafant fweetnesse, is made that wine which the Vintners call **Alligant**: but we haue a Vine or Grape come to vs vnder the name of the **Alligant** Grape, as you fhall finde it fet downe hereafter among the Grapes; and therefore it is likely to be but an opinion, and no truth in this, as it may be alfo in the other.

Chap. II.

Ribes rubra, alba, nigra. Currans red, white, and blacke.

THe bufhes that beare thofe berries, which are vfually called red Currans, are not thofe Currans either blew or red, that are fold at the Grocers, nor any kind thereof; for that they are the grapes of a certaine Vine, as fhall be fhewed by and by: but a farre differing kinde of berry, whereof there are three forts, red, white, and blacke.

The red Curran bufh is of two forts, and groweth to the height of a man, hauing fometimes a ftemme of two inches thicknesse, and diuers armes and branches, couered with a fmooth, darke, brownifh barke, without anie pricke or thorne at all vpon anie part thereof, whereon doe growe large cornered blackifh greene leaues, cut in on the edges, feeming to be made of fiue parts, almoft like a Vine leafe, the ends a little pointing out, and ftanding one aboue another on both fides of the branches: the flowers are little and hollow, comming forth at the ioynts of the leaues, growing many together on a long ftalke, hanging downe aboue a fingers length, and of an herbie colour: after which come fmall round fruit or berries, greene at the firft, and red as a Cherry when they are ripe, of a pleafant and tart tafte: the other differeth not in anie other thing then in the berries, being twice as bigge as the former: the roote is wooddy, and fpreadeth diuerfly.

The white Curran bufh rifeth vfually both higher then the red, and ftraighter or more vpright, bigger alfo in the ftemme, and couered with a whiter barke: the leaues are cornered, fomewhat like the former, but not fo large: the flowers are fmall and hollow like the other, hanging downe in the fame manner on long ftalkes, being of a whiter colour: the berries likewife growe on the long ftalkes, fomewhat thicker fet together, and of a cleare white colour, with a little blacke head, fo tranfparent that the feedes may be eafily feene thorough them, and of a more pleafant winie tafte then the red by much.

The blacke Curran bufh rifeth higher then the white, with more plentifull branches, and more pliant and twiggie: the ftemme and the elder branches being couered with a brownifh barke, and the younger with a paler: the flowers are alfo like vnto little bottles as the others be, of a greenifh purple colour, which turne into blacke berries, of the bignesse of the fmaller red Currans: the leaues are fomewhat like vnto the leaues of the red Currans, but not fo large: both branches, leaues, and fruit haue a kind of ftinking fent with them, yet they are not vnwholfome, but the berries are eaten of many, without offending either tafte or fmell.

The Vfe of Currans.

The red Currans are vfually eaten when they are ripe, as a refrefhing to an

hot

1 *Rubus Idæus.* The Raſpis.   2 *Ribes fructu rubro vel albo.* White or red Currans.   3 *Groſſularia vulgaris.* The ordinary Gooſeberry
4 *Groſſularia fructu rubro.* The great red Gooſeberry.   5 *Groſſularia aculeata.* The prickly Gooſeberry. 6 *Oxyacantha ſeu Berberis,* The
Barbary buſh. 7 *Avellana Byzantina.* The Filberd of Conſtantinople. 8 *Avellana rubra noſtras.* The beſt red Filberd.

hot ſtomacke in the heate of the yeare, which by the tartneſſe is much delighted. Some preſerue them, and conſerue them alſo as other fruits, and ſpend them at neede.

The white Currans, by reaſon of the more pleaſant winie taſte, are more accepted and deſired, as alſo becauſe they are more daintie, and leſſe common.

Some vſe both the leaues and berries of the blacke Currans in ſawces, and other meates, and are well pleaſed both with the ſauour and taſte thereof, although many miſlike it.

Chap. III.

Vva Criſpa ſiue Groſſularia. Gooſeberries or Feaberries.

WEe haue diuers ſorts of Gooſeberries, beſides the common kinde, which is of three ſorts, ſmall, great, and long. For wee haue three red Gooſeberries, a blew and a greene.

The common Gooſeberrie, or Feaberrie buſh, as it is called in diuers Countries of England, hath oftentimes a great ſtemme, couered with a ſmooth darke coloured bark, without anie thorne thereon, but the elder branches haue here and there ſome on them, and the younger are whitiſh, armed with verie ſharpe and cruell crooked thorns, which no mans hand can well auoide that doth handle them, whereon are ſet verie greene and ſmall cornered leaues cut in, of the faſhion almoſt of Smallage, or Hawthorne leaues, but broad at the ſtalke : the flowers come forth ſingle, at euerie ioynt of the leafe one or two, of a purpliſh greene colour, hollow and turning vp the brims a little : the berries follow, bearing the flowers on the heads of them, which are of a pale greene at the firſt, and of a greeniſh yellow colour when they are ripe, ſtriped in diuers places, and cleare, almoſt tranſparent, in which the ſeede lyeth. In ſome theſe berries are ſmall and round ; in others much greater ; a third is great, but longer then the other : all of them haue a pleaſant winie taſte, acceptable to the ſtomacke of anie (but the long kinde hath both the thicker skin, and the worſer taſte of the other) and none haue been diſtempered by the eating of them, that euer I could heare of.

The firſt of the red Gooſeberries is better knowne I thinke then the reſt, and by reaſon of the ſmall bearing not much regarded ; the ſtemme is ſomewhat bigge, and couered with a ſmooth darke coloured barke, the younger branches are whiter, and without anie thorne or pricke at all, ſo long, weake, ſmall, and ſlender, that they lye vpon the ground, and will there roote againe : the leaues are like vnto the former Gooſeberries, but larger : the flowers and berries ſtand ſingle, and not manie to bee found anie yeare vpon them, but are ſomewhat long, and are as great as the ordinarie Gooſeberry, of a darke browniſh red colour, almoſt blackiſh when they are ripe, and of a ſweetiſh taſte, but without any great delight.

The ſecond red Gooſeberry riſeth vp with a more ſtraight ſtemme, couered with a browniſh barke ; the young branches are ſtraight likewiſe, and whitiſh, and grow not ſo thicke vpon it as the former red kinde, and without any thorne alſo vpon them : the leaues are like vnto the former red, but ſmaller : the berries ſtand ſingly at the leaues as Gooſeberries doe, and are of a fine red colour when they are ripe, but change with ſtanding to be of a darker red colour, of the bigneſſe of the ſmall ordinary Gooſeberry, of a pretty tart taſte, and ſomewhat ſweete withall.

The third red Gooſeberry which is the greateſt, and knowne but vnto few, is ſo like vnto the common great Gooſeberry, that it is hardly diſtinguiſhed : the fruit or berries grow as plentifully on the branches as the ordinary, and are as great & round as the great ordinary kinde, but reddiſh, and ſome of them paler, with red ſtripes.

The blew Gooſeberry riſeth vp to bee a buſh like vnto the red Curran, and of the ſame bigneſſe and height, with broader and redder leaues at the firſt ſhooting out, then the ſecond red Gooſeberry : the berries are more ſparingly ſet on the branches, then on the ſmall red, and much about the ſame bigneſſe, or rather leſſer, of the colour of a Damſon, with an ouerſhadowing of a blewiſh colour vpon them, as the Damſon hath, before it be handled or wiped away.

The greene prickly Gooſeberry is very like vnto the ordinary Gooſeberry in ſtemme and branches, but that they are not ſtored with ſo many ſharpe prickles; but the young ſhootes are more plentifull in ſmall prickles about, and the greene leaſe is a little ſmaller: the flowers are alike, and ſo are the berries, being of a middle ſize, and not very great, greene when they are thorough ripe as well as before, but mellower, and hauing a few ſmall ſhort prickles, like ſmall ſhort haires vpon them, which are harmleſſe, and without danger to anie the moſt dainty and tender palate that is, and of a verie good pleaſant taſte. The ſeede hereof hath produced buſhes bearing berries, hauing few or no prickles vpon them.

The Vſe of Gooſeberries.

The berries of the ordinary Gooſeberries, while they are ſmall, greene, and hard, are much vſed to bee boyled or ſcalded to make ſawce, both for fiſh and fleſh of diuers ſorts, for the ſicke ſometimes as well as the ſound, as alſo before they be neere ripe, to bake into tarts, or otherwiſe, after manie faſhions, as the cunning of the Cooke, or the pleaſure of his commanders will appoint. They are a fit diſh for women with childe to ſtay their longings, and to procure an appetite vnto meate.

The other ſorts are not vſed in Cookery that I know, but ſerue to bee eaten at pleaſure; but in regard they are not ſo tart before maturity as the former, they are not put to thoſe vſes they be.

CHAP. IIII.

Oxyacantha, ſed potius Berberis. Barberries.

THe Barberry buſh groweth oftentimes with very high ſtemmes, almoſt two mens height, but vſually ſomewhat lower, with manie ſhootes from the roote, couered with a whitiſh rinde or barke, and yellow vnderneath, the wood being white and pithy in the middle: the leaues are ſmall, long, and very greene, nicked or finely dented about the edges, with three ſmall white ſharpe thornes, for the moſt part ſet together at the ſetting on of the leaues: the flowers doe growe vpon long cluſtering ſtalkes, ſmall, round, and yellow, ſweete in ſmell while they are freſh, which turne into ſmall, long, and round berries, white at the firſt, and very red when they are ripe, of a ſharpe ſowre taſte, fit to ſet their teeth on edge that eate them: the roote is yellow, ſpreading far vnder the vpper part of the ground, but not very deepe.

There is (as it is thought) another kinde, whoſe berries are thrice as bigge as the former, which I confeſſe I haue not ſeene, and know not whether it be true or no: for it may peraduenture be but the ſame, the goodneſſe of the ground and ayre where they growe, and the youngneſſe of the buſhes cauſing that largeneſſe, as I haue obſerued in the ſame kinde, to yeeld greater berries.

There is ſaid to be alſo another kinde, whoſe berries ſhould be without ſtones or ſeede within them, not differing elſe in anie thing from the former: but becauſe I haue long heard of it, and cannot vnderſtand by all the inquirie I haue made, that any hath ſeene ſuch a fruit, I reſt doubtfull of it.

The Vſe of Barberries.

Some doe vſe the leaues of Barberries in the ſtead of Sorrell, to make ſawce for meate, and by reaſon of their ſowreneſſe are of the ſame quality.

The berries are vſed to be pickled, to ſerue to trimme or ſet out diſhes of fiſh and fleſh in broth, or otherwiſe, as alſo ſometime to bee boyled in the broth, to giue it a ſharpe relliſh, and many other wayes, as a Maſter Cooke can better tell then my ſelfe.

The

The berries are preſerued and conſerued to giue to ſicke bodies, to helpe to coole any heate in the ſtomacke or mouth, and quicken the appetite.

The depurate iuyce is a fine menſtrue to diſſolue many things, and to verie good purpoſe, if it be cunningly handled by an Artiſt.

The yellow inner barke of the branches, or of the rootes, are vſed to be boyled in Ale, or other drinkes, to be giuen to thoſe that haue the yellow iaundiſe: As alſo for them that haue anie fluxes of choller, to helpe to ſtay and binde.

Cluſius ſetteth downe a ſecret that hee had of a friend, of a cleane differing propertie, which was, that if the yellow barke were laid in ſteepe in white wine for the ſpace of three houres, and afterwards drunke, it would purge one very wonderfully.

Chap. V.

Nux Auellana. The Filberd.

THe Filberd tree that is planted in Orchards, is very like vnto the Haſell nut tree that groweth wilde in the woods, growing vpright, parted into many boughes and tough plyable twigges, without knots, couered with a browniſh, ſpeckled, ſmooth, thinne rinde, and greene vnderneath: the leaues are broad, large, wrinkled, and full of veines, cut in on the edges into deepe dents, but not into any gaſhes, of a darke greene colour on the vpperſide, and of a grayiſh aſh colour vnderneath: it hath ſmall and long catkins in ſtead of flowers, that come forth in the Winter, when as they are firme and cloſe, and in the Spring open themſelues ſomewhat more, growing longer, and of a browniſh yellow colour: the nuts come not vpon thoſe ſtalkes that bore thoſe catkins, but by themſelues, and are wholly incloſed in long, thicke, rough huskes, bearded as it were at the vpper ends, or cut into diuers long iagges, much more then the wood nut: the nut hath a thinne and ſomewhat hard ſhell, but not ſo thicke and hard as the wood nut, in ſome longer then in other, and in the long kinde, one hath the skinne white that couereth the kernels, and another red.

There is another ſort of the round kinde that came from Conſtantinople, whoſe huske is more cut, torne, or iagged, both aboue and belowe, then any of our country; the barke alſo is whiter, and more rugged then ours, and the leaues ſomewhat larger.

We haue had from Virginia Haſell nuts, that haue beene ſmaller, rounder, browner, thinner ſheld, and more pointed at the end then ours: I know not if any hath planted of them, or if they differ in leafe or any thing elſe.

The Vſe of Filberds.

Filberds are eaten as the beſt kinde of Haſell nuts, at bankets among other dainty fruits, according to the ſeaſon of the yeare, or otherwiſe, as euery one pleaſe: But Macer hath a Verſe, expreſſing prettily the nature of theſe nuts, which is,

Ex minimis nucibus nulli datur eſca ſalubris.

that is, There is no wholſome food or nouriſhment had from theſe ſmall kinde of nuts.

Yet they are vſed ſometime phyſically to be roſted, and made into a Lohoc or Electuary, that is vſed for the cough or cold. And it is thought of ſome, that Mithridates meant the kernels of theſe nuts, to be vſed with Figs and Rue for his Antidote, and not of Walnuts.

Chap. VI.

Vitis. The Vine.

THere is fo great diuerfities of Grapes, and fo confequently of Vines that beare them, that I cannot giue you names to all that here grow with vs : for Iohn Tradefcante my verie good friend, fo often before remembred, hath affured me, that he hath twentie forts growing with him, that hee neuer knew how or by what name to call them. One defcription therefore fhall ferue (as I vfe to doe in fuch varieties) for all the reft, with the names afterwards, of as many as we can giue, and the feuerall formes, colours and proportions of the grapes.

The manured Vine, in the places where it hath abiden long time, groweth to haue a great bodie, ftemme or trunke, fometimes of the bigneffe of a mans arme, fleeue and all, fpreading branches if it bee fuffered without end or meafure, but vfually ftored with many armes or branches, both old and new, but weake, and therefore muft bee fuftained; whereof the old are couered with a thin fcaly rinde, which will often chap and peele off of it felfe; the youngeft being of a reddifh colour, fmooth and firme, with a hollowneffe or pith in the middle : from the ioints of the young branches, and fometimes from the bodie of the elder, breake out on euerie fide broade greene leaues, cut on the edges into fiue diuifions for the moft part, and befides notched or dented about : right againft the leafe, and likewife at other ioynts of the branches, come forth long twining or clafping tendrels, winding themfelues about any thing ftandeth next vnto them : at the bottome of thefe leaues come forth clufters of fmall greenifh yellow bloomes or flowers, and after them the berries, growing in the fame manner in clufters, but of diuers formes, colours, taftes and greatneffe. For fome grapes are great, others leffe, fome very fmall (as the Currans that the Grocers fell) fome white, fome red, blew, blacke, or partie-coloured, fome are are as it were fquare, others round : fome the clufters are clofe, others open, fome are fweete, others fower or harfh, or of fome other mixed tafte; euerie one differing from others, verie notably either in tafte, colour or forme; within euerie one of which grapes, (and yet there is a grape without ftones) are contained one, two, or more kernels or ftones, fome of them being fmal, others greater : the rootes fpread far and deepe. They that keepe their Vines in the beft order, doe cut them low, not fuffering them to grow high, or with too many branches, whereby they grow the better, take vp the leffer roome, and bring their grapes fairer and fweeter.

The kinds of Vines and Grapes.

Our ordinarie Grape both white and red, which excelleth Crabs for veriuice, and is not fit for wine with vs.

The white Mufcadine Grape is a verie great Grape, fweete and firme, fome of the bunches haue weighed fixe pound, and fome of the grapes halfe an ounce.

The redde Mufcadine is as great as the white, and chiefly differeth in colour.

The Burlet is a very great white Grape, but fitter for veriuice then wine for the moft part; yet when a hot yeare happeneth fit for it, the Grape is pleafant.

The little blacke Grape that is ripe very early.

The Raifin of the Sunne Grape is a very great Grape, and very great clufters, of a reddifh colour when it is ripe with vs, yet in an extraordinarie hot yeare, it hath got a little blewneffe caft ouer it by the heat: but naturally verie blew.

The Curran Grape (or the Grape of Corinth) is the leaft Grape of all, and beareth both few, and verie feldome with vs, but in reafonable great clufters, and of a blackifh blew colour, when they are ripe with vs, and very

fweet

fweete. There is another fort of them that are red or browne, and of a fow-er tafte, nothing fo fweete.

The Greeke wine Grape is a blackifh Grape, and very fweete.

The Frontignack is a white Grape, of a verie fweete and delicate tafte, as the wine declareth, that fmelleth as it were of Muske.

The fquare Grape is reported to beare a Grape not fully round, but fided, or as it were fquare, whereby it became fo called.

The Damafco Grape is a great white grape, very fweete, and is the true *Vva Zibeba*, that the Apothecaries fhould vfe in the *Trochifc: Ciphi:* and fuch wee haue had in former times come ouer vnto vs in great, long and round white boxes, containing halfe an hundred weight a peece.

The Ruffet Grape is a reafonable faire grape, exceeding fweet and whi-tifh, with a thicke skinne, crufted ouer with a fhew of afh colour.

The white long Grape is like vnto a Pigeons egge, or as it were pointed pendent like a Pearle.

The partie-coloured Grape is a reafonable great Grape, and difcoloured when it is ripe, fometimes the whole bunches, and fometimes but fome of the grapes being parted whitifh, and blacke halfe through, verie variably.

The Rhenifh wine Grape is a white Grape, and endureth the cold of win-ter when it commeth earely, more then the Mufcadine before fet downe, and is nothing fo fweete.

The White wine Grape is verie like vnto the Rhine Grape, the foile only and climate adding more fweetneffe vnto the one then to the other.

The Claret wine Grape is altogether like the white Grape, but that it is not white, but of a reddifh colour, which lying bruifed vpon the skins be-fore they are preffed, giue that Claret tincture to the wine.

The Teint is a Grape of a deeper or darker colour, whofe iuice is of fo deepe a colour, that it ferueth to colonr other wine.

The Burfarobe is a faire fweete white Grape of much efteeme about Paris.

The Alligant is a verie fweete Grape, giuing fo deep and liuely a coloured red wine, that no other whatfoeuer is comparable to it, and therfore vfually called Spaniards blood.

The blew or blacke Grape of Orleans is another blacke Grape, giuing a darke coloured fweete wine much commended in thofe parts.

The Grape without ftones is alfo a kinde by it felfe, and groweth natural-ly neere Afcalon, as Brochard affirmeth, the wine whereof is redde, and of a good tafte.

The Virginia Vine, whereof I muft needes make mention among other Vines, beareth fmall Grapes without any great ftore of iuice therein, and the ftone within it bigger then in any other Grape : naturally it runneth on the ground, and beareth little.

The Vfe of Vines, Grapes, and other parts that come of them.

The greene leaues of the Vine are cooling and binding, and therefore good to put among other herbes that make gargles and lotions for fore mouthes.

And alfo to put into the broths and drinke of thofe that haue hot burning feauers, or any other inflammation.

They may (as it is held for true) womens longings, if they be either taken inwardly, or applyed outwardly.

Wine is vfually taken both for drinke and medicine, and is often put into fawces, broths, cawdles, and gellies that are giuen to the ficke. As al-fo into diuers Phyficall drinkes, to be as a *vehiculum* for the properties of the ingredients.

It is diftilled likewife after diuers manners, with diuers things, for diuers & fundry waters to drinke, & for diuers purpofes both inward and outward.

As

1 *Vuæ nigræ minores.* The small blacke Grape. 2 *Vuæ cæruleæ maiores.* The great blew Grape. 3 *Vuæ Moschatellinæ.* The Muscadine Grape. 4 *Vuæ Burutenses.* The Burlet Grape. 5 *Vuæ insolatæ.* The Raysins of the sunne Grape. 6 *Ficus.* The Figge tree:

Also distilled of it selfe, is called Spirit of wine, which serueth to dissolue, and to draw out the tincture of diuers things, and for many other purposes.

The iuice or veriuice that is made of greene hard grapes, before they be ripe, is vsed of the Apothecaries to be made into a Syrupe, that is very good to coole and refresh a faint stomacke.

And being made of the riper grapes is the best veriuice, farre exceeding that which is made of crabs, to be kept all the yeare, to be put both into meates and medicines.

The grapes of the best sorts of Vines are pressed into wine by some in these dayes with vs, and much more as I verily beleeue in times past, as by the name of Vineyard giuen to many places in this Kingdome, especially where Abbies and Monasteries stood, may bee coniectured: but the wine of late made hath beene but small, and not durable, like that which commeth from beyond Sea, whether our vnkindly yeares, or the want of skill, or a conuenient place for a Vineyard be the cause, I cannot well tell you.

Grapes of all sorts are familiarly eaten when they are ripe, of the sicke sometimes as well as the sound.

The dryed grapes which we call great Rayfins, and the Currans which we call small Rayfins, are much vsed both for meates, broths, and sawces, in diuers manners, as this Countrey in generall aboue any other, wherein many thousands of Frailes full, Pipes, Hogs-heads, and Buts full are spent yearly, that it breedeth a wonder in them of those parts where they growe and prouide them, how we could spend so many.

The Rayfins of the Sunne are the best dryed grapes, next vnto the Damasco, and are very wholsome to eate fasting, both to nourish, and to helpe to loosen the belly.

The dryed Lees of wine called Argoll or Tartar, is put to the vse of the Goldsmith, Dyer, and Apothecary, who doe all vse it in seuerall manners, uery one in his art.

Of it the Apothecaries make *Cremor Tartari,* a fine medicine to bee vsed, as the Physitian can best appoint, and doth helpe to purge humours by the stoole.

Thereof likewise they make a kinde of water or oyle, fit to bee vsed, to take away freckles, spots, or any such deformities of the face or skinne, and to make it smooth. It causeth likewise haire to growe more aboundantly in those places where it naturally should growe.

The liquor of the Vine that runneth forth when it is cut, is commended to be good against the stone wheresoeuer it be; but that liquor that is taken from the end of the branches when they are burnt, is most effectuall to take away spots and markes, ring-wormes and tetters in any place.

Chap. VII.

Ficus. The Figge tree.

THe Figge trees that are noursed vp in our country are of three forts, whereof two are high; the one bearing against a wall goodly sweete and delicate Figs, called Figs of Algarua, and is blewish when it is ripe: the other tall kinde is nothing so good, neither doth beare ripe Figges so kindly and well, and peraduenture may be the white ordinary kinde that commeth from Spaine. The third is a dwarfe kinde of Figge tree, not growing much higher then to a mans body or shoulders, bearing excellent good Figges and blew, but not so large as the first kinde.

The Figge trees of all these three kindes are in leaues and growing one like vnto another, sauing for their height, colour, and sweetnesse of the fruit, hauing many armes or branches, hollow or pithy in the middle, bearing very large leaues, and somewhat thicke, diuided sometimes into three, but vsually into fiue sections, of a darke greene colour on the vpperside, and whitish vnderneath, yeelding a milkie iayce when it is

broken,

broken, as the branches also or the figges when they are greene : the fruit breaketh out from the branches without anie blossome, contrary to all other trees of our Orchard, being round and long, fashioned very like vnto a small Peare, full of small white grains or kernels within it, of a very sweete taste when it is ripe, and very mellow or soft, that it can hardly be carried farre without bruising.

The other two sorts you may easily know and vnderstand, by so much as hath been said of them. Take only this more of the Figge tree, That if you plant it not against a bricke wall, or the wall of an house, &c. it will not ripen so kindly. The dwarfe Figge tree is more tender, and is therefore planted in great square tubs, to be remoued into the sunne in the Summer time, and into the house in Winter.

The Vse of Figges.

Figges are serued to the table with Rayfins of the Sunne, and blanched Almonds, for a Lenten dish.

The Figs that growe with vs when they are ripe, and fresh gathered, are eaten of diuers with a little salt and pepper, as a dainty banquet to entertaine a freind, which seldome passeth without a cup of wine to wash them downe.

In Italy (as I haue beene enformed by diuers Gentlemen that haue liued there to study physicke) they eate them in the same manner, but dare not eate many for feare of a feuer to follow, they doe account them to be such breeders of bloud, and heaters of it likewise.

The Figges that are brought vs from Spaine, are vsed to make Ptisan drinkes, and diuers other things, that are giuen them that haue coughes or colds.

It is one of the ingredients also with Nuts and Rice, into Mithridates counterpoison.

The small Figges that growe with vs, and will not ripen, are preserued by the Comfitmakers, and candid also, to serue as other moist or candid banquetting stuffe.

Chap. VIII.

Sorbus. The Seruice tree.

THere are two kindes of Seruice trees that are planted in Orchards with vs, and there is also a wilde kinde like vnto the later of them, with Ashen leaues, found in the woods growing of it selfe, whose fruit is not gathered, nor vsed to bee eaten of any but birds. And there is another kinde also growing wilde abroad in many places, taken by the Country people where it groweth, to be a Seruice tree, and is called in Latine, *Aria Theophrasti,* whose leaues are large, somewhat like Nut tree leaues, but greene aboue, and grayish vnderneath : some doe vse the fruit as Seruices, and for the same purposes to good effect, yet both of these wilde kindes wee leaue for another worke, and here declare vnto you onely those two sorts are nourced vp in our Orchards.

The more common or ordinary Seruice tree with vs, is a reasonable great tree, couered with a smooth barke, spread into many great armes, whereon are set large leaues, very much cut in on the edges, almost like vnto a Vine leafe, or rather like vnto that kind of Maple, that is vsually called the Sycomore tree with vs : the flowers are white, and growe many clustering together, which after bring forth small browne berries when they are ripe, of the bignesse almost of Hasell nuts, with a small tuft, as if it were a crowne on the head, wherein are small blacke kernels.

The other kinde, which is more rare with vs, and brought into this Land by Iohn Tradescante, heretofore often remembred, hath diuers winged leaues, many set together like vnto an Ashen leafe, but smaller, and euery one endented about the edges : the flowers growe in long clusters, but nothing so many, or so close set as the wilde kinde : the fruit of this tree is in some round like an Apple, and in others a little longer

like

like a Peare, but of a more pleasant taste then the ordinarie kinde, when they are ripe and mellowed, as they vse to doe with both these kindes, and with Medlars.

The Vse of Seruices.

They are gathered when they growe to be neare ripe (and that is neuer before they haue felt some frosts) and being tyed together, are either hung vp in some warme roome, to ripen them thoroughly, that they may bee eaten, or (as some vse to doe) lay them in strawe, chaffe, or branne, to ripen them.

They are binding, fit to be taken of them that haue any scouring or laske, to helpe to stay the fluxe; but take heed, lest if you binde too much, more paine and danger may come thereof then of the scouring.

Chap. IX.

Mespilus. The Medlar tree.

THere are three sorts of Medlers: The greater and the lesser English, and the Neapolitan.

The great and the small English Medlar differ not one from the other in any thing, but in the size of the fruit, except that the small kinde hath some prickes or thornes vpon it, which the great one hath not, bearing diuers boughes or armes, from whence breake forth diuers branches, whereon are set long and somewhat narrow leaues, many standing together; in the middle whereof, at the end of the branch, commeth the flower, which is great and white, made of fiue leaues, broad at the ends, with a nicke in the middle of euery one; after which commeth the fruit, being round, and of a pale brownish colour, bearing a crowne of those small leaues at the toppe, which were the huske of the flower before, the middle thereof being somewhat hollow, and is harsh, able to choake any that shall eate it before it be made mellow, wherein there are certaine flat and hard kernels.

The Medlar of Naples groweth likewise to bee a reasonable great tree, spreading forth armes and branches, whereon are set many gashed leaues, somewhat like vnto Hawthorne leaues, but greater, and likewise diuers thornes in many places: the flowers are of an herbie greene colour, and small, which turne into smaller fruit then the former, and rounder also, but with a small head or crowne at the toppe like vnto it, and is of a more sweete and pleasant taste then the other, with three seeds only therein ordinarily.

The Vse of Medlars.

Medlars are vsed in the same manner that Seruices are, that is, to be eaten when they are mellowed, and are for the same purposes to binde the body when there is a cause: yet they as well as the Seruices, are often eaten by them that haue no neede of binding, and but onely for the pleasant sweetnesse of them when they are made mellow, and sometimes come as a dish of ripe fruit at their fit season, to be serued with other sorts to the table.

Chap. X.

Lotus. The Lote or Nettle tree.

THe first kinde of Lote tree, whereof Dioscorides maketh mention, is but of one kinde; but there are some other trees spoken of by Theophrastus, that may be referred thereunto, which may bee accounted as bastard kindes thereof, of which I meane to entreate in this Chapter, hauing giuen you before the description

of

1 *Sorbus legitima.* The true Seruice tree. 2 *Sorbus vulgaris siue Torminalis.* The ordinary Seruice tree. 3 *Mespilus vulgaris.* The common Medlar tree. 4 *Mespilus Aronia.* The Medlar of Naples. 5 *Lotus arbor.* The Nettle tree. 6 *Lotus Virginiana.* The Pishamin or Virginia Plumme. 7 *Cornus mas.* The Cornell Cherry tree.

Bbb 3

of another kinde hereof (by the opinion of good Authors) vnder the name of *Laurocerasus.*

The firſt or true Lote tree groweth to be a tree of a great height, whoſe bodie and elder branches are couered with a ſmooth darke greene barke, the leaues are ſomewhat rough in handling, of a darke greene colour, long pointed, and ſomewhat deepe dented about the edges, ſomewhat like vnto a Nettle leafe, and oftentimes growe yellow toward Autumne : the flowers ſtand here and there ſcattered vpon the branches; after which come round berries like vnto Cherries, hanging downewards vpon long foot-ſtalkes, greene at the firſt, and whitiſh afterwards; but when they are ripe they become reddiſh, and if they be ſuffered to hang too long on the branches, they grow blackiſh, of a pleaſant auſtere taſte, not to be miſliked, wherein is a hard round ſtone.

The ſecond, which is a baſtard kinde, and called *Guaiacum Patauinum*, groweth to bee a faire tree, with a ſmooth darke greene barke, ſhooting out many faire great boughes, and alſo ſlender greene branches, beſet with faire broad greene leaues, almoſt like vnto the leaues of the Cornell tree, but larger : the flowers growe along the branches cloſe vnto them, without any or with a very ſhort foote-ſtalke conſiſting of foure greene leaues, which are as the huſke, containing within it a purpliſh flower, made of foure leaues ſomewhat reddiſh : the fruit ſtandeth in the middle of the green huſke, greene at the firſt, and very harſh, but red and round when it is ripe, and ſomewhat like a Plumme, with a ſmall point or pricke at the head thereof, and of a reaſonable pleaſant taſte or relliſh, wherein are contained flat and thicke browne ſeeds or kernels, like vnto the kernels of *Caſſia Fiſtula*, ſomewhat hard, and not ſo ſtonie, but that it may ſomewhat eaſily be cut with a knife.

The third is called in Virginia *Piſhamin*, The Virginia Plumme (if it be not all one with the former Guaiacana, whereof I am more then halfe perſwaded) hath growne with vs of the kernels that were ſent out of Virginia, into great trees, whoſe wood is very hard and brittle, and ſomewhat white withall : the branches are many, and grow ſlender to the end, couered with a very thinne greeniſh bark, whereon doe grow many faire broad greene leaues, without dent or notch on the edges, and ſo like vnto the former *Guaiacum*, that I verily thinke it (as I before ſaid) to bee the ſame. It hath not yet borne flower or fruit in our Countrey that I can vnderſtand : but the fruit, as it was ſent to vs, is in forme and bigneſſe like vnto a Date, couered with a blackiſh skinne, ſet in a huſke of foure hard leaues, very firme like vnto a Date, and almoſt as ſweete, with great flat and thicke kernels within them, very like vnto the former, but larger.

The Vſe of theſe Lote trees.

The firſt ſort is eaten as an helper to coole and binde the body : the laſt, as Captaine Smith relateth in the diſcouery of Virginia, if the fruit be eaten while it is greene, and not ripe, is able by the harſh and binding taſte and quality to draw ones mouth awry (euen as it is ſaid of the former Guaiacana) but when it is thorough ripe it is pleaſant, as I ſaid before.

Chap. XI.

Cornus mas. The Cornell tree.

THe Cornell tree that is planted in Orchards, being the male (for the female is an hedge buſh) is of two ſorts, the one bearing red, the other whiter berries, which is very rare yet in our country, and not differing elſe.

It groweth to a reaſonable bigneſſe and height, yet neuer to any great tree, the wood whereof is very hard, like vnto horne, and thereof it obtained the name : the body and branches are couered with a rugged barke, and ſpreadeth reaſonable well, hauing ſomewhat ſmooth leaues, full of veines, plaine, and not dented on the edges: the flowers are many ſmall yellow tufts, as it were of ſhort haires or threads ſet together, which come forth before any leafe, and fall away likewiſe before any leafe bee much open : the fruit are long and round berries, of the bigneſſe of ſmall Oliues, with an

hard

hard round ftone within them, like vnto an Oliue ftone, and are of a yellowifh red when they are ripe, of a reafonable pleafant tafte, yet fomewhat auftere withall.

The white (as I faid) is like vnto the red, but onely that his fruit is more white when it is ripe.

The Vfe of the Cornelles.

They helpe to binde the body, and to ftay laskes, and by reafon of the pleafantneffe in them when they are ripe, they are much defired.

They are alfo preferued and eaten, both for rarity and delight, and for the purpofe aforefaid.

Chap. XII.

Cerafus. The Cherry tree.

THere are fo many varieties and differences of Cherries, that I know not well how to expreffe them vnto you, without a large relation of their feuerall formes. I will therefore endeauour after one generall defcription (as my cuftome is in many other the like variable fruits) to giue as briefe and fhort notes vpon all the reft, as I can both for leafe and fruit, that fo you may the better know what the fruit is, when you haue the name.

The Englifh Cherrie tree groweth in time to be of a reafonable bigneffe and height, fpreading great armes, and alfo fmall twiggy branches plentifully ; the leaues whereof are not verie large or long, but nicked or dented about the edges : the flowers come forth two or three or foure at the moft together, at a knot or ioynt, euerie one by it felfe, vpon his owne fmall and long footeftalke, confifting of fiue white leaues, with fome threds in the middle ; after which come round berries, greene at the firft, and red when they are through ripe, of a meane bigneffe, and of a pleafant fweete tafte, fomewhat tart withall, with a hard white ftone within it, whofe kernell is fomewhat bitter, but not vnpleafant.

The Flanders Cherrie differeth not from the Englifh, but that it is fomewhat larger, and the Cherry fomewhat greater and fweeter, and not fo fower.

The early Flanders Cherry is more rathe or early ripe, almoft as foone as the May Cherry, efpecially planted againft a wall, and of many falfe knaues or Gardiners are fold for May Cherrie trees.

The May Cherrie in a ftandard beareth ripe fruite later then planted againft a wall, where the berries will be red in the verie beginning of May fometimes.

The Arch-Dukes Cherrie is one of the faireft and beft cherries wee haue, being of a very red colour when it is ripe, and a little long more then round, and fomewhat pointed at the end, of the beft rellifh of any Cherrie whatfoeuer, and of a firme fubftance; fcarce one of twentie of our Nurferie men doe fell the right, but giue one for another: for it is an inherent qualitie almoft hereditarie with moft of them, to fell any man an ordinary fruit for whatfoeuer rare fruit he fhall aske for : fo little they are to be trufted.

The ounce Cherrie hath the greateft and broadeft leafe of any other cherrie, but beareth the fmalleft ftore of cherries euerie yeare that any doth, and yet bloffometh well : the fruit alfo is nothing anfwerable to the name being not verie great, of a pale yellowifh red, neere the colour of Amber, and therefore fome haue called it, the Amber Cherrie.

The great leafed Cherrie is thought of diuers to bee the Ounce Cherrie, becaufe it hath almoft as great a leafe as the former : but the fruit of this alfo doth not anfwer the expectation of fo great a leafe, being but of a meane bigneffe, and a fmall bearer, yet of a pale reddifh colour.

The true Gafcoign Cherry is known but to a few; for our Nurfery men do fo change the names of moft fruits they fell, that they deliuer but very few true names to any : In former times before our wilde blacke Cherrie was found to grow plentifully in our owne woods in many places of this Land, the French continually ftored vs with wilde ftockes to graft vpon, which then were called Gafcoigne ftocks, but fince they haue fo

termed

termed another red Cherrie, and obtruded it vpon their cuſtomers: but the true is one of our late ripe white Cherries, euen as Gerard ſaith, it is a great cherrie and ſpotted: and this is that Cherrie I ſo commend to be a fit ſtocke to graft May cherries vpon.

The Morello Cherrie is of a reaſonable bigneſſe, of a darke red colour when they are full ripe, and hang long on, of a ſweetiſh ſower taſte, the pulpe or ſubſtance is red, and ſomewhat firme: if they be dryed they will haue a fine ſharpe or ſower taſte very delectable.

The Hartlippe Cherrie is ſo called of the place where the beſt of this kinde is nourſed vp, being betweene Sittingbourne and Chattam in Kent, and is the biggeſt of our Engliſh kindes.

The ſmaller Lacure or Hart Cherrie is a reaſonable faire Cherrie, full aboue, and a little pointing downward, after the faſhion of an heart, as it is vſually painted, blackiſh when it is full ripe, and leſſer then the next.

The great Lacure or Hart Cherrie differeth not in forme, but in greatneſſe, being vſually twice as great as the former, and of a reddiſh blacke colour alſo: both of them are of a firme ſubſtance, and reaſonable ſweete. Some doe call the white cherrie, the White hart cherrie.

The Luke Wardes Cherrie hath a reaſonable large leafe, and a larger flower then many other: the cherries grow with long ſtalkes, and a ſtone of a meane ſize within them, of a darke reddiſh colour when they are full ripe, of a reaſonable good relliſh, and beareth well.

The Corone Cherrie hath a leafe little differing from the Luke Wardes cherrie, the fruit when it is ripe, is of a faire deepe red colour, of a good bigneſſe, and of a verie good taſte, neither verie ſweete or ſower: the pulpe or iuice will ſtaine the hands.

The Vrinall Cherrie in a moſt fruitfull yeare is a ſmall bearer, hauing many yeares none, and the beſt but a few; yet doth bloſſome plentifully euery yeare for the moſt part: the cherrie is long and round, like vnto an Vrinall, from whence it tooke his name; reddiſh when it is full ripe, and of an indifferent ſweete relliſh.

The Agriot Cherrie is but a ſmall Cherrie, of a deepe redde colour when it is ripe, which is late; of a fine ſharpe taſte, moſt pleaſant and wholſome to the ſtomacke of all other cherries, as well while they are freſh as being dryed, which manner they much vſe in France, and keepe them for the vſe both of the ſicke and ſound at all times.

The Biguarre Cherrie is a fair cherrie, much ſpotted with white ſpots vpon the pale red berry, and ſometimes diſcoloured halfe white and halfe reddiſh, of a reaſonable good relliſh.

The Morocco Cherrie hath a large white bloſſome, and an indifferent big berrie, long and round, with a long ſtalke of a darke reddiſh purple colour, a little tending to a blew when it is full ripe, of a firme ſubſtance: the iuice is of a blackiſh red, diſcolouring the hands or lips, and of a pleaſant taſte: Some doe thinke that this and the Morello be both one.

The Naples Cherrie is alſo thought to bee all one with the Morello or Morocco.

The white Spaniſh Cherrie is an indifferent good bearer, the leafe and bloſſome ſomewhat large, and like the Luke Wardes cherrie: the cherries are reaſonable faire berries, with long ſtalkes and great ſtones, white on the outſide, with ſome redneſſe, on the one ſide of a firme ſubſtance, and reaſonable ſweet, but with a little aciditie, and is one of the late ripe ones: But there is another late ripe white Cherry, which ſome call the Gaſcoigne, before remembred.

The Flanders cluſter Cherrie is of two ſorts, one greater then another: the greater kinde hath an indifferent large leafe; the bloſſomes haue many threds within them, ſhewing as it were many parts, which after turne into cluſters of berries, foure, fiue or ſixe together, and but with one ſtalke vnder them, as if they grew one out of another, and ſometimes they will beare but two or three, and moſt of them but one cherry on a ſtalke, which are red when they are ripe, very tender, and watriſh ſweete in eating.

The leſſer is in all things like the greater, but ſmaller, which maketh the difference.

The wilde cluſter or birds cluſter Cherry beareth many bloſſomes ſet all along the ſtalkes, and cherries after them in the ſame maner, like a long thinne bunch of grapes, and therefore called of ſome the Grape cherry: there are of them both red and blacke.

The

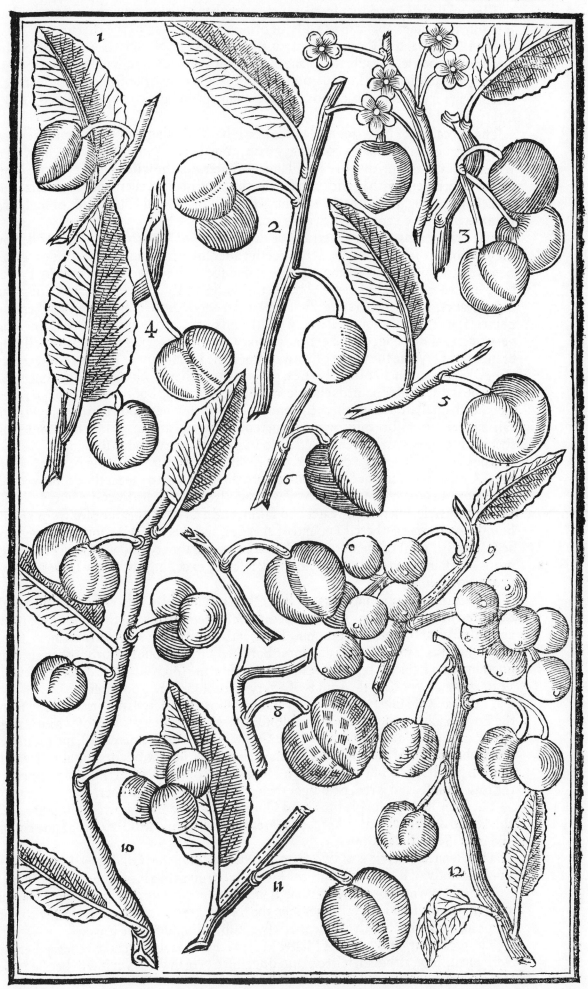

1 *Cerasus præcox*. The May Cherry. 2 *Cerasus Batauica*. The Flanders Cherry. 3 *Cerasus Hispanica siue alba*. The white Cherry. 4 *Cerasus plate-phyllos*. The great leafed Cherry. 5 *Cerasus Luca Wardi*. Luke Wards Cherry. 6 *Cerasus Neapolitana*. The Naples Cherry. 7 *Cerasus Cordata*. The Heart Cherry. 8 *Cerasus maculata*. The bignarre or spotted Cherry. 9 *Cerasus autumracemosa*. The wilde cluster Cherry. 10 *Cerasus Corymbifera*. The Flanders cluster Cherry. 11 *Cerasus Archiducis*. The Archdukes Cherry. 12 *Chamacerasus*. The dwarfe Cherry.

The soft sheld Cherrie is a small red cherrie when it is ripe, hauing the stone within it so soft and tender, that it may easily be broken in the eating of the cherrie.

Iohn Tradescantes Cherrie is most vsually sold by our Nursery Gardiners, for the Archdukes cherrie, because they haue more plenty thereof, and will better be increased, and because it is so faire and good a cherrie that it may be obtruded without much discontent: it is a reasonable good bearer, a faire great berrie, deepe coloured, and a little pointed.

The Baccalaos or New-found-land Cherrie hath a shining long leafe, most like vnto a Peach leafe, the blossomes come very many together as it were in an vmbell, which is such a cluster as is neither like the Flanders cluster, nor the wilde cluster cherrie blossome: it bringeth forth berries standing in the same manner euerie one vpon his own footestalke, being no bigger then the largest berrie of the red Curran tree or bush, of a pale or waterish red colour when it is ripe.

The strange long cluster Cherrie, or *Padus Theophrasti Dalechampio* is reckoned by the Author of that great Herball that goeth vnder his name, among the sorts of cherries ; and so must I vntill a fitter place be found for it. It groweth in time to be a great tree, with a sad coloured barke both on the bodie and branches, whereon doe grow many leaues, somewhat broade, shorter, harder, and a little more crumpled then any cherrie leafe: the blossomes are very small, and of a pale or whitish colour, smelling very sweete and strong, or rather heady, like Orenge flowers, growing on small long branches, very like the toppe of flowers vpon the Laburnum or Beane trefoile trees: after which come small blacke berries, growing together all along the long stalke, like vnto the wilde cluster or birds cherrie mentioned before, but not much bigger then tares, with small stones within them, and little or no sustance vpon them : the French call the tree *Putier*, because the wood thereof stinketh, and make it to be wonderfull that the blossomes of the tree should be so sweete, and the wood so stinking.

The Cullen Cherrie is a darke red cherrie like the Agriot, which they of those parts neere Cullen and Vtrecht &c. vse to put into their drinke, to giue it the deeper colour.

The great Hungarian Cherrie of Zwerts is like both in leafe and fruite vnto the Morello cherrie, but much greater and fairer, and a far better bearer: for from a small branch hath beene gathered a pound of cherries, and this is vsuall continually, and not accidentally, most of them foure inches in compasse about, and very many of them more of a faire deepe red colour, and very sweete, excelling the Arch-Dukes cherry, or any other whatsoeuer.

The Cameleon or strange changeable Cherry deseruedly hath this name, although of mine owne imposition, not only because it beareth vsually both blossomes, greene and ripe fruit at one time thereupon, but that the fruit will be of many formes ; some round, some as it were square, and some bunched forth on one side or another, abiding constant in no fashion, but for the most part shewing forth all these diuersities euerie yeare growing vpon it: the fruit is of a very red colour, and good taste.

The great Rose Cherry, or double blossomd Cherry differeth not in any thing from the English Cherrie, but only in the blossomes, which are very thicke of white leaues, as great and double as the double white Crowfoote, before remembred, and somtimes out of the middle of them will spring another smaller flower, but double also ; this seldome beareth fruit, but when it doth I suppose it commeth from those blossomes are the least double, and is red, no bigger then our ordinary English cherrie.

The lesser Rose or double blossomd Cherrie beareth double flowers also, but not so thicke and double as the former ; but beareth fruit more plentifully, of the same colour and bignesse with the former.

The Dwarfe Cherrie is of two sorts ; one whose branches fall downe low, round about the body of it, with small greene leaues, and fruit as small, of a deep red colour.

The other, whose branches, although small, grow more vpright, hauing greener shining leaues: the fruit is little bigger then the former, red also when it is ripe, with a little point at the end: both of them of a sweetish rellish, but more sower.

The great bearing Cherry of Master Millen is a reasonable great red cherry, bearing very plentifully, although it bee planted against a North wall, yet it will bee late ripe, but of an indifferent sweet and good rellish.

The long finger Cherry is another small long red one, being long & round like a finger, wherof it tooke the name: this is not the Vrinall cherry before, but differing from it.

The

The Vfe of Cherries.

All thefe forts of Cherries ferue wholly to pleafe the palate, and are eaten at all times, both before and after meales.

All Cherries are cold, yet the fower more then the fweete ; and although the fweete doe moft pleafe, yet the fower are more wholfome, if there bee regard taken in the vfing.

The Agriot or fower Cherries are in France much vfed to bee dryed (as is faid before) as Pruines are, and fo ferue to miniftred to be the fick in all hot difeafes, as feuers &c. being both boyled in their drinkes, and taken now and then of themfelues, which by reafon of their tartneffe, doe pleafe the ftomacke paffing well.

The Gum of the Cherrie tree is commended to bee good for thofe are troubled with the grauell or ftone. It is alfo good for the cough being diffolued in liquour, and ftirreth vp an appetite. The diftilled water of the blacke Cherries, the ftones being broken among them, is vfed for the fame purpofe, for the grauell, ftone, and winde.

CHAP. XIII.

Prunus. The Plumme tree.

THere are many more varieties of Plummes then of Cherries, fo that I muft follow the fame order with thefe that I did with them, euen giue you their names apart, with briefe notes vpon them, and one defcription to ferue for all the reft. And in this recitall I fhall leaue out the Apricockes which are certainly a kind of Plum, of an efpeciall difference, and not of a Peach, as Galen and fome others haue thought, and fet them in a chapter by themfelues, and only in this fet down thofe fruits are vfually called Plums.

The Plum tree (efpecially diuers of them) rifeth in time to bee a reafonable tall and great tree, whofe bodie and greater armes are couered with a more rugged barke, yet in fome more or leffe, the younger branches being fmooth in all, the leaues are fomewhat rounder then thofe of the Cherrie tree, and much differing among themfelues, fome being longer, or larger, or rounder then others, and many that are exercifed herein, can tell by the leafe what Plum the tree beareth (I fpeake this of many, not of all) as in many Cherries they can doe the like : the flowers are white, confifting of fiue leaues: the fruit is as variable in forme, as in tafte or colour, fome being ovall, or Peare fafhion or Almond like, or fphericall or round, fome firme, fome foft and waterifh, fome fweete, fome fower or harfh, or differing from all thefe taftes : and fome white, others blacke, fome red, others yellow, fome purple, others blew, as they fhall bee briefly fet downe vnto you in the following lines, where I meane not to infert any the wilde or hedge fruit, but thofe only are fit for an Orchard, to be ftored with good fruit: and of all which forts, the choyfeft for goodneffe, and rareft for knowledge, are to be had of my very good friend Mafter Iohn Tradefcante, who hath wonderfully laboured to obtaine all the rareft fruits hee can heare off in any place of Chriftendome, Turky, yea or the whole world ; as alfo with Mafter Iohn Millen, dwelling in Olde ftreete, who from Iohn Tradefcante and all others that haue had good fruit, hath ftored himfelfe with the beft only, and he can fufficiently furnifh any.

The Amber Primordian Plumme is an indifferent faire Plumme, early ripe, of a pale yellowifh colour, and of a waterifh tafte, not pleafing.

The red Primordian Plumme is of a reafonable fize, long and round, reddifh on the outfide, of a more dry tafte, and ripe with the firft forts in the beginning of Auguft.

The blew Primordian is a fmall plumme, almoft like the Damafcene, and is fubiect to drop off from the tree before it be ripe.

The white Date Plum is no very good plum.

The

The red Date plumme is a great long red pointed plumme, and late ripe, little better then the white.

The blacke Muſſell plumme is a good plumme, reaſonable drye, and taſteth well.

The red Muſſell Plumme is ſomewhat flat as well as round, of a very good taſte, and is ripe about the middle of Auguſt.

The white Muſſell plumme is like the redde, but ſomewhat ſmaller, and of a whitiſh greene colour, but not ſo well taſted.

The Imperiall plum is a great long reddiſh plum, very wateriſh, and ripeneth ſomewhat late.

The Gaunt plum is a great round reddiſh plum, ripe ſomewhat late, and eateth wateriſh.

The red Peſcod plum is a reaſonable good plum.

The white Peſcod plum is a reaſonable good relliſhed plumme, but ſomewhat wateriſh.

The greene Peſcod plum is a reaſonable big and long pointed plum, and ripe in the beginning of September.

The Orenge plum is a yellowiſh plum, moiſt, and ſomewhat ſweetiſh.

The Morocco plumme is blacke like a Damſon, well taſted, and ſomewhat drye in eating.

The Dine plum is a late ripe plum, great and whitiſh, ſpeckled all ouer.

The Turkie plum is a large long blackiſh plum, and ſomewhat flat like the Muſſell plum, a well relliſhed dry plum.

The Nutmeg plumme is no bigger then a Damſon, and is of a greeniſh yellow colour when it is ripe, which is with vs about Bartholmew tide, and is a good plum.

The Perdigon plumme is a dainty good plumme, early blackiſh, and well relliſhed.

The Verdoch plum is a great fine greene ſhining plum fit to preſerue.

The Ienua plum is the white Date plum, before remembred.

The Barberry plum is a great early blacke plum, and well taſted.

The Pruneola plum is a ſmall white plum, of a fine tart taſte : it was wont to bee vſually brought ouer in ſmall round boxes, and ſold moſt commonly at the Comfitmakers, (cut in twaine, the ſtone caſt away) at a very deere rate: the tree groweth and beareth well with vs.

The Shepway Bulleis is of a darke blewiſh brown colour, of a larger ſize then the ordinary, and of a ſharpe taſte, but not ſo good as the common.

The white and the blacke Bulleis are common in moſt Countries, being ſmall round plums, leſſer then Damſons, ſharper in taſte, and later ripe.

The Fluſhing Bulleis groweth with his fruite thicke cluſtring together like grapes.

The Winter Creke is the lateſt ripe plum of all ſorts, it groweth plentifully about Biſhops Hatfield.

The white Peare plum early ripe, is of a pale yellowiſh greene colour.

The late ripe white Peare plum is a greater and longer plum, greeniſh white, and is not ripe vntill it be neere the end of September, both wateriſh plums.

The blacke Peare plum is like vnto the white Peare plumme, but that the colour is blackiſh when it is ripe, and is of a very good relliſh, more firme and drye then the other.

The red Peare plumme is of the ſame faſhion and goodneſſe, but is the worſt of the three.

The white Wheate plum is a wateriſh fulſome plum.

The red Wheate plum is like the other for taſte.

The Bowle plum is flat and round, yet flatter on the one ſide then on the other, which cauſed the name, and is a very good relliſhed blacke plum.

The Friars plumme is a very good plum, well taſted, and comming cleane from the ſtone, being blacke when it is ripe, and ſome whitiſh ſpots vpon it.

The Catalonia plum is a very good plum.

The don Alteza is alſo a very good plum.

The Muſcadine plum, ſome call the Queene mother plumme, and ſome the Cherry plum, is a faire red plum, of a reaſonable bigneſſe, and ripe about Bartholmew tide.

The Chriſtian plum, called alſo the Nutmeg plum ; the tree groweth very ſhrubby,

and

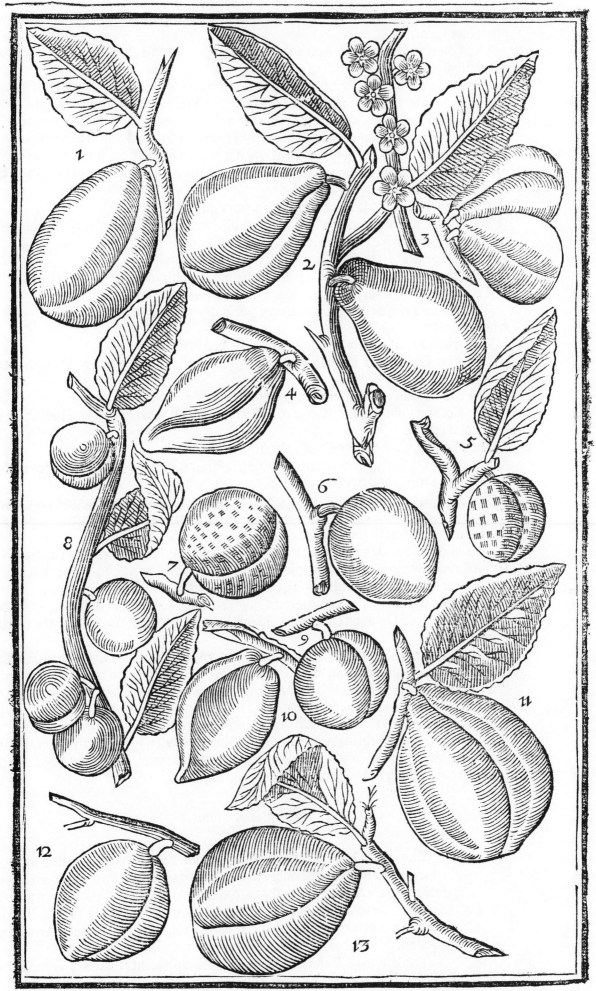

1 *Prunum Imperiale.* The Imperiall Plum. 2 *Prunum Turcicum.* The Turkey Plum. 3 *Prunum præcox rubrum.* The red Primordian Plum. 4 *Prunum Mytellinum.* The Mussell Plum. 5 *Prunum Ambarinum.* The Amber Plum. 6 *Prunum Regineum.* The Queene mother Plum. 7 *Prunum viride.* Tho green Oysterly Plum. 8 *Prunum Arantiacum.* The Orenge Plum. 9 *Prunum Myristicum.* The Nutmeg Plum. 10 *Prunum Siliquosum.* The Pescod Plum. 11 *Prunum Gandauense.* The Gaunt Plum. 12 *Prunum Dactylites.* The Date Plum. 13 *Prunum Pyrinum præcox.* The early Peare Plum.

and will abide good for six weekes at the least after it is gathered , and after all other plums are spent.

The Cherry plum remembred before, speaking of the Muscadine plum , is a very good plum, but small.

The Amber plum is a round plum, as yellow on the outside almost as yellow waxe, of a sowre vnpleasant taste that which I tasted, but I thinke it was not the right ; for I haue seene and tasted another of the same bignesse, of a paler colour, farre better relli-shed, and a firmer substance, comming cleane from the stone like an Apricocke.

The Apricocke plum is a good plum when it is in its perfection, but that is seldome; for it doth most vsually cracke , thereby diminishing much of its goodnesse , and be-sides yeeldeth gumme at the crackes.

The Eason plum is a little red plum, but very good in taste.

The Violet plum is a small and long blackish blew plum , ripe about Bartholmew tide, a very good dry eating fruit.

The Grape plum is the Flushing Bulleis before remembred

The Dennie plum is called also the Chefton, or the Friars plum before remembred.

The Damaske Violet plum, or Queen mother plum spoken of before.

The blacke Damascene plum is a very good dry plum , and of a darke blew colour when it is ripe.

The white Damson is nothing so well rellished as the other.

The great Damson or Damaske plum is greater then the ordinary Damson, and sweeter in taste.

The blew Damson well knowne, a good fruit.

The Coferers plum is flat, like vnto a Peare plum , it is early ripe and blacke, of a very good rellish.

The Margate plum the worst of an hundred.

The green Oysterly plum is a reasonable great plum, of a whitish green colour when it is ripe, of a moist and sweete taste, reasonable good.

The red Mirobalane plum groweth to be a great tree quickly, spreading very thicke and farre, very like the blacke Thorne or Sloe bush : the fruit is red, earlier ripe , and of a better taste then the white.

The white Mirobalane plum is in most things like the former red, but the fruit is of a whitish yellow colour, and very pleasant, especially if it be not ouer ripe : both these had need to be plashed against a wall, or else they will hardly beare ripe fruit.

The Oliue plum is very like a greene Oliue, both for colour and bignesse, and grow-eth lowe on a small bushing tree , and ripeneth late , but is the best of all the sorts of greene plums.

The white diapred plum of Malta, scarce knowne to any in our Land but Iohn Trade-scante, is a very good plum, and striped all ouer like diaper, and thereby so called.

The blacke diapred plum is like the Damascene plum, being blacke with spots, as small as pins points vpon it, of a very good rellish.

The Peake plum is a long whitish plum, and very good.

The Pishamin or Virginia plum is called a plum, but vtterly differeth from all forts of plums , the description whereof may truely enforme you, as it is set downe in the tenth Chapter going before, whereunto I referre you.

The Vse of Plums.

The great Damaske or Damson Plummes are dryed in France in great quantities, and brought ouer vnto vs in Hogs-heads, and other great ves-sels, and are those Prunes that are vsually sold at the Grocers, vnder the name of Damaske Prunes : the blacke Bulleis also are those (being dryed in the same manner) that they call French Prunes, and by their tartnesse are thought to binde, as the other, being sweet, to loosen the body.

The Bruneola Plumme, by reason of his pleasant tartnesse, is much ac-counted of, and being dryed, the stones taken from them, are brought ouer to vs in small boxes, and sold deere at the Comfitmakers, where they very often accompany all other sorts of banquetting stuffes.

Some

Some of these Plums, becaufe of their firmneffe, are vndoubtedly more wholfome then others that are fweete and waterifh, and caufe leffe offence in their ftomackes that eate them; and therefore are preferued with Sugar, to be kept all the yeare. None of them all is vfed in medicines fo much as the great Damfon or Damaske Prune, although all of them for the moft part doe coole, lenifie, and draw forth choller, and thereby are fitteft to be vfed of fuch as haue chollericke Agues.

CHAP. XIIII.

Mala Armeniaca fiue Præcocia. Apricockes.

THe Apricocke (as I faid) is without queftion a kinde of Plumme, rather then a Peach, both the flower being white, and the ftone of the fruit fmooth alfo, like a Plumme, and yet becaufe of the excellencie of the fruit, and the difference therein from all other Plummes, I haue thought it meete to entreate thereof by it felfe, and fhew you the varieties haue been obferued in thefe times.

The Apricocke tree rifeth vp to a very great height, either ftanding by it felfe (where it beareth not fo kindly, and very little in our country) or planted againft a wall, as it is moft vfuall, hauing a great ftemme or body, and likewife many great armes or branches, couered with a fmooth barke: the leaues are large, broad, and almoft round, but pointed at the ends, and finely dented about the edges: the flowers are white, as the Plumme tree bloffomes, but fomewhat larger, and rounder fet: the fruit is round, with a cleft on the one fide, fomewhat like vnto a Peach, being of a yellowifh colour as well on the infide as outfide, of a firme or faft fubftance, and dry, not ouer-moift in the eating, and very pleafant in tafte, containing within it a broad and flat ftone, fomewhat round and fmooth, not rugged as the Peach ftone, with a pleafant fweete kernell (yet fome haue reported, that there is fuch as haue their kernels bitter, which I did neuer fee or know) and is ripe almoft with our firft or earlieft Plummes, and thereof it tooke the name of *Præcox*; and it may bee was the earlieft of all others was then knowne, when that name was giuen.

The great Apricocke, which fome call the long Apricocke, is the greateft and faireft of all the reft.

The fmaller Apricocke, which fome call the fmall round Apricocke, is thought to be fmall, becaufe it firft fprang from a ftone: but that is not fo; for the kinde it felfe being inoculated, will bee alwaies fmall, and neuer halfe fo faire and great as the former.

The white Apricocke hath his leaues more folded together, as if it were halfe double: it beareth but feldome, and very few, which differ not from the ordinary, but in being more white, without any red when it is ripe.

The Mafcoline Apricocke hath a finer greene leafe, and thinner then the former, and beareth very feldome any ftore of fruit, which differeth in nothing from the firft, but that it is a little more delicate.

The long Mafcoline Apricocke hath his fruit growing a little longer then the former, and differeth in nothing elfe.

The Argier Apricocke is a fmaller fruit then any of the other, and yellow, but as fweete and delicate as any of them, hauing a blackifh ftone within it, little bigger then a Lacure Cherry ftone: this with many other forts Iohn Tradefcante brought with him returning from the Argier voyage, whither hee went voluntary with the Fleete, that went againft the Pyrates in the yeare 1620.

The Vfe of Apricockes.

Apricockes are eaten oftentimes in the fame manner that other dainty Plummes are, betweene meales of themfelues, or among other fruit at banquets.

They

They are alſo preſerued and candid, as it pleaſeth Gentlewomen to beſtowe their time and charge, or the Comfitmaker to ſort among other candid fruits.

Some likewiſe dry them, like vnto Peares, Apples, Damſons, and other
Plummes.

Matthiolus doth wonderfully commend the oyle drawne from the kernels of the ſtones, to annoint the inflamed *hæmorrhoides* or piles, the ſwellings of vlcers, the roughneſſe of the tongue and throate, and likewiſe the
paines of the eares.

Chap. XV.

Mala Perſica. Peaches.

AS I ordered the Cherries and Plummes, ſo I intend to deale with Peaches, becauſe their varieties are many, and more knowne in theſe dayes then in former
times: but becauſe the Nectorin is a differing kinde of Peach, I muſt deale
with it as I did with the Apricocke among the Plummes, that is, place it in a Chapter
by it ſelfe.

The Peach tree of it ſelfe groweth not vſually altogether ſo great, or high as the A-
pricocke, becauſe it is leſſe durable, but yet ſpreadeth with faire great branches, from
whence ſpring ſmaller and ſlenderer reddiſh twigges, whereon are ſet long narrow
greene leaues, dented about the edges: the bloſſomes are greater then of any Plumme,
of a deepe bluſh or light purple colour: after which commeth the fruit, which is round,
and ſometimes as great as a reaſonable Apple or Pippin (I ſpeake of ſome ſorts; for
there be ſome kindes that are much ſmaller) with a furrow or cleft on the one ſide, and
couered with a freeſe or cotton on the outſide, of colour either ruſſet, or red, or yellow, or of a blackiſh red colour; of differing ſubſtances and taſtes alſo, ſome being
firme, others wateriſh, ſome cleauing faſt to the ſtone on the inſide, others parting
from it more or leſſe eaſily, one excelling another very farre, wherein is contained a
rugged ſtone, with many chinkes or clefts in it, the kernell whereof is bitter: the roots
growe neither deepe nor farre; and therefore are ſubiect to the winds, ſtanding alone,
and not againſt a wall. It ſooner waxeth old and decayeth, being ſprung of a ſtone,
then being inoculated on a Plumme ſtocke, whereby it is more durable.

The great white Peach is white on the outſide as the meate is alſo, and is a good well
relliſhed fruit.

The ſmall white Peach is all one with the greater, but differeth in ſize.

The Carnation Peach is of three ſorts, two are round, and the third long; they are
all of a whitiſh colour, ſhadowed ouer with red, and more red on the ſide is next the
ſunne: the leſſer round is the more common, and the later ripe.

The grand Carnation Peach is like the former round Peach, but greater, and is as
late ripe, that is, in the beginning of September.

The red Peach is an exceeding well relliſhed fruit.

The ruſſet Peach is one of the moſt ordinary Peaches in the Kingdome, being of a
ruſſet colour on the outſide, and but of a reaſonable relliſh, farre meaner then many
other.

The Iſland Peach is a faire Peach, and of a very good relliſh.

The Newington Peach is a very good Peach, and of an excellent good relliſh, being
of a whitiſh greene colour on the outſide, yet halfe reddiſh, and is ripe about Bartholmew tide.

The yellow Peach is of a deepe yellow colour; there be hereof diuers ſorts, ſome
good and ſome bad.

The St. Iames Peach is the ſame with the Queenes Peach, here belowe ſet downe,
although ſome would make them differing.

The Melocotone Peach is a yellow faire Peach, but differing from the former yellow both in forme and taſte, in that this hath a ſmall crooked end or point for the moſt
part, it is ripe before them, and better relliſhed then any of them.

The

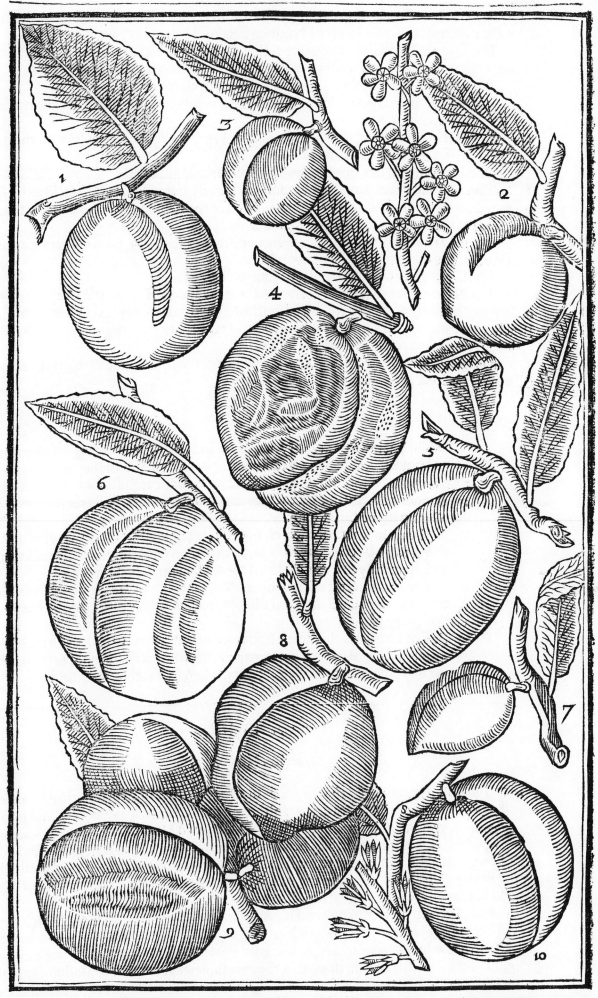

1 *Malus Armeniaca siue Precocia.* The Apricocke. 2 *Malus Persica Melocotonea.* The Melocotone Peach. 3 *Persica Moschatellina.* The Nutmeg Peach
4 *Persica nigra.* The blacke Peach. 5 *Persica Carnea longa.* The long Carnation Peach. 6 *Persica Reginea.* The Queenes Peach. 7 *Amygdalus.* The Almond. 8 *Persica du Troas.* The Peach du Troas. 9 *Nucipersica rubra optima.* The best Romane red Nectorin. 10 *Nucipersica rubra altera.* The bastard
red Nectorin with a pincking blossome.

The Peach *du Troas* is a long and great whitish yellow Peach, red on the outside, early ripe, and is another kinde of Nutmeg Peach.

The Queenes Peach is a faire great yellowish browne Peach, shadowed as it were ouer with deepe red, and is ripe at Bartholmew tide, of a very pleasant good taste.

The Romane Peach is a very good Peach, and well rellished.

The Durasme or Spanish Peach is of a darke yellowish red colour on the outside, and white within.

The blacke Peach is a great large Peach, of a very darke browne colour on the outside, it is of a waterish taste, and late ripe.

The Alberza Peach is late ripe, and of a reasonable good taste.

The Almond Peach, so called, because the kernell of the stone is sweete, like the Almond, and the fruit also somewhat pointed like the Almond in the huske ; it is early ripe, and like the Newington Peach, but lesser.

The Man Peach is of two sorts, the one longer then the other, both of them are good Peaches, but the shorter is the better rellished.

The Cherry Peach is a small Peach, but well tasted.

The Nutmeg Peach is of two sorts, one that will be hard when it is ripe, and eateth not so pleasantly as the other, which will bee soft and mellow ; they are both small Peaches, hauing very little or no resemblance at all to a Nutmeg, except in being a little longer then round, and are early ripe.

Many other sorts of Peaches there are, whereunto wee can giue no especiall name ; and therefore I passe them ouer in silence.

The Vse of Peaches.

Those Peaches that are very moist and waterish (as many of them are) and not firme, doe soone putrefie in the stomacke, causing surfeits oftentimes ; and therefore euery one had neede bee carefull, what and in what manner they eate them : yet they are much and often well accepted with all the Gentry of the Kingdome.

The leaues, because of their bitternesse, serue well being boyled in Ale or Milke, to be giuen vnto children that haue wormes, to help to kill them and doe gently open the belly, if there be a sufficient quantity vsed.

The flowers haue the like operation, that is, to purge the body somewhat more forceably then Damaske Roses ; a Syrupe therefore made of the flowers is very good.

The kernels of the Peach stones are oftentimes vsed to be giuen to them that cannot well make water, or are troubled with the stone ; for it openeth the stoppings of the vritory passages, whereby much ease ensueth.

Chap. XVI.

Nucipersica. Nectorins.

I Presume that the name *Nucipersica* doth most rightly belong vnto that kinde of Peach, which we call Nectorins, and although they haue beene with vs not many yeares, yet haue they beene knowne both in Italy to Matthiolus, and others before him, who it seemeth knew no other then the yellow Nectorin, as Dalechampius also : But we at this day doe know fiue seuerall sorts of Nectorins, as they shall be presently set downe ; and as in the former fruits, so in this, I will giue you the description of one, and briefe notes of the rest.

The Nectorin is a tree of no great bignesse, most vsually lesser then the Peach tree, his body and elder boughes being whitish, the younger branches very red, whereon grow narrow long greene leaues, so like vnto Peach leaues, that none can well distinguish them, vnlesse it be in this, that they are somewhat lesser : the blossomes are all reddish, as the Peach, but one of a differing fashion from all the other, as I shall shew you by and by : the fruit that followeth is smaller, rounder, and smoother then Peaches, without any cleft on the side, and without any douny cotton or freeze at all ; and

herein

herein is like vnto the outer greene rinde of the Wallnut, whereof as I am perswaded it tooke the name, of a faſt and firme meate, and very delicate in taſte, eſpecially the beſt kindes, with a rugged ſtone within it, and a bitter kernell.

The Muske Nectorin, ſo called, becauſe it being a kinde of the beſt red Nectorins, both ſmelleth and eateth as if the fruit were ſteeped in Muske : ſome thinke that this and the next Romane Nectorin are all one.

The Romane red Nectorin, or cluſter Nectorin, hath a large or great purpliſh bloſſome, like vnto a Peach, reddiſh at the bottome on the outſide, and greeniſh within : the fruit is of a fine red colour on the outſide, and groweth in cluſters, two or three at a ioynt together, of an excellent good taſte.

The baſtard red Nectorin hath a ſmaller or pincking bloſſome, more like threads then leaues, neither ſo large nor open as the former, and yellowiſh within at the bottome : the fruit is red on the outſide, and groweth neuer but one at a ioynt ; it is a good fruit, but eateth a little more rawiſh then the other, euen when it is full ripe.

The yellow Nectorin is of two ſorts, the one an excellent fruit, mellow, and of a very good relliſh ; the other hard, and no way comparable to it.

The greene Nectorin, great and ſmall ; for ſuch I haue ſeene abiding conſtant, although both planted in one ground : they are both of one goodneſſe, and accounted with moſt to be the beſt relliſhed Nectorin of all others.

The white Nectorin is ſaid to bee differing from the other, in that it will bee more white on the outſide when it is ripe, then either the yellow or greene : but I haue not yet ſeene it.

The Vſe of Nectorins.

The fruit is more firme then the Peach, and more delectable in taſte ; and is therefore of more eſteeme, and that worthily.

Chap. XVII.

Amygdala. Almonds.

THe Almond alſo may be reckoned vnto the ſtock or kindred of the Peaches, it is ſo like both in leafe and bloſſome, and ſomewhat alſo in the fruit, for the outward forme, although it hath onely a dry skinne, and no pulpe or meate to bee eaten : but the kernell of the ſtone or ſhell, which is called the Almond, maketh recompenſe of that defect, whereof ſome are ſweete, ſome bitter, ſome great, ſome ſmall, ſome long, and ſome ſhort.

The Almond tree groweth vpright, higher and greater then any Peach ; and is therefore vſually planted by it ſelfe, and not againſt a wall, whoſe body ſometime exceedeth any mans fadome, whereby it ſheweth to be of longer continuance, bearing large armes, and ſmaller branches alſo, but brittle, whereon are ſet long and narrow leaues, like vnto the Peach tree : the bloſſomes are purpliſh, like vnto Peach bloſſoms, but paler : the fruit is ſomewhat like a Peach for the forme of the skinne or outſide, which is rough, but not with any ſuch cleft therein, or with any pulpe or meate fit to bee eaten, but is a thicke dry skinne when it is ripe, couering the ſtone or ſhell, which is ſmooth and not rugged, and is either long and great, or ſmall, or thicke and ſhort, according as the nut or kernell within it is, which is ſweete both in the greater and ſmaller, and onely one ſmaller kinde which is bitter : yet this I haue obſerued, that all the Almond trees that I haue ſeene growe in England, both of the ſweete and bitter kindes, beare Almonds thicke and ſhort, and not long, as that ſort which is called the Iorden Almond.

The Vſe of Almonds.

They are vſed many wayes, and for many purpoſes, either eaten alone with Figges, or Rayſins of the Sunne, or made into paſte with Sugar and Roſewater for Marchpanes, or put among Floure, Egges, and Sugar, to make

make Mackerons, or crufted ouer with Sugar, to make Comfits, or mixed with Rofewater and Sugar, to make Butter, or with Barley water, to make Milke, and many other waies, as euery one lift, that hath skill in fuch things.

The oyle alfo of Almonds is vfed many waies, both inwardly and outwardly, for many purpofes; as the oyle of fweete Almonds mixt with poudered white Sugar Candy, for coughes and hoarfeneffe, and to be drunk alone, or with fome other thing (as the Syrupe of Marfh Mallowes) for the ftone, to open and lenifie the paffages, and make them flipperie, that the ftone may paffe the eafier. And alfo for women in Child bed after their fore trauell. And outwardly either by it felfe, or with oyle of Tartar to make a creame, to lenifie the skin, parched with the winde or otherwife, or to annoint the ftomacke either alone, or with other things to helpe a cold.

The oyle of bitter Almonds is much vfed to be dropped into their eares that are hard of hearing, to helpe to open them. And as it is thought, doth more fcoure and cleanfe the skin then the fweet oyle doth, and is therefore more vfed of many for that purpofe, as the Almonds themfelues are.

<h2 style="text-align:center">CHAP. XVIII.</h2>

Mala Arantia. Orenges.

I Bring here to your confideration, as you fee, the Orenge tree alone, without mentioning the Citron or Lemmon trees, in regard of the experience we haue feen made of them in diuers places: For the Orenge tree hath abiden with fome extraordinary looking and tending of it, when as neither of the other would by any meanes be preferued any long time. If therefore any be defirous to keepe this tree, he muft fo prouide for it, that it be preferued from any cold, either in the winter or fpring, and expofed to the comfort of the funne in fummer. And for that purpofe fome keepe them in great fquare boxes, and lift them to and fro by iron hooks on the fides, or caufe them to be rowled by trundels, or fmall wheeles vnder them, to place them in an houfe, or clofe gallerie for the winter time: others plant them againft a bricke wall in the ground, and defend them by a fhed of boardes, couered ouer with feare-cloth in the winter, and by the warmth of a ftoue, or other fuch thing, giue them fome comfort in the colder times: but no tent or meane prouifion will preferue them.

The Orenge tree in the warme Countries groweth very high, but with vs (or elfe it is a dwarfe kinde thereof) rifeth not very high: the barke of the elder ftemmes being of a darke colour, and the young branches very greene, whereon grow here and there fome few thornes: the leaues are faire, large, and very greene, in forme almoft like a Bay leafe, but that it hath a fmall eare, or peece of a leafe, fafhioned like vnto an heart vnder euery one of them, with many fmall holes to be feene in them, if you hold them vp betweene you and the light, of a fweet but ftrong fmell, naturally not falling away, but alwaies abiding on, or vntill new be come vp, bearing greene leaues continually: the flowers are whitifh, of a very ftrong and heady fent; after which come fmall round fruit, greene at the firft, while they are fmall, and not neere maturitie, but being grown and ripe, are (as all men know) red on the out fide, fome more pale then others, and fome kindes of a deeper yellowifh red, according to the climate, and as it receiueth the heate of the funne, wherein is contained fower or fweete iuice, and thicke white kernels among it: it beareth in the warme Countries both bloffomes and greene fruit continually vpon it, and ripe fruit alfo with them for the beft part of the yeare, but efpecially in Autumne and Winter.

The Vfe of Orenges.

Orenges are vfed as fawce for many forts of meates, in refpect of their fweete fowerneffe, giuing a rellifh of delight, whereinfoeuer they are vfed.

The inner pulpe or iuice doth ferue in agues and hot difeafes, and in Summer to coole the heate of deiected ftomackes, or fainting fpirits.

The

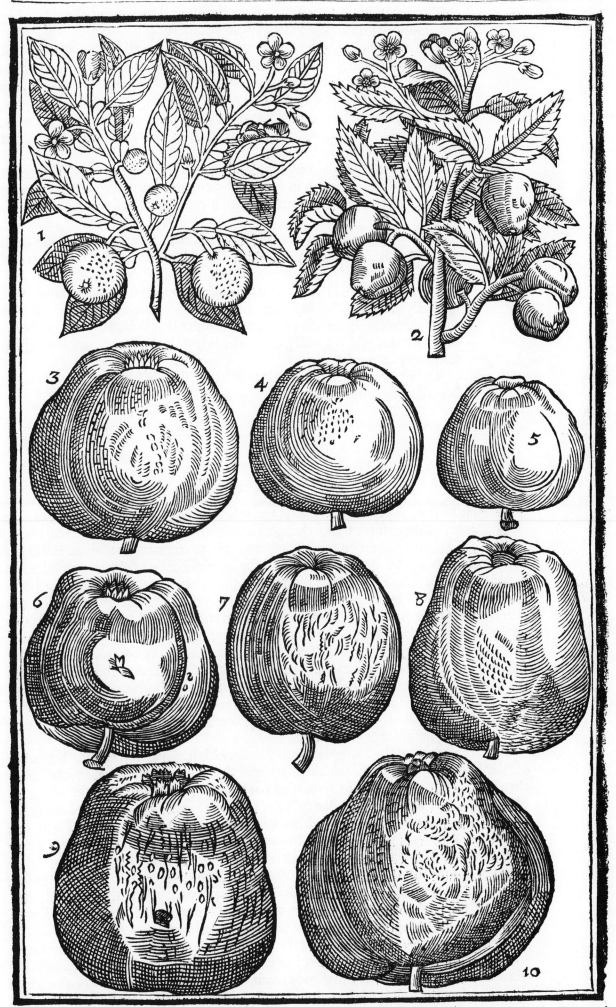

1 *Malus Arantia.* The Orenge tree. 2 *Malus.* The Apple tree. 3 *Malum Carbonarium.* The Pomewater. 4 *Malum Curtipendulum.* The golden Pippin.
5 *Melapium.* The Pearemaine. 6 *Malum Regineum.* The Queene Apple. 7 *Malum primò maturum.* The Genneting. 8 *Malum Regale.* The pound Royall. 9 *Malum Kentij ad servescendum.* The Kentish Codlin. 10 *Malum Regineum spurium.* The Bardfield Quining.

The dryed rinde, by reafon of the fweete and ftrong fent, ferueth to bee put among other things to make fweet pouthers.

The outer rindes, when they are clenfed from all the inner pulpe and skins, are preferued in Sugar, after the bitterneffe by often fteepings hath been taken away, & do ferue either as Succots, and banquetting ftuffes, or as ornaments to fet out difhes for the table, or to giue a rellifh vnto meats, whether baked or boyled : Phyfically they helpe to warme a cold ftomack, and to digeft or breake winde therein : or they are candid with Sugar, and ferue with other dryed Iunquets.

The water of Orange flowers is oftentimes vfed as a great perfume for gloues, to wafhe them, or in ftead of Rofe-water to mixe with other things.

It is vfed to bee drunke by fome, to preuent or to helpe any peftilentiall feuer.

The oyntment that is made of the flowers, is very comfortable both for the ftomache, againft the could or cough, or for the head, for paines and difineffe.

The kernels or feede beeing caft into the ground in the fpring time, will quickely grow vp, (but will not abide the winter with vs, to bee kept for growing trees) and when they are of a finger length high, being pluckt vp, and put among fallats, will giue them a maruellous fine aromaticke or fpicy tafte, very acceptable.

The feed or kernels are a little cordiall, although nothing fo much as the kernels of the Pomecitron.

Chap. XIX.

Poma. Apples.

THe forts of Apples are fo many, and infinite almoft as I may fay, that I cannot giue you the names of all, though I haue endeauoured to giue a great many, and I thinke it almoft impoffible for any one, to attaine to the full perfection of knowledge herein, not onely in regard of the multiplicitie of fafhions, colours and taftes, but in that fome are more familiar to one Countrey then to another, being of a better or worfe tafte in one place then in another, and therefore diuerfly called : I will therefore as I haue done before, giue you the defcription of the Tree in generall, as alfo of the Paradife or dwarfe Apple, becaufe of fome efpeciall difference, and afterwards the names of as many, with their fafhions, as haue come to my knowledge, either by fight or relation : for I doe confeffe I haue not feene all that I here fet downe, but vfe the helpe of fome friends, and therefore if it happen that the feuerall names doe not anfwer vnto feuerall forts, but that the fame fruit may bee called by one name in one Country, that is called by another elfewhere, excufe it I pray you ; for in fuch a number, fuch a fault may efcape vnknowne.

The Apple tree for the moft part is neyther very high, great or ftraight, but rather vfually boweth and fpreadeth (although in fome places it groweth fairer and ftraighter then in others) hauing long and great armes or boughes, and from them fmaller branches, whereon doe grow fomewhat broade, and long greene leaues, nicked about the edges : the flowers are large and white, with blufh coloured fides, confifting of fiue leaues : the fruit (as I faid) is of diuers formes, colours and taftes, and likewife of a very variable durabilitie; for fome muft be eaten prefently after they are gathered, and they are for the moft part the earlieft ripe ; others will abide longer vpon the trees, before they bee fit to be gathered ; fome alfo will be fo hard when others are gathered, that they will not be fit to be eaten, for one, two or three months after they bee gathered ; and fome will abide good but one, two or three moneths, and no more ; and fome will be beft, after a quarter or halfe a yeares lying, vnto the end of that yeare or the next.

The Paradife or dwarfe Apple tree groweth nothing fo high as the former, and many times not much higher then a man may reach, hauing leaues and flowers altogether like the other, the fruit is a faire yellow Apple, and reafonable great, but very light and fpongy or loofe, and of a bitterifh fweet tafte, nothing pleafant. And thefe faults alſo are incident vnto this tree, that both bodie and branches are much ſubieƈt vnto cancker, which will quickely eate it round, and kill it ; beſides it will haue many bunches, or tuberous ſwellings in many places, which grow as it were ſcabby or rough, and will ſoone cauſe it to periſh : the roote ſendeth forth many ſhootes and ſuckers, whereby it may be much increaſed. But this benefit may be had of it, to recompence the former faults, That being a dwarfe Tree, whatſoeuer fruit ſhall bee grafted on it, will keepe the graft low like vnto it ſelfe, and yet beare fruit reaſonable well. And this is a pretty way to haue Pippins, Pomewaters, or any other ſort of Apples (as I haue had my ſelfe, and alſo ſeene with others) growing low, that if any will, they may make a hedge rowe of theſe low fruits, planted in an Orchard all along by a walke ſide : but take this Caueat, if you will auoide the danger of the cancker and knots, which ſpoile the tree, to graft it hard vnto the ground, that therby you may giue as little of the nature of the ſtock thereunto as poſſibly you can, which wil vndoubtedly help it very much.

The kindes or ſorts of Apples.

The Summer pippin is a very good apple firſt ripe, and therefore to bee firſt ſpent, becauſe it will not abide ſo long as the other.

The French pippin is alſo a good fruit and yellow.

The Golding pippin is the greateſt and beſt of all ſorts of pippins.

The Ruſſet pippin is as good an apple as moſt of the other ſorts of pippins.

The ſpotted pippin is the moſt durable pippin of all the other ſorts.

The ordinary yellow pippin is like the other, and as good; for indeed I know no ſort of pippins but are excellent good well relliſhed fruites.

The great pearemaine differeth little either in taſte or durabilitie from the pippin, and therefore next vnto it is accounted the beſt of all apples.

The ſummer pearemaine is of equall goodneſſe with the former, or rather a little more pleaſing, eſpecially for the time of its eating, which will not bee ſo long laſting, but is ſpent and gone when the other beginneth to be good to eate.

The Ruſſetting is alſo a firme and a very good apple, not ſo wateriſh as the pippin or pearemaine, and will laſt the beſt part of the year, but will be very mellow at the laſt, or rather halfe dryed.

The Broading is a very good apple.

The Pomewater is an excellent good and great whitiſh apple, full of ſap or moiſture, ſomewhat pleaſant ſharpe, but a little bitter withall : it will not laſt long, the winter froſts ſoone cauſing it to rot, and periſh.

The Flower of Kent is a faire yellowiſh greene apple both good and great.

The Gilloflower apple is a fine apple, and finely ſpotted.

The Marligo is the ſame, that is called the Marigold apple, it is a middle ſized apple, very yellow on the outſide, ſhadowed ouer as it were with red, and more red on one ſide, a reaſonable well relliſhed fruit.

The Blandrill is a good apple.

The Dauie Gentle is a very good apple

The Gruntlin is ſomewhat a long apple, ſmaller at the crowne then at the ſtalke, and is a reaſonable good apple.

The gray Coſterd is a good great apple, ſomewhat whitiſh on the outſide, and abideth the winter.

The greene Coſterd is like the other, but greener on the outſide continually.

The Haruy apple is a faire great goodly apple, and very well relliſhed.

The Dowſe apple is a ſweetiſh apple not much accounted of.

The Pome-paris is a very good apple.

The Belle boon of two ſorts winter and ſummer, both of them good apples, and fair fruit to look on, being yellow and of a meane bigneſſe.

The pound Royall is a very great apple, of a very good and ſharpe taſte.

The Doues Bill a ſmall apple.

The Deuſan or apple Iohn is a delicate fine fruit, well relliſhed when it beginneth to be fit to be eaten, and endureth good longer then any other apple.

The Maſter William is greater then a pippin, but of no very good relliſh.

The Maſter Iohn is a better taſted apple then the other by much.

The

The Spicing is a well tasted fruite.

Pome de Rambures
Pome de Capanda   }all faire and good apples brought from France.
Pome de Calual

The Queene apple is of two sorts, both of them great faire red apples, and well rellished, but the greater is the best.

The Bastard Queene apple is like the other for forme and colour, but not so good in taste : some call this the bardfield Queening.

The Boughton or greening is a very good and well tasted apple.

The Leathercoate apple is a good winter apple, of no great bignesse, but of a very good and sharpe taste.

The Pot apple is a plaine Country apple.

The Cowsnout is no very good fruit.

The Gildiling apple is a yellow one, not much accounted.

The Cats head apple tooke the name of the likenesse, and is a reasonable good apple and great.

The Kentish Codlin is a faire great greenish apple, very good to eate when it is ripe; but the best to coddle of all other apples.

The Stoken apple is a reasonable good apple.

The Geneting apple is a very pleasant and good apple.

The Worcester apple is a very good apple, as bigge as a Pomewater.

Dosime Couadis is a French apple, and of a good rellish.

The French Goodwin is a very good apple.

The old wife is a very good, and well rellished apple.

The towne Crab is an hard apple, not so good to be eaten rawe as roasted, but excellent to make Cider.

· The Virgilling apple is a reasonable good apple.

The Crowes egge is no good rellished fruit, but nourced vp in some places of the common people.

The Sugar apple is so called of the sweetnesse.

Sops in wine is so named both of the pleasantnesse of the fruit, and beautie of the apple.

The womans breast apple is a great apple.

The blacke apple or pippin is a very good eating apple, and very like a Pearemaine, both for forme and bignesse, but of a blacke sooty colour.

Tweenty sorts of Sweetings and none good.

The Peare apple is a small fruit, but well rellished being ripe, and is for shape very like vnto a small short Peare, and greene.

The Paradise apple is a faire goodly yellow apple, but light and spongy, and of a bitterish sweet taste, not to be commended.

The apple without blossome, so called because although it haue a small shew of a blossome, yet they are but small threds rather than leaues, neuer shewing to bee like a flower, and therefore termed without blossome : the apple is neyther good eating nor baking fruit.

Wildings and Crabs are without number or vse in our Orchard, being to be had out of the woods, fields and hedges rather then any where else.

The Vse of Apples.

The best sorts of Apples serue at the last course for the table, in most mens houses of account, where, if there grow any rare or excellent fruit, it is then set forth to be seene and tasted.

Diuers other sorts serue to bake, either for the Masters Table, or the meynes sustenance, either in pyes or pans, or else stewed in dishes with Rosewater and Sugar, and Cinamon or Ginger cast vpon.

Some kinds are fittest to roast in the winter time, to warme a cup of wine, ale or beere ; or to be eaten alone, for the nature of some fruit is neuer so good, or worth the eating, as when they are roasted.

                                                                Some

Some forts are fitteft to fcald for Codlins, and are taken to coole the ftomacke, as well as to pleafe the tafte, hauing Rofewater and Sugar put to them.

Some forts are beft to make Cider of, as in the Weft Countrey of England great quantities, yea many Hogfheads and Tunnes full are made, efpecially to bee carried to the Sea in long voyages, and is found by experience to bee of excellent vfe, to mixe with water for beuerage. It is vfually feene that thofe fruits that are neither fit to eate raw, roafted, nor baked, are fitteft for Cider, and make the beft.

The iuice of Apples likewife, as of pippins, and pearemaines, is of very good vfe in Melancholicke difeafes, helping to procure mirth, and to expell heauineffe.

The diftilled water of the fame Apples is of the like effect.

There is a fine fweet oyntment made of Apples called *Pomatum*, which is much vfed to helpe chapt lips, or hands, or for the face, or any other part of the skinne that is rough with winde, or any other accident, to fupple them, and make them fmooth.

Chap. XX.

Cydonia. Quinces.

WEe haue fome diuerfities of Quinces, although not many, yet more then our elder times were acquainted with, which fhall be here expreffed.

The Quince tree groweth oftentimes to the height and bigneffe of a good Apple tree, but more vfually lower, with crooked and fpreading armes and branches farre abroad, the leaues are fomewhat round, and like the leaues of the Apple tree, but thicker, harder, fuller of veines, and white on the vnderfide: the bloffomes or flowers are white, now and then dafht ouer with blufh, being large and open, like vnto a fingle Rofe: the fruit followeth, which when it is ripe is yellow, and couered with a white cotton or freeze, which in the younger is thicker and more plentifull, but waxeth leffe and leffe, as the fruit ripeneth, being bunched out many times in feuerall places, and round, efpecially about the head, fome greater, others fmaller, fome round like an Apple, others long like a Peare, of a ftrong heady fent, accounted not wholfome or long to be endured, and of no durabilitie to keepe, in the middle whereof is a core, with many blackifh feedes or kernels therein, lying clofe together in cels, and compaffed with a kinde of cleare gelly, which is eafier feene in the fcalded fruit, then in the raw.

The Englifh Quince is the ordinarie Apple Quince, fet downe before, and is of fo harfh a tafte being greene, that no man can endure to eate it rawe, but eyther boyled, ftewed, roafted or baked; all which waies it is very good.

The Portingall Apple Quince is a great yellow Quince, feldome comming to bee whole and faire without chapping; this is fo pleafant being frefh gathered, that it may be eaten like vnto an Apple without offence.

The Portingall Peare Quince is not fit to be eaten rawe like the former, but muft be vfed after fome of the waies the Englifh Quince is appointed, and fo it will make more dainty difhes then the Englifh, becaufe it is leffe harfh, will bee more tender, and take leffe fugar for the ordering then the Englifh kinde.

The Barbary Quince is like in goodneffe vnto the Portingall Quince laft fpoken of, but leffer in bigneffe.

The Lyons Quince.

The Brunfwicke Quince.

The Vfe of Quinces.

There is no fruit growing in this Land that is of fo many excellent vfes as this, feruing as well to make many difhes of meate for the table, as for ban-

banquets, and much more for the Phyſicall vertues, whereof to write at large is neither conuenient for mee, nor for this worke : I will onely briefly recite ſome, as it were to giue you a taſte of that plenty remaineth therein, to bee conuerted into ſundry formes : as firſt for the table, while they are freſh (and all the yeare long after being pickled vp) to be baked, as a dainty diſh, being well and orderly cookt. And being preſerued whole in Sugar, either white or red, ſerue likewiſe, not onely as an after diſh to cloſe vp the ſtomacke, but is placed among other Preſerues by Ladies and Gentlewomen, and beſtowed on their friends to entertaine them, and among other ſorts of Preſerues at Banquets. Codiniacke alſo and Marmilade, Ielly and Paſte, are all made of Quinces, chiefly for delight and pleaſure, although they haue alſo with them ſome phyſicall properties.

We haue for the vſe of phyſicke, both Iuyce and Syrupe, both Conſerue and Condite, both binding and looſening medicines, both inward and outward, and all made of Quinces.

The Ielly or Muccilage of the ſeedes, is often vſed to be laid vpon womens breaſts, to heale them being ſore or rawe, by their childrens default giuing them ſucke.

Athenæus reciteth in his third booke, that one Philarchus found, that the ſmell of Quinces tooke away the ſtrength of a certaine poiſon, called *Phariacum.* And the Spaniards haue alſo found, that the ſtrength of the iuyce of white Ellebor (which the Hunters vſe as a poyſon to dippe their arrow heads in, that they ſhoote at wilde beaſts to kill them) is quite taken away, if it ſtand within the compaſſe of the ſmell of Quinces. And alſo that Grapes, being hung vp to bee kept, and ſpent in Winter, doe quickly rot with the ſmell of a Quince.

Chap. XXI.

Pyra. Peares.

THe variety of peares is as much or more then of apples, and I thinke it is as hard in this, as before in apples, for any to be ſo exquiſite, as that hee could number vp all the ſorts that are to be had : for wee haue in our country ſo manie, as I ſhall giue you the names of by and by, and are hitherto come to our knowledge : but I verily beleeue that there be many, both in our country, and in others, that we haue not yet knowne or heard of ; for euery yeare almoſt wee attaine to the knowledge of ſome, we knew not of before. Take therefore, according to the manner before held, the deſcription of one, with the ſeuerall names of the reſt, vntill a more exact diſcourſe be had of them, euery one apart.

The Peare tree groweth more ſlowly, but higher, and more vpright then the apple tree, and not leſſe in the bulke of the body : his branches ſpread not ſo farre or wide, but growe vprighter and cloſer : the leaues are ſomewhat broader and rounder, greene aboue, and whiter vnderneath then thoſe of the apple tree : the flowers are whiter and greater : the fruit is longer then round for the moſt part, ſmaller at the ſtalke, and greater at the head, of ſo many differing formes, colours, and taſtes, that hardly can one diſtinguiſh rightly between them, the times alſo being as variable in the gathering and ſpending of them, as in apples : the roote groweth deeper then the apple tree, and therefore abideth longer, and giueth a faſter, cloſer, & ſmoother gentle wood, eaſie to be wrought vpon.

The kindes of Peares.

The Summer bon Chretien is ſomewhat a long peare, with a greene and yellow ruſſetiſh coate, and will haue ſometimes red ſides ; it is ripe at Michaelmas : ſome vſe to dry them as they doe Prunes, and keepe them all the yeare after. I haue not ſeene or heard any more Summer kindes hereof then this one, and needeth no wall to nourſe it as the other.

The

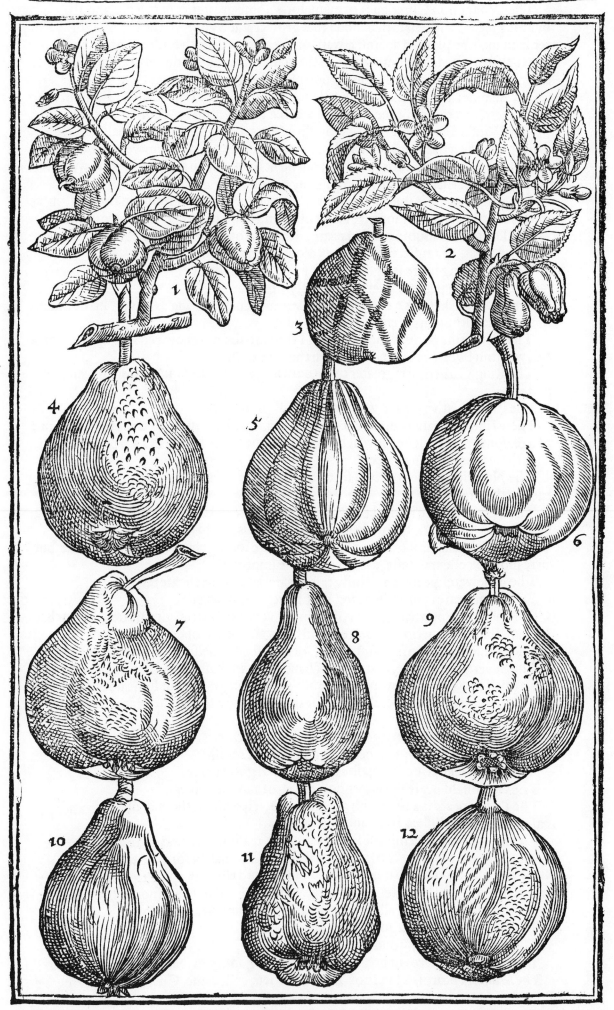

1 *Malus Cotonea*. The Quince tree. 2 *Cydonium Lusitanicum*. The Portingall Quince. 3 *Pyrus*. The Peare tree. 4 *Pyrum Pompeianum, sive Cucumerinum hyemale*. The Winter Bon Chretien. 5 *Pyrum pictum vel striatum*. The painted or striped Peare of Ierusalem. 6 *Pyrum Palatinale*. The Burgomot Peare. 7 *Pyrum Cucumerinum sive Pompeianum aestivum*. The Summer Bon Chretien. 8 *Pyrum Volemum*. The best Warden. 9 *Pyrum Librale*. The pound Peare. 10 *Pyrum Windsorianum*. The Windsor Peare. 11 *Pyrum Cucumerinum*. The Gratiola Peare. 12 *Pyrum Caryophyllatum*. The Gilloflower Peare.

The Winter bon Chretien is of many forts, fome greater, others leffer, and all good; but the greateft and beft is that kinde that groweth at Syon: All the kinds of this Winter fruit muft be planted againft a wall, or elfe they will both feldome beare, and bring fewer alfo to ripeneffe, comparable to the wall fruit: the kindes alfo are according to their lafting; for fome will endure good much longer then others.

The Summer Bergomot is an excellent well rellifhed peare, flattifh, & fhort, not long like others, of a meane bigneffe, and of a darke yellowifh greene colour on the outfide.

The Winter Bergomot is of two or three forts, being all of them fmall fruit, fome-what greener on the outfide then the Summer kindes; all of them very delicate and good in their due time: for fome will not be fit to bee eaten when others are will-nigh fpent, euery of them outlafting another by a moneth or more.

The Diego peare is but a fmall peare, but an excellent well rellifhed fruit, tafting as if Muske had been put among it; many of them growe together, as it were in clufters.

The Duetete or double headed peare, fo called of the forme, is a very good peare, not very great, of a ruffettifh browne colour on the outfide.

The Primating peare is a good moift peare, and early ripe.

The Geneting peare is a very good early ripe peare.

The greene Chefill is a delicate mellow peare, euen melting as it were in the mouth of the eater, although greenifh on the outfide.

The Catherine peare is knowne to all I thinke to be a yellow red fided peare, of a full waterifh fweete tafte, and ripe with the foremoft.

The King Catherine is greater then the other, and of the fame goodneffe, or rather better.

The Ruffet Catherine is a very good middle fized peare.

The Windfor peare is an excellent good peare, well knowne to moft perfons, and of a reafonable greatneffe: it will beare fruit fome times twice in a yeare (and as it is faid) three times in fome places.

The Norwich peare is of two forts, Summer and Winter, both of them good fruit, each in their feafon.

The Worfter peare is blackifh, a farre better peare to bake (when as it will be like a Warden, and as good) then to eate rawe; yet fo it is not to be mifliked.

The Muske peare is like vnto a Catherine peare for bigneffe, colour, and forme, but farre more excellent in tafte, as the very name importeth.

The Rofewater peare is a goodly faire peare, and of a delicate tafte.

The Sugar peare is an early peare, very fweete, but waterifh.

The Summer Popperin ⎱ both of them are very good firme dry peares, fomewhat
The Winter Popperin ⎰ fpotted, and brownifh on the outfide.

The greene Popperin is a winter fruit, of equall goodneffe with the former.

The Soueraingne peare, that which I haue feene and tafted, and fo termed vnto me, was a fmall brownifh yellow peare, but of a moft dainty tafte; but fome doe take a kind of Bon Chretien, called the Elizabeth peare, to be the Soueraigne peare; how truely let others iudge.

The Kings peare is a very good and well tafted peare.

The peare Royall is a great peare, and of a good rellifh.

The Warwicke peare is a reafonable faire and good peare.

The Greenfield peare is a very good peare, of a middle fize.

The Lewes peare is a brownifh greene peare, ripe about the end of September, a reafonable well rellifhed fruit, and very moift.

The Bifhop peare is a middle fized peare, of a reafonable good tafte, not very waterifh; but this property is oftentimes feene in it, that before the fruit is gathered, (but more vfually thofe that fall of themfelues, and the reft within a while after they are gathered) it will be rotten at the core, when there wil not be a fpot or blemifh to be feene on the outfide, or in all the peare, vntill you come neare the core.

The Wilford peare is a good and a faire peare.

The Bell peare a very good greene peare.

The Portingall peare is a great peare, but more goodly in fhew then good indeed.

The Gratiola peare is a kinde of Bon Chretien, called the Cowcumber peare, or Spinola's peare.

The Rowling peare is a good peare, but hard, and not good before it bee a little rowled or bruifed, to make it eate the more mellow.

The

The Pimpe peare is as great as the Windfor peare, but rounder, and of a very good rellifh.

The Turnep peare is a hard winter peare, not fo good to eate rawe, as it is to bake.

The Arundell peare is moft plentifull in Suffolke, and there commended to be a verie good peare.

The Berry peare is a Summer peare, reafonable faire and great, and of fo good and wholfome a tafte, that few or none take harme by eating neuer fo many of them.

The Sand peare is a reafonable good peare, but fmall.

The Morley peare is a very good peare, like in forme and colour vnto the Windfor, but fomewhat grayer.

The peare pricke is very like vnto the Greenfield peare, being both faire, great, and good.

The good Rewell is a reafonable great peare, as good to bake as to eate rawe, and both wayes it is a good fruit.

The Hawkes bill peare is of a middle fize, fomewhat like vnto the Rowling peare.

The Petworth peare is a winter peare, and is great, fomewhat long, faire, and good.

The Slipper peare is a reafonable good peare.

The Robert peare is a very good peare, plentifull in Suffolke and Norfolke.

The pound peare is a reafonable good peare, both to eate rawe, and to bake.

The ten pound peare, or the hundred pound peare, the trueft and beft, is the beft Bon Chretien of Syon, fo called, becaufe the grafts coft the Mafter fo much the fetching by the meffengers expences, when he brought nothing elfe.

The Gilloflower peare is a winter peare, faire in fhew, but hard, and not fit to bee eaten rawe, but very good to bake.

The peare Couteau is neither good one way nor other.

The Binfce peare is a reafonable good winter peare, of a ruffetifh colour, and a fmall fruit : but will abide good a long while.

The Pucell is a greene peare, of an indifferent good tafte.

The blacke Sorrell is a reafonable great long peare, of a darke red colour on the outfide.

The red Sorrell is of a redder colour, elfe like the other.

The Surrine is no very good peare.

The Summer Hafting is a little greene peare, of an indifferent good rellifh.

Peare Gergonell is an early peare, fomewhat long, and of a very pleafant tafte.

The white Genneting is a reafonable good peare, yet not equall to the other.

The Sweater is fomewhat like the Windfor for colour and bigneffe, but nothing neare of fo good a tafte.

The bloud red peare is of a darke red colour on the outfide, but piercing very little into the inner pulpe.

The Hony peare is a long greene Summer peare.

The Winter peare is of many forts, but this is onely fo called, to bee diftinguifhed from all other Winter peares, which haue feuerall names giuen them, and is a very good peare.

The Warden or Luke Wards peare of two forts, both white and red, both great and fmall.

The Spanifh Warden is greater then either of both the former, and better alfo.

The peare of Ierufalem, or the ftript peare, whofe barke while it is young, is as plainly feene to be ftript with greene, red, and yellow, as the fruit it felfe is alfo, and is of a very good tafte : being baked alfo, it is as red as the beft Warden, whereof Mafter William Ward of Effex hath affured mee, who is the chiefe keeper of the Kings Granary at Whitehall.

Hereof likewife there is a wilde kinde no bigger then ones thumbe, and ftriped in the like manner, but much more.

The Choke peares, and other wilde peares, both great and fmall, as they are not to furnifh our Orchard, but the Woods, Forrefts, Fields, and Hedges, fo wee leaue them to their naturall places, and to them that keep them, and make good vfe of them.

The Vfe of Peares.

The moft excellent forts of Peares, ferue (as I faid before of Apples) to

make

make an after-courſe for their maſters table, where the goodneſſe of his Orchard is tryed. They are dryed alſo, and ſo are an excellent repaſte, if they
be of the beſt kindes, fit for the purpoſe.

They are eaten familiarly of all ſorts of people, of ſome for delight, and
of others for nouriſhment, being baked, ſtewed, or ſcalded.

The red Warden and the Spaniſh Warden are reckoned among the moſt
excellent of Peares, either to bake or to roaſt, for the ſicke or for the ſound:
And indeede, the Quince and the Warden are the two onely fruits are permitted to the ſicke, to eate at any time.

Perry, which is the iuyce of Peares preſſed out, is a drinke much eſteemed
as well as Cyder, to be both drunke at home, and carried to the Sea, and
found to be of good vſe in long voyages.

The Perry made of Choke Peares, notwithſtanding the harſhneſſe, and
euill taſte, both of the fruit when it is greene, as alſo of the iuyce when it is
new made, doth yet after a few moneths become as milde and pleaſant as
wine, and will hardly bee knowne by the ſight or taſte from it : this hath
beene found true by often experience; and therefore wee may admire the
goodneſſe of God, that hath giuen ſuch facility to ſo wilde fruits, altogether thought vſeleſſe, to become vſefull, and apply the benefit thereof both
to the comfort of our ſoules and bodies.

For the Phyſicall properties, if we doe as Galen teacheth vs, *in ſecundo
Alimentorum,* referre the qualities of Peares to their ſeuerall taſtes, as before he had done in Apples, we ſhall not neede to make a new worke; thoſe
that are harſh and ſowre doe coole and binde, ſweet do nouriſh and warme,
and thoſe betweene theſe, to haue middle vertues, anſwerable to their tem-
peratures, &c.

Much more might be ſaid, both of this and the other kinds of fruits; but
let this ſuffice for this place and worke, vntill a more exact be accompliſhed.

Chap. XXII.

Nux Iuglans. The Wallnut.

Although the Wallnut tree bee often planted in the middle of great Court-
yards, where by reaſon of his great ſpreading armes it taketh vp a great deale of
roome, his ſhadow reaching farre, ſo that ſcarce any thing can well grow neare
it; yet becauſe it is likewiſe planted in fit places or corners of Orchards, and that it
beareth fruit or nuts, often brought to the table, eſpecially while they are freſheſt,
ſweeteſt, and fitteſt to be eaten, let not my Orchard want his company, or you the
knowledge of it. Some doe thinke that there are many ſorts of them, becauſe ſome
are much greater then others, and ſome longer then others, and ſome haue a more fran-
gible ſhell then others; but I am certainly perſwaded, that the ſoyle and climate
where they grow, are the whole and onely cauſe of the varieties and differences. Indeed Virginia hath ſent vnto vs two ſorts of Wallnuts, the one blacke, the other white,
whereof as yet wee haue no further knowledge. And I know that Cluſius reporteth,
he tooke vp at a banquet a long Wallnut, differing in forme and tenderneſſe of ſhell
from others, which being ſet, grew and bore farre tenderer leaues then the other, and
a little ſnipt about the edges, which (as I ſaid) might alter with the ſoyle and climate:
and beſides you may obſerue, that many of Cluſius differences are very nice, and ſo I
leaue it.

The Wallnut tree groweth very high and great, with a large and thicke body or
trunke, couered with a thicke clouen whitiſh greene barke, tending to an aſh-colour;
the armes are great, and ſpread farre, breaking out into ſmaller branches, whereon doe
grow long & large leaues, fiue or ſeuen ſet together one againſt another, with an odde
one at the end, ſomewhat like vnto Aſhen leaues, but farre larger, and not ſo many on a
ſtalke, ſmooth, and ſomewhat reddiſh at the firſt ſpringing, and tender alſo, of a reaſonable good ſent, but more ſtrong and headie when they growe old: the fruit or nut is
great and round, growing cloſe to the ſtalkes of the leaues, either by couples or by
three

three set together, couered with a double shell, that is to say, with a greene thicke and soft outer rinde, and an inner hard shell, within which the white kernell is contained, couered with a thinne yellow rinde or peeling, which is more easily peeled away while it is greene then afterwards, and is as it were parted into foure quarters, with a thinne wooddy peece parting it at the head, very sweete and pleasant while it is fresh, and for a while after the gathering; but the elder they growe, the harder and more oily : the catkins or blowings are long and yellow, made of many scaly leaues set close together, which come forth early in the Spring, and when they open and fall away, vpon their stalkes arise certaine small flowers, which turne into so many nuts.

The Vse of Wallnuts.

They are often serued to the table with other fruits while they abide fresh and sweete ; and therefore many to keepe them fresh a long time haue deuised many wayes, as to put them into great pots, and bury them in the ground, and so take them out as they spend them, which is a very good way, and will keepe them long.

The small young nuts while they are tender, being preserued or candid, are vsed among other sorts of candid fruits, that serue at banquets.

The iuyce of the outer greene huskes are held to be a soueraigne remedy against either poyson, or plague, or pestilentiall feuer.

The distilled water of the huskes drunke with a little vinegar, if the fits growe hot and tedious, is an approued remedy for the same.

The water distilled from the leaues, is effectuall to be applyed to fluent or running vlcers, to dry and binde the humours.

Some haue vsed the pouder of the catkins in white wine, for the suffocation or strangling of the mother.

The oyle of Wallnuts is vsed to varnish Ioyners workes. As also is accounted farre to excell Linseede oyle, to mixe a white colour withall, that the colour bee not dimmed. It is of excellent vse for the coldnesse, hardnesse and contracting of the sinewes and ioynts, to warme, supple, and to extend them.

Chap. XXIII.

Castanea Equina. The Horse Chesnut.

ALthough the ordinary Chesnut is not a tree planted in Orchards, but left to Woods, Parkes, and other such like places ; yet wee haue another sort which wee haue noursed vp from the nuts sent vs from Turky, of a greater and more pleasant aspect for the faire leaues, and of as good vse for the fruit. It groweth in time to be a great tree, spreading with great armes and branches, whereon are set at seuerall distances goodly faire great greene leaues, diuided into six, seuen, or nine parts or leaues, euery one of them nicked about the edges, very like vnto the leaues of *Ricnus,* or *Palma Christi,* and almost as great : it beareth at the ends of the branches many flowers set together vpon a long stalke, consisting of foure white leaues a peece, with many threads in the middle, which afterwards turne into nuts, like vnto the ordinary Chesnuts, but set in rougher and more prickly huskes : the nuts themselues being rounder and blacker, with a white spot at the head of each, formed somewhat like an heart, and of a little sweeter taste.

The Vse of this Chesnut.

It serueth to binde and stop any maner of fluxe, be it of bloud or humours, either of the belly or stomacke; as also the much spitting of bloud. They are roasted and eaten as the ordinary sort, to make them taste the better.

They are vsually in Turkie giuen to horses in their prouender, to cure them of coughes, and helpe them being broken winded.

Chap.

Chap. XXIIII.

Morus. The Mulberrie.

There are two forts of Mulberries fufficiently known to moft, the blackifh and the white : but wee haue had brought vs from Virginia another fort, which is of greater refpect then eyther of the other two, not onely in regard of the rauitie, but of the vfe, as you fhall prefently vnderftand.

1. *Morus nigra.* The blacke Mulberrie.

The blacke Mulberrie tree groweth oftentimes tall and great, and oftentimes alfo crooked, and fpreading abroade, rather then high; for it is fubiect to abide what forme you will conforme it vnto : if by fuffering it to grow, it will mount vp, and if you will binde it, or plafh the boughes, they will fo abide, and be carried ouer arbours, or other things as you will haue it. The bodie groweth in time to bee very great, couered with a rugged or thicke barke, the armes or branches being fmoother, whereon doe grow round thicke leaues pointed at the ends, and nicked about the edges, and in fome there are to be feene deep gafhes, making it feeme fomewhat like the Vine leafe: the flowers are certaine fhort dounie catkings, which turne into greene berries at the firft, afterwards red, and when they are full ripe blacke, made of many graines fet together, like vnto the blacke berrie, but longer and greater : before they are ripe, they haue an auftere and harfh tafte, but when they are full ripe, they are more fweete and pleafant; the iuice whereof is fo red, that it will ftaine the hands of them that handle and eate them.

2. *Morus alba.* The white Mulberrie.

The white Mulberrie tree groweth not with vs to that greatneffe or bulke of bodie that the blacke doth, but runneth vp higher, flenderer, more knotty, hard and brittle, with thinner fpreade armes and branches : the leaues are like the former, but not fo thicke fet on the branches, nor fo hard in handling, a little paler alfo, hauing fomewhat longer ftalkes: the fruit is fmaller and clofer fet together, greene, and fomewhat harfh before they be ripe, but of a wonderfull fweetneffe, almoft ready to procure loathing when they are thorough ripe, and white, with fuch like feede in them as in the former, but fmaller.

3. *Morus Virginiana.* The Virginia Mulberrie.

The Virginia Mulberry tree groweth quickely with vs to be a very great tree, fpreading many armes and branches, whereon grow faire great leaues, very like vnto the leaues of the white Mulberrie tree : the berry or fruit is longer and redder then either of the other, and of a very pleafant tafte.

The Vfe of Mulberries.

The greateft and moft efpeciall vfe of the planting of white Mulberries, is for the feeding of Silke wormes, for which purpofe all the Eafterne Countries, as Perfia, Syria, Armenia, Arabia &c. and alfo the hither part of Turkie, Spaine alfo and Italie, and many other hot Countries doe nourifh them, becaufe it is beft for that purpofe, the wormes feeding thereon, giuing the fineft and beft filke; yet fome are confident that the leaues of the blacke will doe as much good as the white : but that refpect muft be had to change your feede, becaufe therein lyeth the greateft myfterie. But there is a Booke or Tractate printed, declaring the whole vfe of whatfoeuer can belong vnto them : I will therefore referre them thereunto, that

would

1 *Nux Inglans.* The Wallnut. 2 *Castanea equina.* The horse Chesnut. 3 *Morus nigra vel alba.* The Mulberry. 4 *Morus Virginiana.* The Virginia Mulberry. 5 *Laurus vulgaris.* The ordinary Bay tree. 6 *Laurea Cerasus Virginiana.* The Virginia Cherry Bay.

would further vnderstand of that matter.

Mulberries are not much desired to be eaten, although they be somewhat pleasant, both for that they staine their fingers and lips that eate them, and doe quickly putrefie in the stomacke, if they bee not taken before meate.

They haue yet a Physicall vse, which is by reason of the astringent quality while they are red, and before they bee ripe, for sore mouthes and throats, or the like, whereunto also the Syrup, called Diamoron, is effectuall.

Corollarium.
A COROLLARIE
To this Orchard.

Here are certaine other trees that beare no fruit fit to bee eaten, which yet are often seene planted in Orchards, and other fit and conuenient places bout an house, whereof some are of especiall vse, as the Bay tree &c. others for their beauty and shadow are fit for walkes or arbours; some being euer green are most fit for hedge-rowes; and some others more for their raritie then for any other great vse, wherof I thought good to entreat apart by themselues, and bring them after the fruit trees of this Orchard, as an ornament to accomplish the same.

1. *Laurus.* The Bay tree.

There are to bee reckoned vp fiue kindes of Bay trees, three whereof haue been entreated of in the first part, a fourth wee will only bring hereto your consideration, which is that kinde that is vsually planted in euery mans yard or orchard, for their vse throughout the whole land, the other we will leaue to bee considered of in that place is fit for it.

The Bay tree riseth vp oftentimes to carry the face of a tree of a meane bignesse in our Countrey (although much greater in the hoter) and oftentimes shooteth vp with many suckers from the roote, shewing it selfe more like to a tall shrubbe or hedge-bush, then a tree, bauing many branches, the young ones whereof are sometimes reddish, but most vsually of a light or fresh greene colour, when the stemme and elder boughes are couered with a darke greene barke : the leaues are somewhat broad, and long pointed as it were at both the ends, hard and sometimes crumpled on the edges, of a darke greene colour aboue, and of a yellowish greene vnderneath, in smell sweet, in taste bitter, and abiding euer greene : the flowers are yellow and mossie, which turne into berries that are a little long as well as round, whose shell or outermost peele is greene at the first, and blacke when it is ripe; wherein is contained an hard bitter kernell, which cleaueth in two parts.

The Vse of Bayes.

The Bay leaues are of as necessary vse as any other in Garden or Orchard, for they serue both for pleasure and profit, both for ornament and for vse, both for honest Ciuill vses, and for Physicke, yea both for the sicke and for the sound, both for the liuing and for the dead : And so much might be said of this one tree, that if it were all told, would as well weary the Reader, as the Relater : but to explaine my selfe ; It serueth to adorne the house of God as well as of man : to procure warmth, comfort and strength to the limmes of men and women, by bathings and annoyntings outward, and by drinkes &c. inward to the stomacke, and other parts : to season vessels &c. wherein are preserued our meates, as well as our drinkes : to crowne or encircle

circle as with a garland, the heads of the liuing, and to fticke and decke forth the bodies of the dead : fo that from the cradle to the graue we haue ftill vfe of it, we haue ftill neede of it.

The berries likewife ferue for ftitches inward, and for paines outward, that come of cold eyther in the ioynts, finewes, or other places.

2. *Laurea Cerafus, fiue Laurus Virginiana*. The Virginian Bay, or Cherry Baye.

THis Virginian (whether you will call it a Baye, or a Cherrie, or a Cherrie Bay, I leaue it to euery ones free will and iudgement, but yet I thinke I may as well call it a Bay as others a Cherrie, neither of them being anfwerable to the tree, which neyther beareth fuch berries as are like Cherries, neither beareth euer greene leaues like the Bay : if it may therefore bee called the Virginia Cherry Bay, for a diftinction from the former Bay Cherry that beareth faire blacke Cherries, it will more fitly agree thereunto, vntill a more proper may be impofed) rifeth vp to be a tree of a reafonable height, the ftemme or bodie thereof being almoft as great as a mans legge, fpreading forth into diuers armes or boughes, and they againe into diuers fmall branches, whereon are fet without order diuers faire broade greene leaues, fomewhat like vnto the former Bay leaues, but more limber and gentle, and not fo hard in handling, broader alfo, and for the moft part ending in a point, but in many fomewhat round pointed, very finely notched or toothed about the edges, of a bitter tafte, very neere refembling the tafte of the Bay leafe, but of little or no fent at all, either greene or dryed, which fall away euery autumne, and fpring afrefh euery yeare : the bloffomes are fmall and white, many growing together vpon a long ftalke, fomewhat like the Bird Cherry bloffomes, but fmaller, and come forth at the ends of the young branches, which after turne into fmall berries, euery one fet in a fmall cup or hufke, greene at the firft, and blacke when they are ripe, of the bigneffe of a fmall peafe, of a ftrong bitter tafte, and fomewhat aromaticall withall, but without any flefhy fubftance like a Cherry at all vpon it ; for it is altogether like a berry.

The Vfe of this Virginia Cherry Bay.

Being a ftranger in our Land, and poffeffed but of a very few, I doe not heare that there hath beene any triall made thereof what properties are in it : let this therefore fuffice for this prefent, to haue fhewed you the defcription and forme thereof, vntill we can learne further of his vfes.

3. *Pinus*. The Pine tree.

MY purpofe in this place is not to fhew you all the diuerfities of Pine trees, or of the reft that follow, but of that one kinde is planted in many places of our Land for ornament and delight, and there doth reafonably well abide : take it therefore into this Orchard, for the raritie and beautie of it, though we haue little other vfe of it.

The Pine tree groweth with vs, though flowely, to a very great height in many places, with a great ftraight bodie, couered with a grayifh greene barke, the younger branches are fet round about, with very narrow long whitifh greene leaues, which fall away from the elder, but abide on the younger, being both winter and fummer alwaies greene. It hath growing in fundry places on the branches, certaine great hard wooddy clogs (called of fome apples, of others nuts) compofed of many hard wooddy fcales, or tuberous knobs, which abide for the moft part alwaies greene in our Countrey, and hardly become brownifh, as in other Countries, where they haue more heat and comfort of the Sun, and where the fcales open themfelues ; wherein are contained white long and round kernels, very fweete while they are frefh, but quickely growing oylely and rancide.

The

The Vfe of the Pine apples and kernels.

The Cones or Apples are vfed of diuers Vintners in this City, being painted, to expreffe a bunch of grapes, whereunto they are very like, and are hung vp in their bufhes, as alfo to faften keyes vnto them, as is feene in many places.

The kernels within the hard fhels, while they are frefh or newly taken out, are vfed many waies, both with Apothecaries, Comfit-makers, and Cookes : for of them are made medicines, good to lenifie the pipes and paffages of the lungs and throate, when it is hoarfe. Of them are made Comfits, Paftes, Marchpanes, and diuers other fuch like : And with them a cunning Cooke can make diuers Keck fhofes for his Mafters table.

Matthiolus commendeth the water of the greene apples diftilled, to take away the wrinkles in the face, to abate the ouer-fwelling breafts of Maidens, by fomenting them after with linnen clothes, wet in the water; and to reftore fuch as are rauifht into better termes.

4. *Abies.* The Firre tree.

THe Firre tree groweth naturally higher then any other tree in thefe parts of Chriftendome where no Cedars grow, and euen equalling or ouer-topping the Pine : the ftemme or bodie is bare without branches for a great height, if they bee elder trees, and then branching forth at one place of the bodie foure wayes in manner of a croffe, thofe boughes againe hauing two branches at euery ioynt, on which are fet on all fides very thicke together many fmall narrow long hard whitifh greene leaues, and while they are young tending to yellowneffe, but nothing fo long or hard or fharpe pointed as the Pine tree leaues, growing fmaller and fhorter to the end of the branches : the bloomings are certaine fmall long fcaly catkins, of a yellowifh colour, comming forth at the ioynts of the branches, which fall away : the cones are fmaller and longer then of the Pine tree, wherein are fmall three fquare feede contained, not halfe fo big as the Pine kernels.

The Vfe of the Firre tree.

The vfe of this tree is growne with vs of late daies to bee more frequent for the building of houfes then euer before : for hereof (namely of Deale timber and Deale boords) are framed many houfes, and their floores, without the helpe of any other timber or boord of any other tree almoft ; as alfo for many other workes and purpofes. The yellow Roffen that is vfed as well to make falues as for many other common vfes, is taken from this tree, as the Pitch is both from the Pitch and Pine trees, and is boyled to make it to bee hard, but was at the firft a yellow thin cleere Turpentine, and is that beft fort of common Turpentine is altogether in vfe with vs, as alfo another more thicke, whitifh, and troubled, both which are vfed in falues, both for man and beaft (but not inwardly as the cleere white Venice Turpentine is) and ferueth both to draw, cleanfe and heale. Dodonæus feemeth to fay, that the cleere white Turpentine, called Venice Turpentine, is drawn from the Firre : but Matthiolus confuteth that opinion, which Fulfius alfo held before him.

5. *Ilex arbor.* The euer-greene Oake.

THe *Ilex* or euer-greene Oake rifeth in time to be a very great tree, but very long and flow in growing (as is to be feene in the Kings priuy Garden at Whitehall, growing iuft againft the backe gate that openeth into the way going to Weftminfter, and in fome other places) fpreading many fair large great armes and branches, whereon are fet fmall and hard greene leaues, fomewhat endented or cornered, and

prickly

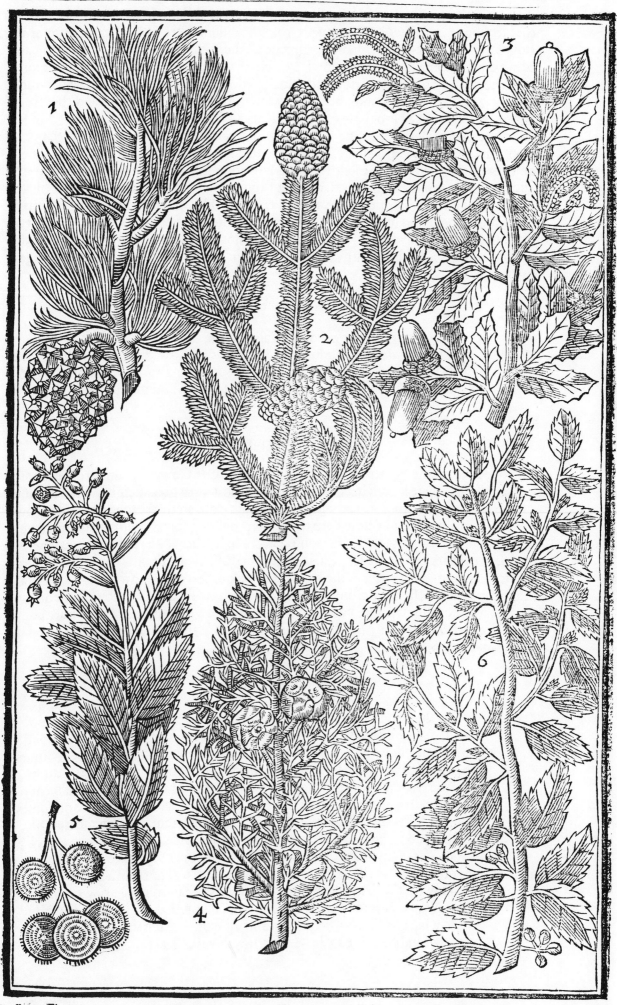

1 *Pinus* The Pine tree. 2 *Abies*. The Firre tree. 3 *Ilex*. The euer greene Oake. 4 *Cupreſſus*, The Cipreſſe tree. 5 *Arbutus*. The Strawberry tree. 6 *Alaternus*. The euer greene Priuet.

prickly on the edges, especially in the young trees, and sometimes on those branches that are young and newly sprung forth from the elder rootes, but else in a manner all smooth in the elder growne, abiding greene all the winter as well as summer, and are of a grayish greene on the vnderside. It beareth in the spring time certaine slender long branches (like as other Okes doe) with small yellowish mossie flowers on them, which fall away, and are vnprofitable, the acornes not growing from those places, but from others which are like vnto those of our ordinary Oake, but smaller and blacker, and set in a more rugged huske or cuppe. This and no other kinde of *Ilex* doe I know to grow in all our land in any Garden or Orchard: for that kind with long and narrower leaues, and not prickly, growing so plentifully as Matthiolus saith in Tuscane, I haue not seen: and it is very probable to bee the same that Plinie remembreth to haue the leafe of an Oliue ; but not as some would haue it, that *Smilax* Theophrastus maketh mention of in his third Booke and sixteenth Chapter of his Historie of Plants, which the Arcadians so called, and had the leafe of the *Ilex*, but not prickly : for Theophrastus saith, the timber of *Smilax* is smooth and soft, and this of the *Ilex* is harder, and stronger then an Oake.

The Vse of the *Ilex* or euer-greene Oake.

Seeing this is to be accounted among the kindes of Oake (and all Oakes by Dioscorides his opinion are binding) it is also of the same qualitie, but a little weaker, and may serue to strengthen weake members. The young tops and leaues are also vsed in gargles for the mouth and throate.

---

6 *Cupressus*. The Cypresse tree.

THe Cypresse tree that is nourfed vp by vs, in our Country, doth grow in those places where it hath beene long planted, to a very great height, whose bodie and boughes are couered with a reddish ash-coloured bark ; the branches grow not spreading, but vpright close vnto the bodie, bushing thicke below, and small vpwards, spire fashion, those below reaching neere halfe the way to them aboue, whereon doe grow euer greene leaues, small, long and flat, of a resinous sweete smell, and strong taste, somewhat bitter : the fruit, which are called nuts, grow here and there among the boughes, sticking close vnto them, which are small, and clouen into diuers parts, but close while they are young, of a russetish browne colour ; wherein are contained small browne seede, but not so small as motes in the Sunne, as Matthiolus and others make them to be.

The Vse of the Cypresse tree.

For the goodly proportion this tree beareth, as also for his euer-greene head, it is and hath beene of great account with all Princes, both beyond, and on this side of the Sea, to plant them in rowes on both sides of some spatious walke, which by reason of their high growing, and little spreading, must be planted the thicker together, and so they giue a goodly, pleasant and sweet shadow : or else alone, if they haue not many, in the middle of some quarter, or as they thinke meete. The wood thereof is firme and durable, or neuer decaying, of a brown yellow colour, and of a strong sweete smell, whereof Chests or Boxes are made to keepe apparell, linnen, furres, and other things, to preserue them from moths, and to giue them a good smell.

Many Physicall properties, both wood, leaues and nuts haue, which here is not my purpose to vnfold, but only to tell you, that the leaues being boyled in wine, and drunke, helpe the difficultie of making vrine, and that the nuts are binding, fit to bee vsed to stay fluxes or laskes, and good also for ruptures.

7. *Arbutus*. The Strawberry tree.

THe Strawberry tree groweth but flowly, and rifeth not to the height of any great tree, no not in France, Italy, or Spaine: and with vs the coldneffe of our country doth the more abate his vigour, fo that it feldome rifeth to the height of a man: the barke of the body is rough, and fmooth in the younger branches : the leaues are faire and greene, very like vnto Baye leaues, finely dented or fnipped about the edges, abiding alwayes greene thereon both Winter and Summer : the flowers come forth at the end of the branches vpon long ftalkes, not cluftering thicke together, but in long bunches, and are fmall, white, and hollow, like a little bottle, or the flower of Lilly Conually, which after turne into rough or rugged berries, moft like vnto Strawberries (which hath giuen the name to the tree) fomewhat reddifh when they are ripe, of a harfh tafte, nothing pleafant, wherein are contained many fmall feedes : It hardly bringeth his fruit to ripeneffe in our countrey; for in their naturall places they ripen not vntill Winter, which there is much milder then with vs.

The Vfe of the Strawberry tree.

Amatus Lufitanus I thinke is the firft that euer recorded, that the water diftilled from the leaues and flowers hereof, fhould bee very powerfull againft the plague and poyfons : for all the ancient Writers doe report, that the fruit hereof being eaten, is an enemy to the ftomacke and head. And Clufius likewife fetteth downe, that at Lifhbone, and other places in Portingall where they are frequent, they are chiefly eaten, but of the poorer fort, women and boyes. They are fomewhat aftringent or binding, and therefore may well ferue for fluxes. It is chiefly nourfed with vs for the beauty and rareneffe of the tree; for that it beareth his leaues alwayes green.

8. *Alaternus*. The euer greene Priuet.

THe tree which we haue growing in our country called *Alaternus*, groweth not to be a tree of any height; but abiding lowe, fpreadeth forth many branches, whereon are fet diuers fmall and hard greene leaues, fomewhat round for the forme, and endented a little by the edges : it beareth many fmall whitifh greene flowers at the ioynts of the ftalkes, and fetting on of the lower leaues cluftering thicke together, which after turne into fmall blacke berries, wherein are contained many fmall graines or feedes : the beauty and verdure of thefe leaues abiding fo frefh all the yeare, doth caufe it to be of the greater refpect; and therefore findeth place in their Gardens onely, that are curious conferuers of all natures beauties.

The Vfe of the euer greene Priuet.

It is feldome vfed for any Phyficall property, neither with vs, nor in the places where it is naturall and plentifull : but as Clufius reporteth, hee learned that the Portingall Fifhermen do dye their nets red with the decoction of the barke hereof, and that the Dyers in thofe parts doe vfe the fmall peeces of the wood to ftrike a blackifh blew colour.

9. *Celaftrus Theophrafti Clufio*. Clufius his Celaftrus.

ALthough the Collectour (who is thought to be Ioannes Molineus of the great Herball or Hiftory of plants, and generally bearing Dalefchampius name, becaufe the finding and relation of diuers herbes therein expreffed, is appropiate to him, and printed at Lyons) of all our moderne Writers doth firft of all others appoint the *Celaftrus*, whereof Theophraftus onely among all the ancient Writers of plants maketh mention, to be the first *Alaternus* that Clusius hath set forth in his History of rarer plants : yet I finde, that Clusius himselfe before his death doth appropiate that *Celastrus* of Theophrastus to another plant, growing in the Garden at Leyden, which formerly of diuers had beene taken to be a kinde of *Laurus Tinus*, or the wilde Baye ; but he impugning that opinion for diuers respects, decyphreth out that Leyden tree in the same manner that I doe : and because it is not onely faire, in bearing his leaues alwayes greene, but rare also, being nourced vp in our Land in very few places, but principally with a good old Lady, the widow of Sir Iohn Leuson, dwelling neere Rochester in Kent ; I thought it fit to commend it for an ornament, to adorne this our Garden and Orchard. It groweth vp to the height of a reasonable tree, the body whereof is couered with a darke coloured barke, as the elder branches are in like manner ; the younger branches being greene, whereon are set diuers leaues thicke together, two alwayes at a ioynt, one against another, of a sad but faire greene colour on the vpperside, and paler vnderneath, which are little or nothing at all snipped about the edges, as large as the leaues of the *Laurus Tinus*, or wilde Baye tree : at the end of the young branches breake forth between the leaues diuers small stalkes, with foure or fiue flowers on each of them, of a yellowish greene colour, which turne into small berries, of the bignesse of blacke Cherries, greene at the first, and red when they begin to be ripe, but growing blacke if they hang too long vpon the branches, wherein is contained a hard shell, and a white hard kernell within it, couered with a yellowish skin. This abideth (as I said before) with greene leaues as well Winter as Summer ; and therefore fittest to be planted among other of the same nature, to make an euer greene hedge.

The Vse of Clusius his Celastrus.

Being so great a stranger in this part of the Christian world, I know none hath made tryall of what property it is, but that the taste of the leaues is somewhat bitter.

---

10. *Pyracantha.* The euer greene Hawthorne, or prickly Corall tree.

THis euer greene shrubbe is so fine an ornament to a Garden or Orchard, either to be nourced vp into a small tree by it selfe, by pruining and taking away the suckers and vnder branches, or by suffering it to grow with suckers, thicke and plashing the branches into a hedge, for that it is plyable to be ordered either way ; that I could not but giue you the knowledge thereof, with the description in this manner. The younger branches are couered with a smooth darke blewish greene barke, and the elder with a more ash coloured, thicke set with leaues without order, some greater and others smaller, somewhat like both in forme and bignesse vnto the leaues of the Barberry tree, but somewhat larger, and more snipt about the edges, of a deeper green colour also, and with small long thornes scattered here & there vpon the branches : the flowers come forth as well at the ends of the branches, as at diuers places at the ioynts of the leaues, standing thicke together, of a pale whitish colour, a little dasht ouer with a shew of blush, consisting of fiue leaues a peece, with some small threads in the middle, which turne into berries, very like vnto Hawthorne berries, but much redder and dryer, almost like polished Corall, wherein are contained foure or fiue small yellowish white three square seede, somewhat shining. It is thought to be the *Oxyacantha* of Dioscorides ; but seeing Dioscorides doth explaine the forme of the leafe in his Chapter of Medlars, which he concealed in the Chapter of *Oxyacantha*, it cannot be the same : for *Mespilus Anthedon* of Theophrastus, or *Aronia* of Dioscorides, hath the leafe of *Oxyacantha*, as Dioscorides saith, or of Smalladge, as Theophrastus, which cannot agree to this Thorne ; but doth most liuely delineate out our white Thorne or Hawthorne, that now there is no doubt, but that *Oxyacantha* of Dioscorides is the Hawthorne tree or bush.

The Vse of this Corall tree.

Although Lobel maketh mention of this tree to grow both in Italy, and

Prouince

1 *Celastrus Theophrasti Clusio*. Clusius his Celastrus. 2 *Pyracantha*. The euer green prickly Corall tree. 3 *Taxus*. The Yewe tree. 4 *Buxus arbor*. The Boxe tree 5 *Buxus humilis*. The lowe or dwarfe Boxe. 6 *Sabina*. The Sauine tree. 7 *Paliurus* Christs thorne. 8 *Larix*. The Larch tree.

Prouence in France, in fome of their hedges, yet he faith it is neglected in
the naturall places, and to be of no vfe with them : neither doe I heare, that
it is applyed to any Phyficall vfe with vs , but (as I before faid) it is preferued with diuers as an ornament to a Garden or Orchard , by reafon of his
euer greene leaues, and red berries among them , being a pleafant fpectacle,
and fit to be brought into the forme of an hedge, as one pleafe to lead it.

---

11. *Taxus*. The Yewe tree.

THe Yewe tree groweth with vs in many places to bee a reafonable great tree,
but in hoter countries much bigger, couered with a reddifh gray fcaly barke ;
the younger branches are reddifh likewife , whereon grow many winged
leaues, that is, many narrow long darke greene leaues, fet on both fides of a long ftalke
or branch, neuer dying or falling away, but abiding on perpetually, except it be on the
elder boughes : the flowers are fmall, growing by the leaues , which turne into round
red berries, like vnto red Afparagus berries, in tafte fweetifh , with a little bitterneffe,
and caufing no harme to them for any thing hath been knowne in our country,

The Vfe of the Yewe tree.

It is found planted both in the corners of Orchards, and againft the windowes of Houfes, to be both a fhadow and an ornament , in being alwayes
greene, and to decke vp Houfes in Winter : but ancient Writers haue euer
reckoned it to be dangerous at the leaft, if not deadly.

---

12. *Buxus*. The Boxe tree.

THe Boxe tree in fome places is a reafonable tall tree , yet growing flowly ; the
trunke or body whereof is of the bigneffe of a mans thigh, which is the biggeft
that euer I faw : but fometimes, and in other places it groweth much lower,
vfually not aboue a yard, or a yard and a halfe high , on the backe fides of many Houfes, and in the Orchards likewife : the leaues are fmall, thicke and hard, and ftill the
greater or leffer the tree is. the greater or leffer are the leaues, round pointed, and of a
frefh fhining greene colour : the flowers are fmall and greenifh , which turne into
heads or berries, with foure hornes, whittifh on the outfide, and with reddifh feede
within them.

Buxus aureus.
Gilded Boxe. There is another kinde hereof but lately come to our knowledge , which differeth
not in any thing from the former , but onely that all the leaues haue a yellow lift or
gard about the edge of them on the vpperfide, and none on the lower , which maketh
it feeme very beautifull ; and is therefore called gilded Boxe.

Buxus humilis.
Dwarfe Boxe. We haue yet another kinde of Boxe, growing fmall and lowe, not aboue halfe a
foote, or a foote high at the moft, vnleffe it be neglected , which then doth grow a little more fhrubby, bearing the like leaues, but fmaller, according to the growth, and of
a deeper greene colour : I could neuer know that this kinde euer bore flower or feede,
but is propagated by flipping the roote, which encreafeth very much.

The Vfe of Boxe.

The wood of the Boxe tree is vfed in many kindes of fmall works among
Turners, becaufe it is hard, clofe, and firme, and as fome haue faid, the roots
much more, in regard of the diuers waues and crooked veines running
through it. It hath no Phyficall vfe among the moft and beft Phyfitians, although fome haue reported it to ftay fluxes, and to be as good as the wood
of *Guaiacum*, or *Lignum vitæ* for the French difeafe. The leaues and branches ferue both Summer and Winter to decke vp houfes ; and are many
times giuen to horfes for the bots.
The lowe or dwarfe Boxe is of excellent vfe to border vp a knot, or the

long beds in a Garden, being a maruailous fine ornament thereunto, in regard it both groweth lowe, is euer greene, and by cutting may bee kept in what maner euery one pleafe, as I haue before fpoken more largely.

13. *Sabina.* The Sauine tree or bufh.

THe Sauine tree or bufh that is moft vfuall in our country, is a fmall lowe bufh, not fo high as a man in any place, nor fo bigge in the ftemme or trunke as a mans arme, with many crooked bending boughes and branches, whereon are fet many fmall, fhort, hard, and prickly leaues, of a darke green colour, frefh and green both Winter and Summer: it is reported, that in the naturall places it beareth fmall blacke berries, like vnto Iuniper, but with vs it was neuer knowne to beare any.

The Vfe of Sauine.

It is planted in out-yards, backfides, or voide places of Orchards, as well to caft clothes thereon to dry, as for medicines both for men and horfes: being made into an oyle, it is good to annoint childrens bellies for to kill the Wormes: and the powder thereof mixed with Hogs greafe, to annoint the running fores or fcabs in their heads; but beware how you giue it inwardly to men, women, or children. It is often put into horfes drenches, to helpe to cure them of the bots, and other difeafes.

14. *Paliurus.* Chrifts thorne.

THis thorny fhrubbe (wherewith as it is thought, our Sauiour Chrift was crowned, becaufe as thofe that haue trauelled through Paleftina and Iudæa, doe report no other thorne doth grow therein fo frequent, or fo apt to be wrethed) rifeth in fome places to a reafonable height, but in our country feldome exceedeth the height of a man, bearing many flender branches, full of leaues, fet on either fide thereof one by one, which are fomewhat broad and round, yet pointed, and full of veines, thicke fet alfo with fmall thornes, euen at the foote of euery branch, and at the foote of euery leafe one or two, fome ftanding vpright, others a little bending downe: the flowers are fmall and yellow, ftanding for the moft part at the end of the branches, many growing vpon a long ftalke, which after turne into round, flat, and hard fhelly fruit, yet couered with a foft flefhy skinne, within which are included two or three hard, fmall, and browne flat feeds, lying in feuerall partitions. The leaues hereof fall away euery yeare, and fpring forth afrefh againe the next May following. The rarity and beauty of this fhrubbe, but chiefly (as I thinke) the name hath caufed this to be much accounted of with all louers of plants.

The Vfe of Chrifts thorne.

Wee haue fo few of thefe fhrubbes growing in our country, and thofe that are, doe, for any thing I can vnderftand, neuer beare fruit with vs; that there is no other vfe made hereof then to delight the owners: but this is certainly receiued for the *Paliurus* of Diofcorides and Theophraftus, and thought alfo by Matthiolus to be the very true *Rhamnus tertius* of Diofcorides. Matthiolus alfo feemeth to contradict the opinion is held by the Phyfitians of Mompelier, and others, that it cannot be the *Paliurus* of Theophraftus. It is held to be effectuall to helpe to breake the ftone, both in the bladder, reines, and kidneyes: the leaues and young branches haue an aftringent quality, and good againft poyfons and the bitings of ferpents.

15. *Larix.*

15. *Larix.* The Larch tree.

THe Larch tree, where it naturally groweth, riseth vp to be as tall as the Pine or Firre tree, but in our Land being rare, and nourſed vp but with a few, and thoſe onely louers of rarities, it groweth both ſlowly, and becommeth not high : the baike hereof is very rugged and thicke, the boughes and branches grow one aboue another in a very comely order, hauing diuers ſmall yellowiſh knobs or bunches ſet thereon at ſeuerall diſtances; from whence doe yearely ſhoote forth many ſmall, long, and narrow ſmooth leaues together, both ſhorter and ſmaller, and not ſo hard or ſharpe pointed as either the Pine or Firre tree leaues, which doe not abide the Winter as they doe, but fall away euery yeare, as other trees which ſhed their leaues, and gaine freſh euery Spring : the bloſſomes are very beautifull and delectable, being of an excellent fine crimſon colour, which ſtanding among the greene leaues, allure the eyes of the beholders to regard it with the more deſire : it alſo beareth in the naturall places (but not in our Land that I could heare) ſmall ſoft cones or fruit, ſomewhat like vnto Cypreſſe nuts, when they are greene and cloſe.

The Vſe of the Larch tree.

The coles of the wood hereof (becauſe it is ſo hard and durable as none more) is held to be of moſt force being fired, to cauſe the Iron oare to melt, which none other would doe ſo well. Matthiolus conteſteth againſt Fuchſius, for deeming the Venice Turpentine to be the liquid Roſſen of the Firre tree, which he aſſureth vpon his owne experience and certaine knowledge, to be drawne from this Larch tree, and none other; which cleere Turpintine is altogether vſed inwardly, and no other, except that of the true Turpintine tree, and is very effectuall to cleanſe the reines, kidneyes, and bladder, both of grauell and the ſtone, and to prouoke vrine : it is alſo of eſpeciall property for the *gonorrhæa*, or running of the reines, as it is called, with ſome powder of white Amber mixed therewith, taken for certaine dayes together. Taken alſo in an Electuary, it is ſingular good for to expectorate rotten flegme, and to helpe the conſumption of the lungs. It is vſed in plaiſters and ſalues, as the beſt ſort of Turpintine. The Agaricke that is vſed in phyſicke, is taken from the bodies and armes of this tree. And Matthiolus doth much inſiſt againſt Braſauolus, that thought other trees had produced Agaricke, affirming them to be hard *Fungi*, or Muſhroms (ſuch as wee call Touch-wood) wherwith many vſe to take fire, ſtrooke thereinto from ſteele.

16. *Tilia.* The Line or Linden tree.

THere are two ſorts of Line trees, the male and the female; but becauſe the male is rare to be ſeene, and the female is more familiar, I will onely giue you the deſcription of the female, and leaue the other.

The female Line tree groweth exceeding high and great, like vnto an Elme, with many large ſpreading boughes, couered with a ſmooth barke, the innermoſt being very plyant and bending from whence come ſmaller branches, all of them ſo plyable, that they may bee led or carried into any forme you pleaſe : the leaues thereon are very faire, broad, and round, ſomewhat like vnto Elme leaues, but fairer, ſmoother, and of a freſher greene colour, dented finely about the edges, and ending in a ſharpe point : the flowers are white, and of a good ſmell, many ſtanding together at the top of a ſtalke, which runneth all along the middle ribbe of a ſmall long whitiſh leafe; after which come ſmall round berries, wherein is contained ſmall blackiſh ſeede : this tree is wholly neglected by thoſe that haue them, or dwell neere them, becauſe they ſuppoſe it to be fruitleſſe, in regard it beareth chaffie huskes, which in many places fall away, without giuing ripe ſeede.

The

1 *Tilia fœmina.* The Line or Linden tree. 2 *Tamariscus.* The Tamariske tree. 3 *Acer maius latifolium.* The Sycomore tree. 4 *Staphylodendron.* The bladder nut. 5 *Rhus Myrtifolia.* The Mirtle leafed Sumach. 6 *Rhus Virginiana,* The Bucks horne tree. 7 *Vitis seu Potius Hedera Virginensis.* The Virginia Vine or rather Iuie.

The Vse of the Line tree.

It is planted both to make goodly Arbours, and Summer banquetting houses, either belowe vpon the ground, the boughes seruing very handsomely to plash round about it, or vp higher, for a second aboue it, and a third also: for the more it is depressed, the better it will grow. And I haue seene at Cobham in Kent, a tall or great bodied Line tree, bare withont boughes for eight foote high, and then the branches were spread round about so orderly, as if it were done by art, and brought to compasse that middle Arbour: And from those boughes the body was bare againe for eight or nine foote (wherein might bee placed halfe an hundred men at the least, as there might be likewise in that vnderneath this)& then another rowe of branches to encompasse a third Arbour, with stayres made for the purpose to this and that vnderneath it: vpon the boughes were laid boards to tread vpon, which was the goodliest spectacle mine eyes euer beheld for one tree to carry.

The coles of the wood are the best to make Gunpowder. And being kindled, and quenched in vinegar, are good to dissolue clotted bloud in those that are bruised with a fall. The inner barke being steeped in water yeeldeth a slimie iuyce, which is found by experience, to be very profitable for them that haue been burnt with fire.

17. *Tamarix*. Tamariske tree.

THe Tamariske tree that is common in our country, although in some places it doth not grow great, yet I haue seene it in some other, to be as great as a great apple tree in the body, bearing great arms; from whose smaller branches spring forth young slender red shootes, set with many very fine, small, and short leaues, a little crisped, like vnto the leaues of Sauine, not hard or rough, but soft and greene: the flowers be white mossie threads, which turne into dounie seede, that is carried away with the winde.

Tamariscus folijs ablidis. White Tamariske.
   There is another kinde hereof very beautifull and rare, not to be seene in this Land I thinke, but with M^r. William Ward, the Kings seruant in his Granary, before remembred, who brought me a small twigge to see from his house at Boram in Essex, whose branches are all red while they are young, and all the leaues white, abiding so all the Summer long, without changing into any shew of greene like the other, and so abideth constant yeare after yeare, yet shedding the leaues in Winter like the other.

The Vse of Tamariske.

The greatest vse of Tamariske is for spleneticke diseases, either the leaues or the barke made into drinkes; or the wood made into small Cans or Cups to drinke in.

18. *Acer maius latifolium*. The great Maple or Sycomore tree.

THe Sycomore tree, as we vsually call it (and is the greatest kind of Maple, cherished in our Land onely in Orchards, or elsewhere for shade and walkes, both here in England, and in some other countries also) groweth quickly to bee a faire spreading great tree, with many boughes and branches, whose barke is somewhat smooth: the leaues are very great, large, and smooth, cut into foure or fiue diuisions, and ending into so many corners, euery one standing on a long reddish stalke: the bloomings are of a yellowish greene colour, growing many together on each side of a long stalke, which after turne into long and broad winged seede, two alwaies standing together on a stalke, and bunched out in the middle, where the seed or kernell lyeth, very like vnto the common Maple growing wilde abroad, but many more together, and larger.

The

The Vse of the Sycomore tree.

It is altogether planted for shady walkes, and hath no other vse with vs that I know.

---

19. *Nux Vesicaria.* The bladder Nut.

THis tree groweth not very high, but is of a meane stature, when it is preserued and pruined to grow vpright, or else it shooteth forth many twigges from the rootes, and so is fit to plant in a hedge rowe, as it is vsed in some places: the body and armes are couered with a whitish greene barke: the branches and leaues on them are like vnto the Elder, hauing three or fiue leaues set one against another, with one of them at the end, each whereof is nicked or dented about the edges: the flowers are sweete and white, many growing together on a long stalke, hanging downeward, in forme resembling a small Daffodill, hauing a small round cup in the middle, and leaues about it: after which come the fruit, inclosed in russetish greene bladders, containing one or two brownish nuts, lesser then Hasell nuts, whose outer shell is not hard and woody, like the shell of a nut, but tough, and hard withall, not easie to breake, within which is a greene kernell, sweetish at the first, but lothsome afterwards, ready to procure casting, and yet liked of some people, who can well endure to eate them.

The Vse of the Bladder Nut.

The greatest vse that I know the tree or his fruit is put vnto, is, that it is receiued into an Orchard, either for the rarity of the kinde, being suffered to grow into a tree, or (as I said before) to make an hedge, being let grow into suckers.

Some Quacksaluers haue vsed these nuts as a medicine of rare vertue for the stone, but what good they haue done, I neuer yet could learne.

---

20. *Rhus Myrtifolia.* The Mirtle leafed Sumach.

THis lowe shrubbe groweth seldome to the height of a man, hauing many slender branches, and long winged leaues set thereon, euery one whereof is of the bignesse of the broad or large Mirtle leafe, and set by couples all the length of the ribbe, running through the middle of them. It beareth diuers flowers at the tops of the branches, made of many purple threads, which turne into small blacke berries, wherein are contained small, white, and rough seed, somewhat like vnto Grape kernels or stones. This vseth to dye down to the ground in my Garden euery Winter, and rise vp again euery Spring, whether the nature thereof were so, or the coldnesse of our climate the cause therof, I am not well assured. It is also rare, and to be seen but with a few.

The Vse of this Sumach.

It is vsed to thicken or tanne leather or hides, in the same manner that the ordinary Sumach doth; as also to stay fluxes both in men and women.

---

21. *Rhus Virginiana.* The Virginia Sumach, or Buckes horne tree of Virginia.

THis strange tree becommeth in some places to bee of a reasonable height and bignesse, the wood whereof is white, soft, and pithy in the middle, like vnto an Elder, couered with a darke coloured barke, somewhat smooth: the young branches that are of the last yeares growth are somewhat reddish or browne, very soft and

and fmooth in handling, and fo like vnto the Veluet head of a Deere, that if one were cut off from the tree, and fhewed by it felfe, it might foone deceiue a right good Woodman, and as they grow feeme moft like thereunto, yeelding a yellowifh milke when it is broken, which in a fmall time becommeth thicke like a gumme : the leaues grow without order on the branches, but are themfelues fet in a feemly order on each fide of a middle ribbe, feuen, nine, ten, or more on a fide, and one at the end, each whereof are fomewhat broad and long, of a darke greene colour on the vpperfide, and paler greene vnderneath, finely fnipped or toothed round about the edges : at the ends of the branches come forth long and thicke browne tufts, very foft, and as it were woolly in handling, made all of fhort threads or thrums ; from among which appeare many fmall flowers, much more red or crimfon then the tufts, which turne into a very fmall feede : the roote fhooteth forth young fuckers farre away, and round about, whereby it is mightily encreafed.

The Vfe of this Sumach.

It is onely kept as a rarity and ornament to a Garden or Orchard, no bo-die, that I can heare of, hauing made any tryall of the Phyficall properties.

22. *Vitis, feu potius Hedera Virginenfis.* The Virginia Vine, or rather Iuie.

THis flender, but tall climing Virginia Vine (as it was firft called ; but Iuie, as it doth better refemble) rifeth out of the ground with diuers ftems, none much bigger then a mans thumbe, many leffe ; from whence fhoote forth many long weake branches, not able to ftand vpright, vnleffe they be fuftained: yet planted neere vnto a wall or pale, the branches at feuerall diftances of the leaues will fhoote forth fmall fhort tendrels, not twining themfelues about any thing, but ending into foure, fiue, or fix, or more fmall fhort and fomewhat broad clawes, which will faften like a hand with fingers fo clofe thereunto, that it will bring part of the wall, morter, or board away with it, if it be pulled from it, and thereby ftay it felfe, to climbe vp to the toppe of the higheft chimney of a houfe, being planted thereat : the leaues are crum-pled, or rather folded together at the firft comming forth, and very red, which after growing forth, are very faire, large, and greene, diuided into foure, fiue, fix, or feuen leaues, ftanding together vpon a fmall foote-ftalke, fet without order on the branches, at the ends whereof, as alfo at other places fometime, come forth diuers fhort tufts of buds for flowers ; but we could neuer fee them open themfelues, to fhew what manner of flower it would be, or what fruit would follow in our country : the roote fpreadeth here and there, and not very deepe.

The Vfe of this Virginian.

We know of no other vfe, but to furnifh a Garden, and to encreafe the number of rarities.

And thus haue I finifhed this worke, and furnifhed it with whatfoeuer Art and Na-ture concurring, could effect to bring delight to thofe that liue in our Climate, and take pleafure in fuch things ; which how well or ill done, I muft abide euery ones cenfure : the iudicious and courteous I onely refpect, let Momus bite his lips, and eate his heart ; and fo Farewell.

FINIS.

Index omnium stirpium

quæ in hoc opere continentur.

Fff   Bellis

Index.

Eff 2   Gnaphalium

INDEX.

INDEX.

INDEX.

INDEX,

A

A Table of the English names of such Plants as are contained in this Booke.

THE

A Table of the Vertues and Properties of the Hearbes contained in this Booke.

Faults eſcaped in ſome Copies.

FOlio 8. line 14. for own reade home. f.12.l.27. for trouble reade treble. f.42. l.5. reade, like vnto that of a Lilly f.66.l.42. for χάρις read χρίνα or λείειs. f.73.l.37. for top of the flower, read cup. f.134.l.36. for compoſed reade compaſſed. f.150.l.4. for hath, reade haue, and line 5. for is are. f.173.l.12. put out theſe Wordes, the inſide, in the beginning of the line. f.189.l.38. reade *Biniſflorum ordinibus.* f.218.l.19. reade goulons, and line 28 pratenſis. f.272. line 36. read Pothos. f.276 l.12. Chelidonia. f.281.l.37. for hath, haue. and l.28. Maſtuerzo. f.284 l.15. Vicenza. f.287.l.39. Citrina. f.290.l.39. reade prouoke, and, helpe. f.329.l.37. for Melancholicke, reade Flegmaticke. f.330.331.333. reade Eryngium in all places f.336.l.8. reade, and not very flat. f.356.l.31. Americanum. f.357.l.26. Cervicaria. f.358. l.45. reade, before it can haue. f.372.l.9. blot out, except it. f.389. for ſpockes, reade ſmockes. f.393.l.3. in the margent for cæruleo, read pleno. f.397.l.10. reade dwarfe. f.424.l.45. Hirculus. f.428.l.20. Tarentina. f.431.l.10. Cyprium. l.19. Amomum. f.438.l.17. for Dioſcorides, reade Theophraſtus. f.442.l.3. for caſtings reade purgings. f.509.l.35. reade 2εγγύλη. f.513.l.24. transferre all that clauſe of Onions vnto the other ſide, vnder the vſe of Onions. f.516.l.37. transferre theſe words, [Bauhinus vpon Matthiolus calleth it *Solanum tuberoſum eſculentum*] vnto the former Potatoes of Virginia. f.520.l.13. for ſwelleth, read ſmelleth. f.541.l.51. reade, after your ſtockes rayſed from ſtones. f.566.l.20. for as, read and. and l.29. euery one. f.567.l.24. for Rice, read Rue. f.575.l.8. reade ſerue to be miniſtred to the ſicke. f.588 l.3. Capandu. f.594.l.18. for facility, read faculty. f.595.l.39. reade Ricinus. f.600.l.4. Fuchſius

LONDON,

Printed by HVMFREY LOWNES *and* ROBERT YOVNG

at the ſigne of the Starre on Bread-ſtreet hill.

1629.